Fundamentals of
THE PHYSICAL ENVIRONMENT

'An excellent introductory text, well illustrated with very clear figures which support the text very well. It provides a good cross-section of physical geography and should remain very popular with students.'

David Evans, *University of Glasgow*

'A comprehensive and well written text. The illustrations are excellent, clear and easily understood. The text succeeds in linking the physical processes to the human experience.'

Marilyn Raphael, *University of California, Los Angeles*

'This is an extensively revised edition of a popular undergraduate text in physical geography. The authors have adopted an interesting narrative style which develops environmental systems themes by working from a simple description of a local landscale to the scale of the entire planet. Text boxes provide more detailed description of key points and the work is well illustrated with simple and clearly labelled line diagrams. The book provides foundation studies in most sub-disciplines of physical geography and would be appropriate for first years units in environmental studies.'

Peter Mitchell, *Macquarie University*

David Briggs is Research Director at Nene Centre for Research, Northampton; **Peter Smithson** is Senior Lecturer in Geography at the University of Sheffield; **Kenneth Addison** is Senior Lecturer at the School of Applied Sciences, University of Wolverhampton, and Fellow and Tutor in Physical Geography at St Peter's College, Oxford; **Ken Atkinson** is Senior Lecturer in Geography at the University of Leeds.

Comments on the first edition, Fundamentals of Physical Geography

An excellent comprehensive text for introductory undergraduate courses.

Roger Dackombe, *Wolverhampton University*

Relevant and clear diagrams, lucid and easy to understand text. Good coverage of the subject and a clear layout.

Mike Tuke, *Cambridge Regional College*

Clearly structured with a good breadth of coverage for physical geography. Explanations are comprehensive, avoiding jargon and over-conceptualisation.

Fiona Tweed, *Staffordshire University*

A very good overview of the subject, logical divisions, well illustrated.

Mark Mulligan, *Kings College, London*

A good introductory text. The diagrams and further reading sections are particularly useful for students.

Kegang Wu, *Liverpool University*

A simple, easy to follow yet informative text. This text should fill an enormous gap in literature availability.

Brian Wilson, *Central Lancashire University*

Fundamentals of
THE PHYSICAL ENVIRONMENT

Second Edition

*David Briggs, Peter Smithson,
Kenneth Addison and
Ken Atkinson*

LONDON AND NEW YORK

To Matthew;
to Adam and Cathy
who found the first edition so useful
for their geography exams;
to Charles and Rosanagh
and to Dee, Bruce and Rachel

First edition published by Hutchinson Education in 1985
Fifth impression 1989 published by the Academic Division of
Unwin Hyman

Reprinted 1992, 1993, 1995
by Routledge
11 New Fetter Lane, London EC4P 4EE

Second edition published 1997
Reprinted 1998, 2001

Simultaneously published in the USA and Canada
by Routledge
29 West 35th Street, New York, NY 10001

Routledge is an imprint of the Taylor & Francis Group

© 1997 David Briggs, Peter Smithson, Kenneth Addison and
Ken Atkinson

Typeset by Florencetype Ltd, Stoodleigh, Devon

Figure artwork for chapters 3, 4, 12, 13, 14, 15, 16, 17, 18, 19, 20 21, 22, 23, 24, 25
and 26 redrawn by Florencetype Ltd; figure artwork for chapters 1, 2, 5, 6, 7, 8, 9, 10,
11, 27 and 28 redrawn by Paul Coles, University of Sheffield

Printed and bound in Great Britain by Butler & Tanner Ltd,
Frome and London

British Library Cataloguing in Publication Data
A catalogue record for this book is available from the British Library

Library of Congress Cataloguing in Publication Data
A catalogue record for this book has been requested

ISBN 0–415–10890–X
0–415–10891–8 (pbk)

Contents

Colour plates

1 The Trifid Nebula (M20), situated in the constellation Sagittarius.
2 The Central Cordillera of the Andes in Ecuador.
3 Igneous rock selection.
4 Metamorphic rock selection.
5 Sedimentary rock selection.
6 Peaks near Pasu in the Karakoram Range.
7 Southern Lake District and western Pennines of north-west England.
8 Desert landforms around Gebel Gerf, Southeastern Desert, Egypt.
9 The cuspate, wave-dominated Red River delta, Vietnam, fronted by a barrier coast with offshore bars.
10 A brown earth soil (FAO: Eutric Cambisol) developed under oak woodland in glacial till at Loch Lomond, Scotland.
11 An iron podzol soil (FAO: Ferric Podzol) formed in fluvioglacial sands beneath spruce in the boreal coniferous forest zone of Canada.
12 A humus-ironpan stagnopodzol soil (FAO: Placic Podzol) developed in acidic glacial till in Glen Fiddich, Scotland.
13 Photomicrograph of an horizon formed by clay illuviation (Bt) in an argillic brown earth soil (FAO: Orthic Luvisol).
14 A groundwater gley soil (FAO: Eutric Gleysol) formed in estuarine clay in Stirlingshire, Scotland.
15 A black chernozem soil (FAO: Calcic Chernozem) in the prairie of Alberta, western Canada.
16 A salt crust has formed on the surface of an irrigated field on sands in North Africa.
17 A lateritic soil (FAO: Rhodic Ferralsol) developed under tropical rain forest in Ghana, West Africa.
18 An arctic brown soil (FAO: Gelic Cambisol) developed above permafrost on Devon Island, Canadian Arctic.
19 A red Mediterranean soil or terra rossa (FAO: Chromic Luvisol) developed on limestone in northern Libya.

Black and white plates

Figures

Tables

Boxes

Preface to the second edition

It is over ten years since the first edition of this book, which was then called *Fundamentals of Physical Geography*. It has proved extremely popular in that time, with a Canadian edition appearing as well as many reprints. Although the basic philosophy of the first edition still prevails, we have been very conscious of the need to update the text and its illustrations. Unfortunately severe demands on our time, such as Teaching Quality Assessments, Research Assessment Exercises and the need to bring in money for research groups, have prevented the two original authors revising the first edition. To ensure the book retained its popularity through the appearance of a second edition, we are delighted to welcome Ken Addison of the Universities of Wolverhampton and Oxford and Ken Atkinson of the University of Leeds to the writing team. With their skills and expertise we have been able to prepare a completely new version of the book which we hope will continue to appeal to large numbers of sixth form and University geography students in the way that the previous edition has done.

The second edition presents a total rewrite and a partial restructuring. The first edition aimed at bringing together an appropriate foundation for the various '-ologies', such as climatology, hydrology, geomorphology, etc., so that those interested could step more easily into their selected specialisms. We stressed the systems approach so that the links within the environment could be emphasized. In this edition the -ologies have been condensed slightly, with hydrology being subsumed into chapters on climatology, geomorphology and the ocean, rather than being a separate section. In addition a section has been added on various regional environments to reflect the new title of the book. A thorough knowledge of physical and biological systems and their scientific processes is necessary in order to predict and manage the effects of human development activities. In the modern world it becomes increasingly clear that the physical environment impacts on human activities and in turn, that human activities impact on the world ecosystems. As a result we have taken a problem-solving approach to these environments rather than providing just a comprehensive systematic survey of each one. The choice of problems to emphasize is largely personal but concentrates on our perception of what are the most pressing areas of concern. We trust that they will provide a valuable demonstration of the practicalities of studying the physical environment in an applied manner.

To improve the impact of certain topics we have used the idea of 'boxes'. Many chapters contain a box where a subject is discussed in more detail. Most of them are topics which, we hope, will stimulate debate and discussion. A further change has been the preparation of an Instructor's Manual, which should provide a more effective way of working through the key issues raised in each chapter, as well as including multiple-choice and short-answer questions to test performance.

We hope that these improvements will prove useful enough to ensure that the book continues as a leading reference work on the physical environment into the twenty-first century.

Acknowledgements

There are many people to whom we are indebted for help, advice, encouragement and forbearance in producing this book. We are grateful to Sarah Lloyd and Moira Taylor at Routledge for guiding the publication of this edition and ensuring that delivery of the manuscript fitted their time schedule. Many colleagues willingly provided photographs and data; thanks are thus due to Howard Oliver, Ross Reynolds, Derek Elsom, Liz Rollin, and Abdalaziz Alswilem. Dave Maddison and John Owen of the Geography Department, University of Sheffield, processed many of the negatives to the correct size and tone. We must particularly thank Paul Coles for his work in redrawing so many diagrams that he must have thought they were never going to end. Our greatest debt is to our families for their tolerance and support throughout the long gestation period of the second edition.

Peter Smithson would like to thank W.H. Freeman & Co., New York, for permission to reproduce Figures 2.7 and 2.12 originally from *Understanding the Atmospheric Environment* by Neiburger, Edinger and Bonner; the University of Chicago Press for Figures 2.15, 5.2, 5.7 and 7.27 originally from *Physical Climatology* by Sellers and for Figure 28.8 originally from *Deforestation in the Postwar Philippines* by Kummer; the Royal Meteorological Society for permission to reproduce Figures 8.19, 9.13, 10.12, 11.1 and 28.2 from *Quarterly Journal, Weather*, the *International Journal of Climatology* and *The Global Circulation of the Atmosphere*; the American Meteorological Society for Figure 10.3, reproduced from 'Temporal march of the Chicago heat island' by Ackerman in the *Journal of Climate and Applied Meteorology*, vol. 24, no. 6, 1985, pp. 547–54; the World Meteorological Organization for Figures 11.1 and 11.4 reproduced from Intergovernmental Panel on Climate Change; *Climate Change* by Houghton, Jenkins and Ephraums; *Nature*, vol. 339, 1989, pp. 655–6, Table 1, Valuation of a Amazonian rain forest by C.M. Peters *et al.*; Nick Middleton for Table 27.1 originally from *The Global Casino*, Edward Arnold, and for Tables 27.2 and 27.3 from *Desertification: exploding the myth*, Wiley; Crown copyright governs Figure 7.6 and Figure 9.9 from vols 121 and 114 of the *Meteorological Magazine*, reproduced by permission of the Controller of Her Majesty's Stationery Office. Thanks are also due to Professor Dennis Hartmann for permission to reproduce Figure 2.11a from *Global Physical Climatology* (p. 36), published by Academic Press Ltd.

Kenneth Addison would like to thank the University of Wolverhampton for study leave which supported this project and to the following Oxford University departmental staff for their technical advice and assistance: Martin Barfoot (Photographic Unit, School of Geography), Chris Harris (Teaching Laboratory Technician, Department of Earth Sciences) and Peter Hayward (Cartographer, School of Geography). David Drewry (NERC, Swindon), Mike Fullen (Wolverhampton) and Professor Andrew Goudie (Oxford) very kindly provided a wide range of their photographs from which a selection appear in the book. John Turner (Head of Remote Sensing, British Antarctic Survey), Eugene O'Connor (British Geological Survey), Françoise Hivernel (Librarian, University of Cambridge Committee for Aerial Photography), Ken Gardner (Landform Slides, Lowestoft) and Jason Cowan and David Morgan (Royal Observatory, Edinburgh) were particularly helpful with advice leading to the selection of telescope, satellite and airborne imagery. He would also especially like to thank his wife, Lyn, for all her support and constructive agitation in completing the project and, with Charles and Rosie, for help and companionship in the field.

Ken Atkinson expresses his thanks and appreciation to Bob Eyre, Mike Kirkby and Richard Smith – past and present colleagues at Leeds University – who have been a constant source of ideas and encouragement in an understanding of soils and vegetation, and several of whose ideas doubtless appear unacknowledged in this volume. He would also like to acknowledge a debt of gratitude to his typists, Catherine McKinna and Michelle Byrne.

In addition we acknowledge with thanks the use or adaptation of material from the figures and plates listed below, with every effort being made to trace the holders of individual copyright for permission to reproduce material. Where no reply has been received we have assumed that there was no objection to the use of the material and we apologise for any unwitting omissions.

Academic Press, 12.18, 16.6, 17.3, 17.5, 17.7, 17.11; Addison Wesley Longman, 16.4; American Geophysical Union, 3.11, 14.2, 14.20; Association of American Geographers, 13.7; Basil Blackwell Publisher Ltd, 17.12; Blackie and Son Ltd, 25.4; Blackwell Science Ltd, 3.3, 3.8, 4.1; Blackwell Scientific Publications, 16.2, 16.5, 17.10; British Geological Survey, 12.7; Cambridge University Press, 3.14, 3.15, 3.16, 12.4, 15.6, 15.7, 15.18; Chapman and Hall, 3.16, 12.5, 17.6, 25.2, 25.3; Clarendon Press, 14.14, 14.16, 15.13, 25.7; Edward Arnold, 4.10, 4.12, 14.15, 14.18, 15.3, 15.4, 16.1; Gebrüder Borntraeger, Berlin, 13.19; Geological Institute of the University of Uppsala, 12.11; George Allen & Unwin (Publishers) Ltd, 3.15, 14.21; Harper and Row, 14.11, 16.3; HMSO, 17.20; Houston Geological Society, 17.9; Institute of Mining and Metallurgy, 13.17; International Association of Engineering Geology, 13.10; John Wiley and Sons Ltd, 3.2, 4.2, 14.6, 25.5; John Wiley and Sons, Inc., 12.8, 13.1, 13.2; Longman Group Ltd, 13.3; Macmillan Publishing Company, 4.8, 4.10; Manson Publishing Ltd, 4.15; McGraw-Hill Book Company (UK) Ltd, 14.4, 14.27; Methuen, 25.8, 25.12, 25.14, 25.15; Methuen and Co. Ltd, 14.7; Oliver and Boyd, 17.16, 17.17, 17.18; Oxford University Press, 4.3, 12.20, 13.9, 13.14; Pergamon Press, 14.5; Prentice Hall, Inc. 12.10, 17.22; Keith B. Ronnholm, Plate 12.1; Routledge, 17.21, 25.9, 25.11, 25.13; W.H. Freeman and Co., 25.1; Seismological Society of America, 3.18; The Geographical Association, 15.8, 15.14; The Geological Society of London, 14.25, 25.16; The University of Chicago, 17.15; The University of Washington, 17.19; UCL Press, 3.18; Unwin Paperbacks, 12.6; Zed Books, 25.6.

In the Peak District of Derbyshire, just a few kilometres to the west of Sheffield, there is a small river valley. It is not a spectacular valley like that of the Rhine, the Indus or the Zambezi. It is like thousands of others throughout the world. Grindsbrook, which rises on the flanks of Kinder Scout, tumbles through the gritstone boulders of the Pennines before flowing through the village of Edale and on to join the river Noe. From a few hundred metres north of the village it is possible to stand beside the stream and look up the valley towards the craggy edge of Hartshorn (Plate 1.1). What we can see is a small segment of a most unexceptional Peak District valley.

It is a scene which can tell us a great deal, and it contains the key to understanding the physical world around us. If we could start to understand this valley – if we could begin to appreciate how the individual features which we see are linked together, how they function – then we would have opened a door on our physical environment. For this small valley, like almost any other we might have chosen, demonstrates many of the fundamental processes and factors which make up the environment. If we can in some way recognize and comprehend these things, we would have a basis for answering a vital question: how does our natural world work?

Let us look a little more closely at the view. The river, we can see, splashes over the rocks and rapids which line its channel. It is bounded in front of us by a fairly flat area – the flood plain of the river. On the right, the valley side rises to the bracken-covered flanks of the hills; on the left, there is a small scar in the hillside and, above that, a patch of woodland. Here and there, boulders and outcrops of bedrock protrude through the vegetation and we can see the pale line of the Pennine Way footpath where it follows the valley towards Kinder Scout. It is a scene typical of countless parts of upland Britain. How can we start to learn from it?

Process in Grindsbrook valley

What becomes clear if we inspect this scene is that a great deal is happening. Some of the events are obvious. The air is constantly moving, bending the grass, loosening the leaves from the trees, sweeping clouds across the sky. At times rain or snow falls, and storms unleash their fury on the land. All the time, too, water is flowing though the stream channel, carrying with it logs and twigs which have been washed from the banks and woods, pebbles and boulders which have slid or rolled down the valley side. The boulders can be seen in periods of flood bouncing and shuffling along the channel floor; they come to rest in shoals and bars at the margins of, or within, the channel, and are moved on again as the river rises. Sheep are moving around on the flood plain, grazing the vegetation. Occasionally a hiker walks along the path.

Other events are less apparent. Beneath the soil, the rocks are being weathered and broken down. On the faces of the cliffs, ice freezes in the crevices and loosens small fragments of rock. Plant roots extend

Plate 1.1 *Grindsbrook valley, Edale, Derbyshire: an example of an environmental system.*
Photo: David Briggs.

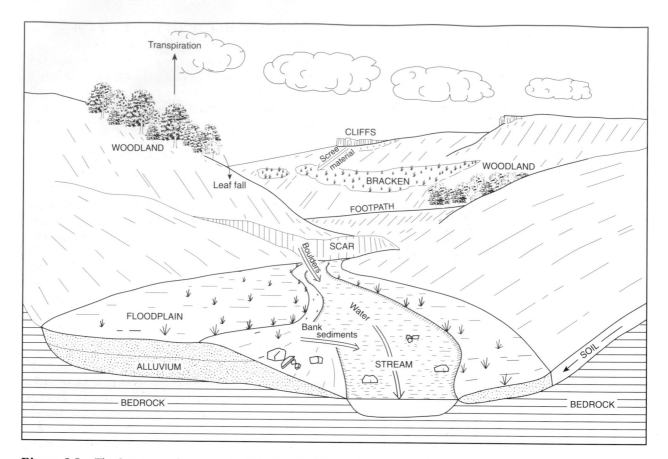

Figure 1.1 *The features and processes in Grindsbrook valley make up an environmental system.*

into larger fissures and open up the cracks. The soil itself is moving, pushed and prodded downslope by the development of ice crystals during the winter, carried by the wash of water over and through the soil, splashed by the impact of raindrops. Water and dissolved substances are being drawn from the soil by the plants; the water is being released to the atmosphere from the leaves. Slowly the vegetation is changing, advancing in some areas, retreating in others. Unseen, energy is being passed through the atmosphere to the land and vegetation, and returned to the air and space.

A model of the Grindsbrook valley

These processes are the lifeblood of the valley. They are the means by which the valley is being shaped and is developing. They show the pathways by which material and energy is being moved through the valley. They thus provide us with a way of understanding the scene. We can start to build up a picture of the way the valley is functioning; we can start to account for the features we see in the valley at the present time and even begin to predict some of the changes which may occur in the future.

We might go even further, for we could try to represent this picture in diagrammatic form. We might use the technique employed in motor car manuals, for example, to show the workings of the valley as we have interpreted it. Just as the car designer explains the position and operation of all the components of the engine by drawing exploded diagrams of the vehicle, so we can construct a similar picture of the Grindsbrook valley. We can show the various features of the valley, use arrows and lines to indicate the main processes and movements affecting them, annotate the diagram to show what is happening (Figure 1.1). What we end up with is a picture – a model – of the valley.

This model, however, is a little limited at present. It helps us to see what is happening within the Grindsbrook valley, but it does not tell us much about the rest of the world. The reason is that the model is at present too specific; it retains too much of the detail of this particular valley. What we need is a more general, a more abstract, model. How can we produce it?

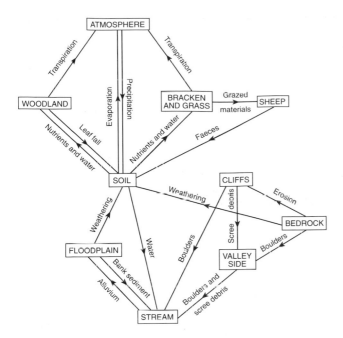

Figure 1.2 *A preliminary model of Grindsbrook valley, showing the general relationships between the main features of the environment*

From Grindsbrook valley to the world!

The first thing we must do is to omit some of the unnecessary detail. After all, it matters little when we are trying to understand the broad aspects of our environment that the river in this valley twists in this way here, or that the slope has a little undulation in it there. We can dispose of much of the local detail and concentrate instead on the most important features of the area. We can show the clouds, the stream, the flood plain, the hillsides, the woods and moorland – in fact all the features that we feel characterize the valley. We can represent all these features as boxes and we can show the processes operating within and between them by arrows (Figure 1.2). As a consequence, we obtain a more abstract, yet simpler and clearer, picture of the valley.

We now have a model which is less specific to the particular view before us. Nevertheless, we still cannot use it to understand the world as a whole. For not every valley or area we may be interested in contains the features we have recognized here. To progress further, to get to the basic structure and character of the scene, we need to take a rather different approach. We need to ask two even more fundamental questions. What does this valley (or indeed any other part of the natural world) consist of? And what is it that links these components together?

The atmosphere, the water, the rocks and land-

forms, the soil, animals and vegetation – these are the basic building blocks of our environment. The mortar between them is provided by the flow of energy and matter; the movement of heat and other forms of energy from the atmosphere to the ground, through the soil, vegetation and landscape and ultimately back to space; the movement of water from the atmosphere to the oceans and back again; the movement of rock debris from the rocks through the landscape; the flow of nutrients from the soil to the vegetation and back to the soil. If we describe the Grindsbrook valley in this way, as we have done in Figure 1.3, we indeed have a model which is relevant not just to this area but to anywhere in the world.

We have made some real progress, therefore. We have discovered what we set out to find: a key to understanding our environment. And we have done more than that, for we have recognized and described a system.

We have described a system. Let us define what we mean by this. A system is simply a set of objects and the relationships between them. Almost anything, at any scale, can be seen as a system, from a small droplet of water on a table top to the whole universe. But the concept of a system gives us a way of looking at those objects. It is an approach which helps us to focus on the way the objects interact.

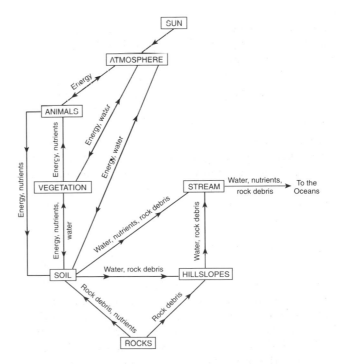

Figure 1.3 *A general model of Grindsbrook valley, showing the major flows of energy, water, rock debris and nutrients.*

This is important because the natural world is highly interactive, and if we are to understand – or, even more critically, to manage – the world effectively we need to be aware of these interactions.

The other important step we have taken is to represent this system in terms of a model. What we have done is to simplify the reality into a more manageable form. The need for this is all too apparent. The world is an extremely complex place, and we cannot hope to comprehend all the details of its complexity. We need, instead, to produce models which reflect the most important components and linkages of the system, but allow us to ignore (at least for a while) the less important details. Subsequently we may wish to extend our models and bring them closer to reality, though there is always a danger that they may become too complex and defeat the purpose of simplification.

There are, in fact, many different sorts of models which we can construct: **hardware models** in which we actually build a physical likeness of the system we are concerned with; or **mathematical models** in which we represent all the components of the system by mathematical symbols and the relationships

between them by equations (Figure 1.4); and more general, **conceptual models** like the one we have built of the Grindsbrook valley.

A systems approach

The structure of systems

Our example of a system in Figure 1.1 leaves a lot to be desired. It is not, in truth, a very clearly defined system; its boundaries are determined simply by the limits of our view. In most cases we are dealing with rather more distinct systems, bounded by more easily definable features. We may, for example, take a whole river valley, the edges of which are marked by the line of highest ground which encircles the valley, and treat it as a system. Or we may take an area covered by a particular vegetation type, a field or a farm, a glacier or a cloud.

Although the boundaries of these systems may appear obvious, they are rarely impermeable. The systems are not self-contained. Instead, matter and energy flow into and out of the system, across its boundaries as **inputs** and **outputs** (Figure 1.5). We can see this in our example: inputs of water, sediment and other materials enter the system from upstream and leave it downstream; energy enters from the sun and is lost by reradiation back to the atmosphere and space. This flow of inputs and outputs is typical of all natural systems (except, perhaps, at the scale of the whole earth or universe). Such systems are called *open systems*. Strictly Earth is not a fully open system, as there are virtually no inputs or outputs of matter, only energy.

Sometimes that is all we know about a system – its boundaries and the inputs and outputs. We can detect a relationship between the inputs and outputs, but we do not understand what goes on inside. In that case we are looking at the system as a **black box** (Figure 1.6). As our knowledge progresses, however, we may be able to discover what lies inside the system. What we find is a series of smaller systems, or **subsystems**, each linked by a series of flows of energy and matter. We then have a view of the system as a **grey box**. But delve a little deeper and we may be able to see the whole internal working of the system: its individual **components**, the pathways by which the energy and matter flow between them, the **storages** where the energy and matter may be held for certain periods of time. We now have a view of the system as a **white box** (Figure 1.6).

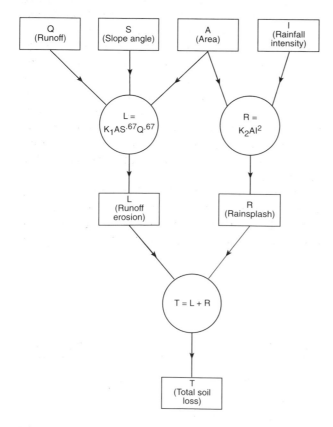

Figure 1.4 *A mathematical model: soil erosion by water. K_1 and K_2 are different coefficients in the equations. Source: Based on Meyer and Wischmeier (1969).*

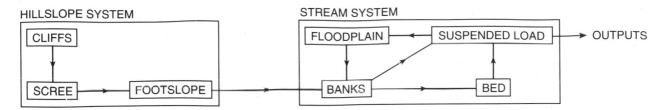

Figure 1.5 *The relationship between two connected systems. Sediment moves through the hillslope system and into the stream system; there it is transferred through the banks, bed, suspended load and flood plain deposits before being lost as outputs. Many examples fit this general structure.*

Types of system

A system, therefore, is a set of components linked by flows of energy and matter. But what does that mean? If we take our example in Plate 1.1 we could, for example, measure all the main features and attributes of the valley and look at the relationships between them. We might see how the size of the trees relates to the soil depth, how soil depth relates to slope angle, how slope angle relates to rock structure, and so on (Figure 1.7). Is that a system?

The answer is yes. It is what is referred to as a **morphological system**. In looking at the system in this way we are concerned not with the dynamics of the interactions and flows, but merely with their morphological expression. Another way of looking at

the system, however, could be to focus attention on the flows of matter and energy through the valley. We might, for example, represent the movement of the sediment through the system as in Figure 1.8. In this case we have described the valley as a **cascading system**: one in which there is a cascade of energy and matter through the environment from one component to another. It is a particularly useful way of dealing with systems and we will use it to look at the flow of solar energy through the atmosphere (Chapter 5) and the cascade of sediment through the landscape (Chapters 13 and 14).

There are other ways of looking at environmental systems. We can combine morphological and cascading systems to define what are called **process–response systems**. Figure 1.9 depicts part of the Grindsbrook valley in these terms. As it shows, the morphology of the system is related to the flows of energy and matter. In other words, the form of the system is a function of the processes operating within it. This is a vital concept and one we shall use repeatedly throughout this book, for it helps us to see the ways in which the environment develops and is maintained.

Finally, we can also define what are referred to as **ecosystems**. These are concerned with the biological relationships within the environment – the interaction between the plants, animals and their physical

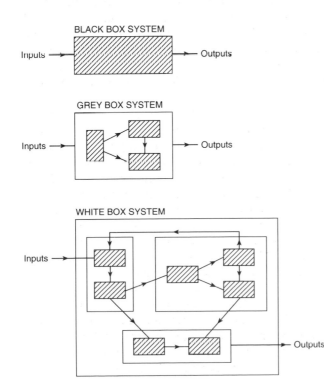

Figure 1.6 *Types of system.*

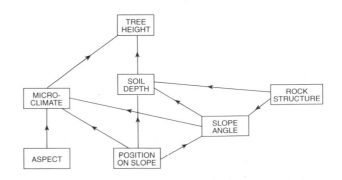

Figure 1.7 *Factors affecting the rate of tree growth in Grindsbrook valley: an example of a morphological system.*

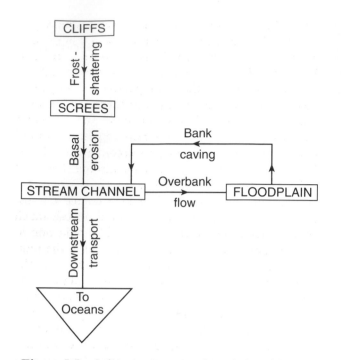

Figure 1.8 *Sediment movement through Grindsbrook valley: an example of a cascading system.*

surroundings. It is an approach we shall adopt in a later part of the book when we look at soil–vegetation–animal relationships.

System change and system stability

The environment is dynamic. It is constantly changing. What causes this change? The immediate answer is energy. As we shall see in Chapter 2, Earth obtains most of its energy from the sun (as **radiant energy**). The energy within the environment tends to be unevenly distributed, however, and so it tends to flow through the environment, in an attempt to achieve a more equitable distribution. In the process it is itself changed but it also carries out work. It alters the environment.

If we were to monitor an environmental system such as that which we have identified in the Grindsbrook valley, we would be able to detect some of these changes. We might see, for example, that the stream changed its course, that the woodland extended as seedlings grew amid the surrounding grassland, that the weather changed from day to day and from season to season. Much of this change would often appear haphazard and random. Changes in one direction at one time might be reversed at the next. The reason is that, although the system is active, it is also in a state of **equilibrium**; over time

it tends to maintain its general structure and character in sympathy with the processes acting upon it. Thus although the channel position varies, on average it follows the same route down the valley. Although the extent of the woodland fluctuates, its mean area and position are constant. Though the weather varies, the long-term average climate is more or less consistent. Systems such as this are said to be in a condition of **steady state equilibrium** (Figure 1.10).

Not all systems behave like this, however. Some are subject to much more marked and often irreversible changes. Such changes may be **triggered** off by certain events which knock the system out of equilibrium. Hill slopes provide an excellent example. Within certain limits they remain stable and their

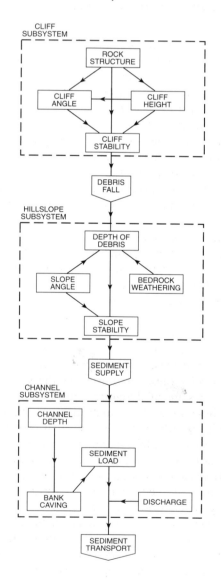

Figure 1.9 *The cliff–hillslope–channel system of Grindsbrook valley: an example of a process–response system.*

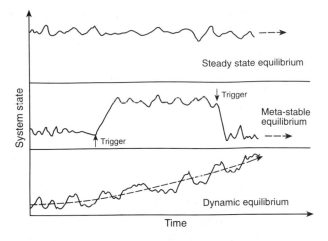

Figure 1.10 *Types of equilibrium.*

form does not change. But if those limits are exceeded – if the ground becomes too wet or the pressures exerted on the slope are too great – slope failure may occur and a new slope form is created. The same is true of many atmospheric systems. The air near the ground may be stable within limits, but if the system is disturbed (for example, if the air is overheated), so that it passes those limits, it may become unstable; the air may start to rise until it reaches a new equilibrium position. Systems of this type are said to be in a condition of **metastable equilibrium**.

In other cases, environmental systems may change more gradually and progressively over time. Short-term, random fluctuations may still be detectable due to the effects of minor variations in environmental conditions. But in the longer term a distinct and consistent trend may be visible. This is characteristic of climatic change. As we shall see in Chapter 11, the present climate shows signs of warming, probably because of increases in carbon dioxide in the atmosphere. Under these circumstances the system is said to be in a state of **dynamic equilibrium**.

All these types of equilibrium are illustrated in Figure 1.10. Although they vary in detail they all reflect a fundamental principle: that the form of any system adjusts towards a state of equilibrium with the processes and conditions to which it is subject. However, these examples also illustrate a further property of environmental systems – the operation within them of feedback processes. **Feedback** refers to the ability of a system to modify earlier links in the chain of interrelationships so that an initial change in the system is either amplified (**positive feedback**) or damped down (**negative feedback**).

We can see how these feedback processes operate with an example. Suppose the Peak District was affected by a prolonged drought (or even a more permanent change in climate) which resulted in the area receiving much less rainfall than it does at present. What would be the effect on the Grindsbrook valley?

In the short term the consequences are fairly obvious. Less rainfall would mean that there was less water available for run-off. Less water would therefore enter the stream channel, and stream discharge would decline. At the same time, less sediment would be brought into the stream by water flowing down the hillside, so the sediment load of the stream would fall. In addition, the stream itself would be unable to transport its load so effectively, and would deposit much of its material on the channel bed. The reduction in the amount of water in the channel and in the sediment load of the stream would, in turn, diminish the erosive capacity of the stream, which would then be unable to erode its banks. That, too, would reduce the amount of sediment entering, or passing down, the channel. Thus the initial decline in the amount of rainfall would have triggered off a cycle of positive feedback which resulted in progressively less sediment being transported by the stream (Figure 1.11).

But the story does not end there, for in the longer run other changes would be taking place. The increased aridity might lead to changes in the vegetation cover, for the trees and heather would start to experience moisture stress. In time, much of the vegetation might wilt and die (as has happened in England during the summer droughts of the 1980s and 1990s). That would reduce the amount of evapotranspiration and would leave areas of bare soil. The rain which did fall, therefore, would not be trapped or taken up by the vegetation, but would be free to run-off across the unprotected surface. That would ultimately help to increase the quantity of sediment being washed into the stream. Moreover, because there would be less vegetation to intercept or slow the water down after rainfall, much of the water would reach the stream channel relatively swiftly, so that the stream would respond more quickly to storm events. At the peak of its flow, it would be able to carry large amounts of material and would have the capacity once more to erode its bed and banks. Thus a longer-term cycle is operating to restabilize the system; to halt the progressive reduction in the rate of sediment movement. This is a process of negative feedback (Figure 1.11).

This example is simple but it illustrates a number of important principles which are worth emphasizing. First, environmental systems are dynamic and intimately interrelated. Changes in one part of the system work through to affect other parts. Second, change tends to be exacerbated by processes of positive

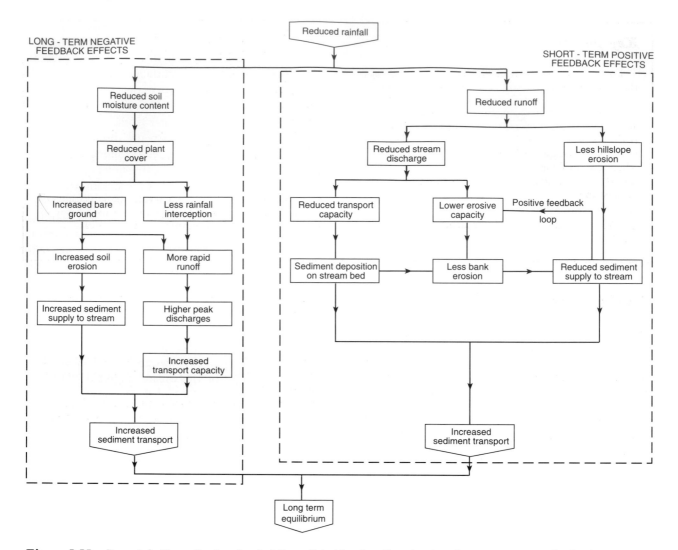

Figure 1.11   *Potential effects of reduced rainfall on Grindsbrook valley, showing the operation of feedback relationships.*

feedback, inhibited by negative feedback, as they change in response to some external impulse then restabilize (this is characteristic of systems which are in a state of metastable equilibrium).

In addition, we may draw two further implications about change and stability in environmental systems. The first is that because conditions change, and because the systems adjust to that change, many of the features we see in the environment at any moment may, at least in part, be inherited from previous times when conditions and processes were different. We can certainly see this in the case of the Grindsbrook valley. Several of the features there are due not to any present-day processes but to the effects of glacial conditions thousands of years ago. The glaciers have gone, but the features do not disappear immediately. They remain and are slowly modified by new processes. Their present form is thus a product of

several superimposed periods of development during which different processes have been at work.

The final point is that, because of their complex interrelationships and their potential for change, environmental systems are difficult to manage. All too easily people may trigger changes in those systems, more far-reaching and destructive perhaps than they intend, simply because they do not appreciate all the ramifications of their actions. For example, the construction of a road across the natural slope can alter the drainage over a large area even if culverts are built. We shall see many examples in the rest of this book. In almost every case the lesson will be the same. We need to understand how environmental systems work, how they change, how they respond to external conditions, if we are to manage them effectively. For that reason if no other a systems approach to the physical environment is useful.

Key points

1 Environmental processes are the means by which the physical landscape is being shaped and is developing. We can depict these processes and their relationships to form a model of the system. There are different levels of understanding of models. There is the black box model where only the inputs and outputs of energy are known, a grey box model where parts of the internal subsystems are known, and the ideal white box model where all elements are understood.

2 There are also different approaches to systems. We can look at the morphological expression of the system in the landscape rather than the dynamic processes. We may consider the flow of energy and matter through the system as a cascade. The two approaches can also be combined to give a process–response system. Most systems will change through time. A change in a single element may trigger changes throughout the system. Whether the system will return to its former state will depend upon its level of equilibrium. In some systems, drastic changes could occur, emphasized by positive feedback.

3 Despite the difficulties, portraying the operation of environmental processes as a system is a useful way of understanding the complexities of the physical environment.

Further reading

Chorley, R.J. and Kennedy, B.A. (1971), *Physical Geography: A systems approach*. London: Prentice-Hall. An advanced and in some ways difficult text, but useful for its methodical treatment of the theory of systems, and for the examples of systems in physical geography.

Bennett, R.J. and Chorley, R.J. (1978), *Environmental Systems: Philosophy, analysis and control*. London: Methuen. Attempts to explore to what extent systems theory provides an interdisciplinary focus for those concerned with environmental matters. Numerous examples provided.

White, I.D., Mottershead, D.N. and Harrison, S.J. (1992), *Environmental systems* (second edition). London: Chapman and Hall. A comprehensive text examining all aspects of systems, then applying them to a wide range of geographical environments, from the atmosphere to the ground surface and the plant kingdom.

CHAPTER 2
Energy and Earth

Earth is one of the smaller of the nine planets forming the solar system. Each planet is distinct in terms of its physical geography. The different distances of the planets from the sun, their different sizes and composition ensure that each world is unique. They all depend for virtually all their energy upon the nearest star, which we call the Sun. A basic understanding of what energy is, how it moves and how it can be transformed is required. In this chapter we examine the nature of energy and how it is emitted by the sun. Variations in sun–Earth relationships create changes in the pattern and distribution of energy at the top of our atmosphere. Finally we study the mechanisms that carry this energy to all parts of Earth's global system. No further reference will be made to the other planets but similar physical laws operate there.

The nature of inputs and outputs of energy

Imagine Earth from 300,000 km into space. An isolated sphere; predominantly blue, patched with brown and green and wreathed in white. A world of water, dotted with land, partly clothed in swirling cloud. This is a view of the global system. Into this system pours the input of solar energy; from it comes reflected energy and reradiated energy, which are its outputs. From our privileged vantage point we could measure the inputs to Earth and, with suitable equipment, monitor the global outputs. We could therefore draw up a simple model of the globe as an energy system (Figure 2.1) showing the inputs and outputs, but that would give us no idea of what happens inside. It would be a picture of the globe as a **black box system**. It would be the simplest view of the system we could obtain, but it would tell us nothing about the internal components or subsystems, or about the relationships between them. We would see only what enters and leaves the globe.

Let us start by looking at those energy flows we *can* examine. Without doubt, the main input of energy to the global system comes from the sun. Compared with the solar contribution, all other inputs are almost negligible. The gravitational effect of the moon and the sun, and reflected and radiant energy from the moon provide some energy, with even smaller inputs from the impact of objects such as meteorites and comets. Many of these extraterrestrial objects burn up on entering the atmosphere and do not reach the surface, although they do act as a minute input of energy to the atmosphere.

Just as the input of energy is dominated by radiant energy from the sun, so the output of energy from Earth is almost entirely radiant energy, although this time with somewhat different properties. Much of the energy has been radiated or emitted by Earth and its atmosphere, but some is solar radiation

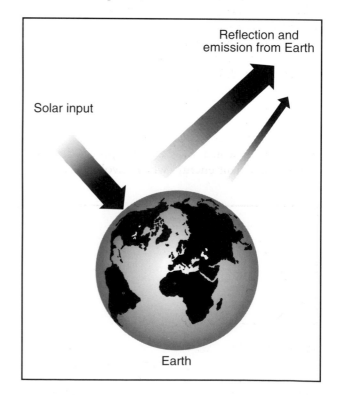

Reflection and emission from Earth

Solar input

Earth

Figure 2.1 *A black box model of Earth's energy system.*

reflected from clouds or from Earth's surface without any major modification. As the overall energy level of Earth is not changing, we can assume that there must be a balance between the energy input to and energy output from the globe as a whole.

Concepts of energy

Before discussing the quantities of energy received by Earth, it is necessary to consider, briefly, the nature of energy and the ways we measure it.

Energy exists in a variety of forms. We are familiar with electrical energy in the home and increasingly with nuclear (or atomic) energy. Neither of these has any great significance with regard to environmental processes. More important as far as environmental processes are concerned are radiant, thermal, kinetic, chemical and potential energy.

Radiant energy is the most relevant to our discussion here, for it is in this form that the sun's energy is transmitted to Earth. The heat from the sun excites or disturbs electrical and magnetic fields, setting up a wave-like activity in space, known as electromagnetic radiation. The length of the waves – that is, their distance apart (Figure 2.2) – varies considerably, so that solar radiation comprises a wide range of electromagnetic wavelengths from 0.2 to 5.0 μm (Figure 2.3). Only a few are visible to the human eye, reaching us as light, but all transmit energy from the sun to Earth. Assuming a mean distance from the sun to Earth of about 15×10^7 km, it takes this energy about 8.3 minutes to reach us.

When this solar radiation reaches Earth's surface, it is converted from radiant to other forms of energy. Much is altered to **thermal (heat) energy**. It warms Earth's surface and the atmosphere by exciting the molecules of which they are composed. In simple terms, the radiant energy (which involves disturbance

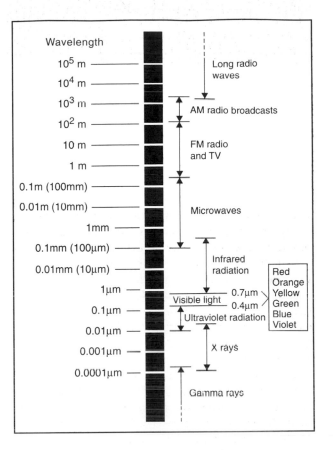

Figure 2.3 *The electromagnetic spectrum.*

of magnetic and electric fields) is transmitted into the molecules making up the ground and atmosphere, with a resulting change in the type of energy.

Thermal energy can therefore be considered as energy involved in the motion of extremely small components of matter. The energy of motion is referred to as **kinetic energy** (and thus thermal energy is sometimes described as the kinetic energy of molecules). Any moving object possesses kinetic energy, and it is through the use of this energy that, for example, a stone thrown into a lake can disturb the water and produce waves. It is also through the exploitation of kinetic energy that turbines and engines are able to produce heat, light and so on (Plate 2.1).

Chemical energy represents a form of electrical energy bound up within the chemical structure of any substance. It is released in the form of thermal or kinetic energy when the substance breaks down. Coal, when it is burnt, releases heat. Food, when it is digested, provides the body with heat and movement.

Potential energy is related to gravity. Because of the apparent pull that Earth exerts upon objects

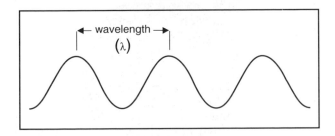

Figure 2.2 *Electromagnetic radiation: the distance from one crest to the next crest or from one trough to the next trough is known as the wavelength (λ). It is an important indicator of the properties of the electromagnetic radiation.*

Plate 2.1 *Power generation by wind on Madeira. The trade winds provide a steady airflow at this high-altitude site on the island.*
Photo: Peter Smithson.

within its gravitational field, material is drawn towards Earth's centre. Thus objects lying at greater distances from its centre (for example, rocks on a hillside, water at the top of a waterfall or the air near a mountain summit) possess more potential energy. This energy is converted to kinetic energy when the rock, the water or the air descends to lower levels; some energy is converted to heat through friction.

Thermal, kinetic, chemical and potential energy are important to Earth's system but operate internally and so cannot be observed directly from space. To understand the results of these different flows of energy, we must look more closely at them, concentrating on the forms of energy that have significance for the physical geography of Earth.

Transfers of energy

The types of energy we have considered so far do not have a uniform distribution over the globe. Both earth and the air experience major inequalities in energy receipts and emissions. As a result of these

differences, spatial transfers of energy take place, for energy is redistributed to minimize the inequalities, or to maintain (or to achieve) an equilibrium.

To understand how energy is transferred we need to consider a little further the principles of energy transformation and modification. We have seen already that energy can exist in a number of forms, and as a general principle energy will be transferred from areas of high energy status to areas of lower energy status in an attempt to eradicate the differences. Thus energy differences expressed by the level of temperature in two bodies, such as the air and the soil, tend to be reduced over time as heat is transferred from the hotter to the cooler body. In this way the soil is heated during the day when the air is warm and loses heat energy back to the air at night when the atmosphere is cool (Figure 2.4).

In the case of thermal energy, three main methods of transfer can be identified: radiation, convection and conduction. **Radiation** is the process by which energy is transmitted through space, mainly in the form of electromagnetic waves. **Convection** involves the physical movement of substances containing heat,

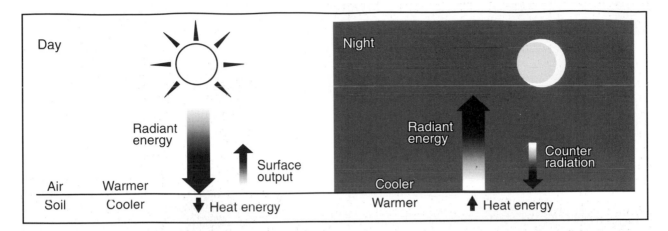

Figure 2.4 *Energy exchange at the soil surface. The arrows are not to scale. More energy is lost by day, when the surface is warmer, than at night.*

such as water or air, and is not possible in a solid. **Conduction** is the transfer of heat through a medium from molecule to molecule.

These three processes of transfer are often closely related. Thus energy may be conducted through the soil to the surface and then radiated or convected into the atmosphere. Similarly, in the air, convection currents may raise warm air masses to higher levels and then conduction to the surrounding cooler atmosphere may occur (see Chapter 5), while condensation of water vapour releases latent heat. Convection is very important as an energy transfer mechanism because it transfers energy in two forms. The first is the **sensible heat** content of the air, which is transferred directly by the rising and mixing of warmed air. It can also be transferred by conduction. The other form of energy transfer by convection is less obvious, as there is no temperature change involved, hence its name, **latent heat**. The evaporation of water into vapour or the melting of ice into water involves a supply of heat to allow the change to take place. When the reverse process operates, from vapour to liquid, or liquid to ice, this heat is released. We shall return to these mechanisms in more detail in Chapters 5 and 6.

Transfers also occur between other forms of energy. If two objects with different kinetic energies are brought together, a transfer takes place between the two which tends to equalize the energy levels. For example, a rapidly flowing stream (high kinetic energy) that comes into contact with a static boulder (no kinetic energy) tends to push the boulder into motion. In doing so, the stream loses energy by friction but imparts some of this energy to the boulder in the form of motion (kinetic energy). Similar principles apply to chemical energy.

One way of looking at these transfers of energy is to regard them as movements down an energy gradient. It is easier to see this principle in the example of heat energy, for we can all appreciate that heat moves from hotter to cooler areas. Heat the end of a metal bar in a fire and the heat will move along the metal until it burns your fingers! Heat energy in this case moves down the energy gradient in the bar. The same general processes operate with other forms of energy (Figure 2.5)

Energy transformations

During these transfers of energy it is clear that the nature of the energy often changes, although the total quantity of energy involved remains constant. Radiant energy heats the objects it meets; it is converted from radiant energy to heat energy. Kinetic energy may similarly be converted to heat energy; the friction of a moving body against another liberates heat, as we can demonstrate by filing or sawing a piece of wood or metal.

Under natural conditions the range of probable transformations is fairly limited. That is, the various forms of energy are normally able to be converted to all other forms, but follow relatively well defined pathways (Figure 2.6) towards the lowest level of energy – that of heat.

General patterns and principles of electromagnetic radiation

Solar energy is transmitted to Earth in the form of radiant energy. How does this energy reach us? Why

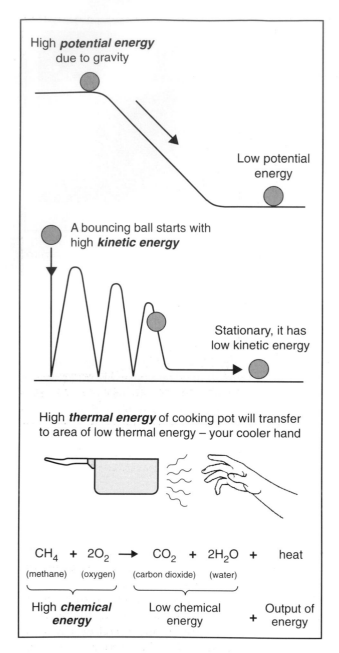

Figure 2.5 *Examples of energy gradients.*

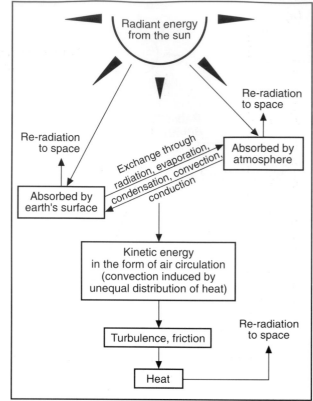

Figure 2.6 *Energy transformations in the atmosphere. Source: After Miller (1966).*

do we receive the amount we do? Why does the energy have the particular properties it has? To answer these questions we need to examine some of the principles of radiation.

Radiant energy consists of electromagnetic waves of varying length. Any object whose temperature is above **absolute zero** (0 K or −273°C) emits radiant energy. The intensity and the character of this radiation depend upon the temperature of the emitting object. As the temperature rises, the radiant energy increases in intensity, but its wavelength decreases; as the temperature falls the intensity decreases and the wavelength increases (Figure 2.7). In addition, the amount of radiation reaching any object is inversely proportional to the square of the distance from the source (Figure 2.8). This distance decay factor accounts for the difference in solar inputs to the various planets in our solar system.

To a certain extent radiation is able to penetrate matter, as, for example, X rays, which can pass through the human body, but most radiant energy is either absorbed or reflected by objects in its path. Absorption occurs when the electromagnetic waves penetrate but do not pass through the object; reflection involves the diversion or deflection of the waves from the surface of objects without any change of wavelength. The ability of an object to absorb or reflect radiant energy depends upon a number of factors, including the detailed physical structure of the material, its colour and surface roughness, the angle of the incident radiation and the wavelength of the radiant energy.

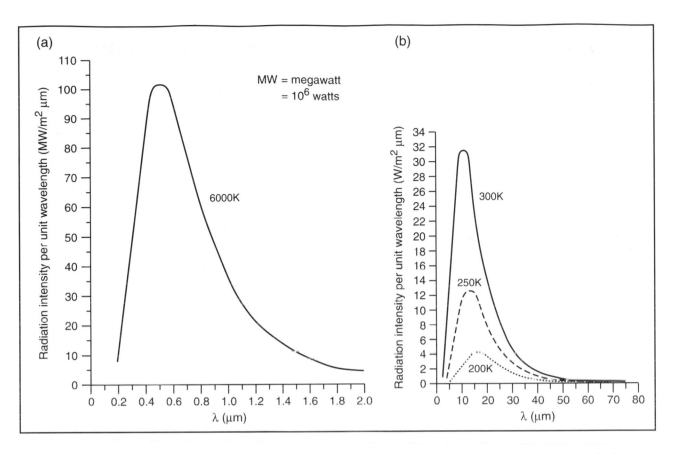

Figure 2.7 *Variation in the intensity of black-body radiation with wavelength. (a) T = 6000 K (approximately the emission temperature of the sun); (b) T = 200 K, 250 K and 300 K (range of Earth emission temperatures). Note the differences in scale.*
Source: After Neiburger et al. (1982).

An object that is able to absorb all the incoming radiation is referred to as a **black body**. Although it has conceptual value, a perfect black body does not exist in reality. All objects absorb a proportion of incoming energy and reflect the remainder. The amount of radiation reflected from a surface is called the **albedo**. The term is most frequently used for the visible part of the spectrum. It is calculated by dividing the amount reflected by the total amount arriving at a surface and is normally expressed as a percentage. The colour of the surface determines the amount reflected. Solar collection panels are matt black to ensure that the maximum amount of short-wave energy is absorbed and converted to heat. Differences also occur according to the wavelength of the energy. Thus snow and sand both absorb **long-wave radiation** (5–50 μm) quite efficiently, but they reflect relatively large proportions of **short-wave radiation** (0.4–0.8 μm). Indeed, under constant conditions, it is possible to define the wavelengths that specific materials selectively absorb, and this knowledge can

be used to characterize or identify materials through remote sensing. It is frequently used in astronomy to determine the gases present in stars.

Whereas solid substances usually absorb most wavelengths of radiation, gases tend to be very selective in their absorption and therefore emission wavelengths (Figure 2.9). This property is very important to the Earth, as it means that the atmosphere absorbs and emits only in certain wavelengths. At other wavelengths, radiation is able to pass right through the atmosphere with little modification. The atmosphere is composed of gas molecules, particles of matter such as dust, water droplets and ice crystals. Light waves striking these obstacles are scattered in all directions, so that radiant energy is scattered back to space as well as down to the surface. There is no change of wavelength in this process, known as **scattering**, simply a change of direction for some of the radiant energy.

Gas molecules are most effective at scattering light in the blue wavelength. Since gas molecules compose

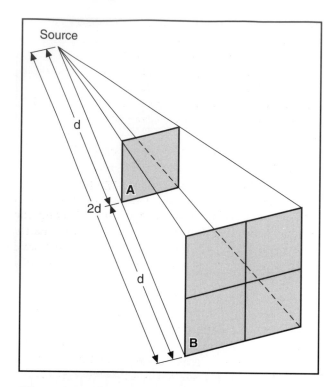

Figure 2.8   *The inverse square law. Intercept area at distance d from source is just one-fourth that at distance 2d; energy passing through area A is spread over an area four times as large at B.*

much of the atmosphere, we see the sky as blue whether we view it from the ground or from space. When the sun is setting or rising the radiant energy passes at a lower angle through the larger particles of dust in the lower atmosphere. The result is that more of the red wavelength is scattered, producing more colourful skies at sunrise and sunset.

Absorption of the radiant energy has more far-reaching consequences than reflection or scattering. As an object absorbs energy its temperature rises, because the radiant energy is converted to heat (thermal energy). Reradiation of this energy tends to occur at a temperature different from that of the initial, radiating object, and thus the radiation emitted is at a different wavelength. The Earth, for example, is considerably cooler than the sun; thus the energy it emits is characteristically of longer wavelengths than the initial solar inputs. We can summarize the radiation laws as follows:

1   All substances emit radiation when their temperature is above absolute zero (−273° C or 0 K).
2   Some substances absorb and emit radiation at certain wavelengths only. This is true mainly of gases.

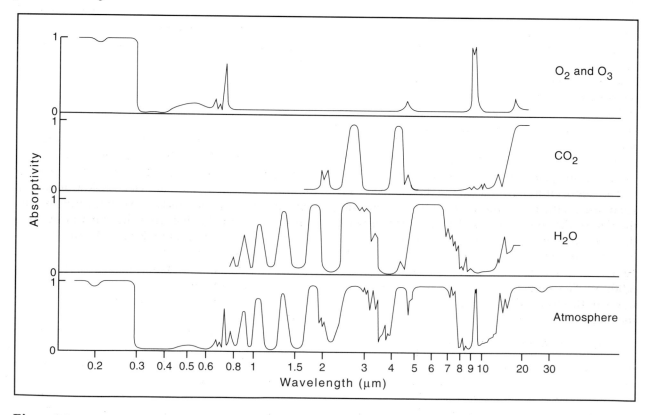

Figure 2.9   *Absorptivity at different wavelengths by selected constituents of the atmosphere and by the atmosphere as a whole. Source: After Fleagle and Businger (1963).*

3 If the substance is an ideal emitter (a black body) the amount of radiation given off is proportional to the fourth power of its absolute temperature. This is known as the Stefan–Boltzmann law and can be represented as $E = \sigma T^4$, where E equals the maximum rate of radiation emitted by each square centimetre of the surface of the object, σ is a constant (the Stefan–Boltzmann constant) with a value of 5.67×10^{-8} W m^{-2} K^{-4}, and T is the absolute temperature.

4 As substances get hotter, the wavelength at which radiation is emitted will become shorter (Figure 2.7). This is called Wien's displacement law, which can be represented as $\lambda_m = \alpha/T$, where λ_m is the wavelength at which the peak occurs in the spectrum, α is a constant with a value of 2898 if λ_m is expressed in micrometres, and T is the absolute temperature of the body.

5 The amount of radiation passing through a particular unit area is inversely proportional to the square of the distance of that area from the source $(1/d^2)$, as shown in Figure 2.8.

Solar radiation input

Now that the principles of radiation have been outlined, we can look at the details of solar radiation input to Earth in a more meaningful manner.

Because we know the mean distance of Earth from the sun we can work out, from law 5 above, how much radiation Earth should receive. This amount is the solar constant and has a value of about 1370 W m^{-2} at the top of the atmosphere. Recent work from satellites shows that the solar constant decreased about 0.1 per cent between 1978 and 1985, with an increase from 1985 to 1988. A 1 per cent increase would be adequate to cause an increase of 0.5° C in global temperature. By measuring how much radiation reaches the top of the atmosphere, and knowing the size of the sun, as well as Earth's mean distance, the emission temperature of the sun can be determined from law 3. For the photosphere, or visible light surface of the sun, this value works out to about 6,000 K. This figure enables us to determine at what wavelength most radiation will be emitted from the sun from law 4, that is,

$$\lambda_m = 2898/6000 = 0.48 \; \mu m$$

From Figure 2.3 we can see that this value is in the middle of the visible part of the spectrum. Note that it is the wavelength of blue light.

From the radiation laws it has been possible to determine how much radiation Earth ought to receive, as well as the amount and properties of solar radiation. Similar calculations can be made for Earth when we are considering outputs.

The input of energy to Earth at its mean distance from the sun is only an average value, for changes are taking place all the time. For example, Earth is rotating on its axis once in twenty-four hours, it is orbiting the sun about once in 365 days and, as its axis of rotation is at an angle of about 23.5° to the vertical, the distribution of radiation at the top of the atmosphere is constantly changing. Over even longer periods of time the nature of Earth's orbit and its angle of tilt also change, thus affecting the amount and distribution of radiation over Earth. These, however, are important only on a time scale of thousands of years and will be discussed more fully in Chapter 11.

The sun also emits energy in what is called corpuscular radiation (sometimes referred to as the **solar wind**), which is composed primarily of ionized particles and magnetic fields. There is a connection between variations in the strength of the solar wind and activity on the surface of the sun. This activity is most clearly seen in the form of sunspots and solar flares. The solar wind interacts with the magnetosphere, the magnetic field that surrounds Earth, and this interaction is visible as the **aurora borealis** in the northern hemisphere and as the **aurora australis** in the southern hemisphere.

Let us look in more detail here at the diurnal and seasonal effects.

Diurnal variation

As Earth rotates on its axis a different portion of the top of the atmosphere will be exposed to the incoming solar radiation (often abbreviated to **insolation**). At dawn the sun will be low in the sky and the amount of radiation passing through a unit area normal to the line from the sun will be spread over a large area (Figure 2.10). As the sun rises in the sky the surface area decreases, and so intensity increases. If our surface is eventually at right-angles to the solar beam it will receive the maximum intensity of radiation – the surface area is at its smallest. As well as the angle between the sun's rays and the top of the atmosphere and Earth's surface, the length of daylight will also affect the amount of radiation received. At the equator the day remains approximately twelve hours long throughout the year. At the poles it varies between zero and twenty-four hours, depending upon the time of year.

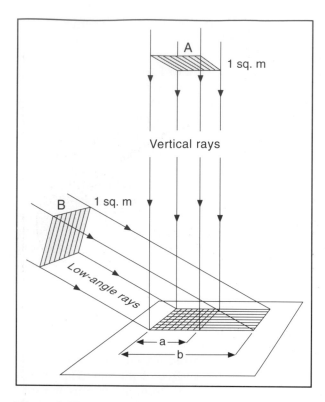

Figure 2.10 *Energy distribution on an intercepting surface depends upon the angle of the incoming energy rays. Energy distribution is more concentrated on a perpendicular surface (A) than on a surface at a lower angle (B).*

Seasonal variation

Seasonal variations in insolation arise from the changing axial tilt of Earth throughout the year (Figure 2.11) and the eccentricity of Earth's orbit. The orbit is an ellipse, not a circle, so that Earth is slightly nearer the sun (147 million km) on 4 January and at its farthest distance (152 million km) on 4 July. The variation in distance means that the amount of energy received also varies. The variation in energy received is ±3.5 per cent, which does make a measurable difference in total insolation received in the two hemispheres (Figure 2.12). Being nearer the sun means that the radiation input will be slightly higher. Earth is closest to the sun (perihelion) in the northern hemisphere winter and farthest away (aphelion) in the southern hemisphere winter at the present time. Because of changes in the shape of Earth's orbit, to be discussed in Chapter 11, these relationships are constantly changing.

As Earth orbits the sun with its axis of rotation pointing in a constant direction, the area that is illuminated by the sun and the angle between the sun's rays and the top of the atmosphere will change. At

the June solstice the sun is above the horizon throughout the twenty-four hours for all latitudes north of the Arctic Circle, while south of the Antarctic Circle the sun would not be visible. Between the autumn equinox (22 September) and the winter solstice (22 December) the latitude at which the midday sun is overhead gradually moves southward from the equator to the Tropic of Capricorn (23.5° S). By 22 December insolation will be at a maximum at that latitude and zero north of the Arctic Circle. Between 22 December and 21 March the sequence is reversed, and in the period leading up to the summer solstice the latitude of the overhead sun moves northward from the equator to the Tropic of Cancer, insolation increases in the northern hemisphere, and the South Pole is thrown progressively into shadow (Figure 2.13).

If you stand with your back to the North Pole, the **altitude** of the sun is the angle between the horizon and the sun at noon. A navigator uses a sextant to measure this angle. Latitude can be calculated with the following formula:

$$\text{Altitude} = 90° - \text{Latitude} \pm \text{Declination}$$

Declination is the latitude at which the sun's rays are vertical at noon. You add declination if you are in the same hemisphere as the sun, subtract if the sun is in the opposite hemisphere. For example, the altitude of the sun at solar noon in Hong Kong (latitude 22° N) would be:

22 December

$$\begin{aligned}\text{Altitude} &= 90° - 22° - 23.5° \\ &= 44.5°\end{aligned}$$

21 March

$$\begin{aligned}\text{Altitude} &= 90° - 22° - 0° \\ &= 68°\end{aligned}$$

21 June

$$\begin{aligned}\text{Altitude} &= 90° - 22° + 23.5° \\ &= 91.5°\end{aligned}$$

(which means the sun is slightly to the north of an overhead sun position).

The presence of Earth's atmosphere has a dramatic effect on the amount of radiant energy which reaches the surface, but we can illustrate the essentially astronomic controls of the input either at the top of the atmosphere or by assuming there was no atmosphere; the result is the same. It is clear from Figure 2.12 that, taking an annual figure, the tropics would receive the most radiation, as the input never falls to low values, unlike the situation at the poles, where

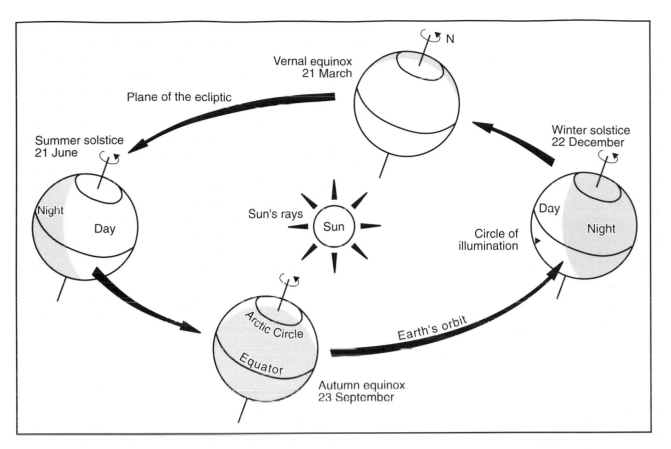

Figure 2.11 *The revolution of Earth around the sun. The seasons result because Earth's tilted axis maintains a constant orientation in space as Earth revolves around the sun.*

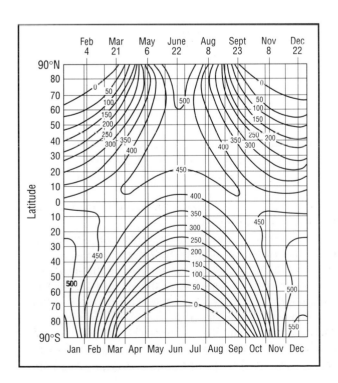

Figure 2.12 *Solar radiation (W m⁻²) falling on a horizontal surface at the outside of the atmosphere. Source: After Neiburger et al. (1982).*

twenty-four hours of daylight in summer becomes twenty-four hours of darkness in winter.

Endogenetic energy

Energy supplied by the sun is known as **exogenetic**, in that it is derived from outside Earth (Greek *exo*, 'outside' + *genos*, 'creation'). It can be argued that, indirectly, almost all Earth's energy is exogenetic and all but a small amount is derived from the sun; the remainder comes from other cosmic bodies, including the attractive force of the moon, which causes tidal activity (a form of kinetic energy). In addition, a minute proportion of the total energy comes from within Earth and is thus **endogenetic** (Greek *endon*, 'within' + *genos*).

The most obvious source of endogenetic energy is the hot interior of Earth. The outer core of the globe

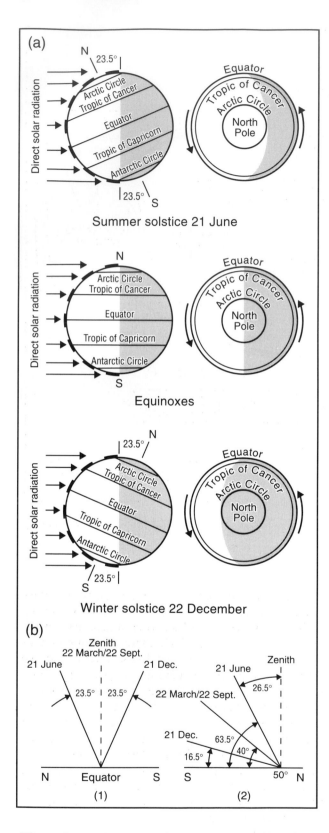

Figure 2.13 (a) Exposure of Earth to the sun's radiation at the solstices and the equinoxes; (b) position of the midday sun (1) at the equator and (2) at 50° N at the solstices and the equinoxes.

consists of molten materials at immense pressures, and at temperatures up to 2600° C. There is an almost immeasurably small and continuous conduction of this heat to the ground surface that adds to the energy inputs acting on the landscape. It can be detected in deep mines and caves. Locally, the decay of radioactive minerals can provide energy to the surface. More dramatic leakages of this endogenetic energy are seen in the form of volcanoes, hot springs and various other tectonic activities. Taken together, all sources of endogenetic energy appear to contribute no more than 0.0001 per cent of the total energy supply averaged over Earth's total surface.

Energy outputs of the globe

The nature of Earth's energy output

The output of energy from Earth is in radiant form, but it is not identical to the input of radiant energy from the sun. Earth has modified the input by a variety of processes. Some of the original solar energy input is reflected by clouds or the ground surface and returned to space with little change in its radiative properties; it is still short-wave radiation. As insolation passes through the atmosphere it is scattered, much of it towards Earth, but a small proportion goes back to space as an output of short-wave energy.

Of much greater importance is the emission of radiant energy from Earth itself. As a result of the absorption of solar energy in the atmosphere and at the surface, Earth will have gained energy that will be converted into heat. In turn, Earth and its atmosphere emit radiation following Wien's law. The average temperature of Earth is about 290 K, while that of the atmosphere is a chilly 250 K. Consequently the energy emission will reach a maximum at a wavelength of 2898/290 K or 2898/250 K, which is 9.99 or 11.59 μm; and overall emission is entirely within the infra-red range (Figure 2.3).

In this form the energy is susceptible to absorption by the atmosphere, so very little escapes directly to space; most is repeatedly reabsorbed and re-emitted before it is able to leave the system. The ability of the globe to trap energy in this way helps to keep the temperature of Earth and atmosphere higher than it would otherwise be. In other words, it promotes energy storage within the system.

At a global scale these processes lead to energy outputs of which about 36 per cent are in the short wavelengths derived from reflected insolation, and about

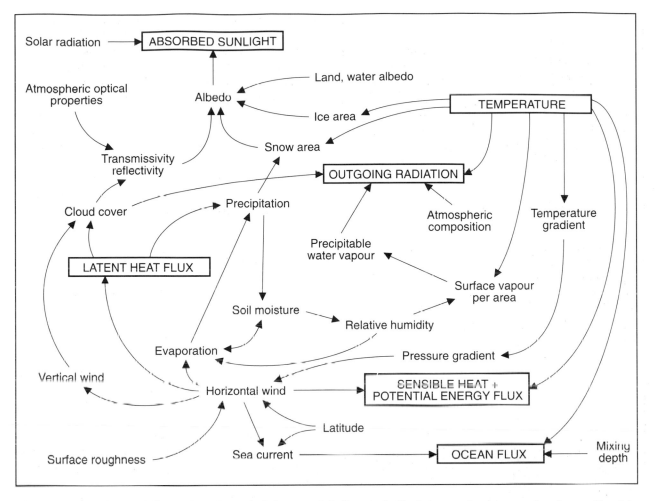

Figure 2.14 *Schematic illustration of many of the potential climatic feedback interactions that need to be considered in a climatic model.*
Source: After Kellogg and Schneider (1974).

64 per cent in the long wavelengths, largely from emission by the atmosphere. Taken together, the difference between the incoming radiation and the outgoing radiation is Earth's **net radiation budget**.

Spatial and temporal variations in outputs

Radiation outputs from the globe vary considerably over time and across the global surface. Spatial fluctuations depend upon a number of factors, including the character of the atmosphere (e.g. its temperature and the degree of cloudiness) and the nature of Earth's surface (e.g. vegetation cover and topography). From the polar regions an output of about 140 W m⁻² compares with 250 W m⁻² from equatorial areas – a ratio of about 2:1 – whereas the ratio for short-wavelength input is about 6:1. These aspects will be covered in Chapter 5.

Over the long term the fluctuations in global energy outputs possibly relate to outside influences; a change in input may lead to an adjustment in the output. The ways in which these adjustments take place are complex, and involve interactions called feedback mechanisms (Figure 2.14). Vegetation cover, atmosphere conditions (including moisture content and cloud cover), the extent of polar and mountain snow cover, the area of the sea surface and even soil cover and roughness may change in response to alterations in energy inputs. Through such changes Earth is able to adjust its energy outputs in the event of any long-term variation in inputs by altering the balance between the absorption, retention, emission and reflection of energy.

The question, however, is whether long-term variations of this kind occur. Certainly over geological time quite marked fluctuations in climate have taken place, as is attested by the evidence of Ice Ages and tropical conditions contained in the rocks of many parts of the world. Some of these changes are due to movement of the continental plates but some may be related to alterations in energy inputs and, if so, it is clear that outputs, too, must have changed. As the snow cover was extended during the Ice Ages, reflection must have increased, while absorption (and hence reradiation) must have been reduced. We will examine some of the possible consequences of this process in Chapter 11. Ultimately, however, a new equilibrium seems to be established as energy outputs decline to match the new, lower levels of input.

It is an intriguing question, also, whether changes in global conditions could arise owing to adjustments in the outputs independently of change in energy inputs. Any event that significantly alters the reflectivity of Earth's surface might trigger such changes. An increase in the extent of the oceans relative to land due, perhaps, to major earth movements; increased snow cover as a result of mountain building; changes in vegetation cover due to these events (or even human activity); or changes in the atmosphere brought about by massive volcanic eruptions – all could lead to significant changes in the global climate and hence in energy outputs. The implications for the world's climate are very important.

What is certain is that marked variations in global energy outputs do occur in the long term. Many of these variations are probably cyclical, related to changes in solar inputs such as those resulting from differences in the tilt and orbit of Earth. It is also apparent that such variations in output are critical if Earth is to adjust to alterations in the energy inputs that are known to occur, and thereby maintain steady-state equilibrium. An unanswered question is: to what extent can humans change these outputs and upset the equilibrium?

Thermodynamics

The laws of thermodynamics

The basic principles of energy are embodied in the laws of thermodynamics. These were initially developed in 1843 by Prescott Joule to explain processes seen in steam engines. Since then it has been appreciated that they have far wider significance, and they now represent basic precepts of science. The first two laws of thermodynamics state that:

1 Energy can be transformed but not destroyed.
2 Heat can never pass spontaneously from a colder to a hotter body; a temperature change can never

occur spontaneously in a body at uniform temperature.

The first law therefore defines the conservation of energy. The second law leads to the principle that energy transfers are a result of inequalities in energy distribution and that energy is always transferred from areas of high energy status to areas of low energy status, that is, down the energy gradient.

The third law of thermodynamics is less easy to understand. In very general terms, it says that systems tend towards equilibrium, that is, a random distribution of energy over time.

Energy and work

The transfer and conversion of energy are associated with the performance of work. The sun performs work in heating Earth through its provision of radiant energy. A river uses kinetic energy to perform the work of moving boulders. The weathering of rocks or the decomposition of plant debris involves work carried out largely by chemical energy. Indeed, it is the work done in these ways that characterizes the myriad processes operating in the environment.

When this work is carried out, therefore, energy is transferred from one body to another, and in some cases it is also converted from one form to another. In the process, the total energy content remains the same, it is changed only in form. When a river or glacier cuts a valley, the energy it uses is not destroyed but transferred or converted to other forms – some to heat energy, some to potential energy, some remaining as kinetic energy. When a plant grows, it takes in energy from the sun, from the air and from the soil and stores it; the energy is not lost, merely transferred and transformed.

Global energy transfers

Every feature and every part of the globe is at some stage or another involved in energy transfers and transformations, and, as conditions change, so the nature of the transfers and conversions operating at any one place also changes. We cannot, therefore, describe the processes operating throughout the entire global system in any detail. We can, however, try to identify the dominant transfers operating at a global scale and indicate, within this general pattern, the roles played by the various subsystems.

We have already noted that the balance between incoming and outgoing radiation is such that marked disparities occur between the energy status of different parts of the globe. The most obvious effect of this is the range in temperature we find when travelling from pole to equator, a range of 30° C to 60° C depending upon the time of year and the hemisphere. In simple terms, it is these differences that drive the global energy circulation. In order to achieve equilibrium, energy is transferred from the warmer to the cooler parts of the globe. If someone turned off the sun these transfers would result eventually in a more or less uniform distribution of energy across the globe; the fact that the sun continues to supply this unequal distribution of energy, however, maintains the imbalance and makes the attempt to achieve uniformity a losing battle. On the other hand, if the battle were not fought, the fact that the equatorial areas are constantly gaining more energy than they lose, while the polar areas are losing more than they gain, would result in a massive accumulation of heat in the lower latitudes and indescribable cold in higher latitudes.

Thus there is a net poleward transfer of energy, and this transfer maintains the existing pattern of energy distribution; it feeds the higher latitudes and drains the lower latitudes (Figure 2.15) This transfer is effected by a variety of processes. Undoubtedly the main transfers occur in the atmosphere. Winds carry warm air and water vapour away from the tropics. The warm air thus transfers heat (thermal energy) to the cooler latitudes. The water vapour carries energy in the form of latent heat. When the water vapour condenses, this energy is released as heat and warms the surrounding atmosphere. The oceans, too, transfer significant amounts of energy poleward. Heating of the sea in equatorial areas creates a temperature gradient between the lower and higher latitudes. Ocean currents carry the warmer waters down this gradient by a process of **lateral convection**, that is, the force created by the difference in temperature between one part of the ocean and another. Equally important, surface winds move water poleward and equatorward, a reflection of the close interaction between atmosphere and ocean.

In both air and sea, however, the transfers are not one-way. If that were the case, we would be faced with a build-up of air and water in the higher latitudes and a slow emptying of the tropical areas. Clearly this does not happen; the warm air and ocean currents that flow towards the poles are replaced by a counter-flow of cooler air and water moving from the poles. In the case of the sea, the flow tends to occur at depth, for the cool water sinks. In the air, the pattern is a little more complex. The transfer of latent heat (that is, energy tied up within water

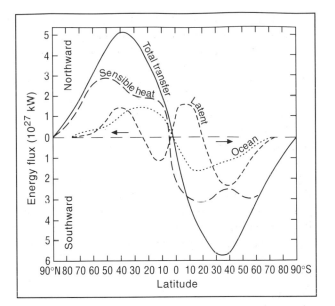

Figure 2.15 *The average annual latitudinal distribution of the components of the poleward energy transfer in the earth–atmosphere system. Source: After Sellers (1965).*

vapour) occurs mainly in the lowest 2 km or 3 km of the atmosphere. It is closely related to the surface wind network. The sensible heat transfer (that is, of warm air masses) occurs both close to the ground surface and also at high altitudes (around 10 km). Both flows, however, are balanced by counter-flows of cooler air from higher latitudes. We will examine these processes more closely in Chapter 5.

Thus the three main processes of energy transfer at a global scale are:

1 The horizontal transfer of sensible heat by warm air masses.
2 The transfer of latent heat in the form of atmospheric moisture
3 The horizontal convection of sensible heat by ocean currents.

Of these three processes, the first is the most important, accounting for about 50 per cent of the total annual energy flow. The other two processes account for about 20 per cent and 30 per cent respectively.

We shall consider the detailed processes involved in these transfers in Chapter 5, and will see there the factors that lead to the spatial distribution of these transfer mechanisms, but it is worth noting here that marked latitudinal variations in the three processes occur. Sensible heat transfers by the atmosphere, for example, are at a maximum between 50° and 60° north and south of the equator, and again at 10°

and 30° north and south (Figure 2.15). This pattern reflects the two types of transfer referred to earlier; the higher-level transfers are dominant in the subtropical zone while surface transfers are most active in middle latitudes.

The transfer of latent heat also shows a complex pattern, related to the distribution of water vapour in the atmosphere and the dominant, lower-air wind patterns. Thus its main effects are seen between 20 and 50 north and south of the equator, where winds blowing outward from the subtropics carry moist air poleward. Nearer the equator the pattern is reversed. Winds created by equatorial low pressures carry this air into the lower latitudes. As we shall see in Chapter 5, this pattern is closely related to the global wind system. Oceanic transfers of energy are most important either side of the equator, reflecting the outward movement of warm water from the tropical region. As the waters move polewards they lose heat to the overlying air. As wind patterns move air eastwards at these higher latitudes, the heat released by the sea eventually warms the western coasts of the continents such as North America and Europe.

In total, the processes of energy transfer maintain a steady-state equilibrium within the global system; they replenish energy losses in areas where outputs exceed inputs (the higher latitudes) and they remove energy from areas where inputs are in excess (the lower latitudes).

Local and regional energy transfers

While these atmospheric and oceanic processes account for the spatial redistribution of energy at a global scale, they are not the only means of energy transfer in the global system. At a more local level, numerous other transfers are taking place.

Atmospheric transfers

Within the atmosphere, local and regional winds, convection currents and air masses carry energy as sensible heat and as latent heat. The uplift of air and the water contained within it transform some of this energy into potential energy, which is released when the air sinks or the water condenses. Small, local transfers of energy to the earth's surface occur owing to friction, while the kinetic energy of the wind is transmitted to soil and rock particles as these are picked up and blown along. Heat energy from the atmosphere is also transferred to soils and plants through conduction and radiation (Figure 2.16).

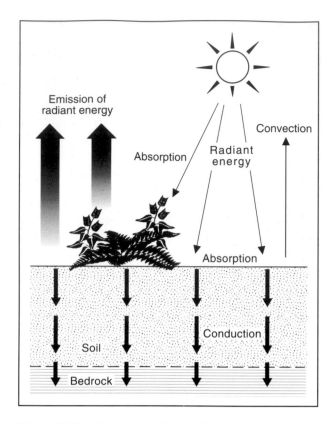

Figure 2.16 *Energy transfers at the surface.*

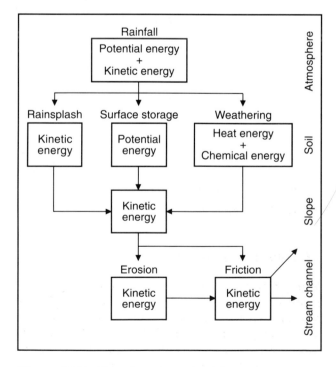

Figure 2.17 *Transfer processes involving water.*

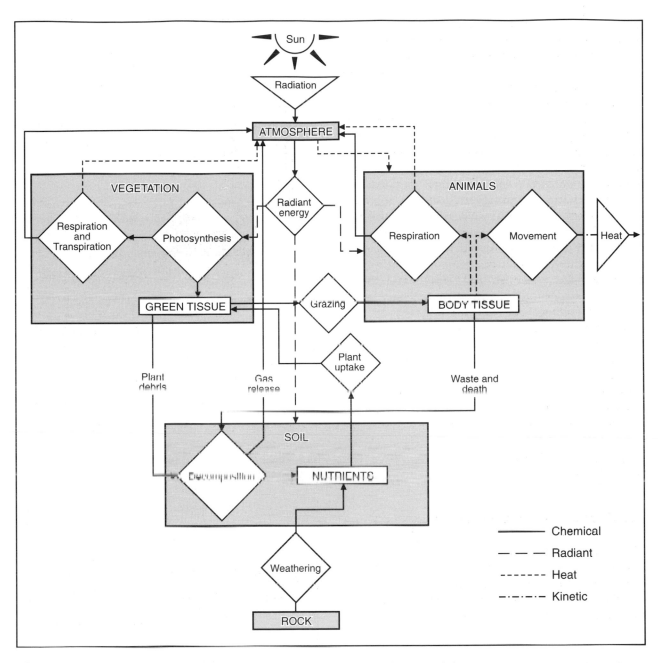

Figure 2.18 *Simplified depiction of energy flows and transformations in terrestrial ecosystems.*

Hydrological transfers

Water similarly takes part in a variety of transfer processes (Figure 2.17). Water condensing in the atmosphere releases latent heat; this warms the surrounding atmosphere. Potential energy derived from the initial uplift of water vapour into the atmosphere is transformed into kinetic energy as raindrops fall, and some of this kinetic energy is transmitted to Earth's surface as rock and soil particles are splashed into motion. Further potential energy is expended and converted to kinetic energy as the water percolates through the soil, runs into streams and flows to the sea. The flowing water again imparts some of its kinetic energy to material that it picks up and carries along.

The water also takes part in chemical processes of weathering and thus chemical energy is transferred to heat energy, given off during the chemical reactions. In the sea, the currents transfer energy laterally, while the upwelling and sinking of water masses leads to vertical transfers. Finally the evaporation of water from the sea, from rivers and lakes, and from the soil involves the conversion of thermal or radiant energy to kinetic and potential energy as the water is again raised from its original position and carried to higher levels in the atmosphere.

Landscape transfers

Many of these transfers influence landscape processes, for the movement of water through the landscape is one of the main ways Earth's surface is altered and moulded. The potential energy possessed, for example, by boulders on a slope is a product of the erosion of the valley by the water and ice. Potential energy is also derived from earth movements, for mountain-building lifts the rock to leave it higher than the surrounding earth surface. Since these mountain-building processes are powered by heat energy within the earth, they represent the transformation of heat energy to kinetic and, ultimately, potential energy. The potential energy is subsequently converted to kinetic energy as the rock particles tumble, sludge or wash downslope. Friction with the surface and between the particles releases further energy in the form of heat.

Ecological transfers

On land the formation of soil, the growth of plants and the support given by this vegetation to animals all reflect further energy transfers and conversions.

In the case of terrestrial ecosystems (Figure 2.18) the development of soil cover involves weathering, which in turn reflects the transfer of chemical energy from rocks to soil. Plants take up substances from the soil and store the chemical energy in their tissue. They also use radiant energy from the sun, and chemical and heat energy from the atmosphere, all three forms being converted to chemical energy by the plant. As the vegetation dies, or animals devour the plant material, this energy is cycled through the environment. Animals convert the chemical energy to heat for bodily warmth and to kinetic energy for motion. They return some energy to the soil and the atmosphere as chemical energy.

Similar processes operate in aquatic ecosystems, although in their case much of the initial input of chemical energy is derived from organic matter washed into the waters from the land.

On a global scale it is impossible to quantify precisely the effects of all these processes. What is clear is that energy transfers create a fabric of relationships that bind the global system together, and provide the motive power for the processes that operate within our world, and which are the very foundation of our existence.

Key points

1 Energy is the driving force of all the processes operating in the global system. It performs the work in processes such as moving rocks, eroding valleys, lifting mountains, and making water flow, the wind blow and plants grow. This work is performed through the transfer and transformation of energy. These transformations tend to follow well defined routes.

2 The work is carried out because of differences in the energy status of different objects or conditions. Inequalities in the distribution of available energy (that which is capable of performing work) lead to energy transfers; in the course of these transfers work is done. The energy involved in the transfers is not destroyed; it merely changes form.

3 The energy transfers that operate in the global system derive from the inequalities in inputs and outputs across the world. On a global scale, they involve the movement of energy as sensible heat and latent heat by the atmosphere and as sensible heat by the oceans.

4 At a more detailed level, these transfers permeate every part of the global system. They involve transfers from rocks to soil, from soil to plants, from plants to animals and the atmosphere; in fact all the components of the world are interconnected by these transfers. They also involve transformations of energy from one state to another.

5 Together these transfers and transformations provide the power for all the processes operating in our environment. They bind the global system into a unified whole. They are the lifeblood of our planet.

Further reading

Ahrens, C.D. (1994) *Meteorology Today* (fifth edition), West Publishing (Chapters 1 and 2).
A modern and very colourful meteorology text at an elementary level.

Neiburger, M., Edinger, J.G. and Bonner, W.D. (1982) *Understanding our Atmospheric Environment*, San Francisco: W.H. Freeman (Chapters 3 and 4).
A good introductory textbook on meteorology which provides a clear account of the principles of radiation and energy.

Peixoto, J.P. and Oort, A.H. (1992) *Physics of Climate*, New York: American Institute of Physics.
An advanced text for those with an understanding of mathematics and physics. It sets out to explain the principles of climate in physical terms as a basis for atmospheric modelling.

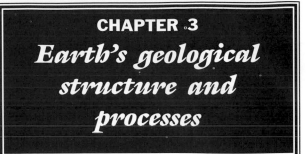

Rock material forms Earth's largest constituent by far and has undergone continuous transformation since condensing from a contracting interstellar cloud 4.6×10^9 years (4.6 billion, or 4.6 Ba) ago. It has a volume of 1.083×10^{12} km^3 and mass of 5.977×10^{24} kg, compared with 1.4×10^9 km^3, 1.40×10^{21} kg of global water and 5.13×10^{18} kg of atmospheric gases. International travel and orbiting satellites highlight the relatively small size of our planet, just 40,000 km in circumference and averaging 6371 km from surface to centre, equivalent to the London–Chicago distance. Yet the mass, character and age of its rocks can still be hard to comprehend. So, too, are its origins in astro-geological processes which formed our solar system, through the gravity-concentration of matter from a supernova explosion

c. 6 Ba ago. What interest, then, should geographers have in planetary processes dominated by imponderable origins, astronomic time scales, tiny *geothermal* energy flows compared with the atmosphere and vast but almost entirely hidden material reserves? How far do they influence human habitat and lives at Earth's surface? Table 3.1 shows their far-reaching effects and time scales of operation. This chapter explains long term, large scale geological processes which form Earth's dynamic foundations. **Plate tectonics** provides its unifying theme, enabling us to link Earth's early evolution with the geologically recent **neotectonic** emplacement of its principal global landforms. Subsequent shaping by surface **geomorphic** processes is the subject of later chapters.

Table 3.1 *The nature, time scales and relevance of geological and related processes.*

Processes	Time scale (years)
Macro-scale geological processes drive the ever-changing global configuration of ocean basins, continents and mountain ranges, and in turn:	10^{6-8}
● created, and continue to modify the composition of, atmosphere and oceans through outgassing	10^{1-9}
● cycles rock material from its formation, through destruction to reformation	10^{1-8}
● creates global seismic belts which locate most earthquake and volcanic activity	10^{6-7}
● creates random variations in the patterns of solar radiation receipt and absorption which drive global climate	10^{5-7}
● channels global ocean currents and disturbs meridional and zonal atmospheric circulation, with major impacts on climate systems and weather events	10^{1-7}
● drives geomorphological processes through displacement of continental crust and sea levels	10^{1-7}
Meso-scale geomorphological processes etch and sculpt the continental crust into distinct landforms and landform assemblages or landsystems.	10^{1-7}
Micro-scale pedological processes drive the formation of soils and, through them:	10^{1-4}
● provision of physical support and attachment sites for flora and fauna	
● the principal inorganic source and cycling system of the biosphere's nutrient cycle	
Geological, geomorphological and pedological processes collectively also create:	
● all our nuclear and fossil fuels, metal and non-metalliferous ores, building stones and aggregates	
● the variety of surface landsystems and foundation of landscape	
● the substrate on which we construct our buildings, urban regions, farmland and other economic and recreation systems	

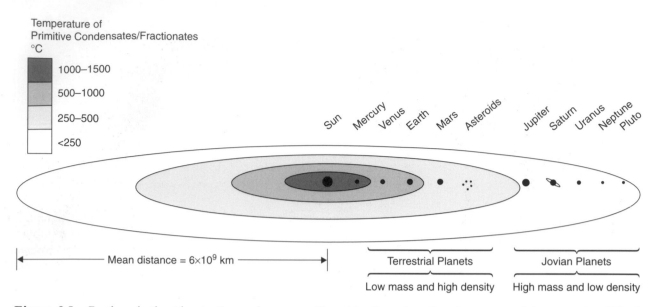

Figure 3.1   *Earth and other planets of our solar system. Planetisimals condensed and concentrated in a rotating disk of planetary matter around our Sun, with early fractionation according to temperature zones within the disk. (Not to scale.)*

Origin and dynamics

The Sun formed at the centre of a rotating nebula of planetary matter (Figure 3.1). Condensing around tiny planetesimals, gaseous planets (Jupiter, Neptune, etc.) formed in outer, cooler parts of the nebula. Solid, terrestrial planets (Earth, Mars, etc.) formed in its inner, hotter zone. This occurred through **fractionation** or segregation of the elements composing our solar system into distinct assemblages, determined by their physical properties, which we will see throughout planetary geology. Infant Earth would be unrecognizable today, dominated then by heat-generating accretion around a dense core. Outer cooling after an initial hot phase formed a crust, violently pockmarked by out-gassing of volatile gases through continuing fractionation and pulverized by planetesimals and other space debris. This dramatic *Hadean* aeon, named after Hades – the underworld in Greek mythology – was short-lived. Earth's essential structure was in place 4.4 Ba ago. Meteorite impacts, occasionally large enough to form craters, and cooling accompanied by volcanic activity still occur on a reduced scale. Most geological activity is now confined to the crust and upper mantle within approximately 150 km of the surface. The present form of the continents and oceans is less than 200 Ma old, which allows us to concentrate on just 7 per cent of global rock mass and 4 per cent of Earth history.

In the intervening 4.2 Ba continuous but uneven cooling developed a process of crustal evolution which acts as the radiator to Earth's internal engine. New crust forms over hot-spots and old crust sinks and is recycled elsewhere, actively venting geothermal energy as well as passively emitting it to space. Mobile crust in transit between these zones takes the form of semi-rigid plates, and their boundaries coincide with global-scale landforms, earthquake and volcanic belts. Crustal formation differentiates between lighter, granitic continental rafts 'floating' on heavier, basaltic oceanic crust which acts as the conveyor belt in plate motion. The persistence of plate tectonics over geological time accounts for the extreme youth of ocean crust, with a mean age of 55 Ma and none more than 200 Ma old; and the relative youth of Earth's crust as a whole, with 98 per cent less than 2.5 Ba and 90 per cent less than 0.6 Ba old. Continental crust has an average age fifteen to twenty times greater than oceanic crust because it is recycled more slowly. Fragments from the early *Archaean* Earth, 3.7–4.3 Ba old, survive in parts of Canada, Greenland, Australia and South Africa. Modern continents are a collage of quite different crustal **terranes**, or fragments of widely dispersed form and origin, which reflect the repeated accretion and break-up of older crust (Figure 3.2).

A **supercontinental cycle** – the **Wilson cycle** – is at work. Continents converge and coalesce during one phase of Earth history and subsequently rift apart through relentless plate motion. Supercontinents would be surrounded by a single global ocean, whereas fragmented continents would isolate several smaller oceans (Figure 3.3). Earth is small enough for rifted fragments to reassemble eventually elsewhere. At 10^1 cm yr^{-1} today – comparable with fingernail

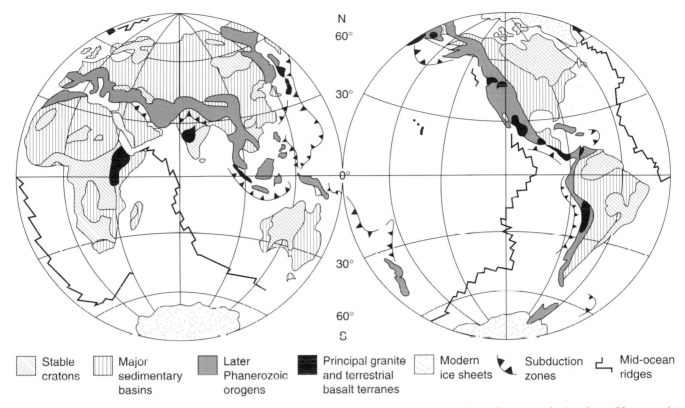

| Stable cratons | Major sedimentary basins | Later Phanerozoic orogens | Principal granite and terrestrial basalt terranes | Modern ice sheets | Subduction zones | Mid-ocean ridges |

Figure 3.2   *Earth's principal surface structures and terranes, showing stable cratons, major sedimentary basins, later Phanerozoic orogens, principal granitic and terrestrial basalt terranes, mid-ocean ridges and subduction zones.*
Source: After Wyllie (1971).

growth rates! – the cycle seems imperceptibly slow but could be completed within 500 Ma. This is short enough to have occurred eight to ten times during Earth's history, especially as greater heat flow may have driven the cycle faster during Archaean time.

Earth's present crustal configuration is half-way through a cycle which commenced with the rifting of the supercontinent **Pangaea** *c*. 200 Ma ago, in the *Mesozoic* era of the *Phanerozoic* aeon. The global *Panthalassic Ocean* has been replaced by the new, equatorially centred basins of the Pacific, Atlantic and Indian oceans, whilst its *Tethys Sea* arm was closed by convergence of Africa with Eurasia. Modern oceans are partially enclosed by the fragmentation of North and South America, Antarctica, Australia and India and the emergence of South East Asia–Pacific Island arcs. Our north polar, landlocked micro-ocean contrasts with a south polar continent surrounded by ocean. We need to think of the global map, therefore, as mobile and dynamic rather than fixed in an 'average' position which we take for granted. Major topographical features which profoundly influence modern global ocean and atmospheric circulation such the Panama isthmus, linking North and South America, and the Tibetan plateau are less than 3 Ma

old. Closer inspection of Earth's crustal 'boundary layer' reveals the global **morphotectonic landforms** of current plate dynamics, clear evidence of past rifts and collisions and the potential sites of future ocean basins and mountain ranges. Plate tectonics provides the framework for understanding the geological evolution of the crust. Its subsidiary *supercontinental cycle* and *rock cycle* drive the formation and destruction of rock material and the landform assemblages involved.

Most geographical references in the text refer to the *modern* location and identity of crustal fragments, which achieved their form and global position only recently. The geological age of events in their evolution is indicated where appropriate.

Earth structure and internal energy

Core, mantle, crust, ocean, atmosphere and biosphere

Plate tectonics continues the geological distillation and fractionation of planetary raw materials which began as the planets condensed from interstellar gases. Controlled by temperature and pressure, dense

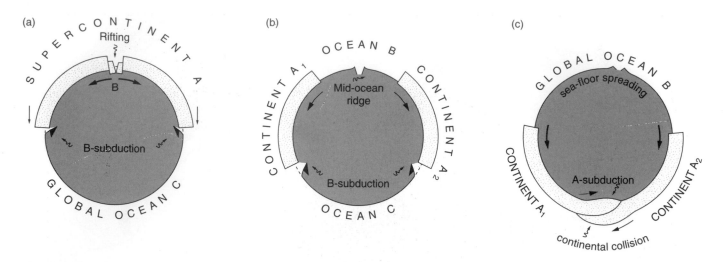

Figure 3.3 *The supercontinental or Wilson cycle: (a) supercontinent A begins to rift at B, causing B-subduction and contraction in the global ocean C; (b) ocean B floods and expands as fragmentary continents A₁ and A₂ drift apart; (c) ocean C closes and continents A₁ and A₂ eventually collide, with B subduction giving way to A-subduction.*
Source: After Kearey and Vine (1996).

refractory or heat-resistant materials such as iron, nickel, silicates and carbonates condensed at higher temperatures nearer the Sun and dominate terrestrial planets, including Earth. Less dense *volatile* materials such as hydrogen, helium, nitrogen and oxygen condensed at low temperatures furthest from the Sun, forming outer planets rich in gas–liquid–ice. Our embryonic Earth gained kinetic energy through the accretion of mass and thermal energy from radio-active decay, creating high temperatures at the core and raising surface temperatures briefly as high as 8000° C–10,000° C. As a result, planetary materials segregated according to their chemical and physical character. Earth's internal structure is a microcosm of the solar system, with a core surrounded by five concentric, progressively cooler and less dense layers or geospheres (Figure 3.4). High-density refractory elements survived at the core whilst volatile gases and fluids were driven off to form the ocean–atmosphere systems or were exhaled to space.

The **core** formed by the separation of a nickel–iron mixture from lighter silicon-rich material and generates Earth's magnetic field. Its mean density of 10.7 gm cm⁻³ rises to almost 14 gm cm⁻³ at Earth's centre, from which the core extends 3460 km, concentrating 32.2 per cent of rock mass in just 16.9 per cent of planetary volume. Seismic evidence described later indicates that the inner core is solid for 1300 km, beyond which the outer core is liquid. Density falls sharply at its boundary with the **mantle**, which extends for a further 2970 km. With a mean density of 4.5 g cm⁻³, the mantle is composed of minerals transitional in weight between the iron of

the core and oxides of silicon and aluminium which comprise 75 per cent of the eventual crust. Like the core, it is not internally homogenous. An inner solid *mesosphere* extends about 2560 km to within 350 km of the crust, overlain by partially melted and viscous *asthenosphere*. Cool solid *lithosphere*, averaging 70 km in thickness, forms the outer mantle and its over-lying, recyclable **crust**. Traditional mantle/crust distinctions are less important than the astheno-sphere/lithosphere boundary, despite differences in mineral composition and density, averaging 3.5 g cm⁻³ in lithospheric mantle compared with less than 3.0 g cm⁻³ in crust. Here mobile rigid crustal plates are decoupled from underlying viscous mantle. They form a distinctive surface architecture of global land-forms, based on mineralogical differences. Denser, heavier **basalt**-rich oceanic crust (2.8–3.4 g cm⁻³) is only 7–10 km thick, compared with less dense, lighter **granite**-rich continental crust (2.7 g cm⁻³) 25–75 km thick with a mean of 35 km.

The outermost planetary layers are quite distinct from the 'solid' mineral earth and not traditionally considered as geological systems. **Hydrosphere** (97 per cent ocean) and **atmosphere** have their own distinct character and behavioural patterns but never-theless originate from the same planetary fractiona-tion processes and continue to exchange and synthesize materials with the lithosphere. The hydros-phere has the greater mass but with ocean (saline) water and ice densities of 1.03 g cm⁻³ and 0.9 g cm⁻³ respectively, accounting for over 99 per cent of mass, it forms an intermittent surface layer averaging only 4.0 km thick. By comparison, atmospheric

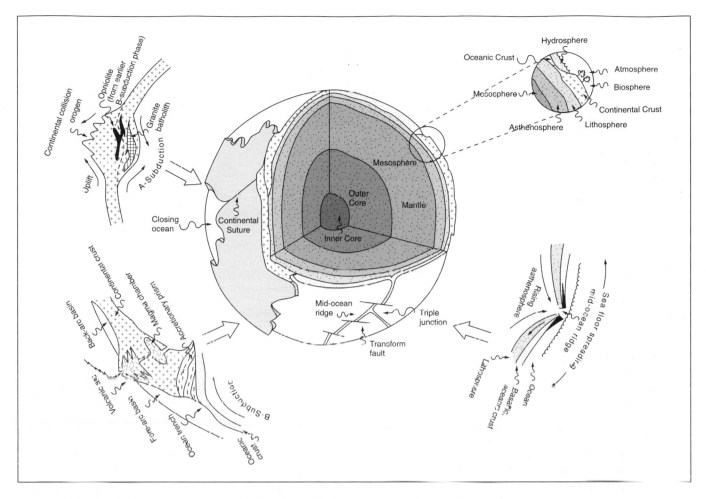

Figure 3.4 *Earth's internal structure and the relationship between outer spheres and crustal processes.*

surface density is only 0.00012 g cm^{-3} falling by two-thirds at 10 km aloft, below which lies 75 per cent of its mass. Both spheres are residues of the lightest, most volatile outgassed elements of the early Earth retained by gravity. Low temperatures in our planetary boundary layer determine that hydrogen, helium, nitrogen, oxygen, methane, carbon dioxide and some trace elements are found primarily as gases and the precise range of surface temperature ensures that H_2O appears commonly as gas, liquid or solid. The tenuous nature of both systems has resulted in the loss of up to 40 per cent of water mass and substantial changes in atmospheric composition since early formation, as the lightest elements were exhaled into space. They continue to be sourced by volcanic outgassing and to evolve compositionally through two-way material transfers driven by crustal recycling, geomorphological, pedological and biological processes. The **biosphere** is a hydrocarbon derivative of geological fractionation where lithosphere, hydrosphere and atmosphere meet. All three systems share

dependence on solar heating, photodissociation or fractionation by sunlight and subsequent chemical reorganization, in addition to geological processes.

Internal energy and heat flow

All planetary processes require energy, and Earth has five energy sources intimately linked with the formation and operation of our solar system; three are concerned with heat energy and two with gravity. We saw the importance of solar radiation as the principal *exogenetic* source of electromagnetic energy in Chapter 2 and explain its significance for atmospheric, geomorphological and biospheric processes in later chapters. Nucleosynthesis of helium from hydrogen in the Sun is the essential energy source of the universe. Similar *endogenetic* nuclear reactions generate heat from the continuing decay of radioactive isotopes of uranium ^{235}U, ^{238}U, to lead, ^{206}Pb, ^{207}Pb, and of potassium, ^{40}K, to argon, ^{40}Ar, etc., which are most abundant in continental crust. Planetary accretion and fractionation also provide a source of residual heat. Kinetic energy from planetesimal and meteorite impacts generated heat and this was augmented by heat generated through friction as core and mantle materials segregated past each other. *Adiabatic* heating through compression of the core and a corresponding decrease in volume is similar to atmospheric processes described in Chapter 6.

The principal effects of endogenetic **thermogenesis** or heat generation, are to establish convection currents within the mantle, which drive plate tectonics, and to cause geological phase transformations, mobilizing rock between its solid–liquid–gas states. This is the key to continuing fractionation and the creation of **magma**, a molten viscous state essential to crustal evolution. Exogenetic heat sources power geomorphological processes which ornament the continental crust, as we shall see in later chapters, and share dependence on the additional role of **gravity** in geological processes. Gravity is the force of mutual attraction between two bodies and is a function of their masses and distance apart. Earth's large mass centred around a dense core provides the primary, endogenetic source of gravity for most geological processes but the gravitational fields of our Sun and Moon influence astro-geological and some surface (especially tidal) processes. **Gravitational energy** describes the potential and/or kinetic energy of a mass dependent upon its displacement away from the Earth's core and is a by-product of fractionation. This is also an important consequence of tectonic uplift and drives surface geomorphological processes. Gravity adds a further twist, quite literally: a centrifugal component, due to Earth's rotation, flattens its spherical shape at the poles into an *oblate spheroid*.

Internal heat sources establish a **geothermal heat flux** from the core towards the cool crust. With a calculated core temperature of some $4000°$ C and a mean surface temperature of $10°$ C, the average thermal gradient would be $0.62°$ C km^{-1}. It is thought that core and mesosphere gradients are slightly lower, owing to the slow release of primordial heat stored from the early accreting Earth, limited mostly to conduction in rigid rocks. However, it is observed in mines that near-surface gradients are up to sixty times greater at 20–$40°$ C km^{-1}, sufficient to be tapped for geothermal power. This is due to crustal radioactive thermogenesis, responsible for some 70 per cent of the continental crust flux, and also to convection aided by the viscous state of the asthenosphere.

Measured as a heat flux in milliwatts, rather than as a thermal gradient in degrees Celsius, the mean surface flux is 82 mW m^{-2} or 0.082 watts m^{-2}. There are, however, a number of interesting systematic variations. Heat flux in the crust decreases with age and, in oceanic crust, away from mid-ocean ridges, where it may reach 200 mW m^{-2}. The oceanic mean flux of 98 mW m^{-2} is 75 per cent higher than the mean flux in continental crust at 56 mW m^{-2}, despite the latter's radioactive source. On the other hand, ocean crust is virtually devoid of radioactive elements, so 95 per cent of its heat flux must come from depth. Gradients are steepest in oceanic lithosphere, which conducts heat twice as efficiently, and continental lithosphere is cooler than ocean lithosphere. Overall, oceanic crust accounts for 75 per cent of global geothermal heat flux by virtue of its larger area and superior rate. Volcanoes and hot spots, not surprisingly, experience the highest fluxes of 200–250 mW m^{-2}. All this indicates great thermal activity in the shallow lithosphere. In particular, persistent contrasts between 'hot' oceanic and 'cool' continental lithosphere show that the sea floor holds the key to crustal evolution via plate tectonics.

Crustal evolution: plate tectonics

From the great voyages of exploration in the *Age of Discovery* between AD 1500 and AD 1900, it was noted that many continental coastlines appeared to fit together, particularly those bordering the Atlantic Ocean, and seemed to have become separated like the dispersed parts of a jigsaw puzzle – perhaps by Noah's Flood! In 1912 Alfred Wegener consolidated emerging theories of dispersal by **continental drift** to propose the prior existence of a single land mass, Pangaea, on the basis of a common geological history culminating in a great Permo-Carboniferous glaciation. This was consistent with Darwin's belief in common biological ancestors, with modern counterparts found in different continents once connected by land bridges or, ironically, natural rafts. No proven mechanism of crustal rifting existed prior to 1960, despite considerable geological evidence of *palaeo-* (past) climatic, palaeoenvironmental and palaeoecological continuity across Pangaea until the Mesozoic era *c.* 200 Ma ago.

Recent advances in the earth sciences confirm that **sea-floor spreading** is the driving mechanism, through the convection of new crust from the asthenosphere. Palaeomagnetic signatures reveal changes in Earth's magnetic field, involving polar wandering and total reversal, and allow us to reassemble the former global location of crustal rocks. Deep-sea drilling into ocean sediments and lithosphere provides evidence of past environments, age-correlated by isotopic dating. Seismology (see

Seismic studies and seismo-volcanic hazards

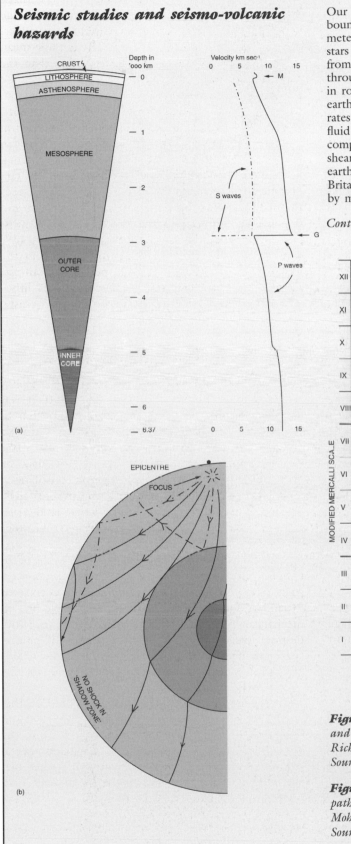

(a)

(b)

Our knowledge of inaccessible internal layers and boundaries is dependent on astro-chemistry, including meteorite mineralogy, the spectral signature of other stars and Earth's **seismic activity**. Earthquakes result from discontinuous crustal movement. Stress applied through plate motion is stored as elastic strain energy in rocks, capable of sudden release. The resultant earthquake transmits shock, or deep body waves, at rates dependent on the rock's density and its solid or fluid state. Faster **Primary** or **P** waves travel by compression and slower **Secondary** or **S waves** by shear. Sensitive seismographs record innumerable daily earth tremors, even in apparently stable zones like Britain. Seismic activity is also triggered deliberately by modest explosions for research purposes or,

Continued on page 35

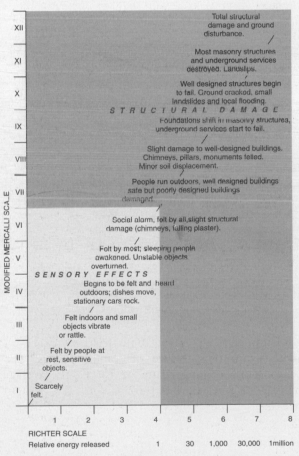

Figure 2 (above)   An outline of earthquake effects and damage associated with the Modified Mercalli and Richter scales of earthquake intensity.
Source: After Bolt et al. (1975); Smith (1982).

Figure 1 (left)   Typical (a) velocities and (b) pathways of earthquake waves; M and G mark the Mohorovičić and Gutenberg discontinuities.
Source: After Duff (1993) and Smith (1982).

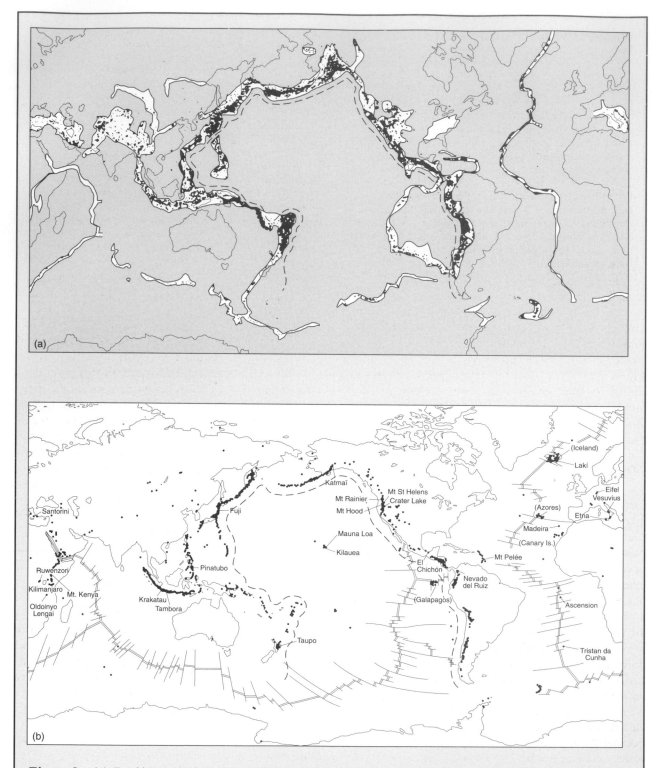

Figure 3   (a) Earth's principal earthquake belts (white, with recent epicentres shown by dots; after Barazangi and Dorman 1969) and (b) active volcanoes (dots; after Duff 1993), including the location of those named in the text. Clusters of island volcanoes, not named individually, are enclosed in brackets. The broken line marks the 'andesite line', separating oceanic basaltic volcanoes from arc–continent andesite, dacite and rhyolite volcanoes.

unintentionally, by nuclear weapon testing and other human actions.

Transmission times between earthquake epicentre and recording seismographs for a given rock layer are a function of distance, allowing for earthquake magnitude, epicentre location and transmission routes to be calculated for a particular shock. Marked changes in the velocity and direction of P and S waves allow us to plot Earth's internal structure by identifying different material densities at boundary discontinuities. Particular interest centres on: the *Mohorovčić* discontinuity between outer crust and lithosphere; wave deceleration between lithosphere and asthenosphere, signifying the partial melt status of the latter; acceleration at the *Gutenberg* discontinuity between mantle and dense core; and the deceleration of P waves and absence of S wave transmission in the outer core, indicative of its fluid condition (Figure 1).

Seismicity directly impacts human life through the destructive power of earthquakes, measured on *Modified Mercalli* (descriptive) and *Richter* (logarithmic energy) scales of severity (Figure 2). These relate chiefly to shallow, surface waves. Mercalli intensities emphasize the human price paid in lives – averaging 10,000 yr^{-1} globally – and property by the direct destruction of housing and other structures. Earthquakes also strike indirectly by triggering landslides, **tsunamis**, or tidal waves, and volcanic eruptions. Indeed, although they may occur independently, **seismo-volcanic** hazards are intimately linked. Major earthquakes in Mexico City (1985), San Francisco (1989) and Kobe (Japan, 1994) share the same global network of subduction zones as the explosive andesitic volcanoes of Mount St Helens and Pinatubo (Figure 3). Minor British earthquakes near Caernarfon (1984) and Shrewsbury (1990, 1996), registering 4–4.6 on the Richter scale, remind us that older tectonic belts are not yet dead.

box) confirms that narrow belts of intense earthquake activity, girdling the earth for over 50,000 km, are located at intraplate boundaries. Bathymetry demonstrates that their mid-ocean segments form submarine ridges. Satellite geo-positioning now provides accurate measurements of the rates and directions of plate motion.

Plates and plate motion

Cool, outer lithosphere does not form unbroken crust but is divided into a mosaic of interlocking rigid **plates** with active boundaries and relatively stable interiors. Each plate consists of lighter continental and/or ocean crust and its underlying denser lithospheric mantle. The global mosaic is dominated by seven major plates, individually over 10^7 km^2 in area, and a further six minor plates plates, each less than 10 per cent of the former's size (in the range 10^{6-7} km^2). A number of microplates assist in articulating the differential movement of plates over Earth's curved surface (Figure 3.5). Their names suggest a series of separate oceanic and continental plates but the reality is more complex. Although American Pacific coastlines closely follow the eastern boundary of the essentially oceanic Pacific, Cocos and Nazca plates, two American plates account for both continents and the western half of the Atlantic Ocean. In contrast, the Eurasian plate is mostly continental but includes the north-east Atlantic and eastern Arctic Oceans, whereas the Eurasian land mass includes continental fragments of the African, Indo-Australian and other small plates.

Continents thus reflect the accretion of terranes from more than one plate and may themselves eventually rift apart at new plate boundaries. Continental collisions **suture**, or weld together, distinct terranes, and parts of the Alpine and Himalayan mountains of the modern Eurasian plate mark such sutures formed on the closure of the Tethys Sea. The closure of the long-gone Iapetus Ocean in the Lower Palaeozoic era *c*. 430 Ma ago formed the Caledonian mountains of northwest Europe and eastern North America. Conversely, the Carboniferous basin of equatorial Pangaea is now divided by the Atlantic Ocean; the Appalachian, Scottish and Norwegian mountains are surviving Caledonian remnants dispersed as the Atlantic Ocean formed; and the East African rift valley system may lead to the formation of a future ocean.

How, then, do convection and gravity forces enable these huge plates to move over distances of 10^{3-4} km? Mantle convection, stirred by local thermogenesis and heat conduction from the core, appears to be the dominant process. Convection in fluids occurs as material is heated, and becomes less dense and more buoyant. Rising to the surface, it spreads, cools and eventually sinks as it becomes denser than surrounding warm plumes. High-temperature rock flows rapidly as molten **lava** only when free of confining pressures at Earth's surface. Pressure increases with temperature as depth increases, raising the melting point of any particular mineral assemblage. The 'solid' nature of the mantle thus reduces normal fluid motion to an extremely slow crystalline creep. However, the pressure–temperature balance in the asthenosphere permits a partial melt of up to 10 per cent of mass, giving it the texture of a stiff, granular slush whose lower viscosity accelerates convection.

Crucially, it also provides the basis of coupling/decoupling at the lithosphere–asthenosphere

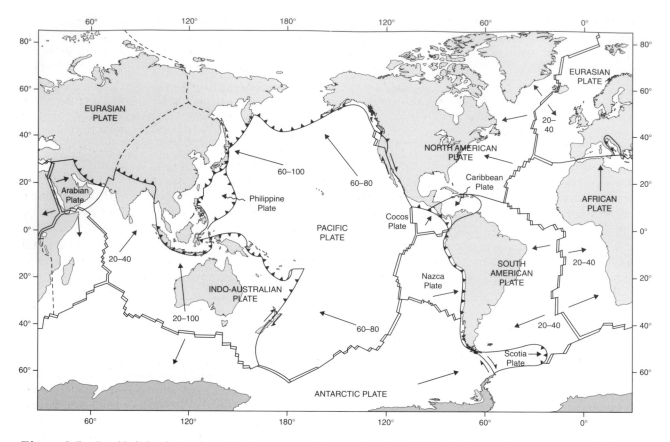

Figure 3.5 *Earth's lithospheric plates, showing constructive margins (parallel lines), destructive margins (toothed lines; teeth point in direction of subduction) and transform margins (single or broken lines). Arrows indicate general direction and velocities of movement in mm yr⁻¹.*

boundary at the depth of the 1400° C isotherm, which represents the minimum melting point of upper mantle rock at the pressures found there. Since continental lithosphere is lighter and cooler than oceanic lithosphere, the boundary is found at mean depths of 100–50 km and 70–80 km below each respectively. It was thought originally that plates were moved by **viscous drag**, coupling the base of the lithosphere to the asthenosphere, as convection cells in the latter spread out over a rising **mantle plume** and rafted crust around the Earth. The cooling limb of the cell would eventually return into the mantle. This convective drive does not account for enough of the thermal energy dissipated at the surface, however, and is augmented by gravitational effects at plate margins. In essence, the ascending convection current develops a **thermal bulge** in lithospheric slab, which then slides away under gravity. Cooling as it moves and ages, it eventually subsides into the subjacent warm asthenosphere, accentuating gravitational pull. In this mechanism, the lithosphere and asthenosphere are clearly decoupled in their boundary zone of melt-driven reduced friction as **ridge push**

or **slab pull** gravity forces are applied at opposite ends of the lithospheric slab (Figure 3.6).

This combination of thermally direct and gravity-induced forces accounts for plate motions and imparts direction and velocity to them. Simultaneously, ascending currents transport the raw material of new crust with them and descending currents recycle older crust back into the mantle. These processes, acting at opposing ends of the 'crustal conveyor belt', must broadly maintain a continuity of mass, since it appears that Earth is neither expanding nor contracting.

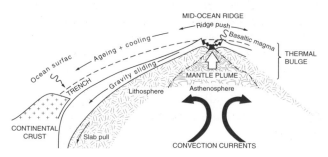

Figure 3.6 *The forces operating on a lithospheric plate, triggered by mantle convection currents.*

If anything, global continental crust is growing extremely slowly at the expense of ocean crust. An estimated 12–15 km³ of crust is recycled annually in this way. Plate motions relative to each other are of great interest to us and occur in three principal ways. Divergent or spreading boundaries are associated with extension of the crust, forming new plates at **constructive margins**. Spreading rates are typically 20–40 mm yr⁻¹ between the American and Eurasian plates in the mid-Atlantic Ocean and 20–100 mm yr⁻¹ between the Antarctic plate and the southern Indo-Australian and Pacific plates (Figure 3.5). Convergent boundaries are associated with crustal compression and plate is consumed by subduction back into the mantle at **destructive margins**. Active slab pull accelerates subduction, with velocities of 60–100 mm yr⁻¹ around the western Pacific margin and 50–120 mm yr⁻¹ on the eastern Pacific ocean-plate boundaries with the Americas. In its absence at the Africa–Eurasia boundary, velocities fall to 10–35 mm yr⁻¹.

Convection and gravity slide are unlikely to be uniform plate-wide, and motion must also accommodate the spherical shape of the earth and drag-resistant, stable continental lithosphere. As a result, plates may articulate internally along **transform faults** or slide past each other at **transform margins**, which are conservative boundaries, since normally plate is neither created nor destroyed, or meet at **triple junctions** (Figure 3.4). Most plates also have an absolute motion about the earth but parts or all of some, especially the smaller plates, still have active margins where they are caught liked fixed 'eddies' rotated by the 'stream' of larger plates. Despite their general rigidity, plates also experience plastic and/or brittle deformation in the form of doming, bulging, subsidence, folding and faulting. Tectonic deformation concentrated at convergent plate margins forms mountains by **orogenesis**. More general uplift/subsidence or **epeirogenesis** is associated with continental plate interiors. It, too, may be generated *thermally* by expansion over isolated hotspots or mantle plumes; or *mechanically* through crustal loading/subsidence by ice sheets, sediment deposition, rising sea level, etc., or unloading/elevation by deglaciation, erosion, falling sea level, etc. This **isostatic adjustment** slowly attempts to restore equilibrium loads to every part of the crust. We shall see later the vertical rates at which plates also move.

Plate architecture and morphotectonic landforms

World-wide tectonic activity, involving the creation and destruction of lithosphere, impresses itself on landforms at all scales. Extremely slow average rates of motion, which persist for 10⁶⁻⁸ years, contrast with violent volcanic eruptions and earthquakes. Even the low vertical amplitude of the lithosphere (10²⁻³ km) and surface relief (20 km spans the deepest ocean to the highest mountain) is not eclipsed by the very large area and horizontal dimensions of plates. The creation of that relief endows geomorphological processes with endogenetic energy. Plate architecture – literally, the style, design and construction of plate structures – is the key to global morphotectonic landforms and the **rock cycle**. Constructive, destructive and conservative styles of plate margins are translated into **mid-ocean ridge**, **subduction zone** and transform faults and related structures in the oceans and continents. The logical place to start is where new lithosphere is created at mid-ocean ridges, but this actually commences with continental rifting in the classic Wilson cycle, which charts the birth and eventual death of the ocean.

Rift formation and development

Rifting involves the splitting and separation of crust and/or lithosphere under high shear stresses. Sustained stress propagates or extends rifts, often along major **lineaments** or existing linear weaknesses such as sutures and faults. Oceanic rifting takes the form of mid-ocean ridges, and continental **rift valleys** form with symmetrical separation on both sides of the rift. **Structural basins** form on asymmetrical crustal extension (Figure 3.7). *Active rifting* occurs over mantle plumes and leads eventually to the emergence of new crust. Rifting may still occur in their absence, in which case the necessary **crustal extension** for *passive rifting* must occur mechanically in various ways. Subduction and associated slab pull on the opposite side of a continent may trigger a corresponding **trench suction force**, and thereby extension and rifting in the continental lithosphere on the near side. Surface erosion may have a similar extensional effect. In that case the reduced mass of upper, brittle crust requires an isostatic adjustment. It is achieved by 'inflow' of lower ductile (pliable) crust which undermines adjacent brittle crust. Extension then causes **faulting** as lithosphere is stretched beyond its brittle strength limits, and the rift, or **graben**, is formed as crust subsides between inward-facing faults. New crust is unlikely to form in the rift unless faulting penetrates the entire lithosphere and/or there is sufficient weakness to induce magmatic flow from underlying asthenosphere (Figure 3.7).

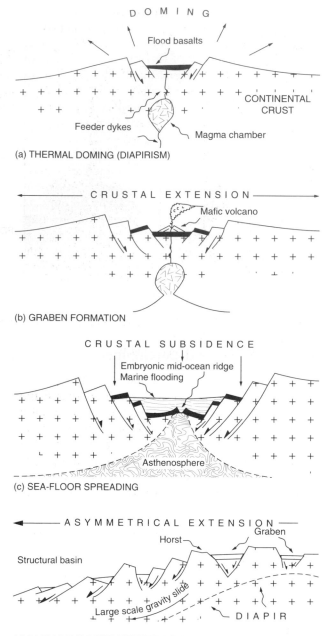

Figure 3.7 *Rift formation and development. Thermal doming or diapirism (a) causes crustal extension and graben formation (b). Continuing subsidence of new, denser basaltic crust leads eventually to marine flooding and the birth of a new ocean (c). Asymmetrical extension frequently leads to basin-range structures (d) with terrestrial sediments (dots) collecting in graben. Faulting is indicated by split arrows on downthrow side.*

Mid-ocean ridges

The **thermal welt** or dome over a rising mantle plume stretches, thins and weakens the lithosphere. The consequent fall in *overburden* pressure lowers the melting point of asthenosphere, which rises faster than it cools. Rock-forming processes are dealt with in detail in Chapter 12. It is sufficient for now to appreciate that the fractionation of different rocks is most intense in the lithosphere and that a more buoyant **gabbro–basalt** mixture segregates and accelerates away from denser asthenosphere **peridotite** at depths of 15–25 km to form sub-surface magma reservoirs. Magma creates new layered oceanic crust where it penetrates the lithosphere and inevitably leaves behind *depleted* peridotite. Gabbro cools to form a sub-surface **intrusive** layer 4–6 km thick whilst the basalt continues to the surface, forming an **extrusive** layer of lava 1–2.5 km thick. Volcanic activity is also associated with mid-ocean ridges and is seen best where the ridge surfaces in the Atlantic Ocean at the volcanic islands of Iceland, the Azores, Ascension and Tristan da Cunha.

The focus of this activity forms a topographical rise or *ridge* 1–3 km high in the sea bed (Figure 3.8). Extension faulting at its heart triggers shallow seismic activity and may create a central rift in slow-spreading ridges. Since the asthenosphere feeds the magma reservoir, and thereby the continuous formation of oceanic lithosphere during the lifetime of the cycle, the ridge system eventually extends to widths of 10^{3-4} km. Why do we speak of oceanic crust and mid-ocean ridges when the process starts by continental rifting? Heat accumulation takes 30–80 Ma to accumulate to the point of rifting, during which time light continental lithosphere is replaced by rising, denser asthenosphere. Isostatic adjustments cause new crust to 'float' at a lower level in the mantle as the continent splits, flooding areas below the contemporary sea level. Continued sea floor spreading gives birth to a new ocean. Hot new crust reacts with sea water. This and other aspects of ocean geochemistry, architecture, associated volcanic activity and the duration of the oceanic stage of the Wilson cycle are described in Chapter 4.

Subduction zones

We know that as the cycle proceeds, and continental lithosphere is rolled away, subduction must be induced elsewhere. It is aided by the cooling and thickening of spreading oceanic lithosphere and enhanced by the basal adhesion of asthenosphere, which itself spreads and cools where it is in contact with cold lithosphere. Where continental and oceanic lithosphere converge, the greater density of the latter ensures that it is always subducted, but cold oceanic crust will also be subducted beneath oceanic crust where regional earth stresses permit. The consequences are a unique global landsystem and hydrothermally driven geochemical reactions on the **resorption** of lithosphere. They may be oceanic or continental in style and location, and can be differentiated further into primary, mechanical/isostatic effects or secondary geochemical reworking of lithosphere and mantle. Indeed, the notion that subduction zones are *destructive* margins is only partially correct. It is not the original oceanic lithosphere which is recycled but a version altered by hydrothermal reaction with sea water on its emplacement, together with sea water itself, ocean sediments and fragments of continental crust and asthenosphere. It perpetuates fractionation processes and is responsible for a wide range of *constructive* materials which form new continental crust, and tectonic landforms such as mountain belts and the majority of surface volcanoes.

Subduction proceeds by the displacement of lithosphere with negative buoyancy through densification and cooling, along a plane inclined in the direction of downgoing oceanic lithosphere. Known as the **Wadati–Benioff** or, commonly, **Benioff** or **B-subduction Zone**, it experiences deep seismic activity. Descending slab is heated by conduction, on contact with hotter lithosphere, and friction in the narrow seismic zone. It undergoes **metamorphism** or physicochemical alteration, and eventually melts at a depth set by the thermal and pressure environment of the subduction zone and its own thickness and geochemistry. Melting will occur at greater depths than at constructive margins, since the critical 1400° C isotherm may be drawn down to over 200 km in subduction zones

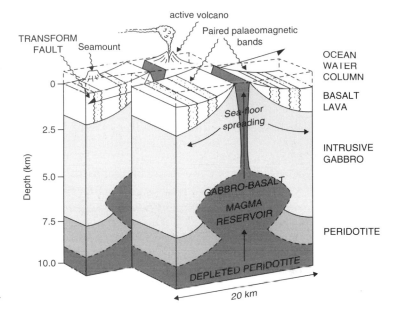

Figure 3.8 *Vertical section through a mid-ocean ridge, drawn at right-angles to the ridge axis.*
Source: After Kearey and Vine (1996).

by cold, descending slab. However, contamination by surface materials reduces the initial melting temperature to less than 650° C; in particular, ocean-wetted basaltic lithosphere begins to melt at about 80 km and the water driven off aids the melting of peridotite in the adjacent continental asthenosphere. Subduction eventually ceases when thermal equilib-

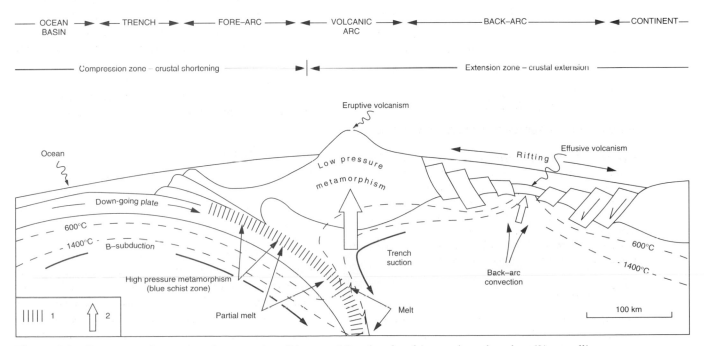

Figure 3.9 *Processes at destructive plate margins: (1) zone of fractional melting and earthquakes, (2) upwelling magma.*

rium is reached with surrounding mantle, at depths of 600–700 km. By then it has created the most globally unstable and complicated surface architecture of marine basins, volcanoes and mountains (Figure 3.9).

Volcanic island arcs and ocean trenches

A glance at an atlas shows that most island systems form curved 'necklaces' strung towards ocean margins. An almost continuous string in the western Pacific extends from the Aleutian Islands of Alaska, south through the Kuril Islands, Japan, the Marianas, the Solomon Islands, the New Hebrides, Samoa and Tonga to New Zealand. Branches also extend through the Philippines and Indonesia into the Indian Ocean. These islands are associated with the evocative names of volcanoes such as Krakatau (erupted 1883), Tambora (erupted 1815), Pinatubo (erupted 1991) and Fuji. Together they form the western half of the circum-Pacific volcano-seismic *Ring of Fire* (Plate 3.1). Volcanic arcs also occur where smaller Caribbean (the Antilles arc, including Mount Pelee, erupted 1902) and Scotia (the Scotia arc) plates oppose the westward-spreading Atlantic Ocean. A small arc in southern Greece and Italy marks the residual thrust of the African plate under Europe. Here the Santorini eruption of *c.* 1625 BC is thought to have ended the Bronze Age *Minoan* civilization of ancient Crete, Vesuvius destroyed Pompeii in AD 79 and Etna entered a new active phase in 1980.

Their three-dimensional architecture consists of three to six concentric, narrow structures invariably convex towards the subducting plate; this is the nature of angled incision into a curved surface, as demonstrated by an oblique knife-cut into an apple!

Plate 3.1 *Mayon, an andesite strato-volcano 2421 m high on the Philippine island of Luzon, 300 km south-east of Manila, which erupted earlier this century. Photo: Landform Slides, Lowestoft.*

The outermost structure is an ocean **trench**, typically 50–100 km wide, over 1000 km long and 5–10 km deep. The Marianas trench is the deepest known, at 11.04 km. Beyond the trench, completing the **fore-arc**, lies an **accretionary prism** and **fore-arc basin**, followed by the **volcanic arc** itself, which may have an outer, inactive zone fringing an inner line of active volcanoes. The **back-arc** completes the full sequence and may contain a marine basin and one or more **remnant arcs**, each representing an extinct volcanic axis. Volcanic arc dynamics and architecture clearly indicate that the entire zone is mobile and its focus of activity may shift (Figure 3.10).

Active subduction keeps the trench open against isostatic forces. It also acts as a *sump* for a potentially major sediment flux of low-density erosion products from the volcanic arc, organic and inorganic **pelagic** debris rain-out from the overlying ocean and **turbidity currents** of **flysch**, or sediment. These eventually compose the bulk of the accretionary prism by *off-scrape* from the upper side of the descending slab against the non-subducting plate, as their low density resists subduction. Successive offscrapes occur on the underside of the prism, which occasionally emerges above sea level. Accretion can be augmented by large slivers of layered oceanic crust sheared off during subduction and known as **ophiolite** in the **mélange**, or chaotic mix of rock material, forming the prism (Figure 3.10). Ophiolite found high in the Alps, including the Matterhorn, and Himalayas is important evidence of the power of plate tectonics. So too are the north-east–south-west parallel bands of rocks, ageing northwards, now forming the Scottish southern uplands, which represent the accretionary prism of the closing Lower Palaeozoic Iapetus Ocean.

Trench–arc distances are determined by the angle and rate of subduction since the volcanic arc overlies the zone of maximum melt. At angles above 25°, slab may be resorbed without vulcanism; 30–60° provides a fore-arc width of approximately 200 km, and steeper angles can halve this distance. Continuous, **effusive** outflows of basaltic lava at mid-ocean ridges contrast with episodic, **explosive** volcanic eruptions in volcanic arcs, exhaling gases and ejecting ash and larger debris or **tephra**. As the arc matures the depth of its magma source increases, causing geochemical compositional changes as different mineral cocktails fractionate from the wet, subducted oceanic crust and overlying melted asthenosphere. Early melts produce silicate-poor, less viscous basalt–andesite volcanoes switching to silicate-rich, more viscous andesite–rhyolite later. Intrusive, non-erupting magmas solidify below the surface and will become exposed only on subsequent erosion. These processes are detailed in Chapter 12.

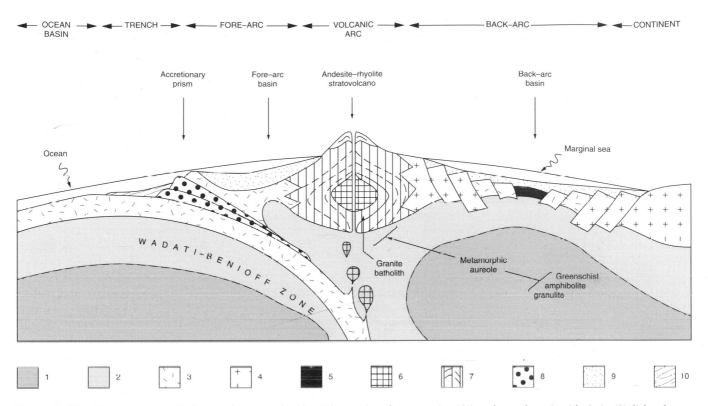

Figure 3.10 *The tectonic morphology and principal rocks of destructive plate margins: (1) asthenosphere (peridotite), (2) lithosphere, (3) oceanic (basaltic) crust, (4) continental (granitic) crust, (5) new basaltic magma, (6) magma diapir, (7) volcanic arc, including metamorphic facies, (8) mélange, including ophiolite, (9) flysch, including volcanoclastic sediment, (10) flysch, with trench turbidite.*

The back arc, which may be 200–600 km wide, develops by extension in contrast with **crustal shortening** or compression in the subduction zone. Extension occurs in the non-descending slab as a result of either active oceanward overriding, movement in the same direction as the descending slab (including trench suction force), rifting of the elevated volcanic arc or even a magma **diapir** creating thermal doming (Figure 3.10). Occasionally the back-arc basin contains a remnant arc, abandoned as the back arc spreads or where a subduction zone migrates *away* from an arc. Subsidence on extension creates a basin which may flood, and most volcanic arcs identified above impound **marginal seas** on oceanic plate between them and adjacent continents. North-west Pacific arcs enclose six such marginal seas but this is a complex zone. East of the Philippines there are two arcs, which meet farther north, as *two* plates subduct beneath Japan simultaneously, one stacked above the other (Figure 3.11).

Marginal arcs and continental collision

So far we have reviewed subduction in the oceanic plate context. In the supercontinental cycle the eventual fate of volcanic arc complexes is to migrate and accrete on to continental plate. Oceanic subduction 'goes onshore' beneath continental lithosphere in arc–continent convergence. This is seen at various stages of completion around the Pacific Ocean and explains its tectonic asymmetry. Intra-oceanic arcs of the western ocean lie well offshore, contrasting with continental-margin orogens to the east, where the Pacific mid-ocean ridge and its branches, defining the Cocos and Nazca oceanic plates, are much nearer the continental subduction zone. Convex arc shapes can be seen in the coastlines of British Columbia, Central America and Ecuador–Peru, completing the Pacific Ring of Fire, with the trench a short distance offshore. The Sunda arc forming the Indonesian islands between Malaysia and northern Australia represents a transitional phase, present on a smaller scale in South Island, New Zealand.

During convergence, B-subduction magmas erupt through continental crust as terrestrial volcanoes. Recent major eruptions along the Pacific coast of the Americas include Katmaï (erupted 1912) and Mount St Helens (1980) in North America and El Chichón (1982) and Nevado del Ruiz (1985) in Central and South America. Alternatively, magma crystallizes as huge granitic **batholiths** from deep magma reservoirs. More than 2 M km³ of granite was intruded

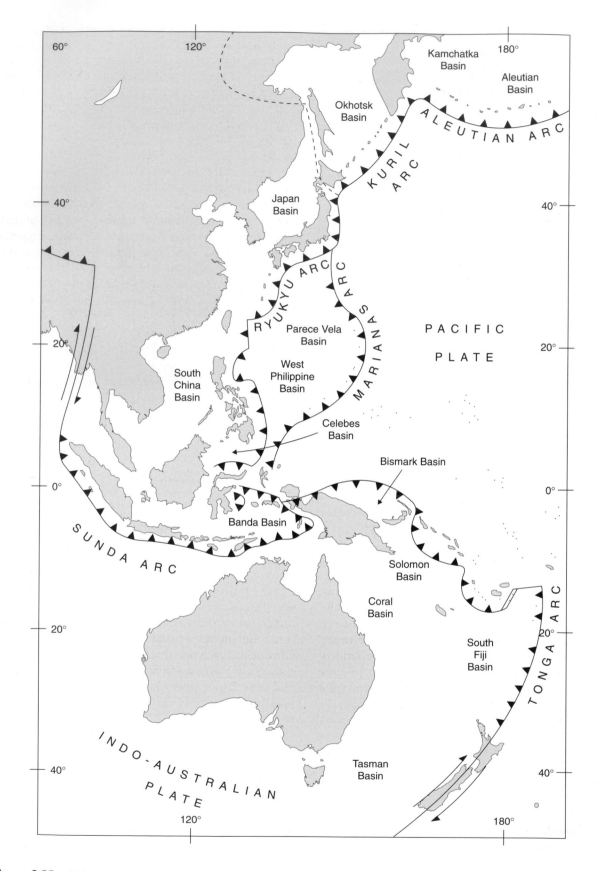

Figure 3.11 *Volcanic arcs, back-arc basins and marginal seas of the western Pacific and Indonesia; key as for Figure 3.5.*

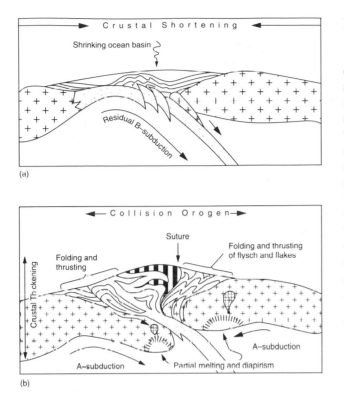

Figure 3.12 *Two stages in continental collision. The convergence of two continents has almost closed the intervening ocean basin (a) leading to continental collision, A-subduction, folding and faulting (b). Ophiolite nappes (stripes) lie along the suture; key otherwise as for Figure 3.10.*

in the Andes, now exposed by erosion over an area of nearly 500 k km².

In due course, subduction of remnant oceanic crust marks the death of the ocean and leads eventually to collision between converging continental plates. Despite the acquisition of denser rocks, lighter continental crust remains buoyant and influences the development of different continental subduction processes. Convergence rates fall sharply along the intra-continental suture but driving forces are still sufficient to cause crustal shortening. Since neither slab of lightweight continental crust is capable of significant B-subduction, crustal shortening must be compensated by **crustal thickening**. This is achieved by complex **thrusting** of flakes or slivers of crust over and under each other, known as **Ampferer-** or **A-subduction**, which does not proceed to sufficient depths/temperatures for crustal recycling (Figure 3.12). Instead, downward displacement of light crust is compensated by an isostatic elevation of the developing pile, forming thick continental plate in which the subducted slivers constitute deep roots. *Intercontinental collision tectonics* in the suture zone are active along almost the entire Alpine-

Himalayan systems, also known collectively as the *Tethyan orogen*, after the ocean which spawned them.

Transform faulting is a variation of general subduction. Different rates and directions of ocean spreading are transmitted onshore as **strike-slip** faulting, with horizontal displacement along the strike or fault axis. Plates may move past each other with nothing more than seismic effects at their conservative margin, or with oblique subduction or spreading. This occurs at the plate boundary itself or, through drag effects, in adjacent lithosphere. New Zealand is transected by active strike-slip B-subduction at the Pacific/Indo-Australian plate boundary. The San Andreas fault system of California, triggered where the East Pacific Rise (mid-ocean ridge) comes onshore, is moving San Francisco and Los Angeles slowly together. The Great Glen fault in Scotland displaced what is now the Scottish north-west highland zone 300 km south-west during the Palaeozoic. The many oblique, as well as head-on, movements at convergent plate boundaries determine that *accretionary tectonics* incorporate slivers and flakes of **suspect** or **displaced terranes**, including ophiolites far removed from their origins.

Orogens

Plate convergence creates major **orogens**, or narrow, linear **cordilleran mountain** systems – literally, 'chords' of sub-parallel ranges and intervening basins reflecting substantial crustal deformation. The origin and dynamic character of major examples are explained in Chapter 25. Modern, Cenozoic orogens represent later stages of the post-Pangaea supercontinental cycle and many experience rapid uplift of 10–20 mm yr⁻¹, stimulating vigorous erosion as endogenetic and exogenetic forces compete (Figure 3.13). Although annual rates seem modest, they amount to several thousand metres if sustained for just 1 Ma and are convincing evidence that great tracts of our highest mountain ranges achieved their present altitude and appearance only during the later Tertiary and Quaternary periods. Older, tectonically quiet and more subdued orogens, such as the Palaeozoic mountains of Appalachia–Scandinavia (the Caledonian orogeny) and the Atlas, Urals and Tien Shan (the Hercynian orogeny), can still be recognized in continental interiors and now-stable Archaean orogens contribute to continental cratons, or shields.

Subduction processes absorb collision and generally convert crustal shortening into thickening along a narrow front. Occasionally, sustained thickening and isostatic uplift elevate broad plateaux rather than narrow orogens. The Tibetan plateau is the most

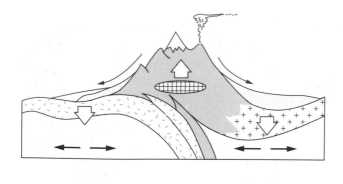

Figure 3.13 *The endogenic (tectonic) and exogenic (geomorphic) processes which shape an orogen. A volcanic arc has 'gone onshore' and the resultant orogen is formed by isostatic uplift (1) of mélange, flysch and continental slivers around a granitic batholith. Lithospheric creep (2), which reinforces uplift, creates subsidence (3) elsewhere, and subsiding basins trap siliciclastic sediment (4) eroded from the orogen. B-subduction magmatic eruptions and alpine glaciation ornament the orogen.*

celebrated example, covering over 2 M km² at a mean altitude of some 5 km and a crustal thickness of 70–80 km. It reached its present height during the past 3–5 Ma and is instrumental in Quaternary changes in global atmospheric circulation and the south-east Asian monsoon. The Bolivian Altiplano, 4 km in mean elevation over 0.4 M km², and the Colorado Plateau (0.5 M km² and 1.9 km high) have similar origins. Uplift of the latter triggered the spectacular 2 km incision of the Grand Canyon by the Colorado river, exposing some 1 Ma of earth history. Plateaux not underlain by thick crustal lithosphere are elevated by epeirogenesis. Occasionally one plate continues its advance into another. The Indian continental sub-plate has achieved this, *indenting* (penetrating) the Eurasian plate by a further 2000 km since initial collision *c.* 40 Ma ago and continuing to promote **indentation tectonics** well into Asia.

Continental formation, evolution and architecture

We have seen that tectonic activity exchanges material between ocean and continental crust, consigning some terrestrial erosion products to the subduction melting pot and accreting oceanic ophiolite on to continental margins. Yet their mean ages indicate that continental crust (1.1 Ba) is largely conserved, whereas oceanic crust (55 Ma) is largely recycled. It is thought now that new continental crust has been added slowly, by the accretion of oceanic plate at

a rate of approximately 1.3 km³ yr⁻¹ over the past 1.0 Ba. Some 50–70 per cent of primary continental crust was formed by *c.* 2.5 Ba ago, during the Archaean. Continental lithosphere – a 'penultimate silicate froth' containing, additionally, some of Earth's least common elements – is the descendant of Archaean crust. Its outer terrestrial fractionates formed an upper layer of volcanic and intrusive granitic, low-melting-temperature products on a metamorphosed granitic base. Only 15 per cent survives, which emphasizes the role of erosional, sedimentary and metamorphic processes in reworking primary continental crust without removal from the continental system; and offscrape, accretion and B-subduction recycling at destructive plate boundaries and remelt of continental lithosphere over hot-spots.

Continental crust is more extensive than is suggested by the ratio of land to sea area. It accounts for 39 per cent of all crust, or 0.6 per cent of Earth's volume, whereas continents cover only 29 per cent of the surface. The difference is accounted for by the presence of **epicontinental seas**, such as the North Sea, the Black Sea, parts of the Mediterranean and Hudson Bay, on continental crust and a portion of the **continental shelf** (see Chapter 4). Continental architecture consists primarily of stable ancient **cratons** swathed in six to eight orogens of various, younger ages (Figure 3.14). Cratons are stable nuclei around which continents form and reform, with only minor 'bruising', contrasting with the high geothermal flux, relief and geomorphic activity of the Cenozoic orogens. Despite their stability and location, usually over cool spots, the cratons have endured sustained erosion since their Archaean or Proterozoic formation and are now areas of modest overall and relative relief. Between stable cratons and unstable orogens, epeirogenesis and isostatic adjustments create lesser, long-term disturbances of continental lithosphere. Their principal effect is to adjust gravitational energy inputs and hence erosion rates, with elevated plateaux forming sediment sources and subsidence zones creating basins which act as terrestrial sediment traps.

What are the consequences of rifting, which initiates the Wilson cycle and eventually forms new oceans, for the continents themselves? Emerging mid-ocean rifts may propagate across ocean–continent boundaries. The East African rift, for example, is the landward extension of the triple junction formed with the Red Sea and Gulf of Aden. If it fails to propagate further and develop into a full mid-ocean ridge system it will join a long list of **aulacogens**. Although classed as failed rifts, they still play a significant role in continental architecture, often forming major

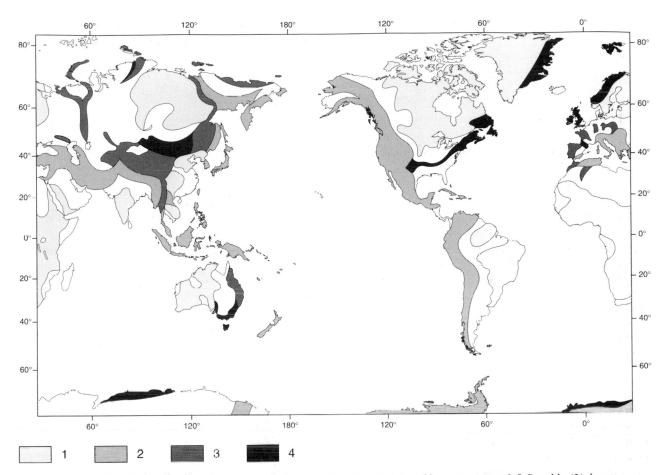

Figure 3.14 *Global distribution of cratons and Phanerozoic orogens: (1) stable cratons over 1.5 Ba old, (2) late Mesozoic–Cenozoic orogens up to 200 Ma old, (3) late Palaeozoic (Variscan) orogens c. 300 Ma old, (4) mid–late Palaeozoic (Caledonian) orogens c. 400 Ma old. Most intervening areas are sedimentary basins. Sources: After Fowler (1990) and Smith (1982).*

topographical depressions which channel world-scale rivers and their sediments; the North Sea and Rhine graben are probable examples (see Figure 4.2). Continental rifting in which significant crustal extension or thinning occurs has other consequences. Asymmetrical rifting may generate **basin and range** topography, best seen between the coastal and Rocky mountain ranges of the western United States, and may be accompanied by basaltic extrusions.

Effusive **flood basalts** inundate the existing landscape and may extend over large areas. The Mesozoic break-up of Pangaea generated large basalt flows near rifted margins which often survive as resistant plateaux, as in parts of India (Deccan, 0.5 M km²), the United States (Columbia, in Washington/Oregon 0.13 M km²) and South Africa (Karroo). The latter flood basalts formerly covered over 5 M km². Up to 1.8 km of Tertiary basalts were extruded in the Irish–Hebridean basin of the British Isles during the formation of the Atlantic Ocean, forming the modern Antrim plateau and Fingal's Cave landmarks (Plate

3.2). *Eruptive* alkali-basalts and other magmas form volcanoes. The largest concentration forms the East African Rift Valley complex, with well known individuals such as Mount Kenya, Kilimanjaro and Ruwenzori over 5 km high. In Europe the Eifel volcanoes of Germany are associated with the Rhine graben.

Sea floor spreading also leaves **passive margins** on the trailing edge of continents which often form major escarpments, upwarped either by the initial crustal elevation producing the rift or a subsequent isostatic or thermal (epeirogenetic) response. Passive margin escarpments, reaching elevations of 1.8–3.5 km, are best developed on the eastern seaboards of southern Brazil, southern Africa (the Drakensburg range) and Australia (the Great Escarpment), both sides of the Red Sea and the west coast of India (the Western Ghats). The Piedmont Fall Line marks a lower persistent passive margin on the south-east coast of the United States. In all cases, escarpments source continental-margin sedimentation on their seaward side.

Plate 3.2 *Coast of the Antrim basalt plateau, Northern Ireland, around Benbane Head and Giant's Causeway. Individual, near-horizontal flood basalts form prominent cliffs.*
Photo: Kenneth Addison.

The geological evolution of Britain

What modern continental area illustrates crustal evolution better than the British Isles? Into their diminutive 150,000 km² are crammed rocks representative of half Earth's history and structures of all three *Phanerozoic* orogens. Origins are traced through rocks which tell a story of fragments lost, gained and surviving as terranes were assembled and dismantled in the long drift across Earth's surface. Britain's familiar coastline is less than 10 ka old and dependent on global sea level. The story is elaborated by three vital strands of the science of **stratigraphy**. *Litho*-stratigraphy and *bio*-stratigraphy reveal the physical and biological character of past environments and *chrono*-stratigraphy provides a time scale, based on the decay of constituent radioactive minerals. The early history is very obscure but we have a clearer view of the past 0.5 Ba, in which fragments originating 60° *south* of the equator were joined by others as 'Britain'

drifted to its modern position at 50°–60° north. *En route*, subtropical Silurian coral reefs were joined by desert sands 'sandwiching' equatorial swamp forests and the whole was subjected most recently to Quaternary glaciation (Figure 3.15).

Conclusion

Earth is of almost unimaginable age and yet its modern character and geological processes can be traced directly to its astronomic origins. Many geological processes are imperceptibly slow, measured against our own life spans, but earthquakes and volcanic eruptions are sharp reminders of Earth's relentless evolution. The accretion of a cocktail of planetary matter over 4.6 Ba ago set in train enduring processes of chemical fractionation, heat generation and cooling which progressively refine our planet's constituent parts. Some processes are necessarily familiar to us – the weathering of simple nutrients from more complex rocks, the evaporation of H_2O from ocean solutions, the abstraction and reformation of carbon, hydrogen and oxygen from a variety of sources to form living cells and even the segregation of different isotopes of these fractionates (^{14}C, ^{16}O, ^{18}O, etc.). All require energy, and the nature and rate of their reactions are also determined by temperature and pressure.

By regarding the familiar atmosphere, hydrosphere and biosphere as Earth's outer fractionates we can more easily understand the mantle, asthenosphere and lithosphere inner fractionates with which they are linked. This lays the foundations for appreciating the morphotectonic and rock cycles which shape our Earth, its principal architecture of oceans, continents, mountains and sedimentary basins, etc., and surface geomorphic processes and landsystems. We see Earth's surface no longer as just a static outline on the world map but as an evolving scene whose contemporary components have come together so recently – to create the world we know – and are already changing towards the world our descendants will inhabit.

Key points

1 Earth acquired its own heat engine and a particular set of planetary raw materials, determined by its position in the solar system, at its formation. The continuous fractionation or segregation and refinement of these raw materials, in conjunction with solar radiation at Earth's surface, has created our atmosphere, hydrosphere and biosphere. Their

reaction with fractionation processes in Earth's geological interior produces familiar surface physical features.

2 Earth's interior is similarly segregated into an inner core surrounded concentrically by a mantle and crust. The outer mantle and crust form a cool, light and brittle lithosphere capable of being dislocated and moved as semi-rigid plates by convection currents in the underlying deformable mantle asthenosphere. This

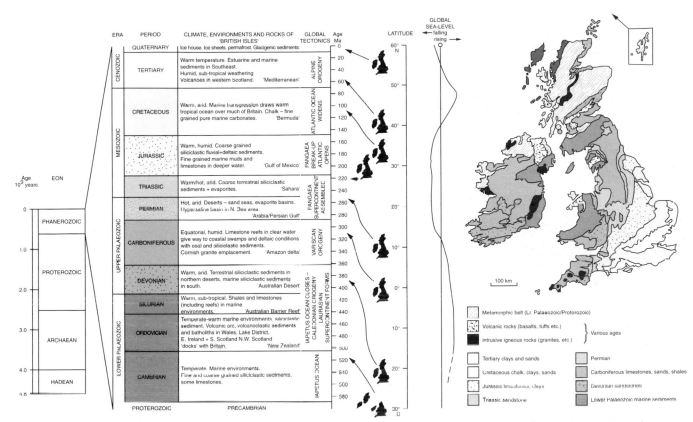

Figure 3.15 *A geological time scale (after Harland et al. 1982), with the principal stratigraphy, regional geology and palaeo-environments of the British Isles (in part, after Lovell 1977). Modern analogues for Britain's past environments are given.*

movement is assisted by partial melting of the asthenosphere in prevailing temperature–pressure conditions.

3 Continental lithosphere thins and rifts apart over rising mantle plumes and new oceans form as sea water floods in. Asthenosphere peridotite rises faster than it cools into the rift and forms new, denser oceanic lithosphere. The ocean enlarges by sea-floor spreading and the divergent plates eventually converge elsewhere with other plates.

4 One plate is subducted below another at convergent boundaries, and crustal thickening in the form of orogenic uplift compensates for the resultant crustal shortening. Denser oceanic crust slides more easily

beneath light continental crust and drags adjacent sea-floor sediments and continental slivers into a remelt zone. The resulting volcanic island arc complex eventually migrates and welds on to the continent. Continental collision orogens are strongly metamorphosed and intruded by granite batholiths.

5 These processes cycle between supercontinent–fragmentary continent and associated single–fragmentary ocean phases and back again over approximately 500 Ma. Plate tectonics creates the fundamental architectural units of Earth's surface, potential energy for denudation through uplift and spatial patterns of rock formation, alteration and destruction.

Further reading

Howell, D.G. (1995) *Principles of Terrane Analysis* (second edition), London: Chapman and Hall.
The title reflects a new generation of research into plate tectonics. A useful summary of general tectonic processes is followed by a continent-by-continent outline of their accretion from plate fragments.

Kearey, P. and Vine, F.J. (1996) *Global Tectonics* (second edition), Oxford and London: Blackwell.

This is an important but readable text on plate tectonics, supported by an unobtrusive level of technical explanation and by simple line drawings rather than photographs.

Smith, D.G. (ed.) (1982) *The Cambridge Encyclopedia of Earth Sciences*, Cambridge and New York: Cambridge University Press.
A comprehensive review of Earth from astronomic origins through geological processes to geological resources, hazards and extraterrestrial geology. The book is well illustrated in colour and follows a concise, chronological and systematic narrative rather than an alphabetical listing.

The global ocean is the 'Cinderella' of physical geography, neglected in or omitted from many textbooks and courses for a variety of reasons. It is perceived as a **black box** of largely invisible parts operating as a homogenous, stable system with monotonous surface expression except at the coastline – scarcely the geographically interesting stuff of rapid spatial and temporal change. Above all, it is not the habitat of the human genus. However, meteorologically it is the largest single moderating influence on extremes of radiation budgets and climate and the source of most precipitation. Geochemically, with the atmosphere, it buffers Earth from the geological extremes of lithosphere and space, and receives and recycles sediments derived from terrestrial erosion. Biologically, it houses some of the world's simpler and locally most productive ecosystems.

Although the marine environment is inimical to human life, the land–sea '*ecotone*' – the coastline – houses Earth's highest concentrations of human population, agriculture and industry by choice or of necessity (Plate 4.1). Many peoples – the British, Japanese, Caribbeans and Indonesians, for example – are islanders; many others occupy the ocean fringes of less hospitable desert, polar or mountainous regions. All have a strong maritime thread to their lives and are becoming aware of threats to coastline integrity, through rising sea level, and water quality, through pollution. Globally, we view oceans increasingly as a potential source of food and minerals as we outgrow or deplete terrestrial resources. Geographers recognize the need to understand our oceans better and this chapter sets out the more important dynamic characteristics of the oceans.

Evolution of Earth's ocean basins

After Pangaea: the formation of modern oceans

Pangaea coalesced as Earth's most recent supercontinent in the late Palaeozoic, *c.* 290 Ma ago, and survived for *c.* 100 Ma before restless tectonic stresses rifted it apart again. Rocks common to its now separated parts reveal that it had a primarily meridional distribution, extending from *Gondwanaland*, centred between the south pole and the equator, well into the northern hemisphere mass of *Laurasia*. Their *Hercynian* collision-suture orogen is now dispersed from the Appalachians, through Cornwall, South Wales and north central Europe, to the Urals and north-eastern Asia. Pangaea's common Carboniferous–Permian–Triassic rocks reflect global palaeo-climates, with a major south-polar ice sheet, subtropical deserts and tropical/equatorial swamp deltas and shallow, carbonate-rich seas (Figure 4.1).

Rifting began as a precursor of the eventual break-up in parts of Pangaea as other parts were sutured together, *c.* 255 Ma ago in the Permian. It commenced in what is now north-west Europe, with a typical suite of early continental rift basalts in a

Plate 4.1 *Cliffs, beaches, dune barriers and a tidal estuary form a typical coastline. Humans seek sites for defence, economic advantage, recreation and aesthetic pleasure.*
Photo: Kenneth Addison.

subtropical continental arid environment of red sands and evaporites. Graben formed around the British Isles, including the midland valley of Scotland and the North Sea, and in the New York–Connecticut–Liberia (West Africa) zone. Extension led to sea-floor spreading, and continental separation commenced in the Jurassic *c.* 180 Ma ago, 40–70 Ma after initial rifts. The early Mesozoic global Panthalassic Ocean covered an entire hemisphere off Pangaea's west coast, with a major arm, the Tethys Ocean, partially enclosed by the more indented east coast. Sea-floor spreading opened new oceans at their expense.

The formation of the Atlantic holds the key to modern ocean evolution and related continental tectonic architecture. It began to open in its modern north central zone, between North America, northwest Africa and South America, followed by the South and then the North Atlantic (Figure 4.2). Sea-floor spreading is progressive, like a slow-motion view of a chick hatching from its egg. Stronger individual cracks develop but weaker cracks cease to propagate and remain as aulacogens as the active rifts become mid-ocean ridges. There is no question of a single, linear rift opening uniformly over the 16,000 km of the Atlantic, still less over the more than 50,000 km extent of modern mid-ocean ridges. Alignment was driven by the initial disposition of hot-spots, either at random or controlled by structural **lineaments** from earlier events. The North Atlantic opened from Florida to Britain, parallel to Caledonian/Hercynian orogens and the Iapetus suture. Spreading puts extensional strains on adjacent areas and can also propagate through intact crust. Adjustment to movements elsewhere and the earth's curved surface create transform faults and **triple junctions**. A number developed in the Atlantic, especially during the final stages between Canada, Greenland and Europe. Some arms failed, forming aulacogens which now guide major continental river basins such as the North Sea–Rhine graben (Europe), the Benue–Niger (Africa), Hudson and Mississippi (North America) and the Orinoco, Amazon and Parana (South America) (Figure 4.2).

American plates moved west as the Atlantic opened, consuming the Panthalassic/Pacific Ocean faster than it was spreading. Their respective ocean basin areas have changed by about +160 per cent and −35 per cent during the Cenozoic (Figure 4.3). Cordilleran systems on their Pacific coasts have elements older than Mesozoic–Cenozoic and include earlier, Palaeozoic subduction terranes. North America has now overrun much of the mid-ocean ridge. The Panthalassic and Tethys Oceans were also under attack from the eastward fragmentation of Gondwanaland and its collision with Eurasia. The

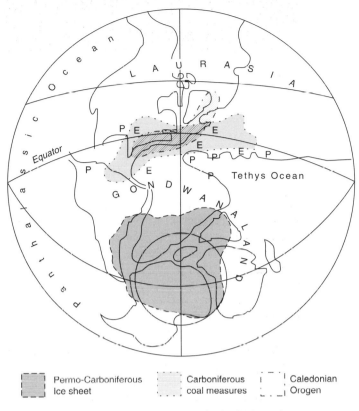

| Permo-Carboniferous Ice sheet | Carboniferous coal measures | Caledonian Orogen |

P, Petroleum deposits   E, Evaporite deposits (rock salt, gypsum)

Figure 4.1   *Reconstruction of the supercontinent Pangaea during the Permo-Triassic c. 250 Ma ago. Note the Caledonian collision orogen, formed on closure of the Iapetus Ocean as Pangaea coalesced, and the concentration of coal and evaporite deposits along the Equator. Petroleum deposits were also formed there later, in the Cretaceous. Source: In part after Kearey and Vine (1996).*

Tethys connected briefly with the developing Atlantic to form a narrow equatorial Cretaceous ocean before closing like a zip. Biogenic sediments in its marginal basins and epicontinental seas now form the world's principal oil reserves. The 'zip' did not close smoothly, and progressive Euro-African collision in the west less than 140 Ma ago opened younger rifts in East Africa and the Red Sea–Gulf of Aden–Persian Gulf region (Figure 4.4).

India and Australia broke away to the north-east, opening up the Indian Ocean behind and closing the eastern Tethys ahead, culminating in the Cenozoic India–Asia collision 40 Ma ago. Continuing indentation of India into Asia nudged south-east Asia and China eastward to aid the consumption of the northwest Panthalassic Ocean. The Pacific Ocean is its diminutive successor, almost completely refloored since break-up. The oldest surviving early Jurassic floor is now subducting in the Marianas trench, compared with younger Cenozoic crust beneath Peru and Quaternary crust beneath California and Mexico

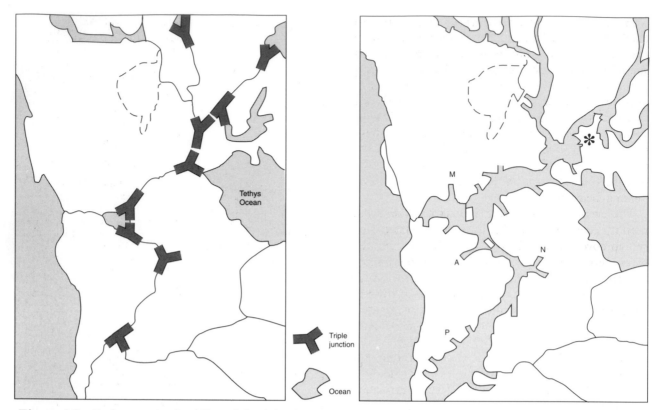

Figure 4.2   *Early stages in the rifting of the Atlantic Ocean from c. 200 Ma ago. (a) Hot-spots generated triple-junction rifts which connected up and initiated sea-floor spreading. (b) Aulacogens (failed rifts) form important structural weaknesses and may direct the line of major river basins such as the Amazon (A), Mississippi (M), Niger (N) and Parana (P). Asterisk (*) shows the position of the embryonic British Isles.*
Source: In part after Windley (1977).

(Figure 4.5). The Tethys survives only in the Mediterranean–Black Sea region. The opening of the Drake Passage and the Scotia arc between South America and the Antarctic Peninsula, and the closure of the Panama seaway linking the Americas, are two other further Cenozoic stages in ocean evolution. The former completed the circumpolar Southern Ocean and the latter closed the tropical Atlantic–Pacific. Both had a major influence on Quaternary global glaciation. The next supercontinent will take shape as Atlantic widening continues before eventually subducting beneath the Americas, whilst Africa rifts apart to create a new ocean.

Ocean architecture

The dynamic architecture of mid-ocean ridges and trenches drives the supercontinental cycle and covers approximately one-third of global ocean area. Ridges account for 95 per cent of this and form Earth's principal continuous relief. They are broadly symmetrical, parallel crest-and-trough structures, 10^{2-3} km wide and 2–4 km high, offset along transform faults and

sloping away from the thermal rise at their central axis, at angles proportional to their spreading rate. Topographic symmetry is matched by a geomagnetic 'bar code' of bands showing normal and reversed magnetic polarity. Magnetic minerals in basalt extruded at the central rift assume Earth's magnetic polarity before cooling and record its reversals over 10^{4-6} years. Paired bands with similar palaeomagnetic direction and radiometric age either side of the axis underpin our reconstruction and timing of ocean evolution (see Figure 3.8). Ridge axes, beneath an average ocean depth of 3 km, are also important minerogenic centres where hydrothermal processes exchange minerals at the ocean–lithosphere interface (see Chapter 13).

Abyssal plains occupy most of the deep ocean floor between the ridges and trenches, covering 42 per cent of the total area, with an average depth of 5–6 km. They are floored by cool, older oceanic crust which has subsided into the lithosphere beyond the spreading ridges and are the flattest places on Earth, broken only by submarine plateaux and **seamounts**. The latter are distributed randomly and form away from ridges, although they may be associated with their positive thermal anomalies. Elsewhere, seamount chains form as ocean crust migrates over fixed hot-spots. Islands such as Hawaii appear where they break surface but most are submarine, with summits levelled by marine planation or post-eruptive subsidence. Known in this form as **guyots**, many provide attachment sites for coral reefs in shallow, clear-water tropical seas. The stark contrast between steeply sloping seamounts ($15°–25°$) and abyssal plains (less than $0.1°$) is provided by a thick carpet of **pelagic sediments**, derived by rain-out or solid precipitation from minero-biogenic sediment sources, infilling bedrock depressions.

The active ocean–continent boundary has a fluctuating coastline and characteristic offshore slope system covering over 20 per cent of ocean area. The hypsometric curve suggests that the **continental shelf** and **coastal plain** are a continuous feature and that the principal boundary occurs at the **continental shelf break** at approximately 200 m depth and an average 70 km from the shore. Below this, the **continental slope** inclines at $3°–6°$ towards the abyssal plain, which it meets at the **continental rise**. Elements of this model may be seen in trench-arc coastlines but it is most applicable to passive continental margins. The entire zone is draped with **terrigenous sediments**, sourced from land, in an assemblage transitional between terrestrial and marine environments because of marine **transgression** (advance) and **regression** (retreat) across the zone. Bedrock channels with sediment infills incise both shelf and slope alike. The former are likely to be the **buried channels** of rivers extending offshore but the latter are invariably **submarine canyons** formed by marine processes alone (Figure 4.6).

Ocean basin geometry and sea levels

The position of the coast is determined by **sea level**, which, in turn, is dependent on ocean water volume and ocean basin geometry. Sea level can fluctuate over the following fundamentally different time scales. *Short* – **tides** and **waves** operate over $10^{-6}–10^{1}$ years (minutes–years), with diurnal and monthly tides being the most regular. *Intermediate* – **eustatic** changes in global water volume and **isostatic** changes in basin geometry driven by localized vertical displacements of crust operate over 10^{1-5} years. *Long*

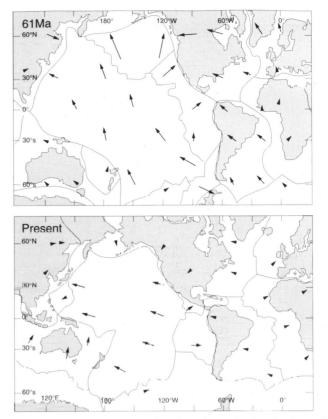

Figure 4.3 *Evolution of plate boundaries during the Cenozoic, showing shrinkage of the Pacific basin, largely at the expense of the spreading Atlantic basin. Arrows show directions and relative rates of plate motion. Source: Open University (1992).*

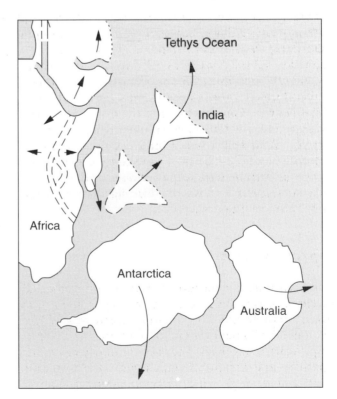

Figure 4.4 *Progressive continental rifting of eastern Gondwanaland, showing composite motions c. 140–40 Ma ago. The Red Sea divergent plate boundary and associated East African–Middle East rifts may mark the birth of a future ocean.*

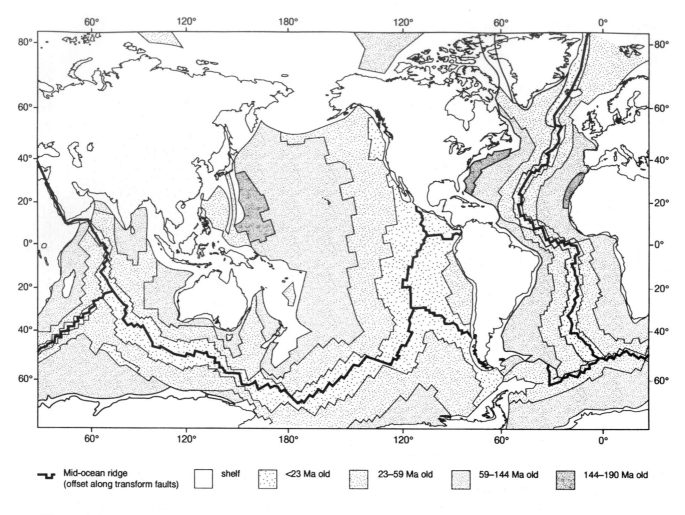

Figure 4.5 *Age of the ocean floor.*
Source: After Scotese et al. (1988).

– **tectonic cycles** alter ocean geometry at the longest time scales (10^{6-8}) discussed earlier. It is also noted that hydrogen outgassing from the early Earth probably reduced initial water volume. The coastline is essentially in equilibrium with tidal variations because of their short periodicity. We are concerned primarily with the average sea level and average position of the coast, susceptible to intermediate fluctuations alone, of considerable significance despite far slower rates and smaller magnitude than tides. Volumetric changes depend on the global hydrological cycle but with over 97 per cent of global water mass held in the oceans they are intrinsically stable. Ocean basin and water mass dimensions are shown in Table 4.1.

Eustatic control of sea level

Eustasy is the control of sea level by water volume. Eustatic change is world-wide and immediate because water effectively finds a common level. Change occurs either by *steric* effects – adjustments to sea-water density via temperature or salinity – or through net mass transfers between *coupled* (linked) stores. Thermal expansion would raise sea level by approximately 0.8 m for a 1° C rise in global temperature before any ice melts. Changes in atmospheric pressure also create measurable changes in sea height, falling in anticyclonic (high-pressure) and rising in cyclonic (low-pressure) conditions at the rate of about 1 cm mb^{-1}. The ocean–lithosphere couple, which cycles water through oceanic crust via subduction and hydrothermal circulation, is assumed to be in equilibrium – partly because it is difficult to assess!

We have a better grasp of ocean–atmosphere–cryosphere coupling and its component terrestrial hydrological cycle. Intermediate-term instability is associated with the Quaternary growth and decay of global ice sheets which form the bulk of the **cryosphere** (ice-bound systems). Evaporated ocean water

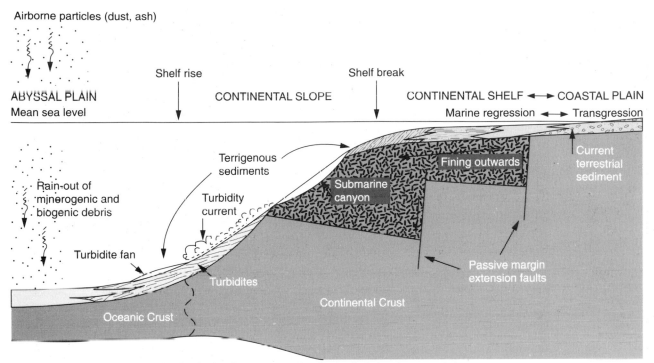

Figure 4.6 *The continental margin landsystem on a 'trailing edge' (passive margin) coast.*

stored in terrestrial glaciers during a cold stage causes a eustatic *fall* in sea level. Deglaciation causes a *rise*, corresponding to the water-equivalent ice mass melted and returned to the oceans. The most recent glacial/interglacial cycle of the past 125 ka recorded overall eustatic changes of about ± 130–165 m. Modern sea level would rise by a further 60–80 m if the remaining ice sheets were to melt, with major coastline implications. The rate of change can be rapid, with a rise from –130 m to –60 m between 15,000 **BP** and 9000 BP (years Before the Present, using AD 1950 as the radiometric index year), and exceeded 20 m ka^{-1} during catastrophic ice sheet collapse *c.* 12,500 BP. Sea-ice formation/melt has a negligible eustatic effect, since it replaces virtually its own volume of water.

Isostatic control of sea level

Isostasy is the gravitational equilibrium between crustal lithosphere of different thickness/density, and therefore 'buoyancy', through vertical or lateral adjustments in adjacent lithosphere. At the largest scale this explains why thin, dense oceanic lithosphere 'floats' lower than thicker, less dense continental lithosphere, hinted at by the bimodal ('twin-peak') nature of the hypsometric curve. Either form of crust can also be loaded/unloaded by the addition or removal of water, ice, rock mass or sediment, causing isostatic depression or uplift. Response is more complicated than eustatic change, since it depends on crustal flow/creep deformation. This is both slow and, by its nature, not confined to the exact area of

Table 4.1 *Principal ocean basin statistics.*

Ocean	Area (10^6 km^2)	% global ocean	% Earth's surface	Volume (10^6 km^3)	Mass (10^{23} kg)	Density (g cm^{-3})	Mean depth (km)
Atlantic	94.3	24.9	17.7	340.3	3.5	>1.03	3.57
Arctic	12.2	0.9	0.6	13.7	0.13	~1.03	1.17
Indian	74.1	21.1	14.9	286.7	3.0	~1.03	3.84
Pacific	181.3	53.1	37.6	717.8	7.4	<1.03	3.94
Global	361.9	100.0	70.8	1358.4	14.03	1.03	3.73

Note: The *Southern Ocean* is counted into the southern areas of the three main oceans; the figures include all epi-continental seas, assessed with their most appropriate ocean.

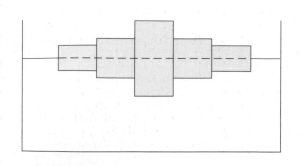

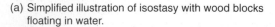

(a) Simplified illustration of isostasy with wood blocks floating in water.

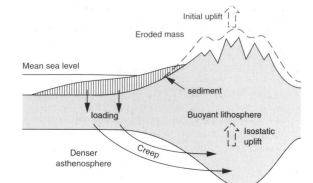

(b) Orogenic uplift, erosion, sediment transfer and isostatic adjustment.

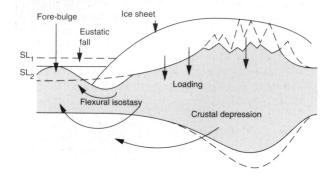

(c) Glaciation, crustal depression, complex margin responses and eustatic sea-level fall.

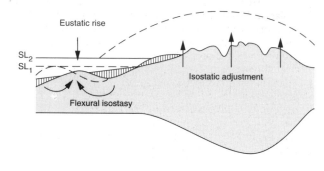

(d) Deglaciation, isostatic recovery, sediment transfer and eustatic sea-level rise.

Figure 4.7 *Some principles and controlling processes of sea level.*

load change. Although neither immediate nor world-wide, it does alter ocean basin geometry. **Flexural isostasy** due to lithosphere creep away from/towards the increase/decrease in load causes fore-bulge or downwarp beyond its margins (Figure 4.7).

For all these reasons, northern Scotland, at the centre of the last British ice sheet, is still in **glacio-isostatic rebound** of 3 mm a^{-1} whereas the Thames/Rhine estuary is subsiding at 1–2 mm a^{-1}. Zero isobase (no change) in Britain passes through Anglesey–Manchester–Middlesbrough. Rebound rates are three and 4.5 times higher in northern Scandinavia and the Hudson's Bay area of eastern Canada, respective former centres of the *Scandinavian* and *Laurentian* ice sheets. Erosion unloads continental crust, and the corresponding transfer of sediment to the deltas of major rivers such as the Mississippi, Ganges and Amazon has depressed sea floors by an average 1–3 mm a^{-1} throughout 10,000 years of Postglacial (*Holocene*) time. Unloading by mining or loading by engineering schemes, such as large reservoir con-struction, introduce a human agency into isostasy but either can occur inland away from coastline impacts. Eustatic/isostatic effects overlap, leading to complex,

fluctuating levels which may be particularly rapid at passive, low-angled continental margins and shallow epicontinental seas – with consequences for ocean circulation.

Composition and structure of ocean waters

Ocean water chemistry

Ocean water is a weak cocktail of nearly 90 per cent of known elements dissolved in 1.4 billion km^3 of water, or carried in suspension largely from terrigen-ous sources. Most elements occur only as traces (less than one part per million) and just eleven account for over 99 per cent of solutes – Cl, Na, Mg, K, Ca, Si, Cu, Zn, Co, Mn and Fe in order of mass. This is reminiscent of, but not identical to, the composi-tional character of the lithosphere. The elements are derived from sea water/crust interactions and rain-fall, as well as the land, and raise water density from 1.0 g cm^{-3} to an average of 1.03 g cm^{-3} with an alkaline pH of 7.8–8.4. The vast bulk of ocean

Oceans and global environmental change

The general effects of oceans on global climate are explained in later chapters. In summary, oceans source about 86 per cent of water cycled through the atmosphere–lithosphere and poleward components of ocean circulation account for 20–40 per cent of global heat transfer. Ocean thermal capacity moderates terrestrial temperature extremes. The global extreme ocean surface range, between the waters of the Persian Gulf and the Arctic Ocean, is about 37.5° C, compared with terrestrial extreme ranges of over 140° C between the Antarctic and the Sahara desert. Annual ranges within oceans are less than 2° C in the tropics, rising to about 4° C and up to 8° C in polar and mid-latitude waters respectively. Ocean currents produce strong regional temperature anomalies, as a comparison of coastal temperatures in both major oceans shows (Figure 1).

It follows that oceans are instrumental in climatic change, leading or responding to changes in other interactive systems. Six important oceanic parameters – geometry, volume, composition (especially salinity), temperature, circulation and sea level – are involved and some are pursued here, commencing with the short-term sub-climatic change referred to as an **El Niño** event. At two to ten-year intervals around Christmas (hence its Hispanic name, 'The Christ Child'), south-east trades relax for some weeks as they cross the Pacific equator, allowing the equatorial counter-current to flow back unhindered to the Peruvian coast. It displaces the cold upwelling Peruvian current, triggering major weather disturbances along the normally dry coastline (and much farther afield), with catastrophic losses in the anchovy fishing industry which the current feeds. The climatic consequences are discussed further in Chapter 8.

Ocean–atmosphere–ice sheet coupling is a principal focus of contemporary research. Tectonics slowly transform ocean geometry, impacting ocean circulation and its climatic effect. We can speculate on the pre-Cenozoic climatic character of land bordering the Atlantic when it was too narrow to support major ocean gyres and the Gulf Stream, or the effects of the equatorial current between the Americas before the late Cenozoic closure of the Panama seaway. What was the impact of the mid-Cenozoic opening of the Drake Passage, connecting up the Antarctic circumpolar current and thus isolating the Southern Ocean and emphasizing the Antarctic convergence?

Far more rapid but smaller-scale impacts on ocean circulation are caused by glacio-eustatic sea-level

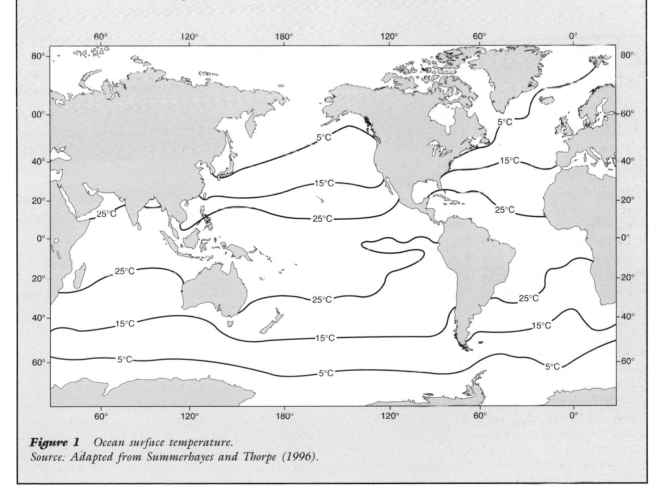

Figure 1 Ocean surface temperature.
Source: Adapted from Summerhayes and Thorpe (1996).

changes, themselves a product of climatic change, and concern us today. Full glacial conditions substantially narrow and shallow submarine sills around the North Atlantic, halving their modern depth of 400–800 m to about 200–600 m. What effect might this have on NADW formation, Arctic climate and the global ocean conveyor? There is now evidence for major pulses of glacial meltwater, or **Heinrich events**, during the past 15 ka. One such occurred when the St Lawrence seaway became ice-free *c.* 11 ka BP and drained meltwater from the Great Lakes region. Fresh water surged across the north Atlantic, retained at the surface by lower density and suppressing the formation of NADW. This may have triggered the British **Loch Lomond stadial** or **Younger Dryas**

(European) ice readvance, lasting a millennium to 10,000 BP.

We conclude with two further possible consequences of an enhanced greenhouse effect. The Arctic sea ice area already halves in summer, owing to its low average thickness of 2.4 m. Global warming is forecast to be at its greatest in the Arctic basin, reaching +7° C. How may this impact Arctic ice cover, planetary albedo and ocean circulation ? What happens when deep ocean circulation of dissolved CO_2, currently buffering the effects of its rise in the atmosphere, reaches saturation after a 10^{2-3} year cycle of the global conveyor? These are serious questions today and may persuade us not to take our oceans quite so much for granted.

water is chemically homogenous and stable, despite substantial active fluxes of water and minerals between adjacent 'spheres' and its own biosphere and sediments. Exceptions occur at particular points of major *influx* or *efflux* of water and/or minerals such as the ocean–atmosphere boundary, estuaries, tidewater glaciers, human-sourced pollution. We note that increased water or decreased mineral flux *dilutes* and decreased water or increased mineral flux *concentrates* the solution. Density varies inversely with temperature but is complicated by changes in salinity, outlined below. In addition to 'solid' minerals, oceans are also reservoirs of dissolved atmospheric gases, incorporated by diffusion from the atmosphere and in sea spray. Concentrations are usually related directly to pressure and inversely to temperature. Average concentrations of N_2 and O_2 are about 1.1 and 0.5 parts per thousand respectively but, at 1.3 parts per thousand, CO_2 is relatively more

abundant. Levels of both oxygen and carbon dioxide vary considerably in the photic zone (see below), owing to biosynthesis.

Salinity

The common expression of solute content is **salinity**, or the quantity of solutes in parts per thousand (‰) by weight and irrespective of composition. NaCl (sodium chloride) is by far the commonest species at over 90 per cent by mass and thus the principal constituent of **brine** or 'salt water'. Global surface values range from about 1 part per thousand in freshwater estuaries to a normal ocean range between 32 and 37.5 parts per thousand. In general, saline water is found at depth, owing to its greater density below a **halocline** isolating less saline surface waters. The halocline may coincide with the **pycnocline**, or zone of greatest density increase (Figure 4.8). However, salinity rises substantially with a weak or negative water balance with the rest of the hydrological cycle through strong evaporation, low rainfall or low freshwater influx. Thus average salinity exceeds 38 parts per thousand in the Mediterranean, 40 parts per thousand in the Red Sea and staggering values of over 350 parts per thousand at depth through density-settling in the latter. The bulk average salinity of 34.5 parts per thousand is approached below the *thermocline* (see below), where surface heating effects are negligible.

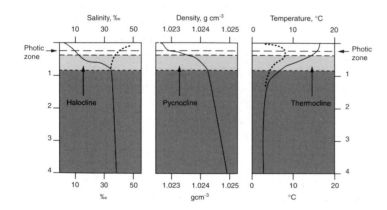

Figure 4.8 *Vertical stratification of the ocean with respect to (a) salinity, (b) density and (c) temperature. Salinity may rise rapidly towards the surface in low-latitude bays (strong evaporation) and temperature may fall towards the surface near freshwater and glacial inflow. Depth in kilometres.*
Source: Modified from Gross (1990).

Temperature and structure

Oceans exhibit both thermal and density stratification or layering, based respectively on temperature and salinity. Solar radiation flux at the ocean surface strongly influences the upper 100–200 m, which has a temperature range of 0–30° C around a global average of 17° C (Figure 4.8). Shallow, local thermoclines may be created diurnally or seasonally in this layer but they are susceptible to turbulent mixing by wind and ocean surface currents. The main **thermocline** is the zone of steepest temperature gradient between 100 m and 500 m, at rates of 0.4–0.7° C per 10 m depth, below which these effects are not felt and ocean water assumes its stable mean temperature of −1 to +5° C. It generally prevents thermal mixing between surface and deep water because the upper layer is warmer and therefore lighter. Modest changes in salinity have a greater impact on density than modest changes in temperature. However, temperature ultimately controls the pycnocline because ocean temperature range greatly exceeds salinity range.

By comparison with surface solar heating, geothermal heat flow through 'average' oceanic crust contributes little to the thermal balance of the oceans; its concentration at mid-ocean ridges has no measurable effect on ocean circulation patterns. Solar radiation flux has two other significant effects. The importance of stored heat is discussed at the end of this chapter. Light penetrates only to shallow depths, owing to high attenuation by sea water. Ocean colour typically varies from a surface blue colour (light scattering) through a green hue (chlorophyll-rich phytoplankton) to brown (muddy or polluted waters). The **photic zone** contains sufficient light for photosynthesis by **phytoplankton** – the oceans' primary producers – and also depends on water **turbidity**, controlled by suspended sediments. The zone varies in depth between 10 m and 200 m, representing the difference between muddy estuaries and the clearest, sunlit water. Below this, marine life depends on chemosynthesis and scavenging of organic rain-out.

Ocean circulation

Fluid properties shared by atmosphere and ocean permit the development of thermally driven circulatory systems, comprising ribbons of air/water currents moving around seasonally mobile cells (atmosphere) and **gyres** (ocean). The stable character of the oceans is indicative of thorough mixing. Ocean motion occurs in two other, non-circulatory forms. Shallow, transient *wave* motion is generated by air flow at the ocean–atmosphere boundary layer and may superficially mimic larger current systems. Tides are an oscillatory response to gravitational mass attractions between earth, sun and moon.

Ocean and atmosphere are coupled by water, energy and momentum transfers and both reflect directional effects of Earth's rotation. Thus we may model a similar, simple overturning cell of tropically heated water moving poleward, cooling and **downwelling** (subsiding) to form a return bottom current which **upwells** at the equator. The Coriolis force (see Chapter 8) draws water to the right of its path in the northern hemisphere and to the left south of the equator. As in the atmosphere, there are areas of divergence and convergence. Broad similarities end there. Oceans are capable of sustaining internal heat and motion for longer, by virtue of their higher specific heat capacity and slower-moving mass, and basin geometry restricts circulation more than continental relief influences the atmosphere. Surface–upper air flow turbulent connection in the atmosphere contrasts with surface–deep water current *dis*connection in the oceans due to the thermocline.

Wind-driven (surface) circulation

Wind imparts a frictional force or **wind stress** on the ocean, proportional to the square of the wind speed, which creates a film of surface waves over a more persistent, slower current. Moving at 3 per cent to 5 per cent of the wind speed, the current extends 50–100 m down from the surface. Once out of the immediate friction zone, successively deeper layers are deflected by the Coriolis force. Warm equatorial waters are driven west across the oceans by atmospheric trade-wind convergence and are deflected poleward by the opposing continental shoreline. The currents accelerate to conserve angular momentum and develop a westerly component in consort with mid-latitude atmospheric westerlies – transporting warm water into high latitudes. Cooled by glacier melt and long-wave radiation loss, the returning cold currents complete each gyre equatorward along eastern continental shorelines. Minor gyres are driven, like a series of gearwheels, within and between the principal currents and coastlines (Figure 4.9).

The importance of ocean geometry now becomes apparent. It supports two hemispherical gyres in both Atlantic and Pacific Oceans and two major northern hemisphere warm mid-latitude currents – the **Gulf Stream** or *North Atlantic Drift* and the **Kuroshio** or *North Pacific Drift*. Atlantic circulation developed only when the ocean was wide enough (probably a minimum of 1500 km) in the Cenozoic. The formation of the Panama isthmus in the late Pliocene shut off its westerly equatorial current and strengthened the Gulf Stream. The Indian Ocean is restricted to a single gyre and is seasonally more varied by atmospheric monsoon circulation. The polar oceans afford fascinating contrasts between the landlocked Arctic

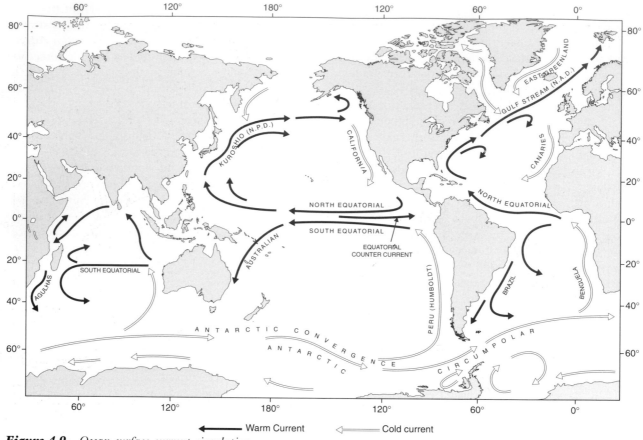

Figure 4.9   *Ocean surface current circulation.*

Ocean and circumpolar Antarctic Ocean. The **Antarctic circumpolar current** in the Southern Ocean attenuates the individual warmth and vigour of the southern hemisphere mid-latitude warm westerly Agulhas, Brazil and Australian currents at the **Antarctic convergence** (Figure 4.9). The Arctic Ocean is fed by the Gulf Stream influx, which circulates beneath polar sea ice and exits via the Denmark, Davis and Bering Straits.

Principal gyres located at about 30° N–S, associated with atmospheric subtropical divergence, aid circulation by pushing water into their cores owing to the Coriolis force (see Chapter 8). This builds very shallow domes 1–2.5 m high in each gyre, enough to add a significant gravity component to circulation as water flows out of the dome. Although currents are shallow and velocities low, their persistence transfers very large quantities of water and heat over time. The Gulf Stream transports a maximum 150×10^6 m³ sec⁻¹ at over 1.5 m sec⁻¹ off Boston and the Kuroshio transports 46×10^6 m³ sec⁻¹ at up to 1.7 m sec⁻¹ off Japan. There is a significant anomaly in the main equatorially driven gyres. Water build-up against the western coast in both oceans creates a rise in sea surface not entirely dissipated by

poleward flow. An **equatorial counter-current** flows back eastward, between the westward limbs of each hemispheric gyre, opposed by only slight winds, between 3° N and 10° N (the doldrum belt) at the thermal equator.

Thermohaline (deep) circulation

Surface ocean circulation obscures the vital slow motion of the deep ocean. The **thermohaline circulation** moves at speeds of 10–50 km yr⁻¹, driven by water masses of different densities determined by temperature and salinity properties. It is isolated from the surface by the thermocline except at points of formation and has an average cycling time of 500–2000 years. Strong radiative cooling of warm currents, as they approach polar oceans, is enhanced by dense *bottom water* formation in cold, saline-enriched sea ice environments at two principal centres. North Atlantic icy straits pour *North Atlantic Deep Water* (NADW) southwards as a tongue penetrating to 60° S, between even denser *Antarctic Bottom Water* (AABW) and less dense *Antarctic Intermediate Water* (AAIW). AABW is generated

(a)

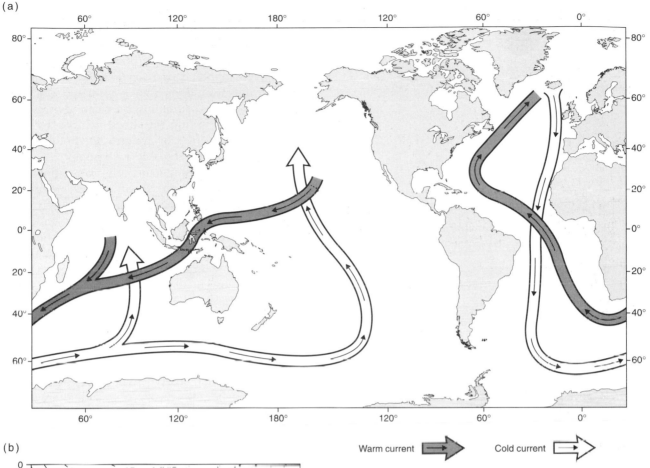

(b)

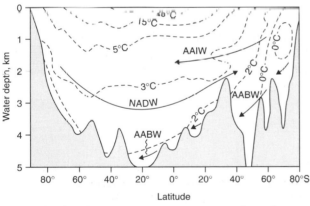

Warm current ➡   Cold current ⇨

Figure 4.10   *Thermohaline deep-water circulation or Global Ocean Conveyor: (a) principal global circulation and (b) vertical section north south through the Atlantic Ocean. See text for initials of stratified flows.*
Source: In part after Gross (1990).

beneath the Antarctic convergence, and undercuts NADW to about 40° N, becoming trapped in deep troughs. The same 'pincer movement' cannot occur in the Indian Ocean, and although AABW also spills into the Pacific there is no equivalent northern Pacific cold outflow. As a result, warmer, less dense *Pacific and Indian Ocean Common Water* (PICW) completes the circulation through a sinuous limb rejoining the Gulf Stream. Involvement of the vast bulk of Earth's 1.4 billion km³ of ocean water compensates for the slowness of thermohaline circulation. Its unimpeded trans-equatorial flow earns the alternative title Global Ocean Conveyor (Figure 4.10).

Tides and waves

Tides

We turn, finally, to the most familiar part of ocean movement – tides and waves, which have the greatest impact at the coastline through their transmission of *energy*. Tides transfer *mass* from one part of the global ocean to another in a regular, oscillating manner by competition between the gravitational fields of Earth, Moon and Sun. Moon and Sun create **tidal bulges** on either side of Earth, extended in the plane of maximum pull (Figure 4.11) but their periodicity is

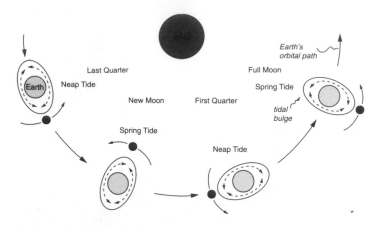

Figure 4.11 *Global tidal bulge and cycles in response to Earth, Sun and Moon configuration.*

not identical. By rotating once in twenty-four hours about its own axis, Earth experiences two tides (periodicity: 12.0 hr) relative to the Sun's 'fixed' position but slightly less than two (periodicity: 12.42 hr) relative to the Moon, which also moves around Earth. *Lunar* tides are stronger than *solar* tides because of the Moon's proximity. The *semi-diurnal* tidal model, of two tides each day, is most applicable in equatorial and mid-latitude waters. Polewards, one tide becomes progressively dominant, giving *mixed* tides or, in high latitudes, a single *diurnal* tide. The Sun adds 47 per cent to tidal pull when both are in line, to form **spring tides** twice during each monthly cycle but reduces it by variable rates when they are not, reaching a lowest or **neap tide** when they pull at right-angles (Figures 4.11 and 4.12).

To understand why tidal levels and cycles are so complex, consider what might happen on a featureless earth covered by a single ocean five to ten times deeper than at present. The Moon would draw a tidal wave some 0.5 m high at the equator at about 1600 km hr^{-1} in its wake – the magnitude of the moon's pull being slight but its rotation around Earth fast. This does *not* happen because the global ocean is comparatively shallow, with greater sea bed friction, and the Moon's orbit 'wobbles' between 28.5° N and 28.5° S of an equatorial plane. Moreover, it is interrupted by large continents and indented coastlines. Thus tides which pass unobserved in mid-ocean are stacked up disharmonically into confined coastal spaces of variable shelf and shoreline slope. This is illustrated by the **Severn bore** in the Bristol Channel, which has one of Earth's largest tidal ranges of 12.3 m at Avonmouth. The Severn estuary is 220 km long, 150 km wide at its mouth between Pembroke and Cornwall but only 1.5 km wide near the Severn bridges. Tides lag by

three hours between its landward and seaward limit and are still rising inland when they begin to fall at sea. Its narrowing shape creates a progressive rise of water, sending a tidal wave or **bore** up-estuary, steepening as it is opposed by river flow. Several million people living around its shores were relieved that calm, anticyclonic conditions prevailed during this century's highest spring tide in September 1993; a **storm surge** reinforced by cyclonic low pressure and strong winds could have been catastrophic.

High tides occur simultaneously at a number of places, linked by **co-tidal lines** as they are drawn across the ocean. The ocean surface tilts as it ebbs and flows, moving away from and towards land, and is also tilted by the Coriolis force. As a result, the water surface in enclosed oceans or seas oscillates from side to side around **amphidromic points** with zero or very low tidal range (Figure 4.13). Coasts also experience a tidal range dependent on their configuration. Open coasts capable of reflecting tides with little complication, or enclosed seas like the Mediterranean with limited scope for tide generation, experience *microtidal* ranges less than 2 m in amplitude. More indented coastlines and wider continental shelves, enhancing wave reflection and retardation, raise tides to *mesotidal* or *macrotidal* ranges of 2–6 m and over 6 m respectively.

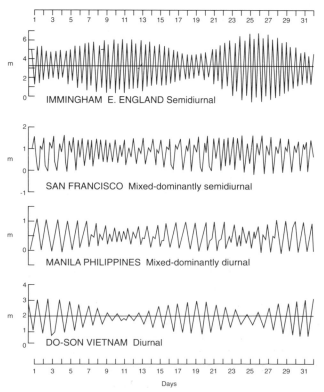

Figure 4.12 *Representative tidal styles and ranges.* Source: *After King (1962).*

Waves

Waves are the smallest disturbances, dependent on the wind and therefore transient and irregular in strength and direction, except in the sense that predominant wind directions excite a similar response in wave direction. It is important to note that waves transmit energy but very little mass – i.e. a wave is an onward transmission of energy from one water particle to the next in which the wave form is created by the rotational rise and fall of each particle in turn (Figure 4.14). This is demonstrated by the rise and fall of a beach ball as each wave passes but otherwise remains in the same general position. In a given wind field, a series of waves, or **wave train**, are separated by their **wavelength** (L) and **wave period** (T). (Plate 4.2). Wave velocity $V = L/T$. Wave height (H) is determined by wind speed, duration and **fetch**, or distance travelled over water. It is also the diameter of the orbital path of individual water particles as they rotate and an important determinant of wave energy, E. $E = \frac{1}{8}\rho H^2$ (where ρ is water density). $L/2$, denoting **wave base**, is the depth beyond which the declining oscillatory motion has effectively ceased and is no longer able to disturb the sea bed. As a wave enters a water depth less than $L/2$ a circular orbit of particles cannot be maintained. Bed friction slows the forward part of the loop first, forcing the still advancing rear to climb, steepening and elevating the wave, and eventually causing it to break. In this way each wave also catches up on decelerating waves ahead, compressing the wave train and so shortening wavelength. This is demonstrated dramatically by **seismic sea waves** or **tsunamis**, triggered by earthquakes. They can travel at over 500 km h^{-1} but at almost imperceptible heights of less than 0.5 m until they decelerate on approaching land, where wave heights build progressively and catastrophically to 30 m or more. The geomorphic impact of wave-related processes is examined in Chapter 17.

Conclusion

Earth's ocean basins are its largest individual surface feature and also collectively its youngest at less than 200 Ma old, owing to the repetitive growth and consumption of dense oceanic crust through the super-continental cycle. In that time a former super-ocean has shrunk considerably (surviving mainly in the Pacific), the Atlantic and Southern Oceans have formed and the Arctic Ocean has become encircled by continents. Oceans play an important but hidden role in most stages of the rock cycle, although their prod-

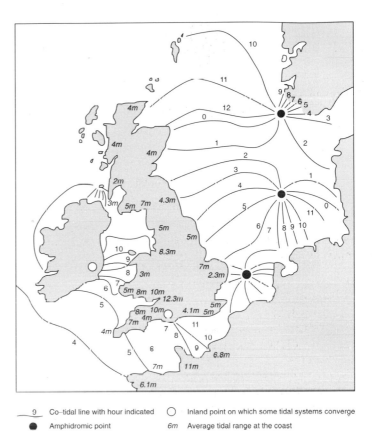

—9— Co–tidal line with hour indicated	◯ Inland point on which some tidal systems converge
● Amphidromic point	6m Average tidal range at the coast

Figure 4.13 *Co-tidal lines, amphidromic points and tidal range around the British coast.*

ucts are concealed until after uplift or sea-level change exposes them. Modern oceanic research reveals the extent to which environmental history is both influenced by, and recorded in, our oceans and warrants a greater future interest in oceans and ocean basins.

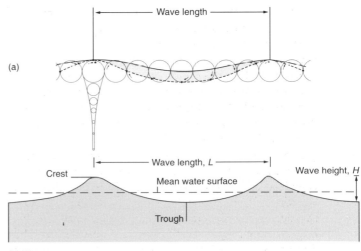

Figure 4.14 *Wave (a) motion and (b) form in open water.*

Plate 4.2 *A wave train entering a bay, showing refraction as the waves are retarded along its flanks before shoaling (breaking), with spilling and surging breakers. Photo: Kenneth Addison.*

Key points

1 Large-scale ocean basin topography reflects its formative tectonic processes. Deep subduction zone trenches contrast with shallow continental shelves, which are broader on passive margins. Abyssal plains are interrupted by mid-ocean ridges and hot-spot submarine volcanoes. All are draped in varying thicknesses of terrigenous or marine sediments, thinning seaward of the continental slope, down which they also slump and flow.

2 Oceans contain almost all planetary water and therefore act, simultaneously, as the principal source and sink of the atmospheric and terrestrial hydrological cycles. Sea water is a weak saline solution of eleven principal elements derived largely from terrigenous sources and lithified in due course as deep-sea chemical precipitates or ocean-margin evaporite rocks.

3 Sea water is stirred superficially by Earth's wind belts and at depth by density and water-temperature differences. In addition, gravitational attraction by Sun and Moon pull tidal waves and currents around the oceans which have a low magnitude in mid-ocean but rise as they encounter coastlines.

4 Sea level fluctuates over geological time scales, determined by a combination of eustatic controls on water volume and isostatic controls on ocean basin geometry. Regular Quaternary iso-eustatic fluctuations of 3–5 per cent of average ocean depth (3.73 km) have significantly altered coastlines and climates.

5 Oceans have a moderating influence on global climate. Surface waters act as a major heat store, having a high thermal capacity which mitigates seasonal temperature fluctuations in maritime climates and reduces meridional temperature extremes. Ocean–atmosphere–ice sheet coupling regulates Earth's energy and moisture balances.

Further reading

Gross, M.G. (1990) *Oceanography*, sixth edition, New York: Macmillan.
A useful, short and concise book which concentrates on sea water, ocean structure, currents, waves and tides between outer chapters on marine geology and resources.

Open University Oceanography Course Team (1989) *The Ocean Basins: Their structure and evolution*, Oxford and New York: Pergamon Press.

The first volume in a series of six thematic books on oceans. This one provides an overview of the origins, form and geology of ocean basins.

Summerhayes, C.P., and Thorpe, S.A., eds (1996) *Oceanography: An illustrated guide*, London: Manson.
This is a timely collection of papers which addresses the whole spectrum of oceanographic science, including the geological, hydrological and biological operation of the oceans. It includes an up-to-date review of research and management problems and is rich in colour illustrations.

The atmospheric energy system

The sun's energy represents the prime source of our climatic system. In this chapter we will look first in some detail at the internal mechanisms of this energy flow, then consider the spatial variability of the flows which give rise to different climates. Perhaps the best way to explain what is happening is to follow the path of sunlight from the top of the atmosphere and describe what affects it on its journey to Earth's surface. Long-wave exchanges can then be described.

Short-wave radiation in the atmosphere

As our beam of sunlight enters the atmosphere it first passes through the thermosphere and the mesosphere with little change. In the stratosphere the density of atmospheric gases increases. There is more oxygen available which reacts with the shortest or ultra-violet wavelengths and effectively removes them, warming the atmosphere and producing ozone in the process (Figure 5.1). About 2 per cent of the original beam is converted to heat at this stage. (See box.)

As we descend into the troposphere the atmosphere becomes rapidly denser and so there is greater interaction between the sunlight and the atmospheric gases. The size of the gas molecules of the air is such that they interact with the insolation, causing some of it to be scattered in many directions. This process depends on wavelength. The shorter waves are scattered more than the longer waves and so we see these scattered waves as blue sky. If the reverse were true the sky would be permanently red, and if there were no atmosphere, as on the moon, the sky would be black. Dust and haze in the atmosphere produce further scattering, but not all of this is lost. Some of the scattered radiation is returned to space, but much is directed downwards towards the surface as diffuse radiation. This is the type of radiation which we also experience during cloudy conditions with no direct sunlight when the solar beam is 'diffused' by the

water droplets or ice particles of the clouds.

Another type of short-wave energy loss is absorption. The gases in the atmosphere absorb some

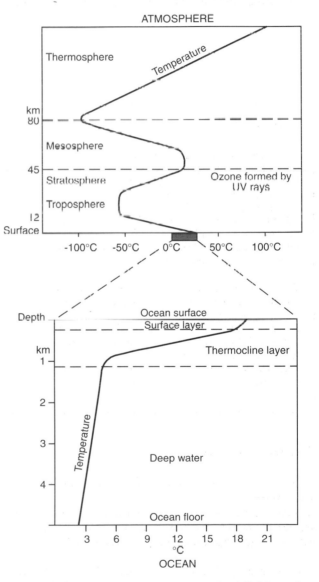

Figure 5.1 *Temperature structure and subdivisions of the atmosphere and ocean.*

Ozone

Ozone is a rare gas made up of three atoms of oxygen. Its concentration rarely exceeds a few parts per billion and yet it is a vital component of our atmosphere. In the stratosphere it is formed through the interaction of the shorter, ultra-violet part of the sun's radiation and the oxygen molecules, which consist of two atoms of oxygen. The reaction for ozone formation is:

$$O_2 + \text{UV light} \rightarrow O + O; \quad O_2 + O \rightarrow O_3$$

The result is the almost total exclusion of the harmful ultra-violet rays of the sun. In the troposphere ozone exists as a by-product of photochemical processes between sunlight and pollutants, particularly nitrogen oxides from car exhausts. It is considered toxic above sixty parts per billion and has harmful effects on plant growth as well as causing respiratory problems.

In the lower part of the atmosphere its concentration has been increasing as a result of higher car pollution, but in the stratosphere its level should remain constant, destruction being balanced by creation.

However, in 1985 scientists working in Antarctica announced that ozone levels in the southern hemisphere stratosphere had fallen by 40 per cent between 1977 and 1984. In the Antarctic spring (October) a hole the size of the United States and about 10 km deep had appeared. By 1995 springtime levels were below 100 Dobson units (parts per billion), less than a third of the natural level. Simple extrapolation of this trend would indicate no ozone in spring by 2005 (see Figure 1).

Subsequent research has shown that the hole has been caused, in part, by the ozone being destroyed by chlorine atoms released from molecules of chloro-fluorocarbon (CFCs) broken down by the ultra-violet radiation from the sun. The simplified reactions are :

$$O_3 + O \rightarrow O_2 + O_2$$

$O_3 + NO \rightarrow NO_2 + O_2$ (nitrous oxide acting as the catalyst)
$O + NO_2 \rightarrow O_2 + NO$

$O_3 + Cl \rightarrow O_2 + ClO$ (with chlorine acting as the catalyst)

$$O + ClO \rightarrow O_2 + Cl$$

Other gases such as nitric oxide and bromine were also implicated. At this time of extreme cold the chlorine atoms are able to destroy ozone faster than it is being formed. The Antarctic stratospheric circulation is effectively isolated from other parts of the atmosphere at this time, so there is little mixing with warmer air richer in ozone.

Background ozone levels have fallen in the northern hemisphere also but only by about 15 per cent. The circulation is less isolated there and the atmosphere less cold. Nonetheless the area of depletion has expanded into the Mediterranean and the southern United States. In their sunny climates more

harmful ultra-violet radiation would have been reaching the surface.

With such a serious decrease in levels of ozone and the consequent reduced protection from ultra-violet radiation, scientists and politicians agreed that something must be done. The first aim was to reduce the production of CFCs, which appeared to be the main culprit. An agreement was reached at Montreal in 1987 to phase out the production and use of CFCs as soon as possible; 'as soon as possible' meant a 50 per cent reduction by 1999. Recognizing that this was too slow, a further agreement in London in 1990 recommended the elimination of the use of CFCs by industrialized countries by 2000. Despite this, ozone levels have continue to decline, though the rate of decrease does appear to have slowed down. Ironically the replacement gases will still contribute to the enhanced greenhouse effect even if they are less damaging to the ozone layer.

The implications of increased ultra-violet light in significant quantities in the southern hemisphere spring are not fully known. Children at school in Australia are encouraged to wear protective hats and to avoid bright sunlight, as large quantities of ultra-violet light are believed to cause skin cancer. There are suspicions that marine plankton may be affected, with unknown effects on the food chain and ecosystems. Measurements of planktonic production have shown that the reduction of photosynthesis induced by ultra-violet radiation increases linearly with the dosage of radiation. The productivity was reduced by a minimum of between 6 per cent and 12 per cent.

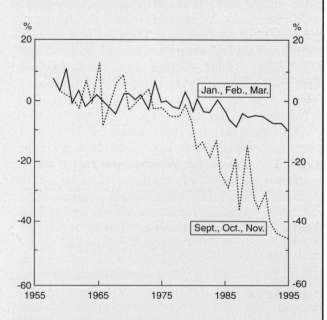

Figure 1   *Seasonal ozone deviations from pre-ozone-hole (1957–78) averages over Antarctica show a rapid decline during the austral spring and less decline during the summer.*
Source: Bojkov (1995).

Even amphibian eggs appear to be affected by ultra-violet radiation, perhaps leading to the current decline in the number of amphibians.

There is still much we do not know about the consequences of using CFCs in aerosol cans and refrigerators. It provides a good example of how human activities can unwittingly have a major impact on a system we do not fully understand.

wavelengths (Figure 2.9), as do clouds. In this way we have a warming of the atmosphere, though the amounts involved are small. The most important loss of short-wave radiation in its path through the atmosphere is by reflection. The water droplets or ice crystals in clouds are very effective in reflecting insolation. Satellite evidence shows that, for Earth, a mean figure of 19 per cent of the original insolation is reflected by clouds. The degree of reflection is called the albedo. The albedo is normally expressed as the ratio of the amount of reflected radiation divided by the incoming radiation. If multiplied by 100, it will be in the form of a percentage. The lowest and thickest clouds tend to reflect most while the thin, high-level ice clouds have an albedo of only about 30 per cent.

By now, the beam has reached the ground surface with, as a global average, about 50 per cent of its original energy. Even then, not all of it is absorbed, as the surface itself has an albedo. The global average albedo represents some 6 per cent of the radiation at the top of the atmosphere, so the loss is not great. However, the figure may seem large when expressed as a percentage of the radiation actually reaching the surface. For example, the albedo of freshly fallen snow may reach as high as 90 per cent (Table 5.1). The greatest variability is over water. When the sun is high in the sky, water has a very low albedo. That is why oceans appear dark on satellite photographs (Plate 5.1). At low angles of the sun, as at dawn or

in midwinter in temperate and sub-polar latitudes, the albedo may reach nearly 80 per cent.

The sunlight reaching Earth's surface which is not reflected by Earth is absorbed and converted into heat energy. The distribution of energy received at the surface is shown in Figure 5.2. Thus incoming radiation can be absorbed (in the atmosphere and at the surface), scattered (in the atmosphere) or reflected (by clouds and at the surface). When reflected, the radiation is returned to space in the short-wave form and becomes part of the outflow of energy from Earth. Similarly, some of the scattered radiation is returned to space to give a short-wave albedo for our planet of 28 per cent. The modifications of the solar beam by the atmosphere are shown diagrammatically on the left-hand side of Figure 5.3.

Long-wave radiation

All substances emit long-wave radiation in proportion to their absolute temperature. Earth's surface receives most short-wave radiation and therefore normally has the highest temperature. It follows that most long-wave emission will be from the ground surface. The atmosphere is much more absorbent of long-wave radiation than of short-wave radiation. Carbon dioxide and water vapour are very effective absorbers of much of the longer part of the spectrum except between 8 μm and 12 μm. As water vapour is concentrated in the lowest layers of the atmosphere, that is where most absorption will take place. Clouds are also very effective at absorbing long-wave radiation and hence their temperature will be higher than otherwise. This cloud effect is more noticeable at night. With clear skies and dry air, radiation is emitted by the surface but little is received from the atmosphere and therefore the temperature falls rapidly. If the sky is cloudy, the clouds will absorb much of the radiation from the surface and, because they are also emitters, more of the radiation will be returned to the ground as counter-radiation than if the sky were clear. It is absorbed by the ground, compensating for the emission of long-wave radiation and so reducing the rate of cooling at the ground. Figure 5.4 compares temperatures on clear and cloudy nights to demonstrate this effect.

Table 5.1 *Albedos for the short-wave part of the spectrum.*

Surface	Albedo (%)
Water (zenith angles above 40°)	2–4
Water (angles less than 40°)	6–80
Fresh snow	75–90
Old snow	40–70
Dry sand	35–45
Dark, wet soil	5–15
Dry concrete	17–27
Black road surface	5–10
Grass	20–30
Deciduous forest	10–20
Coniferous forest	5–15
Crops	15–25
Tundra	15–20

Plate 5.1   *A visible waveband Meteosat image, showing the variation in reflection (albedo) of different surfaces. Where cloud-free the oceans stand out as dark, and the contrast between the vegetated surfaces of West Africa and the Saharan desert is marked.*

Some of the radiation given off by the surface is lost to space but the majority gets caught up in the two-way exchange between the surface and the atmosphere. Figure 5.3 shows the emission and absorption of long-wave radiation as a proportion of incoming energy. Radiation from the atmosphere is emitted spacewards as well as downwards. As there is less water vapour at higher levels, absorption by the atmosphere is less and proportionally more is lost to space.

The Greenhouse effect

The property of the atmosphere that allows the transmission of sunlight, but acts as a partial barrier to the loss of heat from the surface has been called the **Greenhouse effect** because of its analogy with the greenhouse, which appears to produce warming by a similar process. Subsequent work has shown that greenhouses are warmed as much by the removal of

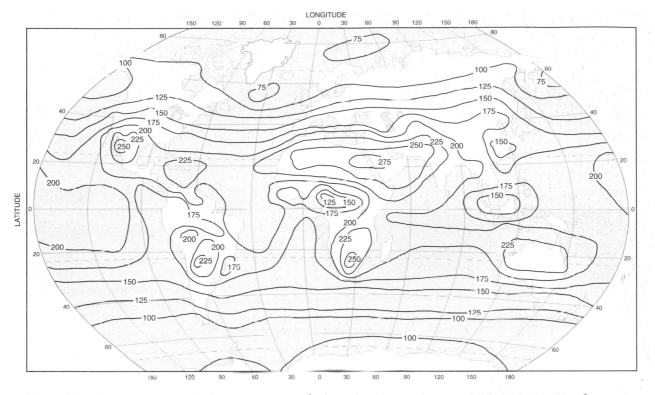

Figure 5.2 *The average annual solar radiation on a horizontal surface at the ground. The units are W m⁻².*
Source: After Sellers (1965).

wind as by any radiational effect but the name of the effect has remained. Without this natural greenhouse effect Earth's equilibrium temperature would be about −19° C and the planet would be almost uninhabitable. We can work this figure out from the amount of long-wave radiation which is lost by the

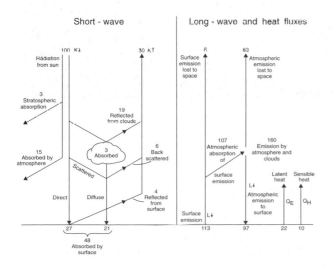

Figure 5.3 *Modification of short- and long-wave radiation by the atmosphere and the surface. Figures are expressed as a percentage of incoming short-wave radiation at the top of the atmosphere based on a global mean.*

planet and now measured by the orbiting satellites.

As a result of the massive consumption of fossil fuels such as coal and oil, the waste products of combustion have been released into the atmosphere, where they slowly accumulate. Changing land use through deforestation, particularly of the temperate and tropical forests, has a similar effect, as less carbon is stored in the replacement crops. From this, the composition of our atmosphere has been changing (Figure 5.5). Since 1720 the concentrations of carbon dioxide have increased from about 280 ppmv (parts per million by volume) to the current levels of over 358 ppmv and those of methane from 0.7 ppmv to 1.7 ppmv. Other minor constituents of the atmosphere with similar effects such as chlorofluorocarbons (CFCs) and nitrous oxide have increased, too. As there is now a greater concentration of gases in the atmosphere which have the capacity to absorb long-wave radiation from Earth's surface, it may be expected that Earth will warm.

Actual changes of climate and their causes will be discussed in Chapter 11, but it is clear that this 'enhanced' greenhouse effect has the potential to warm our planet. It is much more difficult to *prove* that the increase in the concentration of greenhouse gases has been significant in the variations of global temperature in the last 300 years. Mathematical

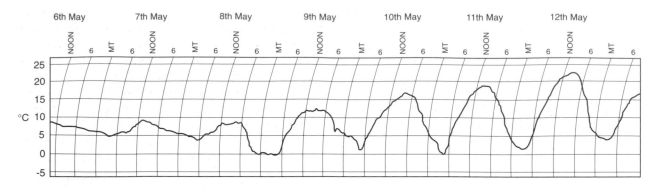

Figure 5.4 *Contrasting diurnal temperature variations on cloudy and clear days at Sheffield, South Yorkshire. Cloudy weather prevailed for the first three days, giving a small diurnal temperature range. As the skies cleared later in the week, daytime temperatures increased but night-time temperatures became lower. There is some indication of a slight, progressive warming of both day and night temperatures as a result of the storage of solar energy.*

models of the climate system demonstrate that a doubling of the proportion of carbon dioxide in our atmosphere should lead to an increase of global temperatures of between 2° C and 4° C, but there is considerable uncertainty about the accuracy of the predictions. We need to know much more about the causes of short-period temperature changes which are known to occur naturally before it is possible to determine the precise role of the 'enhanced' greenhouse effect on our climate.

Global radiation balance

Taking Earth as a whole, we know that no part is getting warmer (apart from the possible 'enhanced' greenhouse effect) or cooler and so there must be an overall balance. More short-wave radiation appears to be absorbed by Earth than leaves it by a mixture of short- and long-wave radiation. The surface seems to be gaining heat (Figure 5.3). Similarly, the atmos-

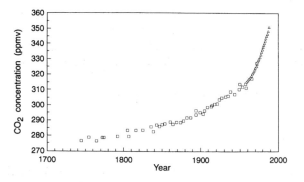

Figure 5.5 *Changes in the atmospheric concentration of carbon dioxide.*
Source: IPCC (1990).

phere seems to be losing heat. If radiation were the only process operating, Earth's surface should be getting warmer and the atmosphere cooler. They do not do so because, in addition to radiation, there are thermal energy transfers in the form of convective heat exchanges. Many of them take place through evaporation and are discussed later in this chapter.

Spatial variability of radiation exchanges

Earth is a large spheroidal body which spins on an axis tilted at 23½° to the vertical and has an elliptical orbit around the sun. These factors alone have a considerable influence on how radiation is distributed at Earth's surface.

In Chapter 2 we described the input of solar energy at the top of the atmosphere and how it was determined by these astronomic controls. Figure 2.12 showed how the radiation would be distributed at the top of the atmosphere. However, if we look at a map of the average annual short-wave radiation reaching the ground, it is appreciably different (Figure 5.2). The general impression of the map is of a decrease in energy input towards the poles, with local anomalies. Most of these are caused by the pattern of clouds. High values are found over the Saharan, Australian and Asian deserts. The lowest values occur in regions of high cloudiness such as Iceland, the Aleutian Islands in the north Pacific, the Zaire basin and parts of West Africa.

The map may be a little misleading. By showing the radiation reaching the ground surface we are omitting an important factor. The surface albedo influences how much radiation is absorbed, and this is not indicated. Oceanic areas absorb a similar total to that shown on the map but for ice-covered surfaces, such as Greenland, and areas with light-coloured, dry surfaces, such as the Sahara, the total radiation absorbed may be significantly lower.

What is it that produces this spatial pattern of radiation? Obviously the astronomic factors have a great effect, giving rise to the poleward decline. But the decrease is far greater than one would expect from the distribution at the top of the atmosphere. We have to look for other reasons. One of the most important is the angle between the sun's rays and Earth's surface. The input is greatest whenever the surface is at right-angles to the sun's rays. If the sun is overhead a horizontal surface will receive the highest intensity of radiation. When the sun is low in the sky, the steeper slopes facing the sun will receive the highest values. As Earth is a sphere at a great distance from the sun, the sun's rays appear parallel and hit the surface at different angles (Figure 2.13).

A secondary effect which further decreases radiation intensity is the longer path through the atmosphere at higher latitudes. Scattering and absorption will be higher, though they increase diffuse radiation at the expense of direct radiation. The amount of scattering and absorption will vary, depending upon the degree of haziness of the atmosphere. Where the atmosphere is very dusty, as in semi-arid or desert areas, more radiation will be absorbed and scattered, preventing it from reaching the ground surface. As the dust particles are much larger than gas molecules, scattering is not dependent upon wavelength and the sky has a whitish hue rather than the deep blue of a clear atmosphere. This effect is also noticeable over urban areas, where pollution produces the same effect.

Latitudinal radiation balance

To see how much radiant energy we have available at any location we must know how much radiation is being lost as well as how much is reaching that location. Long-wave radiation emission is proportional to the absolute temperature of the surface. It is far less variable than the input of solar radiation. The difference between incoming and outgoing radiation is known as net radiation or the radiation balance. For Earth's surface, values are shown in Figure 5.6.

If we include the effects of the atmosphere the picture changes. The atmosphere has a negative balance, even in the tropics (Figure 5.7). In fact, values differ little between equator and poles. For any particular latitude, we can sum the surface and atmospheric radiation balances to find out which areas of Earth have a radiation surplus and which areas have a deficit.

Using satellite data, it is now possible to determine the radiation balance of the surface and atmosphere together, as shown in Figure 5.8. In general there is a surplus of energy between about 38° N and 38° S and a deficit towards the poles. Naturally the magnitude of the surplus is identical to that of the deficit, but it does mean that there must be a steady transfer of energy from the tropics polewards, otherwise the tropics would get hotter and the polar regions cooler. It is the winds of the world, and to a lesser extent the ocean currents (Figure 5.9), which bring about the necessary heat transfer.

The heat balance

Uses of available energy

In the previous section we showed how Earth's surface normally receives a surplus of radiation which leads to warming there. This situation cannot last indefinitely, or the temperature gradient in the air and the soil would become enormous. Energy tends to flow down a gradient, and as radiation is absorbed by the surface, so heat is transmitted into the soil and into the air. This takes place in proportion to the amount of energy originally absorbed. We can express this mathematically as:

$$Q^* - Q_H + Q_G$$

where Q^* is the net radiation, Q_H is the sensible heat transfer into the air and Q_G is the heat flow into or out of the soil. If the surface is damp some of the energy will be used in evaporation. Therefore:

$$Q^* = Q_H + Q_G + Q_E$$

where Q_E is the energy used for evaporation. This is a simplification, as changes in heat storage can take place and a small amount of energy is used in plant growth.

The energy transfer into the atmosphere is the final component of the radiation imbalance between surface and atmosphere. The net radiational loss in the atmosphere is counteracted by this heat transfer from the surface. So, over a long period, the atmosphere gains as much energy as it loses.

Sensible heat

How do these processes which use the net radiational energy operate? One way is by sensible heat transfer.

Sensible heat is the exchange of warm air down the temperature gradient. By day, this will normally be

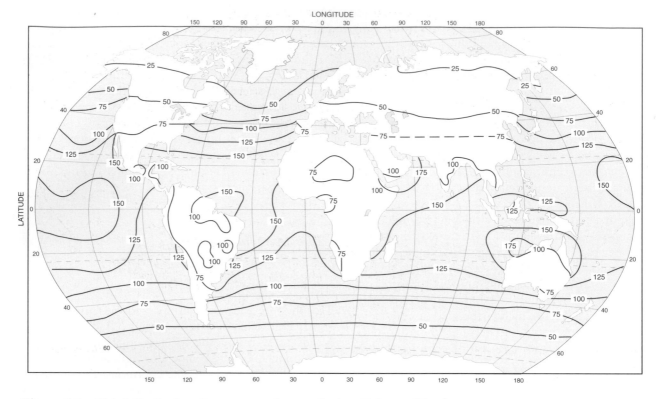

Figure 5.6 *Global distribution of mean annual net radiation. Units are W m⁻².*
Source: After Budyko et al. (1962).

upwards, but at night there may be a weak transfer of sensible heat down to the cooler ground surface. It takes place because the air in contact with the surface becomes warmer through conduction. Being warmer, the air will be less dense than its surroundings and, like a cork in water, will tend to rise until it has the same density (temperature) as its surroundings. Occasionally this process can be seen operating. If the ground is being warmed intensely, the rate of sensible

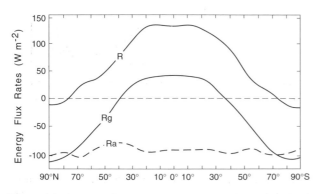

Figure 5.7 *The average annual latitudinal distribution of the radiation balances of the Earth's surface (R), of the atmosphere (R_a) and of the Earth–atmosphere system (R_g) in W m⁻².*
Source: After Sellers (1965).

heat transfer is high. The rising air can be seen as a 'shimmering' of the air layer near the ground due to the variable refractive indices of light through the air of different temperatures. Replacing the rising warm air are pockets of cooler air descending towards the ground.

The significance of sensible heat in the local heat budget depends upon the frequency and intensity of surface heating. Where the surface is usually hotter than the air, values may be high, but where there is little temperature difference, as over most ocean surfaces, sensible heat transfer will be low (Figure 5.10).

Latent heat

The concept of latent heat can best be understood by conducting a small experiment. Start with a large block of ice out of a freezer and measure its temperature; perhaps it may be –15° C. Then place it in a Pyrex glass beaker and heat the beaker at a constant rate, monitoring the temperature of the ice continuously. Keep heating the beaker until all the ice has melted into water; eventually it will reach boiling point and vaporize as steam. If the temperature values are plotted against time, we find a steady increase in

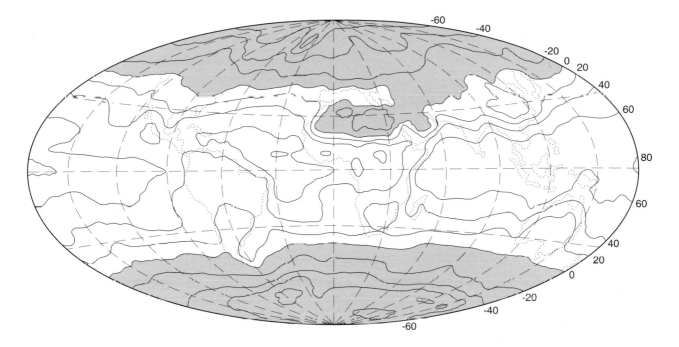

Figure 5.8 *Areas of surplus and deficit (stippled) of annual net radiation at the top of the atmosphere. The figures are W m⁻².*
Source: After Hartmann (1994).

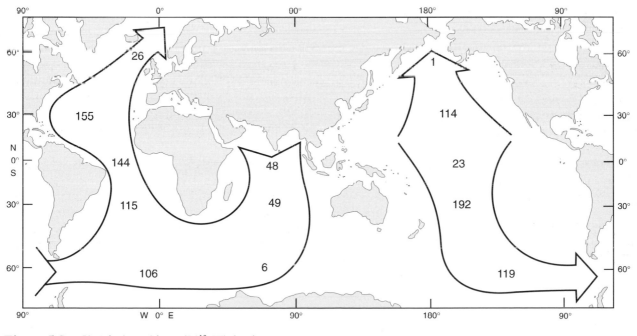

Figure 5.9 *Circulation of heat (10¹³ W) by the oceans.*
Source: After Houghton (1984).

temperature (representing heat input from the heater and some heat flow from the air, which will be warmer than the ice) until melting starts. Despite the steady addition of heat, there is no increase in temperature until the ice melts completely (Figure 5.11). A similar effect is found on vaporization.

Where has the heat that was being added continuously gone? It was being used not to raise the temperature during melting or vaporization but to change the physical state of the water, either from solid to liquid or from liquid to vapour. As the heat appears to be hidden, it is known as latent heat.

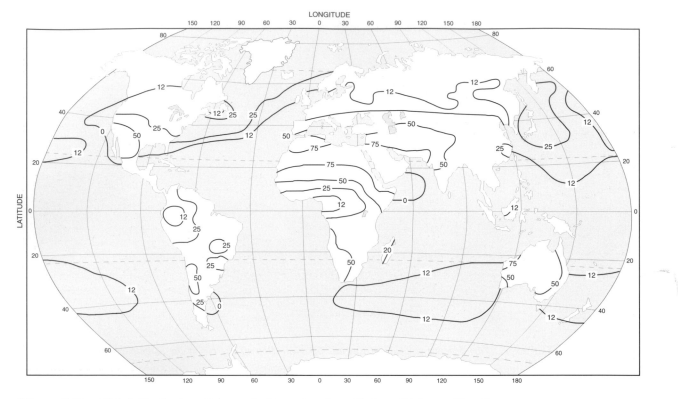

Figure 5.10 *Global distribution of the vertical transfer of sensible heat. Units are W m⁻².*
Source: After Budyko et al. (1962).

Heat consumption

A change of state, from solid to liquid, or from liquid to vapour, involves a considerable use of energy. In the first case we need 3.33×10^5 J kg⁻¹; this quantity of heat is called the latent heat of fusion. In the second, much more energy is needed. At 10° C the latent heat of vaporization is 2.48×10^6 J kg⁻¹ but it falls slightly with increasing temperature. To get a better idea of this large quantity of energy needed for evaporation, the amount consumed in evaporating only 10 g of water is about the same as that needed to raise the temperature of 60 g of water from 0° C to boiling point (100° C). We tend to be most aware of evaporational cooling after swimming. The effect of evaporation leads to the extraction of heat from the skin surface; sweating works in a similar way.

Release of energy

Where the state of water changes to a lower energy level (i.e from vapour to liquid or from liquid to solid) it will release the same quantity of energy that was originally used when it was raised to the higher energy state. This is very important in our atmos-

pheric heat balance. Water that is evaporating from the surface will extract energy from that surface, where there is usually a surplus anyway. Eventually the vapour will condense in the atmosphere, probably as a cloud droplet, releasing latent heat originally extracted from the surface and so helping to warm the atmosphere. This can take place well away from the original evaporation point, so evaporation can transfer heat energy both vertically and horizontally.

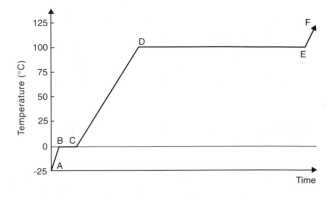

Figure 5.11 *The pattern of temperature and phase changes for water. The temperature remains constant during each phase change as long as pressure remains constant. Differences in specific heat of ice and water give different gradients for the lines A–B and C–D.*

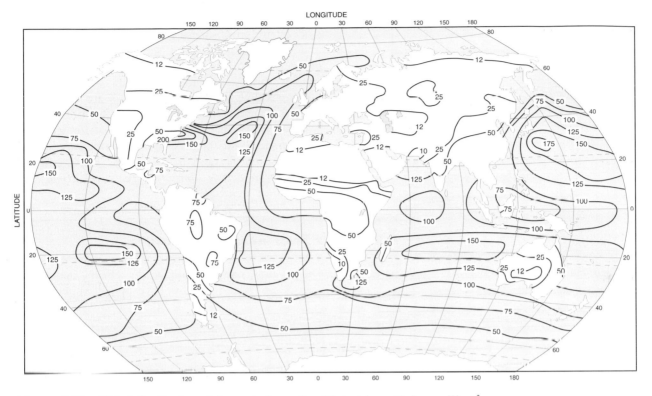

Figure 5.12 *Global distribution of the vertical transfer of latent heat. Units are W m⁻².*
Source: After Budyko et al. (1962).

Much of the earth's surface is covered by oceans, where evaporation takes place continuously. Even a large proportion of the land surface is moist much of the time. Consequently the role of latent heat in balancing the heat budget of Earth is vital. Latent heat transfer by convection carries about one-fifth of the energy of incoming solar radiation back to the atmosphere (Figure 5.12).

The heat used for evaporation over land areas depends upon the availability of moisture and energy. In polar regions it is small, but it increases equatorwards, reaching a maximum in the moist equatorial forests of South America, central Africa and Indonesia. Over the desert areas there is little moisture available and evaporation is insignificant.

Energy transfers and the global circulation

Four main forms of energy exist in atmospheric circulation: latent heat, sensible heat, potential energy and kinetic energy. The total energy of a unit mass of air

(E_t) can therefore be described as follows:

$$E_t = Lq + CpT + gz + V^2/2$$

where Lq = latent heat content, CpT = sensible heat content (specific heat of air × temperature), gz = potential energy (gravitational force × height) and $V^2/2$ = kinetic energy (speed squared divided by two). Latent heat is the quantity of heat released or absorbed, without any change of temperature, during the transformation of a substance from one state to another (e.g. from solid to liquid). Sensible heat can be thought of as the temperature of the atmosphere. More specifically it is the temperature of the air (T) multiplied by the specific heat (Cp)* of the air at a constant pressure. Sensible heat is gained from the ground surface after the absorption of short-wave radiation, or by the release of latent heat through condensation.

The potential energy of the atmosphere is essentially a function of its height above the ground surface (z); gravity (g) is a constant. As air moves in the atmosphere it tends to change its height and alter its energy content. If the air sinks slowly, the potential energy decreases. Normally it is converted to

---

* The specific heat of a substance is the amount of heat required to raise the temperature of 1 g of that substance by 1° C. This is defined at a constant pressure because

adding heat normally alters the volume/pressure relationship of the substance. The specific heat of still air at 10° C is 1.010 J g⁻¹ ° C⁻¹.

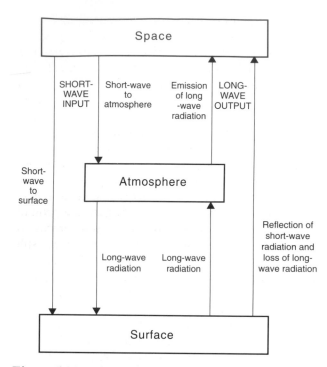

Figure 5.13 *The Earth–atmosphere radiant energy flow system.*

sensible heat, and the air becomes warmer as it subsides. If the air rises, the temperature tends to decline but the potential energy increases.

Kinetic energy is proportional to the square of the velocity of the wind ($V^2/2$). Therefore strong winds have more kinetic energy than gentle winds, as the damage they cause indicates. In fact, on a global scale, hurricanes and other strong winds at the surface are relatively rare, so the quantity of energy in the form of kinetic energy is limited. Even in the regions of strongest winds it probably reaches no more than 0.5 per cent of the total energy content of the atmosphere.

The actual flows of energy can be shown more simply if we consider them as part of a large system in which we distinguish the inputs and outputs with feedback between the different subsystems (Figure 5.13 for radiant energy). The energy is transferred in a variety of forms and, during these transfers, it undergoes numerous transformations.

While the general principles of flow are known, the figures quoted are, in most cases, best estimates. Measurements have been taken at a number of places, but in insufficient quantities to give a reliable global figure. It is little use giving a global average based on a few clustered observations. This has been one of the problems in determining the magnitude of any 'enhanced' greenhouse effect. Satellite observations have helped (and led to appreciable changes in estimates of Earth's albedo) but there are still numerous

flows which are imperfectly known. Long-wave emission by the atmosphere, the separation into direct and diffuse radiation and sensible and latent heat transfer are the main problems, as conditions vary quickly, and, until measurements become more comprehensive, some of the figures are little more than intelligent guesses. The actual value of the flows will depend, in part, on the nature of the assumptions made about them. What we can be sure about, both theoretically and from satellite measurements, is that *what energy comes into the earth/atmosphere system must eventually leave.*

Effects upon temperature

Let us now consider briefly the effects that these energy inputs and outputs have upon temperature.

The daily pattern

If we consider a clear day in the spring in an area in, say, London, sunrise will be at about 6.00 a.m. local time. Temperatures then are low, for during the night the ground has been losing heat by radiational cooling. Slowly, as the sun rises, the ground warms up and, in turn, the air in contact with the surface is heating too (Figure 5.14). By about 2.00 p.m. the ground and air are at their warmest, the maximum temperature at the surface being earlier than that in the air because that is where the heat conversion takes place. From then on, as the sun gradually sinks, the ground surface and the overlying air will cool. The sun sets at about 6.00 p.m.; cooling continues throughout the night until minimum temperatures are reached just before dawn.

This daily variation in insolation and temperature is one of the most basic components of our climate. So obvious is it and so regular that we take it for

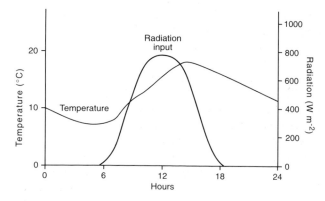

Figure 5.14 *Diurnal changes in short-wave radiation input and temperature on a clear day.*

granted. And yet quite marked differences in atmospheric conditions occur in response to the daily progress of the sun. As we shall see later, the associated changes in temperature may lead to significant changes in humidity, and they often spark off major atmospheric processes such as vertical movements of air and even heavy storms.

It is also apparent that this daily pattern of insolation and temperature change itself varies according to atmospheric conditions. The effects are most obvious when the air is clear and still, for then heating and cooling proceed uninterrupted. If the sky is cloudy or very hazy, however, the daily pattern of temperature is much more variable (Figure 5.4). Similarly, the pattern varies spatially. It is less marked over the sea, for much more of the incoming energy is used to heat up and evaporate the water, and less is returned directly to heat the atmosphere. During the night, the sea cools slowly, with the result that temperatures do not fall so much as on land – one reason why coastal areas are less prone to night-time frost (Figure 5.15). The pattern is most apparent in areas with dry air and ground surfaces, such as deserts. There, incoming radiation is large, and little energy is used for evaporation, so temperatures are high, while radiational cooling at night is intense, giving rise at times to low air temperatures.

The seasonal pattern

A very similar pattern of variation takes place on a seasonal scale. The cause in this case is not Earth's rotation but its changing relationship with the sun: the variation within its orbit that produces the apparent seasonal progress of the sun from the Tropic of Cancer to Capricorn and back.

This change in the position of the sun leads to changes in the angle of the incoming rays and in the duration of daylight. Both factors influence the amount of insolation received by Earth and, therefore, the degree of atmospheric heating. Considering again our area in London, we would find that in the winter the maximum elevation of the sun, at midday, was about 16°, for the sun stands approximately over the Tropic of Capricorn. Thus the rays of the sun still strike the surface at a relatively low angle and the degree of midday heating is limited.

As the sun moves northwards to the equator and thence to the Tropic of Cancer its midday position rises and the rays strike the surface less obliquely. Moreover, the days become longer and the nights shorter. Maximum temperatures increase until, about July, they reach their highest values, slightly after the

maximum radiation in late June. From then until mid-December the sun returns south, its midday position in the sky declines, the quantity of insolation received at the surface is reduced and so temperatures fall.

It is apparent that, in London, the winter months represent a period when incoming radiation is low. Outputs of energy from Earth continue, however, so the area experiences a net radiation deficit. During the spring, as the overhead sun moves north of the equator, radiation inputs rise to match outputs, but the degree of atmospheric warming is restricted because much of the excess energy is used to reheat the ground and the oceans. By August the ground has warmed up; during autumn the sun returns to its position over the equator but now the surface still retains some of the heat gained during the summer. The air, therefore, remains relatively warm compared with spring even through the sun is at the same midday zenith angle.

The seasonal pattern of radiation and associated temperature conditions varies latitudinally. In polar areas the sun never gets high in the sky, but the length of the day varies markedly, so that during summer months these areas experience perpetual daylight. Conversely, in the winter months they are in continuous darkness. The seasonal radiation balance is therefore very variable. At the North Pole, for example, from April to September there is a potential continuous radiation surplus, for the sun would shine for twenty-four hours per day if the sky was cloud-free, so night-time cooling is less. In contrast, for the rest

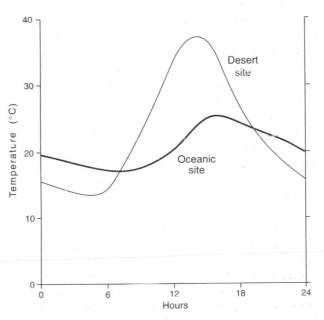

Figure 5.15 *The effect of water on diurnal temperature ranges. Oceanic sites have a small diurnal temperature range, whereas in a desert the range is high.*

of the year a radiation deficit occurs. No insolation is experienced for six months, so radiational cooling continues, interrupted only by the transfer of air from warmer latitudes (Figure 5.16).

The pattern in the tropics is very different. Here the sun never strays far from its overhead position; seasonal variations in radiation are limited and the diurnal variation becomes dominant.

The pattern of energy input to Earth's surface as shown in Figure 5.2 is a vital element in determining the thermal régime. As we shall see, it is not the only factor involved and the actual surface temperatures (Figure 5.17) show many differences from the pattern shown in the figure.

Further reading

Barry, R.G. and Chorley, R.J. (1992) *Atmosphere, Weather and Climate*, sixth edition, London: Routledge (Chapter 1). A popular textbook, now in its sixth edition, which covers the whole field of climatology in considerable detail. Chapter 1 is not always easy to absorb but provides good coverage of atmospheric energy and heat.

Hartmann, D.L. (1994) *Global Physical Climatology*, San Diego: Academic Press (Chapters 2, 3 and 4).
A modern replacement to Sellers' classic *Physical Climatology*. It is pitched at quite an advanced level but includes recent data from satellites.

Hidore, J.J. and Oliver, J.E. (1993) *Climatology: An atmospheric science*, New York: Macmillan (Chapters 2, 3 and 4). An elementary textbook which aims to introduce the processes of climate changes through time. Also looks at human impact on the energy budget.

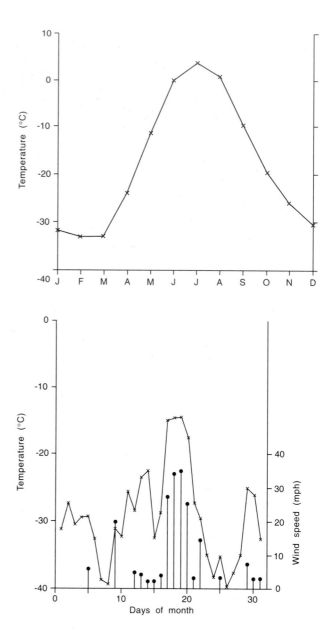

Figure 5.16 *(a) Mean monthly temperatures at Alert (82° N). (b) Temperatures at Alert in winter in relation to wind speed. Under calm conditions temperatures are very low but stronger winds disturb the surface inversion and bring warmer air from more southerly latitudes.*

Key points

The details of the atmospheric energy system may appear complicated but the system is very important as the driving force for our present-day climates. To recap what happens:

1 Energy enters from the sun.

2 It is reflected, scattered, absorbed and reradiated within the system but does not form a uniform distribution. Some areas receive more energy than they lose; in other areas the reverse occurs.

3 If this situation were able to continue for long the areas with an energy surplus would get hotter and those with a deficit would get cooler.

4 This does not happen because the temperature differences produced help to drive the wind and ocean currents of the world. They carry heat with them, either in the sensible or latent forms, and help to counteract the radiation imbalance.

5 Winds from the tropics are therefore normally warm, carrying excess heat with them. Polar winds are blowing from areas with a deficit of heat and so are cold.

6 Acting together, these energy transfer mechanisms help to produce the present climates of the earth.

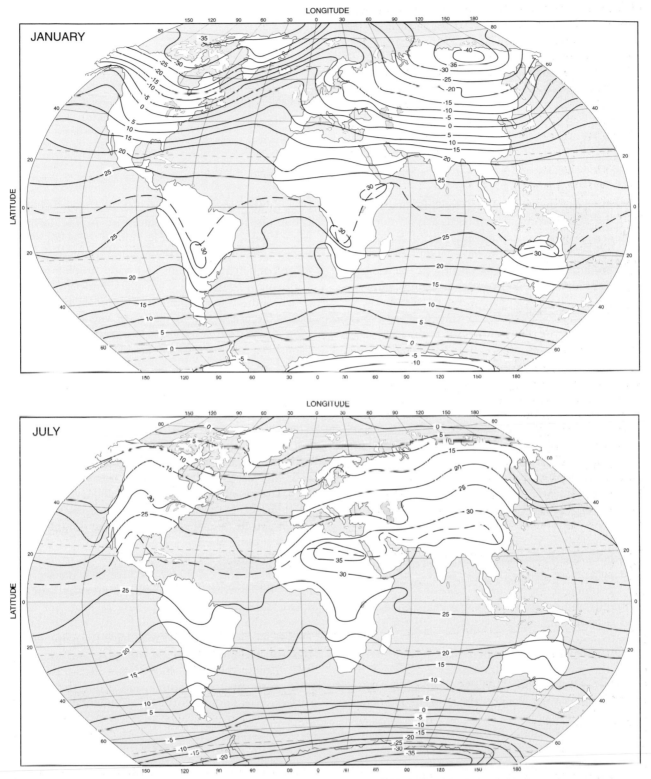

Figure 5.17 *Mean sea-level temperatures in January and July. The approximate positions of the thermal equator are shown by the dashed line.*
Source: After Barry and Chorley (1992).

CHAPTER 6

Moisture in the atmosphere

On a hot day the picture of clouds building up often signifies that a storm is imminent. We do not always appreciate what is happening, but these growing clouds represent one of the most vital processes in the atmosphere – the condensation of water as it is raised to higher levels and cooled within strong updraughts of air. The water, of course, was derived from the surface – evaporated from the oceans, from the soil, or transpired by the vegetation. But within the atmosphere a variety of events combine to convert the water vapour, which is produced by evaporation, to water droplets. The air must rise and cool for condensation to occur. In this chapter we shall be looking at the nature and consequences of these processes.

The effects of heating and cooling in the atmosphere

General effects

The atmosphere is a highly complex system, and the effects of changes in any single property tend to be transmitted to many other properties. Thus heating and cooling of the air cause adjustments in relative humidity and buoyancy; they may cause condensation and evaporation, cloud formation and the development of storms.

What happens, then, when air is heated? To simplify the problem we will consider a parcel of air in contact with the ground. As the ground is heated, the air in contact with it will warm also; its temperature rises and it expands. Gases expand on heating more than liquids or solids, so this effect is quite marked. Moreover, as the air expands its density falls; in simple terms the same mass of air now occupies a larger volume. As its density falls so it becomes lighter than the surrounding air and it tends to rise. Reverse the process, cool the parcel of air, and the opposite occurs. It contracts, its density increases and it sinks.

One effect of the heating and cooling of the surface atmosphere is therefore to cause vertical movements of air. But there are other effects. As the air becomes cooler its ability to hold moisture in the form of water vapour is reduced. If it cools to the point where it can no longer hold the water vapour in vapour form, condensation occurs and water droplets appear. If the air is heated, these droplets tend to evaporate and become water vapour once more. Thus heating and cooling are intimately linked with the processes of evaporation, condensation and precipitation formation. Let us consider these main effects in turn.

Vertical movements

The rising of warm air is a process we can see on a hot day by the shimmering effect of air near the surface; we can see it too if we watch the beautiful and immense towers of a convectional cloud (Plate 6.1). Such vertical movements usually develop if local heating of the atmosphere takes place, so that individual parcels of air become warmer and lighter than the air around them.

Local heating occurs for a number of reasons. Variation in the colour or wetness of a surface may cause differences in atmospheric heating; the air above dark-coloured or dry surfaces heats up more rapidly than that above light or wet surfaces. Differences in slope angle may have the same effect. But there is another factor that plays an important role in these vertical air movements. It is the vertical change in air temperature away from the ground surface. It is known as the environmental lapse rate.

Environmental lapse rate

If we measured air temperatures in the troposphere at different heights under cloudless conditions, we would find that temperature usually falls with height. The reason is quite simple. The incoming radiation heats the ground through absorption. A small proportion of this energy is transmitted downwards

Plate 6.1 *Cumulus clouds developing in an unstable airflow in Slovakia. Individual turrets of rising saturated air can be seen. Parts of the cloud tops can be seen evaporating into the drier air above*
Photo: Peter Smithson.

into the soil by conduction but the majority is returned as either sensible heat or long-wave radiation to heat the atmosphere. Heating is greatest close to the ground surface and declines with height.

The rate at which the temperature falls with increasing altitude is, on average, 6.4° C/1000 m, and this pattern continues as far as the tropopause. This is the environmental lapse rate. It is not constant, however, for it is affected by atmospheric and surface conditions. When the air is turbulent, or is being mixed by strong winds, the environmental lapse rate, at least in the lower layers of the atmosphere, is low; with strong surface heating, it is steep, meaning that air temperature cools rapidly with height. Under still, calm anticyclonic conditions temperatures may even rise with height for short distances. Whatever its value, this rate of temperature change greatly influences air movement.

Stability and instability

We can start to understand the importance of the environmental lapse rate by considering a simple example. Imagine local heating of the air above an island in the sea. The island, because it converts sunlight to heat more effectively than the surrounding water, will act as a thermal source. Above this thermal source the air will be warmed, its density will decrease, its surface pressure will fall and the air will tend to rise. Typically, after the bubble of air has risen a distance equal to about once or twice its own diameter it sinks back. New and larger bubbles form in its wake, however, and each rises a little higher.

What controls this movement? The answer, simply, is relative temperature. If the bubble of air is warmer than its surroundings it will continue to rise; if it is cooler, it will sink. We know already that the general temperature of the air declines upwards – that is the environmental lapse rate. We might imagine, therefore, that once the bubble starts to rise it will continue to do so indefinitely, for the air around it is becoming progressively cooler with height. That does not happen, however, and the reason is that as the air bubble rises it will also cool. The critical factors that determine the height to which the bubble rises are the relative rates of cooling of the bubble and of the surrounding air.

The next question, then, is why does the bubble get cooler? As we move away from the surface, air pressure will decrease. As the air bubble rises, it encounters surrounding air of lower density. The pressure confining the bubble is reduced and it expands. As it does so heat is extracted from the bubble and it becomes cooler. This is in accord with the Gas Laws which state that:

$$PV/T = K \text{ (constant)}$$

In other words, the pressure (P), volume (V) and temperature (T) of a gas are interdependent. A change in any one of these properties tends to cause changes in the others.

The rate at which the air cools with height as a result of this expansion is constant, at 9.8° C for each 1000 m of ascent. It is known as the **dry adiabatic lapse rate** (DALR). *Adiabatic* means that there is no heat exchange between the bubble and its surroundings and, so long as the bubble of air rises rapidly, this condition applies. It is called dry, not because the air does not contain any moisture, but because no condensation has taken place.

We have, therefore, a framework for determining how far a bubble will rise. So long as no condensation occurs, it will cool at the dry adiabatic lapse rate. The surrounding air cools at the environmental lapse rate. The bubble rises until its temperature (and therefore its density) is equal to that of the surrounding air. This is shown diagramatically in Figure 6.1.

As the dry adiabatic lapse rate is a constant, the two variables in this relationship are the environ-mental lapse rate and the initial temperature of the air bubble. The bubble will rise only if it is warmed sufficiently to overcome the restraining effect of the environmental lapse rate. If the bubble cannot rise it is said to be stable. If the temperature is raised far enough, however, or if the environmental lapse rate is great enough, the air bubble can rise a considerable distance. It is in these circumstances that part of the troposphere is said to be unstable.

Condensation

'So long as no condensation occurs . . .' That was the proviso we established when considering the dry adiabatic lapse rate. But all air, even the driest desert atmosphere, does contain some moisture and, if it is cooled sufficiently, will experience condensation. When that happens the processes of uplift are modified.

The ability of the air to hold moisture is dependent upon its temperature. As the temperature of the air increases so its moisture-holding capacity also rises; or, to put it another way, more moisture must be added to reach saturation at a higher temperature.

The amount of moisture which air can hold may be assessed in a number of ways. 'Relative humidity' is the most frequently used term. It is the ratio of the amount of moisture the air contains to the amount of moisture the air could hold when saturated at that air temperature, expressed as a percentage. Relative humidity may be measured indirectly from wet-bulb and dry-bulb temperature readings, using humidity tables. Evaporation of moisture from the wet bulb leads to cooling which is inversely proportional to the relative humidity of the air. If the air is saturated, there will be no evaporation, no cooling and so no difference in temperature between the dry and wet bulbs. Although frequently used, relative humidity does have the disadvantage of being temperature-dependent. For example, as air temperature rises relative humidity will fall, because the air is able to hold more moisture, even though the moisture content of the air has remained constant. An absolute method of measuring moisture content is to determine the vapour pressure, which is that part of the total atmospheric pressure exerted by water vapour. Again it can be obtained indirectly from the wet- and dry-bulb thermometers, using tables and pressure readings. The relationship between temperature and the moisture content at saturation is indicated by the saturation vapour pressure curve (Figure 6.2). Thus as a rising air bubble cools, it approaches the temperature at which condensation occurs. When the air bubble reaches

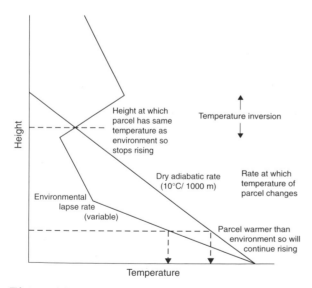

Figure 6.1 *Thermal buoyancy of an air parcel. The parcel will continue to rise as long as it is warmer than the surrounding air.*

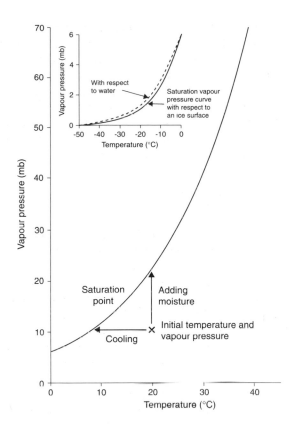

Figure 6.2 *Saturation vapour pressure curve. The curves demonstrate how much moisture the air can hold for any temperature. Below 0° C the curve is slightly different for an ice surface than for a supercooled water droplet.*

that temperature it becomes saturated and condensation takes place.

If condensation was the only thing that happened on saturation then, apart from the extra weight of the droplets, the effect on the air bubble would be small. There is, however, another major effect. As water changes from its vapour state to a liquid it releases latent heat. This heat acts to warm the air and thereby counteracts the cooling resulting from expansion.

We can readily see the implications for our air bubble. Instead of cooling at 9.8° C/1000 m (its dry adiabatic lapse rate), it cools more slowly as it rises. This new, lower rate of cooling is known as the saturated adiabatic lapse rate (SALR). Unlike the dry rate it is not a constant, for, as we can imagine, it depends upon the amount of heat released by condensation, and that, in turn, depends upon the moisture content and hence the temperature of the air. Warm air is able to hold a lot of moisture and thus, on cooling, it releases a lot of latent heat; cold air is able to hold far less moisture, so the heat production during condensation is much less. This

is one reason why some of the world's most severe storms are found in warm climates.

Let us illustrate the effect of condensation by considering a specific example. Figure 6.3 shows the path curve for the bubble. Its initial temperature as a result of surface heating is at 21° C. As it is warmer than its environment, it will cool at 9.8° C/1000 m until saturation point is reached. It is at this level that we first see the visible evidence of our bubble – a small cloud will be seen forming. Above condensation level the rate of cooling slows down to the saturated lapse rate as long as the bubble's temperature is still higher than that of the environment. If it is, we get large convectional clouds building up which will probably give rain (Plate 6.2).

Whether the atmosphere is still stable or not will depend upon the relative rates of cooling of the dry bubbles, the saturated bubbles and the environment. We can summarize this in Figure 6.4. If the environmental lapse rate is cooling more rapidly than the dry adiabatic lapse rate we have absolute instability, as bubbles of air, even if they cool at their maximum rate (the DALR), will be cooling more slowly than the environment. If the environmental lapse rate is cooling more slowly than the saturated adiabatic lapse rate we have absolute stability. If the environmental lapse rate is between the DALR and the SALR we have conditional instability; in other words, instability depends upon the air reaching saturation point.

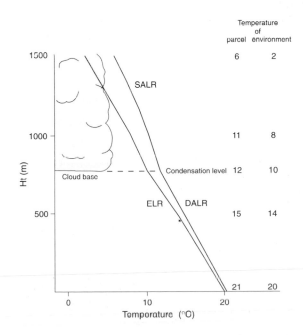

Figure 6.3 *The effect of condensation on the rate of cooling of an air parcel. It is assumed that the parcel starts slightly warmer, owing to localized heating.*

Plate 6.2 *Convectional cloud building towards the tropopause over the Isles of Scilly. The summit of the cloud is more fibrous, indicating that the cloud droplets have frozen.*
Photo: Peter Smithson.

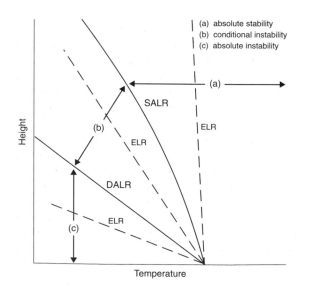

Stability has a considerable effect upon the degree to which convective activity will take place. If the air is unstable it will rise and may produce clouds, whereas if it is stable convection will be reduced. Sometimes, especially under anticyclonic conditions, the temperature will increase with height – a situ-

Figure 6.4 *The relationship between the Environmental Lapse Rate (ELR), the Dry Adiabatic Lapse Rate (DALR) and the Saturated Adiabatic Lapse Rate (SALR). If the ELR is in segment (a) the air is absolutely stable, as, even if the parcel is cooling at its slowest rate (SALR), the ELR indicates the atmosphere is cooling even more slowly. In segment (c) the ELR is cooling more rapidly than the quickest rate of cooling of the parcel, so the parcel will always be warmer than its environment and continue rising. The air is hence absolutely unstable. In segment (b) the ELR is between the DALR and the SALR. The air may become unstable if it reaches the condensation level and can cool at the slower SALR. This situation is called conditional instability, i.e. it depends upon the parcel reaching the condensation level.*

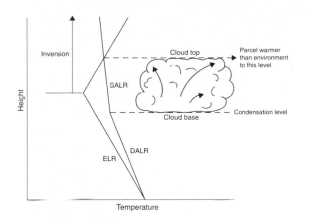

Figure 6.5 *The effect of an inversion of temperature on cloud development with moist air.*

ation known as an inversion of temperature. If the air beneath the inversion is fairly moist, a layer of cloud may develop here. Moist air will have been brought to the inversion by convection and, as it cannot rise further, it spreads out beneath the inversion to form a dense sheet of cloud (Figure 6.5).

Absolute instability in the atmosphere is infrequent except very close to the ground – the convection it initiates helps to transfer heat upwards and so reduces the environmental lapse rate. What is much more common is for the environmental lapse rate to lie between the dry adiabatic lapse rate and the saturated adiabatic lapse rate. In this situation of conditional instability the atmosphere is stable for air which has not reached saturation point, but is unstable for saturated air. If the air can be forced to reach the condensation level, either by ascent over hills or mountains, or by convergence associated with a depression, it will become unstable and assist vertical motion. The former process is one of the mechanisms which leads to higher rainfall over mountains.

Causes of condensation

Clouds are one of the most interesting aspects of the sky. Their shape and form change constantly to reflect the processes of formation and the environment in which they are developing. To produce clouds, we need the air to reach saturation point. It is clear that saturation can be reached either by cooling the air or by adding water to air (Figure 6.2). It is by cooling of the air that the majority of clouds are formed. Orographic lifting, convergent uplift near depressions or within air streams and convection will all produce

vertical motion which may be sufficient to produce clouds. The second process of adding water occurs over warm water surfaces such as the Great Lakes in the autumn, or over the Arctic Ocean, where water will evaporate from the relatively warm sea surface and rapidly condense into the cold air above to form Arctic Sea Smoke. Almost calm conditions are needed to avoid the saturated air being mixed with drier air above.

Radiational or contact cooling at a cold ground surface may also be sufficient to produce saturation, but as these are ground-based processes the resulting condensation is known as fog. It is like cloud in being composed of myriads of water droplets but the detailed mechanisms of formation are different. As there is very little upward movement, the droplets do not increase in size and rain does not fall.

The effects of condensation

Fog

Fogs are a common feature of the climate of some parts of the world. For example, they are frequent on the North Sea coast of Britain in summer, off the Grand Banks of Newfoundland and in coastal Peru. There are two weather situations which can form fog. First, when the ground loses heat at night by long-wave radiational cooling, usually with clear skies of an anticyclone; second, when warm air flows from a warm region to cover a cold surface, particularly a melting snow surface with lots of moisture about. The first type of fog is called radiation fog and the second advection fog.

Fog consists of microscopic droplets of water between 1 μm and 20 μm in diameter. Visibility in a fog will depend upon the size and concentration of droplets in it. When the droplets are small and numerous, visibility is poor, perhaps as little as 5 m. If pollution adds suitable nuclei, condensation of water vapour is favoured. When the droplets are large or sparse, visibility is less affected. The actual formation of radiation fog represents a delicate balance between radiational cooling, air movement and condensation. It forms only when cooling occurs faster than the rate at which latent heat is added by condensation. Because vapour is converted into water droplets, the moisture content and saturation temperature fall, so further cooling is necessary to give saturation. Because of this, fog is more frequent during the long nights of autumn, winter and spring than in summer. If winds are strong, the saturated air near the ground will mix with drier air above and prevent fog forming.

By reducing visibility, fog can be a major environmental hazard. Airports may be closed for several days and road transport is hazardous and slow. Large economic losses can result from such delays, but the potential for artificial fog clearance seems low; too much energy would be needed to warm or dry out the air to prevent condensation.

Clouds

Clouds and fog are the result of similar processes which vary in intensity and duration. Clouds are composed of a mass of water droplets or ice crystals almost microscopic in size. The number of droplets per unit volume of cloud varies considerably, depending upon its origins; smaller concentrations of larger droplets occur in clouds formed in the middle of large oceans, while large concentrations of smaller droplets are found in continental regions. Clearly this is a consequence of the greater availability of nuclei over the dusty continental interiors, but polluted industrial areas may have a similar effect. Studies of such condensation nuclei have shown that there are two broad classes: those with an affinity for water, called hygroscopic particles, like salt; and non-hygroscopic particles, which require relative humidities above 100 per cent before they can act as centres of condensation. The role of sea salt as a source of hygroscopic nuclei has long been debated but recent work suggests that it is not as important as was once thought.

We can find out much about what is happening in the atmosphere by looking at the type of clouds and especially their shape (see box). The low- and medium-level stratiform types are the main rain-bearing clouds of temperate latitudes. Around the centre of a depression we often see a characteristic sequence of clouds as the warm air associated with the depression approaches (Figures 6.6 and 9.6). The first signs of the depression are cirrus clouds which slowly expand in area to become cirrostratus. This cloud sheet thickens and becomes lower, producing altostratus, followed by nimbostratus and, frequently, precipitation. By this final stage there may be a whole complex of cloud sheets which from the ground or even from space are difficult to differentiate.

Cloud types

Clouds are a vital element of the earth's energy budget. They reflect and absorb some of the incoming solar radiation and trap much of the outgoing long-wave radiation. Such is their importance in climate control that when models of the atmosphere are used to predict future climate change the results can vary widely according to the assumptions about the nature of the cloud systems. Cloud top height, thickness, density and spatial distribution are of vital significance in affecting where energy can be absorbed and where it is lost.

Despite this importance, cloud features are one of the least well observed climate variables. Most climate stations will observe the amount of cloud only once a day; it is noted as the proportion of the sky covered by cloud expressed in oktas (eighths) in Europe or in tenths in the United States. Few stations record the type of cloud or its height; observations at night are not easy.

Because clouds develop in an infinite variety of forms and shapes attempts have been made to classify them. The easiest and most widely used way is on the basis of their appearance, a system largely devised by Luke Howard in 1803. Genetic systems based upon the origin of the cloud have been suggested. As it is not always clear exactly how a cloud formed they have been less successful.

The basic division is between clouds which are predominantly layered, known as stratiform, and those where the vertical extent of the cloud is important. These are known as cumuliform (Plate 1). The groups are split up into genera, as shown in the table and then into species, using Latin names in a similar manner to plant classification. Thus we can have *altocumulus lenticularis*, which means a mid-level cloud showing some signs of vertical development (*altocumulus*) which in detail is in the shape of a lens or almond, often elongated and usually with well defined outlines (*lenticularis*). These clouds are usually associated with flow over hills; within them are some moister layers which reach condensation when forced to rise over the hill. There are a large number of these species descriptions because of the variety of cloud forms.

The stratiform types are shown in Figure 6.6 and subdivided according to the height of their formation. In stratiform types of cloud the rate of upward motion is slow, but it may take place over hundreds or even thousands of square kilometres. At low levels these clouds are composed of water droplets, but at higher levels (2000–6000 m) we get a mixture of water droplets and ice crystals. Above about 6000 m stratiform clouds are composed mainly of ice crystals and take the name *cirrus*. Some of the clouds may show signs of convection, even if it is weak. These types have *cumulus* incorporated into their names, such as cirrocumulus or altocumulus.

The other main group of cloud, cumuliform, is the result of local convection or instability. Bubbles of warm air, rising beyond the condensation level (if the air is unstable), are seen as cumulus clouds. The precise shape and form of the cloud will depend upon the degree of instability, the water vapour content of the air and the strength of the horizontal wind (Figure 1). There are many different types of cumulus

Table 1   *Cloud genera.*

Cloud type	Abbreviation	Height level	Cloud base (m)		
			Tropical	Temperate	Polar
Stratiform (*layered*)					
Cirrus	Ci				
Cirrostratus	Cs	High	6000–18000	5000–13000	3000–8000
Cirrocumulus	Cc				
Altostratus	As	Medium	2000–8000	2000–7000	2000–4000
Altocumulus	Ac				
Stratus	St				
Stratocumulus	Sc	Low	Ground to 2000	Ground to 2000	Ground to 2000
Nimbostratus	Nb				
Cumuliform (*vertical development*)					
Cumulus	Cu	Variable	Variable	Variable	Variable
Cumulonimbus	Cb				

cloud, subdivided on the basis of their appearance. Some cumuliform cloud may grow larger and taller. The sharp outlines of the cauliflower-like cumulus become more diffuse and ragged as the upper part of the cloud becomes a fibrous mass of ice crystals. The cumulonimbus stage has then been reached. At this stage of development precipitation is usually occurring, sometimes accompanied by lightning and thunder. As the mass of ice crystals develops, it is often blown downwind by the strong winds of the high troposphere to form an anvil shape (Plate 1), characteristic of cumulonimbus clouds. Convection may initiate other clouds near by or on the flanks of the parent cloud as it gradually decays and evaporates.

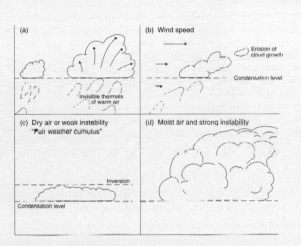

Figure 1   *Atmospheric factors affecting the form of cumuliform clouds.*

Plate 1   *A cumulonimbus cloud where the convection has reached the tropopause and spread out into an anvil shape. Convection of this intensity is associated with unstable air. It is more common in summer, but the snow on the photograph indicates a winter scene.*
Photo: Peter Smithson.

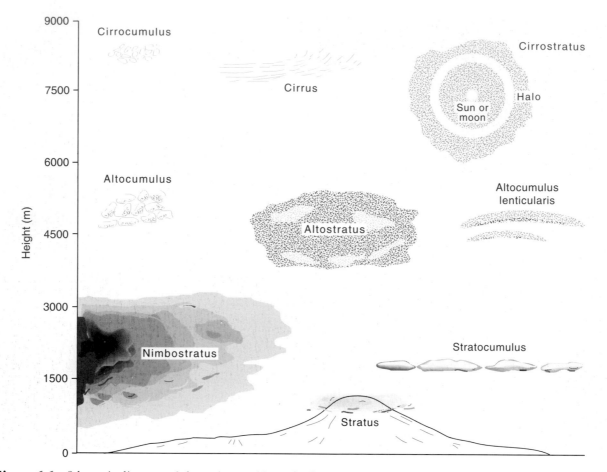

Figure 6.6 *Schematic diagram of the main stratiform cloud types.*

There are many different types of cumulus cloud, subdivided on the basis of their appearance. If an inversion of temperature exists, as often happens with anticyclones, the bubbles will rise to the inversion and then start to level out or descend, evaporating the cloud droplets as they do so. This type of cloud is known as fair-weather cumulus, as it never grows sufficiently to give rain (Plate 6.3). Where there is no inversion to prevent upward growth the cloud may build up as far as the tropopause and give heavy rain.

For rain to fall, we need clouds, but many clouds survive for hours without giving rain. What special circumstances enable some clouds to produce rain whereas others give none?

Precipitation

Formation of precipitation

In parts of Snowdonia average annual precipitation exceeds 4300 mm, and on the summit of Mount Waiaeale, Kauai, Hawaii, it is 11,684 mm.

In terms of the amount of water, this is equivalent to 100,000 t ha⁻¹ y⁻¹ . Without doubt the processes producing precipitation can be very effective when conditions are favourable. But how do these minute cloud droplets (Plate 6.4) grow large enough to fall as rain within as little as twenty minutes of the cloudy air reaching saturation?

To answer this question, we must delve inside a cloud and see what is happening there. In a cloud made up entirely of water droplets there will be a variety of droplet sizes. The air will be rising within the cloud, perhaps at the rate of 10–20 cm per second, though much more rapidly in cumulonimbus clouds. As it rises so the drops get larger through collision and coalescence; some will reach drizzle size. When the uplift is stronger, say 50 cm per second, the downward movement of the drops will be reduced, so there will be more time for them to grow. If the cloud is about 1 km deep small raindrops of 700 μm diameter may be formed.

When temperatures fall below 0° C, because of their small size, the droplets do not freeze immediately but may remain unfrozen in what is said to be a

Plate 6.3 *Fair-weather cumulus clouds over North Wales. The level cloud base and cloud tops can easily be seen.
Photo: Peter Smithson.*

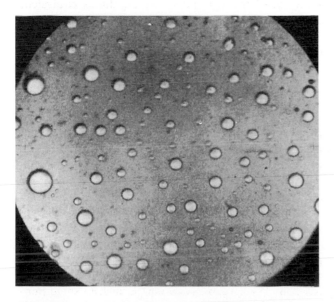

Plate 6.4 *A typical sample of cloud droplets, caught on an oiled slide and photographed under a microscope in the aircraft. The largest droplet has a diameter of about 30 μm. After B.J. Mason.*

supercooled state. With further cooling to –10° C, ice crystals may start to develop among the water droplets. This mixture of water and ice would not be particularly important but for a peculiar property of water. The saturation vapour pressure curve for ice (Figure 6.2) is slightly different from that of water. The air can be saturated for ice when it is not saturated for water. Thus at –10° C, air saturated with respect to liquid water is super-saturated relative to ice by 10 per cent and at –20° by 21 per cent. As a result the ice crystals in the cloud tend to grow and become heavier at the expense of the water droplets.

As the ice crystals sink into lower layers of the cloud where temperatures are only just below freezing, they have a tendency to stick together to form snowflakes. This is brought about by the supercooled droplets of water in the cloud acting as an adhesive. After the snowflakes have melted the resulting drops may grow further by collision with cloud droplets before they reach the ground as rain. This method of producing raindrops is known as the Bergeron–Findeisen process, after the developers of the theory (Figure 6.7). Beneath the base of the cloud, however, evaporation will take place in the drier air and if the drop is small it may be evaporated completely.

Precipitation formation both by collision and coalescence and by the Bergeron–Findeisen process undoubtedly occurs in the atmosphere, though clearly the Bergeron–Findeisen process can operate only when cloud temperatures are well below freezing. The rate at which vapour is converted into water droplets and precipitation depends upon three main factors: the rate of coalescence and ice crystal growth; the cloud thickness; and the strength of the updraughts in the cloud. The total amount of rain will be determined by the life span of the cloud, the height of the cloud above the ground and how long these processes operate. Cloud thickness and updraught speed are largely dependent upon instability and convergence in the atmosphere. Precipitation has been classified in terms of the factor which gives rise to the upward movement, so let us have a look at this in a little more detail.

Types of precipitation

Convectional precipitation

The spontaneous rising of moist air due to instability is known as convection. We have seen that upward-growing clouds are associated with convection. Since the updraughts are usually strong, cooling of the air is rapid and lots of water can be condensed quickly. Collisions and coalescence are likely to be frequent, so the larger droplets rapidly increase in size. Eventually, growing larger and heavier, the droplets overcome the lift provided by the updraught, and they start to fall through the cloud into the clear air beneath. As the volume of water in these big drops is large relative to their surface area, little evaporation takes place in the non-saturated air below the cloud. At the ground there will be a burst of heavy rain as the shower passes.

Unstable air which favours convectional rain is most frequently found in warm and humid areas, but even in the United Kingdom about 20 per cent of the annual rainfall is by convection, with the proportion increasing towards the south and east. This convectional rain may be the result of cold air moving over a warmer ground surface or the result of strong surface heating; both situations will give the steep lapse rates characteristic of instability and convection.

Thunderstorms: When the atmosphere is very unstable, cumulonimbus clouds develop sometimes accompanied by lightning and thunder. At night an intense thunderstorm can be one of the most spectacular displays of the atmosphere. Flashes of lightning shoot from cloud to ground or within the clouds (Plate 6.5), accompanied by great crashes of thunder. How can such dramatic manifestations of energy

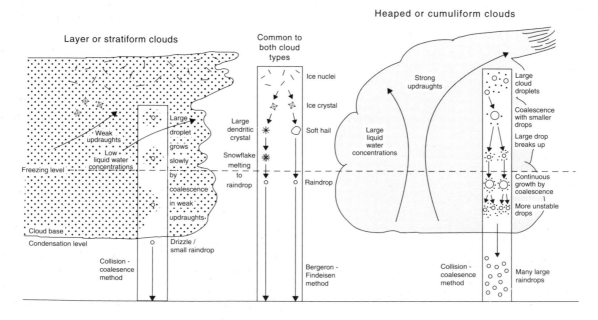

Figure 6.7   *Schematic diagram to demonstrate the processes of precipitation growth in (a) stratiform and (b) cumuliform clouds.*

Plate 6.5 *Lightning strike.*
Photo: *C.J. Richards.*

build up in a cloud? The question has puzzled meteorologists for many years. Recent observations suggest that the main way in which electrical charges are separated is by the formation, growth and electrification of pellets of soft hail, with positive charges being carried into the upper parts of the cloud (Figure 6.8).

As well as electricity, the thunderstorm is often accompanied by squalls of cold wind blowing away from the cloud. They usually originate as downdraughts of air near the main burst of rain (Figure

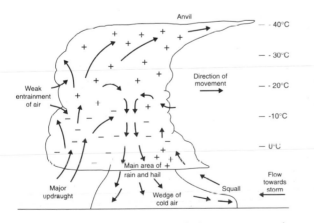

Figure 6.8 *Schematic diagram of air movement and electrical charge concentration in a thunder cloud.*

6.8). Even hail may fall near the centre of the storm, sometimes causing great damage.

Hail: Hail is a form of precipitation composed of spheres or irregular lumps of ice (Plate 6.6). It falls in narrow bands associated with cumulonimbus clouds and so frequently misses the observing stations. However, the destruction it can produce is dramatic. Crops can be torn to shreds, glasshouses ruined and even cars dented by the weight of half a kilogram or more of ice falling from the skies.

Splitting open a large hailstone will show that it is composed of alternating layers of clear and opaque ice (Plate 6.7). It appears that the stone is involved in complex movements within the cloud, being swept up to the higher, colder parts of the cloud several times. When this happens, any moisture condensing on the stone will freeze instantly, including any trapped air, producing opaque ice. At lower levels in the cloud condensed water takes a longer time to freeze. Air bubbles can escape, to leave a layer of clear ice when it eventually freezes. The alternating layers of clear and opaque ice indicate the number of times the hailstone has been swept up by the cloud updraughts.

Cyclonic precipitation

In temperate and subpolar latitudes most of the precipitation comes from cyclones. The temperate cyclone is characterized by areas of rising air associated with convergence. A satellite photograph of a cyclone shows the extensive areas of cloud resulting from this slow but widespread ascent of the air (Plate 6.8).

There are a number of differences from convectional precipitation. The areal extent of rising air associated with a cyclone is much larger, and the rate of upward movement and the rate of condensation in the generally stratiform clouds are much less. Because of this, the droplets grow more slowly and fall out of the cloud sooner. Being small, they can be greatly affected by evaporation in the drier air beneath cloud base. For example, in an atmosphere with a relative humidity of 90 per cent, a droplet of radius 10 μm will fall only 3 cm before evaporating; drops of 100 μm and 1 mm would fall 150 m and 40 km respectively. Despite the relatively small size of the raindrops, the areas affected by rising air are vast. For a particular rain belt, it may take several hours of steady rain before the system has passed, giving a total fall of perhaps between 5 mm and 10 mm. An example of rainfall from a cyclone is shown in Figure 9.9.

Plate 6.6 *Hailstones.*
Photo: M.E. Hardman.

Plate 6.7 *A section through the centre of a large hailstone, taken in reflected light, showing the regions of clear ice which appear black and milky or opaque ice which appears white.*
Photo: By courtesy of Dr K.A. Browning.

In the deep and widespread clouds associated with a cyclone, it is quite common for ice-crystal clouds at higher levels to act as a source of supply to the mixed clouds of ice and water droplets at lower levels – a process known as seeding. The addition of extra ice crystals speeds up the precipitation process and leads to more intense rainfall. Convectional systems may be embedded within the cyclonic circulation to produce more complex patterns of surface precipitation.

Orographic precipitation

Almost all mountain areas are wetter than the surrounding lowlands. To take two examples, Hokitika on the west coast of New Zealand receives an average of 2950 mm per year. At Arthur's Pass, 740 m higher in the New Zealand Alps, the annual average has risen to 3980 mm, compared with less than 670 mm for Christchurch, on the more sheltered lowlands to the east. Even the Ahaggar and Tibesti mountains in the centre of the Sahara receive more rain than do the surrounding lowlands – Asekrem, at 2700 m, has an annual average of about 125 mm, compared with only 13 mm at Silet, 720 m above sea level. Why should this be so?

Where air meets an extensive barrier it is forced to rise. Rising, as we know, leads to cooling of the air, and cooling encourages condensation. On the mountain slopes and above the mountain summits the clouds start to pile up, reflecting the forced ascent of air. Often they reach thicknesses sufficient to give drizzle and rain. From a distance we can sometimes see these dense clouds enveloping the mountains (Plate 6.9).

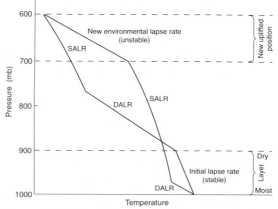

Figure 6.9 *Destabilization of the lower atmosphere through uplift of a layer of air 100 mb in thickness. Between 1000 mb and 900 mb the air is initially stable. It is moist near the base and drier aloft. As a result of uplift, the new environmental lapse rate indicates instability.*

Orographic rain is also produced in another way, due to changes in the stability of the air as it rises. If the air is very moist near the ground surface but much drier above, as it rises the rates of cooling between the top and bottom of the layer will be different (Figure 6.9). The upper part will cool more quickly and so become colder, leading to less stable air. The cloud development associated with instability will increase and rain may fall over the mountains. This situation is known as convective or potential instability.

Hills as well as mountains act as favourable areas for convectional showers. The slopes facing the sun will be warmed more rapidly than flatter areas, because the slopes act as thermal sources. The resulting cloud may produce rainfall which is restricted to the upland area.

The orographic effect is most pronounced when it is already raining upwind of the hills or mountains. Where air is rising – associated with a cyclone, for example – the rate of uplift is increased by the extra ascent forced by the hills. This leads to a greater rate of condensation on the windward side, larger drops of rain being formed, and so a higher rainfall at the surface. Coupled with the slowing down of the rain belt as it passes, owing to increased friction, the net effect is considerably greater rainfall (Figure 6.10).

On the leeward side of the hills, subsidence or descending air begins to dominate, so that the cloud sheet thins or even dissipates and rainfall declines. As the air descends it gets warmer, owing to compression, to give us the rain shadow effect on the leeward slope of the mountains. Here rainfall is far less than on the upslope side and sunshine amounts and daytime temperatures are normally higher.

As much of the precipitation in mountains is due to an intensification of existing rain, it would be wrong to think of orographic precipitation as a truly separate category. It can occur as drizzle or by convective instability, but much more frequently it

Plate 6.8 *A well developed cyclone to the west of Ireland associated with the characteristic spiral of cloud. Shower clouds can be seen over the ocean to the west and north of the low-pressure centre, in contrast to the widespread cloud sheets around the cyclone. Image taken on 12 March 1989 at 13.26 GMT in the infra-red waveband. Photo: University of Dundee.*

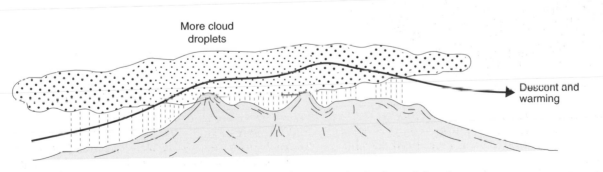

Figure 6.10 *Production of greater rainfall over hills as a result of enhanced forced ascent.*

Plate 6.9 *Orographic cloud near Glencoe, Scotland. The thickest areas of cloud are where uplift is greatest.*
Photo: Peter Smithson.

will depend upon cyclonic or convection processes already operating. Even these two types can occur together in cyclones, so perhaps we should identify convectional, cyclonic and orographic precipitation as interrelated mechanisms of rainfall rather than classifying them into these types.

Conclusion

In this chapter we have followed the exchange and movements of moisture between Earth's surface and the atmosphere. Their effect upon the climate is obvious. But precipitation is also important in other ways. It is a major component of the hydrological cycle which portrays the movement of water around the globe. Rainfall also takes part in many of the processes that build our landscape. And plant and animals are highly dependent upon precipitation. Therefore, in the next chapter we will examine the question of precipitation and follow its consequences on to Earth's surface.

Further reading

Mason, B.J. (1975) *Clouds, Rain and Rain-making,* Cambridge University Press.
A combination of elementary and advanced ideas about rain formation. Only for those who want to go into more detail, especially about the microphysics of clouds and precipitation.

McIlveen, R. (1992) *Fundamentals of Weather and Climate,* second edition, Chapters 5 and 6. London: Chapman and Hall.
Elementary to intermediate level text that provides a useful explanation of stability and moisture with only limited maths and physics.

Schaefer, V.J. and Day, J.A. (1981) *A Field Guide to the Atmosphere,* Boston: Houghton Mifflin.
A very interesting book which stresses the visual approach to the atmosphere, hence the emphasis on clouds. Strikes a nice balance between description and explanation.

Key points

1 The process of evaporation supplies moisture into the lower atmosphere. The prevailing winds then circulate the moisture and mix it with drier air elsewhere. Only if we have a dry surface in areas well away from the oceans and where dry subsiding air is dominant will moisture levels be low.

2 Water vapour is only the first stage of the precipitation chain; the vapour must be converted into liquid form. This is usually achieved by cooling, either rapidly, as in convection, or slowly, as in cyclonic storms; mountains also cause uplift but the rate will depend upon their shape, their height and the direction of the wind.

3 Even this is insufficient, as we can tell from the large number of clouds in the sky which never give precipitation. To produce precipitation, the cloud droplets must become large enough to reach the ground without evaporating. The cloud must possess the right microphysical properties to enable the droplets to grow. It must have ice crystals if the Bergeron–Findeisen process is to operate, or a wide spectrum of drop sizes with plenty of moisture condensing for the collision-coalescence system to work.

4 Even these suitable conditions may be insufficient if the cloud does not last long enough for growth to take place. Clearly precipitation results from a delicate balance of counteracting forces, some leading to droplet growth, others to droplet destruction. Nevertheless, where conditions are basically favourable – where air can rise high enough to produce large vertical developments of cloud – copious amounts of precipitation can occur.

CHAPTER 7
Precipitation and evapotranspiration

Precipitation

For those who live in the humid regions of the world, precipitation is normally so frequent that it is taken for granted. During times of drought the importance of precipitation and its role in feeding the hydrological system and providing water for human use and plant life become all too apparent. In recent years the problems raised by drought in areas such as the Sahel, California and even parts of north-west Europe have become publicized in the media as demand for water has increased at a time when some areas have experienced a decline in precipitation.

Precipitation represents the vital input of water to the surface hydrological system. It is the nature of this input – the character and distribution of precipitation – which we will consider in the first part of this chapter.

Types of precipitation

To most people three types of precipitation come immediately to mind: rain, snow and hail. As we would see if we intercepted these and looked at the raindrops, snowflakes or hailstones more closely, the distinction between them is not always clear, and the terms mean different things in different areas. Moreover, they are not the only forms in which moisture is input to the surface. Dew, fog-drip and rime all transfer water from the atmosphere to the ground (Table 7.1). Their contribution, however, is usually small.

Rain

We have already discussed the main processes of rainfall generation in Chapter 6. Here we are concerned with the nature of the rainfall after it has fallen from the cloud.

Table 7.1 *Types of precipitation.*

Type	Characteristics	Typical amount
Dew	Deposited on surfaces, especially vegetation; hoar frost when frozen	0.1–1.0 mm per night
Fog-drip	Deposited on vegetation and other obstacles from fog; rime when frozen	Up to 4 mm per night
Drizzle	Droplets under 0.5 mm in diameter	0.1–0.5 mm per hour
Rain	Drops over 0.5 mm in diameter, usually 1–2 mm	Light under 2 mm per hour; heavy, over 7 mm per hour
Hail	Roughly spherical lumps of ice 5 to 50 mm or more in diameter, often showing a layered structure of opaque and clear ice in cross-section	Highly variable
Snowflakes	Clusters of ice crystals up to several cm across	Variable
Granular snow	Very small, flat opaque grains of ice; solid equivalent of drizzle	Light, under 1 mm per hour
Snow pellets (graupel or soft hail)	Opaque pellets of ice 2–5 mm in diameter falling in showers	Variable
Ice pellets	Clear ice encasing a snowflake or snow pellet	
Sleet (UK)	Mixture of partly melted snow and rain	
Sleet (USA)	Frozen rain or drizzle drops	

Typically rainfall consists of water droplets that vary in size. Where the rain is produced by thin, low-level stratiform clouds, droplets tend to be small, with a majority in the range from 0.2 mm to 0.5 mm in diameter. Where the clouds are thicker, strong updraughts hold the droplets in the atmosphere longer, so the number of collisions increases and the rain is composed of larger droplets, often several millimetres in diameter. The diameter of a droplet is also affected by events during its fall through the atmosphere. Further collision and coalescence cause droplets to grow in size, while evaporation makes them smaller and may cause the total loss of very tiny droplets. In general, however, the droplets reaching the ground show a logarithmic size distribution, with a large number of small droplets and a much smaller number of large drops.

The size of the droplets has considerable significance, for, together with the strength of updraughts in the air, it controls the fall velocity of the rain. In still air the fall speed of a droplet 0.2 mm in diameter is about 70 cm s^{-1}; for a drop of 2 mm diameter, it is about 650 cm s^{-1}. The momentum of the droplet when it reaches the ground is known as the terminal velocity, and, with the mass of the drop, determines its kinetic energy.

The total kinetic energy of a storm depends upon the number of raindrops reaching the ground. This is a measure of the rainfall intensity and in Figure 7.1 the relationship between the intensity and total kinetic energy of rainfall is illustrated. Rainfall intensity varies considerably both within an individual storm and between storms. Rainfall from thick cumulus-type clouds is particularly variable, owing to spatial differences in cloud thickness and updraught

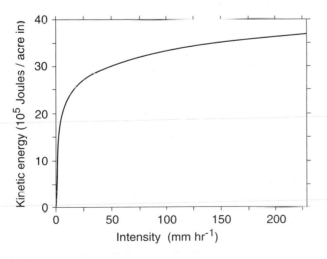

Figure 7.1 *Rainfall intensity against the kinetic energy released by rainstorms.*
Source: After Bennett and Chorley (1978)

Table 7.2 *Water input during a snowstorm in California.*

Period	Duration (hr)	Depth (mm)	Intensity (mm hr^{-1})
Heaviest clock hour	1	5	5
Heaviest 3-hour period	3	11	3.7
Heaviest 6-hour period	6	17	2.8
Heaviest 12-hour period	12	26	2.2
Heaviest 24-hour period	24	38	1.6
Total storm	42	49	1.2

Source: After Miller (1977).

strength, but intensities may be as high as 200 mm hr^{-1} or more for short periods. Precipitation from stratiform clouds is less variable and intensities are usually low – less than 10 mm hr^{-1}.

Snow

In most areas of the world, rainfall is by far the most important input to the surface hydrological system. In some areas significant inputs can fall in the form of snow. Snow occurs mainly in winter, and, despite its thickness and persistence, the quantities of moisture involved are relatively small (Table 7.2). In general 120 mm of freshly fallen snow produces only about 10 mm of water. Where snow is formed in very cold, dry air the moisture equivalent is even smaller, and it may take as much as a metre of snow to produce 10 mm of water. In high mountain and polar regions where temperatures are low throughout the year, the majority of precipitation falls as snow. Even so, because of the low temperatures preventing the atmosphere holding much moisture, many of these areas are, in fact, quite arid. There are no adequate records to provide accurate values, but it seems likely that on a world basis no more than 1 per cent of the total annual precipitation occurs as snow.

Snowfall usually starts in the atmosphere as tiny ice crystals produced at temperatures well below freezing. As the crystals fall they tend to aggregate, particularly where there is sufficient moisture in the air to bind the crystals together. This mainly occurs where temperatures are close to freezing point, and in these conditions large snowflakes may be formed (Plate 7.1). At lower temperatures, moisture is lacking and the crystals do not aggregate (Figure 7.2).

As with rain, the fall velocity of the snowflakes depends on size and, all else being equal, large flakes fall more rapidly than small ones, with maximum speeds of about 100 cm s^{-1}. As we all know, however,

Plate 7.1   *Large flakes of snow falling, with the temperature close to freezing.*
Photo: C.J. Richards

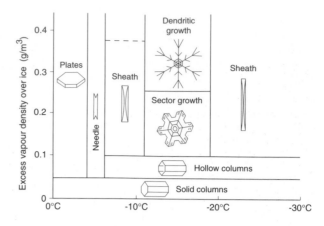

Figure 7.2   *Ice crystal forms in relation to temperature of formation and excess vapour density over ice compared with water.*

snowflakes vary considerably in shape and that too may influence fall speeds. Moreover the density of snowflakes is very low; large flakes often have a density of as little as 100 kg m^{-3} (compared with approximately 1000 kg m^{-3} for raindrops). Consequently, for their size, snowflakes are light and they are readily blown by the wind. For this reason the distribution of snowfall during a storm is greatly influenced by surface wind conditions, and even after reaching the ground the snow may be redistributed to form deep drifts and snow-free areas (Plate 7.2).

Another important feature of moisture inputs in the form of snow is that it is often many weeks or, in polar areas, even years before the water is actually released. Thus in mountain areas, winter snowfall may survive into the spring and so represents a temporary store of water which is released only by melting. In Arctic and Antarctic regions the snow accumulates for centuries, moving with imperceptible slowness in the ice sheets and glaciers before melting, perhaps thousands of years later. Unlike rainfall, therefore, snow is not always an immediate input to the hydrological system.

Hail

The word 'hail' can strike fear into the heart of farmers in many parts of the world. Damage to crops can be severe, though normally the devastation is localized. Hailstorms usually produce a swath of stones as the parent cloud moves across the country. Because of the limited areal extent of the storms, and their relative infrequence, their contribution to water inputs is generally small.

Hailstones vary considerably in size, but are usually less than 1 cm in diameter (Figure 7.3). Stones of

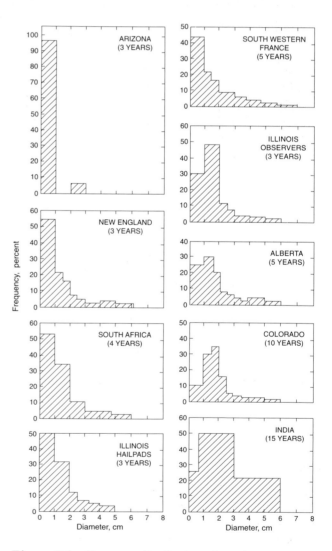

Figure 7.3   *Frequency distributions of maximum hailstone sizes at various locations.*
Source: After Changnon (1971).

Plate 7.2   *Effects of windblow through a hedge following a heavy fall of snow. The snow accumulates where air movement is least and is blown away in windier locations.*
Photo: Peter Smithson.

this size can cause some damage, but it is the larger stones, possessing considerable kinetic energy, that produce spectacular effects, such as damage to cars, greenhouses and vegetation.

Data on the frequency of hailstorms are not entirely reliable. Standard statistics probably underestimate the true frequency of hail because many storms pass between observing stations. For example, in South Africa, where hailstorms are prevalent, the standard network of recording stations gave an average of five storms per year. When the network was increased to one observer per 10 km, eighty days with hail were recorded.

Dew, fog-drip and rime

Walking through grassland on a cold autumn morning after a clear night, we would almost certainly be conscious that the ground surface was wet with dew. Similarly, in a dense, tropical forest with lots of mist or cloud we might see, hear or feel water dripping from the leaves as fog-drip. In these two cases we are dealing with some of the smallest contributors to the precipitation input, although locally they may have some importance. Dew forms on cold surfaces at night when the air is close to saturation. Under these conditions, of course, the air can hold little moisture and, as the atmosphere loses moisture to the ground in the form of dew, it dries out further. Consequently the total amount of dewfall that can occur in a single night in temperate latitudes is normally limited, rarely more than 0.6 mm. In tropical forests the amounts can be greater but they still form an insignificant amount of the annual total. As evaporation rates are high once the sun rises, such small quantities of moisture are soon returned to the atmosphere, so the contribution of dew to the local water budget is likely to be negligible.

Where cloud droplets are blowing continuously across a rough surface such as a forest we get fog-drip. The process results from the deposition of small water droplets moving horizontally by contact with the vegetative surface. Eventually the droplets combine to form larger drops which fall to the ground. The effect is accentuated if the trees increase the turbulent motion of the air as it moves over the

Plate 7.3   *Rime accumulation in the Westerwald, Germany. The direction of air movement can be determined, as rime will grow into the wind when supercooled droplets freeze on contact with the frozen surface.
Photo: Peter Smithson.*

canopy. The vertical motion brings the cloud droplets downwards and impacts them on to the leaves.

Fog-drip is most important in areas where forested mountain ridges extend into persistent cloud sheets which produce little precipitation by normal methods. Examples include California, Hawaii, the Canary Islands and Japan. Studies in all these areas indicate that there is a significant increase in moisture input at the ground which would not otherwise occur. At Berkeley, California, as much as 200–300 kg of water per square metre of surface is found during the summer, when little rain falls. On Hawaii trees planted at 800 m altitude catch trade-wind cloud droplets at the rate of about 4 mm per day; over the year this represents an input of about 750 mm to the island's water budget. Without this additional input it is

unlikely that the forests would be able to survive. Coastal deserts such as in Namibia and Peru also have this feature where mist can blow through shrubs or low vegetation.

If temperatures at the ground are below freezing point, the drifting cloud droplets freeze on the vegetation to form rime. Although this may occasionally produce spectacular scenes (Plate 7.3), the weight of rime can be damaging to trees. The contribution to water input and to tree damage has been investigated in Germany, where coniferous tree growth can be severely hampered by damage to the growth points.

Measurement of precipitation

Rain

It is easy to measure rainfall. Any watertight container sited well away from buildings and trees will act as a rain gauge. How much it collects will depend not only upon the amount of rain and the strength of the wind, but also on the gauge diameter and its height above the ground (Table 7.3). Because of this, rain gauges in the United Kingdom have a standard diameter of 12.7 cm and are set a fixed distance of 30 cm above the ground. Unfortunately the standard varies from country to country, so that in Canada the diameter is 9 cm at a height of 30 cm above the ground, while in the United States the gauges are 20 cm wide and 78 cm high (Figure 7.4). Comparisons of rainfall totals between countries are therefore more difficult than might be expected.

The main reason for these differences in gauge height is snow. We have already mentioned that snowfall is difficult to measure because of its lightness and tendency to drift. In the United States the same gauge is used to measure both snowfall and rain, so it has to be well above the level of drifting snow. In Canada separate gauges are used, while in Britain snowfall is a relatively small component of the annual precipitation.

In normal operation the amount of rainfall collected in a gauge is measured once a day. In the United States an appropriately calibrated stick is used to measure the depth of water which has accumulated in the gauge to obtain the quantity of rainfall. In Canada and the United Kingdom the rainwater in the gauge is poured into a glass measuring cylinder,

Table 7.3   *Variation of rainfall catch with gauge height.*

Height of gauge mouth above ground (cm)	5	10	15	20	30	46	76	152	610
Catch as % of that at 30 cm	105	103	102	101	100	99.2	97.7	95.0	90.0

Source: After Bruce and Clarke (1966).

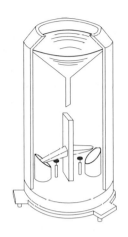
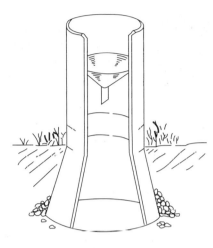
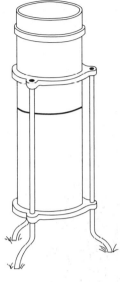

Tipping bucket
rain gauge

British standard
rain gauge

US Weather Bureau
standard rain gauge

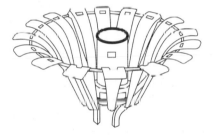

USSR Tretyakov
precipitation gauge

Ground level
rain gauge

Figure 7.4 *Types of standard rain gauges.*

where the rainfall equivalent can be read directly. A standard rain gauge will record only the total rain which has fallen between readings; in many cases it is important to know when the rain fell and at what intensity. For this purpose recording rain gauges are used. The older form of recording gauge uses a natural syphon system of storage with a pen and chart recording the changes in the level of water and hence the rate and amount of rainfall (Figure 7.5). More recent systems use a tipping bucket of known capacity which electronically records the number of times the bucket tips. This information can be stored by a data logger and downloaded directly on to a computer. Some can even be interrogated via telephone or satellite. This is particularly useful in remote areas where

heavy rainstorms may lead to flooding lower in the river basin.

Rain gauges are not the only means of measuring rainfall. Weather radar systems have been developed which can provide quantitative estimates of the rates of rainfall. The method is based on the amount of reflection of the radar signal from falling precipitation. Although there are problems of interpretation of the reflected signal because of scattering from local buildings or hills and when snowflakes melt, it is now possible to produce maps of areas of precipitation and their intensity, as shown in Figure 7.6. In parts of the world, especially over the oceans where there are few or no rain-gauge measurements, satellites have been used to estimate precipitation totals on

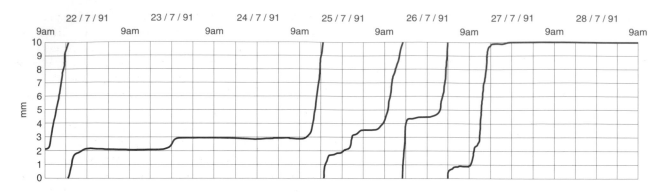

Figure 7.5   *Chart from a recording rain gauge at Hólar, Iceland. The horizontal scale can be a day or a week.*

the basis of the frequency of occurrence of the types of cloud expected to produce rainfall. Even the levels of outgoing long-wave radiation and microwaves have been used in satellite estimation of surface precipitation. Although there are many problems it is possible to obtain an estimate of probable precipitation totals in previously ungauged parts of the world.

Snow

In some countries the water equivalent of snowfall is found by melting the snow which has accumulated in the gauge. Clearly this is not very accurate,

especially during heavy snowfall, when a low gauge may be totally covered. In the United States the tall gauge prevents this happening but the gauge tends to underestimate the amount of snow reaching the ground. In Canada and Russia separate snow gauges are used. Recently there have been experiments in measuring snow depth photogrammetrically, with aerial or satellite photography. Where the snowfall is substantial the depth can be obtained fairly accurately, but without ground observations the water-equivalent of the snow is unknown.

Whichever approach is used, measurements of the water-equivalent of snowfall always entail problems and probable inaccuracies. We have just to accept that, apart from a few areas of intensive observations, we do not know the precise input of water to the ground surface by snow.

Hail

Hail measurement is even more difficult. Hailstones possess considerable kinetic energy and many will bounce out of a conventional gauge, causing underestimation of their water contribution. The size distribution of hailstones can be obtained from a hail pad which measures the degree of impaction made by the stones. If pads are left out for known times, the amount of ice and water-equivalent can be found. Fortunately hail is normally insignificant as a precipitation input to the hydrological cycle, so it is normally recorded in terms of the number of days with hail.

Fog-drip and dewfall

The water content of fog-drip and dew is small, so special measurement techniques have to be used. Fog-drip falls to the surface after contact with leaves or trees, so trough-shaped rain gauges have been

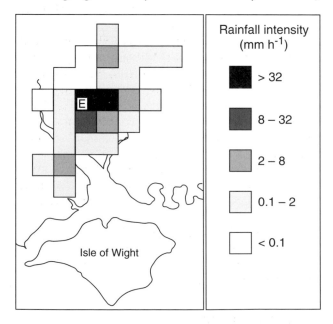

Figure 7.6   *Areal rainfall rates derived from radar information. Data from the UK weather radar at 5 km intervals around Eastleigh (Hampshire) at 11.45 UTC, 29 March 1986.*
Source: Pike (1992).

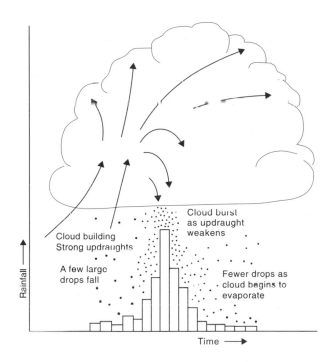

Figure 7.7 *Surface rainfall rates relative to cloud development. In the early stages of cloud build-up, strong updraughts prevent most drops reaching the ground. Later, when the cloud drifts downwind, or as the updraught weakens, a period of much heavier rain may ensue.*

designed to increase the sampling area and make measurements more accurate. In principle they work like an ordinary gauge.

The most commonly used instrument for dewfall is an accurate weighing device. The dewdrops collect on hygroscopic plates which are attached to a balancing system to weigh the amount of water collected. All methods suffer from the basic uncertainty about how accurately the gauges collect dew, compared with natural surfaces. Fortunately, water quantities are minute, so that even large errors are insignificant in relation to the total precipitation input.

Temporal variations of precipitation
Variability in the short term

The variations of rainfall over time are of vital importance to physical geographers and hydrologists. For example, decisions about bridge size, storm sewer construction, culvert dimensions and even flood protection measures must be taken by the engineer on the basis of the expected inputs of precipitation. For this type of decision a single total is not very informative. We need to know not only how much rain is likely to fall but over what period of time.

Twenty-five millimetres of rainfall in a day may not be significant, but if that amount fell in an hour, or even less, there could be drastic consequences. Surface run-off might occur, soil erosion might be initiated, streams might start to swell and flooding might result. Clearly, the intensity of precipitation is extremely important.

If we monitored a storm, we would normally find that precipitation intensity – that is, the amount of rainfall per unit time – varied considerably. Heavy bursts of rain are normally seen to alternate with relatively quiet periods. All types of rainfall show these variations; there is rarely such a thing as steady rain. In convectional storms the variations are often associated with the passage of the main convection zones across the land. Where the updraughts are strong, the raindrops are held in the cloud and prevented from falling, but as the updraughts weaken the drops fall more easily to the ground, giving periods of higher intensity (Figure 7.7). Cyclonic rain, too, shows considerable variability, often associated with temporary zones of instability in the cyclone (Figure 7.8).

In fact, it is only when the source of precipitation is held stationary that we get anything like steady rainfall. One of the most common situations in which this occurs is where moist air is forced to rise over a mountain barrier. If the moist air is blowing from the sea at a constant speed, the air will be fairly uniform and the conversion of vapour to water droplets will proceed at a constant rate. Rainfall then is often prolonged and steady.

The short-term variability of rainfall differs greatly from one area to another. It tends to be greatest in the tropics; at Djakarta (Indonesia), for example, the annual rainfall of 1800 mm falls in only 360 hours on average. By contrast, the average rainfall in London is only 600 mm, yet it falls in about 500 hours. Variability in precipitation is often most important, however, in the more arid parts of the world, for there even quite small storms may be a rare event (Table 7.4); channels that have been dry for months or even years may fill with water, and the baked clay (adobe) used to make houses may crumble and be washed away. Within a matter of hours the rainfall may have ceased and the water almost vanished; within weeks the vegetation will have died down again.

Seasonal variability

In many climates there is a predictable and consistent cycle of rainfall during the course of the year related to the latitudinal migration of the wind and pressure systems. Precipitation areas associated with areas of

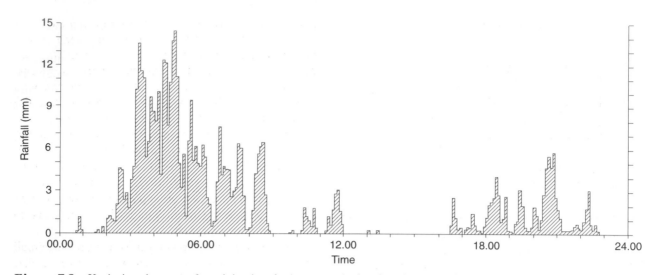

Figure 7.8 *Variations in rates of precipitation during a tropical cyclone in Australia, 20 December 1976. Source: Data after Gilmour and Bonell (1979).*

convergence and uplift tend to shift polewards in summer and equatorwards in winter. Some areas, like the British Isles, remain within the same pressure belt throughout the year and so seasonal variations are subdued. This is also true in the equatorial trough zone, where rainfall can occur at any time throughout the year (Figure 7.9) and in deserts, where rainfall is almost negligible. The brief, rare storms which do occur can come at any time, so monthly rainfall, averaged over the long term, shows little variation (Figure 7.10). Even within the same pressure belt, some seasonal pattern may be evident. In the mid-latitudes, where rainfall is associated with the activity of the rain-bearing cyclones, the winter and autumn are relatively wet, for it is at those periods that the westerlies bring the most intense storms.

In the tropics and subtropics, where convectional rainfall is more important, precipitation tends to be more abundant during the summer months (Figure 7.11). The magnitude of these seasonal variations is even more marked in the monsoonal areas of the

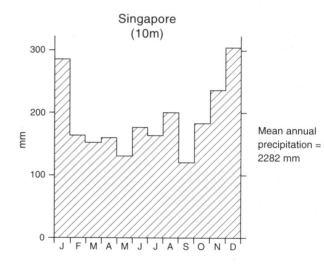

Figure 7.9 *Mean monthly precipitation at Singapore, in the equatorial trough zone.*

Table 7.4 *Rainfall at Yenbo, Saudi Arabia, 24° 8′ N, 38° 03′ E, 1967–90.*

Incidence	No.	%					
Days with no rain	8634	98.5					
Days with rain	78	0.9					
Days with a trace of rain	35	0.4					
Missing data	19	0.2					
Amounts							
Amount (mm)	under 1	1.1–2.0	2.1–3.0	3.1–5.0	5.1–8.0	8.1–15.0	over 15
No. of days in 24 years	27	15	5	14	6	5	6
% of all days	0.31	0.17	0.06	0.16	0.07	0.06	0.07
% of rain days	35	19	6	18	8	6	8

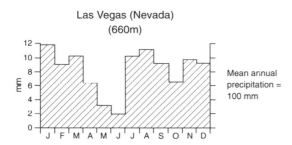

Figure 7.10 *Mean monthly precipitation at Las Vegas, Nevada, in an arid zone.*

world, where the year can be subdivided into a wet and a dry season. At Tilembeya in the Sahel region of Mali, for example, rainfall during August is over 200 mm; from December to February it is almost zero. Similarly, in the monsoon areas of Asia and northern Australia, seasonal differences are great so that hydrological conditions vary considerably throughout the year. During the dry season there is practically no surface run-off, to be followed by a wet season, when run-off is extensive. Vegetation, geomorphological processes and human activities all respond to these changes.

Rainfall frequency

In view of the important consequences of extreme variations in rainfall, it is useful to have some measure of the reliability of precipitation. This may be expressed in a number of different ways. One of the most common is to plot graphs of what are called rainfall recurrence intervals. Using data from a long time period, say fifty years, it is possible to estimate the frequency with which storms of a particular amount or intensity are exceeded. In general, small storms occur most commonly and very heavy storms only rarely. Thus a graph like that in Figure 7.12 is obtained. From this it is possible to tell how

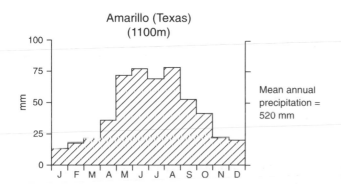

Figure 7.11 *Mean monthly precipitation at Amarillo, Texas, in a subtropical summer rainfall zone.*

frequently a storm giving, for example, 50 mm or less in a day will occur, or how many years it will be on average between storms of 100 mm or more per day. Such information can be very useful in planning bridges or drains, when the aim is normally to produce something that will cope with all but the most extreme events. The frequency of heavy rainfall events may also be of significance in affecting slope processes and landslips. It is important to remember, however, that the figures are only probabilities, derived from average conditions over a specific period. It is quite possible for two storms with an average recurrence interval of fifty years to occur on successive days!

Another way of expressing information on rainfall variation is to plot annual rainfall totals on similar graphs. Thus in Figure 7.13 the frequencies of annual rainfall for Sheffield and Timimoun are shown. We can see that there is a 50 per cent probability of at least 760 mm of rainfall occurring in any year at Sheffield, while at Timimoun the equivalent total is 14 mm. It is also apparent from the graphs that the variability at Sheffield is fairly small compared with that at Timimoun, although the latter is, on average, much drier.

Again this type of data may be very useful. It may be known, for example, that a particular crop may grow satisfactorily only if the annual rainfall exceeds 600 mm. From information on annual rainfall frequencies it is possible to determine the likelihood of receiving that amount of precipitation. If the probability is, say, 90 per cent the farmer may well

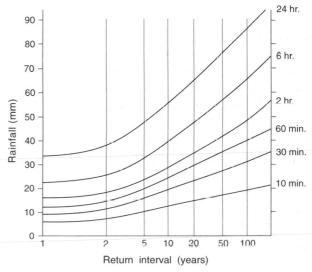

Figure 7.12 *The return period for precipitation totals at Sheffield.*
Source: *Based on methods in Flood Studies Report, NERC, 1976).*

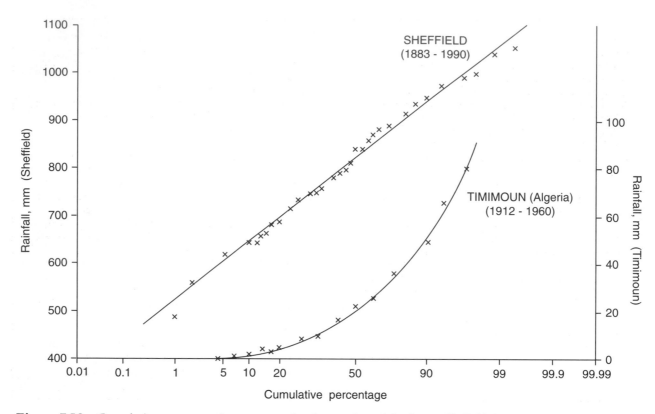

Figure 7.13 *Cumulative percentage frequency graphs of annual precipitation at Sheffield (1883–1990) and Timimoun, Algeria (1912–60).*

think it worth while to grow the crop; if it is only 20 per cent, it is unlikely to be worth the risk. Similarly, it is possible to determine in the same way how often, on average, it will be necessary to irrigate crops.

Rainfall variability may also be expressed statistically by the coefficient of variation (CV). This is calculated from the formula:

$$CV = \frac{s}{\bar{x}} \times 100 \%$$

Drought

In a world of increasing population and development the demand for water is continuously rising. Fresh water can be obtained direct from rivers, by pumping from groundwater or from storage reservoirs. As demand increases, so it is possible, up to certain levels, to increase extraction from each of these systems. Unfortunately, precipitation is not so regular that similar annual totals are received every year. When an area experiences a prolonged period of below-average rainfall, drought conditions may eventually prevail.

Drought is not easy to define. We can think of it as a meteorological drought, and in Britain a drought used to be defined as a period of fifteen consecutive days without rain. A climatological drought would be the result of a longer period with little or no rainfall at a time when rainfall is expected. We would not consider the summer dry period of the Mediterranean

to be a climatological drought because rain is not expected, but dry weather in winter or spring could produce a drought. We might also experience a hydrological drought where water levels in rivers and the ground are well below what would be expected, or an agricultural drought where the impact of reduced precipitation results in crop failure.

Some climatic régimes do have less predictable rainfall with high variability from year to year. As a result agricultural or industrial planning, even the supply of domestic water, became more difficult. With increasing demand the balance with supply may become a problem during periods of below-average rainfall, or even during average conditions. In recent periods many parts of the globe have been affected by drought. Perhaps the best known case is that of the Sahel of West Africa, where rainfall since 1968 has generally been well below the previous levels (Figure 11.7). In addition it is now realized that the periodic changes in sea surface temperatures in the South

Pacific, known as El Niño or El Niño-Southern Oscillation (ENSO), can have a major effect on rainfall levels in Australia, Indonesia and even parts of the western United States. Most of the severe droughts in those areas occur during phases of El Niño. Even areas which normally experience reliable rainfall can occasionally have prolonged periods of below-expected rainfall. In the 1960s much of the north-eastern United States had a dry period, with water levels in reservoirs falling to record values. In western Europe in the 1980s and 1990s there have been a number of dry summers which have caused water supply problems and in some cases a reduction in agricultural production. The worst case for temperate latitudes is when two dry summers are linked by a dry winter and there is no major recharge of the reservoirs or aquifers. This happened in 1975–76 in much of north-west Europe. Over the sixteen-month period from May 1975 to August 1976 less than 50 per cent of average precipitation fell in some areas.

In a developed society, resources can be used to increase the supply of water. Reservoirs can be enlarged, if this is politically acceptable; increased water can be extracted from rivers, if enlargement is environmentally acceptable; an improved distribution system can be achieved by reducing leaks in the pipeline network or by linking water supplies in different parts of the country, if that is economically acceptable. It is assumed that it is less likely that all areas will be suffering drought uniformly. In most cases such measures cost money and take time. In dry areas where energy is cheap, desalination plants can be used to extract fresh water from the sea, as in Saudi Arabia.

In developing areas resources are less readily available. Steps to improve water supplies may be taken only through international action during severe droughts, as in Ethiopia. Short-term measures may be taken to pump groundwater where it is available. This leads to a concentration of population and, in many cases, grazing animals around the new supply, which may cause more problems than it solves. Some countries have tried to increase water storage by constructing large dams to compensate for low river flows during drought. Lake Kariba on the Zambezi and Lake Nasser on the Nile are good examples but they give rise to major environmental problems (see Chapter 27).

Drought is something that affects all parts of the world but its impact varies according to the level of development and the duration of the drought.

where $\bar{x}$ is the average rainfall and s is the standard deviation. This defines the variability relative to the mean. With a standard deviation of 200 mm and a mean annual precipitation of 1600 mm, the coefficient of variation would be 12.5 per cent, but with the same standard deviation and a mean of only 400 mm the coefficient of variation would rise to 50 per cent. This is a useful measure, since it gives an indication of the importance of the variability. In Britain coefficients of variation of annual rainfall range between 10 per cent in north-west Scotland and about 20 per cent in south-east England.

Spatial variations of precipitation

We all know that annual rainfall totals vary from one part of the world to another, even when altitude is allowed for. Locally, however, it seems likely that annual totals will be fairly consistent. It is also clear that, in the short term, quite marked differences in rainfall may occur within short distances, depending upon the route taken by a particular storm or cyclone; indeed, it is sometimes possible for it to be raining on one side of the street and dry on the other.

In order to study spatial variation on a local scale, we need a dense network of recording gauges, for otherwise individual storms may be missed as they pass between the rain gauges. One such investigation was carried out in Illinois, where fifty recording rain gauges were set up in an area of 1400 km² of flat rural land. The experiment was maintained for five years, measuring individual storms, and for thirteen years for monthly and seasonal analyses. Comparisons were made by correlating rainfall at a gauge at the centre of the area with all other gauges. Correlation is a statistical measure which provides an index of the strength of the linear relationship between two variables; a value of +1.0 indicates a perfect positive linear relationship, a value of −1.0 shows a perfect negative linear relationship, and a value of 0.0 shows no relationship (Figure 7.14).

For the shortest time period studied (one minute) the degree of correlation fell rapidly with distance from the central gauge (Figure 7.15). Thus at a distance of only 8 km from the central gauge, the rainfall pattern is different minute by minute; in many cases it may have been raining at the central gauge but not 8 km away. This is what we would expect if rainfall was produced by local summer convection storms, each affecting an area of only a few square kilometres.

At a longer time scale the degree of correlation is better. Taking rainfall totals for whole storms (Figure 7.16a), it is apparent that the gauges close to

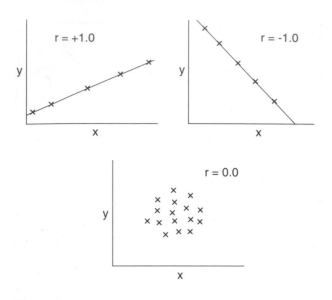

Figure 7.14 *Scatter diagrams demonstrating different correlation coefficients.*

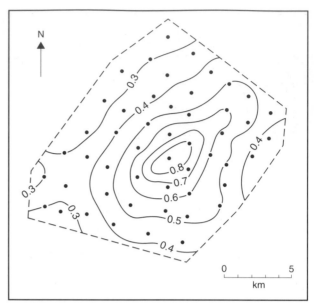

Figure 7.15 *Correlation patterns associated with one-minute rainfall rates in warm season storms in Goose Creek, central Illinois.*
Source: After Huff and Shipp (1969).

the central station are quite strongly correlated. Nevertheless, by the distance of about 20 km the degree of correlation is low. Again, this is probably due to the effect of local variability caused by the passage of small summer convection storms. If, instead, we look at frontal storms or storms associated with low pressure systems, we get a different picture (Figure 7.16b). Now most of the area shows a close correlation with the central gauge – indeed, almost a perfect correlation – indicating that these more general storms affected the whole area equally, despite occasions of variability referred to earlier.

These results indicate some of the atmospheric factors controlling rainfall variability. Convectional storms give high levels of spatial variation, while cyclonic rainfall is spatially much more uniform. In the tropics, where a great proportion of the rainfall comes from convectional storms, the spatial variation is particularly marked. Table 7.5 illustrates this point. It gives rainfall totals at rain gauges only 3.2 km apart in a level area near Dar es Salaam in Tanzania. It is clear that the totals are very different. The reason is that most of the rainfall is derived from individual cumulonimbus clouds which produce intense precipitation over an area of about 2 km^2 to

60 km^2. The storms often build up without any significant movement, so areas just beyond the limits of the cloud may receive no rainfall at all. Sometimes the storms develop over a wider area, perhaps 500 km, but even so they do not give rain everywhere. Using the correlation method, we find that the relationships between rain-gauge totals fall to zero within 100 km and are negative beyond. In other words, if rainfall were high for a particular period in one area, it would tend to be low beyond 100 km. Over the short term these differences may be considerable, but in the long term we would expect them to balance out.

Surface modifications of precipitation

So far we have considered rainfall variability over essentially flat terrain. Few areas of the world are extensively flat, however, and surface irregularities interfere with atmospheric processes to give even

Table 7.5 *Daily rainfall at two stations, 3.2 km apart near Dar es Salaam, Tanzania, April 1967 (mm).*

Station	6th	7–8th	10th	12th	13th	24th	25th	Month
A	54.9	33.3	2.3	13.5	13.5	14.2	21.1	315.2
B	100.3	3.3	31.8	64.8	5.1	8.4	0.8	437.6

Source: After Jackson (1977).

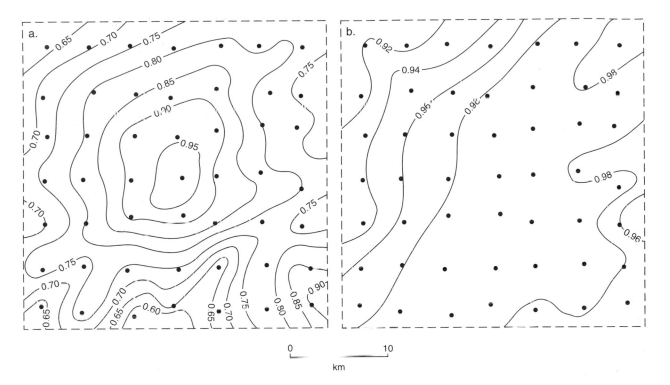

Figure 7.16 *Correlation patterns associated with (a) air mass storms and (b) low-pressure centres, during May–September in Illinois.*
Source: After Huff and Shipp (1969).

more complex spatial patterns of variation in rainfall. Even relatively small hills can have a marked effect. The importance of surface topography on precipitation is indicated at a general scale for the British Isles in Figure 7.17. As can be seen, the general pattern of rainfall is appreciably modified by the Welsh and Scottish mountains. They give rise to higher totals on the western slopes, and a marked rain shadow on the east. The effect of altitude in the leeward areas is less apparent. For example, the Cairngorms in north east Scotland do not stand out as areas of higher rainfall in Figure 7.17, despite their height, because the prevailing winds have lost much of their moisture by the time the eastern side of the country is reached.

Within any climatic region, the relation between rainfall and altitude is generally quite consistent. In most cases, precipitation increases with increasing altitude, even in relatively arid areas. At the Grand Canyon, for example, average annual precipitation increases from less than 250 mm on the canyon floor at 760 m to 400 mm on the southern rim of the canyon at 2100 m. On the forested northern rim, 2600 m above sea level, rainfall totals over 600 mm.

Nevertheless, the progressive increase in rainfall with altitude does not always extend to the summits of the mountains. The Sierra Nevada in California are no wetter on the summit than they are 1200 m lower

(Figure 7.18) In the subtropical trade-wind belt over Hawaii the peaks of Mauna Loa and Mauna Kea receive far less rain (380 mm) than the windward slopes, where maxima between 1000 and 1300 m amount to about 7500 mm yr^{-1}. It is also apparent that the relationship between altitude and precipitation varies from one part of the world to another. In the tropics much of the precipitation is produced by warm clouds whose upper limit is only 3000 m above the ground; thus the effect of altitude is subdued (Figure 7.18a) and the maximum may even be close to sea level. In contrast, in temperate areas a large proportion of the rainfall comes from deep stratiform clouds that extend through a considerable part of the troposphere. Here the effect of altitude on rainfall is more marked, though the increase on windward slopes is usually greater than on leeward slopes. Comparisons are difficult, however, because some of the precipitation on the mountains in temperate areas falls as snow and, as we have seen, snow is impossible to measure accurately.

Evapotranspiration

Evaporation and transpiration form the major flows of moisture away from Earth's surface. Because we

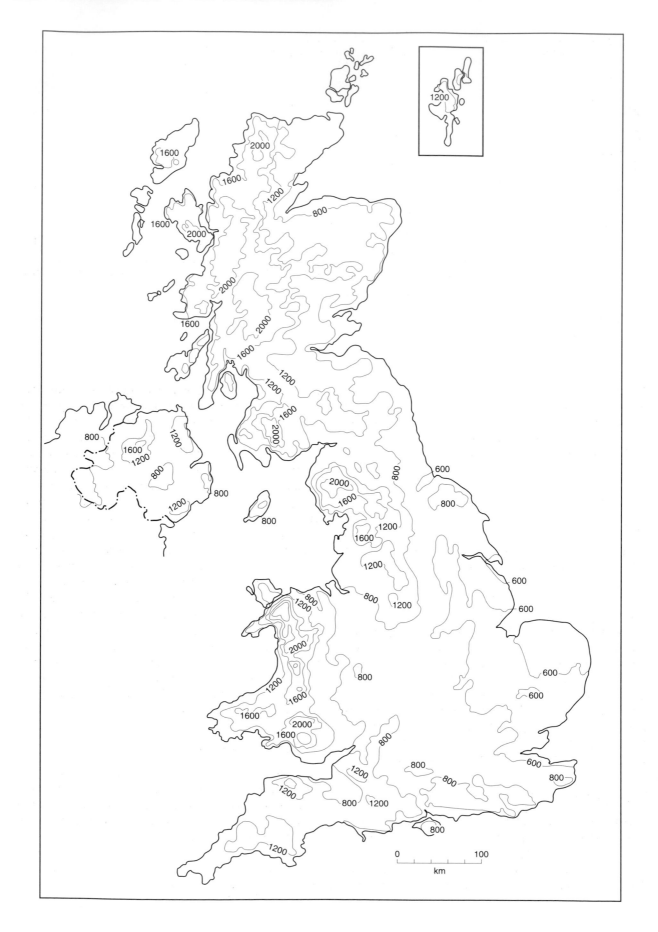

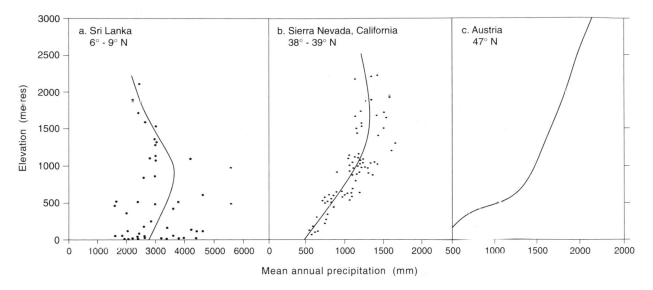

Figure 7.18 *Generalized curves showing the relationship between elevation and mean annual precipitation in different climatic régimes.*

can rarely see the processes taking place it is easy to neglect this component of the hydrological cycle, but it is an extremely important one. It returns moisture to the air, replenishing that lost by precipitation, and it also takes part in the global transfer of energy.

Processes

Evaporation

Evaporation can be defined as the process by which a liquid is converted into a gaseous state. It involves the movement of individual water molecules from the surface of Earth into the atmosphere, a process occurring whenever there is a vapour pressure gradient from the surface to the air. Thus evaporation requires the humidity of the atmosphere to be less than that of the ground. The process also requires energy: 2.48×10^6 Joules to evaporate each kilogram of water at $10°$ C. This energy is normally derived from the sun, although sensible heat from the atmosphere or from the ground may also be significant. However, when the air reaches saturation (100 per cent relative humidity) evaporation cannot take place.

Transpiration

Transpiration is a related process involving water loss from plants. It occurs mainly by day, when small pores, called *stomata*, on the leaves of the plants open up under the influence of sunlight. They expose the moisture in the leaves to the atmosphere and, if the vapour pressure of the air is less than that in the leaf cells, the water is transpired. As a result of this transpiration, the leaf becomes dry and a moisture gradient is set up between the leaf and the base of the plant. Moisture is drawn up through the plant and from the soil into the roots (Figure 7.19).

As far as most plants are concerned, this is a passive process; it is controlled largely by atmospheric and soil conditions, and the plant has little influence over it. Consequently transpiration results in far more water passing through the plant than is needed for growth. Only 1 per cent or so is used directly in the growth process. Nevertheless, the excess movement of moisture through the plant is of great importance, for the water acts as a solvent, transporting vital nutrients from the soil into the roots and carrying them through the cells of the plant. Without this process, plants would die.

Evapotranspiration

In reality it is often difficult to distinguish between evaporation and transpiration. Wherever vegetation is present, both processes tend to be operating together, so the two are normally combined to give the composite term *evapotranspiration*.

Figure 7.17 (opposite) *Mean annual precipitation over the British Isles in millimetres (1961–90).*

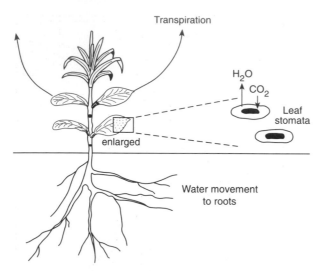

Figure 7.19 *Schematic diagram showing exchanges of water and gases by transpiration in plants.*

Evapotranspiration is governed mainly by atmospheric conditions. Energy is needed to power the process, and wind is necessary to mix the water molecules with the air and transport them away from the surface. In addition, the state of the surface plays an important part, for evaporation can continue only so long as there is a vapour pressure gradient between the ground and the air. Thus as the soil dries out, the rate of evapotranspiration declines. Lack of moisture at the surface often acts as a limiting factor on the process.

We can therefore distinguish between two aspects of evapotranspiration. *Potential evapotranspiration* (PE) is a measure of the ability of the atmosphere to remove water from the surface, assuming no limitation of water supply. *Actual evapotranspiration* (AE) is the amount of water that is actually removed. Except where the surface is continuously moist, actual evapotranspiration is significantly lower than PE.

Potential evapotranspiration

Energy inputs: the sun

The main variable determining potential evapotranspiration is the input of energy from the sun, and it has been estimated that this accounts for about 80 per cent of the variation in PE. The amount of radiant energy available for evapotranspiration depends upon a number of factors, including latitude (and hence the angle of the sun's rays), day length, cloudiness and the amount of atmospheric pollution. Thus PE is at a maximum under the clear skies and hot days of tropical areas, and at a minimum in the cold, cloudy polar regions. In the short term, however,

rates of potential evapotranspiration may vary considerably at any single place. Daily variations in radiation inputs cause marked fluctuations in PE, so that very little evapotranspiration occurs at night. Even subjectively we can get some idea of this by noting how long the ground stays wet after a shower of rain during the night, yet how quickly it dries out during the day. Similar patterns occur seasonally. Potential evapotranspiration reaches a peak during the summer months and declines markedly during the winter (Figure 7.20).

Energy input: wind

The second important factor is the wind. The wind enables the water molecules to be removed from the ground surface by a process known as *eddy diffusion*. This maintains the vapour pressure gradient above the surface. Wind speed is obviously one of the variables determining the efficiency of the wind in removing the water vapour, but it is not the only one. The rate of mixing is also important, and that depends upon the turbulence of the air and the rate of change of wind speed with height.

Energy input: the vapour pressure gradient

Third, evapotranspiration is related to the gradient of vapour pressure between the surface and the air.

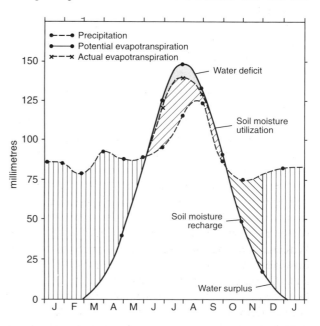

Figure 7.20 *Average moisture budget for Wilmington, Delaware. The difference between potential and actual evapotranspiration is small because precipitation is distributed relatively uniformly throughout the year. Source: After Mather (1974).*

Unfortunately the vapour pressure gradient has proved very difficult to measure precisely in the layer immediately above the surface, so wherever possible, methods of calculating PE use measurements of vapour pressure at one level only.

Actual evapotranspiration

Actual evapotranspiration equals PE only if there is a constant and adequate supply of water to meet the atmospheric demand. Such a situation exists over moist, vegetated surfaces and it is also approximated over water surfaces such as the open sea or large lakes, but most land surfaces experience significant periods when water supply is limited. As a result, actual evapotranspiration falls below PE. We can get some idea of the importance of surface conditions by considering evapotranspiration in a variety of situations. Let us start by examining evapotranspiration from an open water surface.

Evaporation from water surfaces

Because there is an unlimited supply of water to maintain evaporation, and because there is no vegetation to complicate the process, the surface of oceans or large lakes provide the simplest situation in which to study evapotranspiration. Under these conditions, transpiration does not occur and water loss is entirely by evaporation. The main factors determining water loss are therefore the atmospheric conditions, and there is generally a close relationship between actual evaporation and PE.

Nevertheless, the relationship is not perfect, and the main reason is that the water is able to absorb a large amount of energy which is not used in evaporation. This energy is expended in heating the water, and much of it is recirculated through the water body by advection.

Evaporation is greatest when the sea is warm in comparison with the air. In general this is the case, as air temperatures are slightly below those of the sea over much of the globe for much of the time. Where upwelling of cold water from depth occurs, however, the surface temperatures are greatly reduced and the difference between sea and air temperatures becomes small; in some cases the sea may even be cooler than the atmosphere. An example of this phenomenon occurs off the coast of Peru, where the cold Humboldt current brings bottom waters to the surface. As a result, the air is warm relative to the sea, it retains moisture, and so the humidity gradient above the surface is low. This greatly reduces

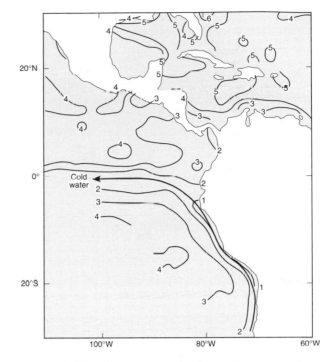

Figure 7.21 *Mean evaporation in millimetres per day from the tropical Atlantic and eastern Pacific Oceans. Source: After Hastenrath and Lamb (1978).*

the rate of evaporation, and, as Figure 7.21 shows, the effect continues some way out into the Pacific.

Evapotranspiration from land surfaces

Because of the importance of the energy and water balance to growing crops, there have been a large number of studies of evapotranspiration from vegetated surfaces. The presence of a vegetated surface complicates the energy exchanges taking place, however, for the plants intercept radiation and rainfall inputs, they affect the temperature and wind profiles near the ground, and they also modify humidity. The degree of these effects varies with the character of the vegetation, so evapotranspiration from a vegetated surface often differs markedly from PE.

Within a mature crop we can identify three layers: the upper layer, or canopy, the main stem zone and the ground surface. During the day most of the incoming radiation is absorbed by the canopy. The air space between the canopy and the ground acts as an insulator, so that it is the top of the vegetation rather than the soil that acts as the active surface. Consequently, transpiration rather than evaporation takes place.

So long as moisture is available in the soil, the plants are able to transpire at or very close to the

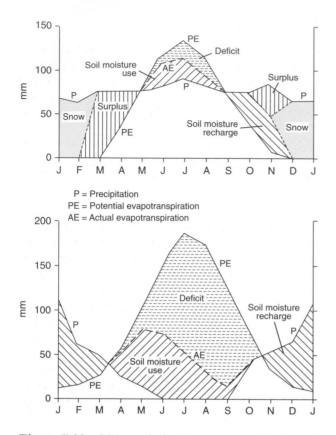

P = Precipitation
PE = Potential evapotranspiration
AE = Actual evapotranspiration

Figure 7.22 *Moisture budget diagrams for (a) Concord, New Hampshire, and (b) Aleppo, Syria. The method assumes that 50 per cent of the soil water surplus runs off in the first month, 50 per cent of the remainder in the next, and so on, unless additional surplus forms. Source: After Barry (1969).*

potential rate. Thus in a moist soil, evapotranspiration proceeds unhindered, water being drawn up the plant from the soil to replace that lost from the leaves. As the soil dries out, however, the plant experiences increasing difficulty in extracting moisture and the rate of transpiration cannot be maintained (Figure 7.22). Several changes take place. The plant starts to suffer from moisture stress and nutrient deficiencies, and in some cases the stomata in the leaves may close, reducing transpiration further. But the drain upon the soil moisture store continues, so the moisture stress gets worse. Progressively the rate of actual evapotranspiration falls below PE to develop a soil moisture deficit.

It now seems clear that the effect of declining moisture availability depends upon a variety of conditions, including vegetation type, rooting depth and density, and soil type. In a heavy clay soil, for example, it seems that evapotranspiration rates fall only slightly as the soil dries out until the point is reached where no more water is available to plants. Evapo-

transpiration then falls rapidly. Conversely, in sandy soil the decline in actual evapotranspiration rates is much more regular (Figure 7.23), as the sandy soil's capacity to retain moisture is less than that of clay.

Plant responses to moisture stress

The reduction in evapotranspiration as the soil dries out has a number of implications. Eventually, of course, the plant experiences severe nutrient deficiencies and the yield is reduced. Thus we often see a close relationship between the degree of moisture stress and crop yields. In addition, the moisture in the plant helps to control its temperature; the energy used in transpiration cannot heat the plant. As transpiration declines, more energy is available for heating and the leaves get warmer. Initially, this may encourage growth, but ultimately, if high temperatures are reached, it may damage the plant. The soil, too, is heated more effectively, so the surface temperature rises. In one study in Wisconsin, with an air temperature of 28° C, the surface of a dry sandy soil was 44° C while the same soil kept moist reached only 32° C. The difference was due to the fact that

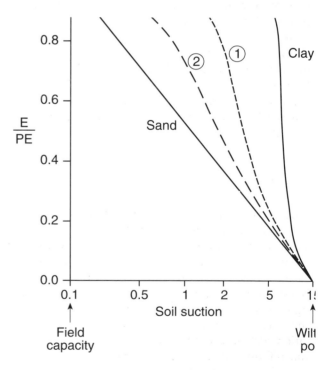

Figure 7.23 *The relationship between the ratio of actual to potential evapotranspiration E/PE and soil moisture. Encircled 1 and 2 are schematic curves for a vegetation-covered clay loam under low evaporation stress and a vegetation-covered sandy soil under high evaporation stress. Source: After Barry (1969).*

more energy was used in evaporation from the wet soil.

The ultimate effect of continued drying of the soil is that plants can no longer obtain any water and transpiration ceases. Since the water in the plant cells is important in keeping the plants rigid (turgid), as the plants dry out they become limp and wilt. At that stage, therefore, the soil is said to be at *wilting point*.

Of course, evapotranspiration does not always continue until wilting point is reached. Instead renewed rainfall generally wets the soil and rejuvenates water uptake by the plants and increases transpiration. Thus the soil acts as an important store of nutrients.

Where the intervals between each period of precipitation are long, the ability of the soil to supply water may be stretched to the limit and moisture stress may be a common occurrence. In such circumstances the vegetation often adapts to the hydrological conditions, by developing deeper roots, or by regulating water use in a variety of ways. For example, the plants may have a dormant period during the dry season, completing their growth cycle during the brief wet period. The sight of deserts blooming after a storm is a most remarkable one (see Plate 27.3) Other plants adapt by controlling their stomata, closing them during periods of dryness in order to reduce water loss. Others are able to alter the orientation of their leaves so that the stomata are more sheltered from the hot sun. Yet others, like the mesquite of the south-western United States, have thick, waxy leaves which protect them from the radiation and slow down water losses.

Measuring evapotranspiration

One of the main needs of the farmer or irrigation engineer is to be able to predict when the plants will suffer from moisture stress and how much water must be applied. This involves being able to measure or calculate the rate of evapotranspiration. Knowledge of evapotranspiration losses is also required by hydrologists who wish to plan water management policies; they need to know what proportion of the precipitation will be available to replenish groundwater or run-off into streams. The measurement of evapotranspiration is therefore important. Unfortunately it is also difficult. Several approaches to measurement have been developed, including *direct measurement* (e.g. with evaporation pans, lysimeters and eddy correlation systems), *meteorological formulas* and *moisture budget* methods.

Direct measurement

Possibly the most widely used method of direct measurement is the *evaporation pan*. This consists of a shallow pan filled with water. The rate at which the water is lost through evaporation is measured with a gauge (Plate 7.4). This procedure measures only potential evaporation, for it does not allow for limitation of moisture supply, nor does it directly determine transpiration losses. In addition, the results seem to vary according to the size, depth, colour, composition and position of the pan, so it is not always easy to compare results from different sites.

In some countries the *atmometer* is used. There are various types but essentially they consist of a porous ceramic or paper evaporating surface and are generally coloured either black or white. The evaporating surface is continuously supplied with water. They are sensitive to variations in wind speed but, properly cared for and calibrated, they can provide reasonable estimates of PE. Their main advantage is cheapness.

An alternative system of direct measurement is the *lysimeter* (Figure 7.24). It may be employed to measure either potential or actual evapotranspiration. To measure PE the column of soil is kept constantly moist so that water deficiencies do not occur. To measure actual evapotranspiration the column is allowed to respond naturally to atmospheric conditions. Regular weighing allows the moisture content to be determined. If the amount of precipitation is known, the moisture loss through evapotranspiration can be calculated (Table 7.6). Commercial lysimeters may weigh up to 60 tonnes, and for really accurate

Plate 7.4 *An evaporation tank and gauge.*
Photo: E.M. Rollin.

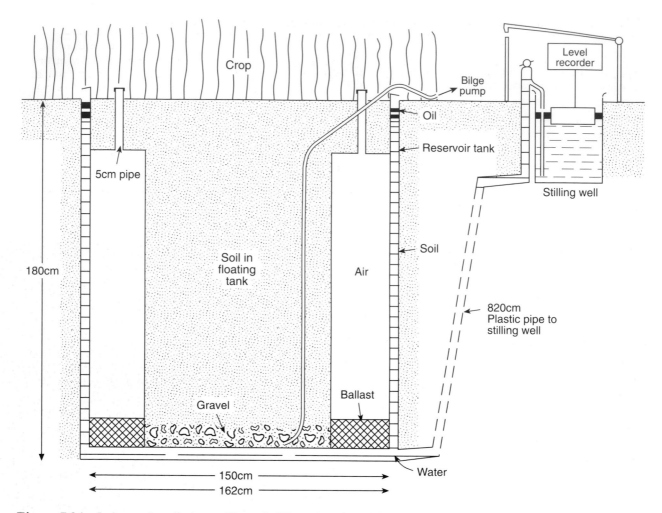

Figure 7.24 *Lysimeter installation at Hancock, Wisconsin. The soil block floats in a tank of water. Changes of water level are recorded instead of weighing the block.*
Source: After Barry (1969).

results large columns and precise yet robust weighing instruments are required. A simple and effective lysimeter can be constructed from old tin cans and a piece of nylon mesh (Figure 7.25).

With improvements in instrumentation and data collection it is technically possible to obtain values for evapotranspiration directly through the measurement of the turbulent fluxes of water vapour, sensible heat and momentum. Sensitive and precise sensors of temperature, vertical and horizontal wind speed and humidity are required, together with a suitable logging system to collect the vast amounts of data generated. Such instruments are now used for research projects (Plate 7.5)

Meteorological formulas

Because evapotranspiration is greatly dependent upon atmospheric conditions, it is possible to derive good estimates of PE from data on meteorological conditions. A wide range of formulae have been developed to do this, some of them so complex that it is almost impossible to use them under normal circumstances; the necessary data are not collected except during special programmes.

This problem is illustrated by what at first seems to be a very simple approach. As we saw in Chapter 5, the energy budget can be expressed as follows:

$$Q^* = Q_H + Q_E + Q_G$$

where Q^* is the net radiation, Q_H is the sensible heat flow, Q_E is the energy use through evaporation and Q_G is the heat flow into the ground. If we could determine all the other components of the equation we could find Q_E by difference, and that would tell us how much evaporation was occurring. Unfortunately we rarely know the value of the other components.

Table 7.6 *Calculation of evapotranspiration through lysimeter moisture measurements.*

1 Date (August 1996)	2 Precipitation (cm)	3 Weight of precipitation (g)	4 Weight of lysimeter (g)	5 Previous weight of lysimeter (g)	6 Change in weight (g) (4–5)	7 Weight trans- pired and evaporated (g) (3–6)	8 Water transpired and evaporated expressed in cm (7 ÷ surface area)
1	0.24	75.36	9110.35	9062.75	+47.60	27.76	0.09
2	–	–	9097.21	9110.35	–13.14	13.14	0.04
3	–	–	9042.94	9097.21	–54.27	54.27	0.17
4	–	–	8986.32	9042.94	–56.62	56.62	0.18
5	0.51	160.14	9124.67	8986.32	+138.35	21.79	0.07

Note: Surface area of lysimeter: 314 cm².

Because of this problem of obtaining data, a large number of simpler, more empirical formulae have been devised. They are much easier to use; they are based, however, not on physical principles but on the observed relationship between evapotranspiration and one or more climatological variables. The relationships have usually been obtained under one particular climatic régime and they may not be applicable elsewhere, hence the number of formulas.

Probably the best known of these empirical equations is that developed by *Thornthwaite* to assess PE. Thornthwaite was trying to devise a climatic classification that went beyond mere description and incorporated indices of heat and water availability and how

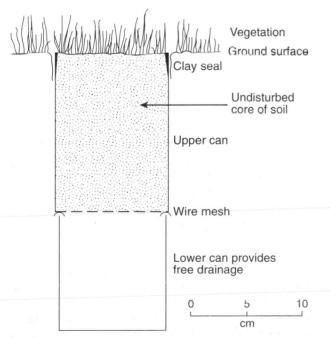

Figure 7.25 *Field weighing lysimeter. The can is removed and weighed in the field.*
Source: After Atkinson (1971).

Plate 7.5 *A data-collecting system called Hydra used to measure evapotranspiration. The sensors measure wind speed and direction, humidity and radiation.*
Photo: H.R. Oliver.

they were related to vegetation. In simple terms, PE is calculated from the formula:

$$PE = 1.6 \, (10T/I)^a$$

where PE is the unadjusted monthly value of potential evapotranspiration, T is the mean monthly temperature in degrees Celsius, I is an annual heat index derived from the sum of twelve monthly index values and a is a constant that varies in relation to I. There are other minor adjustments required to allow for day length variations. Fortunately nomograms and tables have been prepared to simplify the calculations.

In this formula Thornthwaite is using temperature as a substitute for radiation, and it therefore works reasonably well where the two are closely correlated. In the tropics, however, the equation underestimates PE because temperatures lag behind radiation inputs. In addition, the method takes no account of wind, even though it may be locally important. Nevertheless, the relative simplicity of the method makes it popular, and, despite its shortcomings and its inevitable inaccuracies, it is one of the more widely used methods of assessing PE.

In the United Kingdom the *Penman* formula has also been widely used for calculating PE. It is less empirically based than the Thornthwaite method, combining the energy budget and the aerodynamic approaches to the estimation of evaporation, but requires much more meteorological data. Different versions of the basic formula have been developed. They are based on the duration of sunshine, or net radiation if measured, mean air temperature, mean air humidity and mean wind speed. The data are needed at only one level above the ground surface and are recorded at most meteorological stations.

The Penman approach has been modified by Monteith to allow for the nature of vegetation with either optimal or restricted water supply. Although likely to be more realistic, the formula does require information about the resistance of the plants' stomata to water vapour flow. This is not readily available, so the use of the formula has been restricted to research applications.

Moisture budget methods

An alternative method of obtaining actual evapotranspiration is the *moisture balance equation*. At a site the moisture balance can be expressed as:

$$P + I = E + R + D + \Delta S$$

where P is precipitation, I is irrigation water added, E is evapotranspiration, R is run-off, D is drainage to bedrock and ΔS is the change in soil moisture content. As before, if we knew all the other elements in this equation we could calculate E by difference. Several of the components present little problem, for precipitation and irrigation inputs can easily be measured, as can run-off. But drainage to bedrock (D) and changes in soil moisture content (ΔS) are rarely known. As a result, the method can be used only on a large-scale, long-term (e.g. annual) basis where it can be assumed that drainage to bedrock is balanced by release from spring seepage, and where changes in soil moisture content are negligible.

It is clear from what we have said that evapotranspiration remains one of the most difficult aspects of the hydrological system to measure. For this reason, if for no other, our knowledge of the processes involved remains uncertain. For the same reason, it is difficult to give precise figures for the global pattern of evapotranspiration. None the less, it is useful to consider the role of evapotranspiration within the global system.

Evapotranspiration in the hydrological cycle

As we have noted, evapotranspiration provides the main output of moisture from the surface hydrological system, returning water to the atmosphere. So far we have discussed some of the processes involved, but it is important to appreciate that evapotranspiration occurs in many different stages of the hydrological cycle. Thus losses of water to the atmosphere may take place at any point within the system (Figure 7.26).

One of the major losses, for example, occurs during precipitation. Considerable amounts of moisture may be evaporated during rainfall and the small droplets in particular may be totally evaporated before they reach the ground. Similarly, moisture which is intercepted by the vegetation is also susceptible to direct return to the atmosphere by evaporation. The amount of moisture retained on the vegetation during a storm varies according to the character of the storm, the species of plant, the leaf density (and therefore the time of year) and, of course, the vegetation density. In the case of woodlands as much as 50 per cent of the incoming water may be retained in the canopy and returned as evaporation. In the case of more low-lying vegetation, such as grass, the amount of interception is not known with such certainty, partly because of the difficulty of measuring interception in crops. Nevertheless, again it seems likely that interception may reach 20–30 per cent. We only have to walk across a grass field following rain or heavy dewfall to appreciate the quantity of

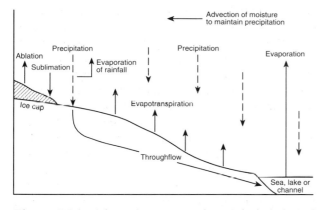

Figure 7.26 *Schematic representation of the hydrological cycle, showing the main losses of water to the atmosphere.*

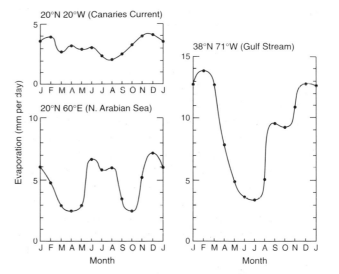

Figure 7.27 *Average annual variation of evaporation at selected oceanic locations.*
Source: After Sellers (1965).

water trapped in the vegetation. Most of it will be lost through evaporation.

A proportion of the water which reaches the soil is also returned by evaporation, for in heavy storms rainfall often collects in surface depressions and these are gradually dried by the sun. Similarly, some of the rainfall flows across the surface as run-off, and further evaporation losses may occur at this stage. Rates are generally low, however, for turbulence mixes the water and disperses the heat from solar radiation through the water body. Thus much greater inputs of energy are needed to heat the water and less is available to carry out evaporation. Nevertheless, losses at this stage are often important for humanity. Open canals and reservoirs may lose considerable quantities of water through evaporation (Table 7.7), and, in arid areas especially, this may represent an irretrievable loss of an important resource.

In vegetated areas, the major process of moisture return is by transpiration. Rates of transpiration vary according to the character of the vegetation and therefore change over both space and time. They are at a maximum when the vegetation is in full leaf and the soil is moist; they decline as the plants lose their leaves or the soil dries out. During the course of a single year, therefore, transpiration losses may show complex fluctuations in response to prevailing conditions. Without doubt, the major evaporative losses occur from the sea – possibly 85 per cent of the

global return to the atmosphere is from the oceans. The reason is not only the great extent of the oceans – some 70 per cent of the world's surface – but also the fact that evaporation can continue at the potential rate. Unlike evapotranspiration from the land, the process is unhindered by water shortage. Even so, seasonal and regional differences in evaporation can be seen, owing to the effect of changing meteorological conditions (Figure 7.27).

Finally, evaporation may occur from the other main storage component of the hydrological cycle – the cryosphere. In general the losses are small, for it requires large amounts of energy to convert ice to water vapour – a process known as sublimation; some 2.83×10^6 J are needed to evaporate 1 kg of ice at 0° C. Sublimation does occur in the marginal areas of glaciers and ice sheets, however, where seasonal inputs of solar radiation may be high; perhaps 2 per cent of the moisture is returned to the atmosphere each year in this way. Moreover, in the past the process was much more important. During the latter parts of the glacial periods, for example, as the ice sheets that had spread into the mid-latitudes began to retreat, sublimation must have been one of the

Table 7.7 *Kempton Park reservoir evaporation (mm).*

					Months						
J	F	M	A	M	J	J	A	S	O	N	D
15	18	28	48	76	94	107	94	74	58	33	18

Source: 'Evaporation from a reservoir near London', Jl.Inst.Water Eng. 19 (1965), 163–81

main processes of stagnation and decay. Warm, turbulent and often relatively dry air masses moved across the ice margins, drawing vast quantities of moisture from the ice sheets and causing them to retreat over areas of thousands of square kilometres.

Conclusion

The precipitation input is probably one of the most important regulators of the hydrological cycle, for it determines the intensity and distribution of many of the processes operating within the system. It is closely related to the rate of evapotranspiration and also influences the pathways of run-off and underground flow and the magnitude of stream flow. Through these processes, and through the direct effects of the impact of rainfall on the ground, it also takes part in many geomorphological processes; it causes rain splash and soil erosion and it plays a vital role in weathering and rock breakdown. The distribution of rainfall across the globe therefore to a large degree controls the operation of the landscape system. Precipitation is similarly a vital input to the ecosystem, and the distribution of vegetation, fauna and population owes much to the pattern of rainfall.

For these reasons, and because of its ultimate importance to human activities, a great deal of attention has been paid to measuring, mapping and predicting precipitation. As we have seen, scarcity of data, particularly in the less accessible parts of the world, limits our ability to gain an accurate picture of precipitation inputs. On the whole, however, rainfall is one of the easiest components of the hydrological cycle to measure. Conversely, evapotranspiration is one of the more difficult. Empirical formulas are the most frequently used ways of obtaining the information.

Key points

1 Precipitation is found in a variety of forms. Which form reaches the ground surface will depend upon many factors: surface temperature, atmospheric moisture, method and rate of cooling and intensity of updraughts, for example. Each type of precipitation has its own characteristics and consequences. The distribution of precipitation varies greatly in time and space, and in quantity.

2 Evaporation and transpiration are more complex processes which return moisture to the atmosphere. The rate of evapotranspiration will depend largely on two factors, (1) how moist the ground is and (2) the capacity of the atmosphere to absorb the moisture. Hence the greatest rates are over the tropical oceans, where moisture is always available and the long hours of sunshine and steady trade winds evaporate vast quantities of water.

Further reading

Jackson, I.J. (1989) *Climate, Water and Agriculture in the Tropics*, Harlow: Longman (Chapters 1–4).
Primarily concerned with the tropics, rainfall is examined in terms of its origins, seasonality, variability and intensity. Intermediate level.

Shaw, E.M. (1994) *Hydrology in Practice*, third edition, London: Chapman and Hall.
A recent edition of a popular book giving a practical approach to the problems of measuring and calculating evapotranspiration. Intermediate level and requires some mathematical expertise.

Sumner, G. (1988) *Precipitation: Process and analysis*, Chichester: Wiley.
An extensive survey of all elements of precipitation from its methods of formation in the atmosphere to how precipitation data can be analysed to draw meaningful conclusions. Intermediate to advanced level but very readable.

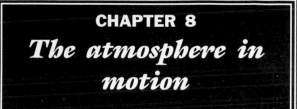
Atmospheric circulation and winds

Earth's atmosphere is in perpetual motion: movement which is striving to eradicate the constant differences in temperature and pressure between different parts of the globe. It is this motion which produces the winds and storms with which we are all familiar. It is this circulation which plays a basic part in maintaining a steady state in the atmosphere and generating the climatic zones which characterize Earth. So far we have considered the upward movements which transfer energy from the surface to the atmosphere. Let us now consider the more obvious horizontal movements that transfer air around the globe.

Sources of information

With modern satellite technology we can watch and monitor these movements. We are no longer dependent solely upon balloons to provide information about the upper atmosphere. Geostationary satellites (e.g. Meteosat, GOES) orbit the globe with the same rate of rotation as Earth, permitting the same portion of Earth to be viewed continuously, using visible light by day and using infra-red photography by day and by night. The images show the main cloud features of the atmosphere. Polar-orbiting satellites with their lower altitudes provide more detailed information about clouds but they pass above a particular part of Earth's surface only twice a day, so wind determination is more difficult.

From the geostationary images, individual cloud patterns can be identified and followed and from successive photographs cloud movements can be calculated and predicted. Unfortunately the satellite photographs show the circulation only in cloudy areas. What happens elsewhere is less clear. For wind speed and direction in these cloudless regions we are dependent upon information supplied by surface or balloon observations. These provide us with a less detailed picture of the pattern of wind circulation over most of the continental areas of the globe (Figure 8.1).

Causes of air movement

Why do we have winds at all? To answer this question it is useful to consider some of the basic principles of motion. Our understanding of these is due in large degree to Isaac Newton. Many people know

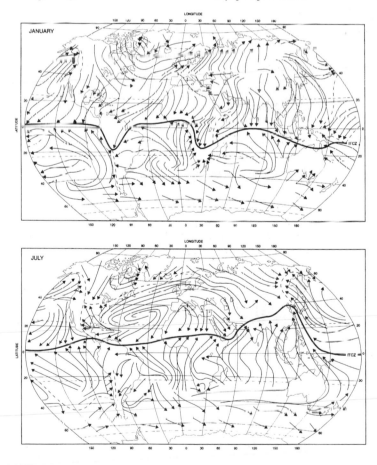

Figure 8.1 *Mean flow patterns of surface winds in (a) January, (b) July.*
Source: After Critchfield (1983).

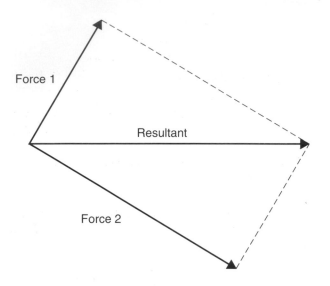

Figure 8.2 *The resultant of two forces acting in different directions. The length of line is proportional to the strength of the force.*

the story of how Isaac Newton 'discovered' gravity when sitting beneath an apple tree. Some will know, too, that he also formulated laws of motion. There are two main laws. The first states that: a particle will remain at rest or in uniform motion unless acted upon by another force. The second law states that: the action of a single force upon a particle causes it to accelerate in the direction of the force. If there is more than one force the particle is accelerated in the direction of the resultant (Figure 8.2).

These forces are particularly important for movement in the atmosphere because forces are continuously acting on particles of air, causing them to accelerate or decelerate and change their direction. The explanation of movement is not unique to Earth, for similar patterns of atmospheric circulation have been identified on other planets with atmospheres.

Forces acting upon the air

Pressure gradient force

Let us imagine that we have a small parcel of air some distance above the ground. What forces will act upon it? The most obvious is the force of gravity, which tends to attract all mass towards Earth's centre. In addition we have the pressure exerted by the air surrounding the parcel (Figure 8.3). If this pressure was the same on all sides of the parcel then its effects would cancel out. But it is not. Pressure decreases upwards in our atmosphere, as we saw in Chapter 5.

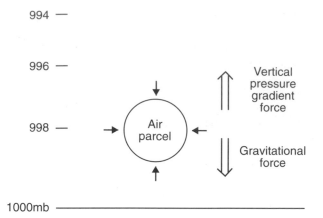

Figure 8.3 *Pressure forces acting upon a parcel of air.*

The force pushing the parcel of air upwards is greater than the downward force from the overlying atmosphere; there is a potential upward acceleration of the parcel. Luckily this vertical force is almost exactly balanced by the force of gravity, otherwise we would have lost our atmosphere long ago. Most of the air movements that we observe are horizontal. Where the atmosphere is denser, the lateral pressure on the parcel of air is great; where the atmosphere has a lower density, the lateral pressure is less. Variations in the density of the atmosphere from one part of the globe to another result in an imbalance of forces and lateral movement of the air (Figure 8.4). The air is 'pushed' from areas of high pressure to areas of low pressure.

This, in fact, is the basic force affecting atmospheric movement. It is called the pressure gradient force. Pressure decreases vertically because, as we move upward through the atmosphere, the weight of overlying air diminishes. It varies laterally because of differences in the intensity of solar heating of the atmosphere. Where solar radiation is intense the air warms up, expands and its density declines; air pressure falls. Where cooling occurs, the air contracts, its density increases and air pressure becomes greater.

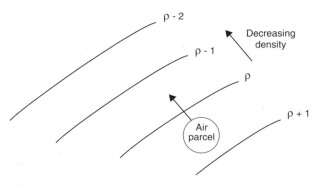

Figure 8.4 *Force exerted on an air parcel produced by density differences.*

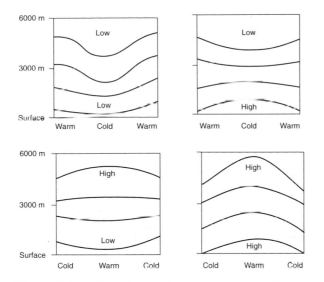

Figure 8.5 *Effect of vertical temperature variations on pressure surfaces.*

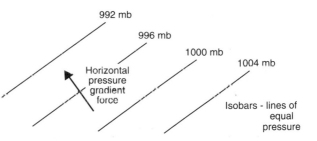

Figure 8.6 *Horizontal pressure gradient force acting at right-angles to the isobars.*

A corollary of this principle is that the pattern of air pressure close to the surface is reversed in the upper atmosphere. Because cold air contracts, the upward decline in pressure is rapid and at any constant height above a zone of cool air the pressure is relatively low. Conversely, warm air expands and rises, so that the vertical pressure gradient is less steep. Above areas of warm air, therefore, the pressure tends to be relatively high (Figure 8.5). The effect upon atmospheric motion is clear. At the surface the air will move from cold to warm zones; at higher altitudes the flow will be from warm to cold.

Differences in air pressure may be mapped by defining lines of equal pressure. These are known as isobars. Air movement occurs at right-angles to the isobars, down the pressure gradient; that is from areas of high pressure to areas of low pressure (Figure 8.6). The magnitude of the force causes movement (the pressure gradient force), and so the speed of the wind is inversely proportional to the distance between the isobars. Thus the closer the isobars are together, and the more rapidly pressure falls with distance, the stronger is the wind.

Mathematically, this relationship can be written as:

$$F = -\frac{1}{\rho} \frac{p2 - p1}{n}$$

where pressure values at points 2 and 1 are p2 and p1, n is the distance separating points 2 and 1, ρ is air density and F is the resulting acceleration. We can use this formula to indicate how quickly the parcel ought to accelerate. The standard isobaric interval on pressure charts is 5 mb and air density is 1.29 kg m^{-3}. Suppose the isobars are 300 km apart on a sea-level chart. What will be the acceleration down the pressure gradient? In uniform units, the formula will become:

$$F = -\frac{1}{1.29} \frac{5 \times 10}{300 \times 10} = 0.00129 \text{ m s}^{-1}$$

If this rate is kept up for 1 hour (3600 seconds) we would have a value of 4.65 m s^{-1} after one hour. As pressure gradients of this size can last for days, we might expect very high wind speeds to develop unless other forces interfered. There are two main forces which prevent this happening. One is friction and the other is Earth's rotation.

If we look at the wind field on a weather map, it will be immediately apparent that air does not flow down the pressure gradient towards areas of low pressure. If it did, the low-pressure areas would fill and the wind movement would stop. Instead we find that the wind is blowing parallel (or almost) to the isobars rather than across them. This is due to the effect of Earth's rotation.

Coriolis force

Although we are not aware of it, Earth is rotating from west to east at 15° longitude per hour. Reference back to Newton's laws shows that if we have a parcel of air moving southwards and there are no forces acting upon it, it will continue to move in the same absolute direction (i.e in a straight line as viewed from space). However, Earth is gradually turning, and so, relative to the ground surface, the parcel will appear to have followed a curved track towards the right in the northern hemisphere and to the left in the southern hemisphere (Figure 8.7). To explain this apparent deflection in Newtonian terms, we have to introduce a force to account for the movement as observed from the ground. This force is called the Coriolis force, after the French mathematician who formalized the concept. The value of the Coriolis force changes with the angle of latitude

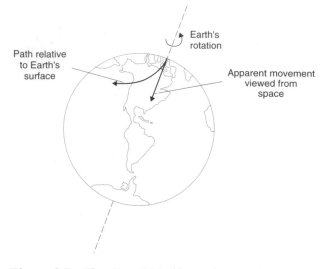

Figure 8.7 *The effect of Earth's rotation on air movement.*

and the speed of the air; mathematically for a unit mass of air it equals $-2\omega V \sin \phi$ (where ω is the rate at which Earth rotates, V is air velocity and ϕ is the angle of latitude). This term ($2\omega V \sin \phi$) is often referred to as the Coriolis parameter. It is greatest at the poles where Earth's surface is at right-angles to the axis of rotation, but it gets progressively less towards the equator where it reaches zero. The reason for this is shown in Figure 8.8. As one proceeds towards the equator so Earth's surface eventually becomes parallel to the axis of rotation. Its effect can be demonstrated by pendulum experiments. Using the Foucault pendulum, which portrays free motion in space as closely as can be achieved at Earth's surface, a disc will rotate under the freely

swinging pendulum in one day at the poles. At latitude 30° it will take two days to rotate (sin 30° = 0.5) and at the equator it does not turn at all.

Geostrophic wind

Let us now return to our parcel of air experiencing a pressure gradient force on the rotating Earth. Initially the parcel will move down the pressure gradient, but as soon as it begins to move it will start to be affected by the Coriolis force, which pulls at 90° to the flow, so that it will be deflected towards the right in the northern hemisphere (Figure 8.9). As the wind accelerates, its speed will increase and, because the Coriolis force is related to speed ($2\omega V \sin\phi$), the two forces pulling together eventually produce an equilibrium flow. This will occur when the two forces are equal and opposite, the resultant wind blowing parallel to the isobars; it is known as the geostrophic wind. Its velocity will be determined primarily by the pressure gradient, though, because the value of the Coriolis force varies with latitude, the geostrophic wind for the same pressure gradient will decrease towards the poles.

Although we have considered only two of the forces acting upon the air parcel, the geostrophic wind is nevertheless a useful approximation. Strictly, it operates only when the isobars are straight – a rare event. Normally isobars are curved and winds are subject to another force termed *centripetal acceleration* which acts towards the centre of rotation. When this rotational component is included, the resultant wind is called the *gradient wind*, which is closer to observed flow in the upper atmosphere (Figure 8.9).

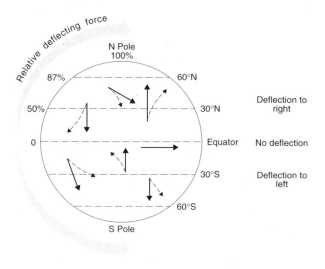

Figure 8.8 *The changing magnitude of the Coriolis force with latitude.*

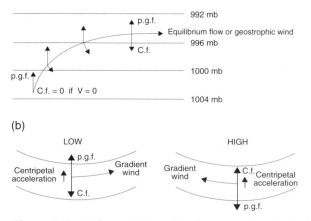

Figure 8.9 *Balance of forces for the geostrophic wind (a) when isobars are straight and (b) for the gradient wind when curvature of the isobars is included. c.f., Coriolis force; p.g.f., pressure gradient force.*

Friction

Inspection of a surface weather map will show that, at ground level, the wind does not blow parallel to the isobars. It blows across the isobars towards the area of lower pressure. The more observant may notice that this angle between the wind flow and the isobars is greater over land areas than over oceans. This may give a clue to the reasons for the change. Land surfaces are rougher than seas; they tend to slow the wind down through friction more effectively. Friction acts as a force pulling against the direction of flow. We can now rearrange our 'balance of forces' to include friction. To achieve balance, the flow will be across the isobars because the Coriolis pull to the right decreases as the air velocity falls (Figure 8.10). From these forces we can now explain equilibrium horizontal flows of air. They are initiated by pressure differences, then modified by the effects of Earth's rotation and friction.

Where flows occur across the isobars in the direction of lower pressure, there will be a transfer of air towards the low-pressure centre, leading to convergence or a net accumulation of air. Where flow is away from a high-pressure centre, there will be a divergence of air away from the surface anticyclone, leading to a net outflow of air. Convergence and divergence can also be found as a result of speed variations within a uniform airflow (Figure 8.11) as well as in ridges and troughs in the upper atmospheric flows (Figure 8.19). If convergence or divergence are maintained for any time, a transfer of mass of air will result and the original pressure gradient will be changed. Convergence will produce an accumulation of air, increase surface pressure and so decrease the pressure gradient and hence the convergence which produced the original air flow. The system will stop. To maintain surface convergence (or divergence), vertical movement is required. In

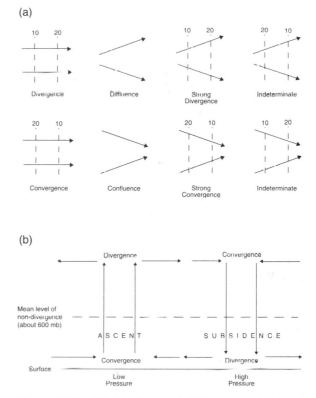

(a)

Divergence Diffluence Strong Divergence Indeterminate

Convergence Confluence Strong Convergence Indeterminate

(b)

Divergence Convergence

Mean level of non-divergence (about 600 mb)

A S C E N T S U B S I D E N C E

Convergence Divergence

Surface

Low Pressure High Pressure

Figure 8.11 *The development of divergence and convergence in (a) horizontal and (b) vertical movements in the atmosphere.*

general, if air is converging at the surface, it must rise, while if it is diverging it is usually associated with subsiding air. Because of these vertical movements resulting from horizontal flows, surface convergence often produces cloud sheets and precipitation whilst surface divergence is associated with clear skies and dry weather. In the middle troposphere there is a level at about 600 mb at which the horizontal convergence and divergence are effectively zero (Figure 8.11b). This link between horizontal and vertical flows in the atmosphere through convergence and divergence is extremely important in determining weather events, as we shall see in Chapter 9.

The global pattern of circulation

With these principles in mind we can try to build up a picture of the global pattern of circulation in Earth's atmosphere. We can start by considering a highly simplified model of the atmospheric system: a uniform, non-rotating, smooth Earth.

As we have seen, the basic force causing atmospheric motion is the pressure gradient; this gradient arises from the unequal heating of the atmosphere by solar

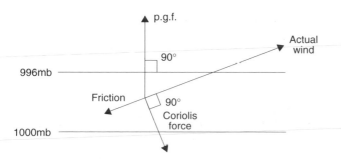

p.g.f.

Actual wind

90°

996mb

Friction

90°
Coriolis force

1000mb

Figure 8.10 *The effect of friction on the geostrophic wind. The Coriolis force is always at right angles to the actual wind. It is smaller than the pressure gradient force because friction has reduced the speed of the wind.*

Plate 8.1 *The characteristic cloud spiral around a mid-latitude cyclone. Image taken in the infra-red waveband on 21 February 1989 at 13.17 GMT.*
Photo: University of Dundee.

radiation. At the equator – the 'firebox' of the circulation, as it has been called – solar radiation is converted into heat. The air expands and rises and flows out towards the poles. Cool, dense air from the poles returns to replace it. We can readily demonstrate the pattern of circulation by heating a dish of water at its centre. Hot water bubbles up above the heat source and flows across the surface to the cold 'polar' areas. At depth the flow is reversed. So long as this unequal heating is continued, the cellular flow is maintained.

In reality, however, the pattern is found to be more complex for, instead of flowing directly to the poles at high altitudes, the warm air from the equator gradually cools and sinks, owing to radiational cooling.

Most of it reaches the surface between about 20° and 30° latitude, and this subsiding air gives rise to zones of high pressure at the tropics – the subtropical high-pressure belts. As the descending air reaches the surface it diverges, some returning towards the equator to complete the cellular circulation of the tropics, the remainder flowing polewards (Figure 8.12).

Various other factors disrupt this pattern further, for Earth is not at rest, nor uniform, as we have so far assumed. It rotates. Its surface is highly variable; it has oceans and continents; it consists of a mosaic of mountains and plains. Moreover, the inputs of solar radiation vary considerably both on a seasonal and on a daily basis.

The effect of Earth's rotation

The rotation of Earth causes the winds to be deflected from the simple pattern just identified. The deflection is towards the right in the northern hemisphere and towards the left in the southern hemisphere. Instead of a direct meridional flow, the Coriolis force produces a surface flow similar to that shown in Figure 8.1

This is not the only effect of Earth's rotation. Air moving towards the poles from the tropics forms a series of irregular eddies, embedded within the generally westerly flow. These can be seen on the satellite photographs as spiralling cloud patterns, similar to the patterns we can see in a turbulent river (Plate 8.1).

Again we can understand the cause of these eddies with the help of a simple experiment. A pan of water is heated at the rim and cooled at the centre. If the pan is slowly rotated it is seen that a simple thermal circulation is produced. If the rate of rotation is increased, however, the flow suddenly becomes unstable. New patterns form like those we see in the atmosphere of the temperate latitudes – eddies and waves. It seems that rapid rotation, like that of Earth, sets up forces which disturb the simple circulation of the atmosphere, particularly near the axis of rotation (i.e. in higher latitudes). These forces destroy the simple pattern and produce more complex circulation (Plate 8.2).

The effect of surface configuration

Even now our picture of atmospheric circulation is far from complete. Earth's surface is not uniform, and the variations in its surface form cause ever more disruption of the pattern of circulation. Friction affects the winds, reducing the effect of the Coriolis force, and, locally, it deflects the surface flow of air to produce highly complicated systems of movement.

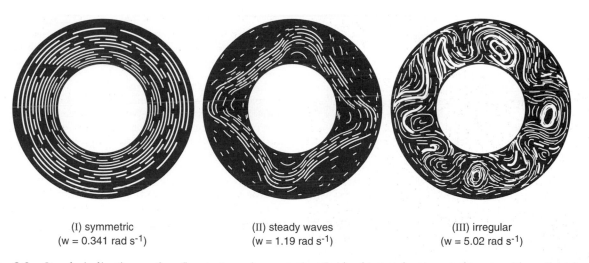

(I) symmetric
(w = 0.341 rad s⁻¹)

(II) steady waves
(w = 1.19 rad s⁻¹)

(III) irregular
(w = 5.02 rad s⁻¹)

Plate 8.2 *Streaks indicating surface flow patterns in a rotating fluid subject to heating at the outer side-wall and cooling at the inner side-wall. At low rates of rotation (left) the flow is symmetrical about the axis of rotation. As the rotation rate increases (centre) the flow develops jet streams and waves. At higher rates (right) the flow is highly irregular, with resemblance to the cyclonic and anticyclonic eddies found in the westerly circulation.
Photo: by courtesy of Dr R. Hide.*

Temperature differences produced by different types of surface, such as land and sea, also have an impact.

It is difficult to model the effects of surface configuration, but a general indication of its influence can be obtained by comparing the northern and southern hemispheres. In the northern hemisphere there are extensive and irregular land masses. Much of the southern hemisphere, by contrast, is ocean, except for the high ice plateau of Antarctica, where very low temperatures are experienced. As we might expect, the pattern is much simpler in the southern hemisphere. A strong westerly flow of cool polar air occurs even in the southern summer. Conversely, in the northern hemisphere the flow is weaker and more irregular, with major meridional air movements and seasonal variations. The temperature differences between pole and equator are less marked (about 30° C compared with 60° C in the southern hemisphere) and so the driving force of the winds – the pressure gradient – is reduced.

Energy transfer in the atmosphere

The pattern of energy transfer in the atmosphere is complex, and we can only consider here some of the general components of the pattern. As a starting point, let us look at the simplified model of what happens in the tropics (Figure 8.12).

The circulation within the tropics consists of two cells. Air blows in towards the low-pressure belt of the equator (the equatorial trough) across the subtropical seas. As it does so evaporation of water from the ocean

utilizes vast quantities of energy so that the sensible heat transfer to the atmosphere is often small (Figure 5.10). The trade winds approaching the equator rise as they meet the equatorial trough, creating a cloudy zone which can often be seen on satellite images (Plate 8.3). The ascent of this air is not a continuous, widespread phenomenon, but occurs mainly in association with localized, often intense and short-lived updraughts such as in thunderstorms. As the air rises and cools, the water vapour condenses and releases latent heat. The increased height of the air also represents an increase of potential energy.

The equatorial air then diverges and flows polewards, so the potential energy is exported to higher latitudes. The cycle is completed as radiational cooling causes subsidence of the air. In the process the air dries and warms as the potential energy is converted to sensible heat. It also checks the rise of convection currents in these subtropical desert areas, producing cloudless skies. Over these desert areas very little evaporation occurs, energy loss is limited and the incoming

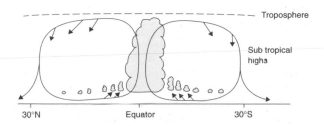

Figure 8.12 *Details of the meridional air circulation or Hadley cell as an energy transfer mechanism in the tropics.*

General circulation models

The rapid development of computing power has made it possible to model the physical processes which operate in the atmosphere and the oceans and to simulate Earth's atmospheric circulation realistically. Such dynamic models are called *general circulation models* (GCMs).

A GCM uses mathematics and the laws of physics to describe the operation of the atmosphere. In summary, the model is started off with a known climatology, usually resembling that of the present Earth. The data are provided for a grid network with horizontal separation of several hundred kilometres (usually 3° latitude by 3° longitude) and information for several heights into the atmosphere for the vertical resolution; more information is obtained about the lower levels of the atmosphere than about the higher levels. The solar input and radiational output are readily known and the main problem is to model the relationship between the surface and the atmosphere. To do so, a number of assumptions and simplifications about the interactions have to be incorporated into the model. An example is the role of clouds, which was discussed in Chapter 6. A major complication is how to link the rapid atmospheric movement with the much slower circulation of the oceans and even slower responses in ice sheets. Early models used fixed sea surface temperatures based on present-day values which were allowed to vary seasonally or incorporated the meridional energy transport of the oceans. The latest models can allow for vertical and horizontal exchanges in the oceans to give more realistic results. Close interaction between the two sub-systems of air and ocean is impossible because the ocean layer needs a much longer time to reach equilibrium from any given change. That is why sea surface temperature anomalies are more persistent than those of the atmosphere.

General circulation models simulate the behaviour of the real atmosphere and reproduce the main circulation features outlined in this chapter. Even individual weather systems are generated by the computer model. The models can either be used for short-period weather prediction extending to about ten days ahead, or they can be modified for climate prediction. In that case the model is run to simulate several decades, to ensure that it reproduces the real atmosphere adequately. Once it is in equilibrium, a variable may be changed. We could alter the concentration of carbon dioxide or the nature of the ground surface to simulate Amazonian deforestation. The model is then run repeatedly with increasing levels of carbon dioxide or reduced areas of forest to see what the effect on the circulation would be. A novel use of GCMs is to attempt to reproduce former circulations. With improved observational techniques world-wide climatic records are being obtained from soils, lake and ocean sediments and ice strata. This wealth of knowledge can be used to infer the nature of the atmosphere and ground surface conditions in the past. It is now possible to allow for changes in Earth's orbit around the sun, to simulate the effect of increased areas of ice at the surface during the last ice age, and even to change the location of the continents to determine what their impact might be.

Although GCMs have a number of limitations they are at present the best way of estimating possible climate change. Developments are taking place in two ways. First, improvements in computer power will allow us to incorporate more information and make calculations even more quickly. In that way the horizontal and vertical resolution of a model can be reduced so that the initial state of the systems can be portrayed more precisely. Second, the modelling of the interaction of air, land, ice and water needs to be improved, perhaps with the incorporation of chemical interactions such as the changes in stratospheric ozone.

Numerical modelling of the global climate system has led to a better understanding of how it works. As human activities may be altering the climate, it is of vital importance to be reasonably certain what the implications are. It will require major resources in computers, observational programmes and scientific research – but at least we are much further along the line of progress than thirty years ago.

radiation heats the ground surface, which then heats the atmosphere. Thus much more of the energy is in the form of sensible heat. During the night this energy is reradiated back to space, for the dry air is unable to intercept much outgoing long-wave radiation. As a result, the net surplus of radiation is fairly small.

In temperate and polar areas the processes of energy transfer are more difficult to decipher. There is no general, cellular circulation of air, as in the tropics, but instead a complicated pattern in which individual, rotating storms play an important part. Within these storms warm air masses rise, releasing latent heat and gaining potential energy. They then become intermixed with descending cold air and gain sensible heat (Figure 8.13). The rotating storms are moving, so the position of this intermixing changes constantly, although there is a tendency towards concentration in certain zones in the northern hemisphere. Labrador, Newfoundland and Greenland are associated with these areas of activity, experiencing cool, southward-moving flows of air (Figure 8.14). Britain and Scandinavia, in contrast, tend to be influenced far more by warm northward-moving air, a phenomenon that greatly improves their climate.

All these transfers of energy through the atmosphere are highly variable, and major differences in the intensity and character of transfers occur over time. Thus the flows of energy represent net increments, often produced by individual, temporary

Plate 8.3 *An infra-red waveband Meteosat image of the cloud patterns on 23 May 1989 at 11.55 GMT. The enhanced cloudy zone of the equatorial trough slightly north of the equator is clearly seen. It expands in East Africa to give a considerable north–south extent. This coincides with one of the rainy seasons in the area. The warmer, lower clouds of the subtropical North and South Atlantic Oceans appear with a greyer tone.*

processes. It is for this reason that it is difficult to detect the nature of energy transfer direct from the general circulation pattern.

Wind patterns

The general circulation of the atmosphere reflects the operation of the atmospheric system as a whole. It is clear, however, that the system is composed of many important subsystems and it is these – the main wind belts of the globe – which provide much of the climatic variation and consistency in the world. We have already indicated that the westerly winds dominate the climate of the temperate latitudes; similarly, the equatorwards movement of air in the regular easterly trade winds has a prevailing influence on tropical climates.

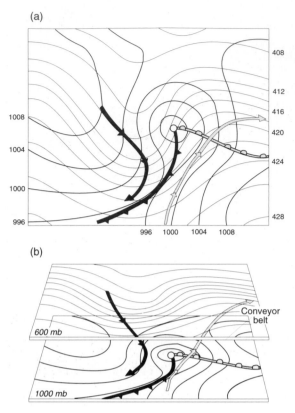

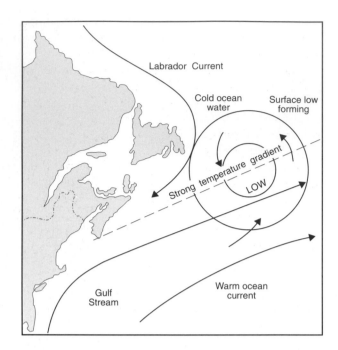

Figure 8.14 *An area of frequent depression formation along the steep temperature gradient between the Labrador current and the Gulf Stream off the north-eastern United States.*

Figure 8.13 *Energy exchanges in a mid-latitude depression: (a) schematic surface and 600 mb contours; (b) a perspective view of the same, with three-dimensional trajectories of air parcels originating in the central part of the cold air and in lower levels in the warm sector. The three portions of each trajectory are for approximately equal time intervals. Source: After Palmén and Newton (1969).*

Surface winds

Four main surface wind belts can be distinguished. Around the equator, in the low-pressure equatorial trough, occurs a zone of convergence where the north-easterly trade winds blowing from the Tropic of Cancer meet the south-easterly trade winds blowing from the Tropic of Capricorn. Either side of the equatorial trough these winds dominate, giving the trade wind belt. Polewards of the tropics, in the temperate latitudes, we find a zone of prevailing westerlies, while around the poles occurs a belt of easterlies. We will examine each of these zones separately, but as we do so it is important to remember that, in reality, these wind belts do not operate in isolation. They are closely interrelated (Plate 8.3).

The equatorial trough

The equatorial trough, or Inter-tropical Convergence Zone, is a shallow trough of low pressure generally situated near the equator. Over the oceans it is fairly static, because seasonal temperature changes are small. In the Pacific, for example, its average position varies by no more than 5° of latitude within the course of a single year (Figure 8.1). The situation is very different over the continents. During summer in continental areas the trough sweeps polewards, reaching 30° or even 40° latitude over eastern China. Behind the trough the winds are predominantly westerly and are the main rain-bearing winds to most of those areas. Where they reach into higher latitudes they are called monsoons (an Arabic word meaning 'season') and they show an almost complete reversal of direction from summer to winter, a change that tends to occur with uncanny regularity about the same dates each year.

With the exception of the monsoon, the winds in the equatorial trough tend to be light and variable and, because sailors often found themselves becalmed there, the area became known as the Doldrums.

The trade winds

The trade wind belts lie between the equatorial trough and the subtropical highs (Figure 8.15). This zone occupies nearly half the globe, much of it ocean, and within that area the steady trade winds provide a stable and relatively constant climate. At the surface the winds have a component towards the equator being from the north-east in the northern and from the south-east in the southern hemisphere. Above the surface friction layer the winds become more easterly.

Viewed from the air, the oceanic trade winds contain innumerable uniform small clouds, all with a similar base and depth (see Plate 9.4). These are the visible expression of the transfer of latent heat from the sea surface, through evaporation, before condensation at higher levels.

As we have noted, the seasonal movement of the equatorial trough is slight over the oceans, so the oceanic tropical areas are dominated by the trades. On the continents the trades are far more restricted in extent, and the equatorial westerlies and monsoons are more important. The two belts interact closely; it is the convergence of moisture in the trade winds that feeds the equatorial trough. The shift in the position of the trough thus determines the relative extent of the easterlies and westerlies. When the trough is farther north, with the overhead sun in July, the trades are restricted in the northern hemisphere, particularly over land. In January the trough is at its most southerly position and the trades extend to the equator. In the southern hemisphere less marked variations occur, for the predominance of ocean means that the southern limit of the trough remains close to the equator (Figure 8.1).

The westerlies

In comparison with the winds of the tropics, the westerlies of the mid-latitudes seem unreliable and fickle. They are westerlies only on average.

Polewards of the subtropical anticyclones, rotating storms are the main mechanism of energy transfer. Unlike hurricanes, these systems cover vast areas and can be seen clearly from space, identified by their characteristic spiral of clouds (Plates 8.1 and 9.2). In the northern hemisphere they tend to move north-eastwards, although directions vary from north to south-east. Typically, they follow an evolutionary pattern which we shall be examining more closely in the next chapter. The storms are initiated in areas of strong temperature gradients, such as off Newfoundland, where the cold Labrador current and warm Gulf Stream are in close proximity, forming roughly circular patterns of low pressure and a rotational movement of winds, which are often strong (Figure 8.14). They are known as lows, cyclones or depressions. As they evolve they become initially more intense – the central pressure has been known to fall as low as 930 mb – before filling as the storm declines. On average the location of maximum intensity is in the areas of Iceland in the Atlantic and the Aleutian

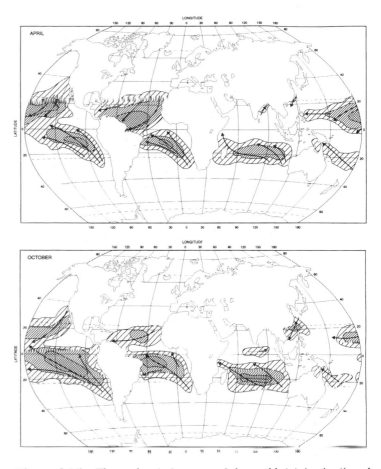

Figure 8.15 The trade wind systems of the world (a) in April and (b) in October. The isopleths are in terms of relative constancy of wind direction and enclose shaded areas where 50 per cent, 70 per cent and 90 per cent of all winds blow from the predominant quadrant, with speeds above 3.3 m s⁻¹.
Source: After Crowe (1971).

Islands in the Pacific. Thus climatologists speak of the Icelandic Low and the Aleutian Low. In the southern hemisphere there are no distinct areas of the genesis of storms, so lows form throughout a wide belt.

As a low approaches, winds increase in strength, initially from a southerly direction, then become westerly and, finally, as the low moves away, they veer to north-westerly or even northerly. The tracks of the lows reach farther poleward in summer than in winter, so the area affected by the storms varies seasonally. Nowhere is this seasonal pattern more clearly seen than over the Mediterranean basin, in California and at equivalent latitudes in the southern hemisphere such as central Chile or parts of West and South Australia. In winter, when the cyclones follow tracks at lower latitudes, they bring rain to these areas. In summer the cyclones move away, to be replaced by the subtropical anticyclones and dry, hot weather. The consequences will be discussed in Chapter 26.

The regular march of cyclones and anticyclones through the temperate latitudes produces a majority of winds between north-west and south in the northern hemisphere and between south-west and north in the southern hemisphere. This pattern is far from invariable, however, and depending upon the precise tracks taken by the lows, and the local topography, winds from any direction are possible (Figure 8.16). The prevailing westerlies, therefore, are anything but prevalent. Moreover, the strong north–south component of winds in these areas allows a more effective transfer of energy between the tropics and the polar regions.

Polar easterlies

Around the poles, beyond the main westerly belt, there is some evidence of prevailing easterlies. The winds are variable and linked with the shallow polar anticyclones. In the northern hemisphere they are often influenced by the circulation around the northern edge of cyclones. As a result they change direction according to the local weather and topography.

In the southern hemisphere the vast Antarctic ice cap controls the atmospheric circulation around the pole. Anticyclones develop frequently over eastern Antarctica, and strong south-easterly winds develop around the margins of the ice plateau with consistencies similar to those of the trades, and occasionally of great strength.

Upper winds
The nature of the upper winds

Looking up at high clouds on a clear day, it is not unusual to find that the direction of their movement is different from that of the surface winds. As this implies, winds in the upper atmosphere can be affected by forces operating in a different direction from those at the surface and may appear to be part of a different system of circulation. If we were to make an ascent by balloon into these upper wind systems we would find that the change from surface to upper atmosphere conditions was not abrupt but transitional. With increasing height, we would discover, the winds tend to follow a gradually more distinct zonal (east–west) direction and they become stronger. The main reason for this change is the disappearance of the frictional influence of the ground surface upon the winds. In other words, the flow more nearly approximates to the geostrophic winds that, it will be remembered, result from the interaction of the pressure gradient and the Coriolis force (Figure 8.17).

The zonal flow of the upper winds can be shown on average as a cross-section from north to south (Figure 8.18). In fact variations around this average picture are slight, except in the monsoon areas of Asia. At each season the same basic pattern exists, with slight shifts in position and intensity. Between about 30° N and 30° S we have a zone of easterly winds which are relatively weak, reaching a maximum speed of 4–5 m s^{-1} (about 17 km hr^{-1}) at about 3 km. On either side of this belt occurs a ring or vortex of much stronger westerly winds.

The upper westerlies

These high-altitude westerly winds are a major feature of our atmosphere. They reach their maximum speed at approximately 12 km between 30° and 40° latitude. The mean speed is as much as 35 m s^{-1} (125 km hr^{-1}) and maximum speeds of several hundred kilometres per hour are not uncommon. It is not surprising that aircraft can travel from the United States to Europe more quickly than on the return journey.

Although these wind patterns are steady, seasonal variations do take place, especially in the northern hemisphere. The upper westerlies are strongest in the winter, when the temperature difference between the tropics and temperate latitudes is at its greatest. From June to August temperatures in the northern hemisphere are relatively warm, even in polar regions, so the pressure gradient is reduced and the upper westerlies decline to speeds of as low as 15 m s^{-1} (55 km hr^{-1}).

As ever, changes in the southern hemisphere are less pronounced, largely owing to the greater thermal stability there. The vast areas of ocean absorb large quantities of heat without any significant increase in temperature. The ice plateau of Antarctica also stays very cool, so the temperature gradients do not change very much from winter to summer.

The position of the boundary between the westerlies and easterlies (of both the upper and the surface winds) varies throughout the year. From December to February the polar vortex of the winter (northern) hemisphere expands, pushing the belts southwards so that, at the surface, the boundaries are at about 30° N and 35° S. As the year progresses, the other polar vortex begins to expand as winter sets in over the southern hemisphere. The boundaries eventually reach about 35° N and 30° S by June to August. The separation between the two systems is not vertical. As a result, some parts of the tropics have easterlies in the lower atmosphere and westerlies above. Only over a small area of the globe do easterlies occur at all levels, whereas westerlies extend

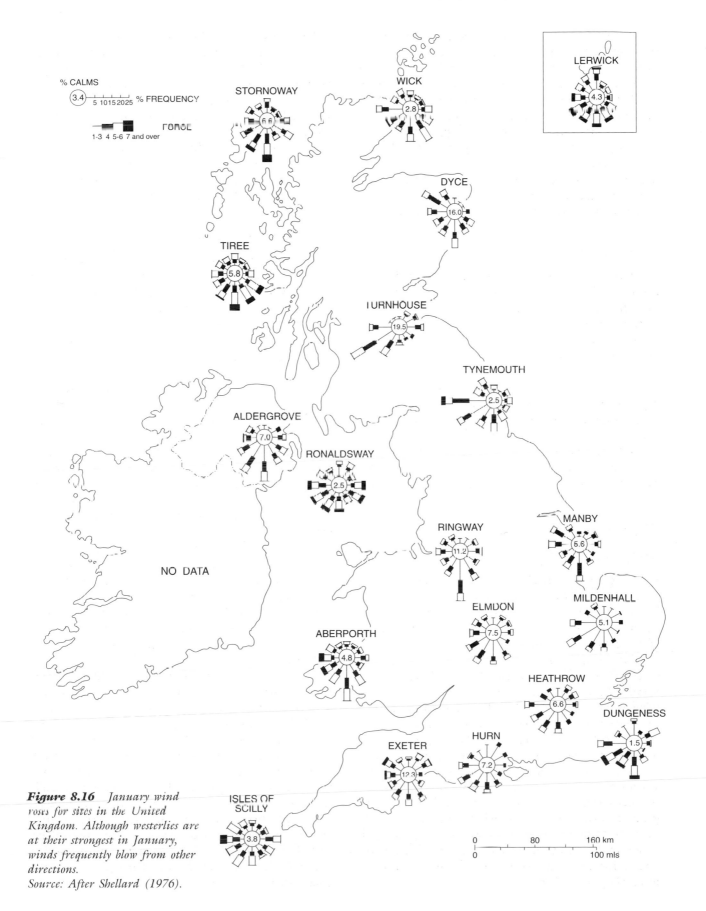

Figure 8.16 *January wind roses for sites in the United Kingdom. Although westerlies are at their strongest in January, winds frequently blow from other directions.*
Source: After Shellard (1976).

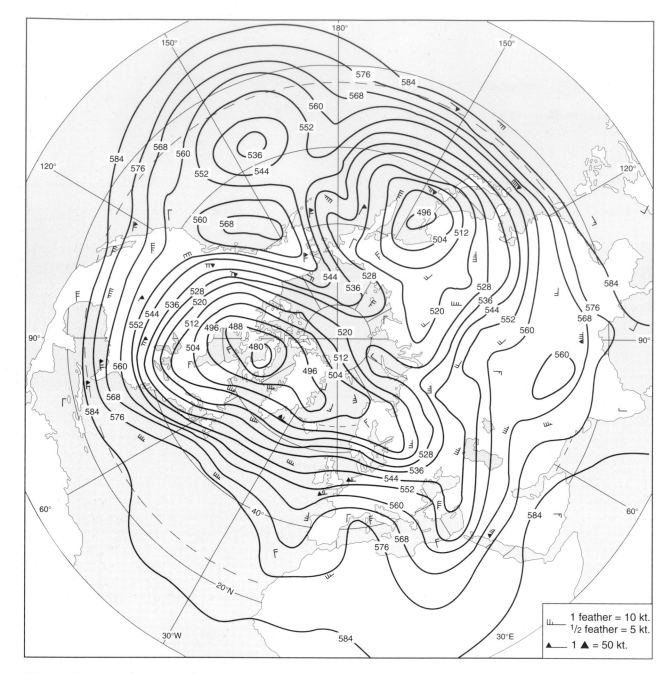

Figure 8.17 *The upper westerlies at 500 mb on 9 February 1981, showing the height of the 500 mb pressure level above a fixed datum near sea level. Winds blow parallel to the contours at a speed proportional to the gradient. Although the flow is predominantly westerly, well marked troughs and ridges can be seen – for example, over the eastern United States and Alaska respectively.*
Source: Reproduced by kind permission of the Deutscher Wetterdienst, Offenbach.

throughout the atmosphere over a large proportion of Earth.

Rossby waves

The pattern of easterlies and westerlies in the upper atmosphere is only part of the total picture. In addi-

tion to the marked zonal flows there are less apparent but none the less important meridional flows. In the circumpolar areas, for example, there occur wave-like patterns of flow called Rossby waves (after C. F. Rossby, a Swedish meteorologist) that play a vital role in the energy exchange between the temperate and polar areas.

It is not easy to detect these meridional flows within the pattern of strong zonal circulation by normal methods of depicting winds. The normal methods usually show average conditions, so that processes which balance each other, flowing northwards for six months, perhaps, then southwards for the next six months, are lost. Yet that is what happens in the case of the Rossby waves. At a particular location southerly flows may last for a few days, to be followed by more northerly winds as the wave progresses eastwards.

In order to see these waves it is necessary to use a rather different technique of presenting atmospheric circulation. Instead of mapping the actual wind directions or speeds, the height at which a particular pressure surface is reached can be plotted. This may seem a strange way of depicting winds, but, as we know, the geostrophic winds blow parallel to the isobars, at a speed inversely proportional to the distance between the isobars. Similarly the winds blow parallel to the contours of the pressure surface. Where the contours are close together, the winds are rapid. Irregularities in the pressure surface indicate local patterns of wind movement.

Figure 8.19 shows a pressure surface (500 mb or hPa) map for January. The projection of the map may make it difficult to appreciate the direction of flow immediately. What is clear is that the flow is not perfectly circular around the north pole. Areas occur, even on this monthly chart, where the mean flow has a northward component, and other areas occur where it is southwards. Effectively the air is flowing in a series of waves around the pole, carrying warmer air northwards on parts of its track and cold air southwards elsewhere. These are the Rossby waves. In January the most prominent features of the waves are the pronounced troughs in the pressure surface near 80° W and 140° E, with a weaker trough between 10° E and 60° E. In July (Figure 8.20), the circulation is less intense and the troughs less well marked.

Many experiments have been conducted to determine the reasons for this pattern. Clearly surface features play an important part, even at this height. The presence of the Rocky Mountains and the Himalayas is believed to 'lock' the troughs at 80° W and 140° E respectively. The distribution of land and sea is also thought to be of importance. In the southern hemisphere there are no mountain ranges of comparable size, nor such marked land and sea temperature contrasts. As a result the mean circulation is much more symmetrical around the South Pole.

Even on a shorter time period, waves may exist in the upper westerlies, though their shape is less regular. The smaller waves tend to be associated with an individual depression and move more rapidly, perhaps

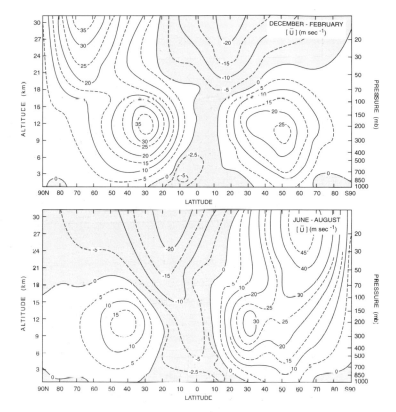

Figure 8.18 *Mean zonal wind for (a) December–February and (b) June–August. Units are m s⁻¹. Positive values denote a westerly wind. Stippling indicates easterly wind.*
Source: After Newell et al. (1969).

up to 15° longitude per day. The longer waves – usually between four and six are apparent – move more slowly and are linked with the major circulation features such as the subtropical highs and Icelandic lows. The long wave flow tends to 'steer' the shorter waves, moving them northwards when ahead of a trough and southward when to the rear of a trough.

The index cycle

It is believed that the Rossby waves do alter their amplitude and wavelength in a roughly cyclical manner over a period of between three and eight weeks. If we follow the pattern of waves over a period of weeks the waves often undergo change, as shown in Figure 8.21. Initially there may be little meridional element in the flow, with a strong westerly circulation – termed a high index flow. Gradually the north–south element becomes more dominant until eventually the flow breaks down into a series of cut-off troughs and ridges like the meandering pattern of a river. This is termed a low index flow, in which north–south exchanges are strong. Low index flow favours the formation of blocking highs and cut-off lows (Chapter 9) which have a

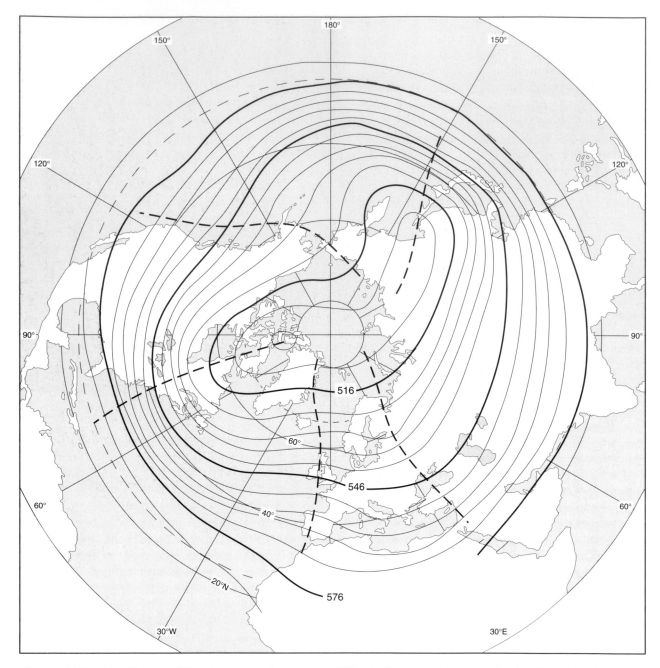

Figure 8.19   *Monthly mean 500 mb contours for January 1951–66. Units are decametres. Dashed lines show the locations of ridges and troughs.*
Source: After Moffitt and Ratcliffe (1972).

marked effect on the weather in some parts of the westerlies of the northern hemisphere.

Jet streams

Within both westerlies and the tropical easterlies, bands of especially strong winds can be found. The existence of these winds or jet streams was appreciated only with the increased use of aircraft during the Second World War. Bombers heading across the Pacific towards Japan reported headwinds so strong that they could hardly advance relative to the ground! More recent investigations have shown that speeds up to 135 m s^{-1} (490 km hr^{-1}) can exist locally in the jet stream maximum. A number of major jets have been found in the troposphere – the polar front jet, the subtropical jet and the tropical easterly

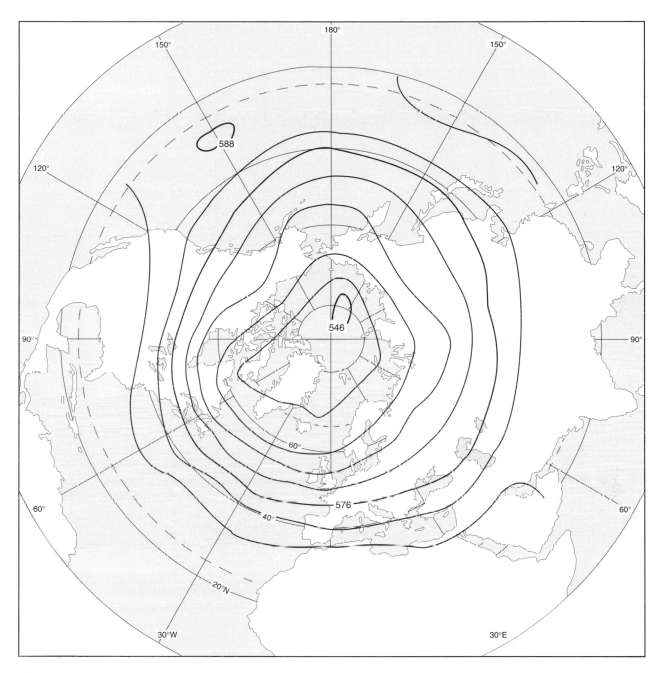

Figure 8.20 *Monthly mean 500 mb contours for July 1951–66. Units are decametres. Note the ridges and troughs are less apparent at this season.*
Source: After Moffitt and Ratcliffe (1972).

jet – and others exist in the stratosphere (Figure 8.22).

What is a jet? Basically it is a very narrow current of air travelling at great speed. Jets are formed in regions of rapid temperature gradient. Typically the westerly jets are connected with the zone of maximum slope or fragmentation of the tropopause, which coincides with the maximum poleward temperature gradient. They can lead to intense accelerations (and decelerations) of air in their vicinity. As we shall

see later, when air is forced to change its rate of flow, tropospheric vertical motion may be started. In turn this may influence events at lower levels.

El Niño–Southern Oscillation

In addition to the major north–south exchanges represented by the Hadley cell, we find a major

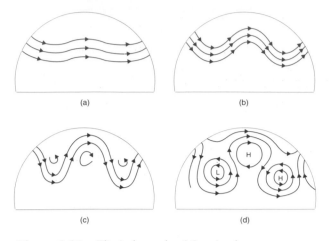

Figure 8.21 *The index cycle of flow in the upper westerlies. The amplitude of the waves increases from (a) to (d) before type (a) becomes re-established.*

east–west exchange in the tropical Pacific Ocean which has been called the Walker circulation. The normal situation is a strong flow from the subtropical high-pressure cells, emphasized by subsidence over the cool ocean currents near South America (Figure 8.23). Over Australia, the archipelago of Indonesia and the warm Pacific Ocean area, air tends to be rising and precipitation is abundant. Periodically this circulation is transformed by the cool Humboldt current off Peru being disrupted and replaced by much warmer waters. As a result, the normally dry areas are wetter as subsidence stops and the wet areas are drier as subsidence zones shift to dominate

these locations. This state of affairs is known as an El Niño or an ENSO (El Niño–Southern Oscillation) event.

Not only does ENSO have a major regional impact in the Pacific, its influence extends to other parts of the world through the interaction of pressure, air flow and temperature effects. During the major El Niño of 1982/3 there were climatic extremes in many parts of the world. Australia, Indonesia, southern India and southern Africa had major droughts. Tropical storms followed anomalous tracks, reducing rainfall in areas which normally experienced the storms, such as north-east Australia, and affecting areas outside the usual range, such as Hawaii. In the northern hemisphere, California suffered major storms as the westerly circulation became more intense in the north Pacific. Increasingly climatologists have begun to realize that anomalies in some parts of the globe can exert an influence on other parts. These effects have been termed atmospheric teleconnections.

Conclusion

Atmospheric movements, together with oceanic circulation, are the main processes by which energy is transferred through the global system. They act to maintain a steady state in the system by transporting excess energy from areas which receive high inputs of solar radiation to areas where inputs are small. Such movements involve two general patterns of flow: the predominantly zonal flow of air within the main wind belts and the less apparent but even more important meridional transfers. Both circulations are controlled by the pressure gradient force, which acts as the driving force of atmospheric motion. Earth's rotation, acting through the Coriolis force, and friction modify the simple pattern of circulation initiated by the pressure gradient force to give the complex systems we find in the atmosphere.

These atmospheric movements are vital for a number of reasons. Many of the features of the world's climates are dependent upon the character of atmospheric circulation, as we shall see in Chapter 9. Seasonal and daily variations in the circulation affect our lives directly, and extreme events may have a dramatic impact on humanity, topics which will be covered in Chapters 24–9.

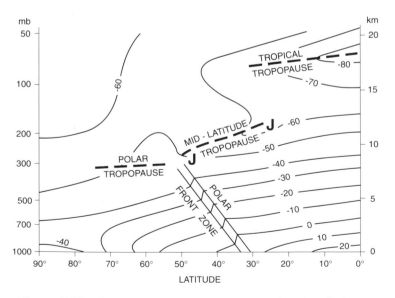

Figure 8.22 *Jet streams (J) in the upper atmosphere in relation to the vertical temperature gradient.*

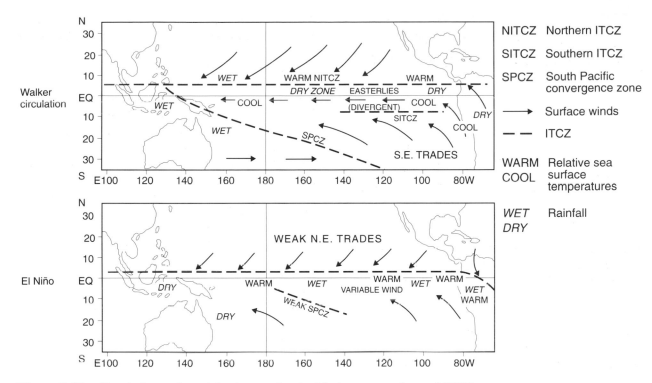

Figure 8.23 *Circulation and precipitation associated with the extreme phases of ENSO.*

Key points

1 Movement of air in the atmosphere is determined by the pressure gradient force and modified by the Coriolis force and, near the ground, by friction.

2 Because of Earth's size and rate of rotation, and the energy imbalance caused by astronomic factors, the large-scale circulation of the globe splits into a series of systems, with easterlies in the Tropics flowing towards the equatorial trough and westerlies in temperate latitudes. Away from Earth's surface the winds strengthen, particularly in temperate and polar latitudes. The upper westerlies flow in a series of waves, called Rossby waves, which have a major effect on surface weather conditions.

3 Long-distance interactions within the atmosphere do occur. Periodic changes in ocean and atmospheric circulations in the south-east Pacific can have an effect on the weather across much of the southern hemisphere.

Further reading

Ahrens, D.L. (1994) *Meteorology Today*, fifth edition, Minneapolis: West Publishing (Chapter 9).
Visual and elementary approach to aspects of winds and factors controlling airflow.

Atkinson, B.W. (1981) *Dynamical Meteorology*, London: Methuen.
Based on articles published in *Weather* by the Royal Meteorological Society, this book describes the complex issues of atmospheric movement in as non-mathematical a way as possible. Still not easy for the non-mathematician.

Barry, R.G. and Chorley, R.J. (1992) *Atmosphere, Weather and Climate*, sixth edition (Chapter 3). London: Routledge.
Serious attempt to explain and inform about the controls of atmospheric motion. Covers a wide range of scales of motion, from micro to global.

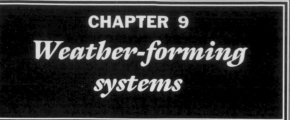

CHAPTER 9
Weather-forming systems

Not a year goes by without weather events some-where in the world causing damage or loss of life. Floods, blizzards, tornadoes, hurricanes or even heat-waves can create problems and generate much economic stress over the areas affected. To be prepared, it is vital to understand the weather, and be able to predict with accuracy, preferably well in advance, events such as these. It is important too that we understand not only the vagaries of day-to-day weather conditions but also longer-term trends. How useful it would be for farmers to know what the weather over the next few weeks or even the whole growing season will be like; they could plan their sowing or ploughing or harvesting far more successfully. How useful it would be to have a clear idea of the weather in the year ahead so that corn harvests could be predicted, plans for winter snow could be made and measures could be taken to deal with drought. Any such detailed under-standing is a long way away. It may come as we gather more knowledge about the medium-term processes operating within the atmosphere, and about the myriad factors that influence those processes. However, some scientists doubt whether it ever will be possible to predict in any detail the long-term movements of a chaotic and turbulent 'fluid' such as our atmosphere.

The key to understanding and predicting short-term weather changes, say up to one week ahead, lies in understanding what we call weather-forming systems. If we look at a satellite photograph showing half the globe it is clear that the distribution of clouds is not random (Plate 9.1). In some areas clouds are abundant, sometimes showing certain patterns which make it possible to identify their origin. Many areas are devoid of cloud altogether and surface features can be seen. Comparing this photograph with a map of surface pressure, we would see that the large spirals of cloud are associated with cyclones in the middle latitudes and the main cloud-free areas with the large anticyclones of the subtropics. Between these areas the cloud patterns are less clear, though

over the south Atlantic Ocean the trade winds have produced some interesting forms and, over the cold Benguela current off south-west Africa, there are extensive layers of low cloud. Viewing this in-stantaneous picture, we can see the way in which different areas of the atmosphere interact, and by using the surface pressure information we can relate these cloud patterns to the weather systems which produce them.

Air masses

An air mass is a large, uniform body of air with no major horizontal gradients of temperature, wind or humidity. In the anticyclonic areas of the world, where air movement is gentle, the air is in contact with the ground surface and gradually acquires the thermal and moisture properties of the ground. We find that the air then has relatively uniform distrib-utions of temperature and humidity over large areas – for example, Siberia in winter. Whether or not the air will fully reach equilibrium with the surface char-acteristics will depend upon how long it remains over the source region.

The character of an air mass is dependent upon conditions in the area in which it forms. Because of this it is possible to classify air masses on the basis of their source area. Four main types are recognized: Arctic (or Antarctic), Polar, Tropical and Equatorial, and these are further subdivided into continental (for those forming over large land masses) and maritime (for those developing over the oceans) (Table 9.1, Figure 9.1).

As the air mass moves away from its source area its character changes, owing to the influence of the new underlying surface. Air moving towards the poles generally comes into contact with cooler surfaces. This causes it to be cooled from below, so that it may become saturated, with the result that low clouds are formed. In addition, the air is made more stable, so rainfall is less likely (Figure 9.2). Conversely, air

Plate 9.1 *Meteosat image taken at 11.55 GMT on 23 May 1989 in the visible waveband. The cloud-free area of the Saharan subtropical anticyclone is evident. To the north we can see the cloud spirals of mid-latitude cyclones. Near the equator, greater cloud development can be seen at the equatorial trough. Comparison with Plate 8.3, which is in the infra-red waveband, determines cloud top heights. High clouds appear white in the infra-red image whilst thick clouds irrespective of height will appear white in the visible image.*

moving towards the equator becomes warmer as it meets warmer surfaces. As we saw in Chapter 6, warming of the lower layers of the air steepens the lapse rate, making the air less stable and convectional showers more likely.

Changes in air masses by these means are particularly marked in the mid-latitudes. Here, cyclones draw in air from several sources; the air is modified by the new surfaces it encounters and is gradually mixed as it rises around the cyclone centre. Its precise thermal properties will depend upon the origin of the air, its track and the speed of its movement from the source area.

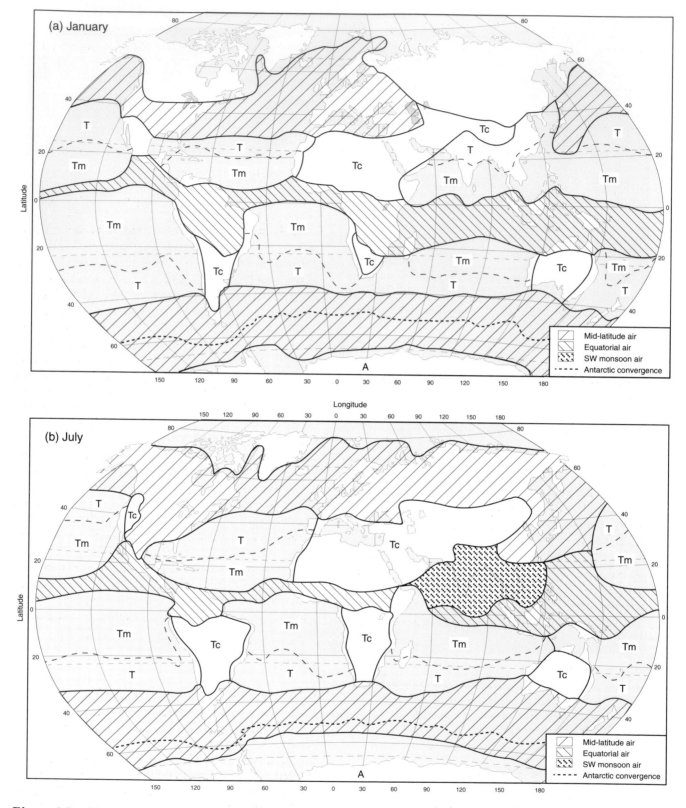

Figure 9.1 *Air mass source regions: (a) January, (b) July.*
Source: *After Crowe (1971).*

Table 9.1 *Average thermal properties of air masses.*

Air mass	Symbol	Properties	Mean temperature (°C)	Specific humidity (g kg^{-1})
Continental Arctic	cA	Very cold, dry	−20	0.1
Continental Polar	cP	Cold, dry (winter)	−10	1.4
Maritime Polar	mP	Cool, moist	5	4.4
Continental Tropical	cT	Warm, dry	25	11
Maritime Tropical	mT	Warm, moist	20	17
Maritime Equatorial	mE	Warm, very moist	27	19

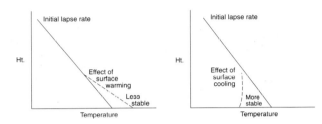

Figure 9.2 *The effect of surface warming and cooling on lapse rates.*

Weather-forming systems of temperate latitudes

Anticyclones

An anticyclone is a mass of relatively high pressure within which the air is subsiding. The major anticyclonic belts are in the subtropics, centred about 30° from the equator. They represent the descending arm of the Hadley cell circulation of the tropics. As air descends it gets warmer and drier (Figure 9.3), but in these regions its descent is restricted by the layer of cool oceanic air below. The result is a semi-permanent inversion. This combination of circumstances gives rise to very stable atmospheric conditions, reducing the possibility of precipitation.

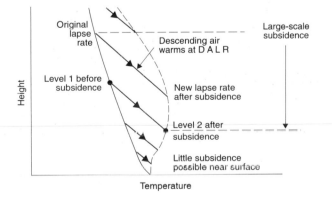

Figure 9.3 *The effect of large-scale subsidence on lapse rates. DALR, dry adiabatic lapse rate.*

Even over heated desert surfaces the effects of subsidence dominate, so the anticyclonic belts are associated with the main dry zones of the world.

In the middle latitudes, anticyclones often develop as a result of convergence in the upper westerlies, particularly where the waves in those westerlies have a large amplitude. The surface anticyclones then build up within the usual depression tracks, diverting the cyclones from their normal routes and giving rise to exceptional patterns of weather. *Blocking anticyclones*, as they are called, are most frequent over north-west Europe and the north Pacific. Blocking in the Atlantic was responsible for the droughts of the late 1980s and 1995 and the severe winter of 1978–9 over north-west Europe. Unfortunately we do not yet know enough about the causes of blocking anticyclones to predict their behaviour.

Anticyclones are normally associated with dry weather and light winds. Clear skies or extensive cloud and very warm or very cold conditions may occur. Which we get depends upon the time of the year, the degree of moistness, the source of the air and the location and intensity of the anticyclone. In Europe, in summer, anticyclones usually bring hot, dry weather if centred over the Mediterranean or central Europe, but in winter cold weather is more usual, especially if the anticyclone is centred over Scandinavia and dry, cold continental air is drawn from the east.

Cyclones

Wind speed rises, pressure falls and the clouds get thicker: a common sequence of events in the mid-latitudes heralding the approach of a cyclone. The cyclone, or depression or low, as it is also known, brings with it conditions very different from those associated with anticyclones. Air pressure is relatively low and the air circulating around the low is rising. Cyclones usually move relatively quickly, in the northern hemisphere normally towards the north-east. They are smaller in size, but within them air is rising more quickly than it descends in an anticyclone. Pressure and temperature gradients are much steeper, so that

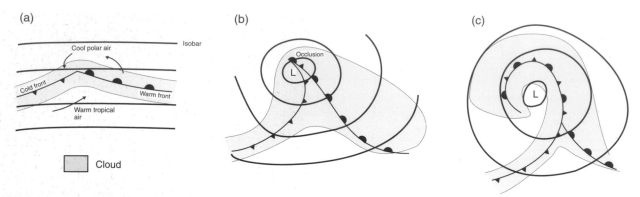

Figure 9.4 *Cloud distribution and pressure changes during the evolution of a mid-latitude cyclone.*

horizontal winds are strong. In essence, they are the main mobile systems of the middle latitudes and they are responsible for the characteristic climates of those regions. Much of the precipitation there comes from this source. Cyclones are also responsible for the sudden swings in temperature from hot to cold or vice versa as air masses change.

For over seventy years the Bergen model of cyclone evolution has dominated our views (Figure 9.4). However, subsequent work, especially that involving the upper atmosphere, has revealed significant deficiencies in the model. For example, it is now clear that depression formation does not need a polar front but rather a zone of strong temperature gradient known as a baroclinic zone. The process of cyclogenesis (or depression formation) actually intensifies the thermal gradients to produce the fronts. In many parts of the world warm fronts, an important part of the Bergen model, are weak or limited in extent. Finally, the classic 'catching-up' occlusion process is difficult to identify and many studies have shown that ideal occluded frontal structures are rarely observed in their entirety. Unfortunately no clear conceptual model of cyclone evolution has replaced the Bergen one, possibly because cyclone development is highly variable, depending upon surface conditions, topography and upper atmospheric flows.

If we follow cyclones over a period of several days we find that many, though by no means all, conform to a general pattern, called the Bergen model. Initially a small wave develops in the front, separating polar and tropical air masses (Figure 9.4a). In some cases no further development takes place and the wave gradually dies out. More often the wave begins to amplify and a small low-pressure centre forms. Gradually the air pressure within this centre falls, the winds strengthen and the area of low pressure expands. Eventually the system starts to fill and the cyclone disappears (Figure 9.4c).

What we see at the surface is only part of the story, however, for the cyclone also extends up into the atmosphere. The low-pressure centre represents a column of rising air – one which is often visible on satellite photographs (Plate 9.2). To understand the cyclone more fully, we need to ascend to the top of the column, to the upper atmosphere, where we find the ridges and troughs in the upper westerlies. The flow around these waves is not always in equilibrium with the pressure gradients. Where air moves out of a trough it accelerates; as it approaches a trough it slows down (Figure 9.5). The air moving away from the trough draws air from the lower atmosphere, causing a reduction in surface pressure. Thus air is seen to converge at the ground within the cyclone, rise upwards into the upper atmosphere, and there diverge as it flows away from the trough. The relative rates of surface convergence and of upper-air divergence control the development of the cyclone. If divergence exceeds convergence the cyclone intensifies as air is drawn out of the system. At that stage we find air pressure at the ground falling. If convergence exceeds divergence the cyclone fills and air pressure at the surface rises. This is what happens in the final stages of the cyclone.

In the northern hemisphere the troughs and ridges of the upper westerlies tend to favour certain locations. There is normally a ridge near the Western Cordillera and a trough near the eastern coast of the United States. This means that cyclone formation is most likely in the area south of Newfoundland (Figure 8.14). The depressions intensify, reach their maximum intensity near Iceland, then decay. The average position of the cyclones shown on mean pressure charts is near Iceland for this reason. It represents the most frequent track of the depressions and where, on average, they reach their lowest pressure. Because the cyclones are areas of rising air, they are almost always accompanied by extensive cloud and precipitation. The steep pressure gradients and rapid falls of pressure which can occur cause problems for the affected areas in terms of gales and heavy rain.

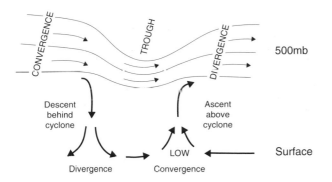

Figure 9.5 *Interaction between surface and upper atmospheric flow near an upper trough.*

Figure 9.6 shows the typical vertical cloud distribution and temperatures associated with a cyclone in mid-latitudes. The details of cloud location and thickness will depend upon the nature of the upper atmospheric divergence and the temperatures, on the time of year and the sources of the air. If we look at the surface pattern of precipitation from the cyclone (Figure 9.7), we can see how the areas of highest rainfall tend to be just on the northern side of the depression track, with amounts decreasing northwards and southwards. The width affected may stretch for about 1200 km but will vary between cyclones. The actual track of the cyclone is determined by air flow in the upper atmosphere and the temperature gradient at the ground.

Fronts

In many cyclones we would find that there is not a gradual change of temperature as the systems pass but several sudden changes. Figure 9.8 shows the trace from a thermograph during the passage of a cyclone. If it has been cold before the storm approaches, temperatures may rise slightly. This is due to cloud and wind stirring up the cold air. If it has been warm, temperatures may fall, because the sun will no longer be shining. Suddenly the temperature starts to rise, perhaps by several degrees within a few hours. It will then remain fairly stable until the arrival of the cold air in the rear of the cyclone. The fall in temperature is usually more sudden than the earlier rise; a fall of 10° C within a few minutes is not unknown.

The sudden change of temperature clearly indicates a change of air mass. The separation surface or zone between air of different origins is call a *front*. Where warm air is replacing cold air we have a warm front, and where cold air is replacing warm air we have a cold front. The typical cloud structure of the fronts is

Plate 9.2 *An example of the typical large spiral cloud pattern associated with the mid-latitude cyclone. The centre of the spiral marks the position of the lowest pressure, with the cold front clearly defined by the band of cloud leading away south-westwards. Further comparison of visible and infra-red images can be made with Plate 8.1. Photo: University of Dundee.*

shown in Figure 9.6. The clouds mark the main areas of rising air produced by divergence in the upper atmosphere. That is why the clouds do not follow the frontal surface as closely as one might expect.

The warm front slopes at a low gradient of about 1 in 300, which means that the first clouds associated with the front can be seen long before the surface front is near. Cirrus clouds are the first indicators of the approach of the front, followed by a sequence of gradual thickening and lowering of the cloud base. Cirrostratus clouds are followed by altostratus, then nimbostratus clouds, by which time rain

Storms

The temperate-latitude cyclones which form a significant feature of the westerly circulation are very varied in their characteristics. Although the majority of them will follow a sequence as outlined in this chapter, occasionally they deepen rapidly and produce much more severe weather than expected. A classic area for the occurrence of such explosive storms is the eastern coast of the United States, though storms of similar origin and intensity also develop to the northeast of Japan. As the storms develop, pressure falls of 10–20 mb over twelve hours are not uncommon. Central pressures may reach as low as 960 mb, with hurricane-force winds over a considerable area. When the storms develop in winter they may be accompanied by large volumes of snow which wreak havoc in coastal cities from Boston to Washington. The size, frequency and intensity of these storms as they affect the coastal areas of the north-east United States make them potentially more dangerous and destructive than hurricanes.

The cyclogenesis takes place about 400 km downstream from a 500 mb trough, and is situated on the cold side of the belt of strongest westerlies where air and sea-surface temperature gradients are steep. The situation is similar in Japan, where explosive cyclogenesis also takes place relatively frequently. Such is the impact of the American storms that some are given names. For example, on 18–20 February 1979 there was the President's Day storm and on 9–10 September 1978 the *Queen Elizabeth II* storm, named because of the damage inflicted on the liner.

Storms of such intensity are rare over populated parts of north-west Europe, but on 15/16 October 1987 explosive deepening of pressure took place over the Bay of Biscay, followed quickly by an even more rapid increase of pressure of over 20 mb in three hours. The centre of the low moved north-eastwards across Brittany, then tracked across southern Britain and out into the North Sea near Norfolk, producing a steep pressure gradient over south-eastern England. Driven by this strong pressure gradient, hurricane-force south to south-westerly winds blew across the south-east to give record wind speeds for many locations (Figure 1). For most of Kent, Sussex and the coastal areas of Essex and Suffolk, the highest gusts were of a speed likely to be exceeded once in 200 years. Even in the built-up area around the London Weather Centre a gust speed of eighty-two knots was recorded, compared with the previous maximum gust of only fifty-seven knots. Heavy rain fell in association with the storm but, unlike the wind speeds, it was not noteworthy.

Wind speeds of this force sweeping across a densely populated and wooded area are likely to have a dramatic effect, and this storm was no exception. More trees were lost in one night than in a decade of Dutch Elm Disease; parkland areas were devastated, forests flattened and many urban trees blown down to block roads. In East Sussex it was estimated that almost 25 per cent of the original standing volume of timber was blown down. In Brittany about 20 per cent of the whole forest area was reported to have been destroyed. In general, conifers appeared more vulnerable to being blown down than deciduous trees, woodland trees were more vulnerable than isolated trees and individual urban trees were more vulnerable than rural trees. Urban trees rarely fell in the direction of the nearest building, perhaps reflecting channelling within the street. Ironically the clearance of trees from many areas has resulted in a resurgence of ground vegetation as increased levels of light and greater nutrient availability from decaying vegetation have changed the local ecosystem.

As well as causing devastation to trees, the storm had an impact on transport, as many trees blocked roads and rail tracks, waves caused major problems along the coast and at sea and more than a hundred flights were cancelled from Heathrow and Gatwick airports. Power lines are always susceptible to damage during storms, and much of south-east England was without electricity for at least six hours between 03.00 and 09.30.

There were a number of other effects as winds damaged buildings, causing scaffolding and even pieces of building fabric to collapse, in some cases causing fatalities. Many glasshouses were destroyed by the force of the wind. Overall the storm proved a financial nightmare to the insurance industry as well as causing major problems for the area affected. Such explosive storms are a rare event in this part of the world though the 'Great Storm' of November 1703 was probably similar in magnitude.

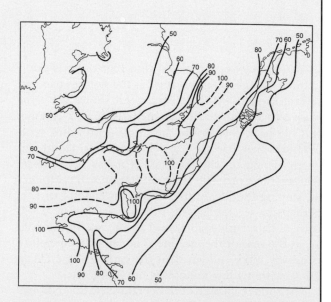

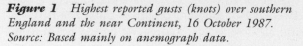

Figure 1 *Highest reported gusts (knots) over southern England and the near Continent, 16 October 1987. Source: Based mainly on anemograph data.*

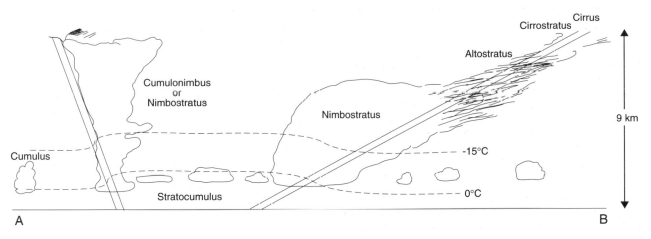

Figure 9.6 *Vertical cloud distribution associated with a model mid-latitude cyclone. Detailed or even major differences occur in most cyclones. Plates 5.8 and 8.2 show cyclonic cloud patterns on satellite images.*

will be falling. In general, the atmosphere is fairly stable at a warm front, but some convection does occur in the middle levels, producing areas of heavier precipitation. Figure 9.9 shows an example of the rainfall patterns associated with a warm front and a cold front.

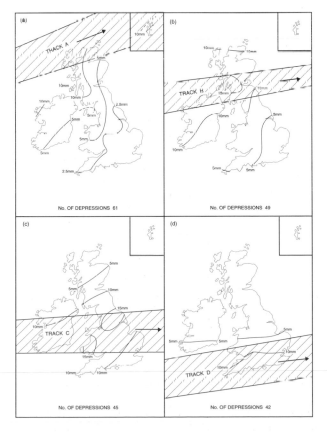

Figure 9.7 *Mean rainfall relative to depression tracks over the British Isles.*
Source: After Sawyer (1956).

The slope of the cold front is much steeper, at about 1 in 50. Weather activity at the cold front can be much more intense than at the warm front. If the warm air is unstable, the effect of uplift at the front generates thunderstorms and even tornadoes. The line of deep cloud may be seen on satellite photographs (Plate 9.3) as a very distinct band. The cold air descending with the heavy rain can intensify the effect of the fall in temperature.

When the air in the warm sector between the fronts is rising, cloud development near the fronts follows the pattern described above; this is known as an *ana-front* (from the Greek word meaning 'up'). Just ahead of the cold front, and at about 1 km above the surface, strong winds develop in the warm sector. This warm, moist flow rises over the warm front and turns south-eastward ahead of it as it merges with the mid-tropospheric flow (Figure 8.13). This flow has been termed the 'conveyor belt', as it conveys large quantities of heat polewards. Convective instability may be produced between this lower, warm, moist air and the cooler, drier air aloft, to produce the typical banded precipitation of the warm front (Figure 9.9).

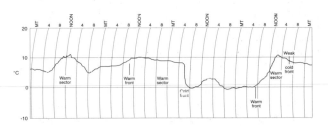

Figure 9.8 *Thermograph trace over a four-day period illustrating the effects on temperature of the passage of two mid-latitude cyclones. The fall of temperature at the cold front was unusually large for the United Kingdom.*

Plate 9.3   *The cloud band associated with a cold front as it crosses England on 4 October 1995. The clearance of cloud at a cold front can be quite sudden, as shown here. Shower clouds form over the relatively warm Atlantic Ocean.*
Photo: University of Dundee.

However, farther away from the cyclone, the intensity of uplift declines and cloud may gradually thin as the front dies out. In this stage of only weakly rising air the front is termed a *kata-front* and the transition zone of temperature is fairly broad (Figure 9.10). Rainfall is slight from kata-fronts, as the clouds are not deep enough and the updraughts are weak.

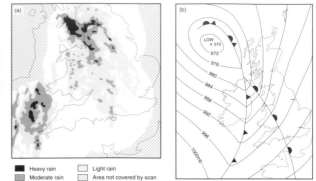

Figure 9.9   *Precipitation patterns obtained by radar associated with a mid-latitude cyclone over the British Isles. Source: After Browning (1985).*

In most cyclones the cold front moves more rapidly than the warm front. The air of the warm sector is raised above the ground surface as the cold front catches up with the warm front. This is known as the stage of *occlusion*, or the *occluded front*. The nature of the front will now depend upon the relative temperatures of the two cold air masses (Figure 9.11). Where the air behind is colder than that ahead, we will have a structure rather like a cold front. If it is warmer than the air ahead, the structure will resemble a warm front.

The detailed air movements and cloud distribution at an occluded front are complex. As fronts represent the mixing of air of different origins, humidities, temperatures and stabilities, it is not surprising that great variation can occur between fronts, or even along the same front. This may also explain why 'true' occluded fronts are relatively rare. Frequently one or more of the frontal components of the Bergen model is missing. It also appears that fronts which seem to have an occluded structure may have been formed by other methods, such as growth northward from the junction of warm and cold fronts, or 'instant' occlusions whereby comma-shaped cloud features behind the cold front join with open frontal waves to produce apparently mature occluded systems over a short period of time. Much remains to be determined about the nature of occluded frontal systems.

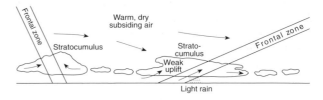

Figure 9.10   *Cloud structure at a kata-warm and a kata-cold front. An ana-frontal structure in shown in Figure 9.6.*

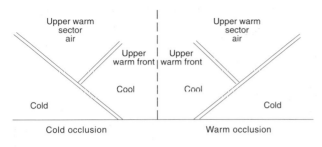

Figure 9.11 *Simplified cross-sections through cold and warm occluded fronts.*

At one time it was believed that it was the air rising along the frontal surface that caused the development of a cyclone. However, the role of divergence in the upper atmosphere is now believed to be the most important factor, the fronts being the result of the rotation of air around the cyclone's centre. From being a cause of the depression, the front has been relegated to a consequence. Nevertheless the weather activity associated with fronts is still a very important aspect of the cyclone. Unfortunately their diversity makes it difficult to generalize about their weather properties.

Weather-forming systems of the tropics

Easterly waves

The weather of the trade wind zone normally shows little variety. It is characterized by small convectional clouds drifting across the sky in response to the prevailing winds (Plate 9.4) and is dominated by the Trade Wind Inversion (Figure 9.12). Showers may develop in the afternoon, and they are likely to be heavier and more frequent in the summer season, but otherwise the weather remains remarkably constant throughout the year.

Occasionally disturbances arise to upset this quiet régime. On a dramatic scale there is the tropical cyclone, which is discussed in the next section, but on a smaller scale there is the easterly wave. As its name implies, this represents weather-forming systems related to wave-like structures in the easterly flow of air. They reach their maximum intensity at about the 700 mb level.

The wave does not necessarily move at the same speed as the easterly flow, and it may even exceed the average wind speed. Preceding the wave, convectional cloud dies down, owing to surface divergence and subsidence of the air, while the wind backs towards the north-east (Figure 9.13). As the main axis of the wave approaches, convergence becomes dominant, causing ascent of the air, cloud formation

Plate 9.4 *Typical trade-wind cumulus clouds. The clouds provide visible evidence of the continuous evaporation from the warm tropical seas. Photo: Peter Smithson.*

and precipitation just ahead of the low-pressure trough. The wind suddenly veers as the wave passes, to be followed fairly quickly by the clearance of the cloud and a return to undisturbed trade wind flow.

The passage of the wave is not dramatic, therefore, but in areas where the weather hardly changes it does at least provide a little variety. Moreover, in areas such as the Caribbean, where the waves are frequent, they are responsible for a significant proportion of the annual precipitation.

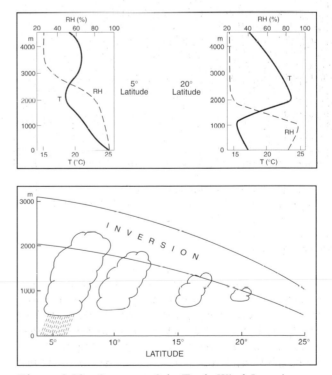

Figure 9.12 *Structure of the Trade Wind Inversion.*

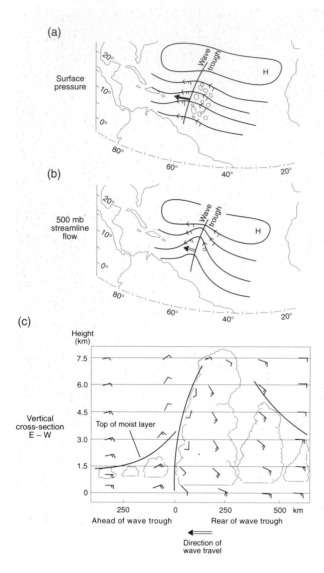

Figure 9.13 (a) Surface pressure, (b) 500 mb streamline flow and (c) vertical structure of an easterly wave. Source: After Malkus (1958).

Tropical cyclones

Throughout much of the tropics one of the other main features of the weather is the tropical cyclone (Plate 9.5). Unlike its counterparts in middle latitudes, this system is not large; instead it consists of a small, intense, revolving storm with cloud bands spiralling away from its centre, usually containing heavy squalls of rain, thunderstorms and even tornadoes (Figure 9.14). It goes under a variety of names: 'hurricane' in the Caribbean and the United States, 'typhoon' in the Pacific or 'cyclone' in the Bay of Bengal. To qualify as a hurricane, the storm must contain winds reaching over 63 knots (32 m s^{-1}). Less intense storms are called tropical cyclones or tropical storms.

If we look at the parts of the globe affected by these cyclones, it is apparent that they develop only over the warmer parts of the oceans (Figure 9.15). In each hemisphere it is during the summer and autumn seasons that cyclones are most likely to strike.

Despite the danger and damage of hurricanes, we know surprisingly little about their origins, except that they all form over the tropical seas where temperatures are above 27° and that they do not form within about 5° latitude of the equator. Once developed, they move towards the west within the trade winds, gradually increasing in intensity. Before dying out the storm usually begins to swing polewards. A few manage to maintain their identity but they gradually decay and acquire the characteristics of a mid-latitude depression. Many September storms and floods in north-west Europe can be traced back to Caribbean hurricanes.

Once started, the development of the hurricane is fairly predictable in terms of flow in the middle levels of the atmosphere and regional temperature patterns. But what starts it off? In order for the cyclonic wind circulation to develop, we must have air converging, which requires some form of initial disturbance. We do find a variety of small disturbances within the tropics where vertical movements and rotation can be started. As they are small in size and found over the seas, their location is difficult to determine. Once the air begins to rise, it cools and, on reaching saturation, large quantities of latent heat are released. It is this process which is believed to be responsible for giving so much energy to the storm. Once the storm moves over land the main source of energy is lost and so it decays.

By understanding the type of atmospheric and surface environment which is most favourable to tropical cyclone development, efforts have been made to predict their occurrence on a seasonal basis. The factors which favour a large number of hurricanes in the Atlantic are: warm sea water off Africa, weaker than normal trade winds, more easterly waves off West Africa, a wet Sahel area of West Africa, and no El Niño in the Pacific. As a result, the number of Atlantic hurricanes and tropical storms is quite variable (Figure 9.16). Researchers in the United States have had considerable success in predicting the number of hurricane-strength storms in the Atlantic, including the near-record year of 1995 with twelve, suggesting that the right factors are being included in the model.

Tornadoes

Tornado! The very word brings alarm in areas such as the Mid West and Mississippi Valley of the United

Plate 9.5 *Hurricane 'Hugo' about to hit the south east coast of the United States, 21 September 1989. The tight spiral of clouds associated with hurricanes can be clearly seen, together with the eye of the storm.*
Photo: By courtesy of Ross Reynolds.

States. It conjures up the vision of a darkening sky, the appearance of a pale cloud, the familiar and frightening tornado funnel. The funnel may descend from the cloud base, getting larger and darker, until it eventually touches the ground, accompanied by a tremendous roaring wind. Debris is caught in the funnel, and as it moves across the countryside it leaves complete devastation in its wake (Plate 9.6).

The tornado is normally narrow, about 0.5 km wide, and seldom does it move more than 20 km. But exceptions do occur, with some being up to 1 km wide and travelling 500 km. How fast the wind blows within the funnel we cannot tell; no recorder has survived its passage. From damage evidence, speeds of over 400 km hr^{-1} are believed to occur.

Tornadoes are not strictly tropical systems, as they occur in most parts of the world, even Britain, but they achieve their greatest strength and frequency over

the continental plains of the United States. The reason for this concentration is the frequent juxtaposition of layers of air with great contrasts in air temperature and moisture. Air ahead of a cold front may be drawn in from the Gulf of Mexico. Behind it, cold air may be sweeping southwards from the Canadian Arctic. Such a situation is ideal for the development of the cumulonimbus clouds needed to spawn tornadoes.

As with hurricanes, the precise mechanism by which a funnel forms is not understood. It is probable that tornadoes are produced by thermal and mechanical effects acting in the cloud. But why some clouds generate tornadoes and others do not is a mystery. Nevertheless, favourable conditions are recognized and tornado warnings are issued by the local American weather services.

Over the sea, similar funnels are termed waterspouts. As convection over the sea tends to be less

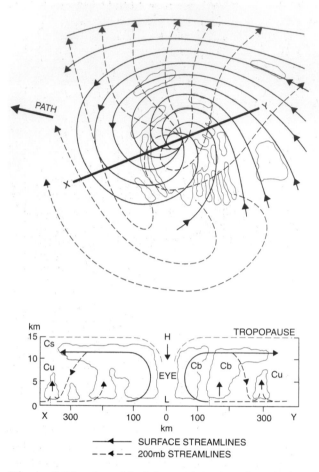

Figure 9.14 *A model of the areal (top) and vertical (bottom) structure of a hurricane.*
Source: After Barry and Chorley (1992).

intense than over land, the waterspout is much weaker than the tornado but may cause some damage to small boats, or to light buildings if it makes landfall.

Weather prediction

We all know from experience how much daily and seasonal variations in weather influence our lives. Clearly it is useful to have an idea of the weather which is in store for us. But how is it possible to foretell the weather? In the past we relied heavily on folklore. 'Red sky at night, shepherd's delight; red sky in the morning, shepherd's warning,' says one country adage. 'When there's sheep-backs (cumulus clouds) in the sky, not long wet, not long dry', goes another. Clearly these sayings sometimes contain a grain of truth – that is presumably why they have survived – but not enough for them to be reliable.

During the course of this century, as we have started to understand atmospheric processes in more detail, methods of forecasting have become more sophisticated. The main approach used today involves understanding the basic processes of weather formation and using these physical laws to predict events. Unfortunately, although we know many of the basic laws, and can express them mathematically, the equations that result are difficult to solve. Only recently, with the development of computers, has it been possible to tackle this mind-stretching task; the first attempt to forecast weather in this way, without computers, in 1921, took several months!

To forecast values of pressure and wind, the globe is subdivided into a grid consisting of about 60,000 squares, each with 217 points from one pole to the other and 288 around most latitude circles. For each point, the upper atmosphere is subdivided into twenty levels and the values of the critical atmospheric properties are determined, mainly from satellite information. The physical equations of motion, continuity and thermodynamics are applied to each grid point at each level to predict the new value a short time period ahead. The new data set then provides the starting point for the next set of predictions, and so on every fifteen minutes until the twenty-four-hour or forty-eight-hour forecast is produced. Clearly a vast amount of calculation is required, though, with the speed of modern computers, six-day forecasts take only about fifteen minutes of computing time. Realistic results are produced in this way, and most meteorological services now use computer methods to predict future weather patterns. For local forecasts up to fifteen hours ahead, the UK Meteorological Office has developed a meso-scale model using a 15 km horizontal grid covering the British Isles. It is particularly good at allowing for hills and the land and sea surface contrasts.

We might get the impression from this technique that we can forecast the weather for the distant future, but that is not true. It appears that small deviations can seriously affect the development of weather-forming systems. New predictions have to be made on a daily basis to incorporate small-scale changes which could become very important. The problem is that we do not have enough information (or large enough computers) to solve the equations accurately.

Efforts are being made to improve our techniques of long-period forecasting but success has been limited. In the United Kingdom, monthly forecasts of temperature and precipitation used to be prepared on the basis of previous weather analogues. For example, if the atmospheric circulation and sea surface temperature patterns in, say, July 1992 were very similar to those of July 1964, it could be assumed that the weather in August 1992 should be the same as that in August 1964. Other controlling factors such as ice cover, the state of the El

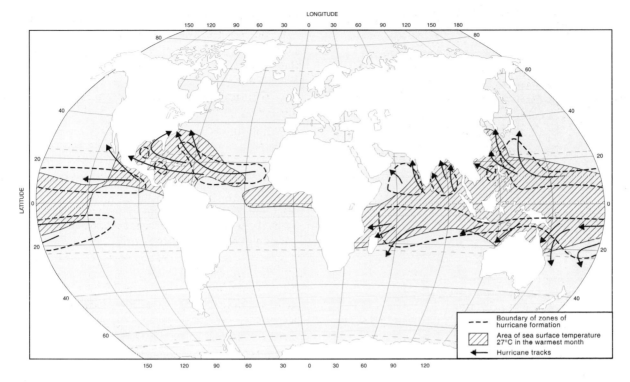

Figure 9.15 *Formation areas and mean tracks of tropical cyclones. Source: After Barry and Chorley (1992).*

Niño–Southern Oscillation (ENSO) and the Quasi-biennial Oscillation (a feature of the wind circulation in the tropical upper troposphere and lower stratosphere) are also included to improve the accuracy of the forecast. However, even if the basic circulation pattern is correctly predicted, slight errors in the tracks of cyclones or anticyclones can produce markedly different weather. Future developments in monthly forecasting are likely to be built around numerical ensemble methods, used to generate probability forecasts. The ensembles are produced as a set of runs from a global numerical model, all slightly different. Their output is interpreted in terms of the probability of the occurrence of particular circulation types.

Weather prediction and hazards

It is possible to predict many of the weather hazards discussed earlier, but we have to distinguish between large-scale and small-scale hazards. The longer is the time scale of development the larger will be the area affected. Major droughts, as the result of a reduction in the number of rain-generating systems, usually affect a large area and take many months to develop, though our techniques of long-term forecasting are not good at predicting when the drought will finish. Tropical and temperate-latitude cyclones can be predicted reasonably well, so that we know approximately

the areas they are likely to affect. On a smaller scale, the warnings of tornado formation are announced for a large area, but precisely where the funnel clouds will touch down on the surface is not known. It is probably impossible to forecast such conditions for more than a few minutes ahead. Flash floods from a single thunderstorm are in a similar category. We have to accept them as one of the microscale features of our atmosphere that occasionally may cause devastation over a small area. The chance of any one site being affected by them is very small.

Further reading

Carlson, T.N. (1991) *Mid-latitude Weather Systems*, London: Harper Collins.
Concerned with all aspects of mid-latitude weather systems. It includes aspects of the development of the systems as well as the weather associated with them. Intermediate to advanced level.

Musk, L.F. (1988) *Weather Systems*, Cambridge University Press.
An elementary text with a clear outline of the types of weather system encountered in temperate latitudes.

O'Hare, G. and Sweeney, J. (1986) *The Atmospheric System*, (Chapter 7), Edinburgh: Oliver & Boyd.
Elementary text with clear description of weather systems at the synoptic scale.

152

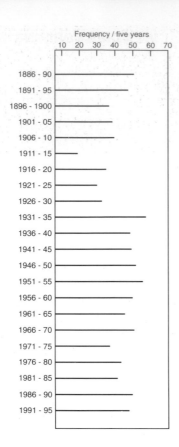

Frequency / five years

10 20 30 40 50 60 70

1886 - 90
1891 - 95
1896 - 1900
1901 - 05
1906 - 10
1911 - 15
1916 - 20
1921 - 25
1926 - 30
1931 - 35
1936 - 40
1941 - 45
1946 - 50
1951 - 55
1956 - 60
1961 - 65
1966 - 70
1971 - 75
1976 - 80
1981 - 85
1986 - 90
1991 - 95

Figure 9.16 *Decadal frequencies of hurricanes and tropical storms in the Atlantic Ocean, 1886–1995. Source: Data from Purdue University hurricane archive.*

Key points

1 Air masses are a feature of the atmospheric circulation. They develop in anticyclonic areas where air movement is less rapid and acquire the characteristics of the underlying surface. As they move away from their source area they transport these thermal characteristics. So we can refer to tropical maritime air masses or tropical continental ones.

2 Within the main circulation flows of the westerlies and the easterlies we can find disturbances which give rise to more unsettled weather. These have a variety of forms and names. In temperate latitudes the cyclone, depression or low and its associated fronts are the main disturbance which brings rain. In tropical latitudes the tropical cyclone is the best known feature but there are other less severe disturbances such as easterly waves.

3 Prediction of weather-forming systems is not easy in any part of the world. For up to about seven days ahead, mathematical and physical models of Earth's atmosphere and surface can be used to determine how the turbulent atmosphere should change from a known starting point. Beyond that length of time, the limitations in knowledge of the original state of the atmosphere and of the models themselves give unrealistic results.

Plate 9.6 *Tornado damage to caravans at Barmouth, West Wales, October 1985. Photo © Torro.*

The climate near the ground

Living as we do in the lowest few metres of the atmosphere, we should have a special interest in the climate of this zone. Unfortunately it turns out to be a very diverse and complicated zone. Climatic differences equivalent to a change in latitude of several degrees can occur in a matter of a few metres. These are examples of microclimatic conditions at the surface.

When the sun is shining, for instance, the ground may become too hot to walk on barefoot, as on a dry, sunny beach in midsummer or on a black road surface. At the ground surface the temperature may exceed 65° C although at head height it may be only 30° C, and in the shade, where most temperature observations are made, it may be as low as 20° C. Similar variations can be found in wind speed and humidity. So, we may ask, what is it about this layer near the ground which produces such major differences – differences that are not repeated anywhere in the free atmosphere?

The main reason for such variability is that we are dealing with the main exchange or activity zone between the ground surface and the atmosphere. Energy is reaching this zone both from the sun and, to a much lesser extent, from the atmosphere. It is absorbed and then returned to the atmosphere in a different form, or is stored in the soil as heat. This absorption process is very sensitive to the nature of the ground surface. Conditions such as surface colour, wetness, vegetation, topography and aspect all affect the interaction between the ground and the atmosphere. We can sometimes see the effects clearly in snowy weather. Clean snow reflects solar radiation and so the surface remains cool and the snow fails to melt. Where the snow is dirty it absorbs more radiation, heats up and is more likely to melt. Vegetation, too, may protect the snow from the heat of the sun, while, even late in spring, snow may be preserved in shaded hollows or on hill slopes facing away from the sun (Plate 10.1).

Let us look at the causes of these differences in more detail. We will start by considering the simplest possible conditions of a horizontal, bare soil surface.

Microclimate over bare soil

Many different properties of the soil influence conditions in the thin layer of atmosphere above it. Soils vary in colour. Darker soils, such as those rich in organic matter, absorb radiant energy more efficiently than do light-coloured soils. Moisture in the soil is also important. Wet soils are normally dark, but water has a large heat capacity; that is to say, it requires a great deal of energy to raise its temperature. A moist soil, therefore, warms up more slowly than a dry one (Figure 10.1).

A further complication with heat transfer into soils is that air is a poor conductor of heat. If there is a large amount of air between the soil particles, heat transfer into the soil is slow. This means that on a hot, sunny day the heat is trapped in the upper layers, so the surface layers warm up more rapidly. Because of this, dry sandy soils can get very hot when the sun shines. Water conducts heat more easily than air, so soils which contain some moisture are able to transmit warmth away from the surface more easily than dry soils. However, if the soil contains a lot of water, the large heat capacity of the water will prevent the soil warming despite heat being conducted from the surface. For most agricultural crops a balance is needed so that soils warm up fairly quickly at depth and are

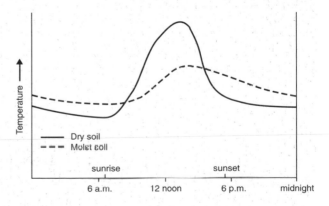

Figure 10.1 *Diurnal temperature changes over a dry soil and a moist soil.*

Plate 10.1 *Aspect effects on snow survival. The south-facing slope has lost its snow through melting, apart from the area in shadow at the base of the slope; the north-facing slope in the foreground is still deeply snow-covered. Photo: Peter Smithson.*

neither too wet nor too dry. This is achieved when the moisture content of the soil is about 20 per cent.

The nature of heat transfer from the soil

As the ground surface gets hotter through absorbing the sun's energy, the layer of air in contact with the ground becomes warm by conduction. If this was the only mechanism of heat transfer, it would take a very long time before even the lowest 1 m of air was warmed. The daytime maximum temperature at that height would not occur until about 9.00 p.m. Clearly this cannot be the only process transferring heat, although it is the most important in the lowest few millimetres, where temperature gradients are extreme. Above that level, the effect of heating the air causes it to become buoyant through being less dense than its surroundings and so it rises, carrying heat with it. Cooler air then moves in to take its place. This air is heated in turn. Consequently we have convection currents rapidly transferring heat to the cooler layers of the lower atmosphere. If there

is a strong wind blowing, the mixing of heat is encouraged and the temperature profile in the lower atmosphere becomes less steep (Figure 10.2).

The radiation balance at night

At night the net radiation at the ground surface is negative (Figure 10.3). More long-wave radiation is lost from the ground than is returned as counter-radiation from the atmosphere. This is especially true with clear skies and dry air, which allow the long-wave radiation to escape to space more easily. In cloudy or humid conditions heat loss is less effective, for water vapour readily absorbs long-wave radiation and re-emits it towards the ground. During the night, therefore, the surface gets cooler, although heat may flow up from lower levels in the soil to reduce the cooling effect of radiation loss. In a sandy soil, with its large air spaces, this process is limited, so the surface becomes particularly cool, as anyone who has slept outdoors on beach sand will know. The air in contact with the ground also gets cooler, making the

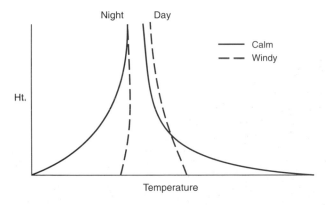

Figure 10.2 *Daytime and night-time temperature profiles on windy and calm days with clear skies.*

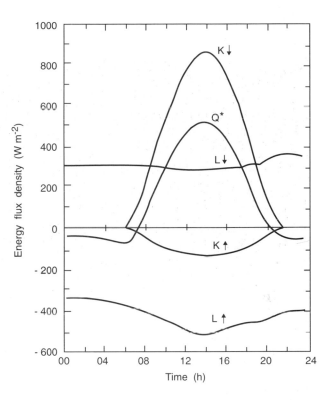

Figure 10.3 *Radiation budget components for 30 July 1971 at Matador, Saskatchewan (50° N) over a 0.2 m stand of native grass. Inputs to the surface have been plotted as positive and outputs as negative to assist interpretation. K is short wave radiation, L is long wave radiation and Q* is net radiation. The arrows indicate the direction of radiation flow. Source: After Oke (1987).*

air denser and preventing any thermals of warmer air rising and mixing with the air above.

The night-time profile of temperature during calm conditions is shown in Figure 10.2, which illustrates the major cooling at the surface. If conditions are windy, the cooler air will be mixed with the warmer air above to give a smaller increase of temperature with height. Clouds are efficient emitters or radiators of long-wave radiation, so low clouds encourage counter-radiation to the surface and the net loss of energy from the surface is reduced. On cloudy and windy nights the cooling at the surface is small and temperatures decrease away from the surface. The factors most favouring low surface temperatures at night are consequently clear skies, dry air, no wind, and sandy, peaty or snow-covered soil. If such conditions occur at the beginning of the growing season in most temperate latitudes then damage to frost-sensitive crops is likely.

Wind near the ground

As we approach the ground, wind speed decreases very rapidly to almost zero in contact with the soil surface (Figure 10.4). This is largely due to the frictional drag exerted on the air by the underlying rigid surface; the rougher the surface is, the more it slows the air down (see Chapter 16). Over a soil surface the effect on the wind is fairly simple, but when we are dealing with a vegetation layer or an urban area, interference is much greater. In addition to friction, buoyancy in the lower layers has an effect on the details of the profile. Rising air will assist mixing and reduce the gradient of wind speed.

The microclimate at a soil surface represents one of the simplest cases of energy exchange at the ground surface. Both the inputs and the outputs of radiation

are changed, and that alters the way energy is used in terms of sensible and latent heat, and heat flow or storage into the soil. This is illustrated in Figure 10.5, where the energy balances over a wet and a dry soil are contrasted, and in Table 10.1, where the thermal differences above such surfaces are demonstrated.

Microclimate above a vegetated surface

The nature of microclimatic conditions and processes becomes far more complex when vegetation cover is present, for not all the energy is absorbed at a single surface. Some is absorbed by the top of the vegetation, some penetrates into the plants, and some may even reach the soil surface. The amount that gets through to the soil depends upon the height of the crop, the density of the leaves and the angle of the sun's rays. As the size of the plants increases, so does the degree of microclimatic modification.

Let us look at some of the detailed effects of plants on the microclimate by considering conditions around

Table 10.1 *Twenty-four-hour diurnal temperature variation in July.*

Height above surface (m)	Hour of day									
	1	5	7	9	11	13	15	17	19	21
	Irrigated oasis									
2	21.4	18.9	20.7	25.4	30.5	33.2	33.9	33.7	30.0	26.4
25	23.8	20.8	21.8	25.3	30.2	33.0	33.9	34.3	30.9	29.7
50	26.2	22.6	22.5	25.5	30.0	33.1	33.5	34.5	31.6	32.7
100	28.6	25.9	23.8	25.9	29.9	33.0	33.3	34.0	31.9	34.2
	Semi-desert									
2	23.0	19.9	23.1	28.4	33.5	37.0	36.7	37.8	33.9	29.4
25	24.5	21.4	23.5	28.1	33.4	35.6	35.3	36.5	33.5	32.2
50	26.2	22.6	23.9	28.2	33.4	35.0	34.7	36.3	33.0	32.9
100	28.6	23.9	25.1	27.8	32.9	34.6	33.9	35.8	32.8	33.1

Source: After Goltsberg (1969).

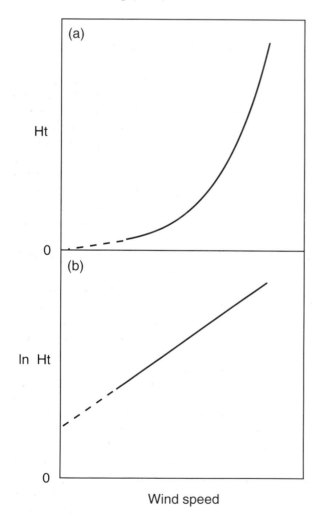

Figure 10.4 *(a) Wind speed profile near the ground. The precise shape of the curve will depend upon the roughness of the surface as well as any buoyancy. (b) The same profile plotted in a semi-logarithmic form, with the height axis converted to the natural logarithm of the value.*

a single leaf. The amount of short-wave radiation absorbed by a leaf depends upon the quantity of radiation reaching its upper surface, the angle between the leaf and the sun's rays, and the colour of the leaf. Through absorption the temperature of the leaf rises and, consequently, the amount of long-wave radiation emitted also increases. Some radiation is transferred downwards towards the soil, and some flows upwards. With a large number of leaves the sun's rays are increasingly obstructed, so the amount of sunlight reaching the ground may be small. The actual quantity depends upon the type and number of leaves (or leaf area index) and the crop height.

Because of its agricultural importance there have been numerous studies of the climate within crops. Agronomists and plant physiologists use the information in order to increase yields from plants best suited to the micro- and macroclimate in which they grow. It is now possible to determine the types of plants growing and to check their health by aircraft or satellite photography. The nature of the radiation reflected and emitted from leaves varies from one species to another, and from healthy to unhealthy plants, owing to alterations in the distribution of pigments in the leaves.

Temperatures in the vegetated layer

If we look at mean profiles of wind speed, temperature and humidity within a plant crop, there is some similarity with those found above a bare soil surface (Figure 10.6). In this instance the main heat exchange zone is found slightly below the canopy top rather than at the soil surface. As a result, daytime temperatures reach their maximum values within the

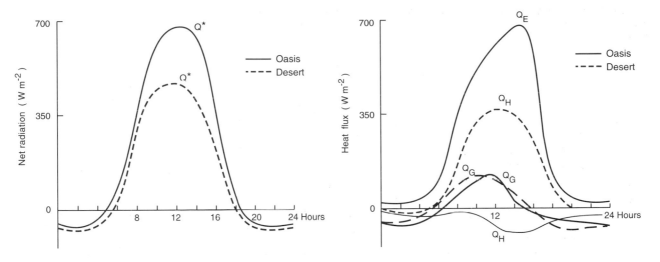

Figure 10.5 (a) Net radiation and (b) the energy budget in an oasis and a desert. Q_H is sensible heat flux, Q_E is latent heat flux and Q_G is soil heat flux.

canopy. The actual location represents a balance between the reduction in sunlight intensity as it penetrates into the crop and the decrease in wind speed and turbulence which would help to remove the heated air. At night, under clear skies and with light winds, long-wave radiation continues to flow from the leaf surfaces, but only that from the upper leaves is able to escape from the plant system. At lower levels in the crop, radiation is trapped and re-emitted, maintaining warmer temperatures. Thus the temperature profile has a minimum value just below the canopy top and gets warmer both upwards and down towards the soil surface. If the crop has a low density, with large gaps between plants, the air cooled by contact with the radiating leaves becomes denser and sinks towards the ground to give the minimum temperatures there. In the soil, temperature changes are much smaller because surface heating and cooling are greatly reduced through shading by the leaves.

Wind in the vegetated layer

The wind speed profile is also more complex, owing to the presence of the crop. Its precise form depends upon the nature of the crop and the prevailing wind speeds. By day, there is normally a progressive decrease in speed as far as the middle canopy. Below that level, most crops have fewer leaves, enabling the wind to blow through the crop. So we get a slightly windier zone before the final decrease towards the soil surface. This effect can be felt behind a hedgerow or windbreak, where the stems are not so effective at reducing wind speed as the leafier branches at higher levels.

Moisture in the vegetated layer

Daytime humidity levels usually show a progressive decrease from the soil surface, through the crop, into the atmosphere. Moisture is evaporated from the soil and transpired by the plant leaves, so that the main moisture sources are within the crop. As wind speeds are low, much of the moisture remains within the vegetation, but that in the upper layers may be carried away by convection and turbulence to mix with the drier air above. At night the shape of the humidity profile is more complicated. Cooling may give rise to dewfall on the upper leaves, producing an inverted profile for a short distance, but normally humidity differences are relatively weak throughout the crop.

Within a plant canopy, moisture exchanges are extensive and of vital importance to the well-being of the crop. In reality these processes are highly complex, but we can get an idea of the exchanges by constructing a simple model of the water balance. Figure 10.7 shows the inputs and outputs of moisture we might expect with an ideal crop. The major input of most climatic régimes is precipitation in the form of either snow or rain, but hail, dewfall, frost and fog can add small amounts. Some of this moisture is intercepted by the leaves. Depending upon the intensity and duration of the precipitation and the nature of the leaf, the water may drop off the leaves, or be directly evaporated without ever having reached the ground surface. This effect is greatest when the rainfall is light and the leaf density high. Small quantities of moisture may flow down the stems of the plants, but with heavy or prolonged rain some droplets will fall right through the crop to moisten the soil surface, eventually reaching the plant roots.

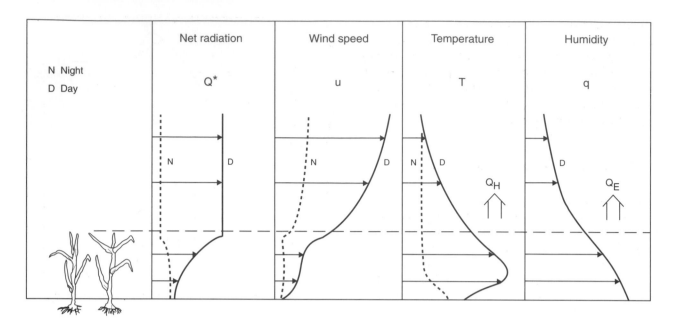

Figure 10.6 *Typical profiles of net radiation, wind speed, temperature and humidity above and within a plant canopy. Dotted lines show the night-time profiles and solid lines those for daytime.*

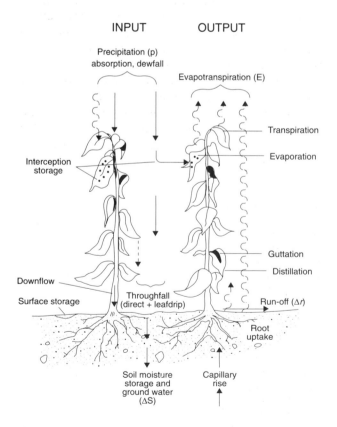

Figure 10.7 *The hydrological cascade in a soil–plant–atmosphere system. Source: After Oke (1987).*

The output of the system is primarily through transpiration from the leaves and evaporation from both soil and leaves. Moisture is extracted from the soil to maintain transpiration, but if the soil becomes too dry during droughts the plants may wilt or even die. During periods of rain, input is usually far higher than evapotranspiration alone. This surplus goes to recharge moisture in the soil or it becomes run-off – the horizontal flow of moisture on the soil – which eventually forms part of the river system.

The microclimate of woodland

So far we have been dealing with the microclimate either on, or at least close to, the ground surface. From it, we have been able to illustrate the processes controlling the climate at that level. As the crop or vegetation gets larger, so the degree of modification increases and the active zone extends from the higher canopy down to the soil surface. The extreme example of this effect is seen in mature forest. So much has been written about the microclimate within a forest that the term *forest climate* is frequently used to indicate the wide variety of conditions that can be experienced.

On a hot summer's day it is noticeable that temperatures in a forest are much lower than outside, providing a respite from the strong glare and baking heat of the sun. Air movement is weak. It feels humid, and the impression is quickly gained of an entirely different climate. This affects plant and animal life as well as people. Quite different ecosystems develop because of the climatic environment produced by the forest. Because of the differences in scale, the microclimates within a forest are more distinct than those in grassland or low crops.

Radiation exchanges in woodland

It is apparent on entering a forest that the forest canopy cuts out much of the incoming radiation. Most of the energy is absorbed by the tree canopy. A significant proportion is reflected – about 5–15 per cent, on average, although in some cases reflection may reach over 30 per cent (Table 10.2). Only a very small proportion reaches the ground directly, normally in the form of small patches of light called sunflecks (Figure 10.8). The remainder penetrates the vegetation indirectly; it is scattered by the atmosphere and arrives as diffuse radiation.

Spectral changes

During the progress of radiation through the forest vegetation, considerable changes in its spectral composition take place, as specific wavelengths are filtered out or scattered by the canopy. The shorter wavelengths (i.e. blue light) are removed preferentially by the leaves, while the amounts of longer-wave red and infra-red radiation increase. This change in the composition of the light is responsible for the characteristic colours that we encounter in woodlands. It also makes the light less suitable for plant growth. As a result, the range of plants that can survive on the forest floor is limited.

The woodland affects not only the inputs of radiation; it similarly affects outputs. The manner of this modification is far more complex, for outgoing long-wave radiation comes from a wide range of sources – from the atmosphere, the top of the canopy, from the leaves and branches of the trees, from the undergrowth and from the soil surface. There is inevitably a great deal of interception, absorption and re-emission of the long-wave radiation, so that little escapes direct to space.

Variations over time

These patterns of microclimate are only averages. Considerable variations occur over time, owing to changes in the inputs of solar radiation and to changes in the woodland itself. If we measured short-wave inputs of radiation throughout the day we

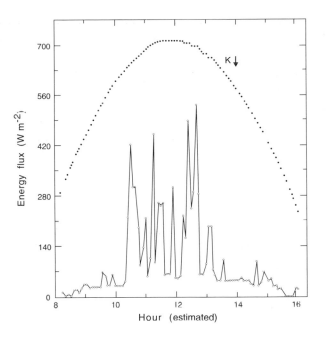

Figure 10.8 *Global solar radiation (K↓) above a pine plantation and at one point on the forest floor on 30 October 1965. Source: After Gay et al. (1971).*

would find that levels remained low with the exception of brief periods associated with the passage of sunflecks; the peak intensity would occur about midday (Figure 10.8). During the night, cooling is slow, for the vegetation traps and returns much of the outgoing long-wave radiation.

This pattern also changes seasonally. In winter the inputs of radiation are low and the effect of the forest on the microclimate diminishes. Moreover, in deciduous woodland, the trees lose their leaves, so that there is much less interception and absorption. If we compare woodland temperatures with those on open

Table 10.2 *Tree albedos (%).*

Aleppo pine	17
Monterey pine	10
Loblolly pine	11
Lodgepole pine	9
Scots pine	9
Oak	
Summer	15
Spring	12
Eucalyptus	19
Sitka spruce	12
Norway spruce	12
Birch and aspen	
Late winter	25
Orange trees	32
Tropical rain forest	13
Cocoa	16

land, therefore, we find a much smaller difference in winter. The effect of the woodland is at a maximum when the trees are in full leaf and radiation inputs are high (Figure 10.9).

The effects of woodland type

The microclimate of woodland depends very much upon the type of woodland we are dealing with. For example, deciduous trees show a strong seasonal change compared with conifers. But considerable variations occur between different species of deciduous trees. Birch leaves, for examples, are small and have a lower density than beech or oak, so that, even when they are in full leaf, birch trees allow more light to reach the ground surface. As a result more plants grow on the woodland floor. Similarly, pine trees give a less dense canopy than do spruce; the dark, unvegetated floor of plantations of Sitka spruce contrast with the much lighter conditions in pine woodland.

In addition, the nature of the understorey is important. An open canopy allows the development of one or more layers of understorey plants, and these, too, intercept both incoming and outgoing radiation. The extreme example is shown by the tropical rain forest. Although radiation inputs are high, the successive layers of trees, bushes and shrubs intercept so much radiation that only about 1 per cent reaches a height 2 m above the ground. Less than 0.1 per cent may reach the forest floor.

Winds in woodland

Patterns of wind in woodland are similar to those in grassland, although the zone of modified flow extends to a much greater height. Above the canopy wind speed may increase slightly, but, as the canopy is approached, velocity falls rapidly. Lowest wind speeds are often found within the leafy canopy and, where the undergrowth is also dense, velocities may remain low. In most cases, however, the main trunk zone is more open, so there is less interference with

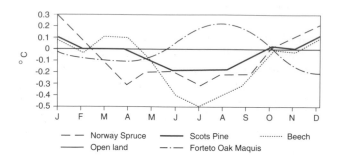

Figure 10.9 *Mean monthly forest temperatures compared with thermal conditions in the open. The woodlands are normally cooler in summer and slightly warmer in winter. The anomalies are beech in spring because of the late opening of the leaves and the very dense oak maquis, which has little transpiration in summer. Source: After Smith (1975).*

air flow and wind speeds increase again. Near the ground, friction and the effect of low-growing plants cause velocity to fall once more. Complex patterns of flow often develop in the forest, with local funnelling and deflection of the wind. We can often see the results of these flow patterns in the distribution of dead leaves on the woodland floor. Sheltered areas trap deep layers of leaves, which, by decay, will add nutrients to the soil, while more exposed zones are swept clear by the wind.

Moisture in woodland

In general, vapour pressure is slightly higher in a forest or in woodland than outside it. This is mainly due to the large amount of leaf area in a forest, transpiring moisture into the atmosphere which is not easily dispersed because of the lighter winds. On the other hand, the interception of moisture by vegetation reduces the amount of water available at the forest floor, so the net effect on humidity levels is small.

As daytime temperatures are cooler than those outside, the relative humidity of the air should also be greater even if the forest atmosphere contained the same absolute amount of water vapour. Experiments suggest values about 5 per cent above those outside, though the precise differences depend upon the type of woodland as well as on the time of year and the weather conditions (Table 10.3).

Table 10.3 *Difference of relative humidity (%) between the inside and outside of a forest.*

Forest	January	April	July	October	Year
Deciduous broad-leaf	3.4	3.2	−0.8	1.1	2.2
Needle tree (conifer)	4.8	4.8	6.5	9.5	6.8
Japanese cedar	1.6	−1.1	1.5	0.5	0.8

Source: After Yoshino (1975).
Note: Positive values indicate that inside the forest is more humid.

Urban climates

The climate modifications found in woodland are small compared with what happens when cities are built. Instead of a mixture of soil and vegetation, Earth is covered with a mosaic of concrete, glass, brick, bitumen and stone surfaces ranging to heights of several hundred metres. Amongst this, grass surfaces and a few trees are scattered to variegate the 'concrete jungle'. The building materials have vastly different physical properties from soil and plants. For example, the warmth of concrete and brick on a summer's evening is due to their high heat capacity. This means that as large quantities of heat are added to the material while the sun is shining it is slowly released during the night, adding warmth to the urban atmosphere. In this way city temperatures are kept relatively high. We notice the effect most in the evening when we travel from the cool of the countryside to the heat of the city (Figure 10.10). It is an effect called the *urban heat island*. Early blooming of flowers and decreased snowfall and frost are both indicators of this effect.

The urban heat island

We can illustrate the different responses of the city and rural areas by comparing their heat budgets as shown in Figure 10.11. It is the change of the heat budget by the urban surface which helps to produce the distinctive urban climate, so let us look in more detail at the way changes are produced. By day, both rural and urban areas experience a radiation surplus. Smoky urban atmospheres may reduce the size of this surplus slightly, but as the quality of urban air

Acid rain

Climatology has a number of misleading terms. We have already mentioned that the *greenhouse effect* of the atmosphere should really be called the 'enhanced greenhouse effect' because it supplements the natural processes operating. Similarly the term *acid rain* is used to indicate precipitation which is more acidic than normal – but even pure rainfall is acidic, with a pH of about 5.5. Neutral water would have a pH of 7.

Pure rainwater is slightly acidic because it absorbs some of the carbon dioxide from the atmosphere to form dilute carbonic acid. Levels of pollutants in the atmosphere have increased greatly recently as a result of human activities, particularly the burning of fossil fuels. Large quantities of sulphur and nitrogen oxides are added to the atmosphere (Figure 1). These gases react with water vapour and sunlight to produce nitrates and sulphates. Some of the pollutants are deposited directly from the atmosphere as dry deposition and some are absorbed into the precipitation process and reach the ground as rain or wet deposition. It is the wet deposition which is measured as acid rain, with pH values as low as 2, but both processes add acidity to the ground surface.

The source of much of this pollution is industry and urban areas. However, it is not just urban areas which receive acid rain. The gases released are dispersed by the winds, and levels of high rainfall acidity extend over considerable areas. Over Europe the highest levels of acidity are to the centre and east, with mean values below 4. It is believed that heavy industry and lignite burning are the main contributors to this peak. As the prevailing winds in Europe are from the west, it is not surprising that the most acid rain occurs towards the east. Wind is not the only factor in dispersal. Atmospheric stability will determine whether pollutants remain concentrated or are dispersed. The westerly winds tend to be relatively unstable and so allow greater vertical mixing; southeasterly winds, though less common, are often relatively stable, and pollution concentration can remain high. Much of Scandinavia's pollution is brought from central Europe in this way. Pollution which may start as a local problem can cross national boundaries to become a regional issue.

The effect of this increased acidity has been debated. It has been argued that the biosphere, human health and building materials can suffer from its effects. What does seem clear is that the ecological effects of acid rain will depend upon the ability of an ecosystem to neutralize the incoming acid. This ability is known as its *buffering capacity* and it depends largely upon the amount of calcium or magnesium in the bedrock. Levels of these minerals are generally low in much of the recently glaciated areas of northern Europe, so lakes in that area have been particularly affected by acid rain. Fish stocks have dropped dramatically, but other aquatic organisms have been affected. The causes are complex. One factor is believed to be the release of aluminium as a result of acid water reacting with heavy metal cations in the soil. Aluminium can affect fish by obstructing their gills.

The other area of impact of acid rain is on vegetation. Acid rain will increase soil acidity, decrease nutrient availability, mobilize toxic metals like aluminium and affect micro-organisms. It is not surprising that it has been held responsible for many changes in the terrestrial ecosystem. Perhaps the most drastic effect has been on forests. Monitoring of forests in Europe since 1986 has shown evidence of increases in the damage, especially to deciduous species. The damage takes the form of thinning of the crown, the shedding of leaves or needles and decreased resistance to disease, drought and frost. Pollution is not the only factor

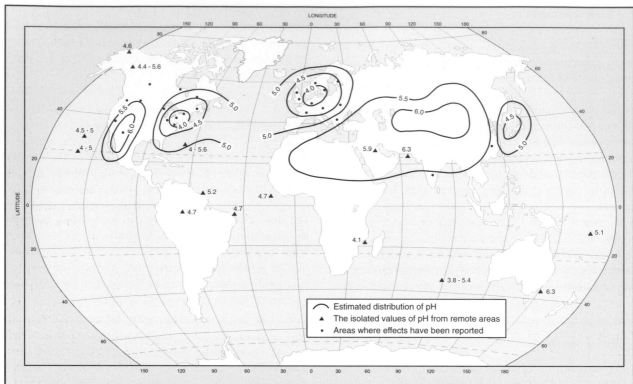

Figure 1 *Schematic representation of the formation, distribution and impact of acid rain.*

involved but is believed to be important. In Britain, Poland and the Czech Republic, it is suggested, more than 50 per cent of trees are showing signs of damage by way of defoliation.

Efforts to combat the problem can be made at source or in the environment affected. Some attempts have been made to reduce sulphur emissions at power stations, such as at Drax in the United Kingdom. Unfortunately they cost money, which makes the electricity more expensive to generate. Greater use is being made of natural gas as a source of energy

though this has more to do with economics than with environmental considerations. Alternatively soil or water acidity can be neutralized by adding lime. It has been argued that it is cheaper to add lime than to adopt expensive systems of reducing sulphur emission. Some success has been achieved in Scandinavia, where over 3000 lakes have been limed. Reduced acidity is followed by recovery of the biota; lower organisms reappear first, succeeded by amphibians and fish. Further liming has to take place every few years as long as the acid rain input continues.

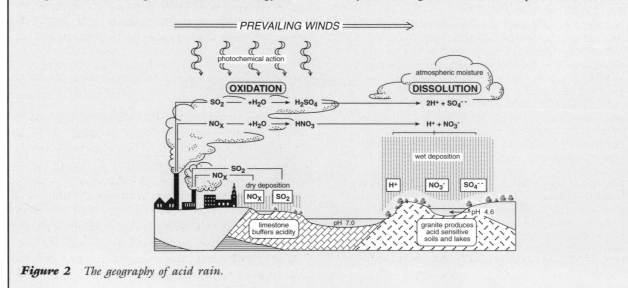

Figure 2 *The geography of acid rain.*

Acid rain is a complex problem which is likely to remain as long as the atmosphere is polluted. It is an international problem; the areas affected are not necessarily the source of the pollution. As with the enhanced greenhouse effect, we are still not certain in detail about the processes at work and hence prediction is difficult.

improves because of pollution controls the differences in inputs have become slight.

At a smaller scale the differences are more significant. Trees and crops allow a certain amount of radiation to pass through them to the ground surface. They transpire moisture and have a low heat capacity. As we saw earlier, this results in cooler temperatures beneath the canopy. In the city, the building materials of concrete, brick and stone all have high heat capacities, enabling them to store large amounts of heat. Shadowing can be important but there are still numerous surfaces exposing large, dry areas to the sun's rays. When the angle between the receptive surface and the sun's rays approaches 90° the heat input will reach its maximum. This effect is likely to occur much more frequently in an urban area, with its vertical walls, than in a rural area. Reflection from light-coloured buildings and glass can also add to the surface heat input.

Of the energy which is available as net radiation, some is used to heat the air, some is used in evaporation and the remainder is absorbed by the soil or buildings and other artificial surfaces. This is where the main contrasts arise. In a city, sewers and drainage systems lead to the rapid removal of water, and actively growing vegetation is infrequent. Surfaces soon become dry once rain has stopped, so the use of energy for evaporation and transpiration is small. This means that more is available for heating the air and the buildings than is being used for evaporation, which is 'non-productive' in terms of heating. A final factor can be significant in the city. Large amounts of fuel are used in industrial processes, to heat build-ings and for transport. Even human activity generates appreciable amounts of heat where population density is high, and all this heat is eventually released into the urban atmosphere (F in Figure 10.11). On Manhattan Island, New York, research has shown that, during the average January, the amount of heat produced from combustion alone is greater than the amount of energy from the sun by a factor of 2.5. In summer that ratio is only about 0.15.

At night the ground surface loses energy, resulting in cooling. In rural areas the ground becomes cooler than the air above, giving an inversion of temperature. There is then a weak transfer of heat to the surface from the soil and from the atmosphere, but these additions do not compensate for the radiational losses and so temperatures fall. In a hot summer this may feel refreshing compared with the sultry warmth of the city. There the buildings continue to give off heat which they have absorbed and stored during the day (Q_G in Figure 10.11) and, coupled with the heat of combustion (F in Figure 10.11), this reduces the rate of cooling. We can see this in Figure 10.12, where the cooling rates in central Birmingham are compared with a rural site. This relative warmth prevents the development of an inversion, so heat transfer and evaporation still take place. Dewfall or condensation is much less frequent than in rural areas. It is this urban heat, especially in the tropics and subtropics, which many city dwellers find so uncomfortable in the summer; it is why they long for the coolness of the countryside; and why, irritated by the conditions, they may tend to react violently.

The effect of winds

If winds were strong, all this surplus heat would be rapidly removed from the city, to be mixed with the cooler air around, and the urban climate would be less distinct. In climates where winds are light and clear skies predominate we find the greatest temperature differences between urban and rural areas (Figure 10.13). The pattern of night-time minimum temperatures usually shows highest values near the high-rise city centre, fairly uniform levels in the low-density suburbs and then a sharp boundary into the cooler rural areas. This is seen most clearly in cities with few relief features. Valleys, hills and parkland

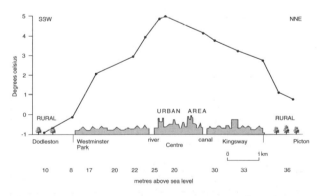

Figure 10.10 *Temperature cross-section of the urban heat island of Chester in relation to built-up area. Source: Nelder (1985).*

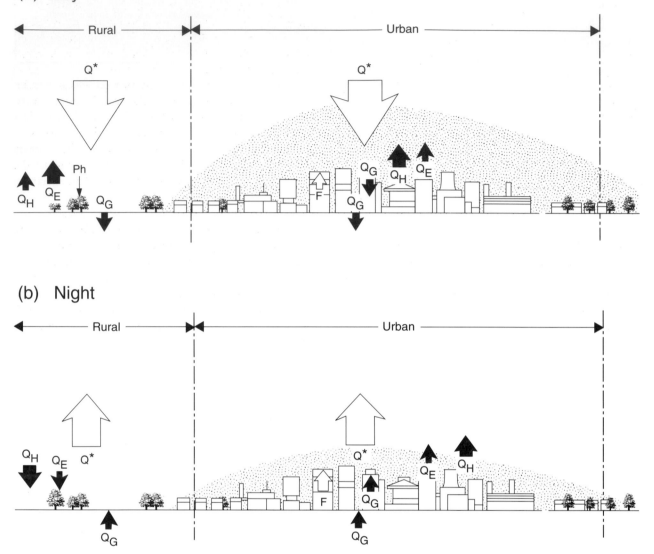

Figure 10.11 *Two-dimensional schematic diagram of the heat balances of urban and rural surfaces (a) by day and (b) by night. Ph is energy use in photosynthesis, F is artificial heat release. The other symbols are as in Figure 10.5.*

within the urban area can produce major changes. The parkland has different heat capacities, albedos, moisture levels and emission temperatures from the surrounding buildings, giving slightly lower day and night-time temperatures. The advantages of these 'urban lungs' extend well beyond their aesthetic appeal, especially during hot summer weather.

Even when winds are not light, the presence of the urban structure tends to slow air movement down. Wind records from city centre sites show lower average speeds than suburban or rural locations near by, although the degree of gustiness may be higher, especially in summer. As the air flows over the very irregular suface of a city, friction with the buildings retards the wind in the lowest layers (Figure 10.14).

The presence of skyscrapers, however, produces eddies which can cause strong local winds. At street level these can become quite unpleasant, raising dust, perhaps even rubbish, and making walking difficult. Quite a few shopping precincts have been unpopular with shoppers until the architects realized that such winds could be a problem and took measures to minimize their effects.

Cloud and precipitation in cities

Most of the climatic changes brought about by urbanization have been well documented. They are summarized in Table 10.4. Some of the changes are

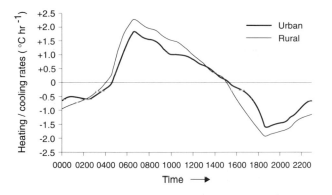

Figure 10.12 *Mean hourly warming and cooling rates in urban and rural areas of Birmingham during clear skies. Source: Johnson (1985).*

appreciable though the decrease in the use of coal as a fuel and power source has led to smaller modifications in insolation, pollutants and fogs. The increase in cloud and precipitation over cities was one aspect which took some time to prove. It was American work, especially on St Louis, which confirmed the urban effect conclusively. There appear to be multiple causes for the increases in cloud cover

and precipitation. Added heating of the air crossing the city, increases in pollutants, the frictional and turbulent effects on air flow and altered moisture all appear to play a role. The confluence zones induced by these urban effects may lead to the preferential development of clouds and rain. Which factors become dominant in a particular storm varies according to the nature of the air circulation over the city on that day. As the effects are less noticeable in winter than in summer, it follows that it is the natural, not the artificial, heating effects which are most important, though the way in which the summer atmosphere responds to the urban surface is also significant.

As the degree of urbanization has increased so an ever greater number of people are affected by an urban climate. Apart from the more obvious effects of pollution, wind and summer heat, few people may realize that their urban area has changed other aspects of the climate. The nature of the urban area represents an extreme example of the way in which human modification can change the climate near the ground.

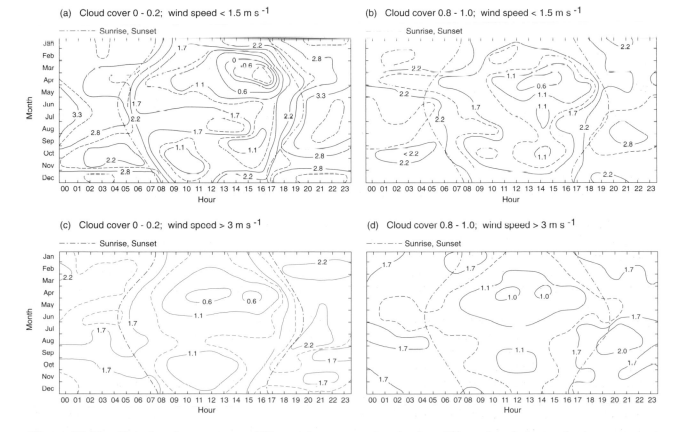

Figure 10.13 *Mean hourly temperature differences between rural and urban Chicago in relation to cloud cover and wind speed as a function of month and time. Source: After Ackerman (1985).*

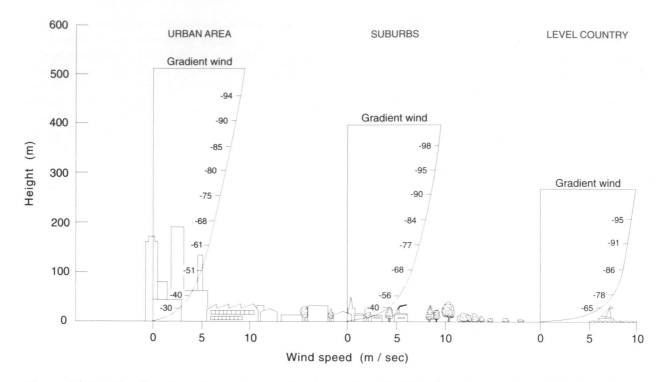

Figure 10.14   *The effect of terrain roughness on the wind speed profile. With decreasing roughness the depth of the modified layer becomes shallower and the profile steeper. Source: After Davenport (1965).*

The microclimate of slopes

So far, all examples quoted have assumed that the ground surface is almost flat. In reality few areas of the world are so level that the effect of topography can be ignored. The reason we need to know more about the topography is that slopes modify how much short-wave radiation reaches the surface. We saw earlier that the maximum intensity of radiation is received when the angle between the surface and the

sun's rays is 90°. If a horizontal surface is tilted so that it is at right-angles to the sun's rays the amount of radiation received increases. This factor is exploited by sunbathers, who can tilt the angle of their reclining seats to achieve maximum heat input. If it was the only factor, calculating the new input for a slope would be easy. However, while the slope remains constant, the sun is continuously changing its position in the sky throughout the day and throughout the year. Slopes, unlike sunbathers, cannot adjust their position. Consequently a slope that receives maximum intensity at one time on a certain day of the year may be in shadow at other times.

Effects on the radiation balance

As the movement of the sun across the sky is known, it is possible to calculate the intensity of short-wave radiation falling on a slope of any combination of gradient and orientation (azimuth) for clear skies. More frequently we are interested in the total radiation rather than the intensity but even this problem has been overcome using computers. A computer program can be devised to calculate the intensity of radiation on the surface for any particular time and slope. So, for the start of the program, radiation intensity is determined for sunrise, depending upon such factors as latitude,

Table 10.4   *Effects of urbanization on climate: average urban climatic differences expressed as a percentage* of rural conditions.*

Measure	Annual	Cold season	Warm season
Pollution	+500	+1000	+250
Solar radiation	−10	−15	−5
Temperature	+2	+3	+1
Humidity	−5	−2	−10
Visibility	−15	−20	−10
Fog	+10	+15	+5
Wind speed	−25	−20	−30
Cloudiness	+8	+5	+10
Rainfall	+5	0	+10
Thunderstorms	+15	+5	+30

* *Note:* Temperature is expressed as a difference only, not as a percentage.

time of year, altitude and atmospheric transmission. Then the computer calculates the sun's position in the sky, say ten minutes later, works out the new radiation intensity and adds its value to the previous total. This is continued until sunset or until the sun drops below the horizon (Figure 10.15). We then have the daily total of short-wave radiation based on intensity values every ten minutes. The contribution from diffuse radiation is assumed to be constant throughout the day and so does not add to the spatial variability of solar receipt at the surface. None the less it is vitally important for slopes with a northerly aspect, which would otherwise receive very little short-wave radiation. Moon explorers are able to see this, for with no atmosphere there is no diffuse radiation and any surface that is not directly in sunlight appears black.

These effects of slopes upon radiation inputs mean that the radiation balance varies locally with topography. In the northern hemisphere, slopes with a southerly aspect receive a greater input of radiation than northerly ones, resulting in larger exchanges in sensible heat and higher temperatures (Table 10.5). In high latitudes this additional energy may be an advantage in sunshine-starved areas, but in more arid countries the increased radiation will evaporate moisture more quickly and may produce moisture stresses in cultivated plants.

Slopes at night

At night, when there is no input of short-wave radiation, the effect of a sloping ground surface on the energy budget is less pronounced. Figure 10.16 shows the exchanges taking place. For slopes between 0° and 30°, emission of long-wave radiation follows the cosine law ($E_{sl} = E_{horiz} \cos \alpha$); at higher angles more radiation is emitted than would be predicted. The only effect of slope direction is in influencing surface heating during the day, which, through heat storage, may affect night-time temperatures and hence emission rates. If the sky is obstructed by trees, other valley slopes or even buildings, much of the long-wave radiation is absorbed and reradiated back to the ground. This reduces the rate of cooling from the surface. The effect can sometimes be seen in frosty weather, when open grassy surfaces are white, but, beneath trees or near buildings where counter-radiation has been greater, there is no sign of frost on the ground (Plate 10.2).

Of much greater importance at night is what happens to the air as it cools through contact with the ground surface. As the air becomes cooler it gets denser. If the surface is flat the cold air remains at

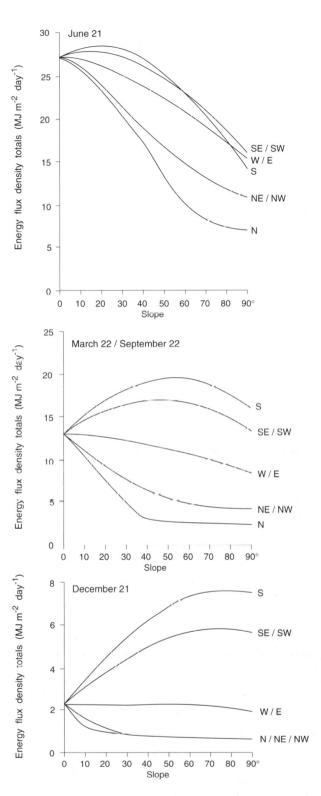

Figure 10.15 *Total daily direct and diffuse solar radiation incident upon slopes of differing angle and aspect at 53° N for (a) 21 June, (b) 22 March–22 September and (c) 21 December. Note different scale for December. Source: Based on a model developed by the Department of Building Science, University of Sheffield.*

Table 10.5 *Influence of slope orientation on microclimate.*

Orientation	After five dry days	After two rain days
	Maximum temperature	
N	−1.9	−1.5
E	−1.3	0.0
S	2.6	1.4
W	0.5	0.2
	Minimum temperature	
N	−0.3	−0.4
E	−0.1	−0.4
S	0.4	0.3
W	0.0	0.5
	Daily mean temperature	
N	−0.9	−0.4
E	0.1	−0.3
S	1.1	0.6
W	−0.4	0.2
	Relative humidity at 13.00 (%)	
N	8	1
E	3	5
S	−13	−3
W	6	−4

Note: Figures are relative to a horizontal surface near by.
Source: Translated from Fuh Baw-Puh (1962).

ground level to give the normal temperature inversion. However, on a slope the cool air may move downslope as a katabatic wind or density current, increasing in strength and volume until it meets a physical barrier, such as a wall or embankment, or until it is no longer colder than its surroundings. Once the cold air stops

Plate 10.2 *Effect of trees on night-time temperatures. The area beneath the tree is free of frost.*
Photo: Peter Smithson.

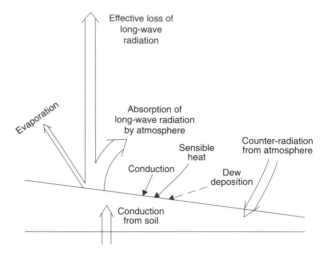

Figure 10.16 *Night-time energy exchanges. The magnitude of the components will vary greatly, depending upon weather conditions such as cloud amounts, wind speed and humidity.*

moving it continues to cool through long-wave radiation emission and may eventually reach very low temperatures. This microclimatological effect can be very pronounced on clear, calm nights which allow radiation cooling to continue at a high rate.

One result of this process is the formation of frost hollows. Farmers should always take care that frost-sensitive crops are not grown where cold air is likely to accumulate and give ground or even air frosts. It is for this reason that, in frost-susceptible areas, fruit orchards are cultivated on valley slopes, allowing the cold air to drain through the trees without accumulating. A classic example of a frost hollow was found in the Austrian Alps. A limestone sink-hole with a steep back wall facing north-east allowed cold air to become stagnant. Figure 10.17 shows temperatures at different levels on one particular night. Towards west-south-west the sink-hole is intersected by a col which allows the stagnant cold air to remain in the lowest 50 m of the hollow. Temperatures as low as −51° have been recorded when the ground was snow-covered. Even coastal Antarctica is usually much warmer than that! The frequent occurrence of frost has affected vegetation, so that few trees grow near the base, to give an inverted vegetation gradient.

Valley-breeze systems

If the katabatic winds, described above, are not prevented from flowing they begin to form an organized system of cold air drainage downslope and down-valley. Speeds are low, perhaps 1 m per second or less,

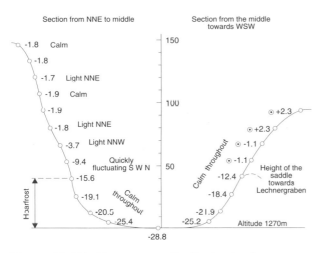

Section from NNE to middle

Section from the middle towards WSW

-1.8 Calm
-1.8
-1.7 Light NNE
-1.9 Calm
-1.9
-1.8 Light NNE
-3.7 Light NNW
-9.4 Quickly fluctuating S W N
-15.6
-19.1
-20.5
-25.4

Calm throughout

-28.8

Hoarfrost

150
100
50

+2.3
+2.3
-1.1
-1.1
-12.4
-18.4
-21.9
-25.2

Height of the saddle towards Lechnergraben

Altitude 1270m

Figure 10.17 *Temperature distribution in the Gstettneralm sinkhole near Lunz, Austria, 21 January 1930. Source: After Schmidt (1930).*

and the movement tends to pulsate with intermittent surges – like that which can be seen in water running down a sloping road surface. The downslope flows eventually combine into a down-valley flow, known as a mountain wind, as it emerges on to the lowlands.

By day this cold air drainage does not occur, except where snow and ice surfaces maintain cooling. Instead it is replaced by anabatic winds upslope. These are produced by heating on the slope which causes the warm air to rise upslope. Cool air from the valley floor flows in to replace the warm air and a valley breeze is generated (Figure 10.18). These valley breeze systems could not last long if the continuity of the flow was not maintained. This is usually found as a counter-wind at higher levels. If the pressure gradient wind is strong it increases local mixing so that major temperature differences are prevented. No cold air is available to sink downslope or warm air to rise upslope, so

the formation of the breeze is stopped. Like so many microclimatological phenomena, valley and mountain breezes require clear skies and light winds for this operation.

Sea breezes

The driving force of the valley and mountain breezes is a temperature gradient. Temperature contrasts develop between slopes and valley floors, between uplands and lowlands, so that the nature and strength of the wind depends upon the precise form of the gradient. This thermal control of winds occurs at all scales, from the general circulation of the atmosphere (Chapter 8) down to the smallest eddy of heat rising from the ground. We have already referred to one wind system which forms at the local scale, but an even more widespread thermally driven wind at this scale is the sea breeze.

Sea breezes are formed by the different responses to heating of water and land. If we have a bright, sunny morning with little wind, the ground surface warms rapidly as it absorbs short-wave radiation. Most of this heat is retained at the surface, although some will be transferred through the soil. As a result, the temperature of the ground surface increases and some of the heat warms the air above. When the sun sets, the surface starts to cool rapidly, as there is little store of heat in the soil. Thus we find that land surfaces are characterized by high day (and summer) temperatures and low night (and winter) temperatures.

The response of the sea is very different. First, sunshine can penetrate through the water to about 30 m, as any skin diver knows. Second, water has a large heat capacity, so a lot of solar energy has to be absorbed to raise its temperature. In addition, the

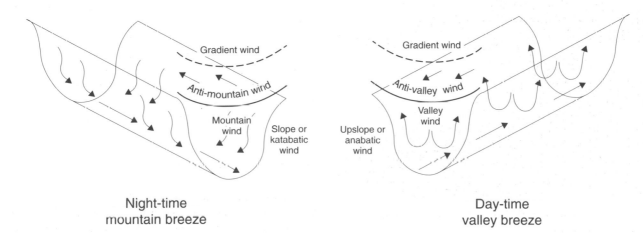

Night-time
mountain breeze

Day-time
valley breeze

Figure 10.18 *Schematic diagram of mountain and valley breezes.*

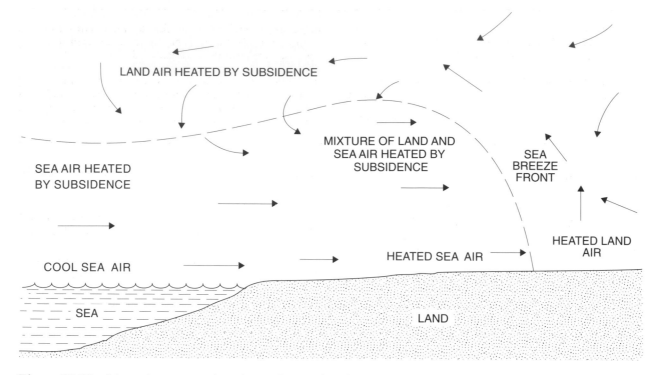

Figure 10.19   Schematic representation of a sea breeze when the geostrophic wind is light.

warming surface water will be mixed with cooler deeper water through wave action and convection. Instead of a thin active layer such as we have in a soil, the top 20 m or so of water forms the active layer; consequently temperature changes are slow. Slight warming occurs during the day and slight cooling at night. This means that the sea is normally cooler than the land by day and warmer by night. (On a longer time scale, the sea is cooler relative to the land in summer and warmer in winter unless there are unusual currents offshore.)

The higher temperature over the land by day generates a weak low-pressure area. As this intensifies during daytime heating, a flow of cool, more humid air spreads inland from the sea, gradually changing in strength and direction during the day (Figure 10.19). At night the reverse circulation evolves, with a flow of air from the cooler land to the warmer sea, though as the temperature difference is usually less, and the atmosphere stable, the land breeze is weak. At higher levels we find a flow in the opposite direction, compensating for the surface land or sea breeze. Even large lakes can show a breeze system of this nature.

In tropical areas the strength and reliability of the sea breeze bring a welcome freshness to the climate of the coastal margin and its effects can extend up to a couple of hundred kilometres inland.

Key points

1 The atmospheric processes of radiation, convection, evaporation and advection interact with themselves and with the variable nature of the ground to produce a mosaic of microclimates. Distinctive effects can be found at a wide variety of scales in increasing size from the microclimate of a single leaf, through crops, forest, valley slopes, urban areas and sea–land breezes. They are the product of the variable interaction between the energy exchanges and the ground surface.

2 In most cases there is no firm boundary between scales: the micro- and local climates form part of a continuum or spectrum from smallest to largest. Certainly within the larger scales like urban climates there would be innumerable microclimates resulting from surface modification. This diversity makes their investigation fascinating.

3 Equally it presents problems of explanation and interpretation, as it is physically impossible to measure the wide variety of possible microclimates and it is easy for so-called understanding to degenerate into a series of case studies. A final understanding (if there is such a thing!) will come only when we appreciate the interactions and links between the myriad of atmospheric processes and surface conditions.

4 The importance of microclimatic modifications goes far beyond the study of climate. It is at this scale that we can see the relationship between climatic processes, landscape and ecosystems. Landforms and vegetation modify the microclimate; the microclimate in turn controls many of the processes involved in landscape and soil development and plant growth. Here, as in so many cases, we need to remember that the world does not fall as conveniently into compartments as students (and authors of textbooks) would sometimes like! It may make the study of geography rather complicated, but it also makes it intriguing.

Further reading

Hanwell, J.D. and Newson, M.D. (1973) *Techniques in Physical Geography*, London: Macmillan (Chapters 2 and 3).
An elementary and practical book demonstrating useful techniques in meteorology and local-scale climatology.

Oke, T.R. (1987) *Boundary Layer Climates*, second edition, London: Methuen.
An intermediate to advanced level book demonstrating the significance of the ground surface in determining micro-climate. Very clearly presented but still needs careful reading.

Rosenburg, N., Bled, B.L. and Verma, S.B. (1983) *Microclimate: The biological environment*, second edition, New York: Wiley.
Looks at microclimate from a biological viewpoint stressing the meteorological factors responsible. Nevertheless it is a clear exposition of the nature and causative factors of microclimate. Particularly good on evapotranspiration and selected environments such as shelter belts.

So far in our discussions concerning the atmosphere it has been the understanding and description of the nature and controls of our *present* climate system that have been stressed. Mathematical and physical models of the atmosphere have been developed which allow us to reproduce the main features of the atmospheric circulation and global climates (Figure 11.1). We know that some climatic controls have changed in the geological past and are likely to change in the future, so we would expect climate to vary in response to these changes. By varying the details of the model for any given time period we should be able to predict what the climate was like or should be on the basis of the model's predictions. These predictions can then be compared with what is known about former climates to see whether there is reasonable accordance between the two. By being more certain about what controlled the climate in the past and at present, we can be more certain about what will happen to climate in the future.

What is meant by a change of climate? If we plot the annual rainfall totals at a particular site through time (Figure 11.2), it is apparent that totals vary markedly from year to year. At some sites the pattern may be entirely random, or we may find oscillations between wetter periods and drier periods, with no long-term trend. Unfortunately the length of instrumental climatic measurements at most sites is short and so it is impossible to reach clear conclusions about whether the recent climate has changed in a statistical sense or whether it merely demonstrates a very variable climate. Over longer periods of time, it is apparent from various lines of evidence that major changes of climate have taken place. In Britain we can find evidence of glaciation and even further back in time tropical forests appear to have thrived during the time the clays of the London basin were being deposited about 60 million years ago.

Initially we will look at the changes which have taken place in the climate, particularly over western Europe, and concentrate on the recent geological period. In earlier times the continents, as plates, have changed their positions on the globe and therefore

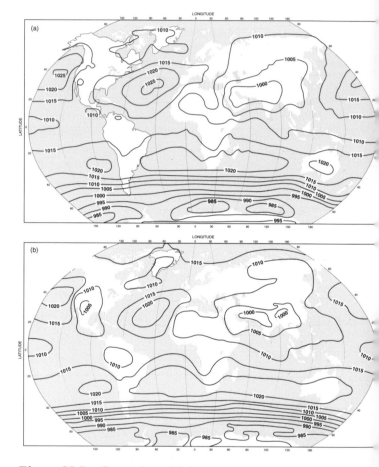

Figure 11.1 *Comparison of (a) summer mean sea-level pressure patterns and (b) model predictions from a general circulation model. Source: After IPCC (1990).*

changes in climate could be caused by this process rather than by atmospheric effects (see Chapter 3).

Evidence for climatic change

Glacial periods

Land and oceanic sediments record clear evidence of numerous alternations between warmer and colder

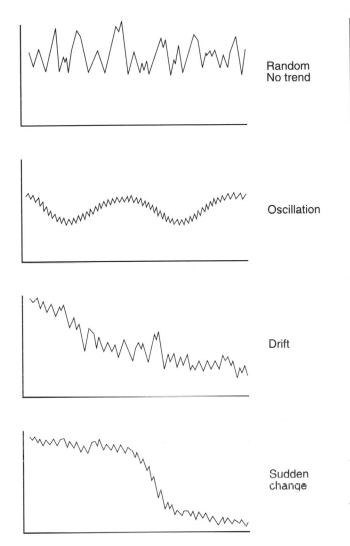

Figure 11.2 *Possible types of precipitation trends.*

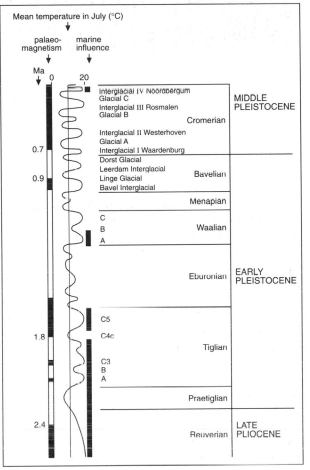

Figure 11.3 *Estimated July temperatures during the early and middle Pleistocene period, showing the oscillations of glacial and interglacial conditions.*
Source: *Jones and Keen (1993).*

conditions over the last two million years. At least eight such cycles have occurred in the last million years, with the warm part of the cycle lasting only a relatively short time (Figure 11.3).

During the cold phases, ice advanced across much of north-west Europe from Scandinavia and across much of North America from centres over northern Canada. It appears that ice occupied these areas for only a short period of time; for the remainder of the cold phase the climate was cold, dry and generally ice-free. In the tropics, lake levels indicate that the glacial periods were generally arid, with sand dune formation in parts of Africa. At the same time, sea level fell by about 100–50 m because so much water would have been locked up in the form of ice in ice caps, ice sheets and glaciers. Oceanic currents almost certainly changed in intensity and even direction to respond to the new conditions.

The last glacial period reached its maximum intensity about 18,000 to 16,000 years Before Present (BP). This term is used whenever dating is based on radiometric methods involving isotopic decay such as Carbon-14 (radiocarbon) or potassium-argon (K-Ar). After this time, much more information is available about changes of climate, based on a variety of types of evidence. As the information is indirect rather than being the result of direct measurements, it is termed *proxy* evidence. Vegetation and animal remains preserved in sediments provide the bulk of information about former climates. The distribution of plants and, to a lesser extent, animals is determined in part by the prevailing climate. Where we find evidence that plants or animals were living in the immediate area it is assumed that the conditions under which those plants or animals live today must have prevailed at that time and location in the past. The present is being used as the key to the past: the *principle of*

Plate 11.1   *Former shoreline on the Ardakan Playa, Yazd, central Iran.*
Photo: D. Mehrshahi.

uniformitarianism is being applied. Pollen is one of the best indicators of former climates, though many assumptions have to be made. In general, more confidence can be placed in a conclusion if it is based on several lines of evidence rather than a single one.

The advances and retreats of glaciers can be used to interpret changes of temperature and moisture régimes. Former lake shorelines may indicate changes in former moisture conditions (Plate 11.1). Ice cores have been taken through the Antarctic and the Greenland ice sheets, where estimates of seasonal climatic conditions may be interpreted from the layers of ice. Even the nature of atmospheric composition may be determined from the content of air inclusions within the ice. Another source of information about former climate is the oxygen isotope composition of some microscopic marine shells called foraminifera.

Information from such proxy data indicates that at about 18,000 BP large parts of the northern hemisphere, especially Europe and North America, were ice-covered (Figure 11.4). Antarctic ice advanced farther equatorwards but, because the southern hemisphere continents have small temperate-zone elements, ice developed only in highland areas, to increase glacier size and frequency. Sea-surface temperatures as estimated from foraminiferal remains and oxygen isotope analysis indicate major decreases in some areas, such as the north-east Atlantic, where the warm oceanic currents changed their position and temperature changes of up to 10° C occur. In other areas, like the South China Sea, the difference was less marked. Temperatures of about 27° C predicted for northern hemisphere summer off Hong Kong (Figure 11.4) compare with about 28.5° C today.

Post-glacial period

The climatic amelioration following the last glacial maximum was rapid though not without fluctuations. In the southern Lake District of Britain, organic mud was being deposited by 14,500 BP, indicating that the ice sheet was beginning to thin rapidly or had disappeared from the area. The rate of warming was so sudden that in many cases the vegetation was out of equilibrium with the climate, as can be deduced by comparing vegetation and insect evidence. It takes years for trees to spread from their refuge areas, but animals and insects can move more quickly in response to a change in climate as long as their food supply moves too. By about 12,200 BP, it is believed, the climate of Britain was similar to that at present (Figure 11.5).

This warmth did not last. Unlike previous glacial/interglacial transitions, as far as we can tell, a major ice advance appears to have occurred in north-west Europe and the former USSR, with possible signs of cooling in other parts of the world. In Britain cirque glaciers became re-established in many parts of the uplands, and in western Scotland ice advanced towards the lowlands near Glasgow. Mean July temperatures fell below 10° C and trees temporarily disappeared from Britain. This brief period of about 800 years is called the *Loch Lomond stadial*. It was too short for extensive glacier growth but the piles of gravelly debris in many mountain cirques show the deposition which took place when the cirque glaciers melted (Plate 11.2).

Following the final retreat of the continental ice sheets from Europe and North America between 10,000 and 7000 BP the climate rapidly ameliorated in middle and high latitudes. A thermal maximum was reached about 5000 years ago, when temperatures are believed to have been 1–2° C higher than those of today. Lake levels in tropical areas indicate moist conditions in the early part of this post-glacial period, with a general decline in levels subsequently. Evaporation and earth movements also affect lake levels, as well as precipitation, so climatic interpretation is not always easy.

The thermal maximum was followed by a period of slowly declining temperatures and fluctuations in precipitation. Within the general cooling there were cooler periods such as around 2000 years ago and warmer ones, for example about AD 1000 to about 1200. At that time there were few severe winter storms in the Atlantic. The Vikings took advantage of this quieter period to colonize Iceland and Greenland and probably visited North America. By AD 1200 cooling began to set in, with increased

Figure 11.4 *Sea-surface temperatures, ice extent (dark stipple) and ice elevation for southern winter, 18,000 years ago, with sea level at −85 m. Source: After CLIMAP Project (1976).*

storminess. In at least four major sea floods of the Dutch and German coasts in the thirteenth century the death toll was estimated at more than 100,000. At the same time, drought was starting to affect Indian settlements in Iowa and South Dakota, but in parts of China moister conditions prevailed. An increase in strength of the westerly circulation in the northern hemisphere has the effect of decreasing precipitation to the lee of the Western Cordillera of the United States but can lead to an increase where jet streams converge after splitting around the Tibetan plateau, so the differences are not contradictory.

The Little Ice Age

After a partial return to more favourable conditions from 1400 to 1550, the climate grew much colder again and for 300 years Europe experienced a cold spell. It is known as the *Little Ice Age*. It began the first period in which instrumental observations could be used to measure climate change. In central England the mean annual temperature in the 1690s was only 8.1° C – about 1.5° C below the current

figure. Agriculture in upland areas became more difficult as the growing season shortened, leading to the abandonment of many farms, whose land often reverted to moorland or rough grazing. An added problem during this period of cooler temperatures

Plate 11.2 *Mounds of glacial moraine deposited during the Loch Lomond stadial at Cwm Idwal, North Wales, in the foreground, with the cirque rockwalls behind.*
Photo: Kenneth Addison.

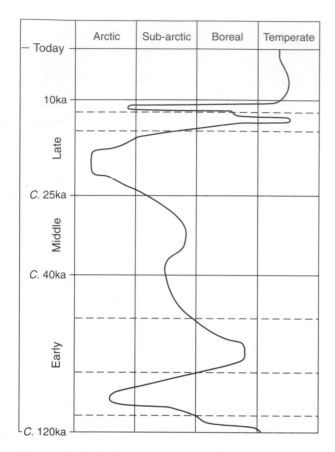

Figure 11.5 *Estimated mean summer temperatures over the British Isles for the last 120,000 years. Source: Jones and Keen (1993).*

appears to have been enhanced variability of temperature. It was not merely a swing from one year to another but a period of several successive years with similar temperature and precipitation before a change to a period of a markedly different character. Other parts of Europe were also affected. Glaciers advanced in the Alps, farms had to be abandoned in Iceland and Scandinavia; in upland Languedoc in southern France there were food shortages and famines associated with severe winters and wet summers. In Spain agricultural difficulties arose through increased aridity and temperature variability. Globally the extent of snow and ice on land and sea seems to have reached its highest levels since the Loch Lomond stadial period, though the timing of its culmination varies. It appears to have been earlier in the United States than in Europe or the southern hemisphere, whilst in China the coldest periods were around 1700 and 1875. Much of the evidence for cooling is based on ice advances. However, it is the combination of temperature and precipitation as well as the 'response time' of the glacier which determines advances rather than temperature alone.

Present climate

By the middle of the nineteenth century the effects of the Little Ice Age were waning in most parts of the world and we begin the steady warming of the last century. The first phase of warming peaked in the 1940s, followed by a slight decline in global mean temperatures. At that time climatologists were predicting the return of cooler conditions and perhaps even another Ice Age. From the mid-1970s the cooling trend reversed and mean temperatures rose suddenly and rapidly through the 1980s into the 1990s (Figure 11.6). Concern became directed towards the effects of global warming and the enhanced greenhouse effect rather than the imminence of the next Ice Age. The impact of a hemispheric mean temperature change of a few tenths of a degree may seem very small but change is not uniform. Some areas experience more significant increases or even decreases of temperature, while rainfall patterns may vary too. The most publicized example of what appears to be a significant recent change in climate has occurred in the Sahel area of Africa. We can see from the rainfall record that at certain times there have been sequences of higher than average rainfall followed by periods with lower than average rainfall (Figure 11.7). From the later 1960s rainfall had nearly always been less than the long-term averages, with years such as 1984 being spectacularly low. These changes may have major human impacts. The role of decreasing rainfall in desertification has been debated, but many of the countries affected by this trend have experienced much political and social upheaval in the last two decades, e.g. Niger, Sudan, Ethiopia and Somalia, to compound the problem.

Causes of climatic change

The summary of climatic history reveals that there are considerable variations of climate at any particular area over time. Many must be the result of natural processes acting on the Earth–atmosphere system, as they occurred well before human activity was sufficient to have an impact on climate. Some of the more recent ones could be the results of human impact on aspects of the system such as changing the composition of the atmosphere or the nature of the ground surface.

What are these processes which might lead to a change in Earth's climate? Why does the climate vary so much over time? These are questions to which we have no easy answer. There are at least four different time scales which require explanation: glacial/interglacial, stadial/interstadial, post-glacial oscillations and fluctuations over the last 150 years. In looking for causes, we can think of mechanisms which are external to Earth, those factors which are purely internal and feedback mechanisms which interact within the atmosphere or between the atmosphere and the Earth (Figure 11.8). Let us look at these in turn.

External factors

The global climate is the product of a complex system involving the atmosphere/hydrosphere/lithosphere/cryosphere. Changes can be forced upon the system by factors which may be either radiative or non-radiative, internal or external.

The most important external radiative factor is the sun. The sun may appear to us as a stable star but satellite observations of the solar beam intensity suggest small variations of output only partly connected with the well known eleven-year sunspot cycle. Long-term observations of sunspot numbers indicate that the cycle is varied in terms of the frequency of sunspots at the peaks of the cycles (Figure 11.9). From 1100 to 1250, 1460 to 1550 and 1645 to 1715 sunspot maxima were very low. One of these periods coincides with the peak of the Little Ice Age and another with a warm period. It seems unlikely that we should expect variations of more than 1 per cent in solar output as a result of the sunspot or other natural changes; simple calculations of Earth's radiative balance suggest that even a 1 per cent difference in output would lead to a change of only 0.6° C in mean annual temperature. Nevertheless, this small figure could be important in climatically marginal areas.

A more certain link between solar variations and long-term climate change has been established through the work of Milankovitch, a Yugoslavian mathematician. He determined the changes in solar radiation reaching Earth's surface as a result of orbital variations. Three interacting variations are known to occur, involving regular changes in (1) the shape of Earth's orbit around the sun; (2) the tilt of Earth's axis of rotation; and (3) the time of year when Earth is closest to the sun. The orbit of Earth around the sun is approximately elliptical. The nearest point of this orbit to the centre of the orbit is known as the *perihelion* (Greek *peri*, 'near' + *helios*, 'sun'), and is about 14.71×10^7 km from the sun. The farthest point is known as the *aphelion* (Greek *ap*, 'far' + *helios*), which is approximately 15.2×10^7 km from the sun.

At present the perihelion occurs on 3 January while the aphelion is on 4 July. The difference in distance

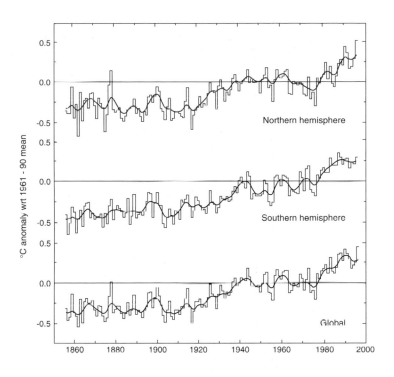

Figure 11.6 *Annual surface temperature anomalies (°C) from the 1961–90 average for combined land and marine records for the two hemispheres and for the whole Earth, 1856–1995. Smooth curve is a ten-year filter. Source: Phil Jones, Climatic Research Unit, University of East Anglia.*

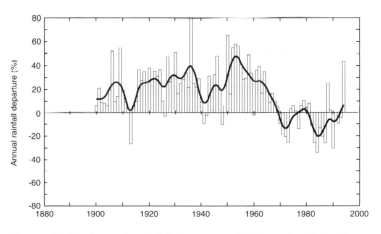

Figure 11.7 *Annual rainfall departures (%) from the 1961–90 average for the African Sahel, 10° N–20° N , 1900–94. Smooth curve is a ten-year filter. Source: Mike Hulme, Climatic Research Unit, University of East Anglia.*

of Earth from the sun at these times affects the amount of solar radiation reaching the atmosphere. At perihelion a maximum of 1400 W m^{-2} is received, whilst at aphelion the value is 1311 W m^{-2}, thus varying by about 7 per cent between perihelion and

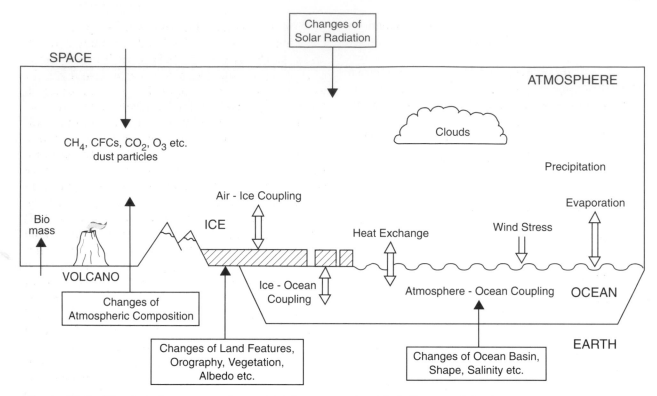

Figure 11.8 *The physical processes and properties that govern the global climate and its changes.*

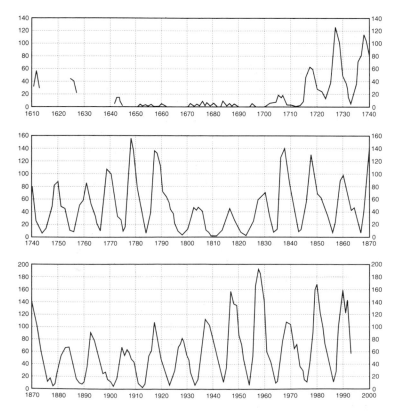

Figure 11.9 *Variations of intensity in sunspot frequencies, 1610–1993. Source: After Zirin (1988) and additional data as compiled by J.A. Eddy.*

aphelion. If Earth is at the perihelion during the northern hemisphere winter it will receive more energy and therefore be warmer than if it were at the aphelion at that time. The time of year at which Earth is nearest the sun does change over time. A complete cycle takes about 21,000 years and is termed the *precession of the equinoxes* (Figure 11.10). It has the effect of changing the relative warmth of winter and summer between the two hemispheres. Aphelion in the northern hemisphere will produce cooler winters but the summer perihelion should give warmer summers, increasing the seasonal difference in temperature. We also find that the degree of ellipticity of Earth's orbit changes through time over a cycle of about 95,000 years; this phenomenon is known is the *eccentricity of the orbit.* At times the orbit is almost circular and there is little difference in input between perihelion and aphelion; 47,500 years later the orbit is at its most elliptical, with a strong difference between perihelion and aphelion. This variation affects the amount of solar radiation intercepted by Earth by a small amount.

The final source of variation in the distribution of solar inputs is the changes in the tilt of Earth's axis of rotation. Although, at present, the tilt is about 23.5°, it can range from 21.8° to 24.4°. This means that the precise latitude of our Tropics will shift slightly. When the axis has a greater tilt, the position

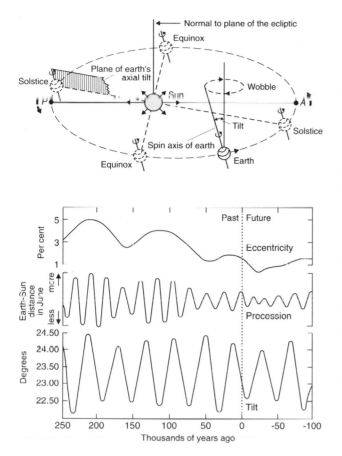

Figure 11.10 *(a) Geometry of the sun–earth system, showing the factors causing variation in radiation receipt by the earth. (b) Changes in eccentricity, tilt and precession for the last 250,000 years and the next 100,000 years. Source: After Imbrie and Imbrie (1979).*

of the overhead sun at midday at the solstices is further polewards by about 2.5° than when the tilt is smallest. This produces greater seasonal contrast with high tilt and less contrast with a small tilt. The variation is sometimes referred to as the *obliquity of the ecliptic*, or more simply as the variation in tilt, and takes place over a full cycle of approximately 42,000 years.

The variations in the solar radiation at different latitudes of Earth's surface due to the orbital changes are shown in Figure 11.11. In high latitudes it is the 42,000 year cycle which dominates, whilst at lower latitudes the 21,000 year cycle is dominant. From the amounts of incoming radiation, with an allowance for ice cover, calculations of Earth's energy budget indicate that the orbital variations have the correct timing and size to start the succession of major advances and retreats of the ice sheets during the last 300,000 years. This is seen most clearly in some of the ocean cores, where undisturbed sediments have accumulated over

thousands of years. Fluctuations in temperature are determined from their fossil and carbonate contents and tie in closely with the Milankovitch cycles (Figure 11.12). However, the graph shows that it is the longest cycle – the eccentricity – which produces the dominant signal in the sediments, despite the fact that it causes only small variations in global solar radiation compared with the other two factors. Other factors than orbital ones must operate. Another problem in climatic change is that orbital processes change slowly, whilst there is frequent evidence from sediments that changes of climate can take place rapidly. For example, in one deposit near Birmingham (UK) a typical northern assemblage of beetles was found dated to 10,025 ± 100 years BP. Ten centimetres higher no Arctic fauna survived at an age of 9970 ± 110 years BP. Conversely the rapid cooling at about 10,900 BP brought a catastrophic readvance of the ice, which destroyed fully grown forests, and caused desiccation in Colombia and a marked cooling in Antarctica within a time span of only 200–300 years. It seems highly unlikely that the orbital variations could have been responsible for such sharp climatic fluctuations as described here. For such changes we must look to other mechanisms.

Internal forcing

Figure 11.8 shows the internal mechanisms which can also generate climate change. Some are likely to operate only very slowly. For example, changes in Earth's orography take millions of years to become significant in terms of their effects on the atmosphere, though they can be extremely important. It is interesting to speculate what the climate of the northern hemisphere would be like without the Western Cordillera of the United States or the Tibetan plateau. The westerly circulation would certainly be different, probably with fewer meridional exchanges and a different pattern of precipitation as areas favouring convergence and divergence changed. The ideas could be tested by using climate models with different surface topography, though the validity of the conclusions, would depend upon the reliability of the model. Similarly, ocean basin shape may vary, as it did to some extent when sea level dropped markedly during glacial phases. Shape changes may affect ocean current patterns and – something which has been appreciated only recently – they may affect salinity levels and interactions between surface and deep-sea waters (Figure 11.13).

Changes in surface features may have drastic effects on climate over rather shorter time periods than orography and ocean shape. Deforestation is a clear

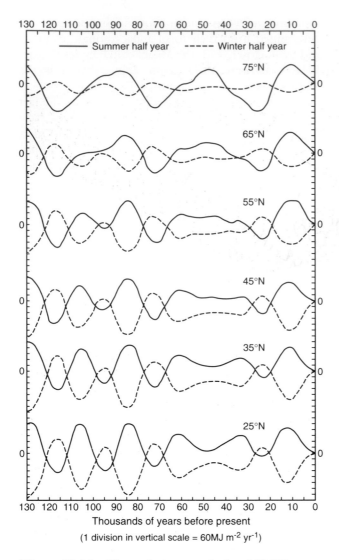

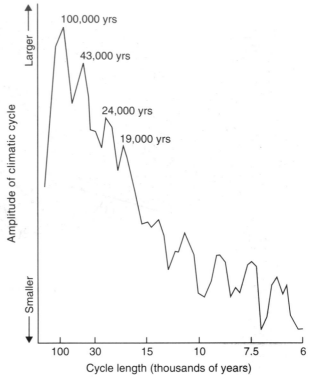

Figure 11.12 *Spectrum of climatic variation over the past half-million years. This graph – showing the relative importance of different climatic cycles in the isotope record of two Indian Ocean cores – confirmed many predictions of the Milankovitch theory. Source: After Imbrie and Imbrie (1979).*

Figure 11.11 *The variation over the last 130,000 years of the radiative flux between latitudes 25° and 75° N in the summer and winter half-years, based on calculations by Milankovitch. Source: After Mason (1976).*

case of a change in surface properties which, by changing surface albedo, could affect climate. Clearance of temperate forests in Europe, and currently of tropical forest in South America and East Asia, is believed to have the potential to modify climate. Some model estimates predict that a change of Brazilian rain forest to savannah would lead to a decrease of evapotranspiration of up to 40 per cent, an increase of run-off from 14 per cent of rainfall to 43 per cent, and an average increase in soil temperature from 27° C to 32° C, but the precise figures depend upon the assumptions embodied in the models. It has also been argued that overgrazing of vegetation can change surface albedo and lead to a decrease in precipitation. There are no clear examples

of how this effect can be determined from climatic records, though climate models do indicate that such effects should occur.

Another surface change which would have clearer effects is snow and ice. If a surface is **deglacierized**, its albedo will decrease from relatively high values to much lower ones; as a result more solar energy will be available to warm the surface. With snow or ice on the ground much energy will be consumed in melting or ablating which will become available for heating when the surface changes. All these factors produce a marked increase of temperature at the surface. Conversely, a change to snow and ice at the surface would trigger a positive feedback to enhance cooling. It has been suggested that the regional cooling over Europe between 11,000 and 10,000 years BP may have resulted from a breakdown of heat release in the North Atlantic due to a sudden massive surge of fresh pro-glacial meltwater from the St Lawrence and the break-up of sea-ice from northern Canada (a Heinrich event).

It is well known that atmospheric composition can and does change through time, though precise levels

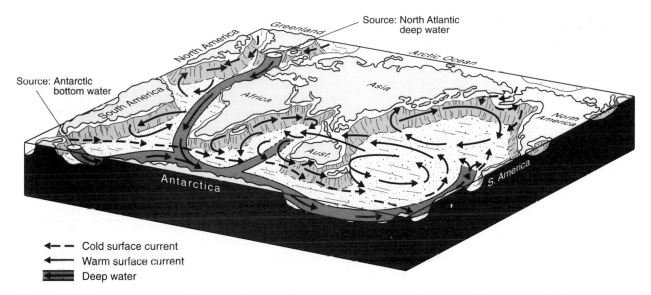

Figure 11.13 *Schematic block diagram of major deep-water oceanic flows believed to influence the global climate. Source: After Bradshaw and Weaver (1993).*

of measurement may not be available. One major influence is volcanic activity. When a volcano erupts it may expel vast quantities of dust and gases such as carbon dioxide and sulphur dioxide into the atmosphere. How significant the eruption is for climatic conditions depends particularly on how much material is ejected into the stratosphere and how long it is able to survive there as well as the latitude of eruption. If dust and sulphate particles can survive in the stratosphere they are able to reduce by reflection the amount of solar radiation reaching the ground surface. The longer they survive the greater will be the impact of the eruption. Major eruptions can result in surface cooling of about 0.2° C for a few years after the event. The eruption of Mount Pinatubo in the Philippines in 1991 was expected to lead to a reduction in the recent trend of increasing global mean temperatures. Although large amounts of smoke were released from the Kuwaiti oilfield fires, none of it was able to penetrate into the stratosphere and its climatic effects were largely local. Volcanic eruptions at high latitudes send dust into the circumpolar vortex which tends to get trapped rather than being dispersed globally. The impact is less likely to be world-wide. A single major eruption can have an immediate impact on global climate. As volcanic activity appears to follow a random pattern, it is possible that several eruptions could occur over a short period of time to increase the particulate matter in the stratosphere and increase the potential climatic impact. This effect may have contributed to the severity of the Little Ice Age, with major eruptions in 1750–70 and 1810–35.

As well as volcanic dust, the atmosphere contains dust blown up from the surface. Occasionally falls of red dust occur in Britain as Saharan dust has been carried northwards. Adding to the effects of natural dust blow, agricultural activities may expose bare soil and lead to major losses of topsoil through wind action. A classic example of this was during the 1930s drought on the Great Plains of the central United States when vast quantities of soil were swept up from the fields and deposited as far afield as New York and Washington. As soil particles are usually heavier and are not ejected into the atmosphere with the force of a volcanic explosion, their climatic impact tends to be more local. However, they do backscatter sunlight whilst absorbing some long-wave radiation from the ground. The precise effect depends upon the albedo of the surface. Man-made particles cause net warming over snow and ice and most land surfaces but cooling over the oceans, with their low albedo.

Another major change in atmospheric composition which is known to occur is in the proportion of carbon dioxide (Figure 11.14). Carbon dioxide is one of the natural atmospheric gases which contribute to the greenhouse effect. We would expect an increase in the proportion of the gas to trap more long-wave radiation emission from the surface and increase the mean temperature of the globe. Investigations have shown a close relationship on a long-term basis between global temperatures and CO_2 levels, though it is not clear which increases first or whether the changes are synchronous. Since 1958 precise measurements of CO_2 levels have been taken at the Mauna Loa Observatory on Hawaii. They show an increase from

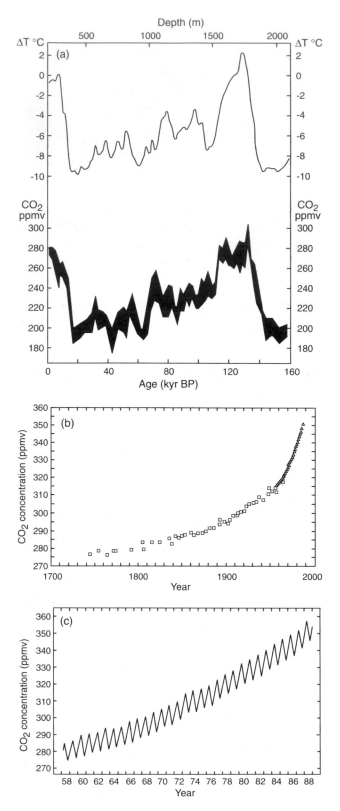

315 ppm by volume to about 360 ppm today, an increase of over 10 per cent in only thirty-five years. The increase is largely the effect of fossil fuel burning, though deforestation and other land use changes have an impact. Other gases such as methane, nitrogen oxides and chlorofluorocarbons (CFCs) are even more effective at absorbing long-wave radiation. At present their concentrations in the atmosphere are low, though they are all increasing as a result of human activities (Figure 11.15). Theory and climate models predict that their increasing concentrations will have an impact on global temperatures.

Although less publicized, the increase in methane (CH_4) is causing concern as it is a by-product of both energy consumption and agricultural activity. It is believed that a large portion of the methane increase may be the result of anaerobic decomposition of organic matter associated with rice paddy cultivation and the digestive processes of ruminants such as cattle. As both rice area and ruminant numbers have increased with the rising human population over the last two centuries, the present annual increase is about 1 per cent per year compared with 0.48 per cent for carbon dioxide.

Conversely the emission of sulphur by-products from biomass and fossil fuel burning into the atmosphere leads to the formation of sulphate aerosols which have a strong regional effect on climate. The suspended particles increase scattering and reflection of insolation at a greater rate than they absorb outgoing long-wave radiation. It is believed that their presence may help to offset the enhanced greenhouse effect to some extent but they will sustain the acidity of precipitation.

The increase in CFCs has had another disturbing impact which was never foreseen. CFCs are extremely stable molecules which gradually disperse throughout the atmosphere. They are destroyed by the action of ultra-violet light in the stratosphere, yielding free chlorine atoms. The highly reactive chlorine reacts with ozone to produce chlorine monoxide and oxygen. Chlorine monoxide is unstable, reacting with free oxygen atoms to form a further oxygen molecule and releasing another free chlorine atom which can react and destroy more ozone. Although the details are still not fully understood, the levels of ozone in the stratosphere have been declining, especially over Antarctica in spring, when very cold temperatures prevail. Concentrations have fallen by over 50 per cent within the last decade (see Box, Chapter 5). As well

Figure 11.14   (a) *Estimated temperature changes as determined from the Vostok (Antarctica) ice core and derived CO_2 levels; (b) atmospheric CO_2 levels as indicated by measurements on air trapped in ice from Siple (Antarctica); (c) measurements of CO_2 concentrations in the atmosphere at Mauna Loa, Hawaii. Source: After IPCC (1990).*

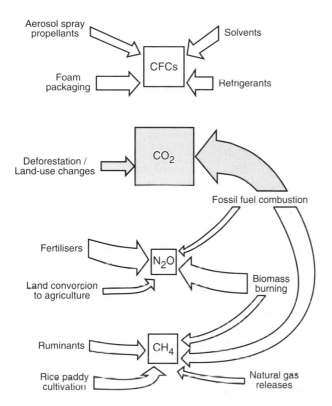

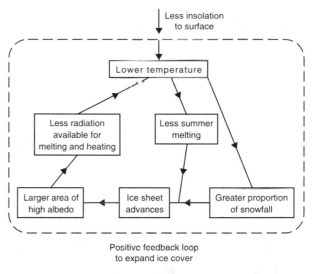

Figure 11.16 *A positive feedback loop, demonstrating how a decrease in insolation and lower surface temperatures may generate further cooling and perhaps even an ice age.*

Figure 11.15 *Main anthropogenic sources of greenhouse gases. Sizes of the boxes are proportional to the contribution to radiative warming. For each gas, the size of the arrows indicates the relative importance of each source to the total concentration change. Source: Warrick et al. (1990).*

as affecting ultra violet levels at the surface and the implications for skin cancer, the degree of heating in the stratosphere will decrease, changing its temperature structure and perhaps its circulation.

Direct warming of the atmosphere by waste heat also affects atmospheric temperatures. Estimates of global energy production have indicated that 8×10^6 MW are generated annually, most of it in densely populated urban and industrial areas. Long-period temperature records at city-centre sites usually show an increase of temperature through time because of this effect coupled with heat storage by buildings.

Feedback effects

Whilst external and internal forcing systems may give rise to changes of climate, the results are not always as straightforward as we might expect because of the complex interactions which take place within the Earth–atmosphere system. Our climatic system consists of several subsystems, such as the atmosphere, the oceans, the ice sheets and the land surfaces. They are all closely related as a system, and changes

in one may affect the others. Moreover, changes within one of the components may act as positive or negative feedback, ultimately influencing inputs of solar radiation to the ground surface. These feedback mechanisms may be responsible for many of the more rapid fluctuations in climate that have occurred throughout Earth's history.

Positive feedback leads to more dramatic and far-reaching changes. The initial effect is magnified, so that quite small changes in the environment produce major adjustments in the system. Perhaps that is why the climate sometimes changes abruptly without any evidence of a clear change in external conditions. Figure 11.16 shows such an effect which has been proposed as a cause of the Ice Ages. A quite small cooling of temperature at the poles of only 1–2° C delays the summer melting of the Arctic ice cap. Because the ice survives for longer, the albedo of the surface stays high longer. More incoming short-wave radiation is reflected back to space. Reduced heating of the surface therefore occurs, allowing the ice caps to survive even longer, which increases reflection further, which lowers temperatures further ... The cycle is self-perpetuating. Once they have been initiated, positive feedback processes magnify the effect of the initial change and cause major adjustments in the system – possibly even an Ice Age.

Another factor which may affect the state of the climate system results from the complex non-linear behaviour of the atmospheric circulation. In a transitive system there would be one normal state of circulation and any disturbances in the circulation, would

Implications of global warming

Although it is not shared world-wide, there is growing concern that the observed increase in carbon dioxide and other greenhouse gas levels is likely to lead to a global increase in temperature of between 2° and 4° over the next century. Global circulation models predict that the increase is not likely to be uniform; some areas will have a slightly lower increase and other parts considerably more. Unfortunately this does not mean that we can look at areas which already experience such temperature levels and assume that conditions there indicate the future climate. In addition to mean annual temperature change, it is likely that minimum temperatures will increase more than maximum temperatures, so that the temperature range will decline. Precipitation is very difficult to predict; it may increase in some areas and decrease in others, but the effect of increased evaporation through higher temperatures will be to decrease moisture levels. In Britain, it has been argued, winter precipitation should increase but there will be a decline in summer precipitation. Storminess may increase in winter in temperate latitudes and there could be an increase in tropical storms as sea surface temperatures increase. All these predictions are conditional: no one is really sure what other climatic implications may result from a rise in global mean temperature.

Other predictions are more certain. If there is an increase in temperature then the oceans will get warmer and expand, so mean sea level will rise. In addition, with warmer temperatures, some of the glaciers and ice caps will melt. The effect is expected to be greatest in mountainous areas but even Greenland and Antarctica will be affected. Recently some ice shelves have been breaking up on the Antarctic Peninsula, which may be an early sign of this warming effect, though even that is debated. The net effect is to add water to the oceans and raise their level. The combined effects of expansion and melting are expected to lead to an increase of sea level of between 0.3 m and 1 m by AD 2100.

As climate changes so all the other environmental elements dependent upon climate will experience an impact. The most obvious and immediate effects will be upon the plant and animal kingdoms. An increased frequency of higher temperatures and lower growing-season rainfall could cause plants to die and any animals dependent upon those species. If the change happens slowly and with no human interference, species may migrate to more suitable areas and other species will move in to replace them. After all, it has happened many times in the last 2 million years. The main difference this time is that it is happening so quickly in geological terms and there is a vast human population which requires feeding. Species may not be able to move and so may become extinct.

Perhaps even more serious, it is likely that pests and diseases will proliferate in the higher temperatures. Many pests are restricted in their poleward distribution by winter cold, so warmer winters would allow an expansion of their habitat. The mosquito could re-establish itself in southern Britain and northern Europe. Even more seriously, crop diseases could become more prevalent to affect food supplies.

Even geomorphologically there would be consequences. The changing moisture régime would affect river levels and channel form. Increased storminess may generate more floods and aid run-off and erosion. Even mass movement may become more important as dry and wet spells alternate. Decreased frost frequency would change freeze–thaw activity in tundra areas, and over time the permafrost would melt. This would lead to greater microbial activity and, as organic matter decomposed, large quantities of methane would be released. Methane is another greenhouse gas which is even more efficient than carbon dioxide in absorbing long-wave radiation.

Over longer periods of time the rates of chemical weathering would speed up as temperatures increased, as long as moisture was available to assist the process. Soil formation should increase but we are still dealing with hundreds of years rather than an instant change. This presents one problem for species migration. Although the climate of a more poleward location might be suitable, the type of soil formation which has developed under present climatic conditions may be unsuitable for large-scale agriculture or forest. It is unlikely that the wheat belt of Canada could simply move northwards in response to climate change.

Taking all environmental factors together, the impact of the predicted rapid increase in temperature over the next century is likely to be dramatic.

be expected to revert to the norm. In an intransitive system there are two equally acceptable outcomes, depending upon the initial state. However, mathematicians have found that some systems can be almost intransitive, i.e. the circulation resembles a transitive state for an indeterminate length of time and then suddenly switches to an alternative resultant state. With such a circulation it is impossible to know which is the normal state and when a switch may take place. Attempts to model such a system with any confidence would be very difficult. Unfortunately geological and historical data are insufficiently detailed to determine which of these circulation types is typical of Earth, but they could account for the known sudden changes. We do not necessarily have to look to external or internal forcings for rapid change.

Conclusion

It is clear that in the medium to long term, over the time scale of tens to thousands of years, our climate

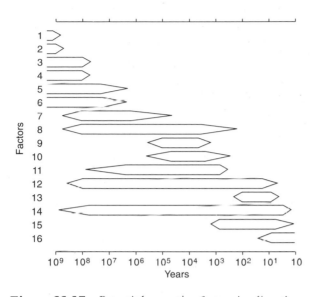

Figure 11.17 *Potential causative factors in climatic change and the probable range of the time-scale of change attributable to each. Factors: (1) evolution of the sun; (2) gravitational waves in the universe; (3) galactic dust: (4) mass and composition of the air (except CO_2, H_2O and O_3; (5) polar wandering; (6) plate tectonic movement; (7) orogenic and continental uplift; (8) CO_2 in air; (9) earth-orbital variations; (10) air–sea–ice cap feedback; (11) abyssal oceanic circulation; (12) solar variability; (13) CO_2 additions by fossil fuel combustion; (14) volcanic dust in the stratosphere; (15) ocean–atmosphere autovariation; (16) atmospheric autovariation. Source: After Mitchell (1968).*

varies, not randomly but systematically. Broad, consistent fluctuations occur, giving periods of relative warmth and periods of coldness, years of aridity and years of wetness. The reasons for the fluctuations are not clear; variations in Earth's orbit and rotation, changes in solar output, internal adjustments to the vegetation, topography and atmosphere, all may be contributory factors. The time scale of possible causative factors is shown in Figure 11.17 but we must not forget that most of the factors are interactive; we cannot isolate a single process and describe its consequences with much confidence. In recent centuries human activity may have begun to have an impact on climate.

Two questions remain. What is the effect of these climatic fluctuations? And where is our climate going now?

Some of the effects are all too apparent to us. In those areas which are marginal to agriculture, like parts of the Sahel, minor changes in climate may have appalling consequences, bring crop failure, soil erosion and famine. Some of the effects are more subtle, but none the less significant. As the pattern

of climate changes people tend to move if they are able; new areas become favourable, others may become unfavourable. It has been suggested that the stimulus for the Viking invasions and settlement of Iceland, Greenland and Britain was climatic deterioration in Scandinavia. Nomadic tribes today respond to similar stimuli as grazing levels vary.

The effects of climatic change are not confined to agriculture. As we shall see in later chapters, fluctuations also influence landscape processes. Throughout the temperate regions of the world the imprint of past climatic change is clear within the landscape. Glacial landforms lie hundreds of kilometres beyond the limits of the present ice caps; lakes which were at one stage huge inland seas are now small pools in comparison; river valleys that once carried vast torrents of water are now occupied by small, generally placid streams; fine, wind-blown silt and former sand dunes in currently moist areas testify to the former strength of winds and aridity. Effects of similar magnitude can be detected in the vegetation. In many areas the range of plants that we find today is a result of the migration and mixing of vegetation in response to climatic changes. The global system, as we have noted before, is intricately interrelated. Changes in one part affect others, and the effect is nowhere more apparent than in the influence of climatic change.

So to our last question. What will be the climate of the future? Numerous predictions have been made. Climatic models have been used to investigate the effects of known or highly likely changes in the near future. They would include the effects of an atmosphere with more greenhouse gases in its composition, together with the orbital variations that we know will take place over the next 100,000 years (Figure 11.18). Output from the orbital models indicates that climates as warm as those of today are relatively rare and suggests that, other things being equal, the global climate should start to deteriorate. Unfortunately because of the very different time scales of operation it is difficult to incorporate both effects into the same model. At present any discussion of their joint impact relies on informed judgement and speculation. Our uncertainty about the future climate is based on the many different forcing factors, which operate over many different time scales, all superimposed on each other and each operating over a different time cycle. Consequently, it is almost impossible to tell how long any trend we identify will persist. We can only guess at what even the immediate future holds.

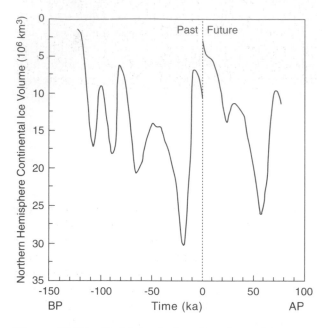

Figure 11.18 *Model simulation of continental ice volume for the last 120,000 years and the next 80,000. The sudden decrease in ice volume at present is due to the enhanced greenhouse effect. Source: After Goodess et al. (1992).*

Further reading

Goudie, A. (1992) *Environmental Change*, third edition, Oxford: Clarendon Press.
Elementary to intermediate text which includes sections on the causes of climatic change as well as a variety of examples through time and across the globe. Especially useful on tropical areas.

Goodess, C.M., Palutikof, J. P. and Davies, T.D. (1992) *The Nature and Causes of Climate Change*, London: Belhaven.
Intermediate to advanced text about the causes of climatic change. It also offers suggestions on future climate, taking natural and anthropogenic factors into account.

Imbrie, J. and Imbrie, K.P. (1979) *Ice Ages: Solving the mystery*, London: Macmillan.
A history of ideas about the origins of recent ice ages. Provides a clear account of the Milankovitch effect, though it gives the impression that it must be the dominant mechanism and other possibilities are subordinate, hence a little dated.

Schnieder, S.H. (1989) *Global Warming*, Cambridge: Lutterworth Press.
Written by one of the leading scientists involved in general circulation modelling of the past, present and future, this is a very readable account about the factors behind global warming and its possible impacts.

Earth's surface evolution has focused so far on the global mosaic of crustal plates and morphotectonic landforms. These great structures undergo detailed geological changes during their global voyage, analogous to a ship picking up commodities and crew at one port and leaving them at another, processing raw and waste materials *en route* and undergoing repairs and structural refits as job requirements change. A ship's plates, rivets and fittings as its working life ends are not all those of its maiden voyage! Rock-forming and recycling processes, integrated within supercontinental/Wilson tectonic cycles, constitute a cycle in their own right. Plate tectonics and architecture determine the general location of each stage and the cycle runs via two interconnected loops (Figure 12.1).

The *primary* or **igneous** loop is the shortest, concerned only with cycling oceanic lithosphere between magma formation and its resorption. The *secondary*, or **sedimentary**, loop exposes magmas, retained in continental environments, to exogenetic **weathering**, **erosion**, transportation and **deposition** to form terrestrial or marine sediments. This sequence may be interrupted at any time, with material returned to an earlier stage before eventually re-entering the loop via subduction and remagmatization. At any point in either loop, rock material may also be subjected to irreversible metamorphism by significant increases in temperature or pressure, usually through volcano-tectonic activity. These processes represent a **geochemical cycle** of continued fractionation and a rock cycle of particular *lithological* styles and masses of material. Both are important for geographers and are amalgamated here. Concern with rocks *and* their geochemistry underpins our ability

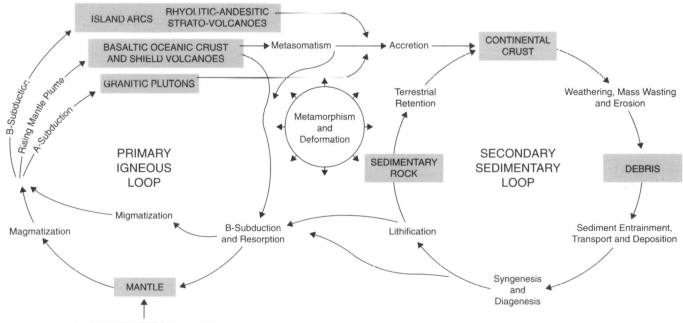

Figure 12.1 *The rock cycle, following Primary (igneous) and Secondary (sedimentary) loops. Rock material assemblages are highlighted in boxes between the operational processes of the cycle. Metamorphism and deformation can occur in any part of the loop.*

to understand not only the formation of rocks but also soils, nutrient cycles, the biosphere, the atmosphere and the oceans – all of which are dynamic, evolving material systems sourced by Earth's rocks.

Rock-forming minerals and processes

Rock-forming minerals

Rock-forming minerals are the crucial link between broadly *homo*genous magmas derived from the upper mantle and particular *hetero*genous assemblages which form a distinct **lithology** or rock type based on geochemical and textural character – rather like the general stock of ingredients in a restaurant and a particular dish. We start with upper mantle geochemistry and the formation of a suite of igneous fractionates, determined largely by temperature/pressure environments. Each *mineral species* has unique chemical and physical properties, related to its elemental composition, and a crystalline structure. The existence of well over 2000 known minerals makes their study a formidable proposition until it is appreciated that just two elements, oxygen (O) and silicon (Si) form 75 per cent of the lithosphere by mass, with a further 24 per cent formed (in declining abundance) by aluminium (Al), iron (Fe), calcium (Ca), sodium (Na), potassium (K) magnesium (Mg) and titanium (Ti); oxygen alone forms 95 per cent by volume. Lithospheric minerals reflect these concentrations.

Mineral structure

We need to consider atomic structures in order to understand minerals further. Atoms comprise a nucleus of protons and neutrons inside an electron shell. Protons and electrons have positive and negative electrical charges respectively and, when balanced, give atoms electrical neutrality. However, atoms exchanging electrons create an electrical imbalance and become known as **ions**. Net loss of negatively charged electrons leaves a smaller, positive **cation** and net gain leaves a larger, negative **anion**. Mineral structure is created by three-dimensional arrangements of suitable anions and cations in a repeated geometric pattern, held together by *electrostatic* attraction in *ionic bonds* or the sharing of electrons by two atoms by *covalent bonds*. Precise patterns depend largely on the size and internal packing of the anions and the presence of suitably sized cations in the spaces between. A compact, snug fit restores electrical equilibrium at the points of contact or electron-sharing provides a symmetrical framework which gives the mineral a distinct, crystalline structure. The geometric arrangement of cations and anions is imparted faithfully to each crystal, providing a three-dimensional symmetry and crystal *lattice*.

When minerals can grow freely from a melt or solution, rather than competing with surrounding minerals for space, crystal lattices become visible to the eye, whereas naturally their atomic structure is not. Each lattice is unique to a particular mineral, although **ionic substitution** occurs in some minerals and others are **polymorphic**. The former refers to substitution of an ion by one of similar size and charge, therefore fitting physically and electrically into the atomic structure without altering the crystal structure. Iron and magnesium, for example, may substitute for each other in *olivine* and its crystals are referred to as a **solid solution**. Polymorphism, on the other hand, refers to identical chemical composition expressed in two mineral and crystalline forms and is indicative of adjustments in crystal structure between lithospheric and denser, deep mantle minerals. Calcite/aragonite and graphite/diamond are alternative polymorphs of calcium carbonate and carbon respectively, forming under different temperature and pressure conditions. The great hardness of diamond, which is formed at depths of approximately 150 km, is indicative of depth/pressure influence on crystal structure.

Mineral chemistry

Minerals such as lead, gold and carbon are single elements but the majority are compounds. Fortunately, a shortlist of mineral groups from our cast of 2000 accounts for virtually all rock mass. One group – the **silicates** – provide not only by far the greatest bulk and diverse range of minerals but also an outstanding illustration of linkage between textural and chemical properties. This centres on SiO_4^{4-} the **silicate tetrahedron**, an anion with four negative charges composed of four large oxygen anions ($4 \times O^{2-}$) clustered around a small silicon cation (Si^{4+}). Covalent bonding leaves four surplus oxygen electrons, giving the silicate compound a negative charge. Tetrahedral structure provides a large range of potential linkages in which electrical equilibrium is reached, either by **polymerization** through replication by sharing oxygen atoms or by the addition of other suitable cations, principally from the above shortlist of elements. Aluminium may also replace silicon, converting Si_4 to $AlSi_3$ or Al_2Si_2 to

SILICATE STRUCTURE AND FORMULA		MINERAL EXAMPLES, CRYSTAL AND CLEAVAGE CHARACTER, SPECIFIC GRAVITY
SILICATE TETRAHEDRON  simplified as	None SiO_4	Olivine, garnet, zircon. Dense, equidimensional crystals Specific gravity = 3.5–4.0
RING SILICATES 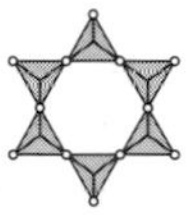	2 Si_6O_{18}	Beryl, tourmaline. Columnar (prismatic) crystals; cleavage between rings and across columns. 2.7–3.2
CHAIN SILICATES (a) Single chain (b) Double chain	2 SiO_3 2–3 Si_4O_{11}	Pyroxenes. Dense, equidimensional crystals; cation bonding inhibits cleavage. 3.0–4.0 Amphiboles. Well-developed cleavage, aided by weak cation bonds *between* chains. 2.7–3.6
SHEET SILICATES 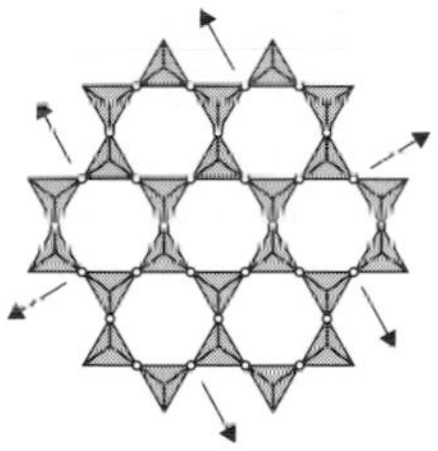	3 Si_2O_5	Micas, clay minerals, talc, serpentine. Low density, excellent cleavage *between* sheets. 2.6–3.3
FRAMEWORK SILICATES Complex, 3-dimensional structures	4 SiO_2	Quartz, feldspars, zeolite. Less dense but strong, three-dimensional bonding. Cleavage absent in quartz, present in feldspars. 2.5–2.8

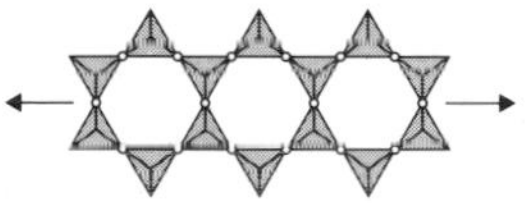

Figure 12.2 *The structure and characteristics of silicates and some representative silicate minerals. The silicate tetrahedron of four large oxygen anions and a single small silicon cation and its simplified form are shown in the first panel. Subsequent structures are shown with the number of shared oxygen anions and their silicon–oxygen formulae.*

form **alumino-silicates** requiring further, balancing ions.

The principal silicate tetrahedra and minerals they construct are illustrated in Figure 12.2. The *valency*, or combining potential based on the number of electrons lost/gained by the atoms, is the key to interpreting their electrical and elemental associations. For example, each tetrahedron shares two common oxygen atoms to create the basic formula $(SiO_2)_n^{2-}$ of *chain silicates*. A divalent cation, for example Mg^{2+} or Fe^{2+}, would then form the pyroxene-group mineral hypersthene $(Mg,Fe)SiO_3$ – which also happens to be a solid solution! Mineral density and hardness are greatest in the *single* and *ring silicates*, decreasing as the number of shared oxygens increases and the framework enlarges by incorporating large balancing cations. Since the tetrahedral complex is based on strong, covalent Si–O (anion) bonds and weaker ionic (cation) bonds with the associated elements, crystals possess strength anisotropy – they are not uniformly strong in all directions and cleave more readily through the ionic bonds. **Cleavage** develops across the columnar crystals in ring silicates and between the bands and sheets of double-chain and sheet silicates. Mica, for example, has a particularly flaky structure. Generally, the strongest common minerals are three-dimensional tetrahedra or *framework silicates* such as quartz and feldspar. Quartz is the purer silicate, formed solely of SiO_2 with all oxygen anions shared, but aluminium replaces some silicon in feldspar and is balanced by potassium (K^{1+}), sodium (Na^{1+}) or calcium (Ca^{2+}) to form orthoclase, albite or anorthite solid solution feldspars. Quartz and feldspars form over 70 per cent of continental lithosphere.

Other principal mineral groups are built on simpler oxides and anion complexes (Table 12.1). Oxides and sulphides are important metallic minerals, with sulphur replacing oxygen as the anion in the latter, and salt cations form halides with fluorine and chlorine. Oxygen in association with the carbonate, sulphate, phosphate and hydroxyl anion complexes $(CO_3)^{2-}$, $(SO_4)^{2-}$, $(PO_4)^{3-}$ and OH^- forms carbonates, sulphates, phosphates and hydroxides respectively. Most of these minerals do not form directly from melts and are dependent instead on metamorphic, metasomatic and sedimentary processes covered later.

How do various mineral combinations come together and develop into distinct lithologies? Magma derived from partial melting of the asthenosphere or continental crust is converted from a hot melt to cold, solid lithified rock. The rate and location of cooling determine its mineralogical evolution and eventual rock character. The initial melt, at temperatures of 900° C–1200° C, does not cool uniformly and its homogenous, minero-elemental composition changes *en route* by **fractional crystallization** as solid minerals with successively lower melt temperatures form and settle out through the rising magma. Fractionation proceeds in several ways – the melt becomes depleted of higher-temperature

products and therefore enriched in remaining elements; denser minerals settle out faster through the viscous melt, although they may still react with it chemically; and further mineral *speciation* occurs as more subtle changes alter element ratios in solid solutions (Table 12.2).

Three classes of magma are recognized. Fractionation of asthenosphere peridotite proceeds through basaltic–andesitic–granitic stages, although andesitic–granitic magmas are also derived from continental crust *wet* melts over subduction zones. Basaltic magmas crystallize first from asthenosphere peridotite, commencing with olivine, followed by plagioclase feldspar (anorthite–albite solid solutions) and pyroxene. Their denser, dark minerals – Mg,Fe-rich and relatively silicate-poor – form *ultrabasic–basic* or **ultramafic–mafic** rock. They are also dry melts, with less than 0.2 per cent water, and therefore anhydrous. Andesitic magmas are *intermediate* in nature, with more albite plagioclase, amphibole (hornblende) and biotite, and are close in composition to average continental crust. They solidify within a temperature range of 900–1000° C. Low-temperature (500–600° C) granitic or *rhyolitic* magma is dominated by the lighter, less dense minerals orthoclase feldspar (potassium-rich), quartz and biotite–muscovite to form Fe,Si-rich, *acid* or **felsic** rocks.

Other important physico-chemical properties change during fractionation. Silicate percentage rises steadily from 45–54 per cent (ultramafic) to 55–64 per cent (intermediate) and 65–78 per cent in felsic rocks, causing a corresponding increase in viscosity and progressively slower flow rates. Feldspar minerals also become less alkaline as the plagioclase–orthoclase series shows. Early settling of denser, ultrabasic minerals and the continued rise of lighter fractions combine to create a *layering* effect in igneous rocks. We have seen this already in layered oceanic crust and mineral distinctions between upper (felsic) and lower (mafic) continental crust but it can also be

Table 12.1   *The principal non-silicate groups of rock-forming minerals.*

Oxides		Sulphides		Halides		Carbonates	
haematite	Fe_2O_3	pyrrhotite	FeS	sylvite	KCl	siderite	$FeCO_3$
magnetite	Fe_3O_4	pyrite	FeS_2	halite	NaCl	magnesite	$MgCO_3$
ilmenite	$FeTiO_3$	galena	PbS	fluorite	CaF_2	calcite	$CaCO_3$
rutile	TiO_2	chalcopyrite	$CuFeS_2$			aragonite	$CaCO_3$
chromite	$(Mg,Fe)Cr_2O_4$					dolomite	$CaMg(CO_3)_2$

Sulphates		Phosphates		Hydroxides	
gypsum	$CaSO_4.2H_2O$	apatite	$Ca_5(PO_4)_3(F,Cl,OH)$	goethite	$Fe_2O_3.H_2O$
anhydrite	$CaSO_4$			gibbsite	$Al(OH)_3$
barite	$BaSO_4$				

Table 12.2 *Comparative composition of selected magmas and rocks (% by weight).*

Mineral	C'tal crust	Canadian Craton	Upper Mantle	Basaltic magma	Basalt	Andesitic magma	Rhyolitic magma	Granite	LR Tuff	Green-schist
SiO_2	58.0	66.1	45.16	50.3	49.1	62.5	75.1	70.9	75.6	42.1
Al_2O_3	18.0	16.1	3.54	20.3	18.2	15.7	12.2	14.5	12.4	15.7
CaO	7.7	3.5	3.08	11.0	11.1	4.3	1.6	1.8	0.1	3.2
FeO	7.7	1.4	8.04	3.0	6.0	3.2	0.9	1.8	2.9	5.5
Fe_2O_3		3.1	0.46	5.5	3.2	1.3	0.8	1.6		11.6
MgO	3.5	2.2	37.49	4.2	7.6	3.4	0.3	0.9	0.5	11.7
Na_2O	3.6	3.9	0.57	3.3	2.5	4.2	4.2	3.3	0.2	0.8
K_2O	1.5	2.9	0.12	0.4	0.9	2.7	3.2	4.0	7.9	0.1
TiO_2		0.5	0.71	1.0	1.0	0.6	0.3	0.4	0.2	2.2
Cr_2O_3			0.43							
NiO			0.2							
MnO		0.1	0.14	0.2		0.1	0.1		0.1	0.2
P_2O_5		0.2	0.06	0.1		0.2	0.1		0.1	
H_2O				0.7	0.4	1.8	1.0	0.8		6.9

C'tal crust, 'average' continental crust; LR Tuff, Lower Rhyolitic Tuff, from the Snowdon caldera eruptions. Greenschist is metamorphosed oceanic basalt (see below).

present in sub-surface **plutons**. The ability of mineral–magma reactions to form new mineral species, formerly set out in *Bowen's Reaction Series*, is now recognized in general fractional crystallization, ionic substitution and solid solution processes. Table 12.2 shows a representative range of magma and rock compositions.

Cooling creates textural as well as chemical properties in igneous rocks, based on the cooling time and silicate content. Rapid cooling – especially of high temperature rocks suddenly exposed to low temperatures – pre-empts the growth of large crystals and forms fine-grained rocks such as basalt. Even within a specific magma, some minerals will cool faster than others, leaving others to enlarge. Larger crystals in finer matrices create a **porphyritic** texture which becomes **pegmatitic** if they grow very large. Slow cooling gives the characteristic coarse-grained texture to granite associated with subsurface environments in low-temperature magmas. Volatiles driven off during cooling leave empty vesicles which create **vesicular** texture if they survive or **amygdaloidal** texture when infused with slower-cooling melt products or gases.

The rock cycle

The rock cycle: (1) Igneous processes and landsystems

Magma mineralogy and specific temperature/pressure environments, found at a predictable and limited

Geological resources

Human prehistory is defined by the currently fashionable geological materials used by early societies. The *Palaeolithic–Mesolithic–Neolithic* progression of 'Stone Ages' takes us from the earliest known humans to just 4000 years ago, witnessing the slow improvement in stone tools to the later development of clay-using ceramic pots. The first use of metals in the subsequent Bronze and Iron Ages extended well into the historical period, 1500 years ago. Only 250 years ago our ancestors started the new 'Iron and Steel Age' of the industrial revolution. It is estimated that 50–100 tonnes of rock material is now consumed each year for every person living in advanced techno-industrial societies. Higher standards of living and rapid industrialization in other parts of the world create an inexorable rise in global consumption. We recycle some materials or extend their useful life but the slow rate of operation of most geological processes makes it almost inevitable that human consumption exceeds geological renewal. It is apparent that our rapidly improving understanding of global geological cycling processes assists the economic, political and moral solution of this dilemma.

Systematic, dynamic links between geological environments and their representative rocks and structures have been the focus of this chapter, together with Chapters 3 and 4. The products of past geological environments in Britain are now outlined in terms of their resource potential (Figure 1). Few geological resources are 'consumed' literally, and almost none is used directly from the ground. Instead they are segregated, concentrated, cleaned or refined to be fit for use in the required form, quantity and quality. All these processes consume energy, much of

Quaternary A thin veneer of glacial, fluvial and aeolian sediments and frost weathered debris. Widely distributed, especially in lowlands and shallow marine basins. Sand and gravel aggregates – construction, glass industries.

Jurassic, Cretaceous and Tertiary Limestones, clays, chalk, sandstone and ironstones in south-eastern lowland Britain; offshore oil and natural gas. Building stone; cement, brick and iron and steel and – formerly – chemical industries. Power and petrochemical industries. Groundwater.

Permo-Triassic Magnesian limestone, marl, sandstones and evaporites in English midland basins. Some building stone and aggregates. Salt and related chemical industries, including phosphates. Groundwater.

Carboniferous Limestone, grits and sandstones and coal measures in Variscan orogen upland and fringes; some hydrothermal mineralization (lead, zinc). Crushed limestone aggregates and cement; building stones. Coal. Former metalliferous mining and smelting industry.

Devonian Sandstone, slate in English borderlands, southern Ireland and Scottish lowlands. Some building and roofing stone. Groundwater.

Later Precambrian and Lower Palaeozoic Volcanic, intrusive and metamorphic rocks of former subduction zones and Caledonian orogens forming highland and upland Britain; slates, grits and sandstones; schists and gneiss. Hydrothermal mineralization – gold, silver, lead, zinc, copper, tin; semi-precious stones. Hard crushed rock aggregates; building and roofing stone. Former extensive metalliferous and smelting industries.

Other igneous rocks (mainly Devonian and Tertiary age) Granites and basalts in Cornwall (Devonian) and Tertiary volcanic province of Northern Ireland and western Scotland. Hard crushed rock aggregates; kaolin (china clay) from weathered granite.

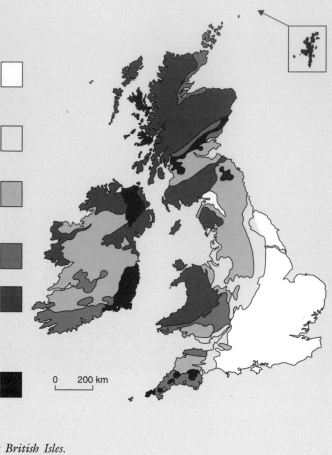

0   200 km

Figure 1 *An outline of the geological resources of the British Isles.*

it to produce high-temperature melts. Wastes are discarded at every stage, from mining to manufacture and after their useful life. In every sense we take Earth's fractionates and fractionate them further still before consigning them to new geological fates in the atmosphere, hydrosphere or lithosphere. We drive the ultimate stage in the rock cycle!

range of sites in the global tectonic framework, determine the style, location and lithology of igneous activity. Magma which solidifies before reaching the surface is intrusive in style and plutonic in location; magma which reaches the surface is extrusive in style, with an effusive (flowing) or explosive (eruptive) nature. The characteristic mineralogy, magma class, texture, viscosity and surface/sub-surface formation of the principal igneous rocks are identified in Figure 12.3. These distinctions extend to the styles of eruptive activity and resultant igneous landforms.

Intrusive activity

Intrusions form a plumbing system of underground magma reservoirs and pipework which connects asthenosphere/lithosphere melts with diapirs and surface extrusions. Two-thirds of new magma is thought to be intrusive. Large igneous intrusions are formed mostly by the viscous flow of low-temperature, silicate-rich granitic magma or more 'solid' lithosphere under relatively thick continental crust by subduction or thermal diapirism. The material permeates the *country* (surrounding) rock at depth via existing voids, either as magmatized continental granitic crust or as the residual part of a less silicic melt. Sustained flow, magma buoyancy and **anatexis** or melting of country rock enlarge them into substantial **plutons** or underground reservoirs. Batholiths are the largest plutons, found in the cores of island arc complexes, subduction zones and some extension orogens. Individual Cenozoic batholiths of 10^{2-5} km^3 contribute to intrusive assemblages of 10^6 km^3 in the Andes. The largest North American Mesozoic batholiths extend for some 1600 km in the Yukon–

VOLCANIC (fine-grained or porphyritic)		BASALT	ANDESITE	DACITE	RHYOLITE
HYPABYSSAL (medium-grained)		DOLERITE			GRANOPHYRE MICROGRANITE
PLUTONIC (coarse-grained)	PERIDOTITE	GABBRO	DIORITE	GRANODIORITE	GRANITE

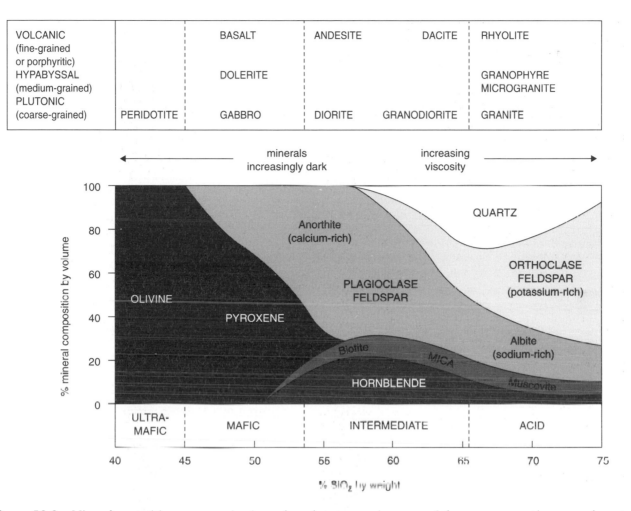

Figure 12.3 *Mineral composition, texture, viscosity and emplacement environment of the more common igneous rocks. Volcanic rocks are eruptive or extruded at the surface; hypabyssal and plutonic rocks intrude existing rocks at intermediate and greater depth.*

British Columbia–Washington coast range. Magma conduits which feed batholiths or surface extrusions intrude country rocks *accordantly* as **sills,** thickening locally into sizeable **laccoliths,** and *discordantly* as **dykes** or fault-directed **cone sheets.** The latter intrude rocks overlying the advancing diapir and may feed volcanoes. Intermediate or basic magmas develop as the original ground surface is approached. Solidified plugs of volcanoes or extrusive lava flows mark the intrusion/extrusion transition (Figure 12.4).

The full nature of intrusion becomes apparent when it is exhumed by erosion of the *overburden* rocks. The intrusion usually forms prominent surface relief by virtue of high mineralogical strength. The mineral species and radioactive isotopes also provide good estimates of the extent and rate of erosion since emplacement. Tens of kilometres of crust stripped off since the late Palaeozoic have partially exhumed the *Cornubian* granite batholith of south-west

England. This diapiric batholith, 250 km long and over 55,000 km^3 in volume, is a residual part of the north European Hercynian orogen and now forms high ground on Dartmoor, other Cornish uplands and the Isles of Scilly. By comparison, erosion has exposed a shallow Tertiary sub-volcanic landscape with cone sheets, ring dykes and volcanic plugs in the Ardnamurchan ring complex of the Scottish Highlands (Figure 12.5).

Extrusive and eruptive activity

Volcanoes are the superstars of the rock cycle. However, only about 6 per cent of new magma is erupted annually from terrestrial volcanoes and we know that basaltic lava effusion at mid-ocean ridges accounts for most extrusive activity (Figure 12.4b) The progression from effusive lava flows to explosive volcanic activity closely follows the fall in viscosity

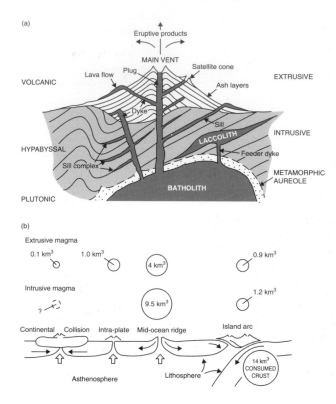

Figure 12.4 *(a) Principal forms of igneous intrusion and extrusion and (b) modern annual rates of global magma production. Source: In part after Smith (1989).*

with increased silicate content, falling melt temperature and parallel increases in crustal contaminates and gas-water volatiles. Consequently, there are broad *volcano-orogenic* associations. B-subduction shows an evolutionary progression, from basalt–andesite–dacite–rhyolite volcanoes with increasing age and distance from the trench and A-subduction volcanoes are usually of dacite–rhyolite composition. Dacite is intermediate between andesite–rhyolite, with 63–69 per cent silica.

The presence of dissolved gases and cooler, more viscous magma is the recipe for spectacular eruptions. Water (over 90 per cent), together with smaller proportions of sulphur, hydrogen, chlorine and carbon gases, SO_2, H_2S, H_2, HCl, CO_2, vaporizes in the near-surface lower-pressure environment and forms an explosive mixture – rather like the decompressive effects of uncorking a champagne bottle! In shallow marine environments, sea water may enter the vent and create a *hydro-volcanic* effect, with an explosive expansion of steam; Surtsey developed via this mode in Iceland in 1963. The existence of a solid, rhyolitic plug in the vent simply enhances explosivity and may cause a lateral blast of the sort displayed by Mount St Helens in 1980 (Plate 12.1).

This violent exhalation of gases, magma and fragmented rock dramatically transforms volcanic products from lava flows over surrounding landscapes into airborne tephra, rained out over a large area. Debris

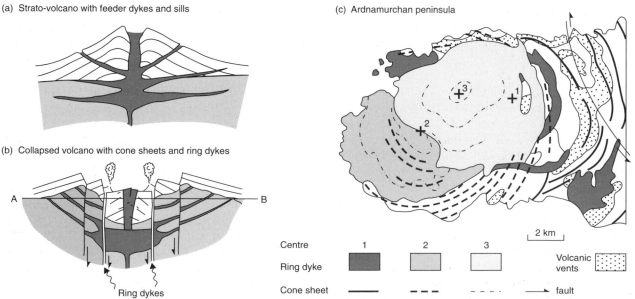

Figure 12.5 *Mode of formation and the modern landsystem of the Ardnamurchan ring complex, Scottish highlands: (a) an active strato-volcano with associated dyke and sill complexes, (b) post-eruptive subsidence above the former diapir forms concentric cones of the original sills whilst minor renewed eruption intrudes ring dykes along fault lines (c) the Ardnamurchan landscape is dominated by three such phases of Tertiary eruption and collapse eroded down to the line A–B in (b). Source: In part after Whittow (1992).*

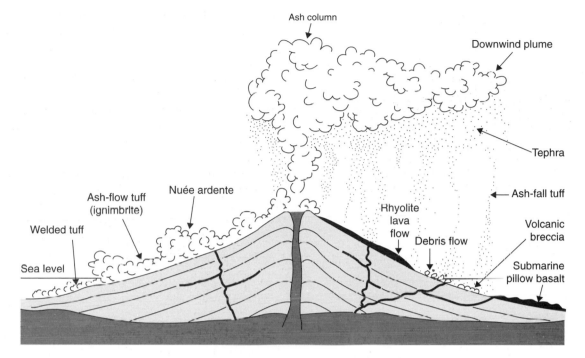

Figure 12.6 *General processes, rocks and landforms of strato-volcano eruptions. Source: After Howells et al. (1981).*

extends as an **ash** column several kilometres into the atmosphere and a *plume* often thousands of kilometres downwind. Column collapse of heavier ash, **lapilli** and **tephra bombs** back towards the volcano develops into fast-moving, incandescent (fiery) ash flows or **nuées ardentes** at ground level, compared with finer **ash fall**. Sedimentation from the atmosphere and through water, from flows or plumes extending seawards, forms **pyroclastic rocks**. Heat often welds fine-grained **ash-fall tuffs**, even under water, and coarse debris forms **volcanic breccia**. Lightweight, highly vesicular and vitreous (glassy) magma cools as **cinders** and **pumice**. Nuées ardentes, with temperatures between 250–700° C, form **ash-flow tuffs** or **ignimbrites** on cooling (Figure 12.6). Widespread distribution of mineralogically distinctive tephra from single events gives **tephrochronology** a powerful role in geological dating.

Explosive activity from andesite–dacite–rhyolite magmas is concentrated in 'mature' arc-margin and island-arc orogens at **strato-volcanoes**. They are the largest terrestrial forms and epitomize the classic view of volcanoes as steep-sided composite cones of stratified (layered) lava flows, tuffs and breccias (Figure 12.6). Several generations of satellite cones surround a main vent, each with its feeder dykes. Strato-volcanoes commonly vent 10^{1-2} km^3 of ash and magma in each eruption and eventually construct volcanoes 1–3 km high with 10^{2-3} km^3 of magma products on base areas of 10^{3-4} km^2 in a few million years. When eruption has

emptied the near-surface magma chamber, the main vent collapses to form a **caldera**, varying in size from the well known Crater Lake in Oregon, some 8 km in diameter, to Lake Taupo in North Island, New Zealand, over 70 km diameter.

For every modern strato-volcano there are hundreds disguised by subsequent events in older continental crust. They reveal important clues to earlier volcano-tectonic cycles. Their volcanic arc land-systems can still be identified in the modern landscape which they help to shape. Wales tells as clearly as anywhere the volcanic-arc/collision orogen story of the closing Iapetus Ocean. A 10 km thick rock pile forms the *Welsh Basin*, occupying most of Wales and consisting of Lower Palaeozoic marine sediments and island basalts. As the ocean subducted south-eastwards, they were increasingly intruded and coated by arc-collision magmas and ashes, seen most extensively in Snowdonia, North Wales. Two eruptive cycles produced over 200 km^3 of *sub-aerial* and *subaqueous* igneous rocks, up to 2 km thick, in just 3–5 Ma. 600 m of tuffs, breccias and basalts form the thickest single formation of Lower Rhyolitic Tuffs, associated with a caldera 30–40 km across centred on Snowdon (Figure 12.7, Table 12.2). Caldera subsidence was probably caused by extensional rifting, adjusting to crustal compression, creating **fissure** rather than central vent eruptions. Today there are some 600 active/dormant strato-volcanoes worldwide, generally of late Cenozoic age. Eighty per cent

Plate 12.1   *Explosive eruption of Mount St Helens on 18 May 1980. The lateral blast sends large clouds of ash and tephra into the atmosphere, with heavier particles beginning to fall out.*
Photo: Keith B. Ronnholm.

are located within the Pacific Ring of Fire, lying outside the **Andesite Line** which divides the Pacific into strato-volcano and shield volcano provinces (see Figure 3 in the box on p. 34).

Subduction magmatism drives plate-boundary volcanoes but isolated hot-spots or sub-rift diapirism are the key to intra-plate volcanoes. Magmas are usually basaltic and generate **shield** volcanoes in ocean intra-plate settings. There are thousands 1–2 km high, extinct or dormant, which do not break surface. Those which do often form long volcanic island chains, best seen in the Hawaiian and other Pacific islands. Slow, intermittent sea-floor effusion of magma over a 'geostationary' hot-spot has studded the main Pacific plate with a 6000 km line of volcanoes in its journey north-west during the past 70 Ma. Moreover a shift in plate direction away from the East Pacific Rise *c*. 40 Ma ago, adjusting to neighbouring plates, realigned islands in the Hawaiian and adjacent chains. Hawaii itself is at the active point in the chain. Its shield volcanoes of Mauna Loa, Kilauea and three others rise 10 km from the sea floor at shallow angles, due to low-viscosity magmas, with the uppermost 4 km above sea level. The Galapagos

(Pacific Ocean), Azores, Madeira and Canary island groups (Atlantic Ocean) are other well-known island shield volcanoes. Basalt effusion rates are more easily assessed on land, where they appear as narrow, linear fissures or vast flood basalts. The 100 km-long Laki fissure, where the Atlantic mid-ocean ridge goes onshore in Iceland, discharged over 5000 m^3 per second for several weeks in 1793 and its sulphurous exhalations upset Benjamin Franklin's visit to Paris, 2000 km away. More impressive still, the Late Cenozoic Roza eruption in Oregon probably contributed a peak of 1500 km^3 per week to the Columbia basalt plateau.

Volcanic eruptions have major environmental and human impacts. *Plinian*-style strato-volcano eruptions, named after Pliny the Younger's graphic account of the eruption of Vesuvius in AD 79, are the most violent. Blast flattened a large area of forest on Mount St Helens in 1980 and slope disturbance disrupted hydrological and vegetation systems, generating landslides and debris avalanches aided by catastrophic ice melt. Ash from 1 km^3 of erupted magma spread over 600 km^2. Pinatubo erupted ten times this volume of ash and a further 3 km^3 of dacite magma

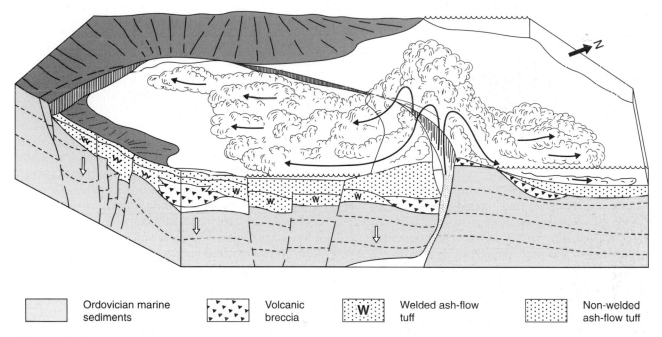

Figure 12.7 *A reconstruction of Ordovician fissure eruptions and caldera collapse, Snowdonia, North Wales. Source: After Howells et al. (1991).*

in nuées ardentes during 1991. **Lahars** of liquefied ash and other debris, named from Indonesia, where they are a major hazard, were particularly destructive around Nevado del Ruiz in 1985. Area impacts and other destructive effects from the blast of these and other recent volcanoes covers some $10^{3\,1}$ km². Herculaneum and Pompeii were destroyed by ignimbrites and ash falls during the AD 79 eruption of Vesuvius with over 25,000 casualties, and nuées ardentes travelling at up to 500 km h⁻¹ from Mount Pelée in the Caribbean killed 26,000 in 1902. Spectacular post-eruptive sunsets hint at gaseous and particulate inputs to the atmosphere, with short-term impacts on radiation and moisture balances. Annual average yields of some 1–3 km³ of new magma and some 20 million tonnes of related sulphur dioxide can be matched or exceeded in single large eruptions considered capable of a small but significant role in global climatic change. Much Quaternary volcanism may be triggered by coastal stress responding to rapid, climate-driven sea level changes.

The rock cycle: (2) Metamorphism

Rock material is subjected, during formation, to immediate **syngenesis** or progressive **diagenesis**. These are changes in form commonly involving low-magnitude compaction by rock or water overburden pressures, *dewatering* (dehydration), *degassing* and even small thermal effects. They create textural and chemical changes generally regarded as the final stages of lithification of previously unconsolidated rock material. Higher temperature/pressure levels such as those experienced in tectonic activity, however, trigger more substantial changes. Magmatism is an extreme response to temperature/pressure changes in the lithosphere. Similarly, tremendous mechanical forces generate wholesale crustal reorganization through subduction or uplift. Both interlinked processes transform rock beyond recognition. **Metamorphism** alters texture or mineralogy permanently, without a liquid phase – i.e. short of melting. **Metasomatism** is minero-chemical change through infusion of high-temperature fluids. They may occur together but metasomatism is particularly important in oceanic crust. **Migmatization** represents the extreme range of metamorphism at the boundary with magmatization. All are further distinguishable from rock **deformation**, which is mostly a mechanical effect on the geometry of entire rock structures rather than on their internal lithological nature.

Metamorphism takes two forms and is graded in intensity. *Contact* metamorphism bakes or recrystallizes rock in a localized **metamorphic aureole** around a magma intrusion. *Regional* metamorphism deforms and recrystallizes rock on a large scale by compression and heating in crustal shortening/subduction zones. We know from magmas that temperature/pressure effects are difficult to separate but our familiarity with some forms of metamorphism may help. Wrought iron is hardened by hammering (compression). Simultaneous heating greatly assists by making the minerals more ductile, capable of adjusting to the required shape, such as a sword blade. When hot, it can also be bent without losing strength. Similarly, fragile snowflakes may assume a dense, crystalline form (ice) by compression, accompanied by *pressure-melting* at crystal edges. This does not break the rule about not melting! Tiny films of water are essential catalysts and permit realignment of constituent minerals in all forms of metamorphism. The snowflakes have *not* melted and refrozen; like the iron, their original constituents have merely recrystallized into a new, stronger configuration.

Proceeding beyond diagenesis, in which a given rock will have assumed textural and chemical equilibrium, regional metamorphism may first compress the material. This causes shortening in the direction of compressive stress by squeezing any remaining plate-shaped grains or minerals into line, enhancing its foliated or planar structure. Poorly cemented *clastic* (granular) sediments deform most readily and their grain-size and relative abundance of phyllo- or sheet silicates (see Figure 12.2) determine the quality of foliation. Thus mudstones with weak bedding – horizontal diagenetic structures – acquired during deposition were converted into shale and then slate by progressive compression. Increasing foliation or cleavage converts flagstone suitable for pavement into strong slate suitable for roofs. Coarser silty clays and sands are transformed into phyllites and schists respectively, the latter developing coarser **schistosity** rather than closely spaced cleavage.

The role of phyllosilicates in pressure metamorphism is a vital clue to an important function of thermal metamorphism. Parallel rise of temperature with pressure in regional metamorphism converts minerals species stable in a less compressed state to mineral species more comfortable in a more compressed state by **solid-state recrystallization**. Foliation is conditioned therefore not by initial textures alone but by the extent to which new crystalline textures can develop. This, in turn, depends largely on the available minerals and is exemplified by granite and its metamorphic 'twin', gneiss. Random quartz–feldspar–biotite crystal assemblage in granite contrasts with marked foliations in gneiss in which the phyllosilicate mineral biotite forms bands. Absence of phyllosilicates leads to massive rather than foliated structure. Marble, quartzite and amphibolite represent metamorphic forms of carbonate, quartz and basalt.

As with magmas, high temperature allows recrystallizing minerals to grow with time, and the right mineral assemblage can replicate fine or coarse textures over short or long time scales. Small amounts of pore fluids enhance recrystallization by diffusing dissolved minerals through the rock, or depositing others as they are driven off at higher temperatures, often as vein minerals in fractures. Some minerals are dependent on particular temperature/pressure environments, exclusive to metamorphic rocks. They form facies or families diagnostic of those environments and reflect the grade or severity of metamorphism. Grade is, in part, a trade-off between temperature and pressure. The principal metamorphic facies and environments are shown in Figure 12.8.

It becomes apparent that metamorphic zones must cover large tracts of continental crust through their formative association with tectonic belts. The cumulative effects of six to eight global orogenic episodes have created complex metamorphic belts around the cores of contemporary and older orogens. Together with ancient gneissose and schistose zones of the cratons, they form up to 70 per cent of continental crust. By substantially strengthening this crust, and in particular the flysch, mélange and pelagic soft sediments of ocean basins accreted via subduction, metamorphism has a major impact on continental architecture. Its reorganization of rock material extends and refines geological fractionation and, in the process, creates distinctive suites of site-specific geological resources used in the human domain (see box).

The rock cycle: (3) Sediments

Sediment source

Sediments are the unconsolidated detrital and dissolved remains of other rocks and organisms.

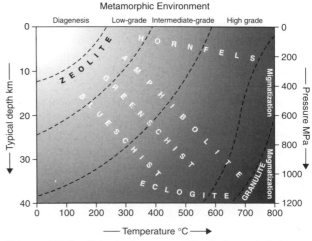

Figure 12.8 *Principal metamorphic facies and their environments of formation. Source: In part after Skinner and Porter (1995).*

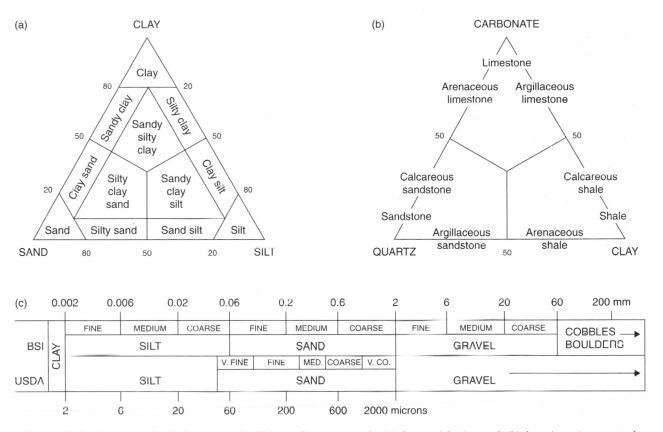

Figure 12.9 *Some standard schemes for classifying sedimentary rocks (a) by particle size and (b) by mineral content; the numbers are percentages. (c) Below these ternary diagrams the British Standards Institute and US Department of Agriculture define particle size and their millimetre and micron size ranges.*

Weathering and erosion source the *sediment cascade* from continental denudation to eventual deposition in local (*autochthonous*) or distant (*allochthonous*) sedimentary basins. The variety of processes by which this occurs and the transient sedimentary landforms produced are covered in later chapters. First, we need to establish the general character of sediment bodies and their depositional environments. Sediments retain characteristics of their source area and environment and acquire new ones through refinement in transit, transported by gravity, water, ice and wind. They become lithified as sedimentary rocks and continue through the rock cycle, reforming continental crust when retained as **terrestrial sediment** in continental basins. Most sediments, however, eventually reach the oceans as land-sourced, **terrigenous sediment**. All are then recycled by deformation, magmatization and metamorphism at plate boundaries. Sediments of pelagic (surface) and **benthic** (deep-water) marine origin, formed by biogeochemical processes in the oceans themselves, share a similar fate.

Denudation transforms rocks into disaggregated minerals and lithic fragments, lacking cohesion, and solutions which are readily removed from their source. At this stage they possess textural (particle size, shape) and chemical properties which are clearly diagnostic of parent rocks and, to some extent, the denudation process responsible. Three principal styles of sediment are recognized. **Clastic sediments** (from the Greek word for 'broken') are formed by particles broken off parent rocks and initially reflect fragment size and shape. A distinction is normally drawn between larger fragments or individual **clasts**, larger than sand size, and finer grains (sand, silt and clay sizes) which form a **matrix** (Figure 12.9). **Chemical sediments** are precipitated primarily from dissolved salts, silicates and carbonates and form **biogenic sediments** when they have an organic origin, shared with 'clastic' (shell and bone beds) and carboniferous (peat and coal) sediments. Figure 12.10 shows the relationship between sourcing process, product and the eventual form of these sediments.

Sediment transport and deposition

Once in transit, the character of the original denudation product weakens and the sediment begins to

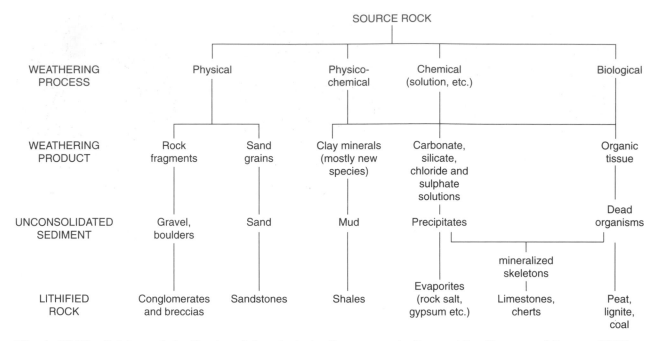

Figure 12.10 *Origins and classification of the principal sedimentary rocks. Source: After Newton and Laporte (1989).*

take on the signature of the transport environment. This occurs through particle attrition and sorting, in response to transportational energy, and further chemical weathering or fractionation as less stable minerals are taken in or out of solution. Particle edge roundness or sphericity tends to increase with time in transit, reflecting clast-to-clast collision. Comparisons between compact and platy grain shapes indicate lithological distinctions between massive and schistose/laminated parent rocks. Attrition also reduces individual particle size, whereas sorting alters the mean size and particle distribution of the sediment body through a close correspondence between particle mass and the velocity and viscosity of the transporting medium. Figure 12.11 shows the particle size/velocity relationships associated with the entrainment, transport and deposition of single spherical grains in water. Water and wind produce well sorted sediments, i.e. with small standard deviations about mean particle size, indicating the energy environment of transport (in-transit sediment) or deposition (deposited sediment). Low-energy lacustrine, lagoonal or deep-water sediments are fine-grained, and high-energy torrent or storm beach sediments are coarse-grained. The much higher viscosity of glaciers, debris flows and basal elements of turbidites (see below) – where ice or clasts themselves provide buoyancy – markedly reduces the degree of sorting, and particle size distributions may be bi- or multi-modal (Figure 12.12). Textural changes occur in transit and are retained at the point

of deposition, where the sediment compactness is measured by the *packing density* of grains and the *void ratio* of remaining spaces. This is determined by particle size, shape and sorting and also reflects syn- or post-depositional **overburden pressure** – exerted by overlying sediments, water or ice bodies.

Sediments also acquire directional and structural properties. Particles deposited by anything other than precipitation through still water or air are subject to an *anistropic force field* and elongated particles tend to

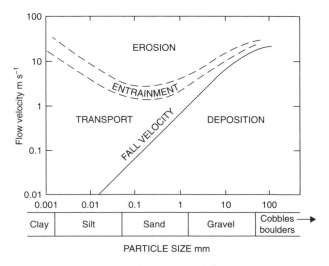

Figure 12.11 *Flow velocity thresholds for the entrainment, transport and deposition of granular particles, using BSI particle size classification. Source: After Hjulström (1935).*

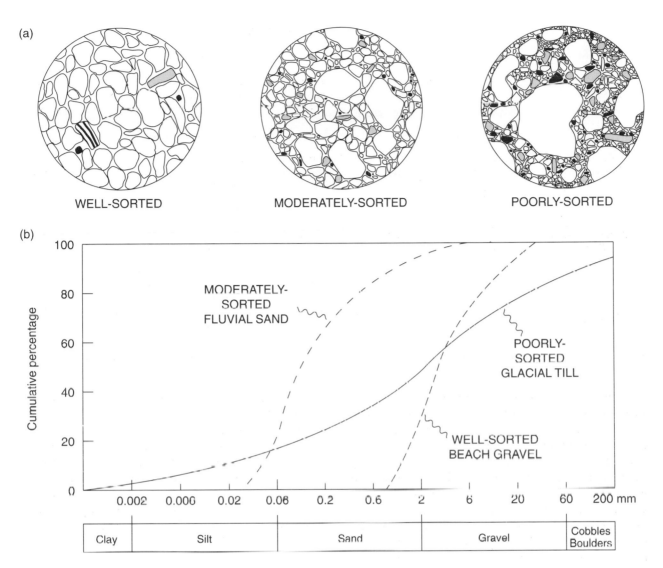

Figure 12.12   *Sorting and BSI particle size characteristics of sediments: (a) three samples shown at the same scale but not ascribed to specific environments; (b) cumulative percentage curves, showing typical particle size distributions, which measure sorting visually by the relative horizontal span of each curve and statistically by standard deviation.*

move either parallel, or transverse (normal), to the flow or slope. This is often preserved as particle orientation in sediments and, together with imbrication – the stacking of particles dipping (sloping) towards or away from the flow direction – provide **palaeocurrent** information of former in-transit material. This extends to include **syndepositional sedimentary structures**, formed during or immediately after deposition. Parcels of sediment are laid down as a layer or stratum determined by the underlying topographical surface, sediment properties and the geometry of the transporting medium. Its lower surface or **bedding plane** usually distinguishes it from the previous (older) parcel by subtle changes in texture or colour in a **conformable sequence**, or dramatically so at an **unconformity** which marks a pause, erosive event, etc. Within each

stratum a series of subsidiary structures reflect the direction and energy of transport (Figure 12.13).

Sedimentary facies, environments and tectonic basins

So far we have reviewed some general, descriptive aspects of sediments. We must also remember that each parcel is the product of specific environmental processes in a broader landsystem and constitutes a **sedimentary facies**. A facies may vary as much internally as it does from adjacent facies. It is said to be a *litho*facies or a *bio*facies, depending on whether minerogenic or organic constituents predominate, and description of its characteristics is the first stage towards reconstructing its environmental origins.

Plate 12.2   *Coarse sandstone – the Millstone Grit of Carboniferous (Namurian) age – now exposed on the summit of Ingleborough, North Yorkshire. Near-horizontal bedding planes separate cross-bedded and rippled sands flowing from left to right into a former delta.*
Photo: Kenneth Addison.

Each facies may be the stratigraphic equivalent of a sedimentary landform and may represent a single event. Contemporary geologic or geomorphical processes are present at the surface and older ones in the stratigraphic record (Plate 12.2).

Facies assemblages linked by common genetic origins constitute a **sedimentary environment** whose principal terrestrial forms correspond very closely with the *geomorphic landsystems* described in later chapters (Figure 12.14). Moreover, the global location of major sedimentary environments is determined largely by plate tectonics, which create **sedimentary basins** at specific locations, collecting the sedimentary signature of the volcano-tectonic, denudation, climatic and biological processes operating there (Figure 12.15). The principal endogenetic basins are subduction trench and arc basin systems; continental, mid-ocean and aulacogen rifts; passive-margin shelf and continental subsidence basins. Exogenetic processes cut valley floors and fill lake basins on a smaller scale. Sea-level movement driven by tectonic processes or climate change alters the size of terrestrial and oceanic basins and influences erosion rates. Sedimentary onlap and offlap sequences may indicate episodes of marine *transgression* or *regression* respectively and/or changes in sediment supply.

Particular reference is made here to biogenic and chemical rocks, which feature less in the review of clastic sedimentation in later chapters. Living organisms form facies and landforms such as wetlands, salt marshes and coral reefs, and burrowing organisms disturb soft sediments (**bioturbation**). Dead organic matter retains much of its life form as *fossils* in partially decomposed sediments such as the peat–lignite–coal sequence, shell and bone beds or whole individuals, or as completely decomposed organic precipitates in carbonate and bio-silicate rocks.

Dissolved minerals are precipitated in most environments but especially in the terrestrial warm arid zone, in tropical tidal zones and in deep oceans. Evaporation leaves dissolved minerals to recrystallize from shallow lakes, fluids surfacing through capillary action to form surface concretions and tidal lagoons and mud flats (**sabkhas** and **salinas**). The most common **evaporites** and their total solids percentage in sea water are halite (NaCl, 78 per cent), potash salts (K_2SO_4, 18 per cent), gypsum/anhydrite ($CaSO_4$/$CaSO_4.2H_2O$, 3.6 per cent) and calcite ($CaCO_3$, 0.3 per cent). Fractional recrystallization occurs in the sequence carbonates–anhydrite/gypsum–halite–potash and is common in tropical epicontinental seas (e.g. the Persian Gulf), inter-

(a)

(b)

(c)

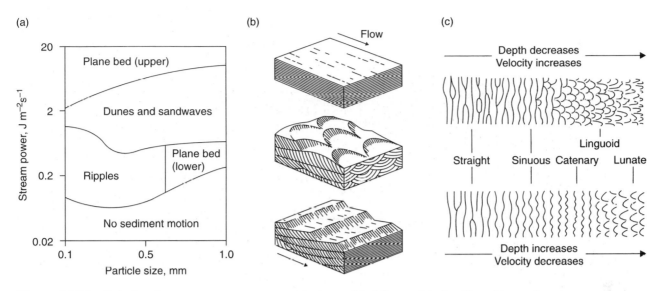

Figure 12.13 *Relations between stream power, particle size and bed forms: (a) the effect of increasing stream power on bedform; (b) upper plane bedform (top), dune (middle) and ripple bedforms (bottom); (c) the form and depth/velocity relationship of ripples. Source: In part after Allen (1968).*

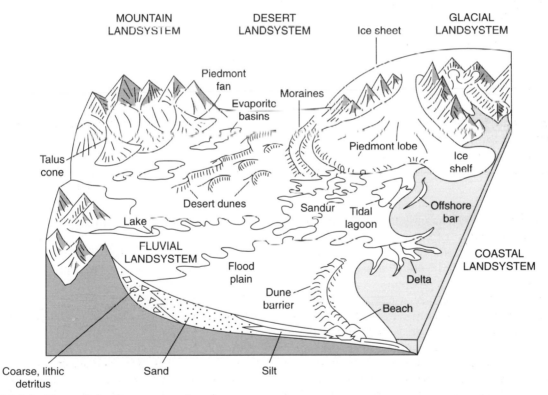

Figure 12.14 *Terrestrial sedimentary environments.*

montane basins (e.g. the Great Salt Lake, Utah, and Salar de Atacama, Chile) and continental basins (Lake Eyre, Australia) today. They were extensive in Permian tropical Pangaea when the Zechstein Sea and Delaware basins generated the large evaporite deposits of modern northern Europe and the south-western United States. Capillary action due to high evaporation in arid environments also draws dissolved minerals to the surface, forming crystalline **duricrusts** of calcrete, silicrete or ferricrete (Figure 12.16). Deep marine precipitation is covered below.

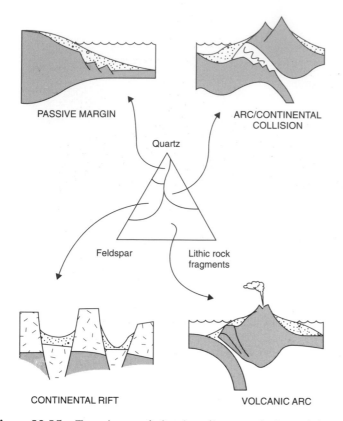

Figure 12.15 *Tectonic control of major sedimentary basins and the character of their siliciclastic sediments (stippled). More mature, quartz-rich sediments are associated with older, lower-energy basins and raw, lithic and immature feldspathic sediments with younger higher-energy basins.*

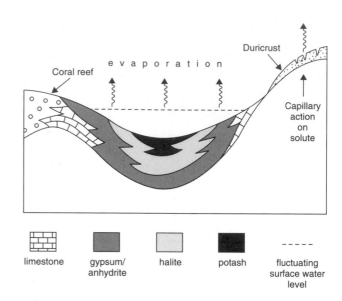

Figure 12.16 *General representations of evaporite rocks in marine and/or terrestrial environments.*

Diagenesis and lithification

Diagenesis refers to transformations which sediment may undergo – either *syn*genetically as facies accumulate or at any time thereafter – up to the point of metamorphism, weathering or erosion. **Lithification** describes only those diagenetic mechanical and chemical processes which transform unconsolidated and invariably wet sediment to hard 'dry' rock. This distinguishes between diagenetic changes which alter the character of rock without substantially altering its strength, such as the development of **load casts** and **bioturbation structures**, and wholesale decomposition, solution, leaching and replacement of buried soft body parts and less stable aragonite skeletons converted to calcite.

Overburden pressure from accumulating sediment piles and overlying ice or water bodies stimulates consolidation, by reducing the void ratio to a level determined by sediment texture and pressure, and dewatering by expelling pore fluids. Mass shrinkage, by up to 50 per cent, establishes orthogonal joint or discontinuity systems in which bedding planes form the

principal horizontal set. Dewatering (dehydration) may also create load casts as material of varying saturation and competence is squeezed, slumps or intrudes other layers. In addition to these largely mechanical effects, dewatering through high pore pressures and dehydration through surface evaporation precipitates dissolved minerals, especially carbonates, silicates and iron compounds. These cement the grains and give **cohesive strength** to the developing rock mass.

The oceanic rock cycle

It is reckoned that the entire global ocean is cycled through oceanic crust every few million years. This process is driven by the same thermal convection responsible for the mid-ocean ridges and is known as **hydrothermal circulation**. It alters the condition of sea-floor rocks and is capable of generating new minerals. Oceans are also a major route for recycling crustal lithosphere by reprocessing terrigenous sediments via subduction or accretion, albeit on far longer time scales. Together these processes form a major part of the rock cycle and influence the quality (geochemistry and turbidity), performance and biochemical processes of ocean water.

The rock cycle: (4) Hydrothermal circulation and metasomatism

Ocean water not only comes into contact with the sea bed but can also circulate to depths of several kilometres in oceanic crust. It therefore penetrates the basalt and gabbro layers and may also reach upper-asthenosphere peridotite. It gains access to the lithosphere via faults and fractures generated by mid-ocean ridge rifting, post-formational cooling and subsidence. It occurs over a very wide area – perhaps over 30 per cent of the ocean floor – driven by mantle convection typified by heat fluxes exceeding 50 mW m^{-2} for 'new' crust up to some 50 Ma old and 200–250 mW m^{-2} at the ridges. This draws water in over a wider area and pumps it in concentration as **hydrothermal plumes** through axial vents. Although the process is known simply as *hydrothermal circulation*, it is clear that water–rock–magma reactions occur in a number of different environments and styles, determined – not surprisingly – by the thermal environment. Even without mantle convection, **cold sea-water weathering** occurs at the sea bed. As may be expected, this is mostly by hydration, which produces hydrated aluminosilicate clay minerals, but oxidation also occurs, forming oxide films on Fe and Mn minerals relevant to red clays considered below.

At elevated temperatures above 200° C hydration may form the lowest-grade metamorphic facies, **zeolite**. It is appropriate to include precipitation of solutes by, in effect, **reverse weathering** here, since it involves hydrothermal minerals reviewed presently along with minerals sourced elsewhere. High temperature/low water volume reactions drive metamorphism, particularly in subduction zones. This is quite distinct from mineralization associated with high water volumes circulating at mid-ocean ridges with water:rock ratios of 10–100:1. Metasomatism takes large volumes of basaltic and other minerals into solution, infusing new species into the crust, depositing others around and beyond the vents and taking yet others out of solution. Three associations are found, reflecting the peridotite–gabbro–basalt layering towards the ocean floor and parallel temperature/pressure decline. Serpentine – $Mg_3Si_2O_5(OH)_4$) – forms on the hydration of olivine in peridotite above 400° C. Hornblende – $(Na,Ca)_2(Mg,Fe,Al)_5O_{22}(OH)_2$ – forms in gabbro at 200–400° C, and a whole cluster of new minerals at similar temperatures in basalt, including albite – $NaAlSi_3O_8$ – chlorite – $(Mg,Fe,Al)_3(Si,Al)_2O_5(OH)$ – and epidote – $Ca_2(Al,Fe)_3Si_3O_{12}(OH)$ (see Figure 12.8).

They are key minerals of the metasomatic facies *Serpentinite*, *Amphibolite* and *Greenschists* and these reactions portray *par excellence* the vital importance of hydration (OH), the magmatic minerals Fe, Mg, Al, Si, Ca and Na, and solid solutions to ocean and oceanic crust geochemistry. Sea water becomes enriched by chloride and the soluble minerals Mn and

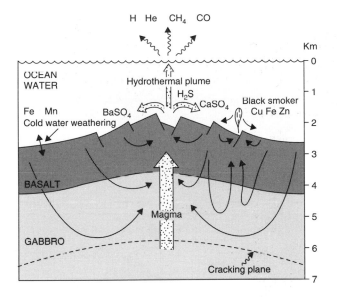

Figure 12.17 *Hydrothermal circulation around a mid-ocean ridge, indicated by solid arrows. Ocean water circulates as far as the cracking plane, above which layered ocean crust is stretched and cracked.*

Fe but depleted of Mg. Hydrogen sulphide (H_2S), barium sulphate ($BaSO_4$), anhydrite ($CaSO_4$) and insoluble metal sulphides of Cu, Fe and Zn are precipitated – particularly by black smokers or hot plumes with suspended and dissolved minerals – to form important mineral and ore deposits. Vents also stimulate their own chemosynthesizing ecosystems and hydrothermal circulation also exhales magmatic gases, including hydrogen (H), helium (He), methane (CH_4) and carbon monoxide (CO) (Figure 12.17).

The rock cycle: (5) Marine sedimentation

Marine sediments vary according to their composition, source of materials and location in either the offshore slope system or the abyssal plain. Terrigenous sources of eroded and transported continental erosion products are mainly **minerogenic** in nature, and their abundance and *calibre* (particle size) diminish seawards. They are mostly more stable minerals like quartz, potassium feldspars and biotite flushed, with dissolved minerals, on to the continental shelf by rivers at rates of 1.8×10^{13} kg solids and 4×10^{12} kg of solutes each year, or derived from marine erosion of the coast itself. Heavy minerals such as gold (Au), copper (Cu) and tin (Sn) form **placer** deposits among the shelf sediments. **Biogenic** rain-out of dead marine organisms and their chemical derivatives increases in importance away from the shore to become the predominant source of abyssal

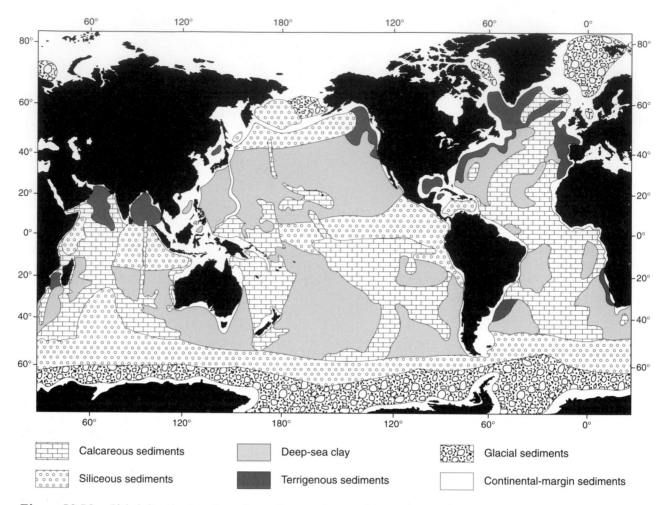

Figure 12.18 *Global distribution of sea floor sediments. Source: After Davies and Gorsline (1986).*

sediments, flooring 62 per cent of deep-ocean basins. They form *calcareous* (calcite/aragonite) and *siliceous* (silicate) muds derived from the skeletal parts of plankton. Carbonate mud is three to four times more abundant and referred to as an **ooze** if over 30 per cent is derived from diatoms and coccoliths (marine micro-organisms). Silicate solution is less depth- and temperature-sensitive than carbonates and hence is found in deeper and colder ocean areas. Carbonate rocks are associated with tropical/warm–temperate oceans today, with a contrasting silicate mud belt in the Southern (Antarctic) Ocean (Figure 12.18). Silurian and Carboniferous limestones and Cretaceous chalk represent former Laurasian and Tethys carbonate oceans in the geological record. Hydrothermal circulation adds a suite of **hydrogenic** sea bed sediments which become buried by younger sediments as they rift away from their mid-ocean ridge source. Manganese (Mn) nodules are widespread and of increasing resource potential for, additionally, Fe, Cu, Co and Ni.

This relatively simple arrangement becomes complicated by intercalation or lateral intergrowth of different sediment bodies through sea-level change, pulsed sediment supply and the rain-out of **red clays** and **glaciomarine sediments**. The former are iron oxide-coated mixes of clay minerals from terrigenous sources carried far into the ocean by suspension in sea water, or atmospheric plumes of fine volcanic ash or desert dust, and biogenic material. Glaciomarine deposition is restricted by the fluctuating survival range of icebergs or grounded marine glaciers during Quaternary cold and temperate stages. Sediments are decanted as icebergs melt, or emerge from **tidewater glaciers** as waterlain till, or as powerful plumes forming extensive *mud drapes*. The Yakataga formation, in the Gulf of Alaska, is a spectacular example of ocean–glacier–tectonic interaction. Pacific Plate subduction and continental plate elevation have caused a vertical accretion of 5 km of Plio-Pleistocene glaciomarine sediments in the coast ranges.

With these sediment sources and characters in mind, we can now consider their depositional envi-

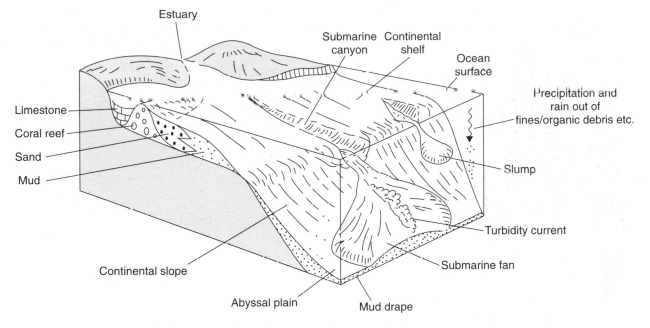

Figure 12.19 *The general range of marine sedimentary environments.*

ronments (Figure 12.19). Shallow-water continental shelves are generally high-energy, unstable environments. Large-calibre gravel and coarse sands are deposited inshore in the *surf belt* and grade progressively into fine sands and muds towards the outer shelf. The shelf is constantly reworked by waves, tides, storms and currents which leave positive (sediment accumulation) or negative (scour) **bedforms** to reflect current velocity and direction. Giant sand waves and offshore sand/shingle bars form the mobile shelf morphology. Mud drapes represent transient fine sediment fluxes in these environments. They are extensive only in sheltered estuaries and epicontinental seas, or at depth. Shell debris accumulates in high-energy onshore and offshore ridge structures, and carbonates derived from them may form cement in sediment diagenesis. **Reefs** build more permanent bio-morphological structures. This entire assemblage is susceptible to fluctuating sea levels, and onshore/offshore zones often display relict features from both environments. The inshore continuum with the coast is developed in Chapter 17.

Continental slopes are probably the most dynamic oceanic sedimentary environments, responsible for transferring shelf sediments to the abyssal plains. Extended efflux from large deltas develops fan deposits on the slope, more commonly during times of low sea level (i.e. cold stages) when much of the shelf becomes an extension of the coastal plain and is thus truncated. Few deltas are large enough to feed such fans today. The Ganges-Brahmaputra fan is the most impressive and extends over 1500 km into the Bay of Bengal. Mass movement by slumping or liquefaction and flow is a more widespread transfer process. Turbidity currents are low-density debris flows, mobilized by top-loading of sediment, currents or earthquakes. They etch canyons in the slope and discharge large volumes of sediment at speeds of 20–50 km h^{-1} for 10^{1-2} km beyond the continental rise. **Turbidite** sequences form the continental rise through the coalescence of coarser, basal fans and assist in levelling the abyssal plain by long-distance transport of suspended fine particles. Each turbidite can be recognized through a fining-upwards progression, gravel–sand–mud, as the respective particles are decanted out over time. Sediment sequences several kilometres thick form in the slope/rise area in this way.

Abyssal plain sediments are usually isolated from adjacent slopes in the 'quietest' parts of the ocean and accumulate extremely slowly. They are primarily undisturbed biogenic muds 0.5–1.0 km thick, forming at rates of 0.1–4 cm ka^{-1}, illustrating the great stability of these areas. Fractionation of the stable isotope composition of oxygen in sea water and organic carbonate provides some of the best long-term records of recent Earth history. The ratio of ^{18}O to ^{16}O in water is temperature-dependent. The lighter isotope ^{16}O is taken up preferentially when sea water evaporates, enriching the remaining water in heavier ^{18}O. Water retention in ice sheets sustains the difference and leads to enrichment of

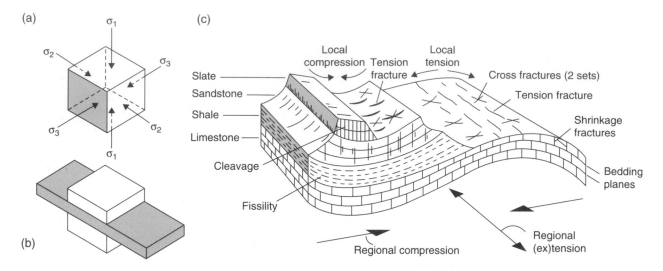

Figure 12.20   *Principal stresses and their application to geological structures. Principal stresses operating on the cube (a) are equal. Cube (b) has been compressed vertically and has responded by extending in one horizontal plane whilst conserving its volume. (c) The application of these forces to a sedimentary rock sequence – which also possesses lithological fractures. Source: In part after Selby (1993).*

^{18}O in marine carbonates and ^{16}O in ice sheets. Both environments thus record global temperature and ice conditions.

Rock deformation: folding and faulting

Tectonic activity sets up huge stresses in rocks which result in strain or deformation. This is quite different from denudation and the associated processes of rock disintegration, which are the subject of later chapters. Deformation occurs along planar structures (fold, faults, etc.) and the vast bulk of intervening rock mass may remain intact, even if relocated *en masse*. If we focus on any small cube of rock within the crust we can measure the force applied by the surrounding rock to each of its six faces. A three-dimensional *orthogonal* or right-angled pattern of forces emerges, with a pair of opposing forces in each **principal stress** direction. Assuming that internal constituents are packed as tightly as possible, the cube will retain its shape provided that the confining forces (σ_2, σ_3) are equal and resist the compressive force (σ_1), creating an *isotropic* force field (Figure 12.20). Deformation occurs when the forces are *anisotropic* (unequal) and the rock is able to deform. This depends on its **rheologic properties**, or ability to flow, and occurs – as we have seen – in crustal convection as slow granular creep, strongly influenced by temperature and fluid content. Fluids at high pore

pressures improve the plastic response. Rock may respond to progressive increases in stress by **elastic** and **plastic** deformation and **brittle failure** (Figure 12.21a). Elastic strain is recoverable once the stress is removed, whereas plastic strain is permanent, as in the different response on stretching an 'elastic' band and bending a soft metal rod. Rock may behave elastically at very low pressures and temperatures and plastically at higher pressures and temperatures (i.e. at greater depth), especially if the strain rate or deformation speed is slow, typically changing length at 1 per cent per 10^4 a! Brittle failure, or fracturing, occurs if the strain rate exceeds the plastic deformation rate. Earthquakes occur when this happens abruptly.

The principal forms of plastic deformation are **foliations** (including cleavage) and **folds**, which produce banded and wave-like structures respectively. Foliation tends to occur at small–intermediate, microscopic–hand specimen, scales whereas folds occur at all scales with wavelengths up to 10^{1-2} km and in a variety of wave forms (Figure 12.21b, c). Material between each foliation or fold plane remains intact but is realigned in sympathy with the deformation. This creates strength anisotropy, which is exploited by denudation. We have seen that foliation is low-grade metamorphism but folding, by comparison, represents a shortening of rock mass and the degree of folding reflects stress intensity and rock deformability. **Nappes** bridge the difference between folds and faults and, arguably, are the most impressive deformation. Rocks are folded well past the upright

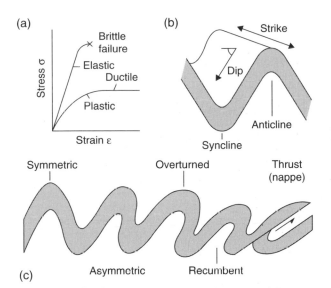

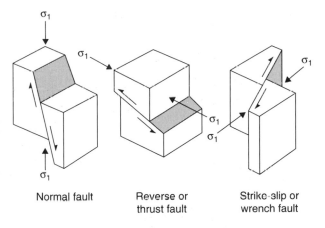

Figure 12.22 *Principal types of fault.*

Figure 12.21 *Styles of rock deformation and folding: (a) stress–strain relationships, (b) the terminology of simple folds, (c) types of fold, showing increasing deformation to the right.*

and assume a recumbent form, but shortening does not stop there. The lower limb continues to deform by thrust faulting, which transports the entire fold in the thrust direction. In continental collision, slivers of cratonic basement may also be thrust forward. However, displacement may occur along a **décollement** or detachment surface instead, requiring – or creating – a zone of significant change in material properties, rather than a fault *sensu stricto*. Such low-angled surfaces may also lead to tectonic-scale gravity sliding in the opposite direction. The Alpine thrust zone (see Figure 25.5), Canadian Rockies and Appalachians show nappe/décollement structures and the basin-range system of the south-western United States exemplifies gravity sliding (see Figure 25.3). Deformation is also associated with *diapirism*, when overlying rock is intruded by less dense magma (hot) or salt and mud (cold) diapirs.

Brittle failure fractures rock without disturbing the intervening rock mass, at a strain rate and magnitude which locally exceed its intact strength. **Joints** are incipient faults, generally restricted to an individual facies, and occur initially in response to contraction brought about by cooling (igneous) and dewatering (sedimentary) processes. **Faults** are joints with differential movement on opposite sides and are identified by the principal displacement direction (Figure 12.22). Faults and joints are organized in geometric (usually orthogonal) patterns and each plane is defined by a **dip** and **strike**. They differ in scale, with joint spacing at 10^{1-3} cm and faults two or three orders of magnitude larger. Fractures may be 'clean

breaks' or occasionally smoothed, where movement has abraded opposing faces to form **slickensides**. However, both forms may be lined with a 'fill' composed of coarse rock fragments (**fault breccia**), fine debris (**gouge**) or cement – which may be weaker or stronger than the fractured rock mass itself.

Deformation accompanies small-scale rock-forming processes, when they are lithological in origin, and the creation of large-scale tectonic structures. Subsequent stress episodes are likely to be accommodated along existing structures first and, in that way, principal mobile belts often drive younger plate motions and orogens. In addition to their primary function, they are of vital importance to denudation.

Conclusion

We occupy the landsurface of a living Earth and, in a minor way, contribute to its rock cycle by using and discarding geological resources and interfering with the energy flows and components of endogenetic processes. The landscapes around us are strongly influenced by their geological foundations and, armed with sufficient knowledge and the right techniques, we are able to reconstruct their palaeo-environmental history. Long time scales should not deflect us from the need to understand our geological environment and heritage. Large human populations living along plate boundaries are only too aware of how dramatically abrupt geological events can be. Every stage of the rock cycle can be located within the global mosaic of moving plates and morphotectonic landforms and permits a better understanding of Earth's dynamic evolution and assessment of geological resources and better hazards.

Key points

1 Over 2000 minerals, occurring as single or compound elements, are the basic rock-forming units. Mineral species, which grow from melts or solutions, are distinguishable by their composition, cation and anion bonds, crystal structure and other properties. Silicates are the most common minerals, constructed of silicate tetrahedral anions in various combinations or with other cations. Mineral assemblages are formed by and later separated, refined or reorganized in a global rock cycle, at specific sites and times defined by plate tectonics.

2 The cycle commences with the fractional crystallization of solid minerals from rising partially melted asthenosphere peridotite (magma). Iron–magnesium-rich minerals forming at high temperatures (1000–1200° C) are replaced by increasing proportions of silicate minerals as magma cools below 1000° C. Magma may intrude, cool and solidify in older rocks to form plutons below the landsurface or erupt as lava flows, or as effusive or explosive volcanoes, to create new surface landforms.

3 Exposure to significantly different hydrothermal and mechanical conditions at the landsurface triggers denudation. This leads to a cascade of weathering and erosion products which are deposited as sediments, in the short term on the landsurface or continental shelf, but they may eventually reach the sea floor before being recycled. Biogenic sediments form by the accumulation of dead organisms or the precipitation of their dissolved derivatives.

4 All rocks undergo mild diagenesis during and after formation, usually as chemical and textural properties stabilize. They can also be altered geochemically to progressively greater extents by metamorphism, metasomatism or migmatization in higher temperature and pressure conditions. These processes stop short of remelt but, where prevailing conditions exceed the melting point of rock and other surface materials in subduction zones, the cycle is complete.

5 Rock is also deformed mechanically under high stresses, especially those associated with moving plates. Crustal shortening, extension, uplift and subsidence are accompanied by folding, faulting and thrusting and the large-scale displacement of original terranes.

Further reading

Bell, F.G. (1993) *Engineering Geology*, Oxford: Blackwell. The inclusion of a book on applied geology may seem surprising but the first three chapters give an incisive account of rock types, stratigraphy, structures and surface processes. Their application to a wide range of human environments and actions in six further chapters brings geological processes into sharp focus and is a worthwhile bonus.

Park, R.B. (1983) *Foundations of Structural Geology*, Glasgow: Blackie.
This book focuses on deformation and geological structures with good, if rather technical, explanations of folding, faulting, etc. It concludes by relating deformation processes and zones to plate tectonics.

Reading, H.G., ed. (1996) *Sedimentary Environments and Facies*, third edition, Oxford and London: Blackwell.
This is a widely respected text which sets out the environmental character and stratigraphy of sedimentary environments. Links between past and present are well made and extensively illustrated, although there is more on palaeoenvironments than non-geologists need or may understand.

Scarth, A. (1994) *Volcanoes*, London: UCL Press.
Volcanoes are one of the most popular rock- and land-forming processes and this book provides a clearly written, structured and illustrated account of both. It tells the story of developing eruptions at celebrated sites and describes their human impact.

Rocks are stable in the environment in which they form and inherently unstable in any other. As endogenetic processes elevate continental crust into the sub-aerial environment and expose new land surfaces to the atmosphere, geomorphic (exogenetic) processes commence their attack and **denudation** begins. Denudation describes the overall destruction and levelling of continental land mass and is achieved by three sets of processes – weathering, mass wasting and erosion. The first two occur in sequence and are found everywhere, although their form and rates vary. Erosion removes the debris in turn but it can also bypass these stages and remove substantial layers of fresh, unweathered rock, especially during glaciation. This chapter focuses on why and how weathering occurs and its products are redistributed downslope by gravity. Weathering has been described as the 'static attack' of meteorological elements, and the role of gravity on slopes as 'passive', to distinguish it from the *dynamic* role of flowing water, ice and wind. This has some merit but weathering may continue in transit, all mass wasting involves movement (sometimes over long distances at high velocity) and gravity also mobilizes rivers, glaciers, air and coastal currents. Erosional and depositional processes in these four distinct domains are examined in subsequent chapters. We start with an overview of denudation.

Denudation

Tectonic uplift mobilizes potential energy in the landscape; the higher the slopes, the greater the potential energy. Sea level forms a general base level for the terrestrial environment, although we know that terrigenous sediments continue across the continental shelf and slope to the ocean floor. Sea-level change independent of tectonic activity further disturbs denudation rates and, through ice sheet coupling, is a regular feature of the Quaternary Earth. Valley floors, lakes, etc., act as temporary, local bases for adjacent slopes. Denudation also requires the impact of solar-powered systems at Earth's surface, determining hydrothermal and biological conditions of weathering, applying force through wind circulation and raindrops and raising water through the hydrological cycle to provide the kinetic energy of rivers and glaciers.

Denudation rates

Denudation rates interest geographers for a variety of reasons. Measurements of suspended sediment load in streams, changes in coastline location, landslide volume or glacial excavation lead us to extrapolate the magnitude of whole-landscape change and the time scales of events. Armed with a growing database on rates of sea-floor spreading, uplift, new magma formation and ocean sedimentation, we are naturally inquisitive about their dynamic balance with the continents. Knowledge of rates and processes also permits the prediction and management of environmental change, and some global generalizations set the scene. A measure of whole-earth denudation over 4.6 Ba is the extent of reworking given by a *planetary evolution index*, which is the ratio of reworked to original and/or cratered planetary surface. A value of 6.2 for Earth, compared with 0.2 for our Moon and 0.7 for Mars, emphasizes the impact of endogenetic and geomorphic processes here. This is confirmed by the survival of only 15 per cent of Archaean crust, modern mountain systems belonging to three principal orogens less than 450 Ma old and the geological youth of ocean basins.

Detailed assessment of denudation rates, however, is complicated and takes various forms. We can start with tectonics and follow the sediment cascade. Since plutonic rocks and minerals such as granite and diamond are formed at depths between 10 km and 150 km, their surface exposure in the core of orogens testifies to severe denudation since emplacement. Granitic rocks 8.8 km high on Mount Everest required the removal of 30–35 km of overburden, and a similar order of magnitude applies to Alpine, Andean and North American cordilleran batholiths. Average denudation rates calculated from the age of batholiths by K–Ar (potassium–argon) dating, emplacement depth and degree of exhumation are in the range 1–10 km Ma^{-1} (orogens) and 0.1–1 km Ma^{-1} (epeirogens). The highest average rates are often associated with late Cenozoic, even Quaternary, orogens, suggesting that high erosion rates are unsustainable over long periods; this point is explained below.

Tectonic evidence also comes from comparing rates of uplift, the elevation of mountain systems and the conformity of their summit altitudes. Uplift measured over short periods at rates of 5–20 mm a^{-1} would raise another Everest in 1 Ma! This does not happen, owing to a combination of denudation and changing uplift rates which peg Earth's highest summits to 8 km (Himalayas), 7 km (Andes) and 4–6 km (North America). Characteristic saw-tooth profiles of these and lesser ranges shows a remarkable accordance of summits, broken only occasionally by 'one that got away'. Once interpreted as dissected erosion surfaces, these are now considered to represent approximations to **steady state** mountain systems where uplift = denudation. Initial orogenic uplift stimulates denudation, which then triggers *isostatic* uplift (thought to be up to 80 per cent of gross lowering) – the tectonic system 'jacks up' the crust to compensate for the removal of light, continental rocks!

Since *net* denudation does occur, denudation and isostatic compensation must decline over time as the available relief is progressively lowered. Figure 13.1 shows the evolution of a hypothetical orogen uplifted at an average rate of 1.0 mm a^{-1} for 5 Ma. Denudation reduces its potential elevation of 5 km to 4 km during that time and, thereafter (from time zero), the net rate of lowering declines in exact proportion to the remaining relief. Denudation rates therefore decay exponentially, with 50 per cent of initial elevation lost by 10 Ma but about 3 per cent surviving after 50 Ma. This example also assumes a constant, moist climate generating active fluvial and/or glacial erosion, otherwise rates may change, and no **rejuvenation** or new uplift. Isolated higher peaks mark higher local rates of uplift, or immature connection with steep slope/valley systems where denudation is concentrated. The latter exerts general influence on height, with erosion reaching range crests later in broad orogens such as the Himalayas. This may explain why its peaks are 1 km higher than anywhere else and the young Tibetan plateau is not yet dissected. Quaternary glaciation has undoubtedly

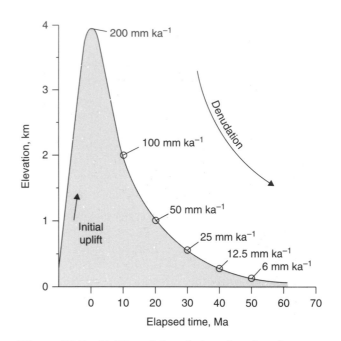

Figure 13.1 *Uplift and denudation of an imaginary orogen; see text for explanation. Source: After Strahler and Strahler (1978).*

played a major role, creating what have been called 'climate-carved' mountains from broad, high plateaux.

Denudation rates *per se* are measured by sediment transfers, particularly in rivers, or sediment accumulation in either transient lake basins or marine basins. Measurements come in various forms, from t km^{-1} a^{-1} or kg m^{-1} a^{-1} (suspended stream load), ppm concentration or electrical conductivity (dissolved load) to mm a^{-1} (thickness) or km^3 a^{-1} (volume) of accumulating sediment. Assuming rock density of 2.7 g cm^{-3}, they are converted into rates of **average surface lowering** (or rock wall retreat on slopes) in mm a^{-1} This is unsatisfactory, because it does not equate easily with our perception of landscape dissection and the creation of relative relief, which concentrates erosion in valleys. It is a consistent standard, however, and is directly comparable with uplift rates. More important caveats concern the short time scales and other uncertainties of measuring terrestrial processes. Sediments in transit at one point of measurement may be detained elsewhere. This is particularly pertinent, for example, in the case of the Karakoram and Southern Alps, New Zealand, where river terrace sediments are uplifted and recycled before being evacuated to the coast! Dissolved rock is more difficult to track and biogenic processes extract minerals from one part of the sediment cascade and redeposit them elsewhere later. Sediment loads in northern Europe

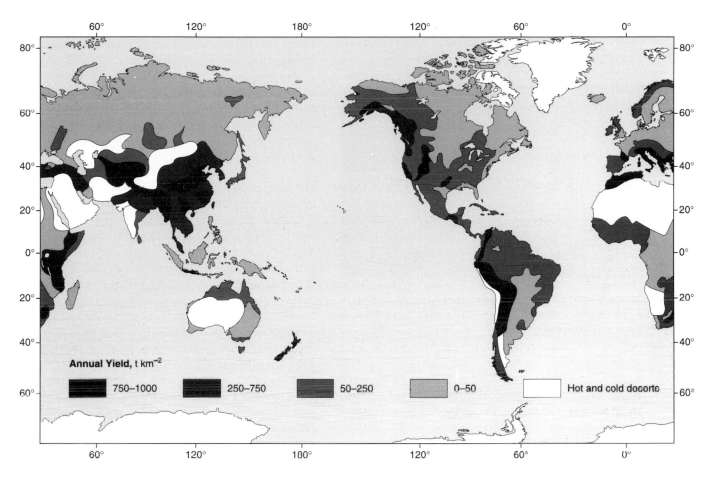

Figure 13.2 *Global suspended sediment yield. Source: After Walling and Webb; in Gregory (1983).*

may be more indicative of the reworking of Pleistocene glacial sediments than of current lowering. The restless state of Quaternary environments and the emergence of hominids with an increasing **anthropogenic** impact complicate assessments of long-term rates. Only deep-ocean undisturbed terrigenous sedimentation produces reliable figures (of *c.* 10 m Ma^{-1}) over long, radiometrically secure time scales. None the less, some gross figures emerge from all this uncertainty. An average of recent estimates suggests that some $25–28 \times 10^9$ t a^{-1} of terrigenous sediment is delivered to the oceans, in a solid:dissolved ratio of 6:1, equivalent to some 62–70 mm ka^{-1} of surface lowering. Allowing for isostatic adjustment, net lowering of terrestrial surfaces would be some 12–14 mm ka^{-1}. Global solid (suspended) sediment transfers is shown in Figure 13.2.

Denudation cycles and chronology

Episodic uplift and the presence of large continental areas of low relief, even at moderate altitudes

(1–2 km) prompted the formulation of model **denudation cycles** and **denudation chronologies** to chart landsurface development. The most enduring models envisaged elevated landsurfaces wearing *down* (W.M. Davis, 1890s) or wearing *back* (W. Penck, 1920s) to an eventual **peneplain**, or through parallel slope *retreat* to a **pediplain** (L.C. King, 1950s). The cycle was rejuvenated through renewed uplift. Davis's model had regions like the Appalachians in mind and, in common with those which followed, employed concepts of *youth* (waxing slopes), *maturity* and *old age* (waning slopes). In this way, for example, much of upland Britain and individual regions such as Wales, the Weald and the Downlands of south-east England were thought to reflect multiple peneplanation and rejuvenation cycles with accordant remnant summits and plateaux surfaces. Plate 13.1 shows the '3000 ft' erosion surface in North Wales, one of the highest and on Davis's terms therefore one of the oldest in Britain. Volcano-tectonic evidence suggests that it is a lowland surface uplifted during the mid to late Tertiary and subsequently breached by Quaternary glaciation. This region is still seismically active.

Plate 13.1 *The '3000 ft' remnant summit plateau of Y Glyderau, North Wales, which was probably a lowland plain elevated tectonically during the mid- to late Tertiary.*
Photo: Kenneth Addison.

Denudation cycles and chronologies have received a bad press subsequently but they were formulated long before sea-floor spreading, volcano-tectonic activity and the dramatic events of the Quaternary were understood. These clarify the complex nature and rates of uplift and denudation, the speed and frequency of disturbances to denudational 'rhythms' and means of dating events. Older models projected whole landsystem effects without comprehending the total landsystem and were rather parochial. Davis's 'normal cycle' assumed fluvial conditions in temperate environments, glacial events were sometimes seen as climatic 'accidents', and King worked largely in semi-arid southern Africa. However, Davis recognized the progressive transformation of potential to kinetic energy as his cycle proceeded and his *graded* river profile reflects exponential energy decay and the progressive reduction of elevation, relief and slope angles.

Force and resistance

Debate shifted away from denudation cycles and explored the validity of **morphogenetic** regions as a basis for modern geomorphology. Its pervasive presumption of climatic control on landform assemblages led to the creation of morphogenetic maps (Figure 13.3). At first sight they are plausible, especially since glacial and hot desert environments equate with modern climatic zones. Climate inconstancy – and tectonic displacement of crust across climatic

zones – cannot be ignored, however, and many landforms are *polygenetic*. Close climatic affinity recedes in one sense, when we consider that many Pleistocene glacier landsystems now lie outside the glacial zone, although in another they are fossil remains of fluctuating climate patterns. A tendency to regard everywhere else as dominated by fluvial landsystems ignores major regional variations in precipitation amounts and régimes and carries echoes of the 'normal cycle'. Many landforms are not specific to any one climatic zone. Slopes and fluvial channels evolve in all climates and coastlines and mountains are distinctly azonal. Details of form or intensity which reflect climatic nuances do not override this principle. No-one doubts the role of climate in contributing materials (water, ice, air, pollutants) in particular amounts and régimes to the geomorphic environment, or of temperature and moisture, etc. in influencing processes. To regard it as dominant or controlling is inappropriate, however: it is to ignore the role of geological factors and to misunderstand the nature of *process*.

In essence, *geomorphological processes* transform earth materials and energy from one state or condition to another. They involve rock in any form (intact, or disaggregated as debris, sediment or soil) and any combination of water, ice, atmospheric gas(es) and organic matter. Energy may be exogenic (light, heat), endogenic (geothermal heat, gravity), chemical (mineral bonds, etc.) or combined in the potential and kinetic energy of moving materials at Earth's surface. **Force** and **resistance** hold the key to the progressive breakdown of intact rock and therefore denudation. Forces applied to rock mass originate *internally* through volumetric changes associated with heating, cooling, chemical reactions and the circulation of fluids (including air). They determine **rock weathering** processes. Forces are also applied *externally*, through the emplacement and/or removal of static (*in situ*) or dynamic (moving) loads of rock, sediment, water, ice, wind or anthropogenic structures. This is the essence of erosion. The latter include tectonic forces whose effects are represented by elastic strain, released in earthquakes, deformation structures and continuing rock deformation. Resistance can be measured as hardness, resisting **abrasion**, or as the sum of internal strength properties capable of resisting **tensile stress** (pull apart), **compressive stress** (crushing) and **shear stress** (sliding rupture) (Figure 13.4). Abrasion occurs when one rock scratches another of lower hardness and is a minor process in slope and glacial environments (**striation**) and in-transit sediments, where **attrition** polishes grains jostled together. Some forms

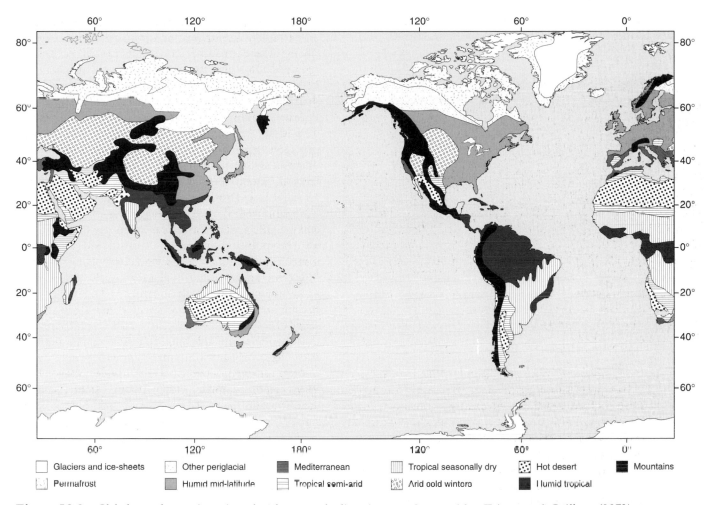

Figure 13.3 *Global morphogenetic regions, based on morphoclimatic zones. Source: After Tricart and Cailleux (1972).*

of weathering establish internal tensile stress where individual minerals expand.

Tensile and compressive stresses are more directly important in tectonic than in geomorphic processes. Together with shear stress, they drive deformation and uplift and therefore are instrumental in initiating denudation. However, the inherited effect of deformation in the form of rock structures is of profound geomorphic significance. All forms of structure – folds, fractures, faults, thrusts, joints, laminations, etc. – regardless of their lithological or tectonic origin, are **planar discontinuities** which render rock mass mechanically and hydraulically defective. That is, they individually represent two-dimensional, planar partings where the homogenous character of the rock mass, and especially its cohesive or **intact strength**, is momentarily interrupted or lost, with a number of key geomorphic consequences. They control the **permeability** of rock, or its capacity to circulate water and air, which are important weathering agents, and

are more important than **porosity** in that respect. By reducing the strength of rock, discontinuities make its removal easier and also provide **release surfaces** along which the rock comes apart. They are found at all scales (Figure 13.5) and those of lithological and tectonic origin form regular, three-dimensional geometric patterns. This reflects compressive and tensile stress, and therefore the strain history of the rock, and directs denudation processes (Figure 12.20). In addition, faulting juxtaposes rocks of different strengths. **Structural control** can be seen at every scale from the shape and slope angles of individual landforms (Plate 13.2), regional patterns of hills, valleys and drainage networks (Figure 13.6) and in continental architecture. All this makes a strong case for a **morphotectonic** approach to geomorphology.

Shear stress is predominant in the operation of geomorphic processes. The balance between force and resistance – the **limiting equilibrium** – marks the

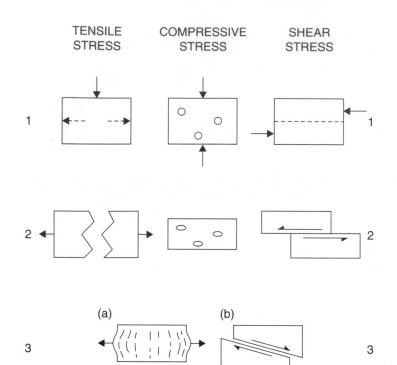

Figure 13.4 *Principal types of stress and failure. Line 1 shows the initial shape of identical blocks prior to failure. Line 2 shows the failed state; the block subject to compressive stress has deformed by compression of void space alone. Line 3 shows alternatives: (a) reveals a tensile component, short of rupture whilst (b) has sheared.*

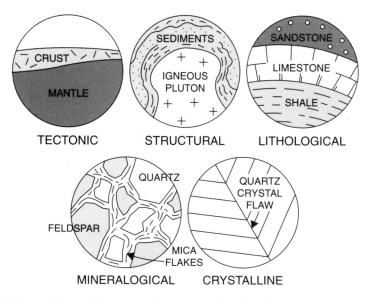

Figure 13.5 *Discontinuities at all geological scales.*

point of imminent rock mass failure, and its criteria are summarized in a **Mohr-Coulomb** equation:

$$\tau = c + \sigma.\tan\phi$$

where τ = shear stress, c = cohesion, σ = normal stress and ϕ = the angle of internal friction. This is an extremely useful way of summarizing, and quantifying where possible, the force mobilized against the rock mass and the **shear strength** which resists it, with three principal components. **Cohesion** is provided by electrostatic and magnetic bonds between minerals, intergranular cement and water. The first two are acquired during the various processes of rock formation. Pore water in moderate quantities exerts a 'suction' force, through surface tension, and its presence depends on rock porosity. **Normal stress** is the anchoring weight of rock overlying a point in the mass and depends on rock density and gravity. Internal **friction strength** is dependent on the point contacts of constituent grains and minerals, determined by their size, shape and packing arrangement. It is additional to cohesion, and is given as an angle along which shearing would take place; the higher the angle, the greater the strength. Its significance is seen clearly in a dry scree or pile of loose sand, where the 'angle of rest' equals the friction angle. Mohr-Coulomb criteria will help us to understand some principal failure conditions and forms later; meanwhile, typical component strengths are shown in Table 13.1

Weathering

Weathering is the preliminary etching of landsurfaces which eases the task of the main sculptors, mass movement and erosion, and is everywhere around us. Informal indications of its processes and rates can be seen on every building, from the corrosion of fine-carved monuments to the cracking of artificial 'stone'. The control of 'lithology', and processes of chemical and mechanical weathering, are self-evident at home! (Plate 13.3). Outside, the natural world reveals their full range, from discoloured and often friable **weathering rinds** of rock walls and pebbles, or staining by the weathering of other minerals, to the complete disintegration represented by soil. Earth's surface environment is alien to most rocks, which display varying degrees and forms of susceptibility on exposure to the atmosphere and biosphere. Hard rock is mechanically stable but chemically unstable in this environment; weathering converts it to a mechanically unstable (disaggregated) but chemically more stable residue. Fractionation is then complete; the long journey of alumino-silicate and clay minerals through the rock cycle sees them finally segregated.

Weathering must overcome the tensile strength of rock mass. It is controlled therefore by geochemistry and texture (*lithological* properties) and by dis-

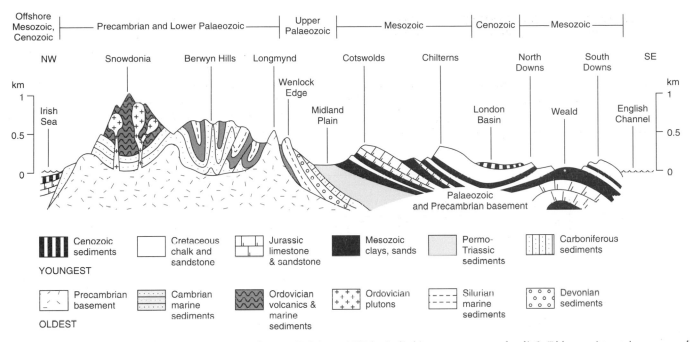

Figure 13.6 *Geological cross-section from Anglesey to Brighton (450 km), linking structure and relief. Older, resistant igneous and metamorphic rocks form higher ground in Wales than younger, less resistant sedimentary rocks in central and south-east England. Mesozoic sediments in the Irish Sea basin suggest that Wales may have been exhumed from beneath a Mesozoic cover.*

continuity geometry and the assemblage of different lithologies (*structural* properties). The former determine the **specific susceptibility** of mineral species and bonds, and porosity; the latter determine the pattern of circulating groundwater and air (permeability). Continuing access to rock is essential and weathering enhances it further by enlarging fractures and voids, **leaching** (flushing) solutes and by the **eluviation** (wash-out) and mass wasting of debris. Initial fragmentation by **mechanical weathering** may be important because it greatly enlarges exposed surface area as a prelude to **chemical weathering**. However, climate stimulates **hydrothermal alteration** of rock and temperature, and moisture régimes exert a strong influence on weathering rates and styles. Climate restricts chemical weathering by

aridity whilst enhancing it at higher temperatures (Figure 13.7).

Physical weathering

Overcoming tensile strength is central to weathering, and one of the more spectacular ways in which it occurs is through **elastic strain release**. This is sometimes described as pressure release and causes much confusion. Deep glacial erosion was ascribed to repeated ice advances followed by interludes of relief, relaxation and pressure release for the unfortunate rocks, only for a readvance to sweep out the newly fractured debris and repeat the cycle. As well as mistaking process, this notion committed the cardinal

Table 13.1 *Some typical geotechnical parameters of principal rock types.*

Type	Typical lithology	Cohesion (MN m⁻²)	Friction angle (°)	Residual angle (°)	Strength (MN m⁻²)			Porosity (%)
					compressive	tensile	shear	
Plutonic	Granite	56.1	45	35	146.4	20.6	31.5	0.5–2.0
Volcanic	Tuff	42.2	35	31	123.9	25.2	37.9	0.5–1.5
Metamorphic	Slate	22.9	27	25	79.6	13.3	22.5	0.1–0.6
Clastic sediment	Sandstone	31.7	29	25	96.3	12.7	32.2	12.0–25.0

Note: MN = mega-newtons; one newton is an SI unit of force, equivalent to 1 kg m⁻¹ sec⁻²; the residual (friction) angle applies once shearing has started and abraded any surface roughness (asperities) along the failure plane.

Plate 13.2   A 15 m high rockwall dominated by columnar 'organ pipe' jointing and cross-joints, in Tertiary basalt near the Giant's Causeway, Northern Ireland.
Photo: Kenneth Addison.

sin of inventing climatic changes and events to suit the story! There is much evidence of **dilation** or expansive enlargement, so what *is* going on? Stress patterns in buried rock mass attest to huge overburden and confining stresses during *em*placement. Deformation of the mass, including elastic strain and brittle failure, is likely to create incipient discontinuities, **orthogonal** (at right-angles) to the principal stress directions. As the structures are exhumed, released strains appear to fracture rock parallel to an erosion surface, creating **sheeting structures** or

exfoliation surfaces. In reality, their pre-existence controlled denudation rather than vice versa. Plate 13.4 shows steep glacially excavated granite rock walls. Apparent sheeting structures control the wall, but it is also clearly penetrated by other fractures, orthogonal to any erosion surface. The geomechanics of these processes are explained below. We should avoid previous pitfalls and appreciate that rock mass fails at the large scale along *existing* fractures long before new ones are needed, and form only in their absence. This is central to the issue of physical (mechanical) weathering and erosion.

Open or incipient fracture systems are exploited by a number of weathering processes which generate tensile stress. Most involve cyclical application and relaxation of stress through thermal change and wetting and drying – hydrothermal changes stimulated by weather and climate. **Frost shattering**, also known as cryofracture and even as hydrofracture, occurs in response to the 9 per cent expansion of water on freezing. In open fractures, this is taken up by voids or ice itself. However, if initial freezing of surface and pore water seals the fracture, subsurface water is now confined and further freezing expansion applies a force of some 20 MN m^{-2} to the rock. Repeated freezing and thawing generate **fatigue** failure in due course. It is thought that diurnal or other high-frequency cycles in more maritime cold climates are far more effective than seasonal freeze–thaw cycles with large temperature ranges (in continental cold climates), provided they range between $-10°$ C and $+10°$ C. Large temperature oscillations are also behind

Plate 13.3   'Geological controls' in a garden retaining wall. Sub-zero air temperatures draw water through porous bricks to a freezing plane at their outer surface, causing frost weathering. Impervious engineering bricks are undamaged and mortar courses exert an additional, structural control.
Photo: Kenneth Addison.

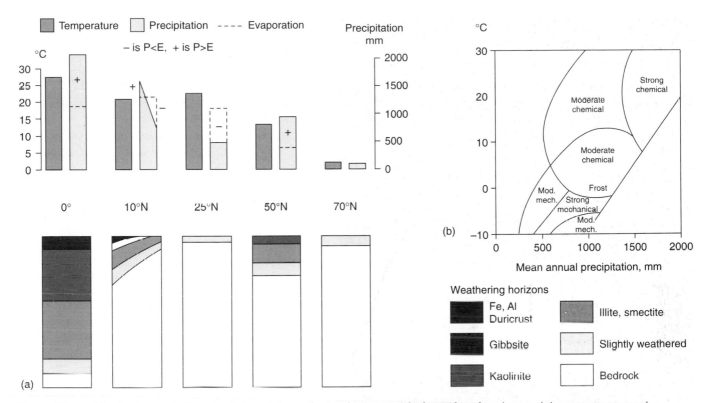

Figure 13.7 *Weathering and climatic régime: (a) weathering horizons with thermal and moisture régimes at representative latitudes and may be viewed in conjunction with Figures 13.3 and 14.6; (b) identifies principal global weathering régimes. Source: After Peltier (1950).*

insolation weathering, this time in hot climates, with diurnal temperature ranges of some 30° C. The process is more effective than chemical weathering in arid climates, although water is probably required to enhance cooling and hence fatigue.

Similar tensile stresses found in other processes emphasize a moisture, rather than thermal, régime. **Slaking** involves cyclical hydration of rock, either by pore water or by the much more expansive effects of swelling clay minerals. **Salt weathering** does not necessarily require cyclical changes, relying instead on crystallization in voids. Water may be sourced externally or drawn to the surface by capillary action, following on from a phase of solution weathering. Salt weathering is more likely to be found in arid climates, although **salt efflorescence** in bricks and concrete is common in cool, temperate climates. Slaking requires a moist climate. The only physical weathering process not directly influenced by climate is bioturbation by plant roots and burrowing animals, which also facilitate the ingress of chemical weathering agents. Of all these processes, **cryofracture** and bioturbation are the most likely to exploit rock fractures directly. Others will operate at intergranular or intercrystalline discontinuities first, although larger-scale weathering fronts open up along fractures.

Chemical weathering

Two important stages in the rock cycle haunt rocks at the landsurface. Mineral stability should be inversely proportional to the temperature at which they formed. By and large this is so, with minerals crystallized from high-temperature melts in the greatest discomfort at the low-temperature land-surface (Figure 13.8). Siliciclastic rocks, formed from rock residues in conditions of closest equilibrium to the landsurface, are the most stable. Water, which is so essential to – but eventually expelled from – B-subduction magmatization, evaporites and most sedimentary rocks, is the principal agent of chemical weathering. The H^+ (hydrogen) and OH^- (hydroxyl) ions in water react with other minerals, creating new species which are readily removed (Figure 13.9); it is therefore both **reagent** and **solvent**. Hydrogen ion concentration is expressed as the **pH** of water, which is neutral when pH = 7.0, **acid** at less than 7.0 and **alkaline** at over 7.0. Chemical weathering potential increases inversely with pH and proportionally with increasing equilibrium solubility of minerals and temperature, until saturation is reached.

Most minerals dissolve slowly in water but those dominated by ionic bonds, e.g. mafic minerals (Mg^{++},

Plate 13.4 *Massive granite sheeting structures in a fjord rockwall, 800 m high, south-west Norway.*
Photo: Kenneth Addison.

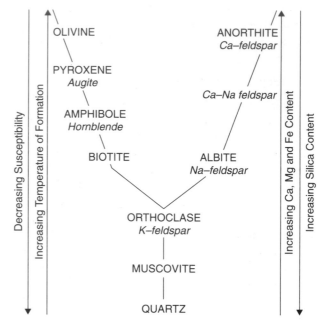

Figure 13.8 *Mineral susceptibility to chemical weathering, closely linked with magma geochemistry and melt temperatures.*

Fe^{++}), calc-plagioclase ($CaAl_2Si_2O_8$) potassium (K^+), sodium (Na^+) and calcium (Ca^+) are more susceptible to **solution** than felsic minerals dominated by covalent bonds, e.g. orthoclase ($KAlSiO_3O_8$), muscovite ($KAl_2(AlSi_3O_{10})(OH,F)_4$) and quartz ($SiO_2$). Alumina ($Al_2SO_3$) is soluble in very acid (pH < 4.0) or alkaline (pH > 9.0) water.

Solution potential is enhanced through **carbonation** – the incorporation of dissolved atmospheric CO_2 during precipitation – forming dilute carbonic acid, H_2CO_3. This stage in the process is a rare instance of a chemical reaction rate inversely related to temperature. Other dilute acids may form in association with atmospheric constituents, including sulphuric acid (H_2SO_4) as *acid rain*. Carbonation is enhanced after passing through soil rich in biogenic carbon dioxide and is a common form of chemical weathering, particularly of limestones forming 15 per cent of global sedimentary rocks. After hydration

equilibrium between H_2O and CO_2 has been reached, the carbonic acid is dissociated to HCO_3^- and H^+ ions. Calcium carbonate, $CaCO_3$, in contact with the solution is dissociated into Ca^{++} and CO_3^{--} ions, combining with H^+ to form more HCO_3^- ions. This complex, multiphase reaction is often simplified to the form

$$CaCO_3 + CO_2 + H_2O \rightarrow Ca(HCO_3)_2$$

with calcium bicarbonate removed in solution. Each process is reversible, and reprecipitation of calcium carbonate may occur eventually in more resistant, crystalline form as *travertine*. In the carbonation of orthoclase feldspar we see the formation of a new clay mineral species, *illite*, the production of potassium and bicarbonate ions and the removal of dissolved silicate.

$$6KAlSi_3O_8 + 4H_2O + 4CO_2 \rightarrow$$
$$K_2AL_4(Si_6Al_2O_{20})(OH_4) + 12SiO_2 + 4K^+$$
$$+ 4HCO_3^-$$

Hydrolysis involves H^+ and OH^- reactions with minerals and is important, for example, in the decomposition of granite containing plagioclase feldspar, with the clay mineral *kaolinite* the principal product.

$$4NaAlSi_3O_8 + 6H_2O_3 \rightarrow Al_4Si_4O_{10}(OH_8)$$
$$+ 8SiO_2 + 4Na^+ + 4OH^-$$

Hydration occurs when minerals absorb water into their crystal lattice and establish tensile stress in addition to chemical alteration. Hydration of iron oxides to the form *limonite*, an important weathering process in mafic rocks, illustrates the latter and is reversible.

$$2Fe_2O_3 + 3H_2O \rightleftharpoons 2Fe_2O_3.3H_2O$$

Two sets of '*redox*' reactions involve oxygen either by combination (**oxidation**) or by removal (**reduction**), occurring in aerobic or anaerobic environments and changing valency in a positive or negative direction respectively. Oxidation promotes weathering in mafic minerals, changing ferrous iron oxide (FeO) to its ferric form (Fe_2O_3), destabilizing the crystal lattice and requiring the compensating loss of another cation. Finally, chemical weathering is also enhanced by certain biological processes. Decomposition inflates soil concentrations of CO_2, potentially increasing solution rates, and secretes biochemical **chelating agents** which aid the solution of otherwise stable cations.

The chief products of chemical weathering are stable quartz – the principal ingredient of siliciclastic sediments – clay minerals and dissolved products. Clay minerals are capable of further weathering, enhancing silica removal, particularly in warm, humid environments. Residual or redeposited minerals create economically viable sources of alumina (bauxite) and iron (laterite), etc. Gibbsite and to a lesser extent kaolinite, found in Wales and Cornwall, are assumed to be relic Tertiary humid-tropical weathering clay minerals. Residual, **core-stone** granite tors on Dartmoor are thought to have been exhumed by Quaternary cold-stage solifluction. The survival of thick weathered profiles and core stones below glacigenic sediments indicates that removal by glacial erosion down to the weathering front or **etch front**, does not have to occur and cannot be taken as evidence of ice-free areas (Plate 13.5).

Weathering landforms

Landforms of physical and chemical weathering range from the survival of residual, less or non-weathered bedrock to deep, *in situ* weathered debris sheets depending on the balance between weathering and removal by mass wasting or erosion (Figure 13.10). **Saprolite** is a fine-grained or amorphous chemical weathering product with no direct physical equivalent. General rock debris or **regolith** is non-specific but **felsenmeer** or **blockfield**, found on low-angled surfaces, is more closely associated with cryofracture. At a grand scale, calcareous (limestone) solution-carbonation weathering is so distinctive as to justify

Plate 13.5   *Core stones separated by rotted, saprolitic material believed to reflect Tertiary humid weathering in a dolerite dyke near Caenarfon, North Wales. The site survived Quaternary glaciation. Photo: Kenneth Addison.*

its own subdiscipline of **karst geomorphology**. Its principal landforms are shown in Figure 13.11.

Mass wasting

Weathering products seldom remain *in situ* for long and products must be removed for vigorous weathering to continue. Their original rock mass strength destroyed, those which are not washed out become a prime target of mass wasting. **Mass wasting** is a

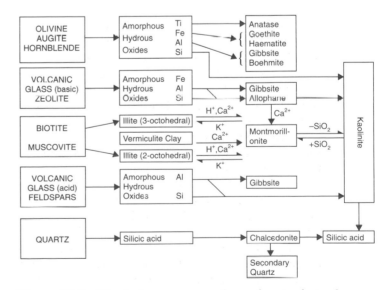

Figure 13.9   *Weathering sequence and secondary products of primary rock-forming minerals. Source: After Selby (1993).*

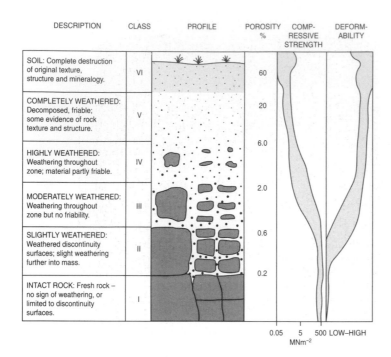

DESCRIPTION	CLASS	PROFILE	POROSITY %	COMP-RESSIVE STRENGTH	DEFORM-ABILITY
SOIL: Complete destruction of original texture, structure and mineralogy.	VI		60		
COMPLETELY WEATHERED: Decomposed, friable; some evidence of rock texture and structure.	V		20		
HIGHLY WEATHERED: Weathering throughout zone; material partly friable.	IV		6.0		
MODERATELY WEATHERED: Weathering throughout zone but no friability.	III		2.0		
SLIGHTLY WEATHERED: Weathered discontinuity surfaces; slight weathering further into mass.	II		0.6		
INTACT ROCK: Fresh rock – no sign of weathering, or limited to discontinuity surfaces.	I		0.2		

0.05   5   500 LOW–HIGH
MNm^{-2}

Figure 13.10 *Weathering profile and some geotechnical properties. Source: In part after Dearman (1974).*

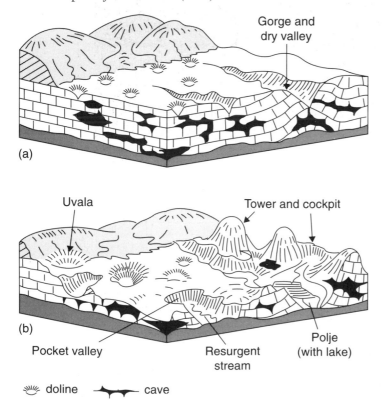

(a)

Gorge and dry valley

(b)

Uvala   Tower and cockpit

Pocket valley   Resurgent stream   Polje (with lake)

〰 doline   ◆ cave

Figure 13.11 *Evolution of a karst landscape. (a) surface solution dolines drain water into the limestone where it develops cave systems before resurgence along an aquiclude. (b) Continuing surface solution weathering and cavern collapse progressively corrode the landscape down to poljes and alluvial plains; residual towers and cockpits are common in tropical karst.*

general term for a variety of slope denudation processes operating under *static* gravity load, rather than by water and ice moving as discrete bodies. It is preferred here to **mass movement**, which implies the coherent movement of rock and soil *en masse*. Most movements occur through **compound translation failure** – with material properties and forms of failure changing radically during a single event. This may be apparent in the resultant landform and is important if we are to understand and manage slope failure. Mass wasting and slopes are not restricted to areas lacking formal designations such as *fluvial, glacial, periglacial, coastal* environments, etc. They are found everywhere and rules governing their behaviour are invariable. Mass wasting also occurs in more formally recognized environments and every landform is a composite of slope forms. It is not surprising that producing schemes of slope classification and evolution is a busy industry!

Slope stability

In its simplest form, slope stability depends on the ratio of stabilizing forces resisting movement to destabilizing forces encouraging it and is reflected in the engineering term **Factor of Safety, F**. When $F = 1$ a state of *limiting equilibrium* or incipient failure exists, as we saw in the Mohr–Coulomb equation earlier, and can be triggered by any further deterioration. The natural angle of rest in granular slope materials approximates $F = 1$, which is too close for comfort in the engineering world. A compromise is struck between increased stability and cost. Hence $F \geq 1$ is measured in expensive stabilization schemes or loss of space/ resources in road cuttings or quarries by reducing slope angles; or $F = 1$ by the cost of failure during the design life of a structure in human lives, resource value, etc. We simplify the balance of forces on natural slopes to the relationship between shear *strength*, shear *stress* and slope angle – and revisit Mohr–Coulomb.

The parameters of the equation are now shown in Figure 13.12. Any point on the solid line shows the shear stress needed to exceed the shear strength related to normal stress, σ, the value of cohesion, c, and a given friction angle, ϕ. This line, however, shows 'ideal' intact rock mass strength (IRS) and the presence of discontinuities and water on slopes substantially reduces shear strength. Discontinuities destroy cohesion between intact blocks. As water infiltrates the mass, any downslope component increases shear stress, v, provides an uplifting force u which reduces normal stress to *effective* stress and may reduce the friction angle to a residual value. Their collective impact is shown by **discontinuous rock**

Debris flow and debris slide hazard

Debris flows are among the most unpredictable, fast-moving forms of mass wasting process, capable of limited self-regeneration in transit. For these reasons they pose particular hazards to human settlements and structures. They are initiated by a wide range of stimuli – landslide, seismic activity, intense rainfall, rapid snow or glacier melt, water eruption (e.g. a bursting pipe), volcanic eruption, etc. – depending on the critical condition of slope materials at the time of potential failure. Their essential requirement is rapid fluidization of granular debris, hence the rapid water delivery and/or high-porosity debris components listed. They are common on arctic and alpine slopes (see Chapter 25). Explosive eruptions on ice-capped strato-volcanoes generate some of the most spectacular, destructive debris flows and provide their own granular debris in the form of ash falls.

Debris flows and lahars (volcanic mudflows) triggered by the Nevado del Ruiz (Colombia) eruption in 1985 swept over tens of kilometres and claimed more than 25,000 lives. The 1980 Mount St Helens (Washington State) eruption claimed few lives but debris flows travelled over 20 km along the Toutle river valley. The typical sequence of events started with an earthquake, triggering the eruption via a major landslide which opened up the blast vent. Unconsolidated slope materials were disturbed by both shocks and fluidized by ice melt, surface drainage disruption and lake burst.

We rarely experience such catastrophic debris flows in Britain, although the 1966 Aberfan disaster claimed 144 lives and was caused by the fluidization of dumped coal waste and debris avalanche/flow. Debris flows are a previously underestimated and probably increasing hazard. Imagine driving along the A5 Euro-route through Snowdonia or around the Great Orme Marine Drive, Llandudno – or any other upland road – and confronting a wall of boulders up to 2 m high, bouncing and jostling along at 20–30 km hr⁻¹. Intense summer rainfall of some 120 mm in three to five hours in September 1983 and June 1996 triggered just such debris flows in complex colluvial deposits of glacial, periglacial and talus materials. Shallow initial slides were rapidly fluidized and transformed to debris flows, travelling downslope as a series of turbulent pulses.

Each flow gouged a track 3–10 m deep (Plate 1), displacing debris thrown out to form parallel **levées** (banks) by violent boulder collisions. The same collisions contribute to the buoyancy of the flow but large 'grains' move faster than others and eventually form the boulder front through which water drains. This debris 'slug' then grinds to a halt but the water may continue and repeat the process several times before dying out. The largest A5 pulsed debris flow travelled approximately 600 m down slopes of 22°–35° before cutting the carriageway, and over twenty flows devastated the Marine Drive. One flow travelled over 900 m along the Drive, unable to drain

Plate 1 *Debris flow and structural damage on the Great Orme, North Wales. (1) Initial gullying sourced downslope debris flow; (2) debris slugs display meandering flow on the Marine Drive below the source, leading to (3) structural damage to the carriageway and retaining walls.*
Photos: Kenneth Addison.

through its impervious surface, leaving a meandering trail of debris slugs (Plate 2). Finally punching several large holes in the Drive and retaining walls, water formed new flows before plunging into the sea (Plate 3).

These are among several dozen similar events in North Wales alone since 1980. The cost of immediate clearance, longer-term repair and remedial work and the loss of revenue amounted to some £500,000. Over 150,000 km² of upland slopes in Britain may be susceptible to debris flow hazard, and their sensitivity may increase with changing land management practices and global warming.

Plate 2   **Plate 3**

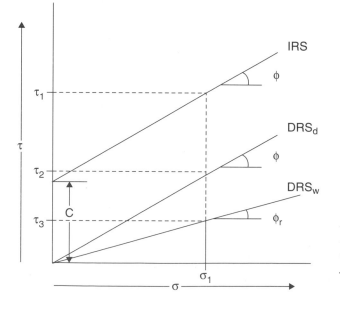

mass strength in dry (DRS_d) and wet (DRS_w) states and the Mohr–Coulomb equation is modified to

$$\tau + v = c + (\sigma - u) \cdot \tan (\sigma - x) \text{ or simplified}$$
$$\text{to } \tau' = c + \sigma'_n \cdot \tan \sigma_r$$

where τ' = effective shear stress, increased by the weight of water. This restates the balance of shear stress and strength but has not yet formally introduced the significance of slope angle and still appears to relate only to rock mass. However, the modified equation applied to discontinuous rock mass is also a reasonable approximation for debris and soil on

Figure 13.12 *Mohr–Coulomb relationship between shear stress (τ), normal stress (σ) and cohesion (c) in rock mass; see text for explanation. The shear stress required for failure in a rock mass with normal stress σ_1 falls from τ_1 in intact (non-fractured) rock to τ_2 in dry, discontinuous rock and to τ_3 in wet, discontinuous rock respectively.*

Figure 13.13 *Forces acting on a block on a rock slope; see text for explanation. In wet conditions, water in discontinuities (dark area) reduces sliding resistance by exerting a buoyancy or uplift force, u, and adding the weight of water behind the block, V, to sliding stress.*

slopes, with a little qualification. Cohesion = 0 in discontinuous rock mass and applies equally to large, loose blocks or uncohesive soil grains on a sloping surface. We can apply remaining Mohr–Coulomb criteria to such a block or grain and see its **sliding resistance**, R, on a slope of known angle, given as

$$R = cA + W. \cos B \tan \phi$$

where c = cohesion (zero), A is the block/grain-to-slope contact area, W = block/grain weight, B = slope angle and ϕ = friction angle. However, part of the mass is mobilized downslope by gravity and limiting equilibrium reflects a balance between *perpendicular* and *tangential* forces. In the case shown in Figure 13.13, water behind and beneath the block, respectively, exerts shear stress and reduces effective stress.

In granular materials, moderate pore water levels generate cohesion or negative pore water pressure. Positive pore water pressures develop as water content rises, transferring some of the normal stress from grains to pore water and comprehensively reducing shear strength. The **plastic** and **liquid limits** define thresholds of increasing saturation at which the mass deforms plastically or as a viscous fluid (Figure 13.14). They are also dependent on particle size character, void ratio and proportion of clay minerals, but plastic and fluid behaviour is not restricted to fine-grained soils. Under exceptional water pressures, large boulders and grains may become fluidized and move rapidly downslope as **debris flows** when grain collisions enhance buoyancy forces. Water content also fluctuates with the weather and the seasons. Precipitation or spring melt may induce pore water pressures to rise rapidly. Steep slope angles and the proximity of rock walls generate rapid run-off. This readily infiltrates slope colluvium but drainage is then inhibited by marked reductions in permeability at the colluvium/bedrock boundary or by perched water tables associated with *indurated* (cemented) or clay-rich horizons. Other significant clay–water interactions include the behaviour of **sensitive** or **quick clays** (e.g. illite), which can lose structure and shear at relatively low moisture contents, or **swelling clays** (e.g. montmorillonite), which can establish heaving (uplifting) forces as they absorb water.

Figure 13.14 *The influence of water content on soil material properties, measured by (a) Atterberg indices or (b) limits. Source: After Selby (1993).*

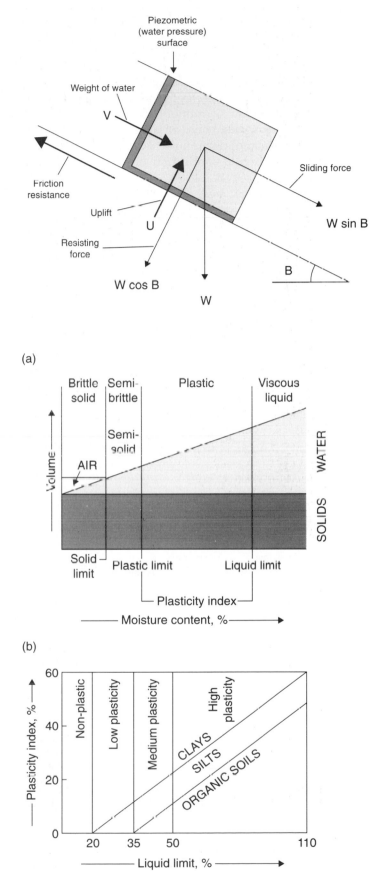

(a)

(b)

Slope failure: materials and modes

There have been many attempts to classify mass wasting comprehensively, mostly by civil engineers, with varying degrees of success and acceptance. It is difficult to pigeonhole events in which material properties and style of motion may change in transit, leading to a wide range of landforms. Tidy schemes on paper cannot discriminate easily between processes operating over a **continuum** of properties. Criteria widely used are as follows: *type of material* – rock–rock debris–soil; *mode of failure* – fall–slide–flow; *rate of movement* – slow–rapid; *water content* – wet–dry; *mass geometry* – sheet–block; *morphology* – scar–lobe; *failure history* – new–reactivation. Table 13.2 reflects classifications in use without preferring any particular scheme. Rates of movement are shown in Figure 13.15. No single criterion unlocks the whole scheme and failure modes are reviewed below according to whether they are in hard rock or debris slopes.

Failure modes on rock slopes

Sliding and toppling failures dominate rock slopes, taking one or more of four forms determined by relations between discontinuity geometry and slope (Figure 13.16). Failure rarely needs to shear intact rock and would find this more difficult (Plate 13.6). The sole requirement for sliding in discontinuous rock mass is one fracture surface (F) steeper than the friction angle (D) but at a lower angle than the rock wall slope (S), which allows it to appear or *daylight* in the rockwall. Single fractures provide **translation** or **release** surfaces in **planar slides** (Plate 13.7a) but most rock walls fail by **wedge slides** (Plate 13.7b), where the release surface is the intersection plane (I) between two fractures. **Toppling failure** (Plate 13.7c) occurs where a primary fracture surface (D_1) dips into the rock wall and appears stable but is intersected by more widely spaced fractures (D_2). Columnar blocks, defined by height (h) and breadth (b), topple when their centre of gravity overhangs a pivot. Fracture geometry, failure modes and rockwall height relationships are shown in Figure 13.17.

Circular sliding is uncommon in hard rock, where the 'massive' nature of orthogonal fracture sets provides sufficient release surfaces, but may be seen in densely fractured material or when hard rock overlies less resistant strata. In the former, initially steep slides are released along less steep fractures as they approach the rock wall foot, leaving circular scars. In the latter, the *incompetent* stratum (e.g. clays, shales) slumps, and may subsequently flow, carrying away *competent* overburden in such celebrated cases as the Black Ven cliffs in Dorset (Plate 13.8). **Cambering**

Table 13.2 *Classification of mass wasting by materials, rates and processes.*

Mode of failure	Rate of movement	Water content	Snow/ice	Rock	Type of material Rock debris	Colluvium	Soil
Flow	Very fast	Very high	Slush avalanche	←	Debris flow —————————————————————→		
	Very fast	Low	Snow avalanche	Rock avalanche	←————— lahar (+ volcanic debris) ————→ Debris avalanche		
	Moderate-fast	High					Mudflow
	Moderate	Moderate				Sand run	
	Very slow	Very high (Frozen)	Rock glacier				
Fall	Fast	Low		Rockfall	←————— Cohesive blockfall —————→		
					←——Individual block/grain fall/topple ——→		
	Fast	Low		Topple	←————— Cohesive topple —————→		
Translation slide	Slow-fast	Low-moderate		Plane, wedge slide	Block glide ←————————→ Grain slide		
Rotation slide	Slow	Low-moderate		Circular slide		Debris slump ←——→ Soil slump	
Creep/heave	Very slow	Low		Rock creep Cambering		Talus creep ←——→ Solifluction Grain creep and rain splash	
Compound translation failure			Combination of two or more, changing along a continuum of components, modes and rates during movement				

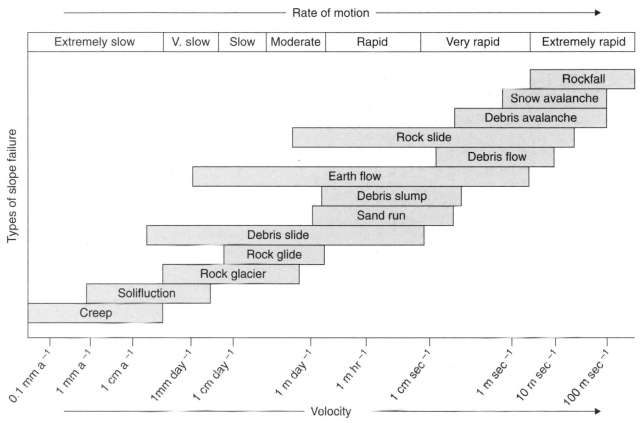

Figure 13.15 *Rates and types of slope failure.*

Plate 13.6 *(a) Quarry surfaces during and after blasting, showing (left) structural control on the release of blast energy and the surviving rockwall; and (right) the tectonic fracture geometry determines the alignment and dip of every rock face, major or minor. Photos. Kenneth Addison.*

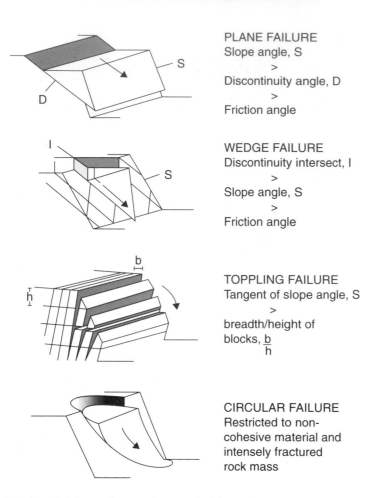

PLANE FAILURE
Slope angle, S
>
Discontinuity angle, D
>
Friction angle

WEDGE FAILURE
Discontinuity intersect, I
>
Slope angle, S
>
Friction angle

TOPPLING FAILURE
Tangent of slope angle, S
>
breadth/height of
blocks, $\frac{b}{h}$

CIRCULAR FAILURE
Restricted to non-
cohesive material and
intensely fractured
rock mass

Figure 13.16   Failure modes in rock slopes, related to discontinuity geometry and friction strength.

occurs as the rigid overburden creeps forward before shearing over a steepening slope. Debris may bend underlying strata during **rock creep**.

Slides release small rock volumes and spectacular failures. Some 36,000 m³ of rock failed in a single wedge slide during glaciation at the site in Plate 13.7b but quantities two to three orders of magnitude larger were moved in catastrophic events such as the Franks Slide, Alberta, in 1903. **Rockfalls** involve smaller volumes, the rock falling off an overhang or being prised off by frost weathering. All rock failures contribute blocky debris either to downslope scree or to avalanches, which continue as collision-buoyed flows. The special case of rock glaciers is covered in Chapter 15.

Failure modes on debris slopes

Various terms are used for the forms of denudation 'debris' or 'detritus'. Several stages in the progressive breakdown of rock are recognized. **Rock debris** is blocky material derived from rockwalls and retains its angular character with little or no further breakdown. **Scree** and **talus** are slope deposits of rock debris with equilibrium angles in the range 35°–45°, depending on lithology and block shape. **Colluvium** is a general term for reworked rock debris derived from slope or other (especially glacial) sediments. It has undergone further breakdown and chemical weathering, with a wider range of particle sizes, and acquired a less blocky, more grain/matrix form. It is unlikely to be well sorted or stratified, except over short distances, reflecting the episodic downslope movement of pulses of debris. Pedogenesis will have begun but burial also testifies to the influx of new material. Equilibrium slope angles range from 32° to 37°, rising above 40° towards an upslope boundary with rock debris. **Soil** refers to material showing substantial pedogenesis and pedological character, normally supporting vegetation.

Debris is generally characterized by low or zero cohesion except at low water content, high void ratios and weak, irregular potential shear surfaces. Dry movements occur when tangential forces exceed shear resistance, assisted by **ground heave** through hydration or ice formation, or rain splash. At low velocities, and in the absence of clearly defined failure surfaces, this amounts to the relentless **creep** of particles downslope either as single blocks or grains or *en masse* by **solifluction** as spreads or sheets. The latter term is used both generally and also in the restricted sense of the permafrost environment. Debris/soil slumping develops on well defined failure surfaces by mass rotation. Cohesive debris, with suction forces or cementation, may behave initially like rock mass and can slide, fall or topple before disintegrating (Plate 13.9 on p. 233). Wet movements are promoted by mass fluidization and the resultant slurry velocity is usually moderate to fast as **mudflows** or very fast as **debris flows**, where grain collisions augment buoyancy. They feature in the box (pp. 223–4).

Slope landforms and slope development

Slope landforms are either residual scars from which material has moved or constructional debris deposits downslope. These components relate to each other and constitute features of a slope landsystem of recognizable, recurring elements (Figure 13.18). Both are transient, since scars represent local oversteepening above an unsustainable shear stress ($F = \leq 1$) and deposits are probably connected with the fluvial landsystem. As with mass movements, slope landform and development models abound, and one of the most enduring and widely used is the *hypothetical nine-unit landsurface model* (Figure 13.19). By no means all elements are found on every slope but it is instantly recognizable, for example, in the *alpine landsystem* described in Chapter 25.

Conclusion

It seems scarcely appropriate to review rock destruction so soon after its formation. Yet destruction is what happens as soon as rock is elevated above sea level or brought within range of percolating water or air by the removal of overlying rocks. In essence, a new set of environmental conditions have replaced those in which the rock formed, and its susceptibility is roughly proportional to the degree of change. Weathering may be seen as a means to an end, 'softening up' and further reducing the strength of earth materials and thereby facilitating subsequent mass wasting and erosion. An appreciation of the components and origins of rock-mass strength is a prerequisite to understanding how the environment mobilizes hydrothermal alteration, mechanical processes and gravity to overcome rock strength. Denudation, weathering and mass wasting produce unstable and transient surface materials but thereby also access to soil parent materials, nutrients and economic mineral deposits vital to the biosphere and human societies.

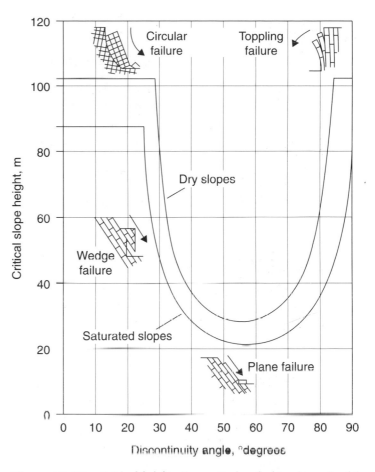

Figure 13.17 *Critical height of a vertical rock slope determined by fracture geometry and condition. Note the major reduction caused by saturated intermediate–high-angled discontinuities. Source: After Hoek and Bray (1977).*

Further reading

Duff, P.M.D., ed. (1993) *Holmes' Principles of Physical Geology*, fourth edition, London, Glasgow and New York: Chapman and Hall.
This volume of nearly 800 pages is a major reference text for any student of the physical environment. Although it ranges over the entire geological environment, a large portion deals generally and then specifically with the processes which denude and sculpture Earth's surface. Well written and concise text is supported by copious illustration.

Goudie, A.S. (1995) *The Changing Earth: Rates of geomorphological processes*, Oxford and Cambridge, Mass.: Blackwell.
This is an essential companion to standard texts on denudation and geomorphic processes. It provides concise definitions and brief explanations of geomorphic processes before addressing its main interest in the rates at which they operate and denude the continents. Acknowledging that the data are incomplete and often experimental, the book is nevertheless comprehensive in its selection of case studies and illustrations.

Selby, M.J. (1993) *Hillslope Materials and Processes*, Oxford and New York: Oxford University Press.
This is an outstanding text which commences with a review of rock and soil properties and strength. It proceeds via weathering to geomorphic processes, focusing on slopes, slope failure and mass wasting before concluding with hill slope models and denudation rates. The book is exceptionally well illustrated, with case studies, line drawings and photographs.

(b)

(c)

Plate 13.7 *The principal styles of rockwall failure, in Cambrian Grits, North Wales. (a) Planar slides down tilted bedding planes determine the rockwall slope angle. (b) A large wedge slide (200 m high) occurred at this intersection of the bedding plane and near-vertical cross fractures. (c) Toppling failure occurs where the latter fractures form cliffs, overhanging bedding planes.*
Photos: Kenneth Addison

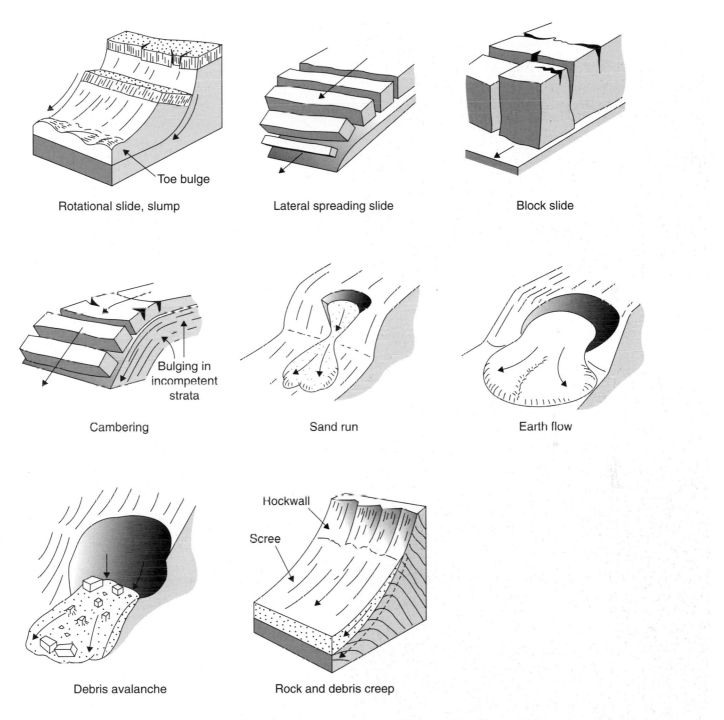

Figure 13.18 *Failure modes in unconsolidated earth materials. Source: In part after Varnes (1958).*

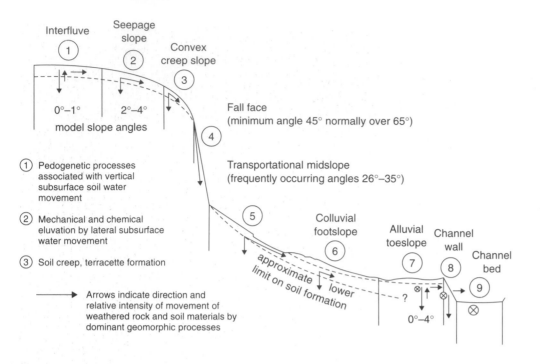

Interfluve

Seepage slope

① Pedogenetic processes associated with vertical subsurface soil water movement

② Mechanical and chemical eluviation by lateral subsurface water movement

③ Soil creep, terracette formation

⟶ Arrows indicate direction and relative intensity of movement of weathered rock and soil materials by dominant geomorphic processes

Convex creep slope

0°–1°   2°–4°
model slope angles

Fall face (minimum angle 45° normally over 65°)

Transportational midslope (frequently occurring angles 26°–35°)

Colluvial footslope

Alluvial toeslope

Channel wall

Channel bed

approximate limit on soil formation   lower

0°–4°

④ Fall, slide, chemical and physical weathering

⑤ Transportation of material by mass movement (flow, slide, slump, creep), terracette formation, surface and subsurface water action

⑥ Redeposition of material by mass movement and some surface wash fan formation. Transportation of material, creep, subsurface water action

⑦ Alluvial deposition, processes resulting from subsurface water movement

⑧ Corrasion, slumping, fall

⑨ Transportation of material downvalley by surface water action, periodic aggradation and corrasion

⊗ Indicates movement in a downvalley direction

Figure 13.19   *A hypothetical nine-unit landsurface model. Source: After Dalrymple et al. (1968).*

Plate 13.8   *Rotational landsliding, slumping and mudflows at Black Ven on the Dorset coast, seen from the air in 1966. Less competent Jurassic and Cretaceous clays trigger rotational failure in overlying resistant cherts and marl. Liquefaction moves fine debris downslope as mudflows.*
Photo: University of Cambridge Committee for Aerial Photography.

Key points

1 Denudation is the total effect of all processes which wear away earth materials and thereby lower the landsurface. It normally commences with rock weathering on exposure to the atmosphere and biosphere. Weathering products are subsequently transferred downslope under gravity by a range of dynamic forces mobilized by water, wind and ice in a variety of geomorphic environments.

2 Denudation exploits rock susceptibility in Earth's surface environment, which is alien to that in which most rocks form. The strength of rock, acquired during formation and diagenesis, depends on cohesive ionic bonds, inter-particulate friction and its own mass. If material resistance is exceeded by static and dynamic gravitational forces mobilized by tectonic and geomorphic processes, earth material fails in a manner reflecting its own properties and the nature of the forces applied.

3 Weathering reduces rock mass strength by the generation of internal stresses or the alteration of geochemical properties. Physical or mechanical processes are initiated by positive internal pressures through heating, hydration or the growth of salt or ice crystals, or elastic strain release caused by the removal of the confining pressures of adjacent rock. Chemical weathering processes may be summarized as hydrothermal alteration, since they involve solution, carbonation, hydration, hydrolysis and oxidation and are generally most effective in warm, moist conditions.

4 Mass wasting occurs when residual rock or earth material strength after weathering is insufficient to withstand the downslope component of gravity. Slope material properties, including water and air in voids, and slope angle determine the rate and mode of mass wasting, for which several classifications exist. However, properties often change in transit and most mass wasting proceeds as compound translation failure.

5 A variety of large-scale denudation chronologies and associated large-scale schemes for slope evolution have been proposed over the years. These reflected slope evolution in particular climate régimes and crude estimated rates of denudation. Modern research focuses on more detailed slope models and a materials science approach. Denudation rates, calculated from sediment yields and laboratory experiments, show that surface lowering rates are ultimately attuned to tectonic cycles and rates of plate motion.

*Plate 13.9 Toppling of cohesive, glaciofluvial gravels in a quarry face, after tension cracking behind the working face (right). The figure is standing on a toppled block.
Photo: Kenneth Addison.*

CHAPTER 14

Flowing water in the landscape

Water is Earth's 'proud setter up and puller down of kings'. Shakespeare's description of Richard, Earl of Warwick – the 'Kingmaker' during the Wars of the Roses – is most appropriate. Water stimulates the low-temperature melts vital to subduction orogens and is then a principal agent in their sub-aerial weathering and mass wasting. Nowhere is its ability to destroy, as well as create, mountain systems and continents more obvious than in running water at Earth's surface. Precipitation is widely distributed on arrival at the landsurface but markedly concentrated at points of **discharge** or departure in liquid (or solid) form through *trunk* rivers (or glaciers). The **catchment**, or landsurface unit generating stream flow, is a fundamental geomorphic and accounting unit. Catchment slope processes are intimately linked with their water and sediment transfers, and their collective yields then drive fluvial processes in the channel, although channel–slope links are rarely in equilibrium. Water–sediment stores and transfers are measured or estimated, leading to calculation of **water** and **sediment balances**. This is not done solely in the interests of geomorphology, for the catchment determines vital options for human occupation and land use. Water is required simultaneously for the essential but conflicting purposes of water supply and waste stream disposal. Data gathered for both hydrological management and geomorphological investigation considerably aid our appreciation of water flow through the landscape.

This chapter follows the cascade of water and sediment, from the generation of channel flow and the behaviour of water and sediment in channels to their creation of fluvial landforms. At first sight the *terrestrial* component of the **global hydrological cycle** appears to be negligible. Surface freshwater rivers, lakes and swamps account for only 110,000 km³ or less than 0.01 per cent of the global water balance of approximately 1.35×10^9 km³ (Figure 14.1). Almost 175 times this amount is stored in terrestrial glaciers, ice sheets and permafrost (see Chapter 15). However, the very short cycling time of surface waters by high-energy, rapid transit over the landsurface (less than twenty days or 0.05 year via rivers, compared with over 10^4 years via ice sheets) compensates for its diminutive mass (Figure 14.2). Rivers are the principal route for water, sediment and energy transmission from the continents.

Generation of channel flow

The catchment and water cascade

The catchment or **drainage basin** converts water, snow and ice input (precipitation + *in*fluent groundwater) to output as stream flow, evapotranspiration and *ef*fluent groundwater flow. Catchment climate and landsystems (soils, geology, slopes, ecosystems, and human structures and activities) determine the volume, routing and time scale of these transfers. This terrestrial hydrological system is the vital link in the global cycle between its principal atmospheric and ocean components. It processes water evaporated from oceanic and terrestrial sources in the ratio of approximately one to two. The importance of net landward advection of evaporated ocean water to drainage basins cannot be overstated – it is approximately equal to global stream flow to the oceans. From the moment

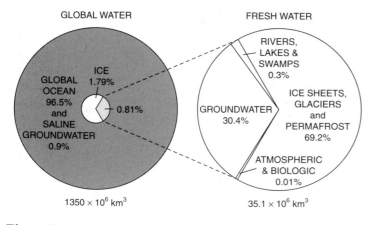

Figure 14.1 *Global water resources.*

precipitation arrives in the system, water may be transmitted through a series of in-line stores or exit any one of them directly or indirectly to a channel. The hydrological *catchment* defines the geographical surface area and geological sub-surface structure which delivers water to each trunk river. This three-dimensional landsystem is bounded by a **watershed** (Figure 14.3). The hydrological *system* defines the structure of component stores, transfer routes and processes whose individual character and spatial location are catchment variables (Figure 14.4).

Hydrological studies focus ultimately on river channels, although they cover less than 1 per cent of catchment area, but most water flow commences underground and the role of the hydrological system is emphasized by a hypothetical case. If all precipitation fell directly into channels, stream flow exiting the catchment would be very rapid, although not instantaneous – a function of catchment shape, stream connectivity, average slope angles and channel friction, discussed below. Catchment storage creates **lag time** (delay) in water transmission which moderates the episodic nature of precipitation and sustains stream flow during drier spells. It also buys time for other water-using systems (e.g. the biosphere, humans) but thereby diverts some water away from the channels and reduces overall stream flow. This is summarized in a basic water balance equation:

$$Q = (P - E) + (\Delta S + \Delta T)$$

where Q = stream flow, P = precipitation, E = evapotranspiration, ΔS = net change in storage and ΔT = net underground (influent – effluent) transfers. Where storage increases and effluence exceeds influence, ΔS and ΔT are negative (and vice versa) and Q correspondingly falls (or rises). The catchment may range in scale from single *first-order* streams (see below) less than 1 km² in area to major trunk rivers, such as the Thames (9950 km²) or Mississippi (3,270,000 km²). Stream flow is measured from single precipitation event responses, in hours or days, to the annual *water balance year*. The extent to which actual stream flow differs from the 'instantaneous' rate and the precipitation pattern is a function of lag time and routing towards or away from the channel system. The hydrological system ought to be one of the most easily understood open systems of energy and material transfers in the physical environment. In essence, water passes from available store to store, in sequence via a set transfer route as the capacity of each is reached. In practice the catchment disguises much of the system; we can measure quantities and rates at a limited number of visible points but depend mostly on estimates or extrapolations.

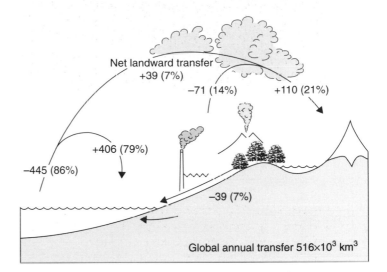

Figure 14.2 *Global hydrological cycle, showing sources of the annual total of 516,000 km³ of evaporated water and its distribution and eventual return to the oceans. Glacier growth or shrinkage respectively reduces or increases surface run-off. Source: Data based in part on L'vovich (1979).*

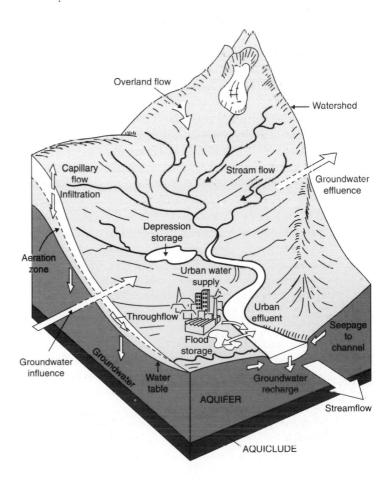

Figure 14.3 *The catchment landsystem.*

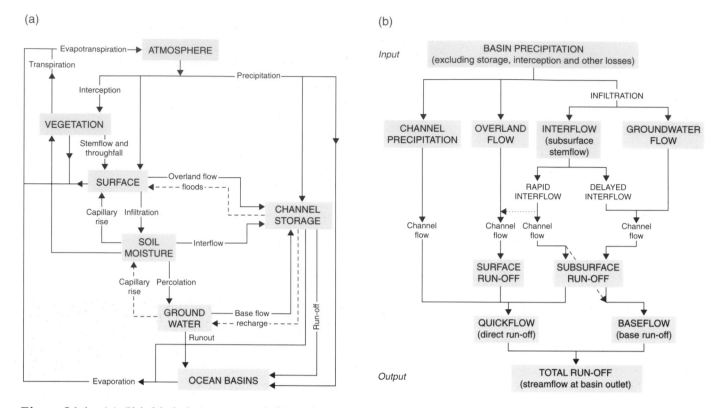

Figure 14.4   (a) Global hydrologic system and (b) catchment streamflow generating system. Source: After Ward (1975).

Hydrometeorological transfers

The (P − E) portion of the water balance deals with primary **hydrometeorological** transfers between atmosphere and catchment (precipitation inputs and evapotranspiration outputs). **Precipitation** is the total atmospheric input of water or water-equivalent mass (snow, ice) and varies according to volume, type, intensity, frequency and annual régime. Specific parameters are measured at *points* in the landscape by rain gauges and *area* estimates are provided by radar assessment of rainfall intensity or gauge-and-radar combinations, i.e with ground truth from gauges refining radar estimates. Data expressed as millimetres of water depth or rates in mm hr^{-1} are directly applicable only to the point or area concerned. However, total catchment data are required for most purposes, and areal estimates are made using statistical weightings, which reflect catchment character such as area and altitude. The spatial density of gauges, in particular, is low – e.g. a minimum of one per 100 km^2 in the United Kingdom for general water balance purposes – and usually reflects a country's remoteness and economic development. Table 14.1 shows key water balance data for the continents, excluding glacial Antarctica

and Greenland. Percentage data for global land area and stream flow (Q) reveal lower shares of stream flow for continents with substantial deserts. The mean value of 69.4 per cent discharged rapidly after rainfall or melt events (Q$_f$) underlines the volatility of stream flow and the maintenance of a substantially lower **base flow** for much of the year.

Precipitation *volume* determines the gross water balance and is composed of *falling* precipitation (rain, snow, hail) and **direct deposition** when moist air is cooled to dew-point temperature on contact with cold surfaces (dew, hoar frost and rime ice). This also identifies precipitation by *type* as a function of temperature and creates patterns determined by climatic zone, altitude, exposure and synoptic situation. All precipitation is measured in water-equivalent terms. This accommodates substantial density differences between water (1 g cc^{-3}), fresh snow (0.01–0.06 g cc^{-3}) and the subsequent evolution of snowpack by densification to **firn** (0.4–0.5 g cc^{-3}) and glacier ice (0.85–0.9 g cc^{-3}) (Plate 14.1). The proportion of precipitation in water and solid forms strongly influences subsequent water transfers and dependent catchment processes. Broad distinctions exist between humid-temperate catchments with transient snow-cover, cool-temperate/cold or alpine catchments with

Table 14.1 *Continental water balances (km³ yr⁻¹)*

Continent	P	E	Stream flow Total Q	Stream flow Surface (flood)	Stream flow Stable (base flow)	% land area	% global Q	Q_E	Q_F
Europe	7.17	4.06	3.11	2.05	1.06	7.7	6.5	43.2	65.8
Asia	32.69	19.50	13.19	9.78	3.41	32.0	30.0	40.3	74.1
N. America	13.91	7.95	5.96	4.22	1.74	18.6	12.6	42.8	70.1
S. America	29.36	18.98	10.38	6.64	3.74	12.9	26.5	35.3	63.9
Africa	20.78	16.56	4.22	2.76	1.46	22.3	18.7	20.4	65.2
Australasia	6.41	4.44	1.97	1.50	0.47	6.5	5.7	30.7	76.1
Total	110.32	71.49	38.83	26.95	11.88	100.0	100.0× =	35.5× =	69.4

Source: In part after M.I. L'vovich *World Water Resources and the Future*, Chelsea, Mich.: American Geophysical Union, 1979.

Note: P precipitation, E evaporation, Q stream flow, Q_E per cent of precipitation discharged as stream flow, Q_F per cent stream flow discharged as *surface flood* or *quickflow*.

enduring seasonal snowpack and rapid spring melt, and glacial catchments where fluvial processes are severely disrupted (Figure 14.5).

The temporal aspect of precipitation relates its duration, intensity and frequency to an overall régime which is crucial in determining the balance between storage and onward transfer set out below. Intensity is inversely proportional to duration and frequency and determines the time-based delivery of water to the catchment within a single event or annual balance. Evapotranspiration reduces the amount of water available for catchment processes by its direct return to the atmosphere. It is, properly, two separate processes subject to different controls beyond the principal requirement of free energy capable of liberating water molecules from a surface (see Chapter 7). *Evaporation* occurs from intercepted water *films* (precipitation coating plants, soil, impervious rock and artificial surfaces) and standing water *bodies* (rivers, lakes). It may also tap sub-surface water where **vapour pressure deficit** and capillary action overcome gravity and surface tension. *Transpiration* from plants results from their maintenance of internal nutrient flow and cell turgidity but has to overcome vascular resistance designed to prevent excessive loss and physiological drought. Losses are more easily estimated as potential evapotranspiration than measured by evaporation pans, evaporimeters and lysimeters owing to the great diversity of surfaces and rates involved. Since they assume a continuous

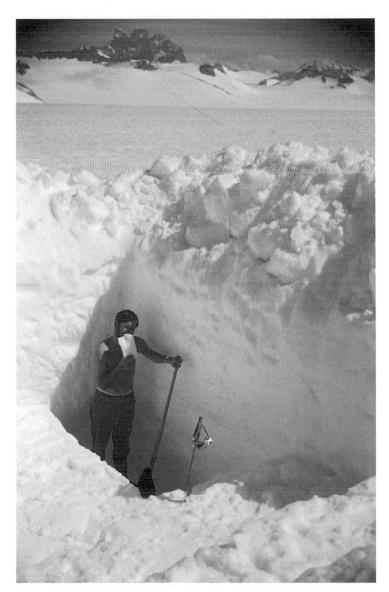

Plate 14.1 *Accumulation on the Taku glacier, Alaska, is calculated as water-equivalent mass, melted from cores from the annual snow-pack. Density increases with depth and includes ice storm layers (darker bands).*
Photo: Kenneth Addison.

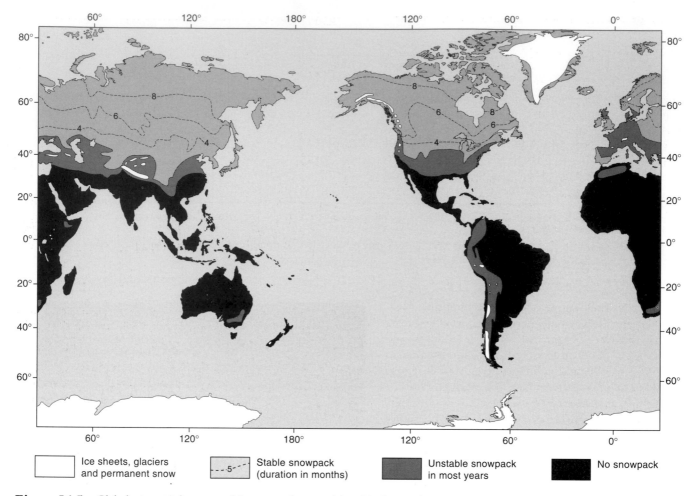

| | Ice sheets, glaciers and permanent snow | | Stable snowpack (duration in months) | | Unstable snowpack in most years | | No snowpack |

Figure 14.5   *Global terrestrial snow and ice cover. Source: After Mackay and Gray (1981).*

evaporable water supply, both methods underestimate water available for transfer and stream flow. This is evident in global water balance and stream flow estimates (Figure 14.6), emphasizing the irregularity of arid region precipitation and stream flow régimes where P < PE (potential evapotranspiration) but **ephemeral stream flow** occurs.

Hydrogeological transfers

Losses to the atmosphere can occur at any time but what happens next after precipitation input depends on hydrogeological ($\Delta S + \Delta T$) responses to input quantity, type and régime. Interception and absorption rates by the principal stores (vegetation, soils and bedrock) are determined by their **interception capacity** or **infiltration capacity**. Onward transfer occurs when either is exceeded, diverting surplus water to the next available store, and through gravity **draw-down** of water through each store. Water

draw-*back* to the surface by evapotranspiration depends on capillary and osmotic pressures, water potential gradients and void connectivity. Storage capacity depends on the total void space and the proportion left unfilled by antecedent events. These are broad definitions and each store has its own character.

Vegetation intercepts and stores water temporarily in foliage, from where it can be evaporated and small quantities absorbed. This is quite distinct from plant transpiration of cell water, with the exception of rainfed epiphytes. Interception rates depend on plant **leaf area index** (leaf area per unit ground area). They vary with vegetation maturity, season in deciduous vegetation, and are inversely proportional to precipitation intensity and duration. Water also reaches the ground by direct **throughfall**, or indirectly by **stemflow** and **leaf drip**. Table 14.2 shows some typical values.

Soil and, to a lesser extent, bare rock receive water direct or transmitted from the vegetation cover.

Shared properties of porosity (void space) and permeability (ability to transmit water) provide both storage capacity and transmission routes. They vary substantially over short distances in soils, at soil horizon boundaries and owing to other pedological processes (see Chapter 18). Intense rain can reduce infiltration rates in bare soil by raindrop compaction and can cause surface **crusting** on drying. Cultivation provides aeration through seedbed preparation (ploughing and harrowing) but compaction by vehicle wheels, which drastically reduces infiltration on cultivated sandy soils and induces gully erosion (Plate 14.2). Soil fissures (millimetre scale), **macropores** (millimetres) and **pipes** (centimetres or metres) greatly increase downward **percolation** and lateral **throughflow**, emphasizing that soils are rarely uniformly porous. Fissures develop seasonally through desiccation or shallow mass wasting. Macropores may be single or connected pores, often reflecting soil structure, and develop into pipes by enlargement during percolation. Pipes also develop through animal burrows and in the presence of swelling clay minerals.

During rainfall, water percolation advances along a *wetting front* in unsaturated soil, filling voids and hydratable particles. If sustained infiltration exceeds onward drainage the soil above this front becomes saturated, with implications for **overland flow**. Gravitational drainage, taking one to two days, does not leave soil dry, however, and water molecules adhering to soil particles and each other develop a **matric force** of 1×10^9 Pa (approximately 10,000 times atmospheric pressure). This binds thin water films (< 0.06 mm) around particles, invulnerable to gravity but available to plant roots and evaporation by capillary suction as high matric force draws molecules from wetter to drier areas. Soil moisture storage, measured by tensiometers and neutron probes, is at **field capacity** when **capillary** and **hygroscopic** water is at a maximum after gravitational drainage and at **wilting point** when accessible capillary water has been removed. Soil water tables, measured in dipping wells, mark fluctuating satura-

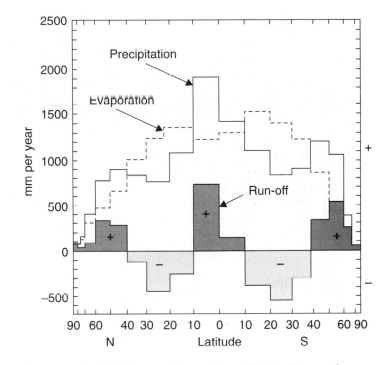

Figure 14.6 *Global water balance. Run off is P E; note the subtropical areas of large water deficit (hot deserts) and very small polar water surplus (cold deserts). Source: After Strahler and Strahler (1978).*

tion zones. Where drainage is inhibited by impermeable soil horizons or bedrock, it is diverted laterally towards channels as throughflow. Global soil moisture stores of 16,500 km³ account for only some 0.0012 per cent of global water, or 0.07 per cent of non-frozen terrestrial water. Despite this, its short cycling time (0.04–1 yr) and rapid fluctuation related to weather, land use and hydrogeology give it a major influence on stream flow (see Figure 7.22).

The **groundwater** store is far larger and more stable, with a global value of 23.4 million km³ or 0.17 per cent of the global balance and forms stable **delayed** or base flow stream flow components. Fifty-five per cent is saline but the remaining 10.5 million km³ accounts for 97 per cent of non-frozen terrestrial fresh water, maintains some 30 per cent of global river

Table 14.2 *Interception as percentage of rainfall.*

Vegetation type	Condition	Intercept (%)	Crop type	Condition	Intercept (%)
Tundra	Dwarf shrubs	45–55	Larch	Plantation	20–25
Pine	Woodland	35–42	Spruce	Plantation	24–32
Spruce	Forest	36–45	Sown grassland	Full cover	18–23
Deciduous woodland	Winter phase	12–15	Maize	Growing	15–18
Deciduous woodland	Leaf phase	25–40	Maize	Full cover	40–50
Tropical hardwood	Forest	12–22	Wheat	Winter cover	3–8
Grassland	Full cover	22–25	Wheat	Full cover	18–20

Plate 14.2 *Active rill erosion during a rainstorm on bare arable land in Shropshire.*
Photo: M.A. Fullen.

flow and contributes substantially to human fresh-water supplies. Hydrologically, bedrock generally acts either as an **aquifer** (underground reservoir rock) or **aquiclude** (barrier to significant absorption/transmission), although a further role exists. **Aquitards**, with low permeability, retard flow between aquifers whereas some aquifuges can absorb but not transmit water. Geological structure is often as important as individual lithologies in determining catchment groundwater character (Figure 14.3). As with soils, water tables delimit the saturated and overlying aeration zones and move in response to discharge–recharge balances. Aquifers *confined* by impermeable

strata generate high water pressures and force a **piezometric surface** in wells above the general water table of *unconfined* aquifers. **Phreatic water** in the former moves more vigorously than **vadose water** in the latter. Rock mass discontinuity networks usually generate high permeability and can be enlarged by corrosive flow, to such an extent in well structured limestone that drainage may be diverted underground through cavern systems (Plate 14.3).

Storage and transmission capacity is more consistent than in soil and a function of porosity and **hydraulic conductivity** – the ease with which porous rock transmits water. This is a component of **Darcy's law**, which shows groundwater flow Q_g as:

$$Q_g = KA \frac{\Delta h}{l}$$

where k is hydraulic conductivity, A is the discharge cross-sectional area, h is the hydraulic head and l the distance of flow. It is applicable only to homogenous porous media, whereas most soil and rock transmits water through fissures. Typical porosities of principal rock types are given in Table 13.1 but we need to know the impact of groundwater on stream flow. **Specific retention**, reminiscent of field capacity, and **specific yield** show the mobility of groundwater under gravity flow, with losses declining over time (Figure 14.7). A more direct measure is the base flow index, or proportion of stream flow contributed by groundwater, shown in Table 14.3.

Stream flow and hydrographs

Stream flow offers one of the most direct measures of the water balance of part or all of the catchment. **Stream gauging** attempts to quantify the water

Table 14.3 *Baseflow indices (BFI) for typical rock types in Britain.*

Rock type	Principal characteristics		Typical BFI range
	Permeability	*Storage*	
Chalk	Fissure	High storage	0.90–0.98
Oolitic limestone	"	"	0.85–0.95
Carboniferous limestone	"	Low storage	0.20–0.75
Millstone grit	"	"	0.35–0.45
Permo-Triassic sandstone	Intergranular	High storage	0.70–0.80
Coal measures	"	Low storage	0.40–0.55
Cretaceous sands and silts	"	"	0.35–0.50
Lias clays	Impermeable	Low storage at	0.40–0.70
Old Red sandstone	"	shallow depth	0.46–0.54
Silurian/Ordovician shales/slates	"	"	0.30–0.50
Metamorphic and igneous	"	"	0.30–0.50
Oxford, Wealden and London Clay	"	No storage	0.14–0.45

Source: After Institute of Hydrology (1980).

volume discharged over a fixed time period and is normally given in cumecs or cubic metres per second. Since it is virtually impossible to collect and measure total flow, except at very small volume and time scales, it is measured usually by **stage** or depth of water passing a flume or weir of known dimensions and water velocity (Plate 14.4). Other techniques measure the dilution of known quantities of injected salts or dyes passing a downstream point, or estimate discharge based on channel parameters. Continuous records of discharge, provided by automated stage-discharge recorders, are the most useful and permit the construction of a variety of **hydrographs**. *Flood* or *storm* hydrographs measure responses to single meteorological events (although they usually involve neither over-bank floods nor storms!) whereas annual hydrographs chart the discharge régime over a water balance year (Figure 14.8). It is also sometimes useful to consider the **unit** hydrograph response to a fixed precipitation input. Hydrographs are essential to the prediction and management of stream flow and water resources, and we return to this application in the box below.

For present purposes, hydrographs summarize outcomes of the catchment hydrological system and processes which generate stream flow as follows. Assuming dry antecedent conditions, initial precipitation infiltrates spare interception and soil storage. Vegetation intercepts the first ~ 1.0 mm and 20 per cent of subsequent rainfall. Surface water flow occurs wherever precipitation + net inward transfer (input) exceeds evapotranspiration + net onward transfer by percolation, groundwater recharge, seepage or abstraction (output). This occurs as transient overland flow on slopes or permanent and sometimes *intermittent* or *ephemeral* **channel flow**. *Horton* overland flow occurs when precipitation intensity exceeds soil infiltration capacity on non-vegetated surfaces and water moves away downslope at 10^{1-2} mm sec^{-1}, initially as **sheet flow**. This is simulated on impervious urban surfaces when drain capacity is exceeded briefly. Much water is detained in surface **depression storage**, evaporates or drains by percolation. Some will reach channels and contribute to **quickflow** – augmenting direct channel precipitation, riparian vegetation drip and throughflow. Vegetation and soil humus largely preclude Horton flow in humid climates but soil throughflow may emerge as *saturated* overland flow towards valley floors.

The impact of these rapid transfers is dramatic but short-lived, leading to a sharp rise in the constant or slowly declining baseflow of the hydrograph. This **rising limb** has four important parameters (Figure 14.9). The **peak discharge value** (PDV) represents the

Plate 14.3 *Hull Pot swallows a surface stream as it leaves sandstones for Carboniferous limestone, on the western flank of Pen-y-ghent, North Yorkshire. Photo: Kenneth Addison.*

maximum event discharge, prior to its recession down the **falling limb**, and is separated by lag time from the precipitation peak. **Time-of-rise** is determined by the antecedent storage capacity and connectivity of catchment components to the channel, which includes the upstream **channel network** for trunk river hydrographs (Figure 14.10) (discharge/basin character). Therefore, the greater the antecedent stores and connectivity of the system, the shorter the lag time and time-of-rise – and the higher the PDV, since less time has elapsed in which other losses may occur. Delayed flow (slower throughflow and baseflow) continues to

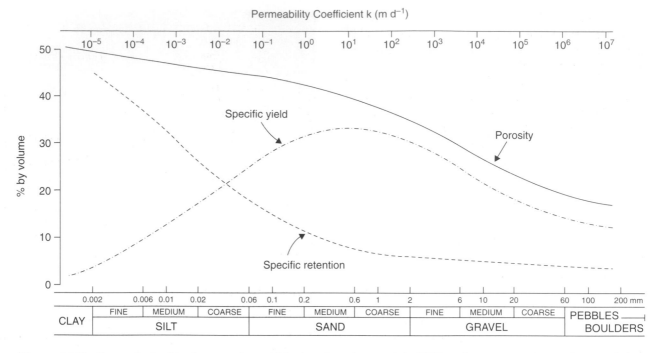

Figure 14.7 *Some relationships between the particle size of earth materials (BSI) and water retention and transmission.*

Plate 14.4 *Flume for stream gauging in a steep catchment on Plynlimon, mid-Wales. Its height, and baffles to dissipate stream energy, reflect the 'flashy' nature of upland stream flow.*
Photo: M.A. Fullen.

rise after the PDV has passed, driven by higher hydrostatic pressures in rising soil and groundwater tables, and meets the falling limb at a higher discharge than at commencement. The **depletion curve** returns to long-term baseflow once the quickflow and flood peak have passed. Where this is insufficient to sustain perennial stream flow, channels may still be maintained by seasonal or intermittent flow. This is typical of small tributaries, channel segments over highly porous strata and ephemeral streams in hot and cold (permafrost) deserts.

Stream flow in channels

Gravity-induced overland flow is both unstable and inefficient, since landsurfaces are neither homogenous nor frictionless. Differences in surface micro-relief and material properties soon concentrate water in parts of the sheet flow at the expense of others. Erosion commences where the **erodibility** of the surface and **erosivity** of the flow combine to exceed shear strength. Conversion of sheet to channel flow is attributed to a *stochastic* process of random events in a time-dependent sequence and can be seen in the smallest **rills**, 10^{1-3} mm deep and wide. Incision is controlled by specific material–energy conditions at every turn but the preferred points of concentration vary randomly from one event to the next. Channels are also initiated where sub-surface throughflow converges at the surface – in slope concavities, for example. Incision creates subsidiary, steeper slopes which then draw adjacent flow into the rill. This increased proportion of *channelled* water reduces overall surface–water friction losses. Rills are transient features and may be infilled by sediment and vegetation, or continue to focus episodic flow and develop at the expense of neighbours into more enduring **gullies**, an order of magnitude larger (Plate 14.5). The ever-increasing downstream focus of catchment flow and energy on the trunk stream is both inevitable and essential, if stream flow and sediment discharge are to overcome the parallel decline in potential energy. The need for **hydraulic efficiency** initiated at the watershed remains with flowing water throughout the catchment to the sea via the **drainage network** of successively larger rills, gullies and river channels.

Stream flow

We return to the network later but first need to understand the movement of water in channels. Stream flow is driven by gravitational energy, inher-

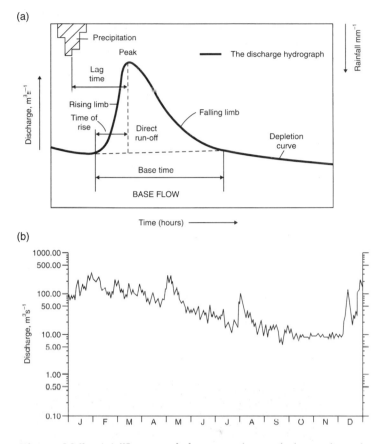

Figure 14.8 (a) The general character of storm hydrographs and (b) the annual hydrograph of the river Thames at Teddington. Source: Compiled from data from the DoE Water Data Unit (1983).

ited from uplift and subject to downstream exponential energy decay (Chapter 13). It is resisted by friction, primarily at the water–channel interface but also between water and solid sediment, individual ribbons of flow within the stream and with the atmosphere. **Dynamic viscosity**, increasing directly with dissolved and suspended sediment load and inversely with temperature, also influences flow resistance. Water moves in one of two ways. **Laminar flow** occurs at low velocities in shallow streams with smooth channels, when the lowest water *lamina* (thin layer) is retarded by channel boundary friction. Overlying laminae move successively faster past each other with a velocity maximum (v_{max}) at the surface (Figure 14.11). This rarely survives for long beyond the immediate boundary layer and breaks down into **turbulent flow** at higher velocities, at greater water depth and in irregular channels. This **eddy viscosity** consumes energy as ribbons of water shear past each other, creating more uniform velocities as faster and slower ribbons mix (Plate 14.6). Turbulence occurs at random at all scales but definite patterns occur

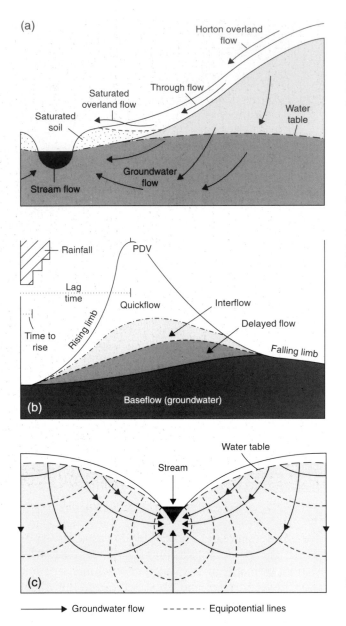

(a)

(b)

(c)

→ Groundwater flow   – – – – – Equipotential lines

Figure 14.9 *(a) and (c) Downslope water pathways and (b) their influence on components of the flood hydrograph.*

within meandering channels (see below), with a spiral or **helical flow** pattern and accompanying transverse currents.

Vital parameters of velocity and discharge are bound up in stream flow and its geomorphic activity. In addition to varying with distance from the stream bed and banks, velocity changes at-a-point with turbulence and discharge but is fairly constant downstream. This reflects downstream increases in hydraulic efficiency, compensating for declining channel slope and potential energy. Discharge varies at-a-point with the hydrographic response to quick-

flow and delayed flow and increases downstream with the contribution of tributaries. However, these parameters cannot be considered in isolation from channel form, which interacts with stream flow. We look first at the water behaviour in channels before considering their geomorphic development.

Stream channels

Channel geometry is defined by width, depth, length and slope, best seen in a short **channel segment** which emphasizes the role of water level in two further, derived channel parameters. The **wetted perimeter** equals $2d + w$ in a rectangular channel and **hydraulic radius** R is the cross-sectional area A ($= d \times w$) divided by wetted perimeter, $A/(2d+w)$

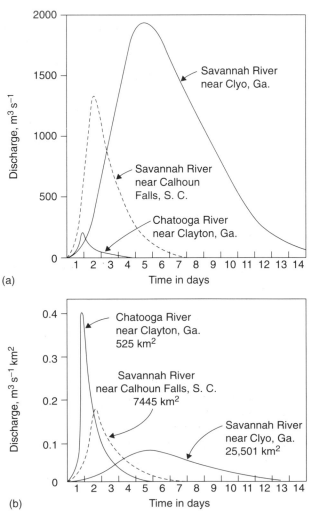

(a)

(b)

Figure 14.10 *Lag times (a) and flood duration (b) extend with catchment area as channel storage capacity increases downstream. Source: Hoyt and Langbein.*

(Figure 14.12). This allows us to measure discharge as:

$$Q = v_{mean} \cdot R$$

where v_{mean} is the mean velocity. Natural channels have irregular wetted perimeters and the magnitude of energy losses at the bed emphasizes the role of bed roughness. This is assessed through the **Manning equation**, which defines the v_{mean} in terms of hydraulic radius, channel slope (S) and a *roughness coefficient* (n)

$$v_{mean} = (R^{2/3} \, S^{1/2}) \, n^{-1}$$

with n ranging from 0.02 for smooth straight channels to 0.1 for rocky channels.

A number of other channel and flow conditions now fall into place. As well as measuring the effect of roughness in flow retardation, the Manning equation also makes the role of slope and hydraulic radius clear. The latter, in particular, is central to hydraulic efficiency and shown by the hypothetical geometry and discharge of three streams shown in Figure 14.12. The shallowest channel has the smallest hydraulic radius and the lowest efficiency. The combined flow of (a) and (b) in (c) demonstrates the normal advantages of trunk over tributary streams and the response to variations in *stage* and discharge. Hydraulic radius and velocity are linked with distinctions between laminar and turbulent flow via the *Reynolds number* (Re)

$$Re = (v_{mean} \, R) \, v^{-1}$$

where v is *kinematic viscosity* (the ratio of dynamic viscosity to density). Laminar flow and turbulent flow are found separately, or both types together, where Re is less than 500, over 2000 and 500–2000 respectively. Velocity, linked this time with stage (d), is also used to distinguish between two types of **flow régime** defined by their **Froude number** (F):

$$F = v_{mean}/\sqrt{gd}$$

where g is the gravitational acceleration. The flow is said to be *critical* when $F = 1$, separating *tranquil* or sub-critical flow régime ($F = \leq 1$) from *rapid* or supercritical flow régime ($F = \geq 1$). Deep water, gliding smoothly down the channel as tranquil flow, may enter a steeper segment over a visible fall or **hydraulic drop** in the water surface. Water depth

Figure 14.11 *(a) and (b) Vertical long-sections and (c) cross-section through a stream, showing styles of water movement and isovels linking points of equal velocity (1.0 marks the v_{max}).*

Plate 14.5 *Gullying through soil and regolith into bedrock on the North York Moors.*
Photo: Kenneth Addison.

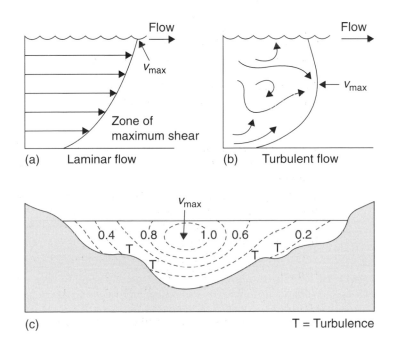

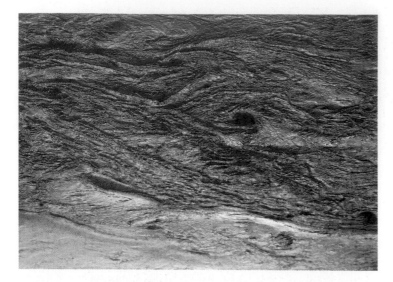

Plate 14.6 *Turbulent flow with eddies in flood conditions on the river Severn, where it is constricted through Ironbridge Gorge, Shropshire. The section is 4 m wide.*
Photo: Kenneth Addison.

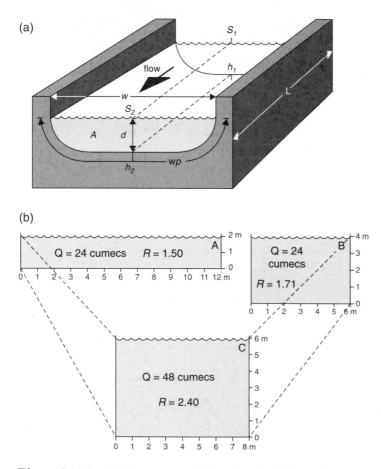

falls as rapid flow develops and the water surface becomes disturbed (Plate 14.7). If the stream then enters another less steep segment, the transition back to deep tranquil flow is marked by a **hydraulic jump** or standing wave. This clarifies the links between the shape of the water surface and bed forms in sandy channels, both of which can change over short distances and time scales. As F increases from values well below 1 small sand bed ripples are transformed to larger dunes out of phase with surface ripples. These are planed off where F approaches 1 and antidunes form where F is more than 1, this time in phase with surface waves (Figure 14.13).

Links between stream flow and channel morphology, which *conserves* discharge from one segment to another in the example given, must also embrace *changes* in discharge over time. It might be argued that ideal channels should accommodate most, but not all, peak discharges and that **bankfull discharge** leading to flooding should be exceeded either annually or only after abnormally wet periods or high rainfall events. We saw earlier how channel efficiency decreases and friction losses increase as discharge falls in fixed-geometry channels. Similarly, wetted perimeter increases and efficiency decreases dramatically in over-bank conditions when the flooded valley briefly becomes part of the channel. Whilst regular flooding is indicative of channels ill adjusted to discharge, there are excessive 'costs' in maintaining channels which never flood. It is easy to see this in the case of large, engineered channels with higher *financial* costs in land and construction, but it is also evident in the *energy* costs of natural channels. In bedrock, streams often cut *channel-in-channel* forms and shrink into the smaller channel during low flows. In soft sediments, stream flow cannot sustain large channels at low flows and **bank caving** effectively reduces channel dimensions (Plate 14.8).

Channels are clearly dynamic landforms, making constant adjustments between form and function to which we return below. Questions still being asked about the way in which water flows, and does work, focus appropriately on very small channel segments or controlled laboratory flumes. For the purpose of summarizing general impacts on the land-surface and fluvial landsystems, we need to appreciate the potential energy available in each channel segment and the shear stress mobilized at the channel

Figure 14.12 *(a) Geometry and (b) hydraulic efficiency in stream channels. Principal parameters are identified in the text; h_1–h_2 and S_1–S_2 represent the height decline in bed and water surface respectively along segment length L. (b) shows the combined flow of two tributary streams, A and B, downstream in stream C, assuming a uniform velocity of 1 m sec⁻¹. C has a greater channel efficiency, reflected by the increase in R; R is still greater than 2.0 even if the stage (depth) falls towards 4 m.*

Plate 14.7 *Riffle and a downstream hydraulic drop, indicated by the broken water surface to the right foreground, in the River Dove, Staffordshire. This riffle reflects slope–channel sediment connectivity below the accelerated (visitor-generated) erosion scar on the far slope.*
Photo: Kenneth Addison.

boundary. **Stream power** Ω, measured in W m^{-2}, is defined as:

$$\Omega = \rho g Q S$$

and **boundary shear stress** τ_o, is defined as:

$$\tau_o = \rho g R S$$

where ρg is the *specific weight* of water (density × gravitational acceleration) and Q, R and S are standard parameters used above. Boundary shear stress multiplied by mean velocity gives the *specific* stream power:

$$\omega = \tau_o v_{mean}$$

and is high (over 1 kW m^{-2}) in steep, high-discharge streams and low (under 100 W m^{-2}) in gentle, low-discharge streams. This confirms the importance of channel efficiency in establishing a **power threshold**, at which available power is just sufficient to overcome friction resistance to mean water and sediment discharge. Below this threshold, the stream deposits a proportion of its sediment load to restore efficiency. Above the threshold, the stream actively erodes its channel.

Channel networks

Principles of channel process and form extend from individual segments to the entire stream and, indeed, the drainage basin. Downstream changes discussed so far are applied later to the geomorphological development of fluvial landsystems. We turn here to their significance for the basin-wide *network*. Channel or drainage networks are organized systems of channels transferring water and sediment, in incremental amounts, through the catchment via a definite sequence (e.g. rills–gullies–river channels–trunk streams). The network possesses measurable order (hierarchy), density and pattern, and reflects the principal catchment attributes, including stage of development and catchment shape.

Stream order recognizes the nature and development of channel hierarchy as more aggressive channels 'capture' neighbours. Having looked earlier at how rills focus flow in a downslope manner, upslope and lateral development extends this simple model. Local convergence of water in the channel, from (ideally)

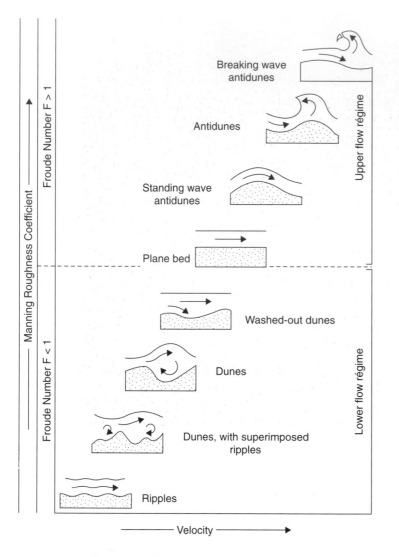

Figure 14.13 *Relationships between bed forms in sandy alluvial channels, channel roughness and stream flow. Compare with Figure 12.13. Source: After Selby (1993).*

isotropic or *equipotential* throughflow or overland flow, causes **spring sapping** and **headward retreat** of the point of initiation. This may continue via branching until the remaining catchment area is too small to feed any additional channels (Figure 14.14). At its simplest, stream ordering identifies all streams lacking tributaries as *first order* streams and each sequential stream by either arithmetic progression (2, 3, etc.) or as the sum of feeder streams. Shrève's scheme is preferred to Strahler's, which fails to recognize the scale of increasing discharge, although it is thought to generate a number of useful correlations between stream connectivity and the catchment (Figure 14.15).

Drainage density, D_d, refers to the channel length (L) draining a unit area (A):

$$D_d = L/A$$

and reflects catchment variables, channel network efficiency and maturity. In hydrometeorological terms, higher drainage densities are associated with higher mean annual precipitation, or highly seasonal régimes, but correlations are far from simple. High densities may also be found in semi-arid areas where lack of vegetation cover compensates for lower rainfall. Drainage density increases during wet spells, as the permanent network expands to incorporate intermittent channels. Geology influences drainage density, which is inversely proportional to permeability, and is central in the creation of **drainage pattern**, which strongly reflects the underlying geological structure (Figure 14.16).

In two respects, network evolution is also time-dependent. The number and density of lower-order streams are likely to increase on immature slopes before stream connectivity improves. This may be indicated by the **bifurcation ratio**

$$B_r = Sn/S(n + 1)$$

Floods and flood control

Flooding is the inundation of land beyond the normal confines of a channel or coastline, either by overflow of excess water or its influx via shallow sub-surface or low-lying routes. Coastal flooding, which occurs through abnormal tidal surges or waves, often enhanced by the coincidence of storms and river discharge, is of less concern here. All other forms of flooding involve water temporarily unable to enter a channel, in the case of **depression storage** and sheetflow, or *re*-enter it after an **over-bank discharge**. Floods originate from extreme meteorological events and their indirect and geophysical consequences (Figure 1). Flooding is the most frequent and widespread form of 'natural' disaster, affecting more humans than any other physical hazard. It appears to be on the increase yet takes a surprisingly small toll of human life. This is not the enigma it first seems. Flooding is so common that people living in flood-susceptible areas rarely do so unawares. Vulnerability is accepted either by choice, in pursuit of economic or aesthetic gain, or of necessity, because of the pressure on, or the particular attributes of, the land. Historically the latter have including flooding itself as a means of natural irrigation and nutrient replenishment (e.g. in farmland along the Nile).

Risk assessments of varying degrees of formality are undertaken and means of evading, avoiding or mitigating flood losses are developed. Unfortunately, human occupation of floodable land – and other activities apparently unconnected with it elsewhere in the

Plate 14.8 *Active bank caving in a 3 m high outer meander bank on the river Dane, Cheshire. Note its fine-grained, floodplain sediments, compared with coarse sediment on the point bar.*
Photo: Kenneth Addison.

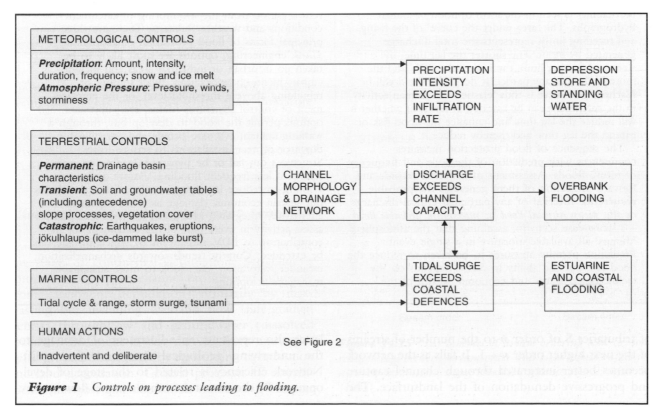

Figure 1 *Controls on processes leading to flooding.*

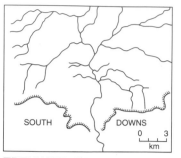

TRELLISED: River Adur,
West Sussex, England

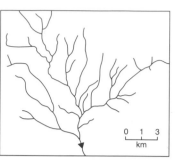

DENDRITIC: Water of Tarf,
Kincardine, Scotland

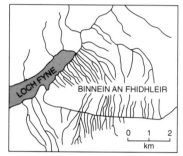

RECTANGULAR:
Adirondack Mountains, USA

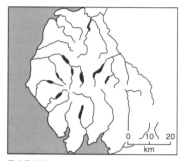

RADIAL:
English Lake District

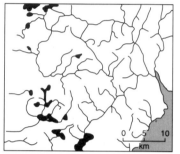

PARALLEL: Glen Fyne,
Argyll, Scotland

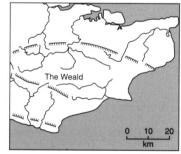

ANNULAR: Wealden
District, south-east England

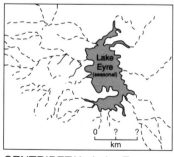

CENTRIPETAL: Lake Eyre
basin, South Australia

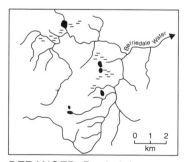

DERANGED: Berriedale
Water, Caithness, Scotland

Figure 14.16 Classified scheme of drainage network patterns.

impact. Corrosion depends on rock susceptibility, water velocity and discharge but most dissolved load is probably acquired from pre-channel processes, since water spends relatively little time in channels. Abrasion depends on bed shear stress, flow turbulence and relative rock hardness. Channel **potholes** containing smoothed pebbles exemplify the general process, known as **evorsion** through its dependence on a fluid vortex (Plate 14.9). However, the ability of large entrained boulders to strike off angular bedrock fragments in turbulent, high-velocity high-discharge flows is probably more effective and maintains angular profiles.

The erosivity by **fluid stressing** of stream flow itself depends on the fluid shear stress directed towards the channel boundary in turbulent flow and bedrock shear strength. This is sufficient to erode softer or well fractured rocks and may be enhanced by **cavitation**, when bubbles or small cavities in rapid turbulent flows implode. Micro-jets of infilling water generate stresses of up to 60 MN m^{-2} when implosion occurs on contact with the channel boundary. Bedrock is less susceptible to fluid erosion but unconsolidated sediments are readily eroded in valley floors, where they form a continuum with the **alluvial toe-slope** of the nine-unit landsurface model, channel wall and bed (see Chapter 13). They are found throughout the catchment but as isolated pockets in upper, steeper segments compared with continuous spreads on the flood plain. Erosion occurs in two ways. At all discharges, subaqueous erosion will occur where bed shear stress exceeds shear strength in cohesionless, granular sediment. At low discharges, bank failure occurs by slumping after both wetting and drying cycles, or toppling where tension cracking is generated by the removal of the supporting effect of water (Plate 14.8). Soft-sediment erosion is a recycling process, as the stream reworks material it previously deposited, but there is net onwards transfer of sediment driven by the influx of mass wasting debris and continuing denudation.

Entrainment, transport and deposition

Entrainment is the incorporation of particles when stream velocity exceeds the *entraining velocity* for a particular particle size or, more accurately, if bed shear stress exceeds particle–bed friction and effective stress. This is a natural extension of erosion and is vital to the movement of stationary particles in changing flow conditions. Conversely, deposition occurs when **stream competence**, or ability to maintain movement as bed load, falls below a given velocity. This applies to large debris particles delivered to the channel by bank caving or landsliding, which simply fall out of the flow, or when stream velocity falls below the **fall velocity** for a particle in transit. All conditions are summarized in a version of Hjülström's diagram in Chapter 12 (Figure 12.11), which shows further important distinctions between *erosion* and *transport* velocities for particles below medium sand size (0.2–0.6 mm). Smaller particles require higher initial velocities because they present smaller, 'streamlined' surfaces to the flow and may also develop weak cohesive strength from surrounding water films. Dissolved load is deposited by precipitation when solutions exceed saturation level.

Between points of entrainment and deposition, particle movement is determined by size, flow conditions and mode of entrainment. Particles above medium–coarse sand size (over 0.2 mm) tend to move by rolling or sliding along the channel bed as *bed* or *traction* **load**. More mobile particles are lifted into the flow if pressure falls in the wake of overlying accelerating and, especially, turbulent flow. Particles move into the partial vacuum and remain in suspension by incorporation into even faster flow paths, or until their weight overcomes buoyancy. Sand particles fall out rapidly and move by **saltation** or repeated bouncing. Silt particles (under 0.06 mm) move as *suspended* load and clay particles (under 0.002 mm), indefinitely, as *wash* load. These modes and overall catchment sediment transfers are summarized in Figure 14.18. Since particles move by different styles and in distinct parts of the channel and stream flow, a considerable amount of **particle sorting** occurs which persists into the depositional environment. Channel and over-bank (flood) sediments form predominantly sand–gravel and clay–silt *facies* respectively. Particle size, grading laterally within the same bed and changing abruptly in a vertical sequence, demonstrates rapid changes in flow conditions (Plate 14.10). The global extent of sediment removal to the coast is shown in Figure 14.19.

Fluvial landsystems

Reference to landforms so far has been incidental and related largely to small-scale channel morphology and bedforms, responding to changes in the power threshold and search for efficiency. Their geomorphic impact is reviewed now at the catchment scale, where fluvial landform assemblages and channel networks form a recognizable fluvial *landsystem*. The transition from erosion-dominated uplands to deposition-

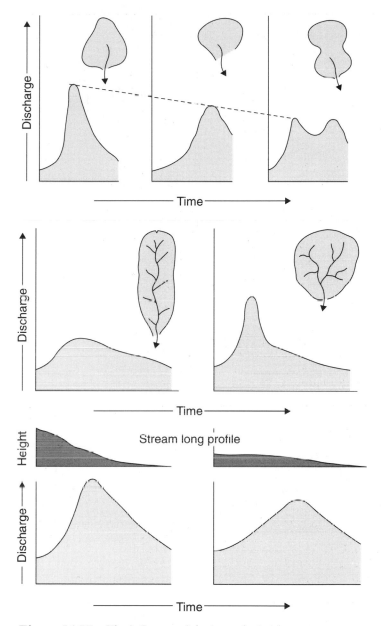

Figure 14.17 *The influence of drainage basin character on individual flood hydrographs. PDV and lag time vary according to shape, network and slope. Note how PDV falls as lag times increase, conserving the 'area under the curve' or total discharge. Source: After Gregory and Walling (1973).*

dominated lowlands is seen by following the trunk stream from source to sea. Its typically concave long profile, along which potential energy increases exponentially towards the source, was mistakenly used to distinguish between vigorous, youthful headwaters and sluggish, senile lowland rivers. Both elements are present in mature, integrated catchments and the profile corresponds to a power curve requiring constant adjustments by the stream to downstream changes in potential energy, discharge, slope and

Plate 14.9 *Potholes in the winter channel of the river Wharfe, North Yorkshire, lined with corroding pebbles (left), contrast with a massive pothole formed by high-pressure subglacial meltwater beneath a glacier moulin (Norway). Photos: Kenneth Addison.*

sediment load. Locally, surplus energy erodes and lowers the profile through **incision** whereas energy deficit leads to its **aggradation** or elevation through deposition. At any time the profile is complicated by different stages of response to tectonic and eustatic changes in base level, climatic, geological and land-use conditions and – increasingly – human regulation of the catchment.

Upper catchments and bedrock channels

Bedrock channels and associated valleys are not the sole preserve of mountains or uplands, although tectonic drive and denudation produce some of Earth's deepest gorges and rock walled valleys there. They are most dramatic in alpine orogens with gorges 10^{2-3} m deep, or in the middle reaches of rivers elevated by recent epeirogenesis in continental interiors and at passive margins. The upper Tsangbo

(Brahmaputra) and Indus trunk rivers and principal tributaries such as the Hunza (Plate 14.11) and Gilgit (Karakoram range), and the Grand Canyon of the Colorado river (Arizona), are excellent examples of the former and latter origins. Gorges tend to be straight or of low sinuosity in orogens, closely directed by geological structure. By contrast, uplift of the lower reaches of established meandering rivers, or their rejuvenation by falling base levels, often conserves sinuous channels as **incised meanders**. The Dee gorge has straight and incised meander sections, with **abandoned meander cores**, 250–300 m deep along a 15 km stretch between Corwen and the English border east of Llangollen. Impressive waterfalls represent either maladjustment between incision and uplift, in which case they form **knick points** marking the upstream limit of adjustment, or between incision and geological structure. The Angel Falls, plunging 1000 m off the Roraima plateau of Venezuela, exemplify the former and the latter

(a)

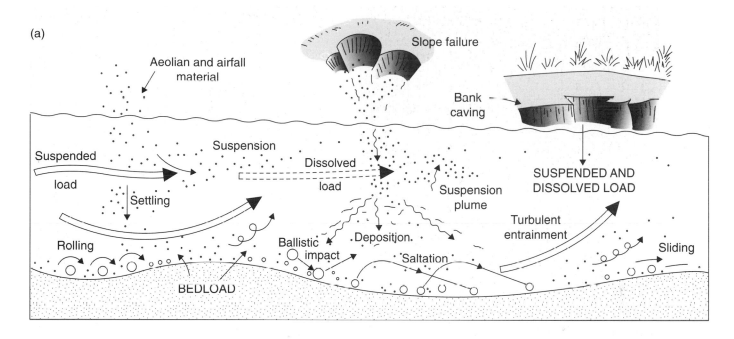

(b)

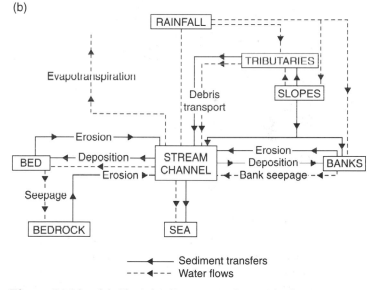

Figure 14.18 *(a) Fluvial sediment entrainment and transport and (b) sediment routing through the catchment.*

include the 52 m high Niagara Falls incising the Niagara escarpment.

Away from gorges in older, lower mountains and uplands generally, fluvial channels dominate the narrow floors of valleys with V-shaped cross-sections and barely concealed rock slopes beneath thin regolith. Valley floors are usually irregular, with steep sections marked by waterfalls and associated plunge pools, rapids, potholed bedrock channels and large angular debris, interrupted by gravel-lined stretches in profile concavities. Tributaries enter higher-order streams across small, steep **debris cones** or more extensive and less steep **alluvial fans** marking a significant reduction in channel slope and stream competence (Plate 14.12). This applies on the grand scale along mountain fronts, where they may coalesce into huge **piedmont fans** 10^{2-4} km^2 in area, or arid-zone bajadas. Much of the later Cenozoic *molasse* south of the Himalayan mountain front was formed in that way.

Most temperate catchments reflect the geomorphic impact of Pleistocene cold stages. Glacial excavation of deep valleys and permafrost ornamentation of slopes in upper catchments disrupted the fluvial landsystem (see Chapters 15 and 25), with glacial diversion of drainage and sedimentation in lowland catchments. High spring permafrost melt discharge eroded enlarged valleys. They survive as **dry valleys** or are occupied by diminutive **misfit streams**, especially in normally porous lithologies such as chalk and limestone, which dominate the uplands of southeast Britain. Altered catchment slope- and sediment-

dynamics still influence fluvial processes in most areas today (Figure 14.20).

Lakes represent a further, transient interruption of fluvial development of the catchment and occur where downstream flow is impeded by rock or debris barriers. They are by no means exclusive to upper catchment areas. Confining valley slopes with glacial excavation, landslide activity and volcano-tectonic processes create favourable sites for impounding water. The huge lowland lake systems of north-west Canada and eastern Scandinavia are largely glacial in origin. Lakes are both fed and drained by rivers but

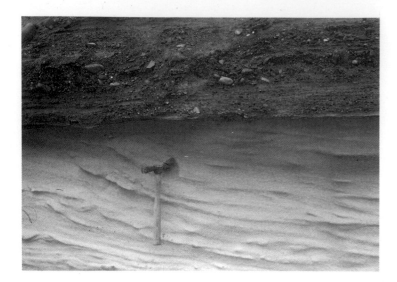

Plate 14.10 *Cross-bedded coarse sand, truncated by fine and medium gravel in fluvial sediments.*
Photo: Kenneth Addison.

they buffer downstream stretches from sediment influx, which progressively infills the basin instead, and also form temporary base levels for upstream reaches (Figure 14.21).

Lower catchments and alluvial channels

Higher potential energy leading to net denudation in upper catchment areas transfers large sediment volumes into the lower catchment, where they line channels and entire valley floors with **alluvium**. This is mostly the sand–gravel fraction (0.06–60.0 mm) and above in straight channels but includes substantial silt (0.002–0.006 mm) and even clay (under 0.002 mm) fractions in sinuous channels. Not all of this material reaches its final, marine destination; the geological record has substantial components of lithified terrestrial sediment in cratons or incorporated in continental collisions. Reference is made in Chapter 25 to the reassimilation of alluvium in New Zealand orogens. For all that, unlithified alluvial sediments are far more easily eroded than bedrock and facilitate channel adjustment to flow régime. The **flood plain** environment also experiences dynamic changes at whole-channel and flood-plain scales, depositing and remobilizing soft sediments.

Straight and meandering channels

Meandering is the natural tendency for alluvial channels, although it was once thought to reflect

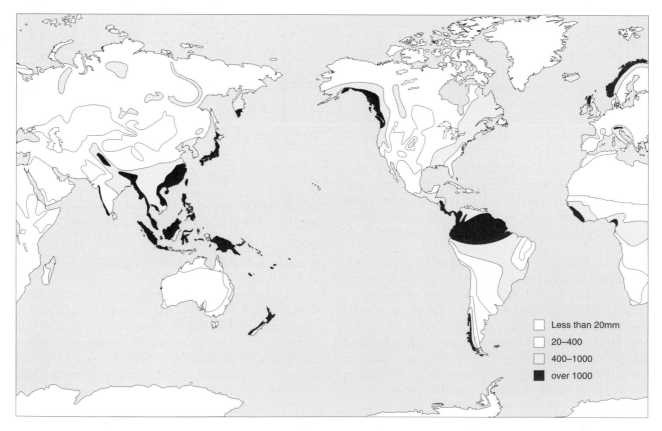

Figure 14.19 *Global mean annual run-off. Source: After L'vovich (1973).*

Legend:
- Less than 20mm
- 20–400
- 400–1000
- over 1000

Plate 1 The Trifid Nebula (M20). Situated in the constellation Sagittarius, this nebula gets its name because it is apparently divided into at least three sections by dark dust lanes. It is believed that new stars are forming in the brightest middle portion of the nebula. The red colouration is characteristic of glowing hydrogen gas, whereas the blue area represents cooler gas which is scattering the light emitted by the star centrally positioned within the blue area. Photography by David Malin of the Anglo-Australian Observatory, and Photolabs Royal Observatory, Edinburgh. Original negatives by UK Schmidt Telescope Unit. Copyright Royal Observatory, Edinburgh.

Plate 2 The Central Cordillera of the Andes in Ecuador. The false colour image shows a belt 240 km wide. Aligned Cordilleran ridges and intervening basins run down the image and are interspersed by the snow-capped strato-volcanoes, shown in red (the false colour image for snow). Vegetation, or bare rocks on higher vents, are shown in green/cyan
Landsat Thematic Mapper image produced by the Remote Sensing Group, British Geological Survey. Copyright NERC.

Plate 3 *Igneous rock selection (clockwise from top left). Basalt with peridotite xenolith (inclusion); surviving mantle peridotite (dark green) contrasts with the very fine grained, dark grey basalt. Vesicular basalt. Rhyolite. Andesite, with white phenocrysts providing porphyritic texture. Ignimbrite or ash-flow tuff, with slower-cooling fragments drawn out in the flow direction. Dartmoor Granite, with white feldspar. Shap Granite, with pink porphyritic feldspar. (Both granites also contain grey/translucent quartz crystals and black biotite (mica) crystals). For scale, Dartmoor Granite is 10 × 10 cm.*

Plate 4 *Metamorphic rock selection (clockwise from top left). Greenschist, with foliation and thrust nappes. Gneiss, with augen (eye-shaped) phenocrysts. Amphibolite, with foliation and micro-folds. Marble, with a sugary semi-crystalline texture. Slate, shown normal to cleavage and with a pale reduction spot showing deformation. Phyllite, with a shiny semi-schistose structure. For scale, amphibolite core is 20 cm long.*

Plate 5 *Sedimentary rock selection (clockwise from top left). Conglomerate, with quartzite and quartz pebble clasts in sand matrix. Sandstsone, with fine climbing ripples. Siltstone, with turbidite structure showing graded bedding with rippled tops. Laterite; a residual, rather than strictly sedimentary, rock representing a highly weathered secondary oxide of iron. Coral, representing the calcareous skeletal remains of colonial, marine invertebrates. Limestone, with fossil marine crinoids. Carbonate mudstone (light bands) with organic stromatolite algal mat (dark bands) from a tidal lagoon. For scale, the sandstone core is 8 × 8 cm.*

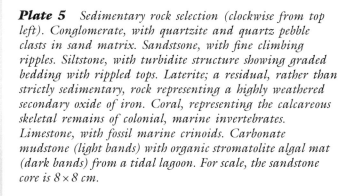

Plate 6 *Peaks near Pasu in the Karakoram Range. Saw-toothed summits and debris cones are the product of uplift and intense denudation.*
Copyright A.S. Goudie.

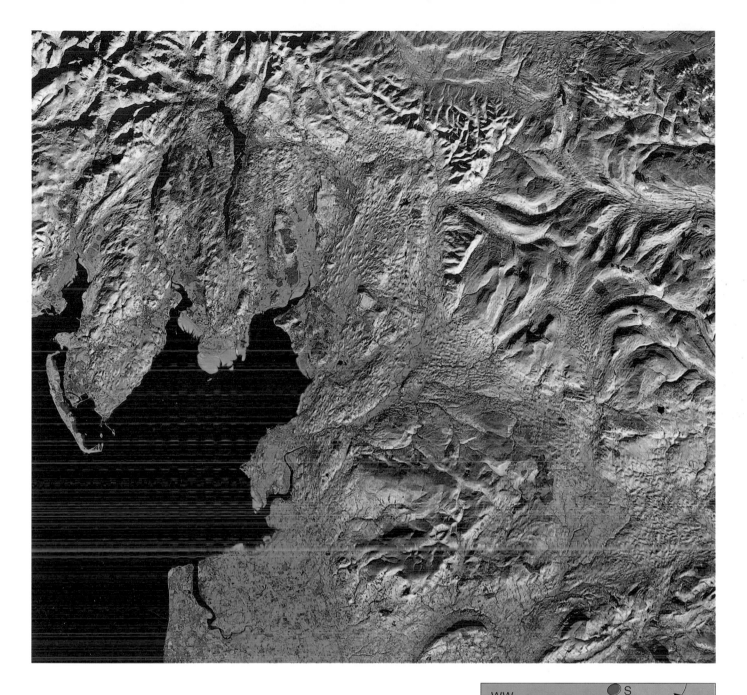

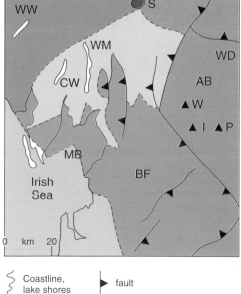

Plate 7 *Southern Lake District and western Pennines of north-west England. Borrowdale Volcanic Group (map – dark blue) and Bannisdale Slate and Coniston Grit formations (light blue) form mountainous terrain. They were domed by thermal diapirism, including the exposed Shap Granite pluton (S), during the Caledonian orogeny. A half-radial pattern of glacial rock basins holding Wastwater (WW), Coniston Water (CW) and Windermere (WM) punctuates the mountains. Upper Palaeozoic strata (brown) form the Variscan orogen Askrigg fault-block (AB) of the western Yorkshire Dales and Bowland Forest (BF). Carboniferous limestone terraces in Wensleydale (WD) and the monadnock peaks Whernside (W), Ingleborough (I) and Pen-y-ghent (P) are visible. Drumlin swarms (dimpled relief) mark the path of Late Quaternary glaciers sweeping out of the Lake District. Barrier and estuarine coasts fringe Morecambe Bay (MB) lowlands on Mesozoic strata (green). Landsat Thematic Mapper image 100 km wide; bands 4, 5 and 7. Moorland and montane vegetation (green) contrasts with plantation, pastoral and arable farmland (brown) and estuarine sediments (cyan). Image produced by the Remote Sensing Group, British Geological Survey. Copyright NERC.*

Plate 8 Desert landforms around Gebel Gerf, South-eastern Desert, Egypt. The mountainous area (an ultramafic ophiolite massif) and other areas of rock desert (grey) source large gravel sheets and fans (reddish brown) which feed sand seas (cream) downwind. Landsat Thematic Mapper image 75 km wide; bands 4, 5 and 7. Image produced by the Remote Sensing Group, British Geological Survey. Copyright NERC.

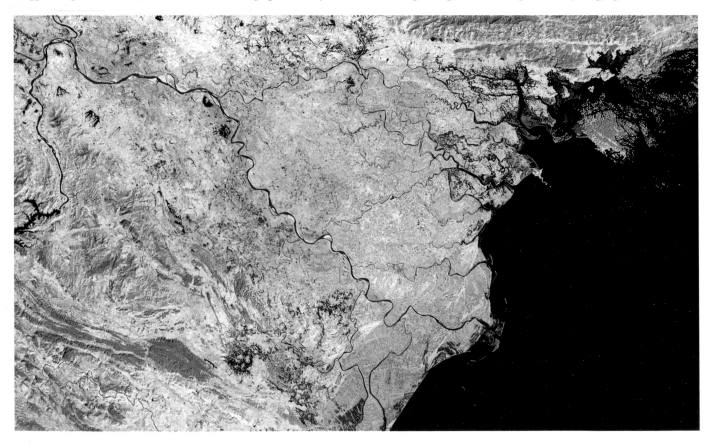

Plate 9 The cuspate, wave-dominated Red River delta, Vietnam, fronted by a barrier coast with offshore bars. A dense network of contemporary and abandoned distributary channels (black; northeast, top) and parallel and en echélon beach ridges (pink; south-east, bottom right) are clearly visible. Hanoi (black) is located where the two principal distributaries part and the intensive farmland of the delta contrasts with flanking, structurally-controlled hills (orange). Landsat Thematic Mapper image 240 km wide; bands 4, 5 and 7. Image produced by the Remote Sensing Group, British Geological Survey. Copyright NERC.

sluggish inability to maintain a more direct line in the 'senile' stage of the river. Straight channels are uncommon, except in heavily regulated and 'channel-ized' rivers where flood evacuation is a required aim. Ironically, this speeds water on to the next down-stream unprotected zone, and new trends in river management include meander *restoration*! Straight channel segments carry low bed loads, compared with meandering forms. Many natural channel segments may appear straight but the **thalweg**, or line of maximum water or channel depth, itself meanders. Flow or channel sinuosity is measured as the ratio between the channel- and straight-line distances between two points and equals 1 in straight channels, more than 1 in sinuous channels. Meandering is, arbitrarily, considered to occur at over 1.5.

Many theories have been advanced for meandering, including chance deflection by obstacles with desta-bilization of stream flow which the channel repeatedly over-corrects. Recent studies emphasize temporal changes in sediment supply and bed shear stress, leading to local erosion/deposition and hence new channel geometry. Since channel efficiency increases downstream to counteract lower potential energy, and velocity remains constant, a downstream increase in discharge must require adjustments to the channel. Meandering consumes surplus energy by lateral erosion and against the larger wetted perimeter implicit in sinuous rather than straight channels (Figure 14.22 and Plate 14.13).

Bars, **riffles** and **pools** are large-scale, dynamic bedforms implicit in the morphology and formation of meanders and it is useful now to assume that the thalweg is also the line of average maximum velocity. *Alternating bars* develop by deposition in slower, less competent flow either side of the sinuous main-stream. Arguably, deposition in a random area of lower bed shear stress could trigger sinuous flow and set up downstream loci of slower flow as the main-stream rebounds off opposing banks. However it

Plate 14.11 *The Minapin glacier valley descending from Mount Rakaposhi (7788 m) into the Hunza valley, Karakoram range, Pakistan. Continuing tectonic uplift and intense glacial and fluvial erosion distinguish this place, together with Nepal, as the steepest on Earth. Photo: Andrew S. Goudie.*

Plate 14.12 *An alluvial fan, located where a steep tributary stream sweeps coarse debris into a Pennine valley. It also provides a dry, elevated site suitable for farm buildings and improved pasture. Photo: Kenneth Addison.*

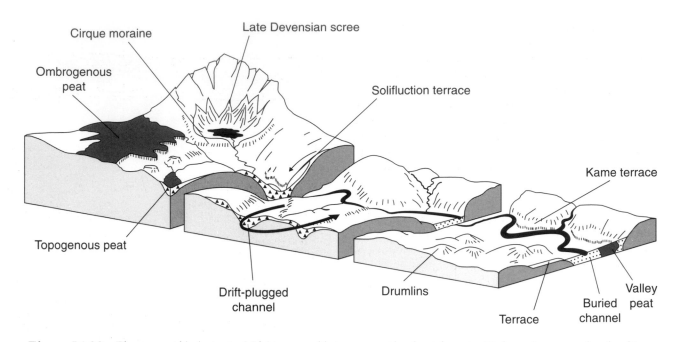

Figure 14.20 *The geomorphic impact of Pleistocene cold stages on upland catchments. Modern rivers rework a landform and sediment system largely not of their making. Source: Newson (1981).*

commences, deposition of coarse bed load locally increases channel roughness, causing further accretion of coarse material and finer sediment on the upstream and downstream ends of the bar respectively. Depleted of bed load, onward-moving water regains competence and increased shear stress which erodes

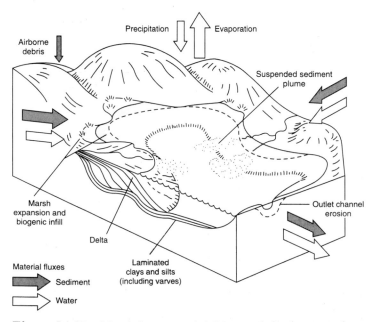

Figure 14.21 *Morphology, material fluxes and development of a lake system. Relative inputs–outputs of the water and sediment fluxes (including dissolved load) are shown by the size of the arrows. More sediment enters than leaves the system, leading to its eventual infill, and outlet lowering progressively lowers the water level.*

a *pool* in the incipient meander – reloading the stream for the next bar. This alternation of erosion and deposition develops until a stable meandering form is reached for most normal water and sediment discharges. Stable pool-and-riffle series develop spacings five to seven times channel width. Meander wavelengths are about one order of magnitude larger than channel width. Alternating bars migrate to form *central bars* or *riffles*, and *point bars* develop in slack flow on the inside of meanders. Sinuosity develops characteristic channel and bank forms with their own terminology and parameters (see Figure 14.22) but their dynamic character is emphasized. New studies suggest that meandering channels are *metastable* and that wholly new forms appear if threshold conditions change significantly. This is apparent with Earth's more dynamic, high-discharge rivers but it is also observed in historical studies of seemingly tame streams in Britain such as the modest River Dane in east Cheshire. A 10 km stretch, now conserved as an SSSI (*Site of Special Scientific Interest*), with a channel width of about 15 m, a mean daily flow of 3 m^3 sec^{-1} and a moderately 'flashy' régime, is known to undergo episodic very rapid rates of meander evolution.

Braided channels

Low flows divide into separate streams around central bars, whilst at higher flows central bars are dissected and point bars cut off by the development of an inside **chute**. The bars are normally resubmerged and single-channel flow is restored at mean flow levels but for a time the channel was *braided*. Persistent braiding represents unstable channel conditions and is usually found in steep, high-discharge, large-calibre, high-sediment-load streams or in glacial or arid environments. More stable high-discharge mountain channels braid on entering the piedmont zone. In effect, stream flow is overloaded with sediment (bed load) beyond the competence of all but the highest discharges and large volumes are shed as soon as discharge or channel slope falls significantly. This drastically increases channel roughness, giving the highest Reynolds number. Flow subsides into a series of *distributary* channels with lower total roughness and individual bars migrate downstream. Rivers draining the Himalayas and New Zealand's Southern Alps may be 5–20 km wide with width-to-depth

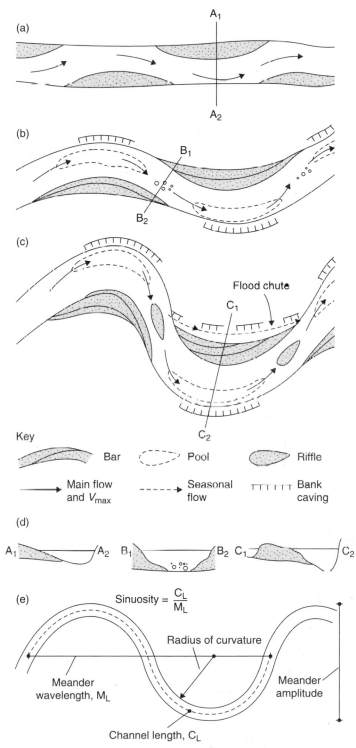

Key

Bar   Pool   Riffle

Main flow   Seasonal   Bank
and V_{max} flow   caving

Figure 14.22 *Development, morphology and geometry of river meanders: (a)–(c) progressive meander development, (d) representative vertical sections, (e) principal geometric elements in the assessment of meander intensity.*

Sinuosity $= \dfrac{C_L}{M_L}$

Plate 14.13 *Stream-flow meandering in an iron quarry drain, even where its banks are not erodible. Photo: Kenneth Addison.*

(a)

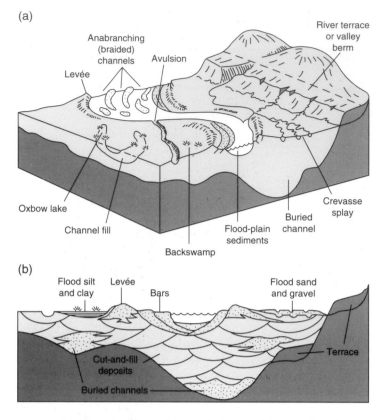

(b)

Figure 14.23 *Floodplain (a) morphology and (b) stratigraphy.*

ratios above 250–300. Vegetation stabilizes bars exposed for long intervals between submergences, redefining them as islands and the braided channel as **anabranching**.

The flood plain

The third and largest scale of dynamic sedimentary environments is the *flood plain*, which loosely describes the valley floor prone to episodic over-bank discharge. In another departure from classic fluvial landsystems, narrow flood plains with chemically and texturally raw sediments occur in pockets within mountain catchments. Lowland flood plains of more mature sediments are far more extensive laterally, developed by lateral accretion by meandering or braided rivers and vertical accretion through over-bank discharge. Both forms usually occur together and developmental history is seen in the array of abandoned channel forms on flood plain surfaces and stratigraphic exposures 10^{1-3} m thick in palaeo-environmental sediments. **Cut-and-fill** structures are indicative of lateral accretion forming a level floor as meandering channels cut into older deposits and

rework them as back-fill into abandoned channels, mostly as bed load. New channels form more dramatically through **avulsion**, where rivers break through old channel segments. Vertical aggradation buries and thereby envelopes older deposits, mostly by suspended sediment, and develops a convex floor *falling* away from the channel. Indeed, raised banks or levées may form as coarser debris is deposited alongside the channel and low-lying areas beyond become poorly drained. Levées 10^{1-3} cm high raise the threshold of the next over-bank flood, which cuts **crevasses** at low or weak points and restarts the process with **crevasse splays**. The composite, three-dimensional floodplain landsystem is shown in Figure 14.23. Flood plains end where the trunk stream enters the sea via an estuary or delta, dealt with in Chapter 17.

Floodplain development may be a product of fluvial processes alone. However, **aggradation** and **terrace** formation are responses to wider environmental changes. Changing catchment climate and vegetation lead to changes in discharge amounts and régimes. Base-level changes alter the catchment energy balances. They are particularly common during the restless changes of the Quaternary era, with linkages between glaciation, sea level, climate and catchment character now altering our whole perception of these fluvial responses. Increased sediment supply from the upper catchment fuels downstream aggradation. Buried channels in flood plains may indicate earlier incision during lower (cold stage) sea levels but terrace aggradation may also reflect a greater sediment flux from glaciated, unvegetated catchments during spring melt in permafrost environments. Unpaired terraces are thought to reflect gradual denudation-led incision as the river meanders across its flood plain. Paired terraces are thought to reflect significant downcutting stages adjusting to falling sea levels (**thalassostatic** terraces). The matter can be made more complex by downstream aggradation on *falling* sea levels as well as *rising* sea levels. Terraces can be cut in bedrock as valley-side berms. Much is now derived from detailed palaeontological and stratigraphic analysis of terrace sediments, of importance in its own right but also because from early prehistory (after *c.* 0.75 Ma BP), hominids chose flood plains as suitable occupation and hunting sites. Dating and correlation are important, and Britain affords useful comparative examples. Terraces in middle sections of England's three main lowland rivers, the Thames, Trent and Severn, possess terraces formed during the past three, two and one temperate (interglacial) stages respectively (Figure 14.24).

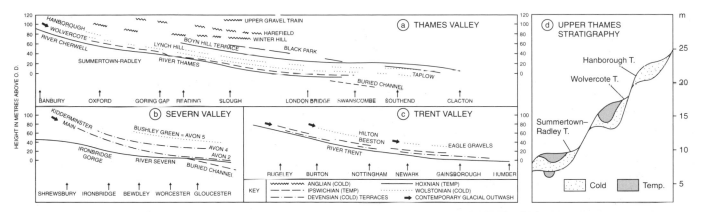

Figure 14.24 *Terraces of the rivers (a) Thames, (b) Severn and (c) Trent and their attribution to cold (glacial) or temperate (interglacial) stages; (d) stratigraphic complexity in the upper Thames around Oxford, where cold stage terraces overlie or are incised by temperate stage channels. (Height in metres above modern flood plain.) Source: Boulton (1992).*

Conclusion

Flowing water and its geomorphic activity have been central to human life since prehistoric hunters exploited their prey concentrations around riparian watering holes. Alluvial sediments are important archaeological sources for their artefacts, succeeded by widespread early to mid-Holocene evidence of the first farmers. Since then, extensive flood plains have become sites of intensive human settlement, agriculture and industry. In exploiting the fertile alluvium, sand and gravel resources, ready access to water and

sites suitable for urban, industrial and communications infrastructure of river corridors we also contend with the fickle behaviour of stream flow and channel instability. Human actions manage and re-route stream flow but often also alter channel dynamics and sediment flux inadvertently. The relative safety of higher river terraces contrasts with direct and human-enhanced flood risks in the contemporary flood plain. Our catchment occupation and actions require an understanding of the dynamics of rivers and fluvial landsystems.

Key points

1 Earth's landsurface is divided into a hierarchical series of drainage basins or catchments which convert precipitation to stream flow. Each catchment contributes surface and sub-surface water to a specific stream or a major trunk river, separated from its neighbours by a watershed. At continental scales, the watershed probably coincides with prominent morphotectonic features. Catchment topography, geology, vegetation and land use systems retain, store and transfer water. They introduce delays and losses to onward water transfer.

2 The fate of precipitation falling on the catchment can be quantified in a water balance equation, with the volume of precipitation generating stream flow known as the discharge. Discharge volume and pattern over time are plotted on a hydrograph and reflect the contribution of the component catchment stores.

3 Gravity-induced overland and sub-surface flow is inefficient when diffuse, encountering high resistance and friction loss. Channel flow is more efficient and is initiated where surface/sub-surface flows converge, initially as intermittent rills and gullies. Efficiency

continues to develop downstream as river channels enlarge with discharge and form compensates for falling gradients and potential energy. Channels connect to form a catchment-wide network, with recognizable patterns and drainage densities determined by catchment hydrometeorology and hydrogeology.

4 Flowing water plays an important role in continental denudation. Erosion occurs through fluid stressing by water itself and/or by water movement of abrasive tools. Stream flow competence and sediment entrainment, transport and deposition are a function of velocity and particle size, summarized by Hjülstrom's curve. Stream flow creates distinctive landforms composed of straight, meandering and braided channels, channel networks and flood plains.

5 Channel segments respond to flow régimes, sediment delivery, slope, etc., and undergo almost continuous change. At whole-landscape scale, upper catchments are dominated by bedrock channels and deep, narrow valleys, etc., with sediment pockets. Lower catchments tend to display extensive flood plains of primary and reworked alluvial sediments. Quaternary sea-level change has driven repeated marine/landward extensions of the flood plain and Holocene delta construction.

Further reading

Knighton, D. (1984) *Fluvial Forms and Processes*, London: Edward Arnold.
An important text covering the origin and forms of stream flow, fluvial channels and fluvial geomorphology. The book depends on line drawings rather than plates for illustration and may be thought a little dry but its appearance in an eighth impression in 1996 attests to its reliability and renown.

Lane, S. (1995) 'The Dynamics of Dynamic River Channels', *Geography* 80 (2), 147–62, Sheffield: Geographical Association.
A state-of-the-art paper on channel dynamics and styles of channel change over short segments. The paper, which is part of a series for geography teachers and students, includes a useful glossary and contemporary bibliography.

Newson, M. (1995) *Hydrology and the River Environment*, Oxford: Clarendon Press.
The book complements Knighton's detailed channel process and geomorphological explanation with a catchment-based approach. It ranges from gauging techniques to river channel and basin management and is well illustrated with plates.

We are witnessing a revolution in glacial science and the study of Earth's intermittent 'Ice Ages', although with popular attention lavished on 'greenhouse' conditions and global warming we sometimes forget that we occupy a brief *temperate* or *interglacial* phase of the Quaternary Ice Age. The inheritance of the last *cold* or *glacial* stage surrounds us. Most mid-latitude farmland and sand and gravel aggregate industry is founded on glacial sediments or their derivatives, and their complex geotechnical character tests civil engineering skills. Engineering problems and opportunities increase in the highlands. Many slopes oversteepened by glacial erosion are still unstable and liable to failure but the same process has created ready-made reservoir sites. Spectacular highland scenery formed during intense Quaternary glacial and frost action in Cenozoic and older orogens and continues to develop in modern alpine glacial areas. Even the full impact of global warming depends on the uncertain reaction of Earth's cryosphere.

Little more than a generation ago glaciation was regarded as a climatic 'accident' and the retreat of glaciers meant a return to 'normal' denudation. Ice was thought simply to have ornamented fluvial valleys and even to have protected the landscape from normal erosion. This was a retreat from the glacial revolution which began a century earlier, when only a brave geologist would have dared propose that glaciers once occupied Britain. Charles Darwin missed the glacial evidence surrounding him in Cwm Idwal (North Wales) in 1831 but made generous amends a decade later, inspired by experience of Andean glaciers on the *Beagle* voyage and a British visit by the pioneer Swiss glaciologist Louis Agassiz. Certain Victorian ancestors preferred theories of icebergs bobbing around on Noah's flood to the land-based ice sheets envisaged by other early geologists which modern research confirms. The modern glacial revolution recognizes world-wide glaciation not only as an important if intermittent feature of Earth's physical environment but as providing an **icehouse** counterpart to long periods of **greenhouse** Earth.

Form, mass and energy balance of ice

Snow and ice accumulate in four different modes. **Ice sheets** and **glaciers** are permanent ice bodies distinguished by their size and thermodynamic activity (see below). They comprise the vast bulk of Earth's 35 million km³ of ice which covers 11 per cent of its landsurface today, with some 98 million km³ covering over 30 per cent at the last global maximum 18 ka ago (Table 15.1). They are responsible almost exclusively for mid- and high-latitude glacial geomorphology and low-latitude alpine mountains (Figure 15.1). Terrestrial ice may enter the sea (or lakes) and remains attached to the parent glacier as **ice shelves** or break off to form floating **icebergs** (Plate 15.1). Modern Antarctic shelf studies stimulate awareness of the impact of glaciomarine environments on shallow continental shelf seas and coastlines around former Pleistocene ice sheets in North America and Europe. **Sea ice** covers some 7 per cent of global ocean area to a mean thickness of up to 2.5 m. Sea water freezes in each hemisphere's winter and shrinks by half in summer. **Ground ice** forms when pore water freezes in terrestrial substrates and accumulates as perennial **permafrost**. It usually occurs where intense cold and aridity stifle glacier development and currently underlies a further 25 per cent of modern landsurfaces.

Glacier ice

Glacier ice is fashioned from snow, hail, sublimation (direct deposition) and rain or dew which subsequently freezes. It eventually forms glacier ice wherever mass *accumulation* exceeds all forms of mass *ablation* or loss through melting, iceberg **calving** and sublimation. Gains or losses of mass may also occur by deflation (wind drifting) of snow and avalanching of snow or ice between the glacier and its surrounds (Figure 15.2). Snowfall is transformed through several recognized stages before becoming mature

Table 15.1   *The size and extent of Late Pleistocene and Modern ice sheets and glaciers.*

Modern ice	Area (10^6 km²)	Volume (km³)	Estimated maximum area of Late Pleistocene ice	(10^6 km²)	Volume (10^6 km³)
Antarctic	13.50	32.0	Antarctic	14.50	37.7
Greenland	1.80	2.6	Greenland	2.35	8.4
Arctic basin	0.24		Laurentide ice sheet	13.40	34.8
Alaska	0.05		North American Cordillera	2.60	1.9
USA (other)	0.03				
Andes	0.03		Andes	0.88	
European Alps	0.004		European Alps	0.04	
Scandinavia	0.004		Scandinavian ice sheet	6.60	14.2
Asia	0.12		Asia	3.90	
Africa	0.0001		Africa	0.003	
Australasia	0.001		Australasia	0.07	
			British ice sheet	0.34	0.8
Total	15.7	35.0		45.15	98.0

Note: The data are compiled from a wide range of sources which do not necessarily measure or estimate ice cover or volume for the same geographical areas. Late Pleistocene (Late Devensian) glacial maximum extent is estimated from geomorphic and other evidence. Global totals may include other small glaciers.

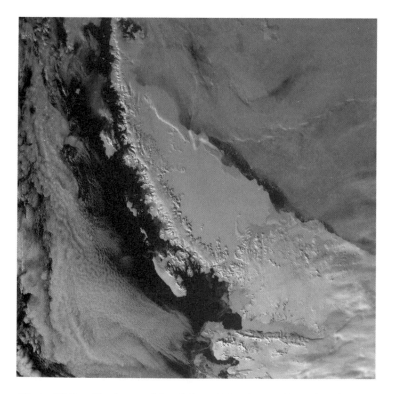

Plate 15.1   *Glaciers and ice shelves of the Antarctic peninsula, flowing from its 2000 km long mountain spine. In 1995 part of the Larsen ice shelf (the largest on the east coast) was seen to have become detached, adding to fears of global warming impacts in the Antarctic. Photo: British Antarctic Survey. (Plates 15.1–2 are visible wavelength images from the AVHRR – Advanced Very High Resolution Radiometer – multispectral scanner of the TIROS-NOAA satellites.)*

glacier ice capable of substantial geomorphic activity. Snowpack is highly porous and held together by a frozen crystal lattice; pores contain air and, depending on temperature, perhaps water vapour and/or water. Mass is measured in water-equivalent terms (against a water density of 1.0 or 1000 kg m³) because of the wide range of densities encountered (Plate 14.1). Initial snowfall densities of 0.04–0.06 increase to some 0.4–0.5 in **firn** or granular snow, beyond which it becomes impermeable, with a density of about 0.6 as bubbly glacier ice and about 0.9 as polycrystalline glacier ice.

In essence, transformation progressively expels air and reduces void space through autocompaction. This process is assisted by the lowering of freezing point under higher pressure. As pressure increases, ice reaches its **pressure-melting point** and supercooled meltwater diffuses to areas of lower pressure in the pack before **regelation** or refreezing/recrystallization occurs. Gravity tends to draw this downhill as regelation creep, which initiates ice flow. All three processes exert a major influence on subsequent glacier behaviour. Snowpack develops internal overburden pressure which exceeds the pressure-melting point at delicate snowflake tips. In addition, localized melt through insolation, advection or geothermal heating assists consolidation if water then regelates in colder parts of the pack. By these processes, which have some affinity with rock crystallization and metamorphism, tiny geometrically complex snowflakes are transformed to assemblages of progressively larger, amorphous ice crystals. Density increases rapidly at first, reaching 0.2 within a few days, but slows considerably thereafter.

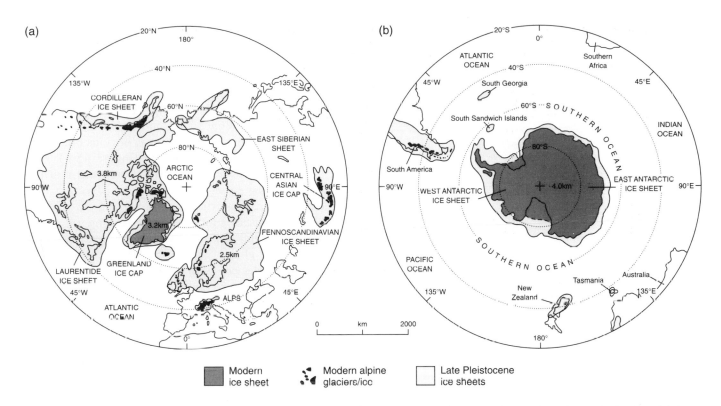

Figure 15.1 *Global distribution of modern and Late Pleistocene ice sheets and glaciers in (a) the northern and (b) the southern hemispheres. The maximum ice surface elevation of each major ice sheet is shown.*

Firn, translated from German, refers literally to snow 'from last year' but its density rarely reaches more than 0.4 so quickly. It increases faster under higher temperatures and accumulation rates, exceeding 0.8 within 10–20 a in alpine glaciers but taking 150–200 a in polar ice.

Glaciologists refer to the difference between a glacier's annual accumulation and ablation as its **mass balance**. Zones of net accumulation in higher, colder parts of the glacier environment and net ablation towards the terminus meet at the **equilibrium line**, whose annual average altitude (ELA) remains fixed in a steady-state glacier. A *positive* mass balance permits the glacier to thicken and extend its terminus, whilst a *negative* mass balance causes thinning and retreat without necessarily halting internal ice flow (Figure 15.3). Ice flow in steady-state glaciers transfers just enough ice from the accumulation zone through the equilibrium line to match ablation zone losses. Ice cut off from its accumulation area by the reappearance of bedrock during thinning becomes stagnant and experiences downwasting *in situ*.

Thermal energy balance complements mass balance to complete our initial glacier profile. Radiative and sensible heat fluxes are the principal energy sources, augmented by latent heat – at a rate of about

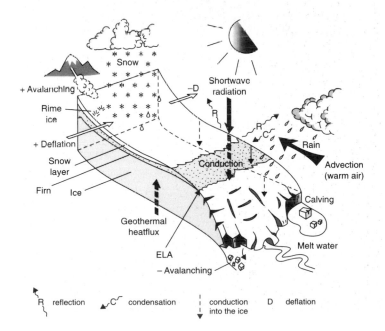

Figure 15.2 *Model glacier with sources and sinks of mass and energy. Prefix + or – indicates positive or negative flux; ELA, equilibrium line altitude.*

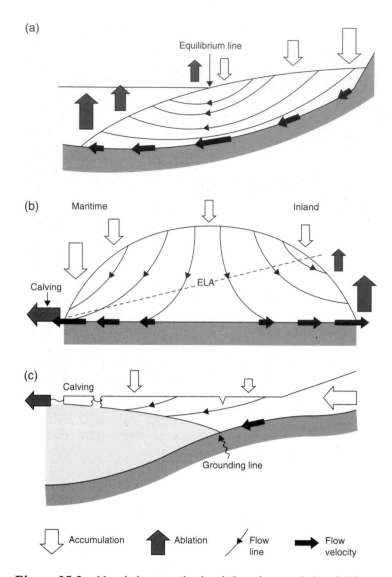

Figure 15.3 *Mass balance and related flow characteristics of (a) a valley glacier, (b) an ice sheet and (c) an ice shelf. Mass transfers and basal velocity are proportional to the size of the arrows. Source: After Sugden and John (1976).*

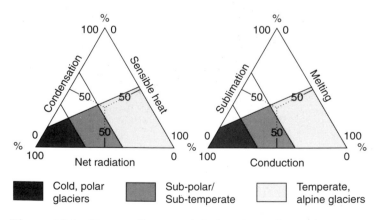

Figure 15.4 *Ternary diagram of glaciers classified according to principal energy sources and sinks. Source: After Andrews (1975).*

2.5×10^6 J kg^{-1} of water – released by condensation and direct deposition at the surface and water freezing at greater depth. Energy is used for sublimation, ice melt – at a rate of 3.33×10^5 J kg^{-1} water equivalent – or is conducted into cold ice, raising its temperature without melting. The relative importance of each energy source and sink is linked closely with glacier type, and one or two obvious but important observations can be made (Figure 15.4). Cold glaciers receive most heat via short-wave radiation flux and least from warm, moist air advection. This dependence, exacerbated by high albedo and the effect of glacial anticyclones in fending off milder winds, explains polar ice sheet development in areas of lowest global radiation receipt. Conversely, temperate glaciers are found in areas of higher radiation flux balanced by orographic effects on moist airstreams. This difference is reinforced by one final distinction. Temperate glaciers are isothermal, i.e. at pressure-melting point throughout their depth and therefore *warm-based*, whereas cold glaciers are polythermal, with basal temperatures below pressure-melting point, and, hence, are *cold-based*.

Ice shelves and sea ice

Ice shelves are extensions of glaciers whose mass balance permits them to reach the coast. Numerous alpine glaciers in Alaska, Chile and New Zealand terminate in *tidewater* – mostly in fjords – but few are thick enough to float. The Ronne–Filchner and Ross ice shelves covering 0.75 million km² and 0.45 million km² respectively are the largest of a suite of ice shelves fringing 30 per cent of the Antarctic coast and accounting for over 7 per cent of Antarctic Ice Sheet area. They develop from **piedmont glaciers** fanning out into coastal lowlands from outlet glaciers and float seawards of a **grounding line** at a water depth about 90 per cent of ice thickness, determined by the greater density of sea water (Plate 15.2). The shelf is pinned to its bed landward of the grounding line and at offshore rises or islands, but otherwise moves unimpeded and may reach velocities of some 1–2 km a^{-1}. It avoids outstripping its supply by thinning seawards, and may even gain snowfall as it moves 10^{2-3} km offshore, but eventually oceanic heat flux and continuous flexing by tides and currents cause icebergs to calve from ice cliffs some 100–200 m high. Bergs over 2,000 km² have recently calved from the Ronne–Filchner and Larsen ice shelves – both of which occupy protected embayments east of the Antarctic Peninsula (Figure 15.5). Shelf ice is implicated in glaciomarine processes outlined later.

Sea water freezes at −1.91° C and, except where rapid freezing occurs, loses most of its salinity on freezing. Initial freezing occurs as small, crystalline platelets known as *frazil ice* which coalesce rapidly. A mean thickness of 2.5 m can develop in a single winter and melt as quickly. Multi-year ice, which is tougher than single-year ice, forms in less extreme conditions by basal accretion and surface melt. Sea ice has a superficial abrading effect on soft coastlines as it is deformed and moved by waves and currents. It is less likely to do so where *fast ice* is literally held fast in fjords or other enclosed locations but is more effective in the *pack ice* along open coasts.

Ground ice (permafrost)

Annual snowpack and glacier ice may insulate their substrates from the full severity of winter cold. In cold arid climates where snowfall is inadequate for glacier growth, however, the ground itself may be perennially frozen to great depths and attention shifts to the forms and effects of ground ice. A cold wave

Plate 15.2   *The Brunt ice shelf, fed by outlet glaciers on the east coast of the Weddell Sea, Antarctica. The change of tone on the otherwise featureless shelf marks its grounding line, and Lyddan Island (50 km long) acts as a pinning point.*
Photo: British Antarctic Survey.

Figure 15.5   *Global sea ice and ice shelf distribution. Polar seas are named in (a) the northern hemisphere and major modern ice shelves in (b) the southern hemisphere.*

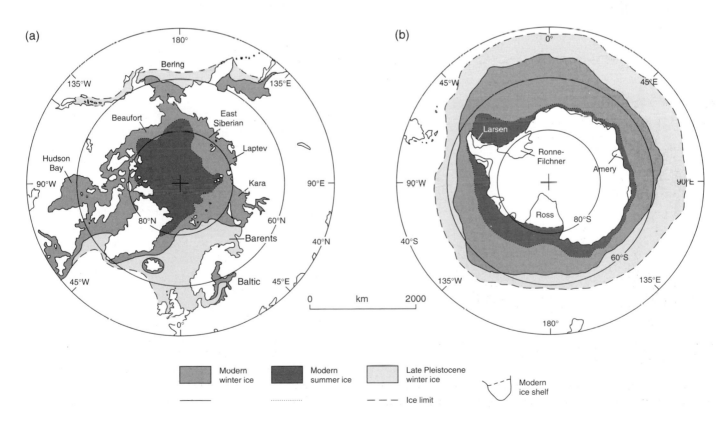

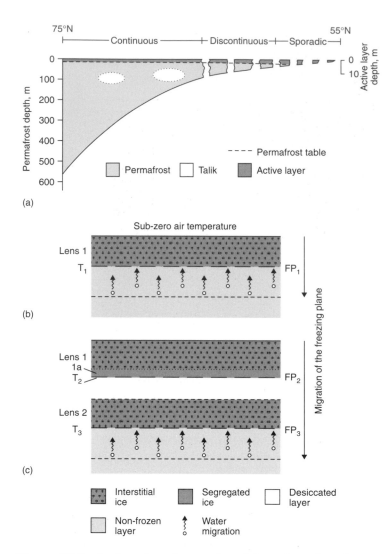

Figure 15.6 *Sections through permafrost: (a) north–south section through principal zones from North West Territories to northern Alberta, Canada; (b) and (c) formation of interstitial and segregated ice lenses (1a) by downward migration of the freezing plane. Note that migration of water to the freezing plane leaves a desiccation layer.*

penetrates the ground once surface temperature falls below 0° C and what happens next depends on the porosity and thermal conductivity of earth materials and the behaviour of pore water, outlined first in Chapter 14. Gravitational water freezes to form **interstitial ice** at 0° C but attractive forces lower the freezing point below 0° C for capillary water and below –20° C for water bonded to soil or rock particles. However, ice crystal growth exerts its own strong attractive force and draws water to the freezing plane. Expansion on freezing generates *heaving* pressures which displace loose soil particles and, in these conditions, **segregated ice lenses** can form at the

freezing front. Its continuing downward penetration passes through the ground layer thus desiccated and into the next zone of pore water beyond initial reach. The process is repeated to depths at which downward penetration of the cold wave ceases or is countered by geothermal heat flow (Figure 15.6). Ground contraction on cooling and local desiccation counter expansion due to frost heave and generate cracks which may become sites of vertical lenses or **ice wedges**. Heaving, contraction and seasonal melt drive permafrost processes, as we shall see below.

Permafrost, or perennially frozen ground, consisting of segregated and interstitial ice zones and desiccated lenses some 400 m thick, is found in the Arctic basin. It forms *continuous* cover on non-glacial polar landsurfaces and cold, arid continental interiors but thins equatorwards and coastward. *Discontinuous* or *sporadic* forms occur as the extent of **talik** or unfrozen ground increases and account for 45 per cent of the approximately 40 million km² of global permafrost (Figure 15.7). Seasonal melt during summer months with temperatures above 0° C develops a saturated surface **active layer** 0.1–3.0 m thick. Meltwater drains laterally but is unable to penetrate the frozen substrate and the layer refreezes in winter. The roles of microclimate, relief, slope, material porosity and near-surface drainage become increasingly important towards the margins. Weak, sporadic permafrost activity may still be present in British mountains for these reasons although in other respects Britain lies outside the circumpolar permafrost zone.

Thermodynamic character of glaciers

There are clear links between glacier climate, mass balance and rates of flow which collectively define a glacier's *thermodynamic state* (Table 15.2). Polar climates are so cold that relatively little snow falls or melts, ice takes longer to accumulate and flow velocity varies from zero (where the glacier is frozen to its bed) to a few tens of metres per year. Low-energy **polar** or **cold glaciers** are consequently large and stable, capable of surviving relatively large climatic fluctuations. The Antarctic Ice Sheet and Greenland Ice Cap are the largest modern polar ice bodies but at global glacial maxima two or more similar ice sheets form over much of North America and Eurasia.

By contrast, warmer alpine climates experience heavier snowfalls and more rapid melt. Ice forms quickly and flow velocities are measured in 10^{1-3} m a^{-1}. However, high-energy **alpine** or **temperate glaciers**

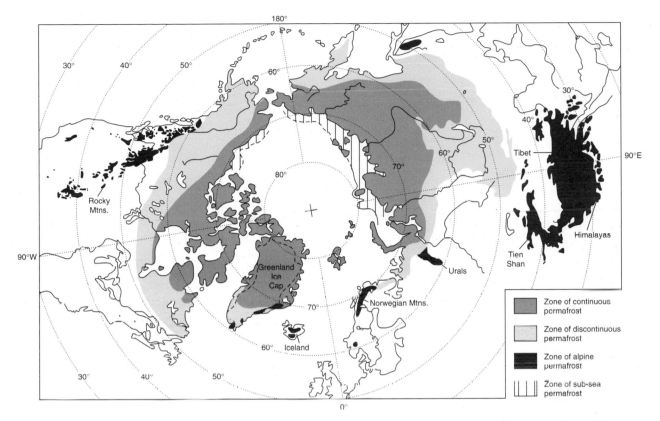

Figure 15.7 *Modern distribution of permafrost in the northern hemisphere. Source: After Ballantyne and Harris (1994).*

are much smaller and far more susceptible to even small climatic changes. They are restricted today to high mountains like the Alps, Himalayas, Andes and North American cordillera, where they are topographically constrained in valleys which they have eroded. Thermodynamic character therefore also determines glacier configuration (Figure 15.8). Larger and thicker ice sheets are less constrained spatially but lower average velocity reduces their geomorphic impact except in the vicinity of **outlet glaciers**, where large ice volumes accelerate towards ice sheet margins and create the most impressive erosional landforms.

Table 15.2 *Mass balance, iceflow and thermodynamic characteristics of principal glacier systems.*

Characteristic	Cold, polar glaciers	Temperate, alpine glaciers
Input (accumulation) and output (ablation)	Low, 10^{4-6} cm^3 m^{-2} a^{-1}	High, 10^{6-7} cm^3 m^{-2} a^{-1}
Mass storage	High, 10^{6-7} km^3	Low, 10^{1-2} km^3
Energy flux to melt accumulation at ELA	Low, 10^{0-1} kcal m^2 a^{-1}	High, 10^{1-3} kcal m^2 a^{-1}
% annual mass turnover	Low, 0.001–0.01 a^{-1}	High, 1–5 a^{-1}
Thermal régime	Polythermal	Isothermal
Basal régime	Cold, frozen	Warm, unfrozen
Principal flow mechanism	Internal deformation	Basal sliding
Secondary flow mechanism	Basal sliding	Internal deformation
Flow velocity	Low, 10^{1-2} m a^{-1}	High, 10^{2-3} m a^{-1}
% area above ELA	Large, approx. 80–90	Moderate, approx. 40–60
Area/thickness configuration	Tabular	Columnar
Channel type	Unconfined	Confined
% land area glacierized	Large, 80–100	Small–moderate. 10–50
System stability	Highly stable–metastable	Unstable–highly unstable

Note: These are general values for glaciers in steady state and demonstrating standard behaviour.

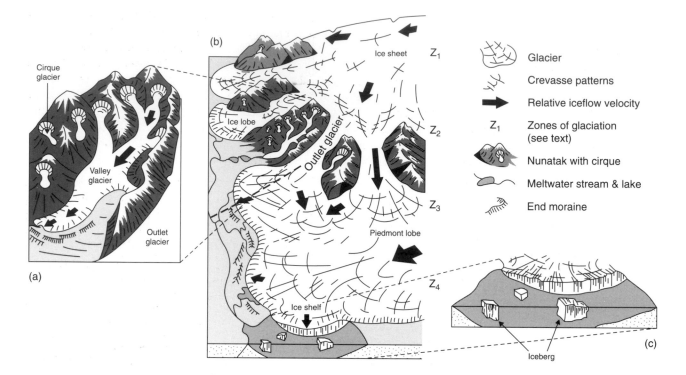

Figure 15.8 *Principal glacier types: (a) temperate (alpine) glaciers, (b) a cold (polar) ice sheet and (c) an ice shelf. Source: After Addison (1983).*

Climate, tectonics and Ice Ages

Links between snowfall and temperature make the ELA a vital measure of glacier climate. It lies close to sea level in polar regions today, rising away from the poles so that, for example, it is at about 1.5 km above sea level in northern Britain, 3 km in the European Alps and 5 km at the Equator. This is why Britain is ice-free but there are small glaciers on equatorial mountains in East Africa (see Figure 25.9). However, we know from geomorphic, biological and oxygen isotope evidence that Quaternary ELAs were often low enough for global or local glaciation in about twenty individual **cold stages**. Why was this and how do we explain earlier Ice Ages?

Radiative forcing of Ice Ages has long been the prime suspect but the magnitude and periodicity of fluctuations in solar activity do not explain adequately the intermittent glacial signature in the geological record. The Quaternary Ice Age is the third in 570 Ma of the Phanerozoic aeon, which commenced shortly after a late Precambrian Ice Age, and earlier Ice Ages are known from lithified glacigenic sediments. The Milankovich mechanism offers exciting prospects of a range of geological processes, with its changing patterns of solar radiation receipt due to Earth's astronomic eccentricities. Although it

probably controls climatic oscillation *within* Ice Ages, its continuing operation between Ice Ages cannot explain the gaps. A range of possible geochemical explanations are being explored, from Earth's passage through clouds of cosmic dust to clear links between atmospheric CO_2 levels and greenhouse–icehouse effects.

Earth's supercontinental cycle is another promising area of research. Quite apart from its influence on CO_2 levels through volcanic outgassing and the extent of landsurface exposed to weathering, which locks up CO_2 in carbonate weathering products, tectonic forcing determines global distributions of sea, land and high relief. Two broad correlations exist – between polar supercontinents, fragmentary oceans and their circulation systems and icehouse conditions; and equatorial supercontinents, well connected oceans and their circulation systems and greenhouse conditions. Tectonic uplift disturbs atmospheric circulation and promotes glaciation in mid to high latitudes and **autocatalysis**, whereby ice sheet growth reinforces icehouse conditions. The high albedo of snow and ice reflects more short-wave radiation and thickens boundary inversion layers. Atmospheric subsidence consequently enhances polar anticyclonic circulation, blocking advection warming, and the associated fall in sea level advances grounding lines.

The reversibility of these effects plays a significant role in glacial–interglacial oscillation but they undoubtedly assist initial ice sheet formation.

Later stages in the break-up of Pangaea were instrumental in initiating the Quaternary Ice Age. Although 'Antarctica' circled the south pole, the Southern Ocean and its isolating circumpolar current could not form until Australia and then South America broke free some 50 Ma and 20 Ma ago. Antarctic glaciation commenced approximately 40 Ma ago during the Eocene and the continent has supported a polar ice sheet ever since. Northern hemisphere glaciation, although doubtless encouraged by Antarctic-driven cooling, had to wait until the Panama isthmus isolated the Atlantic and Pacific Oceans 3 Ma ago and strengthened northern Pacific and Atlantic circulation. The Quaternary Ice Age commenced after 2.4 Ma ago with short, 41 ka cold-temperate stage cycles operating up to the first 1 Ma before settling into a 100 ka rhythm thereafter.

Glacier resources and hazards

Whether Earth reverts to a 'scheduled' icehouse phase in the foreseeable future or maintains the greenhouse trend, human societies will monitor the growth, decay and changing behaviour of glaciers and ice sheets. They convey mixed blessings of resource or resource potential inseparable from glacier hazard and therefore neither their growth nor their decay can be wholly beneficial to us.

At its simplest, glacier resources relate to ice itself, and it is not surprising that nature's refrigerator should have long been used in glaciated regions for food preservation and cooling – sometimes within glacier cavities! Ice is used more obviously and extensively in its melted state for direct human consumption, for irrigation and for hydroelectric power generation. As Earth's largest store of fresh water, its potable and irrigation appeal is inevitable but supplies are restricted geographically and seasonally in the absence of storage schemes. The feasibility of towing Antarctic icebergs to the Middle East has been explored, given the high cost of alternative desalinization strategies. Glacier meltwater is a major global source of hydroelectricity generation but glacier storage does not pre-empt reservoir storage (Plate 1).

Plate 1 *Mauvoisin dam and reservoir, south-west Switzerland. Cirque glaciers above the reservoir would threaten the dam and its hydroelectric installation in the event of regional ice advance. Photo: Kenneth Addison.*

Jökulhlaups demonstrate that it does not always come in manageable quantities. Less directly, glacigenic sediments underpin huge tracts of profitable farmland and glaciers have, over thousands of years, shaped landscapes of high scenic and therefore tourist value in which – gleaming white and blue in sunlight – they are jewels.

All resources are at hazard from inherent glacier variability and climatic change. Discharge, régime and point sources of ice and meltwater vary through continuous internal thermodynamic adjustments. Although meltwater supply is usually out of phase with demand, contemporary use is adjusted to contemporary régimes, vulnerable to change. Farmland takes time to develop but alpine scenery could be degraded rapidly by glacier retreat. In worst-case scenarios, global warming could waste all alpine glaciers on human time scales, or sustained ice *advance* might overrun the very societies they serve. Little Ice Age glacier advances from the mid-fifteenth

to the mid-nineteenth century wrought widespread havoc directly and through their impact on other geophysical processes, clearly evidenced in oral and documentary history and in the landscape.

Perhaps the most serious glacier hazard concerns the West Antarctic Ice Sheet (see Figure 15.1) which holds a water equivalent of 7 m of global sea-level rise. The Ronne–Filchner and Ross ice shelves, which it shares with outlet glaciers breaching the Transantarctic Mountains to the east, comprise a large part of its eastern sector. Much of the remainder is grounded on bedrock below sea level or close to pressure-melting point. It is therefore metastable and its response to global warming and sea-level rise is unpredictable. Catastrophic collapse could be triggered if the initial sea-level rise moved ice shelf grounding lines far enough inland to undermine their pinning effect on the ice sheet. The resulting surge could – literally – put the skids under the ice sheet, leading to draw-down and rapid disintegration.

Ice flow and glacier geomorphic processes

Ice flow mechanisms

Glacier mass and thermal energy balances and general thermodynamic character drive ice flow velocity and style. This, in turn, determines geomorphic activity and subsequent landsystems. Ice behaves as a plastic material and is readily deformable under stress. This is shown by Glen's flow law, defined by:

$$E = A\tau^n$$

where the rate of deformation or strain rate E is determined by the constant A, related to tempera-

ture, shear stress, τ, and the exponent n, which has a mean value of 3. The **basal shear stress**, τ, is given as:

$$\tau = \rho g h \sin a$$

where ρ is ice density, g is gravitational acceleration, h is ice thickness and a is the surface slope of the glacier. Thus basal shear stress increases with glacier thickness and surface slope and the rate of deformation is therefore highly sensitive to an increase in either and to ice temperature. In practice the maximum shear stress ice can exert at its bed before it deforms is about 0.1 MN m^{-2}. These relationships can be appreciated by looking at mass balance, deformation and ice flow in a **cirque glacier** – the smallest glacier type, distinguished from snowpacks by deformation and movement. Annual mass balance adds an incremental wedge of snow above the ELA and, in steady state, melts an identical wedge in the ablation zone (Figure 15.9). The increase in accumulation zone thickness and overall surface slope exceeds the yield stress and equilibrium is restored only by ice flow.

Ice flow style in this or any other glacier is determined by the snow–ice transformation process, basal ice temperature (warm- or cold-based) and the nature and slope of the glacier bed as well as its surface. Intergranular or regelation creep, already identified as initiating ice flow, is one component of **internal deformation**, where parts of the glacier move relative to others. At a larger scale, ice behaves like a viscous fluid in its boundary layer – with lower velocities close to the bed or valley sides – and like a brittle solid where velocities exceed plastic deformation rates

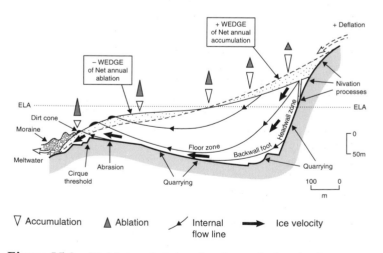

▽ Accumulation   ▲ Ablation   ⚡ Internal flow line   ➡ Ice velocity

Figure 15.9 *Model mass balance of a cirque glacier, showing net accumulation and ablation wedges, ice flow lines, relative velocities and associated geomorphic processes.*

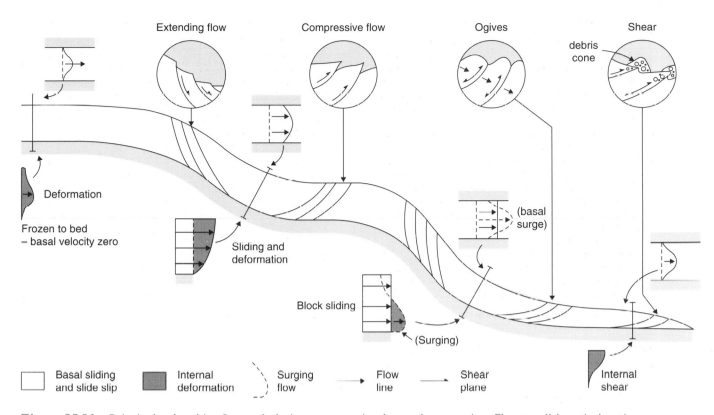

Figure 15.10 *Principal styles of ice flow and glacier movement in plan and cross-section. Flow-parallel vertical sections are shown below the glacier, with flow-parallel surface highlights and transverse plan sections above. Extending flow occurs in icefalls and at the entrance to outlet glaciers. Compressive flow and ogives form at the base of outlet glaciers and icefalls. Shearing occurs over stationary terminus ice.*

to form crevasses and shear planes (Figure 15.10, Plate 15.3). Cold-based glaciers may be frozen to their bed and ice higher in the glacier shears past the stationary basal layer. Warm-based glaciers, and zones of pressure-melting in cold-based glaciers, also slide past their bed and valley sides. **Basal sliding** is facilitated by a thin water film between ice and substrate or by a deformable bed (see below). Surface water (from rainfall or melting) may reach the bed in thin glaciers and average geothermal heat flux is capable of melting some 6 mm a^{-1} of basal ice. The principal source of water, however, is pressure-melting. This occurs quite readily in isothermal ice and is also induced here and in colder ice by basal stress increase on the upstream side of bedrock obstacles. The resultant supercooled water reduces bed friction, encouraging the glacier to slide past or around the obstacle, and regelates as stress falls downstream.

Ice flow patterns and velocities

All glaciers flow via a combination of internal deformation and basal sliding. The proportion of both varies according to thermodynamic character and according to mass balance trends, season and location within the overall ice stream. Cold-based glaciers move primarily by internal deformation, whereas basal sliding is a major component in warm-based glaciers, reflected in their respective velocities. 'Average' velocity lies in the broad range 3–300 m a^{-1}, with cold-based glaciers in the low range of 10^{1-2} m a^{-1} and warm-based glaciers in the high range 10^{2-3} m a^{-1}. Outlet glacier velocities are among the highest at 10^3 m a^{-1}, and the fastest known stable glacier is Jakobshavn Isbrae in west Greenland, moving at about 7.5 km a^{-1} or 20 m *per day*. Extreme velocities encountered in unstable **surging glaciers** (see below) may exceed 50 m day^{-1} or 10^4 m a^{-1}.

Flow mechanism and velocity are inconstant, as *rotational sliding* in the simple glacier illustrates (Figure 15.10). Ice accelerates through the accumulation zone from zero velocity at the ice divide to a velocity maximum beneath the ELA before decelerating to zero again at the snout, accompanied by internal vectors towards and away from the bed respectively. Upstream of the ELA ice therefore experiences divergent or **extending flow**, whereas downstream ice

Plate 15.3 *Iceflow styles. Foliation (internal deformation) parallel to iceflow in the Rhône glacier (left) contrasts with extending flow (due to basal sliding) and associated crevassing in the 0.5 km wide Breithorn glacier icefall (right) in Switzerland.*
Photos: Kenneth Addison.

converges by **compressive flow**. Similar flow patterns develop in glacier long-profiles over bedrock irregularities and generate brittle failure with diagnostic sets of surface crevasses. These processes are seen at their best across steep **icefalls**, with the glacier surface crevassed into huge blocks or *seracs* by extending flow at their crest and fused to form *ogives* in compressive flow at their base (Plate 15.4).

Seasonal changes during the mass balance year and longer-term climatic change influence ice velocity. Accumulation may increase basal shear and accelerate flow in winter, whilst greater atmospheric warmth and meltwater generation can have the same effect in summer. Response does not have to be immediate and it is quite common for several years of increased accumulation to send a pulse or *kinematic wave* of thicker ice down-glacier at velocities three or four times the average flow. Metastable glaciers or zones within ice sheets are susceptible to sudden changes in behaviour. Extreme conditions of ice build-up or sudden failure of the glacier bed lead to surging. This may be catastrophic for the glacier if the snout itself

advances rapidly and the glacier is drawn down faster than new ice accumulates, leading to early downwasting. Surging can also be triggered by earthquakes and landsliding in glaciated orogens and is normally restricted to warm-based glaciers.

Deformation at the glacier bed

Glacier velocity is linked with erosive power but not in the simple relationship applicable to flowing water or wind. The deformability of ice and its bed is worth exploring as the prelude to glacier erosion and deposition, which could be seen as discrete points in a continuum of processes. Glaciers were formerly thought to move over a rigid bed but we now appreciate that the glacier bed is the composite interface between ice, rock, water, sediment and even air. Any one of these materials may locally form the 'bed' over which all or part of the ice moves. Not only is the boundary deformable but the zone of deformation can move from one locus to another and is

Plate 15.4 *Extending and compressive flow in the Vaughan Lewis icefall, Alaska. Crevassing fractures the 1 km-wide glacier surface into building-sized blocks as it enters the icefall (left) compared with compressive ridges and ogives as it exits into the Gilkey glacier (right).*
Photos: Kenneth Addison.

susceptible to changes in any of the material properties. Changes in ice thickness or velocity alter basal shear stress. Changes in the character of granular debris, pressure-melting conditions and pore water pressure alter the strength and deformability of basally lodged or entrained sediments. General relationships between deformation and glacial geomorphic processes are shown in Figure 15.11.

Glacier erosion and entrainment

The low yield stress of ice is scarcely promising for its erosive power, since the shear strength of intact bedrock is two to four orders of magnitude larger, yet glaciers are undoubtedly one of the most powerful erosion agents. Glaciers commonly excavate troughs 1–2 km deep through alpine orogens and beneath ice sheets where the mechanisms responsible are inac-

cessible. This has tested glaciologists' ingenuity, and a variety of more and less acceptable processes have been proposed.

Abrasion, crushing and entrainment

Bedrock striations provide abundant evidence that debris-charged basal ice can abrade a rigid bed provided that the abrading tools are harder than the substrate. Without the supply of larger rock fragments, however, this process is limited by the progressive comminution of debris and smoothing of bedrock. This is unlikely to inflict more than surface ornament (Plate 15.5). and does not occur without first overcoming the technical difficulty posed by the low yield stress of ice. Debris is not moved by traction at the ice–debris–bedrock contact, since the ice readily deforms when shear stress is applied. Pressure-melting continues until the particle is almost wholly

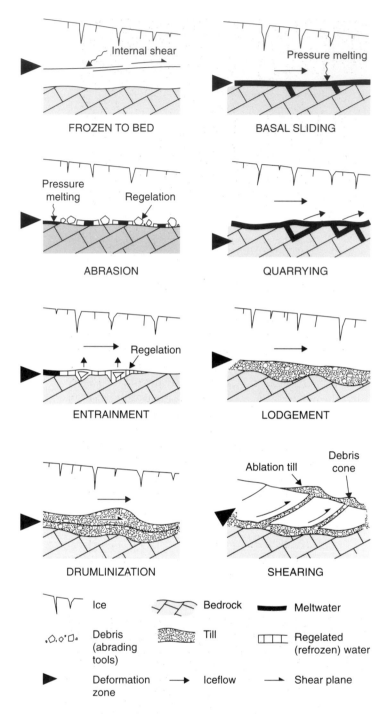

Figure 15.11 *Zones of deformation in glacial geomorphic processes: the ice–bedrock–water–debris interface.*

absorbed by the ice, equalizing bedrock–debris and debris–ice stresses. Gripped by regelation ice in the way that glue grips sand on sandpaper or tarmac grips chippings on road surfaces, the tool is now ready to abrade.

Small tools abrade semi-continuous grooves until they wear out. Larger tools may, through their weight, be held less firmly over small bedrock concavities and strike intermittent chips or *chatter marks* out of the substrate (Plate 15.6). Rates of 1–4 mm a^{-1} of surface abrasion lowering (the low range of which exceeds that of fluvial abrasion) are, in part, a function of ice velocity but through the number of 'abrasive passes' across a point in the substrate rather than higher shear stress. Abraded or crushed debris is flushed away by basal meltwater or becomes entrained like the abrading tools by regelation or a downstream 'freezing-on' process. Abrasion rates are high enough to have cut deep glacial troughs during the Quaternary but the fine-grained **rock flour** less than 0.1 mm in diameter it produces does not account for the abundance of clasts up to very large boulder sizes in glacigenic sediments.

Quarrying and entrainment

In one sense, quarrying presents no problem when we consider that alpine and outlet glaciers confined to bedrock channels undercut and destabilize adjacent rock walls. Repeated mass wasting on to the glacier surface, including large rockfalls and slides, enlarges glacial troughs and entrains the debris as *supraglacial* moraine. The solid character of ice supports debris across the entire range of particle sizes and does not sort them according to velocity as occurs in fluids. *Subglacial* large-scale 'plucking' or joint-block removal has been a harder process to resolve and complex schemes of pre-glacial or subglacial frost shattering and pressure release have been invoked. The principal objection to the latter was outlined in Chapter 13 and, in general, these processes also founder on how ice not only dislodges large blocks of underlying rock (whilst also constraining them) but entrains them in the ice stream. Freezing on seems to defy the pressure-melting effects of very large blocks protruding into the basal ice.

Quarrying is now thought to be particularly effective in the presence of meltwater at the high confining pressures which generate pressure-melting at the glacier bed. Clear traces of high-pressure flow were already known from plastically sculptured **p-forms** and water-scoured **sichelwannen**. Pre-existing rock mass discontinuities pre-empt the need for frost or other fracturing and provide water access. Recalling the Mohr-Coulomb criteria in Chapter 13, high-pressure water provides uplift (u) for individual blocks like a 'jack', powerfully enhances shear stress behind the block (v) and effectively 'fire-hoses' it into the icestream – where it is crushed and/or frozen on in areas of lower temperature or pressure (Plate 15.7).

Glacier transport and deposition

Glaciers transport debris in *sub*glacial, *en*glacial and *supra*glacial positions (Figure 15.12). Debris in transit may remain in one of these positions, from where it is eventually deposited, or may move to another. Glacigenic sediments are deposited by ice directly or may be reworked by meltwater and are not as structureless as was once thought (Plate 15.8 top). The relative importance of ice and water is evident in the sedimentary facies associated with a single glacial event. Modern research draws further distinctions between terrestrial- and tidewater-glacier deposition. Each environment stamps its mark on the character of individual particles and facies.

Subglacial debris entrainment by regelation can build up alternating bands of dirty and cleaner ice several metres thick (Plate 15.7 bottom). In this way debris assumes an englacial position and experiences

Plate 15.5 Large-scale abraded rock surfaces alongside the Findelen glacier, Switzerland (top), with typical abrading tools during (top right) and after extended abrasion (bottom).
Photos: Kenneth Addison.

Plate 15.6 Crescentic chatter marks made by the intermittent contact of a large abrading tool.
Photo: Kenneth Addison.

little attrition compared with material in traction or moving frequently in and out of basal ice as pressure-melting conditions change. Further englacial incorporation occurs where basal debris is squeezed into crevasses which remain open to the bed in thinner ice towards the glacier terminus. In the same zone, moving ice may shear over stationary ice or along debris-rich bands, thrusting debris along the deformation to the glacier surface. Supraglacial debris is sourced primarily by glacier destabilization of adjacent rock walls and their subsequent mass wasting on to the glacier surface and, to a lesser extent, by thrusting and melt-out of englacial debris (Plate 15.9). Glaciers may also receive airborne dust from a variety of extra-glacial sources. Debris enters the ice via crevasses, entrained in meltwater and by pressure- and thermal-melting through its mass or absorption of short-wave radiation. Some is swept from the glacier surface by wind or water.

The character of eventual sediments, facies and landforms evolves through the extent of clast attrition and winnowing of fines from bulk materials in transit. The degree of debris concentration by mass wasting or **glaciotectonic** processes (the development of shear planes, thrusting, etc.) and mode of deposition apply the finishing touches. Poorly sorted, clast-rounded **lodgement till** and well sorted, coarser, angular **ablation till** represent the basal and overlying supraglacial deposits of the same ice advance and, together, demonstrate a degree of bedding. Till is a glacial **diamicton** or poorly sorted sediment in which clasts are embedded in a finer matrix, usually of silt clay but occasionally sand, and replaces the term *boulder clay*. Lodgement occurs through net debris release from moving basal ice and forms till sheets or plains where pressure-melting is widespread or ice flow diminishes and basal shear increases. Both are increasingly common below the ELA and lead to lodgement rates of 10^{1-2} mm yr^{-1}. In effect, the basal deformation zone has shifted from the debris–rock to the debris–ice boundary. It can shift again through changes in the geotechnical properties of till, including **dilatancy** or increase in volume and void ratio, or as a result of ice readvance or surge. Both processes may induce deformation *within* the till as upper layers adhere to basal ice, leading to large-scale till block thrusting and streamlining. This is thought to be a principal mechanism in the formation of **drumlins**, 10^{1-2} m high and 10^{2-3} m long, and fluted moraine bedforms an order of magnitude smaller.

Plate 15.7   *Recent glacier quarrying in granite by the Nigardsbreen glacier, south-west Norway. The side view (top) shows block displacement along vertical and horizontal fractures and the rear view (bottom) reveals the effects of high-pressure water jacking, sufficient to insert water-rounded boulders into the developing fracture.* Photos: Kenneth Addison.

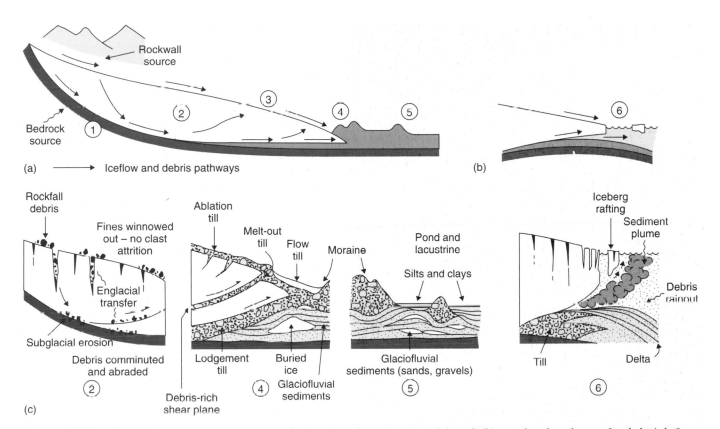

Figure 15.12 *Glacier transport pathways and their depositional environment: (a) and (b) numbered pathways, 1 subglacial, 2 englacial, 3 supraglacial, 4 glacier-marginal, 5 extraglacial, 6 glacio-lacustrine/marine; (c) principal environments and processes typically found in environments 2, 4, 5 and 6.*

Debris is also deposited by **melt-out** from active and, especially, stagnant ice when atmospheric or geothermal heat fluxes are sufficient to melt surface or basal ice respectively. Melt progressively uncovers englacial debris and is enhanced initially at the glacier surface where the lower albedo of debris induces greater heat conduction into the ice. However, the build-up of supraglacial debris eventually insulates the ice, leading to delays in final melt-out and the formation of ice-cored ablation till. In this increasingly water-charged environment, debris slumps and flows off the glacier to form an irregular assemblage of ice-contact and waterlain landforms. The role of meltwater in creating distinct glaciofluvial facies is developed below.

Glacier meltwater

Glacier meltwater is traditionally associated with the distinct roles of erosion of meltwater channels and deposition of a suite of fluvial landforms fed by water and sediment fluxes from the glacial environment. Meltwater processes may occur anywhere in the glacial system but become increasingly important in the ablation zone below the ELA. Sourced primarily by surface melting and with discharge enhanced by rainfall, water proceeds via a glacial plumbing system of surface channels, vertical **moulins** drawing water into the englacial environment and subglacial channels feeding discharge portals at the terminus (Plate 15.10). Water in englacial and subglacial domains is liable to be under high pressure and capable of maintaining ice-walled phreatic channels. Where it incises bedrock, subglacial channels contain uphill segments cut through bed irregularities.

Surface water is inhibited from entering the glacier during winter freezing of the plumbing system in temperate glaciers and more general freezing in cold glaciers. It then cuts lateral ice-marginal channels and becomes ponded in depressions at the glacier margin, often developing overflow channels. Newtondale, for example, marks the spillway of a former ice-marginal lake draining southward across the North York Moors but the recognition of other former ice-dammed lakes from meltwater channels alone may be spurious. Break-out in spring can be catastrophic if sufficient meltwater build-up bursts through still-frozen

Plate 15.8 *Foliation and shear planes in the Tsidjiore Nouve glacier, Switzerland, highlighted by contrasting clean and debris-rich ice (top) and a melt-out version of similar structures in glacier sediments (bottom).*
Photos: Kenneth Addison.

fines which gives meltwater streams a distinctly milky appearance.

Meltwater is increasingly seen as a more subtle influence on processes regarded formerly as glacial in nature. Its presence as a widespread water film at the glacier bed is determined by the glacier's thermo-dynamic character and pressure-melting régime. This facilitates sliding, and thereby glacial abrasion, and exerts a major quarrying influence as a hydraulic jack in bedrock discontinuities. High pore water pressure leads to instability in all forms of glacial debris and promotes deformation, slumping and flowage.

Glacier erosional landsystems

Distinctions between cold and temperate glacier systems allow us to model the erosional and depositional landsystems of contemporary glaciers and, in turn, to reconstruct former glaciers from the landsystem they created (Figure 15.13). Each system develops a recognizable assemblage and pattern from the general range of glacial landforms. Later ice advances obscure or obliterate earlier landforms where there is a sequence of glacial stages. Within a single stage, or glacial event, ice sheet growth at glacial maximum may occur between more localized alpine glaciation phases, with the superimposition of landforms in areas common to both.

Alpine erosional landsystems

Alpine glaciers tend to excavate linear **troughs** fed by in-line and lateral tributary **cirques**. Glaciation commences as regional snowlines fall far enough below mountain summits for sufficient accumulation to generate ice flow. Tributary stream valleys and other sites sheltered from prevailing winds and insolation, enlarged through **nivation** (see Chapter 25), are the first to collect snowbeds. Steep mountain slopes, high precipitation and isothermal warm-based ice combine to generate rapid glacier outflow, which enlarges and deepens existing valley networks. Cirque glaciers coalesce as valley glaciers and concentrate their erosive power below confluences. Troughs develop characteristic parabolic cross-sections typically 0.5–2.0 km deep and with upper rock walls reaching 65°–85° at their steepest above relatively flat floors. Irregular long-profiles reflect extending and compressive flow régimes, with rock basins excavated below ice confluences or in structural weaknesses. **Riegels** or cross-valley barriers separate the basins and rock surfaces are polished and striated by

channels, and the resulting **jökulhlaup** is a powerful erosive agent. Meltwater deposition is subject to normal fluvial 'rules'. It is restricted to ice-walled and bedrock channels throughout the glacial plumbing system but is free to develop unconfined deposits beyond the ice margin. Glaciofluvial sediment budgets are distinguishable by their high point-source concentrations of glacially derived debris, their high spring melt régime and by a high suspended load of vigorous iceflow. Streamlined **roches moutonnées** are abundant.

Valley networks may be rectilinear where glaciers flow away from major morphotectonic watersheds such as the Andes and radial from more isolated eminences, as in the English Lake District. Glaciers remain confined to, and accentuate, rock wall channels even if ice growth intensifies, although some minor transfluence develops as ice sheds are progressively eroded. Erosive intensity over one or more glacial stages is reflected by surviving pyramidal summits and their narrow, precipitous connecting **arêtes**. Less intense glaciation is also recorded by **trimlines**, marking the upper limits of smaller glaciers quarrying their own diminutive troughs into the main valley.

Ice sheet erosional landsystems

Alpine glaciers occupy a greater mountain land area and extend piedmont lobes into surrounding lowlands at glacial maxima. They rarely become the focus of ice sheet growth, as late Pleistocene cordilleran ice limits show (Figure 15.1). Instead, ice sheets envelop large inland areas of lower-lying ground where low mass balance and turnover combine with gentler slopes in sustained glaciation. Plateau ice caps such as Hardangerjökull in Norway, transitional between alpine and ice sheet glaciation, act as embryonic ice sheets early in glacial events.

The size and full range of thermodynamic conditions of continental ice sheets are imprinted on four widely recognized thermodynamic and landsystem zones (Figure 15.14). Zero or low basal velocities and little meltwater in the *ice-shed zone* of cold-based ice sheets severely hamper quarrying and abrasion (Zone I). Such erosion as occurs is inconspicuous and distributed uniformly. Only **nunataks** emerge through almost total ice cover to provide any scope for undercutting and supraglacial debris. Away from ice dispersal centres, abrasive scour is more common and there is evidence that basal ice begins to 'stream' at depth, quarrying bedrock channels in the *selective erosion zone* (Zone II). Local ice thickening increases basal shear stress and pressure-melting, enabling the ice stream to exploit weaker bedrock. Quarrying is self-enhancing, as it draws more ice into the developing rock basin, and this increases dramatically in the *outlet glacier zone* (Zone III).

In a steady state the ELA is located relatively close to ice sheet margins, nourished and melted in advection-driven mass and energy balance conditions. This transforms the ice stream into a temperate, warm-

Plate 15.9 *Supraglacial, ice-cored moraines and debris on the Tsidjiore Nouve glacier, Switzerland.*
Photo: Kenneth Addison.

based state and vigorous outflow draws down adjacent inland areas of the ice sheet through outlet glacier troughs. They are the most impressive of erosional landforms, excavated 1–5 km deep and breached clean through any cordillera in their path regardless of subglacial topography (Plate 15.11). Transfluent ice flow on this scale joins coastal alpine glaciers and excavated **fjords** through the coastal mountains of southern Norway (Plate 15.12), the South Island of New Zealand, British Columbia, Alaska, southern Chile and – to a lesser extent – western Scotland. The Finger Lakes region south of Lake Ontario marks transfluent ice flow towards the southern margin of the Laurentide Ice Sheet. Beyond the constriction of outlet glacier troughs, ice fans out in the *piedmont zone* (Zone IV). Where ice flow is still vigorous it erodes **knock-and-lochan** topography of parallel roches moutonnées interspersed with shallow rock basins which become lake-filled during deglaciation.

Glacier depositional landsystems

Alpine depositional landsystems

Glacier confinement by rock walls also shapes and constrains deposition in the alpine environment. Supraglacial debris derived from the rock walls is usually stacked in lateral ridges or **moraines**, and lateral and medial moraines – the latter marking the confluence of two ice streams – are a prominent feature of most alpine glaciers (Plate 15.13). Lateral

Plate 15.10 *Supraglacial meltwater stream network (left), a single supraglacial stream entering a moulin (bottom left) and a discharge portal on the Taku glacier, Alaska (above).*
Photos: Kenneth Addison.

moraines are linked across the valley where the glacier sheds debris conveyed to its terminus, marking its maximum extent with a chevron of terminal and lateral moraines. Deglaciation may be marked by a sequence of moraine chevrons, each recording the receding ice margin. Retreat moraines from the Little Ice Age advance are still fresh today and most ice-free alpine cirques and valleys also contain visible moraines from the last Pleistocene glaciers (see Plate 11.2). Inside ice limits, sedimentary landforms consist largely of amorphous or hummocky till sheets and fluted moraines interspersed with waterlain facies. Well defined drumlins and **eskers** (subglacial stream beds) are uncommon. Beyond the glacier terminus, glaciofluvial sediments form a **valley train** or braided **outwash plain** confined between the rock walls. Lacustrine sedimentation in rock basins and moraine-dammed depressions is a major feature of deglaciated alpine valleys.

Ice sheet depositional landsystems

Although an ice sheet may advance over 2000 km from its ice shed, and subsequent retreat and deglaciation rework the landscape, the most extensive glacier depositional landsystem is still associated with late Pleistocene ice sheets of North America and Eurasia. The sequence and alignment of landforms, facies and even individual clasts collectively point to the ice-source regions, ice dynamics and ice limits and can be identified using the same zonation as for erosional landforms (see Figure 15.14). Low erosion rates in

Zone I generate little debris, and glacigenic sediment is sparse in ice sheet accumulation areas. Many ice-source regions were formerly unrecognized because of their dearth of conventional glacial landforms. As linear erosion develops in Zone II, deposition also becomes more significant and somewhat streamlined in the form of fluted till. High rates of basal and side-wall erosion in the vicinity of outlet glaciers in Zone III replicate alpine valley glaciers and greatly augment in-transit sediment loads but the landsystem is best developed in Zone IV (Figure 15.15).

Inner, active parts of piedmont glaciers frequently operate over a range of basal shear stresses at which thick lodgement till sheets can form and be deformed, most notably into drumlins parallel to ice flow. This contrasts sharply with transverse deposi-tional landforms in outer, progressively more inactive and even stagnant ice areas at and behind the terminus. The latter commence with *Rogen* moraines, representing transverse crevasse and shear plane melt-out in compression behind a static terminus, and *De Geer* moraines, formed where ice sheds debris beneath a floating tongue in ponded meltwater (Figure 15.16). The terminus may be marked by one or more moraines or a **kame moraine** composed largely of glaciofluvial sediments. The decisive shift from ice-deposited to waterlain facies commences with surviving casts of subglacial meltwater streams, moulin fillings and ice-marginal streams as eskers, **kames** and **kame terraces** with internal structures reflecting ice-wall collapse. Ice melt-out and disinte-gration are marked by chaotic mixtures of till and waterlain sediments with few clear landforms, giving way abruptly beyond the terminus to a **sandur** plain of wholly glaciofluvial sediments. The down-glacier facies sequence reflects changes from subglacial to melt-out/ablation, glaciolacustrine, glaciofluvial and even aeolian environments. The alignment of clasts, trails of **erratics** derived from rocks upstream and fining downstream trends within glacigenic sediments provide palaeocurrent signatures of ice sheet source and flow directions.

The dynamics of glacier sedimentation in British and other piedmont zones are of contemporary interest because of their links with glaciomarine processes. The great variability in sediment type and source, clear evidence of deformation during/after deposition and the occasional presence of arctic marine shells may point to a catastrophic collapse of

Figure 15.13 *Modelled characteristics of the British Late Pleistocene (Late Devensian) ice sheet: (a) general iceflow directions and the intensity of glacial erosion, measured on scale of 1–5 (least to most intense); (b) ice surface, iceflow lines and ice limits. Source: Boulton et al. (1977).*

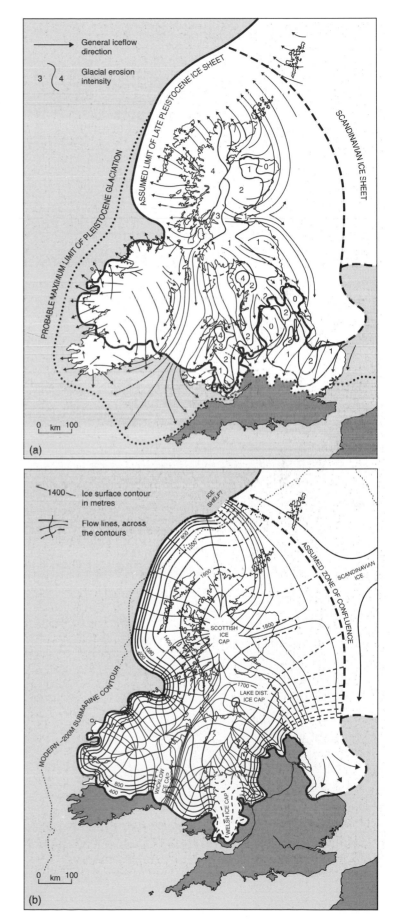

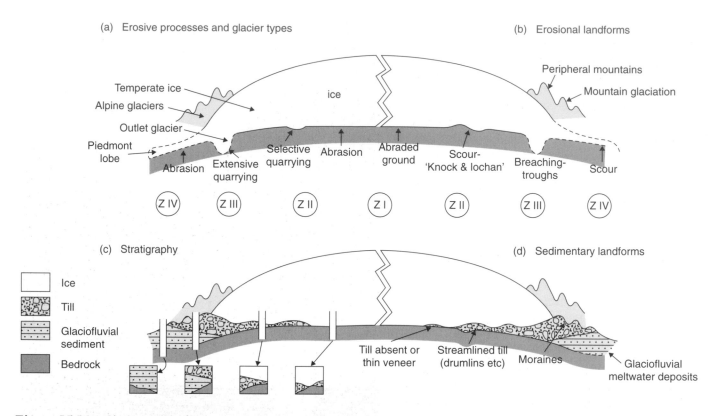

Figure 15.14 *Glacier geomorphic landsystem zones. See text for explanation. Source: After Addison (1983).*

Plate 15.11 *The Ferrar glacier, discharging Antarctic ice through the Transantarctic Mountains into a fjord arm of the Ross Sea.*
Photo: David J. Drewry.

the western British Ice Sheet during deglaciation. It is thought that rising sea level flooded the Irish Sea basin faster than initial ice retreat, turning the ice margin into a floating shelf (Figure 15.17). This may have led first to accelerated iceberg calving and then to ice sheet surging as its frontal support collapsed. Drumlin fields in neighbouring low-lying parts of Ireland, northern England and Wales may be indicative of surging glacier behaviour. This would have reactivated terrestrial ice flow, triggering deformation within the sediment body in the manner required for streamlined bedforms.

Permafrost processes and landsystems

Ground ice processes

Perennially frozen ice lies inert in the ground for the most part, and permafrost includes adjacent earth materials, voids and water. Geomorphic activity depends primarily on how this material system responds to expansion–contraction cycles in that portion of the ice–water mass which regularly changes phase. It is further influenced by the presence of a frozen, impermeable substrate – the ground ice itself – beneath a seasonal active layer in what is commonly described as the **periglacial** environment. The term is misused when the existence of permafrost is equated with the proximity and climatic influence of

Plate 15.12   *A 1000 m rockwall flanking an arm of Sognefjord, south-west Norway.*
Photo: Kenneth Addison.

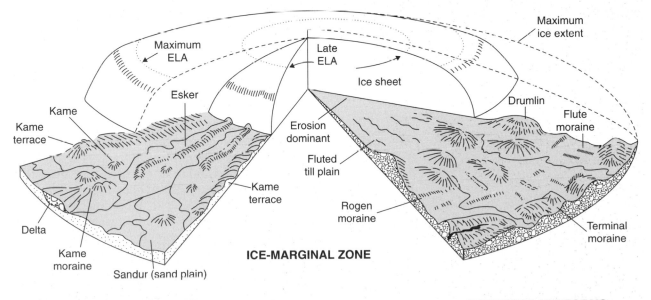

GLACIOFLUVIAL LANDFORMS

MORAINIC LANDFORMS

Figure 15.15   *Ice sheet depositional landsystem. Glaciofluvial (waterlain) and morainic (ice-deposited) subsystems are shown separately in relation to maximum and retreat stages but they may, in practice, be superimposed. Both subsystems are primarily subglacial or ice-marginal in nature and may be draped with supraglacial sediments (flow and ablation till, etc.). Extraglacial landforms must be glaciofluvial but ice and water are both responsible for ice-contact landforms.*

Plate 15.13 *Medial moraine formed by the junction of lateral moraines of the Schmäre and Breithorn glaciers, Switzerland. They mark the Little Ice Age limits of both glaciers and the extent of subsequent retreat.*
Photo: Kenneth Addison.

per cent volumetric increase of freezing water, accentuated by migration of pore water to the freezing plane in porous substrates. This establishes corresponding dewatering contraction elsewhere which accentuates thermal contraction. Thawing effectively reverses these processes, and the combined impact of repetitive stress cycles is responsible for the cryofracture of rock and the cryoturbation of unconsolidated materials (see Chapter 13).

The permafrost landsystem

Erosion and deposition are less easily ascribed to permafrost than other geomorphic environments except for cryofracture (frost weathering). This clearly erodes exposed bedrock (and comminutes clasts in sediments), distributing its products as depositional or residual landforms or feeding them into the mass wasting system. These processes dominate upland permafrost landscapes (see Chapter 25) with a distinctive suite of residual and detrital landforms (Figure 15.18, Plates 25.6–7). Tors represent the surviving bedrock surface on summits and plateau breaks-of-slope, between which the permafrost erosion surface may be represented by an altiplanation terrace influenced by structure or the depth of the active layer. Massive and resistant rocks such as granite, gritstone and tuffs produce the best tors. Rock debris litters the landscape and is crudely stratified on plateau surfaces, with an upper pavement of large angular boulders over a layer of fines washed down by meltwater and sheltered from deflation.

The general form of permafrost slope systems is described in Chapter 25 as a variation on general models of mass wasting and slope evolution (Chapter 13). Cryofracture on summits and slopes is one source of slope debris, together with more specialized processes associated with nivation and rock glaciers. They are indicative of the generally slow and seasonal nature of movement in the permafrost environment and the ability of interstitial ice formation to arrest mobile materials and impart cross-slope turf-banked and stone-banked lobes on solifluction terraces. Ice melt over a frozen substrate is, however, a recipe for rapid movement and summer debris flow is a more hazardous form of *geli*fluction.

Debris joins the supply of superficial deposits susceptible to the most widespread and related permafrost processes of sorting, ice wedging and cryoturbation (Figure 15.19). They do little more than rework or ornament the landsurface but they also contribute indirectly to permafrost mass wasting and denudation. They are most effective in relatively

glaciers, as the increase in permafrost intensity and activity in Alaska *away* from the cordilleran glaciers demonstrates. It is more appropriate to regard glaciation and periglaciation as sharing persistent cold temperatures and permanent ice, albeit in different amounts and forms, than to regard them as sharing geographical proximity.

Freeze–thaw cycles stimulate material expansion and contraction simultaneously. Ground ice styles associated with freezing were described earlier. Initial ground expansion on freezing is equivalent to the 9

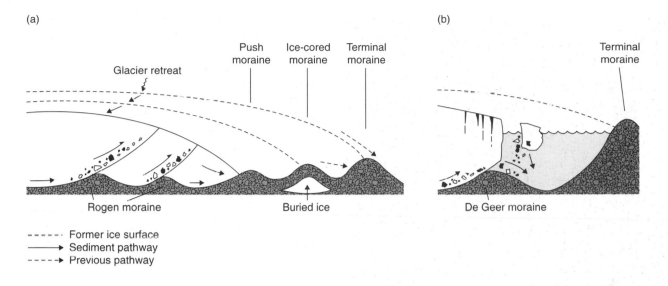

Figure 15.16 *Moraine formation. Terminal and ice-cored moraines form passively by debris melt-out from marginal and buried ice (a). Rogen and push moraines form under active ice by melt-out along debris-rich shear planes and episodic readvance respectively (a). De Geer moraines form by melt-out from active ice decoupled from its bed in ponded water (b).*

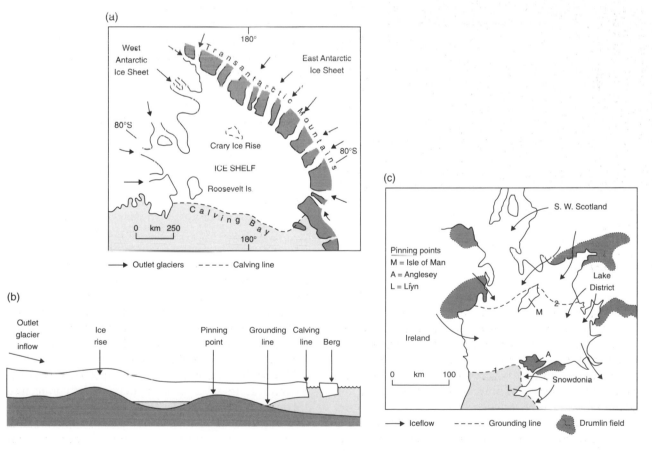

Figure 15.17 *Ice shelves: (a) Ross ice shelf, Antarctica; (b) major shelf features in cross-section; (c) possible late Pleistocene ice shelf, retreating grounding lines (1 and 2) and drumlin fields in the Irish Sea basin.*

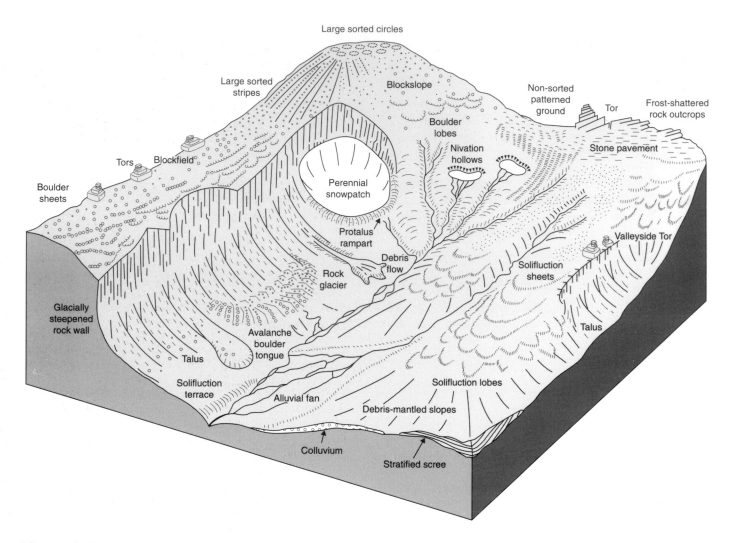

Figure 15.18   *Late Pleistocene (Devensian) permafrost landforms on British mountains. Source: After Ballantyne and Harris (1994).*

thick superficial materials on flat and usually low-lying ground but also develop on plateau detritus. Sorting occurs through the preferential frost heave of stones towards the surface through a finer matrix at rates of 10–50 mm a⁻¹. Small ice lenses freeze first beneath stones with higher thermal conductivity than the surrounding loose matrix, or grip the top of stones during downward migration of the freezing plane. Frost *push* or *pull* is reinforced as further ice crystal growth in the subjacent void maintains the upward progress of the stone, which invariably rotates so that its longest axis is vertical. Sorting is completed as gravity preferentially draws stones down the gentle slopes of frost-heaved ground as they break surface, creating a stone circle around a dome of fines. Ice wedges form in desiccation cracks and effect a cruder form of sorting. Initiated by desiccation, cracks collect water and washed or deflated fine sediment

during melt seasons. They continue to widen and propagate through the active layer during the next freezing cycle. Ice wedges reach 1–10 m depth (exceptionally, an order of magnitude deeper) and 5–50 m apart horizontally, reflecting permafrost duration and the intensity and frost susceptibility of the substrate.

Frost sorting and wedging frequently lead to **patterned ground** development as individual foci make edge contact with each other (Plate 15.14). Constant rates of heaving or wedging in perfectly homogenous material on a flat surface would, in theory, lead to an isomorphic network of connecting polygons or hexagons. Patterned ground is a reasonable approximation but some phenomena are not developed far enough to connect and material inhomogeneity distorts the circles, polygons and nets in those that do. In particular, gravity elongates the

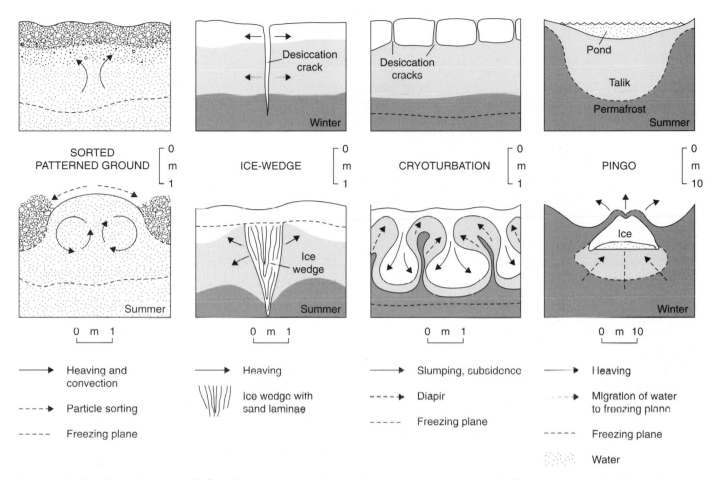

Figure 15.19 *Permafrost ground disturbances; upper and lower levels show initial and concluding stages respectively in each major process.*

downslope component on slopes as low as 2°–5° and stone stripes develop on slopes of 5°–25°. At higher angles, however, the lateral component of thrusting is destroyed and stones develop cross-slope stone-banked lobes or form more extensive screes. Cryoturbation exploits frost-susceptible and hydrological inhomogeneity in a more random manner. Local variations in bulk density, porosity, water saturation and freezing rates generate vertical separation and overturning in the active layer. Cryostatic pressures exerted on a pocket of saturated, unfrozen sediment may cause it to inject a plume into overlying material or erupt at the surface whilst denser blocks may sink and intrude underlying material. In extreme cases, entire unfrozen water bodies may collect in the substrate, especially in valley floor peat or alluvium, and grow by cryostatic and/or artesian pressure as the system starts to freeze. The water eventually forms a segregated ice lens which erupts as a **palsa** up to 10 m high or grows into a perennial **pingo** 10–80 m high (Plate 15.15). The micromorphology of

Plate 15.14 *High-altitude, large-scale patterned ground on Galdhøpiggen (2470 m), central Norway.*
Photo: Kenneth Addison.

Plate 15.15   *Fossil pingos in the Cledlyn valley, mid-Wales. A collapsed core is seen to the left inside its fringing rampart, with another rampart beyond.*
Photo: Kenneth Addison.

frost-disturbed ground creates edaphic variations exploited by plants which often enhance the litho-morphology, creating ground hummocks or *thufur* and string bogs or *muskeg*.

Instability occurs throughout the permafrost environment as the active layer develops during early summer melt. It accelerates slope movement and turns low-lying and level ground into a chaotic mire. The water is unable to drain through underlying permafrost and, for a few weeks, frozen rivers become highly erosive extensions of the fluvial environment. Floating ice is an additional abrasive tool and many valleys become misfits or are abandoned altogether by surface drainage in normally porous rocks on the return of a more temperate climate.

Conclusion

Glaciation is a regular and predictable state and not a freak or 'accident'. It is not merely a by-product of climatic change but also instigates it through ice sheet–ocean–atmosphere coupling. Ice sheet and alpine glacier growth is strongly influenced by tectonic activity and supercontinental cycles as well as Milankovich mechanisms (see Chapter 11). Glaciers, in turn, then drive other forms of glacio-tectonic coupling through the rapid alteration of crustal loading by ice sheet growth and decay, glacial erosion and sediment transfers and glacio-eustasy. All this endows glaciers and permafrost with material and geomorphic systems which are not solely determined by climate but interact with climatic and tectonic processes to set their own rules.

The environmental influence of glaciers and permafrost extends well beyond their current geographical distribution. It matters little to human endeavour that Earth may endure long periods without Ice Ages. The greater part of hominid evolution has occurred during, and continues to be profoundly influenced by, the Quaternary Ice Age. Dramatic human population explosion and almost all our technological innovation have occurred in just 10 ka of the current (Flandrian) interglacial cycle. We need to be as responsive to Earth's icehouse as to its greenhouse modes.

Key points

1   Earth experiences both icehouse and greenhouse extremes of global climate. Over long geological time scales, icehouse conditions appear to coincide with a fragmented continent–ocean stage of the supercontinental cycle, with continents in more polar locations. Each Ice Age lasts millions of years, comprising separate cold or glacial stages interspersed with temperate interglacial stages. During the current Quaternary Ice Age, longer cold stages (10^{4-5} a) and shorter temperate stages (10^{3-4} a) coincide with regular orbital (Milankovich) cycles and contain shorter stadial and interstadial episodes.

2   Large continental ice sheets develop slowly and alpine glacier systems expand during cold stages – particularly in the northern hemisphere – but recede or disappear altogether during temperate stages. The Antarctic Ice Sheet, Earth's largest ice mass, has survived the past 40 Ma. Global changes in albedo and sea level accompany ice sheet growth and initially intensify the cold stage but ice sheet–ocean–atmosphere coupling is sensitive enough to cause rapid deglaciation as the cold stage ends.

3   Regional climate and topography drive glacier mass and energy balances, endowing each glacier with a distinctive thermodynamic character. This determines flow rates and mechanisms and geomorphic activity. Cold, polar ice sheets are stable and cover large areas with generally slow-moving ice, except near their margins, where ice flow replicates the behaviour of temperate, alpine glaciers. The latter are unstable, fast-moving ice streams with considerable geomorphic impact. Floating shelf margins of ice sheets are metastable and hold the key to ice sheet response to global warming.

4 Alpine and Polar glaciers are respectively warm- and cold-based, which influences the deformation zone between moving ice and sediment, water and bedrock at the glacier bed. Deformation can move from one material to another, determining the nature and location of glacial geomorphic processes. Contrasting glacier styles are reflected in their geomorphic landsystems. Alpine glaciers are constrained in their valleys while ice sheets inundate huge land areas, placing different emphases on their scales of operation, spatial variability and supraglacial and subglacial environments. Our knowledge of northern hemisphere Late Pleistocene ice sheets is reconstructed from their residual landsystems.

5 Cold stage climates may be so severe and dry as to prevent glacier growth over large areas, and terrestrial landscapes experience permafrost instead. Surface and underground water is perennially frozen to depths dependent on the severity and duration of the cold stage. The exception is an active layer of seasonal surface melting which houses almost all geomorphic activity, driven by freeze–thaw cycles and the impact of interstitial water over an impermeable substrate. Apart from cryofracture, most processes merely rework and ornament the landscape.

Further reading

Ballantyne, C.K., and Harris, C. (1994) *The Periglaciation of Great Britain*, Cambridge and New York: Cambridge University Press.
A succinct account of global permafrost environments and many timely illustrations of landforms elsewhere justify the British bias of this book. Its excellent account of the extensive range of former and contemporary periglacial processes visible in Britain is most revealing and very well illustrated.

Bennett, M.R., and Glasser, N.F. (1996) *Glacial Geology: Ice sheets and landforms*, Chichester and New York: Wiley.
This book provides a general review of glacial processes. Further reading is placed accessibly at the end of each chapter and 'boxes' are used to explain more detailed processes, although these are sometimes so numerous as to risk disrupting the flow. Its range is similar to Hambrey's earlier book, with which it bears comparison and provides readers with choice.

Dawson, A.G. (1992) *Ice Age Earth: Late Quaternary geology and climate*, London and New York: Routledge.
This book reviews all major aspects of icehouse Earth, including the relevant oceanic, atmospheric and tectonic systems. It places the most recent glacial stages in time and space frameworks and reveals how glacial events are recorded in a wide range of environments.

Hambrey, M. (1994) *Glacial Environments*, London: UCL Press.
A comprehensive and integrative review of glaciological, geological and geomorphic processes uninterrupted by any mathematics, this lavishly illustrated book is supported by an extensive glossary.

CHAPTER 16

The work of the wind

The power of the wind in extreme meteorological events is evident from hurricane, typhoon and tornado damage to property and the resultant human misery. Wind translates this power into geomorphic work indirectly through its ability to drive waves in the coastal zone. Its direct geomorphic impact, however, is much less closely associated with planetary storm belts and is restricted largely to redistributing and ornamenting products of other processes. The predominance of quartz sand and silicic silt in **aeolian** (wind-blown) sediments denotes a final sorting of the final fractionates of other denudation processes. Wind is linked romantically with ever-shifting sand seas of Earth's hot deserts and their nomadic peoples. Common landform terms are often Arabic in origin for this reason but the presence of coastal sand dunes, extensive Pleistocene **loess** (aeolian dust) belts and dustbowls on intensively farmed land in more humid, mid-latitude areas is testimony to its opportunistic attack on vulnerable materials everywhere. Building sites, urban landscapes in general and exposed mountain tops provide additional sources of airborne particles. Wind agency is also important in desertification, enhanced by other prior forms of land degradation (see Chapter 27).

Aeolian processes

Fluid motion of the wind

General dynamics of fluid motion relevant to the entrainment, transport and deposition of earth materials, set out in earlier chapters, show some variation in the wind environment. Although aspects of laminar and turbulent flow and the application of force are broadly similar in aqueous and gaseous fluids, their density differences are very significant. Water is three orders of magnitude more dense than air (1000 kg m^3, compared with 1.22 kg m^3 at sea level) and therefore applies greater force at any given velocity. For example, stream flow maintains particles over 100 mm in diameter (small pebbles) in motion at some 6.0 m s^{-1}. This is the wind velocity required to *entrain* fine sand over 200 μm or maintain coarse sand over 600 μm in motion. However, lower density increases sorting efficiency, with particles falling rapidly out of incompetent flow, and **grain ballistics** are more effective in moving stationary particles on impact. The transmission of force from moving to stationary particles lowers the entrainment threshold for the stationary particle.

Airflow is constrained in the boundary layer with the ground like any other fluid but the nature of the topographical surface is particularly important in controlling its *effective* velocity and patterns of turbulence (see below). Air turbulence may extend through layers 10^{1-4} m thick, unlike turbulent flow in streams and the nearshore zone, which is restricted by much shallower water depths, and airflow is not confined to narrow channels. Aeolian bedforms can develop on a massive scale, provided the sediment source is sufficient. Air temperature also varies over a wider range, which increases its influence on air viscosity. Threshold velocities for sand particles decrease as temperature falls and density rises, enhancing entrainment in cold climates.

Effective wind and deflation

The prime ingredients of aeolian landsystems are effective wind velocities and turbulence with a large supply of uncohesive particles of sand and silt-clay size. In practice this requires that airflow is unimpeded by vegetation, which displaces threshold velocities away from the surface – and therefore the target materials – in the boundary layer (Figure 16.1). Vegetation also retains soil moisture, adds humus and binds grains – all of which increase cohesion. Dry sand is uncohesive and possesses only friction strength under shear stress, whereas clay–silt particles and the presence of moisture develop low cohesive strength. Wind, like water, operates over a wide range of

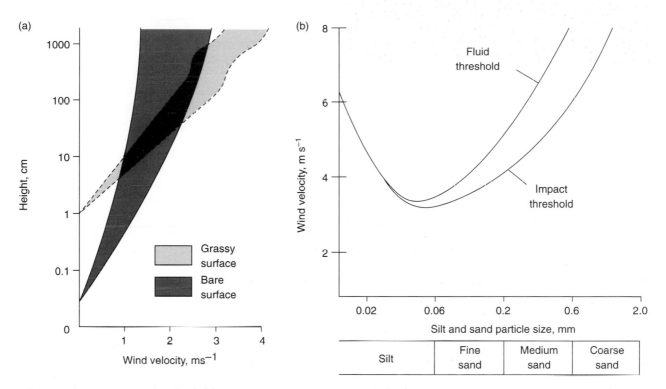

Figure 16.1 *Comparative threshold conditions for deflation: (a) wind velocity over a bare surface and 10 cm tall grass layer and (b) particle size–velocity relationship for entrainment (fluid) and ballistic impact. Source: After Warren (1979).*

velocities but in practice aeolian processes occur only in arid conditions where vegetation and soil are sparse or absent. Airflow also contributes to aridity through its removal of evapotranspired water and steepening of hydrological gradients. Soils are virtually immune to deflation at low matric forces, approximately 1.5×10^3 Pa (permanent wilting point).

Deflation is the entrainment and removal of loose, fine-grained materials from a landsurface by wind which leads to their subsequent deposition downwind. In effect, wind winnows or sifts clay, silt and sand grains and leaves behind rock surfaces and **lag deposits** of coarser particles too large to be deflated. Deflation sets in motion all other processes of the aeolian environment including abrasion and sandblast, the excavation of deflation hollows and creation of mobile bedforms.

Entrainment, transport and deposition

Entrainment occurs when turbulent lift and shear stress, enhanced by ballistic impact, exceed normal stress and friction. A special case can also be made for particulate *injection* by volcanic eruption and sea spray, adding ash, tephra and fine coastal materials to the atmosphere. Although not part of the aeolian

mainstream, they may occur in quantities capable of geomorphic and stratigraphic significance. Average grain size in sand seas lies between 100 μm and 1 mm, with a modal size approximately 300 μm. Entrainment and ballistic thresholds for medium and coarse sand (over 200 μm and 600 μm – 2 mm respectively) are some 5.0 and 4.2 m s^{-1} and 8.0 and 6.0 m s^{-1} (with 5.5 and 4.5 m s^{-1} for the modal size). Shear stress increases exponentially with wind velocity from 2 N m^{-2} at 1.3 m s^{-1} to 100 N m^{-2} at 13 m s^{-1}. As effective wind velocity increases particles may at first roll or creep, without becoming airborne, and non-deflatable particles may continue to move by such means. Lift takes fine particles into suspension, where turbulence may support grains up to 200 μm indefinitely. Larger sand grains have low, short trajectories and move by saltation. As sand grains begin to lift, ballistic impacts set a broader field of particles rolling and lifting in a process known as *avalanching*. This establishes general motion, which interacts with airflow in the development of bedforms discussed later.

Long residence times and incorporation at higher altitudes above 1–2 km can retain particles in transit for 10^{3-4} km and accounts for dust plumes downwind and offshore of major sources. It is calculated that 200–500 Mt a^{-1} is deflated from the Sahara to the Atlantic by the *harmattan*, over 75 per cent of which is deposited within 2000 km of the West African coast but some reaches the Caribbean. The loess deposits of China involve similar distances, and it is not uncommon for Saharan dust to reach northern Europe, circling the western edge of blocking anticyclones. Rain forces wet deposition earlier than dry fall-out. At lower elevations under 1 km, dust storms are common events in the desert environment and are indicative of soil desiccation and the degradation of farmland there and in mid-latitudes. Deposition of deflated material clearly occurs when wind velocity falls below the thresholds needed to maintain motion but is often transient and aeolian sediments are characterized by their relentless progress downwind. Sand and silt are totally segregated *en route*.

Abrasion

Ballistic impacts inevitably bring quartz grains into sharp contact with softer materials which they abrade, at the same time removing their own angular edges where the hardness is very similar or grains are flawed. Fracture may also occur during saltation, and 'collision splash' releases fines for further deflation. Abrasion occurs on larger clasts incapable of deflation and exposed bedrock surfaces which may undergo general surface lowering, at rates varying between 1–10 m ka^{-1} and 10–50 mm ka^{-1} in hot and polar desert respectively. This difference is probably explained by the protection effect of snow cover and the almost perennially frozen state of soil moisture, which resists sublimation and removal. However, deflated ice fragments become hardened as temperature falls and may mimic the hardness of orthoclase feldspar in extreme cold in Siberia and Antarctica below −60° C. Rock structures and differential strength combine with air currents in all environments to produce a series of fluted landforms, which retain palaeocurrent information in their general orientation.

Aeolian landsystems

Tectonic and climatic patterns

Aeolian environments cover some 20 per cent of global landsurfaces associated with hot and cold desert environments in divergent zones of subsiding and generally lighter winds. Large-scale atmospheric subsidence provides a drying influence. It also blows out of continental interiors, isolating them from maritime moisture sources. Hot deserts are located beneath subtropical divergence on the poleward side of the Hadley cell (see Chapter 8). Rather weaker polar divergence is strengthened by the katabatic effects of ice sheets and accounts for mid to high-latitude loess sheets. Aeolian environments are not found, ironically, in the storm belts which double as Earth's principal rain belts and are therefore well vegetated. Ocean circulation contributes to coastal deserts through the upwelling of cold currents on western coasts in the case of the Mojave, Sonora and Atacama deserts of the Americas, the Namibian desert of south-west Africa and the Western Australian desert (Figure 16.2).

Zonal climate does not account wholly for desert location and landsystems and, inevitably, morpho-tectonics superimpose their influence through rain-shadow effects and the active provision of deflatable materials. Coastal cordillera reinforce arid effects of coastal upwelling and extend rain shadow to continental interiors. The latter is most pronounced in the basin-range region of the southwest United States and leeward of the emergent Tibetan plateau, whose elevation of 3.5 km during 2 Ma of the Plio-Pleistocene greatly extended north-central Asian interior deserts. Tectonics also play an active or passive role in hot desert denudation processes which source, and are geomorphically associated with, aeolian landsystems.

Denudation proceeds through mechanical weathering and rock slopes are cut back, leaving residual **buttes** or **inselbergs** above low-angled **pediments**. Rapid evacuation of debris by intense but ephemeral stream flow develops alluvial fans or **bajadas** across the pediments and terminates in mud sheets or **playas** lining basin interiors. Ephemeral lakes add evaporites to the supply of deflatable material (Figure 16.3 and Plate 11.1) Denudation rates are high in syntectonic uplift or basin subsidence. Fluvial facies die out and aeolian facies increase in abundance towards basin centres, where sand seas dominate passive, intercratonic basins. Wetter *pluvial* phases during the Quaternary charged basins with fluvial sediments which are now deflating and the relatively slow development of ergs has not yet obliterated the palaeofluvial landscape.

Desert landsystems

Desert pavements, lags and ventifacts

Deflation initially produces remnant landforms in areas depleted of sand and silt. Non-deflatable,

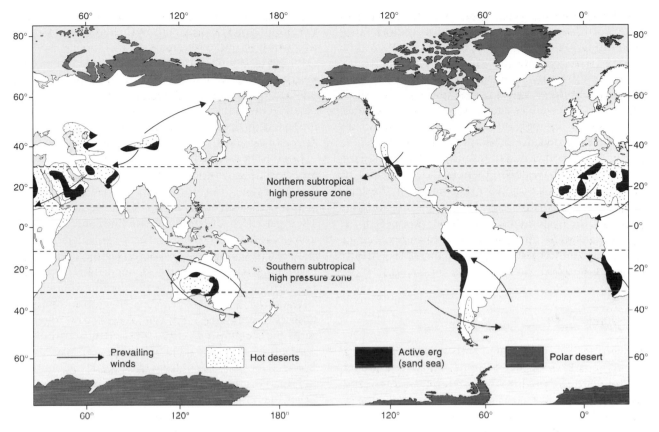

Figure 16.2 *The distribution of hot and polar deserts.*
Source: In part after Collinson (1986).

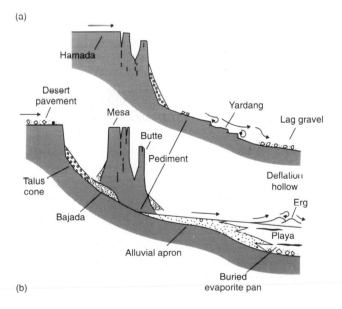

Figure 16.3 *Stages in the development of a hot desert landsystem. Source: After Butzer (1976).*

coarse-grained **lag gravels** loosely protect underlying abraded bedrock surfaces or **hamada** to form **desert pavement**. Residue from the restricted chemical weathering in deserts is rapidly deflated. Lag clasts themselves are abraded *in situ* whilst deflation persists and, if large enough to remained static, sandblasted facets develop on the windward side of these **ventifacts**, often giving a three-faced or *dreikanter* appearance. Desert pavements, known also by their alternate Arabic or aboriginal names **reg** and *gibber*, become sterile unless further fines are introduced to stimulate abrasion.

Yardangs and deflation hollows

Abrasion is likely to pick out lithological weakness and structural discontinuities aligned close to the primary airflow, faceting and fluting the rock into regular or irregular shapes. These in turn channel airflow and create positive velocity anomalies which enhance abrasion locally, offset by lower velocities elsewhere. Fluted channels may run for 10^{2-3} km and are flanked by residual **yardangs** as ridges or pillars of surviving rock. Although sandblast abrades and polishes rock surfaces, the extent to which wind can actively enlarge flutes into large-scale landforms has been in doubt despite the presence in many hot deserts of depressions covering areas of 10^{3-4} km^2. Wind cannot quarry rock, and its capacity to develop **deflation hollows** requires the presence of weak rock, deep water tables and enduring arid conditions. This appears to apply to the suite of large, structurally aligned deflation hollows flooring over 75,000 km^2 of the Egyptian desert west of the Nile. Resistant surface rocks have been penetrated, perhaps by streams in pluvial periods, exposing weak Pliocene shales which now bear clear signs of wind abrasion down to the water table. Deflation products form extensive leeward dune fields or smaller *lunettes*, an Australian counterpart of the ephemeral lake basins of South Australia. Similar forms are found in most deserts.

Sand seas and loess sheets

Coalescence of aeolian sand into a sand sea or **erg** creates a large-scale depositional landsystem whose surface is ornamented by the wind. Single dunes and other bedforms occur wherever deflated material is deposited, but the vast bulk of desert sand is held in active ergs within the desert cores of North Africa, Arabia, Namibia, central Australia and Mexico. Sand volumes of 10^{3-4} km^3 are common and ergs over 25,000 km^2 in area account for some 90 per cent of desert sand, the largest being the Rub'al Khali at some 560,000 km^2 in Saudi Arabia. Erg development generally commences in sheltered topographical depressions where boundary shear stress falls below threshold values, and extends in the direction of effective winds. Sand cover may be incomplete or thin (10^1 m) in peripheral areas but aggradation occurs in erg centres, as active bedforms are superimposed on each other, and may eventually reach thicknesses of 10^{2-3} m. Stabilized relic ergs often fringe active, mobile sand deserts and represent the consequence of regional climatic change.

Loess sheets of silt and clay particles might be expected downwind of sand seas. This happens to a limited extent but loess and related **coversands** almost exclusively form cool/cold desert landsystems. Their particle size – primarily medium and coarse silt from 6 μm to 60 μm – reflects aeolian derivation from glacial sediments, frost deserts and river terraces. Loess rarely develops clearly defined dune bedforms and forms a blanket cover of the existing topography instead over some 10 per cent of global terrestrial surfaces (Figure 16.4). Silt and clay particles, together with the cementing effect of deflated carbonate evaporites, develop stabilizing cohesion. In North America loess deflated from the Rocky Mountains and margins of Laurentian ice lobes in the Great Lakes region blankets much of the high plains (west of the Missouri river) and is funnelled into the Missouri–Ohio–Mississippi basin. Loess forms an extensive apron fringing Fennoscandian Pleistocene ice limits in Europe and the smaller, mountain icefields and frost deserts of Siberia. Southern hemisphere deposits are limited to extensive sheets between the Andes and the river Parana in Argentina and the Southern Alps and east coast in New Zealand. There is no great thickness to any of these loess sheets, which vary generally between about 1 m and 40 m, but they often provide valuable cereal-growing soils, and the designation **brickearth** indicates their local importance to brick manufacture.

British loess deposits scarcely reach 2 m in southeast England, largely through the extent to which glacial retreat was matched by sea-level rise along the North Sea coast and its inundation of loessic source materials. Thin layers are intercalated with Late Pleistocene and Early Holocene palaeosols over much of southern Britain and provide a useful stratigraphic and palaeoclimatic indicator. By comparison, loess exceeds 300 m depth in parts of the great loess sheets of China, which cover over 0.8 million km^2 (Plate 16.1). Basal units date from the Pliocene–Pleistocene boundary 1.8 Ma ago and emphasize the importance of increasing desiccation in the Tibetan plateau rain shadow, leading to deflation of general weathering and erosion products rather than primarily glacial sources. Accumulation reached a peak of 0.5–3.0 m ka^{-1} during

Plate 16.1 *Thick loess overlying late Tertiary clays uncomformably in the Yellow River basin, China.*
Photo: M.A. Fullen.

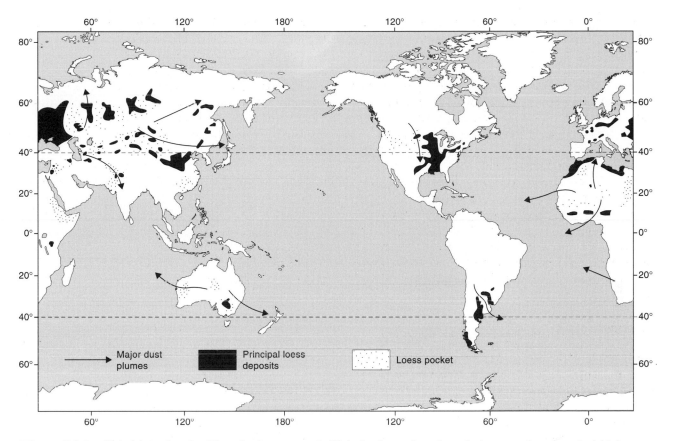

Figure 16.4 *Global loess deposits. Note the focus around 40° latitude north and south, between the subtropical high pressure belt and Pleistocene ice limits. Source: After Livingstone and Warren (1996).*

glacial maxima, falling by 65–80 per cent during temperate stages.

Sand bedforms – dunes and ripples

At first sight, bedforms are very similar to those found on beaches or alluvial river beds but on a far larger scale. As with all bedforms, they reflect particle size, fluid velocity and flow patterns over the sand bed. The most common unit is the sand *dune*, 1–30 m high with wavelengths of 10–500 m, which occurs in a variety of shapes and sizes. Much smaller *ripples* form on its upwind face, 0.1–5 cm high and wavelengths of 0.02–2 m. Wind drives sand forward by creep, saltation and ballistic impact on the windward side of surface irregularities, and grains slide *en masse* down lee slopes oversteepened above the dry sand friction angle. Fast-moving grains catch up with those moving more slowly to form a more prominent ridge transverse to airflow – the ripple – which then tends towards an equilibrium form. At the other end of the scale, **draa** are *megadunes*, i.e. very large dunes or dune complexes 20–400 m high and at wavelengths of 0.3–3 km, on which individual dunes can

be recognized (Plate 16.2). Dune height is limited by the maximum particle size capable of resisting higher (exposed) crestal velocity. *Fixed dunes* develop in the lee of obstacles where airflow is reduced below fall velocity and, although sand is still lost and gained at their perimeter, the landform is metastable. By contrast, *free dunes* develop independently in open flow and are intrinsically unstable and mobile. Dunes and ripples advance, by particle deflation from the windward slope to leeward slopes, more slowly than the movement of individual particles through them. Advance rates are inversely related to bedform size, with barchans fastest at 5–20 m a^{-1} and entire sand seas slowest. At growth rates of 1–10 Ma^{-1}, the modern form and distribution of active ergs fits well into the Plio-Pleistocene global climatic context.

Dune formation commences where sand accumulates on landing, either beneath a slower part of the airflow or where its *effective* velocity is reduced by friction over the embryonic dune. Further growth superimposes zones of faster and slower flow on the general wind field and also initiates vortices as the air tumbles over lee slopes. A relationship develops between dune morphology, regional (primary) and

Plate 16.2   *Dune field (draa) in the Tengger Shamo desert in the Inner Mongolian Autonomous Region, China. Photo: M.A. Fullen.*

local (secondary) airflow which is responsible for the family of distinctive dune shapes. The direction of primary and effective winds may change seasonally, or over longer periods, and complex forms reflect these multidirectional influences.

Where sand is relatively scarce, wind shapes the sides as well as the crest into a classic crescentic dune or **barchan**, but linear dunes form in larger coalescing sand beds. Asymmetrical extension of one *horn* may draw barchans out into longitudinal or **seif dunes**, which can develop in any case where sand is less abundant, or coalesce transversely in **aklé** form. *Draa* are generally developed from longitudinal dunes. Transverse dunes develop where airflow itself acquires wave motion. Troughs in the wave approach the surface and set sand in motion, forming dunes in their lee and below crests where air diverges from the surface. Airflow and vortices rarely stay constant or symmetrical and a number of systematic irregularities readily develop in either form of linear dune. Vortex convergence between longitudinal dunes draws two parallel ridges together into parabolic junctions. Individual **parabolic dunes** form at **blow-outs** in transverse dunes where the windward slope experiences accelerated erosion and breaches the dune crest or, conversely, receives a diminished onward supply. Non-uniform motion of transverse crests may superimpose **barchanoid** (barchan-like) or parabolic elements. Longitudinal dunes subjected to a diverse range of secondary wind directions acquire star-shaped patterns, known as **rhourds** (Figure 16.5).

Coastal dunes

Coastal dunes depend for their formation and nourishment on deflation of sustained sand supplies in the backshore zone and the predominance of effective, onshore winds. Free-draining sands and vigorous coastal airstreams create physiological drought even in humid climates, extending their distribution to mid–high latitudes and more stormy belts on both counts. The seaward edge of the backshore behind a broad beach is most vulnerable to deflation through shoreward sand movement, regular desiccation and minimal vegetation cover. Thereafter the dune system *is* dependent on the development of a biogeomorphic assemblage. Sand would become dispersed as a thin, amorphous layer across the hinterland in the absence of plants and plant succession.

The dune landsystem is mobile until wholly stabilized by woodland or the cessation of sand supply but various forms and degrees of stability characterize different zones. A **psammosere** or pioneer community

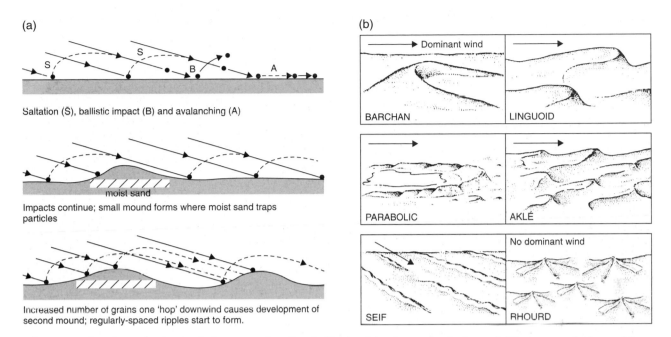

(a)

Saltation (S), ballistic impact (B) and avalanching (A)

moist sand

Impacts continue; small mound forms where moist sand traps particles

Increased number of grains one 'hop' downwind causes development of second mound; regularly-spaced ripples start to form.

(b)

Dominant wind

BARCHAN

LINGUOID

PARABOLIC

AKLÉ

No dominant wind

SEIF

RHOURD

Figure 16.5 *The formation and character of aeolian bedforms: (a) micro-scale ripples and (b) macro-scale dunes. Most of the latter may be superimposed on draa or megadunes. Source: In part after Collinson (1986).*

Medieval storms and coastal dunes

Preoccupation with global environmental change on geological time scales and concern for the short-term environmental future sometimes eclipse the signifi cance of historical environmental change and its human impacts. We have much to learn, for example, from the climatic oscillation in the millennium preceding the industrial revolution and our predecessors' response. Northern hemisphere (perhaps global) warming raised temperature and sea level during three to five centuries of the **Medieval Warm Epoch** – by the same order of magnitude we predict for the next century or two – before falling to below-average levels by a similar margin during the four to five centuries of the **Little Ice Age**. The extent and timing of changes were neither uniform nor synchronous – just as forecast for the twenty-first century. Sea level first rose by about 1 m and then fell by 1–2 m as European temperatures fluctuated by ± 1.5° C. These changes are small by comparison with the Pleistocene–Holocene transition (box, Chapter 17) but still had substantial environmental impacts. The retreat of Arctic sea ice triggered Viking expansion in the north Atlantic region after AD 800 and climatic deterioration led to Europe-wide disease and socio-economic disorder after AD 1300.

Then, as now, the long British coastline was vulnerable to rapid change through the coincidence of eustatic, meteorological and geomorphic shifts. Short-term climatic disturbance is often associated with circulatory and synoptic instability and increased storminess. Marine transgression, however minor, is likely to have pushed sediment shoreward and abandoned it during the subsequent regression. Some of this would have been recovered in storm surges exceeding the extent of transgression and there are many cases of ports and farmland lost to storms, particularly on the European North Sea coast. Elsewhere, onshore gales mobilized large volumes of dry sand in the backshore (see Chapter 17) and formed coastal dunes.

Many sites of rapid dune development and the *ensanding* of settlement arose on the coast of Wales during later medieval times. Approximate age and rates of formation are known from radiocarbon dating, the date of buildings and historical records. Dune slack peats yield ^{14}C dates showing rapid northward growth of a small dune system at Ynys Las in the mid-fifteenth century. The dunes surmount an older shingle ridge which bars the Dyfi estuary at Borth. Llangennith Burrows, a 3 km[2] dune 'wedge' enclosing the northern end of Rhossili Bay in Gower, finally ensanded a Norman church after AD 1200. Smaller dunes also bar Oxwich and Three Cliff Bay in Gower, ensanding Pennard castle, church and vicarage 50 m above sea level by AD 1525 in the latter. Dunes raised in a single storm on 6 December 1330 blocked the harbour and estuary at Aberffraw, on Anglesey, and ensanded valuable arable land.

Aberffraw was the ancestral court of the kings of Gwynedd and Princes of Wales, and the mark of their eventual conqueror, Edward I, helps us to date the impressive dune systems of northern Cardigan Bay.

Two large *morfa* or coastal marshes, each 20–5 km², flank the coast for 20 km between the Mawddach estuary and the Vale of Ffestiniog (see Figure 17.13). Their seaward edge is barred by shingle ridges capped with large dunes. Edward I's military strategy depended on a ring of seventeen castles isolating the granaries of Anglesey from the natural Welsh fortress of Snowdonia and – crucially – stocked and garrisoned by sea. Harlech castle, built after AD 1283 with a water gate and small harbour like the others, was cut off from the sea by AD 1385 through dune formation 1 km seaward of the original rock shore (Plate 1). Dune formation and other coastal impacts of the stormy advent of the Little Ice Age changed the configuration and dynamics of substantial coastal stretches of Britain and Europe. Coastal tourism already threatens dune systems and their protected hinterland through trampling and blow-out along beach access routes, which rising sea level may now exploit.

Plate 1 *The coastal dune system at Morfa Harlech, North Wales, which now isolates Harlech Castle (1 km inland, out of view). Seaward fore-dunes are separated from landward hind-dunes by dune slacks with a species-rich plant community. Blow-outs (bare sand) are closely associated with tourist access routes to the beach. Photo: Kenneth Addison.*

arrests sand movement along a line of embryonic fore-dunes. It is tolerant of physiological drought, nutrient-poor and saline-high embryonic soil and wind shear. *Ammophila arenaria* (marram grass) and *Agropyron sp.* (couch grass) are common in northern hemisphere mid-latitude fore-dunes and have the additional ability to grow through and anchor frequent sand burial. Progressive enlargement and colonization establish a fore- or *yellow-dune* barrier which modifies the airstream and allows a more diverse succession to develop in its lee. Shell fragments deflated from the backshore add calcareous nutrients. The dune is not yet stable and parabolic **blow-outs** occur which advance the dune inland by 1–20 m a⁻¹. In the progressively sheltered conditions of the dune *slack* and *meadow*, soil develops and succession advances towards shrubs and woodland. *Hind-* or *grey-dunes* are more stable although still vulnerable to climatic change and human disturbance (Figure 16.6). Coastal dunes may individually reach over 100 m high and dune landsystems extend for $10^{1–3}$ km inland.

Conclusion

Wind is more dependent on the prior operation of other geomorphic processes than any other agent and is not dominant in Earth's storm belts. Aridity enhances wind power, and the absence of vegetation is a decisive factor in determining the distribution of

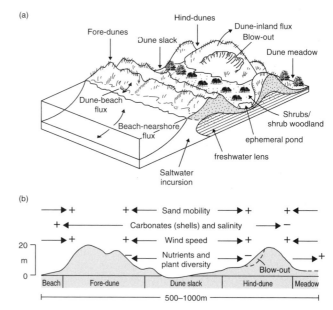

Figure 16.6 *The form and some environmental gradients of a coastal sand dune system, showing the direction in which parameters increase (+) or decrease (−). Source: In part after Carter (1993).*

aeolian landsystems. Wind lacks extensive rock-quarrying power, and its erosive impact is confined largely to the scouring and ornamentation of bedrock surfaces. Dune bedforms reflect the interactive effects of airflow and landforms on each other. Accelerated deflation is one agent of desertification, removing topsoil and burying vegetation incapable of matching sand aggradation rates. However, plant communities tolerant of the harsh aeolian environment form distinctive biogeomorphic dune systems in suitable locations. Aeolian deposits are also common in the geological record. Continental siliciclastic sediments of Devonian (Welsh borderland, Scottish midland valley) and Permo-Triassic (English Midland basin) age show clear dune bedforms and mark Britain's northward passage across the subtropical divergence belts 400–200 Ma ago. They are important sources of long-term palaeoclimate and palaeocurrent information, and modern ergs and loess sheets retain clear evidence of Quaternary climatic change.

Key points

1 Wind is dependent on the prior operation of other geomorphic processes which provide fine-grained products. Wind deflates, i.e. entrains, removes and eventually deposits these materials as metastable aeolian landforms. Armed with the abrasive tools, airflow also scours bedrock surfaces in its path.

2 Aeolian landforms are found primarily in areas of atmospheric subsidence and associated hot and cold arid zones, rather than storm belts. Lack of protective vegetation increases *effective* wind velocity in areas of lower *absolute* velocity and exposes earth materials to desiccation and deflation.

3 Tectonic processes exert secondary control through uplift, basin formation and the creation of rain shadow. Upwelling cold ocean currents suppress rainfall and enhance aridity on adjacent coasts.

4 Residual abraded rock surfaces, supporting lag gravels or scoured into yardangs, and ergs or sand seas with dune and ripple bedforms at all scales, form the aeolian landsystems of hot deserts. Loess sheets derived from glacial and frost desert processes represent cold desert deflation products, especially beyond the margins of Pleistocene ice sheets.

5 Aeolian processes are not restricted to arid climate zones. Physiological drought and the availability of dry, fine-grained materials in coastal, mountain, urban and arable farming environments all promote more localized aeolian activity.

Further reading

Cooke, R.U., Warren, A., and Goudie, A.S. (1993) *Desert Geomorphology*, London: UCL Press.
A well illustrated book which spans the full range of desert geomorphology. Nevertheless, aeolian processes and landforms warrant almost half the text and are complemented by coverage of the fluvial domain and general desert surfaces.

Lancaster, N. (1995) *Geomorphology of Desert Dunes*, London and New York: Routledge.
Sand dunes are the most widespread and evocative form of aeolian deposit and this book focuses on dune processes, landforms and environments. The text is not unduly technical in style and is well illustrated.

Livingstone, I., and Warren, A. (1996) *Aeolian Geomorphology*, Harlow: Addison Wesley Longman.
An excellent and timely book which takes aeolian processes beyond the desert and dune environments. The latter receive due attention but are complemented by chapters/sections on coastal dunes, dust, palaeoenvironments and applied studies. Good line drawings and plates illustrate a readable text.

The work of the sea is focused at the coastline, which is probably the most extensive, active and diverse of Earth's geomorphological features. It stretches for 0.5 M km around the margins of every continent and island – ten times farther than intra-plate boundaries – and for these reasons is familiar to most people. Indeed, 50 per cent of the population of the industrial world and perhaps 60 per cent of global population live within 50 km of the sea. This narrow coastal zone occupies less than 0.05 per cent of Earth's land area but has powerful attractions for agriculture, industry, residence and recreation. Consequently, we want it to stay where it is! Most of us were unwitting geomorphologists in our youth as we built sandcastles doomed by the incoming tide. We may also recall Cnut (Canute), Anglo-Danish King of England in the early eleventh century, and his legendary demonstration that regal power cannot withstand the relentless motion of tides and waves. The coastline is sensitive to rapid geological and biophysical change and we are braced to respond to sea-level rise promoted by global warming.

The *coastal zone* is the region of mutual interaction between terrestrial and marine environments and embraces several components. The **coastline** is the outermost limit of permanent land, which separates the broader coastal hinterland from shore and marine environments. The **backshore** occupies land above modern, normal high tides but is storm-swept. Moving seawards, the **foreshore** (shore) lies between high and low tide limits. Beyond it is the **inshore** zone of breaking or **shoaling waves**, flanked by an **offshore** zone of deeper water occupying inner margins of the continental shelf (Figure 17.1). Foreshore and inshore zones together comprise the **nearshore** wave environment. The coastal zone is a hybrid of terrestrial and oceanic systems at their common boundary, driven by a series of exogenic and morphotectonic processes integrated in some respects and disconnected in others. Wave and tidal (exogenic) energy is at the heart of coastal processes but we also know that land : ocean area and sea levels are intimately connected through tectonic processes and climatic change. They disturb coastal

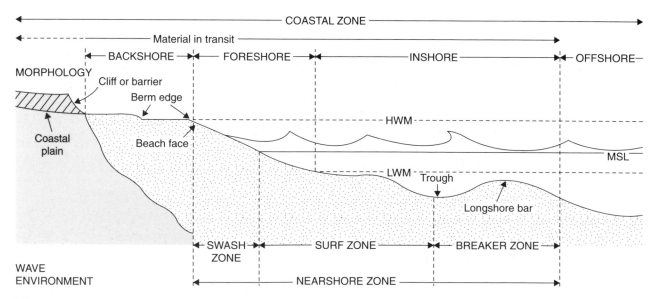

Figure 17.1 *The coastal zone and its component morphology, tide and wave environments.*

equilibrium through uplift and isostatic adjustment and trigger geomorphic responses, further altering the coastline and its supply of terrigenous sediment. Eustatic adjustments have climatic and tectonic origins. Ocean-ice sheet coupling drives Quaternary **glacio-eustatic** responses and is primarily climate-driven but climate itself responds to tectonic as well as radiative forcing.

Wave, current and tidal action

The coast is sculptured primarily by wave action which erodes the landsurface in one place and creates another elsewhere. This dynamic equilibrium between tide–wave energy and earth materials works to maintain an 'average coastline' determined by the average ratio of land to ocean area and average sea level. Coastal geomorphology, thus perceived, classifies the work of the sea into erosional and depositional processes and landforms and allows us to concentrate first on the wave environment.

Wave form and action

The origin and general behaviour of waves in transmitting energy was outlined in Chapter 4. Important relationships exist between *wavelength* (L), *wave base* (L/2) and water depth. Energy is converted to work on meeting the coast, and its geomorphic impact depends on a combination of wave form and coastal material properties. Wave form is the outcome of offshore features of approaching waves, modified by water depth in the inshore zone, coastline geometry and wind-induced wave direction. **Reflected waves** rebound from cliffs terminating in deep water and meet incoming waves to form a **standing wave** which does not break. In all other cases, waves entering water depth less than *L/2* are transformed into **breaking waves** as bed friction destabilizes the orbital path of water particles (Figure 17.2). The retarded wave increases steadily in height to conserve energy and thereby raises the potential energy of the wave, which is released as it breaks. The *starting* height of the wave offshore is determined by the wind environment but its *breaking* height increases with the rate of **shoaling** (shallowing or shelving) of the nearshore zone. The higher the wave, the greater the energy it delivers to the shore.

Waves break when the critical ratio of water depth to wave height lies between 0.6 and 1.2, around a mean value of 0.78. In other words, average waves break in water depths a little less than their own

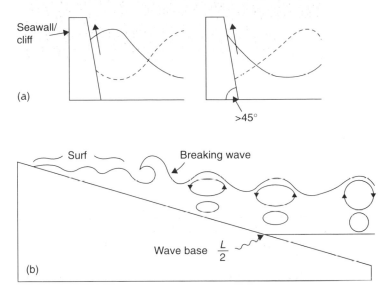

Figure 17.2 *Reflected and standing waves (a) and the effect of shoaling on the orbital path of water particles, leading to breaking waves (b).*

height and so low waves run farther into shallower water than high waves before breaking. The shoaling angle is also important. Waves break close inshore where it is steep and farther out on flat shores, measured by the **breaker coefficient**, B.

$$B = H/LS^2$$

where *H* = wave height, *L* = wavelength and *S* = bed slope. The breaking style influences the way in which wave kinetic energy is used, and four styles are recognized (Figure 17.3). The breaker coefficient falls from *spilling* to *surging* styles as bed slope angles increase.

Most waves do not approach the coastline orthogonally with wave crests parallel to the shore but are driven obliquely onshore or meet an indented coastline. Waves are retarded around headlands but drive on less impeded into bays. Such **refracted waves** alter the pattern of energy flux at the coast, with energy *convergence* around headlands and *divergence* in bays (Plate 17.1).

Wave-generated currents

Breaking waves send pulses of water shorewards until they run out of momentum, whereupon gravity draws water back into the sea. This **surf zone**, between the breaking waves and the point of maximum run-up, is one of turbulent water exchange and maximum geomorphic activity associated with wave-generated currents. Spilling waves dissipate

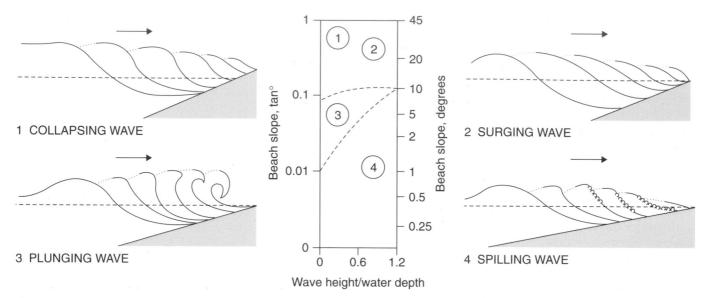

1 COLLAPSING WAVE

3 PLUNGING WAVE

2 SURGING WAVE

4 SPILLING WAVE

Wave height/water depth

Figure 17.3 *Principal types of breaking wave and their associated beach slope and water depth. Source: In part after Carter (1988).*

Plate 17.1 *Wave refraction, surging breakers and a modern rock platform around Old Nab headland, Yorkshire coast. Waves approach the headland almost head-on but are refracted in the bay.*
Photo: Kenneth Addison.

their energy in the surf zone, whereas surging waves are reflected back into following waves. Forward and return pulses – **swash** and **backwash** – move at right-angles to the shore in orthogonal waves, generating shore-normal (right-angle) currents. Some of the swash percolates beach sand and gravel, reducing the immediate backwash volume but leading to later seepage on the receding tide. The swash from refracted waves moves diagonally onshore but back-wash returns normally down the maximum slope, resulting in a net **longshore current**. Swash and backwash interfere with each other to some extent and hold water up ahead of the breaking line. Scarcely visible **edge waves** are established at right angles to breaking waves through their incessant mass shoreward transfer of water. Water level is increased or **set up** when edge wave and incoming wave crests coincide and lowered or **set down** when they are out of phase. **Beach cusps** form where set-up and set-down cause incoming waves to break farther (deeper) and nearer (shallower) the shore respectively, creating a sinuous breaking line. All three processes establish lateral currents in the surf zone which eventually drain seaward as powerful **rip currents**, completing a cellular pattern of water movement (Figure 17.4). Lateral and longshore currents are responsible for **longshore drift** of sediment. Rip currents cut and accelerate through rip channels in soft sediments before dissipating at nearshore rip heads.

Tidal action and currents

Although breaking wave systems appear to dominate coastal geomorphology, tidal waves are also important in a number of respects. Geomorphic activity is concentrated in the surf zone, whose maximum vertical and lateral extent is a product of tidal range as well as of wave height (Figure 17.3). They work together in storm surges which, although relatively infrequent, may have major geomorphic impacts (Figure 17.5). The annual average extent of surges over mean sea level is about +0.6 m in Britain and up to +1.5 m on tropical cyclone coasts. Hurricane Hugo created a storm surge of +11 m on the South Carolina coast of the United States in 1989. Distinctions between micro-, meso- and macro-tidal ranges (see below and Chapter 4) demonstrate the persistent impact of tidal waves in varying the amplitude of the surf zone. Tidal processes dominate coasts, with macro-tidal ranges over 4 m and breaking waves truly dominate only the micro-tidal environment.

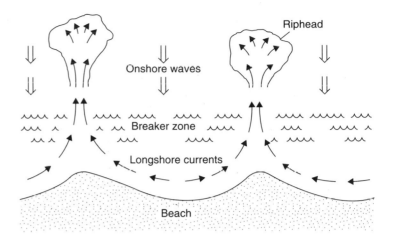

Figure 17.4 *The development of longshore and rip currents.*

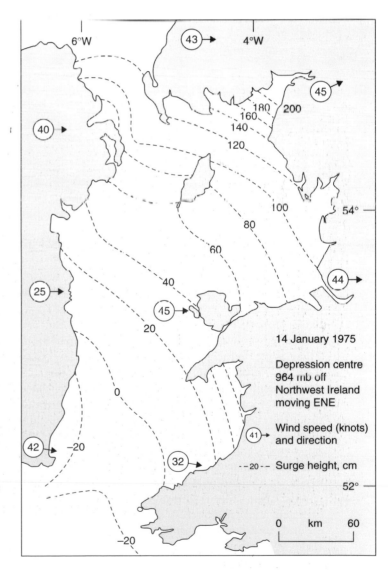

Figure 17.5 *A storm surge in the Irish Sea raised by an intense depression in January 1975. Source: After Carter (1988).*

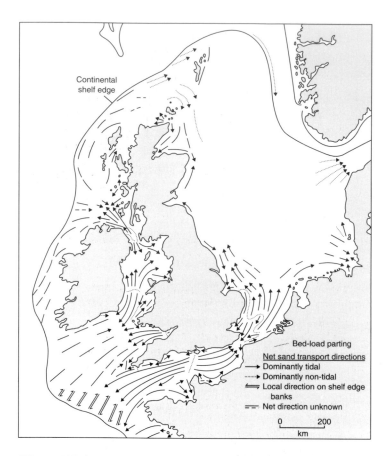

Figure 17.6 *Coastal sediment movement around the British Isles, showing directions of net transfer. Source: Stride (1982).*

Tidal ebb and flow stimulate **tidal currents** with substantial fluxes of water, energy and sediment around the coastline (Figure 17.6). Water density differences based on salinity, and the extent of water body mixing or separation, vary the patterns of circulating currents. Current velocities vary with the size of tidal passes and the frequency of tidal inundation. Tidal wave velocity in open waters is less than 0.05 m s^{-1} but can reach 0.3–3.0 m s^{-1} through passes. Semi-diurnal tides move approximately twice as much water in one day as diurnal tides at about twice the diurnal velocity. The duration of intertidal exposure varies by the same token and, with it, the opportunity for and nature of drying out, weathering and biological activity above the low-water mark. The extent of the **intertidal zone** depends on coastline configuration and tidal range. The foreshore area exposed and re-covered during the tidal cycle expands and contracts during spring and neap tidal cycles respectively.

Coastal geomorphic processes

Coastal energy

General rules governing energy–material interactions, described in Chapters 12 and 13, and flowing water specifically in Chapter 14, underpin coastal processes. We need to understand their particular application in the coastal environment. Waves and tides and their secondary, circulating currents are the principal energy source, although wind energy plays a direct role in the backshore and hinterland. Energy may be *reflected*, without immediate geomorphic consequence, or dissipated in the intertidal zone through turbulence, bed friction and the movement of rock and sediment. Wave energy, normally measured in joules, varies with the square of wave height (see Chapter 4) and can reach levels up to 20 kJ m^{-2} s^{-1}. It is delivered either through hydraulic pressure, capable of compressive stresses reaching 10–100 MN m^{-2}, or water jets and sheets applying bed shear stress. The dynamics of breaking waves determine how energy is likely to be transformed and dissipated.

Prior to breaking, the increase in H/L transfers potential to kinetic energy, with some lost as friction against the bed, at a rate determined by velocity and bed roughness. Dissipation on and after breaking varies according to breaking style. Turbulence commences at the crest of spilling waves and, over several wavelengths, 'spills' down the wave front, leaving little energy for geomorphic work. Collapsing and surging waves may be 'failed' plunging waves but their turbulence pushes a water sheet onshore. Plunging waves are the most dramatic and set up a vortex led by a water jet below its breaking tip. If this is powerful enough, it penetrates the trough ahead of the wave and scours the bed, throwing up a cloud of sediment and trapped air bubbles clearly visible behind the crest (Figure 17.7). Water continues to move onshore as small bores in the surf zone and sheet flow at 1–10 m s^{-2} in swash run-up. The energy available to move sediment depends on the velocity, depth, turbulence (through swash/backwash impedance) and extent of percolation in this zone. Run-up endows backwash and percolated water with potential energy capable of further sediment movement as they evacuate the swash zone. All processes can be observed during *safe* foreshore paddling in appropriate conditions!

Coastal erosion

Erosion occurs through hydraulic action, the mobilization of sediments and their attrition and corrasion, and is most effective under storm wave conditions. Breaking waves apply hydraulic shock or a hammer effect, by trapping water or compressed air ahead of the wave and negative pressure as it retreats. Compressive stress is maximized in plunging waves and when the wave front is vertical, trapping air between crest and trough. The effect of repeated cycles of hydraulic shock and negative pressure depends on the structure and lithology of earth materials. High compressive stress is dispersed along fractures in hard, fractured rock and may generate secondary tangential – i.e. shearing – forces. Together they critically reduce or exceed shear strength and trigger rock-mass failure, quarrying individual blocks or undercutting and destabilizing entire rock walls (see Chapter 13). Much depends on the disposition of fractures relative to the eroding waves. Many coastlines show exemplary structural control on cliff form and the geometry of bays and headlands (Plate 17.2). Similar hydraulic forces accelerate erosion when waves attack coastlines in unconsolidated materials. Soft glacial sediments form many mid-

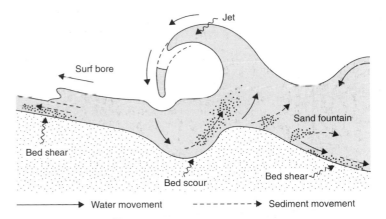

Figure 17.7 *Plunging wave form and associated water and sediment transfers. Source: After Carter (1988).*

high-latitude coastlines, and changing dynamics may convert once constructive or stable coasts into easy erosion targets.

Corrasion and attrition result from the abrasive action of suspended and bed loads in wave and current environments. Corrasion of bedrock or other

Plate 17.2 *Structural control determines cliff angles and alignment in Three Cliff Bay, cut in steeply dipping Carboniferous limestone, Gower, South Wales.*
Photo: Kenneth Addison.

Global warming and rising sea level

Imagine for a moment north-west Europe, lying at one of Earth's most sensitive atmosphere–ocean–ice sheet interactive points, without amelioration by the Gulf Stream. At this latitude our environment would resemble Labrador, with ice-bound seas for several months each year, a cold, arid atmosphere, a more seasonal river régime and a sub-arctic ecosystem. Britons would be *true* Europeans, able to walk to mainland Europe (with lower sea levels), which itself could not have been the birthplace of great civilizations. This was the scene for most of the past 115 ka, since the last or **Eemian** temperate (interglacial) stage. Sea level was over 100 m lower than today and the North Sea area was an extension of the north European Plain. Permafrost and tundra conditions prevailed, with intermittent episodes of Early Devensian mountain glaciation. A late Devensian ice sheet, covering most of Britain and Scandinavia, coincided with worldwide glaciation to drive sea levels down to −130 m at 18 ka BP.

Thereafter, global warming and ice melt led to the Flandrian temperate stage. Global sea levels were restored to 0–3 m above their present level by the Mid-Holocene **hypsithermal** or climatic optimum, *c.* 5 ka BP. Britain's continental shelf shrank progressively and the coastline became far more indented. Orkney and Shetland became islands at *c.* 13 ka BP and land bridges with Ireland and the outer Hebrides were drowned by 12 ka BP. The Loch Lomond Stadial ice re-advance checked further insularization until after 10 ka BP, when the Inner Hebrides, Anglesey and the Isle of Wight were isolated. The low coastal plain connecting the Thames–Rhine estuary as far north as Yorkshire and Sussex–Flanders (northern France) was finally breached by the Flandrian transgression *c.* 8.5 ka BP, which completed the isolation of the British Isles (Figure 1).

Subsequent minor fluctuations may seem insignificant compared with the overall rise of some 130 m. However, the last areas to flood – including the Wash, the inner Severn estuary and the Solway Firth and Morecambe Bay fringe of Lancashire – became the first areas reclaimed naturally during the minor regression (−1–4 m) which accompanied cooling after 5 ka BP. Extensive peat formation in enclosed muddy estuaries, especially the Somerset and Gwent levels (Severn estuary) and Fens (East Anglia), records their subsequent environmental and human history. Similar minor climatic oscillations, such as the *Medieval Warm Epoch* and *Little Ice Age* between *c.* AD 800–1300 and AD 1350–1850 respectively, caused major socio-economic changes in Europe and altered sea level by ±1–2 m. This was enough to create substantial consequences for coastline and hinterland management for our medieval ancestors and provides a warning for the twenty-first century.

The Intergovernmental Panel on Climatic Change (IPCC) confidently predicts a global sea-level rise of 0.13–0.94 m by 2100, made up of three components. Thermal expansion has already begun and will be augmented in due course by accelerated melting of alpine glaciers and then major ice sheets (Figure 2). The behaviour of the West Antarctic Ice Sheet is the most unpredictable component. Sea-level rise beyond an unknown threshold could decouple such a large portion from its bedrock base as to destabilize the whole, leading to catastrophic collapse and a further sea-level rise of up to 7 m (see Chapter 15), threatening huge areas world-wide. Coastal zone management has become a priority. In Britain alone, areas reclaimed last by later Holocene regression would be the first to flood. Large areas of urban settlement and prime agricultural land would be threatened (Plate 1) and necessitate the immediate expenditure of £10 billion to £20 billion to reorganize coastal defences and sewage outflow. This is forcing radical rethinking of coastal zone management. Holistic **coastal cell** studies, embracing all interested agencies, are becoming standard. **Managed retreat** allows natural sand or salt marsh barriers to develop and is a viable alternative to hard defences in suitable locations.

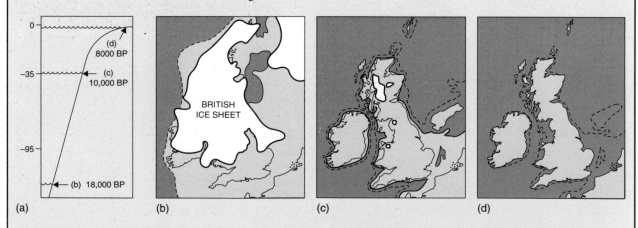

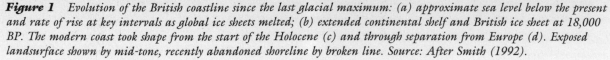

Figure 1 *Evolution of the British coastline since the last glacial maximum: (a) approximate sea level below the present and rate of rise at key intervals as global ice sheets melted; (b) extended continental shelf and British ice sheet at 18,000 BP. The modern coast took shape from the start of the Holocene (c) and through separation from Europe (d). Exposed landsurface shown by mid-tone, recently abandoned shoreline by broken line. Source: After Smith (1992).*

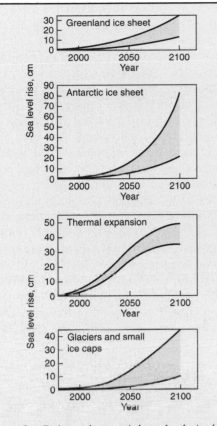

Figure 2 *Estimated potential sea-level rise by AD 2100, attributable to principal sources and showing their 'envelope of uncertainty'. Source: Ince (1990).*

Plate 1 *The expensive infrastructure of Llandarcy oil refinery, requiring hard defences against global sea-level rise, beyond the low dunes of Crymlyn Burrows, Swansea Bay, South Wales.*
Photo: Kenneth Addison.

fixed surfaces is probably restricted to the nearshore and to the backshore within reach of wave spray. Attrition occurs as in any other environment with loose particles in regular, moving contact with each other; beach shingle acquires among the highest mean clast roundness.

Coastal weathering

Sub-aerial weathering processes assist coastal erosion by their progressive reduction of material strength, and particular emphasis is placed on wetting and drying cycles around the intertidal zone and on the presence of salts. The intertidal zone and adjacent areas within reach of sea spray are an area of extremes. Inundation, anaerobic conditions and wetting at high tide are replaced by uncovered surfaces exposed to insolation, airflow and drying at low tide. This leads to **water-layer weathering** by slaking, hydration and salt weathering (see Chapter 13). Rocks must be permeable to water and spray for this to occur, and optimum conditions are found in permeable substrates on tropical coasts with diurnal tides. Carbonate solution is surprisingly active, especially in the tropics, where sea water is often saturated with dissolved carbonates. Biochemical weathering assists, however, by CO_2 respiration in rock pools and through marine boring animals and leachates from algae and marine plants.

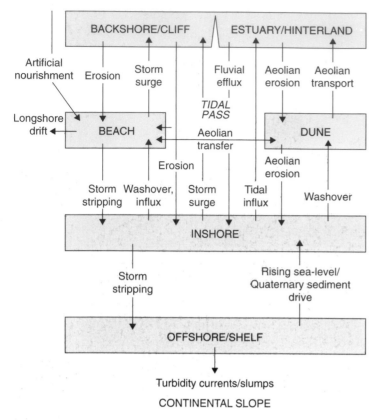

Figure 17.8 *A coastal sediment budget.*

Sediment transport and deposition

Sediment transfers and deposition are major processes in continuously moving coastal waters and account for the greater proportion of landforms. Their principal sources are terrestrial sediments carried seaward by rivers and glaciers, products of coastal erosion, offshore sediments carried landward and *in situ* accumulation of biogenic debris. The continental slope is the ultimate sink for transient coastal sediments. Sediment budgets calculated with increasing attention to coastline management allow us to chart the quantities and transfer routes involved with varying degrees of accuracy (Figure 17.8). Coastal erosion contributes a surprisingly small proportion – about 1 per cent of terrigenous yield – and biogenic sediments make substantial local contributions. Shell and other calcareous debris can be important on wave-dominated coasts, with plant debris more abundant on tide-dominated coasts. Offshore sources are difficult to assess, and temperate high latitudes depend heavily on finite Pleistocene glacigenic supplies. Although the −20 m submarine contour is the maximum limit within which wave motion can move sand shoreward, the offshore source was well stocked

as Late **Devensian** and **Flandrian** sea levels, rising some 130 m between 15 ka and 5 ka BP, drove sediment landward. The mid-Holocene end to the transgression limits future beach nourishment from this source, even with modest further sea-level rise.

All sediment movement is subject to general rules set out in Chapter 14. The breaking-wave zone forms a seaward barrier to coarse sediment transfers except during storm events, when large volumes of sand may extend sand plains seaward. Waves with weak backwash and low swash impedance from previous waves, commonly at 8–10 s intervals, are responsible for net beach construction. By contrast, plunging waves with short (4–5 s) intervals often have a strong backwash and transfer sediment seaward. Longshore drift rates increase with wave power and angle of wave approach and are often the principal transfer process and the only net transport mechanism on equilibrium coastlines. The nature and orientation of bedforms exposed on the foreshore at low tide ('ripples') illustrate tidal sediment mobility.

Coastal landsystems

Deltas

Deltas develop by **progradation** or seaward extension of river flood plains. Most large deltas are located on trailing edge or other passive margin coasts, fed by trunk rivers draining substantial continental areas – often sedimentary or cratonic basins. Many have foundations 10^{5-7} a old but their modern configuration is essentially Mid–Late Holocene in age, post-dating the Flandrian transgression. Low Quaternary sea levels drew flood plains across the continental shelf, delivering sediment closer to shelf-slope margins and its ultimate, oceanic sink. Transgression trimmed flood plains back towards the modern coast, where tides, waves and the requisite high fluvial sediment flux all help to shape the delta landsystem (Figure 17.9).

The delta front advances by sediment deposition as bed- and suspended-load particles enter lower-velocity water at the river mouth. Sediments enter denser (saline) water via a plume from which they are rained out, or less dense water (where high suspended loads raise freshwater density) via sea-bed density currents. Where river and sea-water densities are very similar, a **Gilbert-type** delta forms by successive overlap of sediment packets, providing a classic bottom-set – foreset – topset sequence (Figure 17.10). Wave interaction often forms transverse bars beyond the delta front. The main channel meanders, between natural levées, through older channel and over-bank sediments. A delta plain, reminiscent of the flood plain, develops to either side. Levées are breached periodically, often in river flood or storm conditions, and the resultant crevasse splays and distributary channels form new delta lobes (see Chapter 14).

Deltas extend the coastline most obviously where tide and wave energy are low in protected shelf seas and do little to reshape the delta. The modern Mississippi, entering the hurricane-prone but otherwise sheltered Gulf of Mexico, forms a model elongate, fluvially dominated delta. Increasing wave action arrests the delta front nearer the regional coastline. Transverse bars become more prominent, developing onshore to form an interrupted barrier coastline (see below) and damming tidal lagoons to their rear. Tide-dominated deltas are subject to tidal inundation and low-tide drainage through the distributary channel network and surface saltwater flooding of the delta plain. The higher the tidal range, the more the landward water and sediment fluxes of incoming tides constrain seaward development of the delta.

The delta landsystem is a three-dimensional mosaic of individual channel, plain, lagoon, salt-pan and barrier landforms and sediment calibres. Channel meandering, storm events and fluctuating sea levels create ever-changing surface patterns. Delta plain surfaces are also prone to subsidence through the compaction, dewatering and isostatic depression of sediment under its own weight. Predominantly low-energy, nutrient-fed environments of the delta plain encourage vigorous floral and faunal ecosystems and resource potential. Fossil deltas often contain hydrocarbon deposits, and high bioproductivity today, coupled with regular and sometimes disastrous flooding, leads to high-density but vulnerable human population living at subsistence level. The largest delta (that of the Amazon) covers 0.5 M km²; the Ganges–Brahmaputra (Bangladesh), Mekong (Vietnam) and Yangtze (China) each exceed 50 k km².

Estuaries and lagoons

Estuaries and lagoons are embayments partially enclosing saline and freshwater bodies dominated by fluvial or tidal processes. Estuaries are freshwater-fed, submerged lower reaches of structural, river- or glacier-eroded valleys. They are usually orthogonal to the coastline. Lagoons impounded by barriers are coast-parallel, the direct product of coastal processes and also river- and/or rain-fed. Both are classified by their water chemistry and exchange processes.

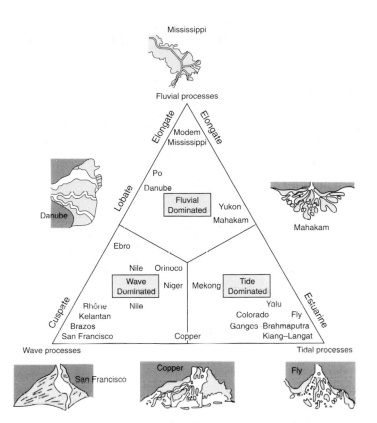

Figure 17.9 *Classification of river deltas according to the importance of fluvial, wave and tidal processes. Source: After Galloway (1975).*

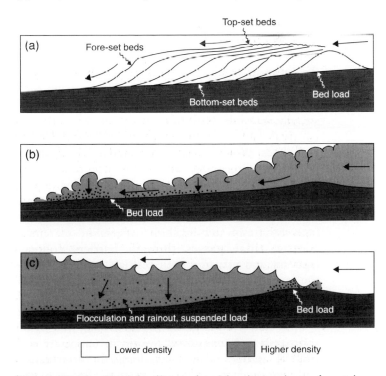

Figure 17.10 *Fluvial sediment deposition in marine or lacustrine basins. Gilbert-type delta formation (a) by rapid deposition of bed load contrasts with sediment plumes entering less-dense (b) or denser water (c). Source: In part after Elliot (1986).*

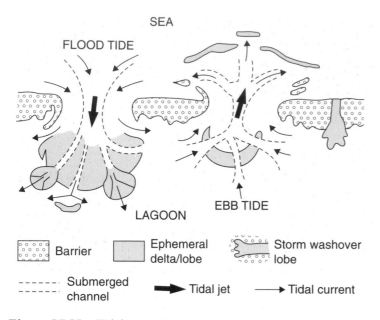

Figure 17.11 *Tidal passes. Separate ebb and flood passes here develop their own ephemeral deltas and lobes but the same pass may be used on both ebb and flood tides. Source: In part after Carter (1988).*

Stratification, or vertical separation, occurs in estuaries lacking significant tidal or current mixing. Freshwater and denser salt water override and undercut each other respectively in river-dominated estuaries where the inland extent and slope of the saltwater wedge is determined by the vigour of river discharge. Their shifting boundary marks a concentrated zone of **flocculation** of clay-silt particles into larger aggregates capable of settling as mud. Tidal mixing produces vertical homogeneity, with salinity increasing steadily seawards. Open lagoons connected to the sea by tidal passes normally maintain balanced tidal exchanges, except in arid-zone sabkhas, where strong evaporation precipitates salts and reduces return flow. Closed lagoons are rain-fed, and, whilst storm washover, lateral seepage through permeable barriers and evaporation may restore some salinity, it varies between freshwater–brackish–hypersaline extremes. Narrow **tidal passes** through barriers connect lagoons, and estuaries with restricted mouths, with the sea. They create their own distinctive water and sediment fluxes which form microcosms of full-scale coastal landsystems (Figure 17.11).

Protected from large waves, **tidal flats**, fringed by **saltmarsh** and **mangrove swamp** in temperate and tropical latitudes respectively, dominate the estuarine and lagoonal intertidal landsystem. Beaches and dunes are rare. Despite complex two-way movements of water and sediment bodies, migrating with river and tidal pulses, patterns of particle size, bedforms

and related vegetation succession emerge. Sand moves as bed load and is deposited most commonly in non-turbid outer and lower parts of the system where currents are at their strongest. Tidal sandflats exhibiting large-scale **megaripple** and **sandwave** bedforms attest to their scour (Plate 17.3). Mud is moved and deposited from turbid suspension in low-energy inner and upper estuarine environments. A **halophyte** (salt-tolerant) vegetation succession responds to, and in turn assists, this zonation. Pioneer algal mats help to arrest and stabilize mud, permitting grasses, sedges and rushes to colonize the upper intertidal zone with its progressively shorter and fewer periods of tidal inundation (Plate 17.4). *Puccinellia maritima* (salt-marsh grass), *Spartina* spp. (cord grass) and *Juncetum* spp. (salt-marsh rush) are the principal members of salt-marsh communities. The halosere may merge inland with a freshwater hydrosere and becomes attractive to reclamation by human intervention. The Ijsselmeer in Holland is the most celebrated European example. Lagoons are floored by muds except where tidal action is strong near passes or they receive terrigenous sediment and sand, blown or washed over the barrier during storms. Carbonate muds form where biological activity is intense and lagoons may house **stromatolites** (algal mats) and **bioherms** or prominent shell beds.

Barriers and barrier islands

Barriers, as their general name suggests, bar advancing waves and buffer or protect the coastline from wave energy. They account for some 15 per cent of the global coastline and assume one or more of four principal forms. Breaking waves and surf create sand, and to a lesser extent shingle, **beaches** in nearshore and foreshore zones described earlier. Parallel **bars** develop in the nearshore and offshore zone and **dunes** develop inland of the backshore with good sand supply and dominant onshore winds. Barriers form at the coastline or offshore and barrier islands develop either by the creation of tidal passes through barriers or by accretion in the sheltered back-barrier zone behind an offshore bar. Offshore systems may migrate onshore, initially trapping tidal lagoons behind **barrier islands** or **spits** open at one end, and eventually become welded to the coastline. Biogenic reefs complete the suite.

Beach morphology is subject to constant short-term change in response to waves, tides, winds and sediment fluxes and may be a composite of several stages at any one time (Figure 17.12). The turbulent nearshore environment segregates sand from gravel which sustain gentler (2°–8°) and steeper slopes (10°–20°) respectively. These profiles and the different permeability of sand and gravel also interact with wave style to create local variations in wave dissipation and morphology. On a larger scale, the swash from spilling breakers constructs a steep-faced **berm** near the high-water mark which survives all but the most destructive storm waves. Winter beach erosion alternates with summer reconstruction on many mid-latitude beaches in response to seasonal storminess, although winter storms may also move large sediment volumes inshore beyond the reach of lesser waves (see box, Chapter 16).

Coast-parallel nearshore bars form on dissipative beaches and offshore at zones of either lower water velocity or higher sediment concentration. They normally form as single or multiple low ridges parallel to the coast, or in crescentic form as linked beach-cusp bars, broken by backwash or rip channels. Bars may be stable in the energy environment at which they form, absorbing 80–100 per cent of the wave energy, but unstable in any other when they wash out or migrate. The bars accentuate as well as respond to longshore currents and are capable of extending across coastal embayments and estuaries as spits in the direction of longshore drift. They also connect islands to mainland with **tombolos** as in the case of Chesil Beach and Portland 'island' on the Dorset coast of southern England.

Dune systems develop by deflation of dry sand from the backshore landwards and, as with estuarine landforms, develop in association with vegetation succession as described in the box in Chapter 16. Biogenic processes assume even greater importance on some tropical coasts. In addition to the role of biogenic debris and bioherms in lagoons, described earlier, reefs form more permanent wave-resistant and biomorphological structures 10^{1-2} m thick, 10^1 km wide and 10^{1-3} km long. Living reef corals, etc., grow on the cemented debris of dead organisms and in that way can contend with slow rates of sea-level change. Despite this, they are sensitive to other conditions and have a predominantly tropical distribution today away from the turbidity of terrigenous sediment fluxes. Two main forms exist: *fringing* reefs weld themselves to the shore, whilst *barrier* reefs parallel the shore beyond an impounded lagoon. The Great Barrier Reef off Queensland (Australia) is the modern equivalent of the Palaeozoic Capitan Reef (Texas/New Mexico) and Wenlock Edge Reef along the Anglo-Welsh border. The latter was formed when 'Britain' was located at 30° S during the Silurian period 420 Ma ago.

Plate 17.3 *Large-scale sand bedforms at low tide near the constricted mouth of the Mawddach estuary, mid-Wales, with a seagull for scale.*
Photo: Kenneth Addison.

Plate 17.4 *Salt-marsh zonation, Mawddach estuary, mid-Wales. The lower marsh commences behind sand banks (top left) and is traversed by creeks which flood and drain the marsh during the tidal cycle. The upper reed marsh (right) is inundated less often and grades inland into a freshwater marsh.*
Photo: Kenneth Addison.

Barrier coasts have developed in mid-Wales, East Anglia and Yorkshire, with substantial longshore drift and associated with low-tidal range coasts (Figure 17.13). Chesil Beach (Dorset) and Dungeness are formed in part by onshore migration of offshore shingle bars. The southern North Sea and Baltic coasts of mainland Europe have long offshore barriers impounding large lagoons. The most impressive

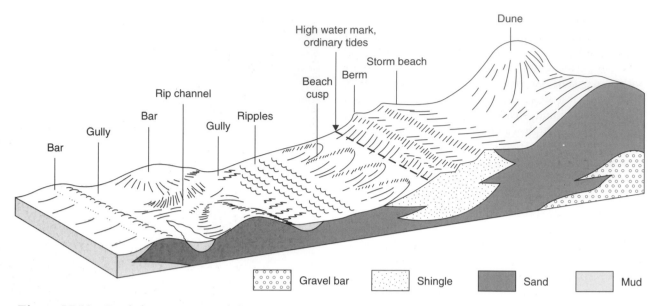

Figure 17.12   Beach-dune system morphology. Source: In part after Goudie (1984).

barrier coastline is found on the North American passive margin. It extends with few interruptions for over 7000 km from the Yucatan peninsula, around the northern Gulf of Mexico and along the Atlantic coast from Florida to Massachusetts.

Rock coasts

The review of coastal landsystems is completed by the least diversified but most extensive, accounting for 75 per cent of all coastlines. Rock coasts, unprotected by barriers, face direct attack by the sea as it mounts the final assault by sub-aerial processes on continental landsurfaces. Wave erosion causes the coastline to retreat at rates determined by effective wave energy and geological resistance. Wave conditions, tidal influence and tectonic settings were described earlier. The erosion front is normally marked by a cliff and a trail of residual landsystem components marking its retreat (Figure 17.14). Assuming constant sea level, marine erosion is concentrated towards the cliff base and may be represented initially by a **wave-cut notch**. This will eventually destabilize overhanging rock to the point of failure and the process continues once wave action has removed the debris. Wave action may account directly for up to 25 per cent of cliff erosion, with mass wasting – triggered indirectly by wave action – and other erosion accounting for the remainder. Net evacuation of eroded debris and any longshore sediment influx is an important requirement for continuing erosion, summarized by reference to *effective* wave energy.

Cliffs retreat across a **rock platform** whose depth is determined by effective wave depth and the extent of water layer and biotic weathering, particularly during exposure at low tide. Platforms may be horizontal, or may slope gently seaward, controlled either by gently dipping structures or by higher tidal ranges in which the seaward margin receives most wave energy. Slow emergence produces a similar effect but entire **raised platforms** and abandoned cliffs are stranded above the contemporary sea level if emergence occurs faster than erosion (Plate 17.5). Incomplete erosion or locally more resistant rock leaves remnant and transient **arches** and **stacks**, with shapes attesting to progressive exploitation of rock structures and decay of rock strength during retreat.

The global coastline

Whatever the nature and controlling mechanisms of individual coastal landforms and landsystems, four universal influences can be recognized in coastline patterns at the global scale. The youth and vigour of sea-floor spreading and continental margin orogens impart morphotectonic control. Global tide and atmospheric circulation establish tidal range and wave energy patterns, despite the complication of continental coastlines. Tectonic and climatic causes of sea-level change, varied temporally and spatially, create patterns of emergent and submergent coastlines.

Morphotectonic coastlines reflect convergent and divergent plate motion and its impact on the continental slope–shelf system and orogens (Figure 17.15). **Leading edge** or convergent-margin coasts (American Pacific, Sunda Arc and New Zealand) typically have narrow shelves and are closely backed by emergent coastal orogens – providing a steep slope continuum between orogen crest and continental slope. Rapid sediment transfer from immature, disconnected terrestrial drainage systems through submarine canyons offsets the opportunity for substantial coastal sedimentation which vigorous erosion in the orogens might otherwise encourage. As a result, leading edge coasts are dominated by wave erosion and cliffs. In direct contrast **trailing edge** passive-margin coasts (American Atlantic, Africa, western Australia and Arctic basin) are generally well served by large, integrated river systems and broad shelves. River-fed wave deposition, with large deltas in humid areas, is dominant. Epicontinental seas (Caribbean and east Asia) accumulate terrigenous and biogenic sediments in relatively sheltered waters. Coastlines are also shaped through their accordance or otherwise with orogenic structures. In addition to island arcs, structurally accordant island archipelagos and peninsulas are found on the Cenozoic orogen coasts of the Alaska 'panhandle' and in British Columbia, southern Chile and Croatia and the Hercynian orogens of south-west Ireland and Brittany.

Tidal cycles refer to the daily number of tides (Figure 17.16). A water-covered Earth with a moon orbiting its equator would experience equal semidiurnal tides. Continents, and a lunar orbit varying between 28.5° N/S of the equator, upset this pattern but semi-diurnal tides are still experienced over large sections of Atlantic, south Pacific and Arctic coastlines. The tidal bulge of one hemisphere dwindles with increasing latitude in the opposite hemisphere, leaving one diurnal tide around Antarctica and parts of the Arctic basin. Mixed tides, predominant in the Pacific and Indian Oceans, show elements of both patterns. *Tidal range* emphasizes the effect of coastline and enclosed seas on the global tide wave (Figure 17.17). The range is lowest on open coasts, which reflect the wave, and increases with increasing width of continental shelf and partial enclosure of marine basins. Tidal range is suppressed by coastal sea ice in polar seas and wholly enclosed seas like the Mediterranean.

Wave energy patterns on a general, global scale are determined by wind speed, duration and fetch superimposed on the land–sea configuration. Two principal storm belts within the general atmospheric circulation – mid-latitude westerlies and tropical cyclone tracks – generate high waves on affected

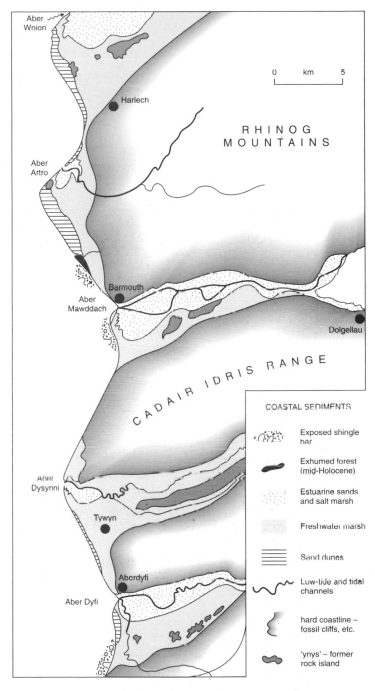

Figure 17.13 *Barrier and estuarine coast, northern Cardigan Bay, Wales. Dominant waves from the south-west help redistribute terrestrial and offshore late Quaternary sediments via longshore drift, diverting or blocking river mouths (Aber-). Most sand dune systems are of medieval age.*

coasts, contrasting with low wave height and energy on equatorial, doldrum belt and circumpolar divergence coasts (Figure 17.18).

Emergent and submergent trends identify relations between global sea level and coastlines over particular

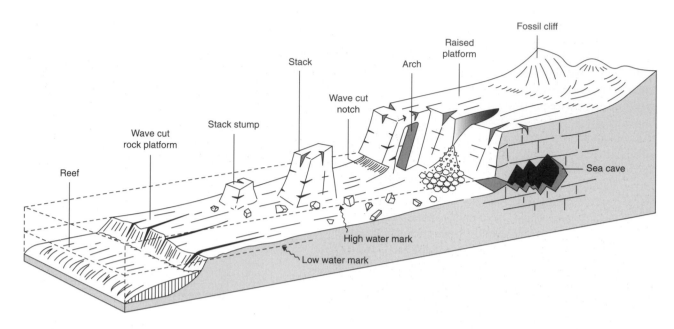

Figure 17.14 *Morphology of a rock coastline.*

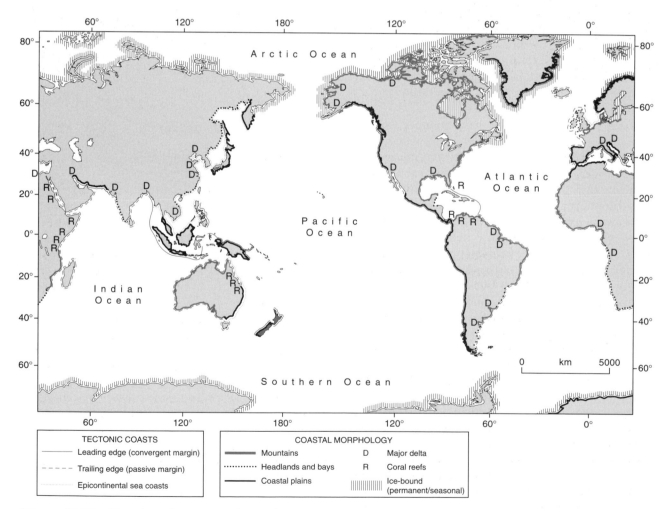

Figure 17.15 *Tectonic and macromorphological character of the global coastline. Source: After Inman and Nordstrom (1971).*

Plate 17.5   *The 61 m raised platform at Rhossili and Worm's Head, Gower, South Wales.*
Photo: Kenneth Addison.

time scales. The most relevant context today is one of Quaternary glacio-eustatic fluctuations superimposed on an overall tectono-eustatic sea-level fall of some 100–200 m. Ocean–ice coupling, capable of + 250 m of eustatic sea-level change if all ice were to melt and reform, has a major impact on continental margin systems of similar relief range. It accounts for marine transgression and regression throughout the Pleistocene, at something less than the full rate, complicated by glacio-isostatic response (see Chapter 4). These mechanisms are evident in global patterns of response during the past 15 ka (Figure 17.19). Submergence flooded river valleys (**rias**) and created structurally discordant fjord coasts in intensely glaciated areas of Norway, west Scotland and the South Island of New Zealand (see Plate 15.12). Emergence leaves coastal landforms, especially **raised beaches**, abandoned cliffs and platforms, stranded above the contemporary coastline (Plate 17.6). The British Isles straddle Zone I–II to provide a microcosm of submergent and emergent coastlines (Figure 17.20). Human-forced global warming is now set to ornament these larger trends through sea-level rise (see box).

Conclusion

Of all dynamic geomorphic landsystems, those subject to the work of the sea are among the most frail. Coastal materials may cross the threshold between terrestrial and marine environments at time scales as short as a breaking wave, tidal flow or storm surge. Coastlines, like flood plains, offer a diverse range of economic opportunities and aesthetic attractions for human occupation but at a price. Low lying barrier coasts and estuaries are particularly unstable and prone to storm surges; rock coasts are more enduring but failure is often dramatic when it occurs. Evidence of the geomorphic and human consequences of historic sea-level change demonstrates the changeable character of the coastline and the problems posed by impending sea-level rise. Increasing appreciation of coastline complexity and sensitivity is leading to more varied, holistic and pragmatic strategies in its management. Cnut's message is being heard!

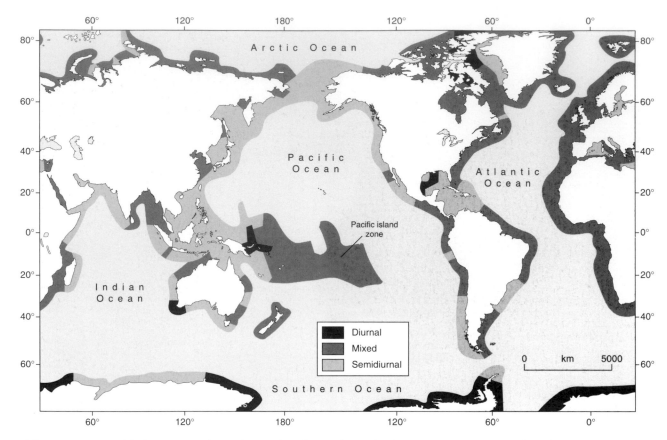

Figure 17.16   *Global pattern of diurnal, mixed and semi-diurnal tides. Source: After Davies (1980).*

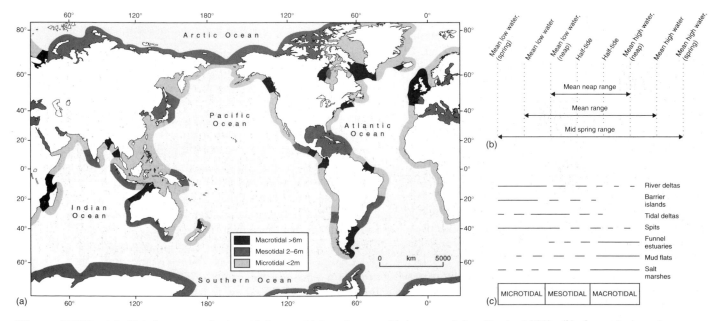

Figure 17.17   *(a) Global pattern of microtidal, mesotidal and macrotidal ranges (after Davies 1980), (b) the variation of range during monthly tidal cycles and (c) the occurrence of coastal landforms associated with tidal range (after Hayes 1976). High frequency is indicated by solid lines.*

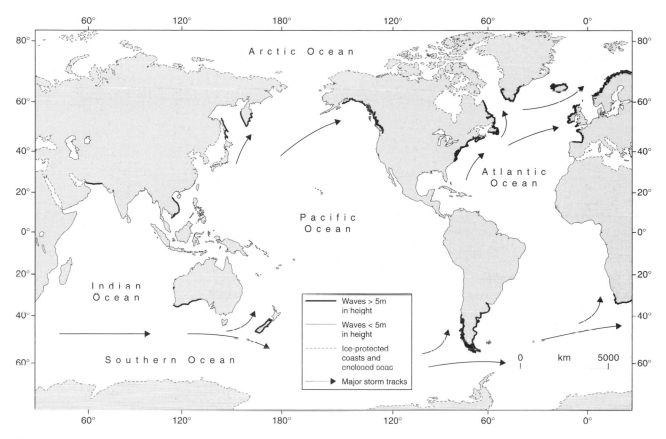

Figure 17.18   *Global wave environments. Source: In part after Davies (1980).*

Plate 17.6   *A raised beach of Ipswichian age (c. 120 ka BP), Gower, South Wales. The beach platform (left) lies 5–10 m above modern sea level and consists of cemented shelly shingle above a rock platform (right).*
Photos: Kenneth Addison.

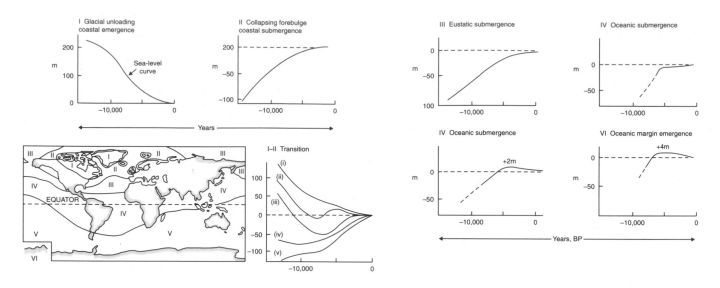

Figure 17.19 *Global zones of probable sea-level variations since 15,000 BP: I, Late Pleistocene ice sheets zone of isostatic recovery; II, Ice sheet periphery zone; I–II transition, complex ice sheet margins experiencing both isostatic and eustatic changes, and particularly appropriate for Britain; III, eustatic submergence beyond ice limits; IV–VI, régimes of oceanic submergence/emergence. Source: After Clark et al. (1978).*

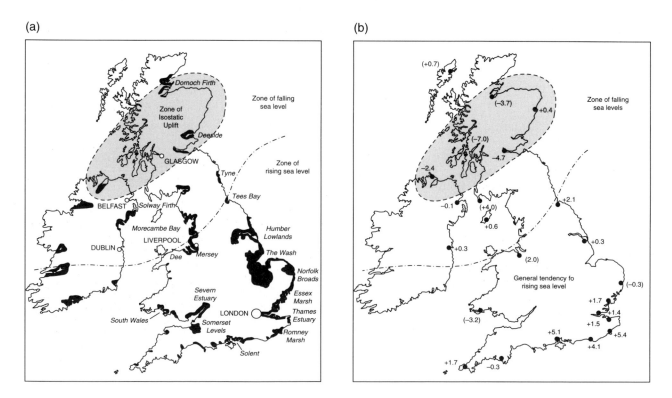

Figure 17.20 *The complex emergent/submergent nature of the British coastline, with residual isostatic recovery in Scotland and submergence in south-east England in mm yr⁻¹, where most areas threatened by rising sea level (black tone) are to be found. Source: In part after Boorman et al. (1989).*

Key points

1 Moving water and air shape the coastline and modify the effects of other sub-aerial processes of weathering and erosion. Water energy is derived from wind-driven breaking waves and, to a lesser extent, tidal waves and currents. It is applied through hydraulic pressure, water jets and sheets. Wind energy deflates sand from the backshore to create and subsequently alter littoral barriers.

2 Breaking waves establish shore-normal and long-shore currents of water and energy. The precise wave-form depends on water depth, the shelving angle of the inshore zone and the direction of approach determined by wind and coastline configuration. Tidal waves and currents also move large bodies of water around the coast. Tidal frequency, range and coastline shape determine their velocity.

3 Coastal marine geomorphic processes are concentrated in the foreshore or intertidal zone and inshore surf zone, influenced by breaking wave style and height. Hydraulic shock, attrition and corrasion erode rock coasts and comminute their debris, aided by water-layer weathering in the intertidal zone. Sediment movement depends on the availability and direction of water currents at or above the entraining velocity for individual particles. Storm waves inevitably cause the greatest amounts of erosion and sediment movement.

4 Soft sediment coastal landsystems develop within partially enclosed estuaries or lagoons and as coast-parallel barriers. Aeolian processes may ornament the barriers with dunes. Deltas form a third landsystem at the seaward extension of fluvial sediments, shaped by the wave and current environment. However, 75 per cent of all coasts are rocky, with their own characteristic rock platform, cliff and cliff remnant landforms. Biogeomorphic processes create important coastal structures in the form of salt marsh, mangrove swamp, reefs, bioherms and vegetated dunes.

5 Coastal patterns are discernible at the global scale. Morphotectonic distinctions between leading and trailing edge coasts are reflected in the respective dominance of rock/erosion dominated coasts and river-fed/depositional coasts. Tidal patterns vary according to diurnal/semi-diurnal frequency and tidal range whilst high to moderate wave energy patterns depend on location within or outside storm belts. Complex isostatic–eustatic adjustment throughout the Holocene has created patterns of emergent and submergent coasts.

Further reading

Carter, R W G (1993) *Coastal Environments*, London and San Diego: Academic Press.
This book provides a detailed account of coastal processes and landforms with an appropriate amount of mathematical explanation. It is an important reference work suitable for more advanced study but still yields plenty of value to the mathematically challenged student. The final third of the book deals with 'cultural systems' of the coastline, its hazards and management

Davis, R.A. (1996) *Coasts*, Englewood Cliffs, N.J.: Prentice-Hall.
Coastal processes and geomorphology without mathematics!

Pethick, J. (1984) *An Introduction to Coastal Geomorphology*, London: Arnold.
A more concise and somewhat easier introduction to the subject, focusing on physical processes, with only a short final chapter on 'applied coastal geomorphology'.

Viles, H., and Spencer, T. (1995) *Coastal Problems*, London, Melbourne and Auckland: Arnold.
Subtitled *Geomorphology, ecology and society at the coast*, this book commences with a short chapter reviewing 'how coasts work'. It is followed by separate accounts of the principal coastal landsystems and an excellent, well illustrated review of the environmental problems and management options they face.

Soil is a dynamic three-phase system. The three phases are: *solid*, which is represented by mineral particles, together with some organic material; *liquid*, consisting of a solution of various salts in water; and a *gas* phase, consisting of air with changing amounts of oxygen, carbon dioxide and nitrogen. The equilibrium of these phases changes continuously as, for example, rainfall fills pores or voids and excludes some of the gases. Soil properties vary greatly from place to place, in line with changes in the nature and the relative content of the three phases. The three phases interact greatly, and the nature of the interactions determines the behaviour of the soil in response to external impacts such as farming, drainage, forestry and engineering.

Physical properties

Texture

Physically the soil is composed of mineral particles of different sizes, with some organic molecules strongly bonded to the minerals and some organic matter physically mixed within it. The mineral particles are classified into groups having definite size limits. Each group is called a **soil separate**, and three basic separates are recognized, namely, sand, silt and clay. The size limits for these are given in Table 18.1, according to the usage of the USDA–FAO (United States Department of Agriculture–Food and Agriculture Organization). The relative proportions of sand, silt and clay determine the **soil texture**, and give the textural name.

Various systems have been used to classify the texture of soils in this way. One of the most commonly used is the USDA–FAO texture triangle shown in Figure 18.1a, giving the names of soils according to different proportions of sand, silt and clay. Figure 18.1b shows the broad grouping of soils into the six most commonly encountered soil textures: sands, light loams, light silts, medium loams, medium silts and clays. The amount of sand, silt or clay in soil samples can be estimated approximately in the field by the simple technique of moistening a handful of soil, working it between the fingers and determining the texture by the 'feel' of the moist soil. Clay is very sticky and hard to work, silt is less sticky but very smooth and greasy. A sandy soil has very little stickiness but a distinctly gritty feel. Accurate determinations using the principles of sedimentation can be performed in the laboratory using a hydrometer or a pipette.

Texture or particle size influences many chemical, physical and biological properties of soil. Larger particles have larger pores between them and therefore allow more rapid **infiltration** and **drainage** of water. Finer particles have finer capillary pores which, in contrast, hold water in the soil and thus improve the soil's **water-holding capacity**. Thus coarse-textured soils are quickly drained of rainfall and are not able to hold much water for plant growth. They are 'droughty' soils, and lack of available water can be a limitation on their productivity and choice of crops. However, because the solid phase has a much lower heat capacity than water, coarse-textured soils heat up much more rapidly in spring, and thus have longer growing seasons. In contrast, fine-textured clays have greater water-holding capacities, and thus show fewer symptoms of drought. Indeed, the farming problem here is often to remove excess water by artificial drainage in order to improve soil aeration. Because

Table 18.1 *Particle-size limits.*

Soil separate	Diameter (mm)
Sand:	
Very coarse	2.0–1.0
Coarse	1.0–0.5
Medium	0.5–0.25
Fine	0.25–0.10
Very fine	0.10–0.05
Silt	0.05–0.002
Clay	< 0.002

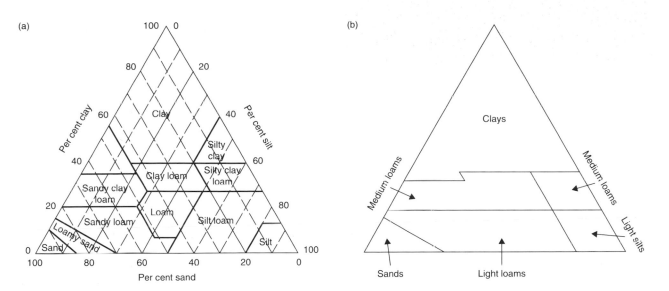

Figure 18.1   *(a) The soil textural triangle according to the US Department of Agriculture; (b) broad groups of textural classes.*

plant roots need oxygen for respiration, soil water-logging can be a serious limitation. Equally, wet clay soils will be slow to heat up, given their high heat capacities. Thus soil texture is important in water capacity and movement, soil temperature and aeration (see Chapter 10).

Chemical properties of soils are also dependent on soil texture. The fine separate of clay determines most of the chemical properties of soils. Particles with a diameter smaller than 0.002 mm are classed as *colloids*, or are said to be in the *colloidal state*. Colloidal properties arise from the very large surface area associated with a small mass. There is an indirect relationship between particle size and the surface area of the particles. Assuming for simplicity that particles are spherical and that the volumes of solid particles are equal, the surface area of soils can be compared by using the value R (ratio of surface area-to-volume for a particle of radius r):

$$R = \frac{4\Pi r^2}{\left(\dfrac{4}{3}\right)\Pi r^3}$$

Thus when r = 1 mm, R = 3 mm^{-1} and when r = 0.001 mm, R = 3000 mm^{-1}. Figure 18.2 illustrates the relationship between surface area-to-volume ratio and particle radius.

Structure

In the field the properties determined by soil texture may be considerably modified by *soil structure*. The solid mineral particles exist in a definite arrangement, and the pore spaces between them are filled partly with water and partly with gases. The arrangement of individual particles into larger **aggregates** or **peds** of various sizes and shapes is the *soil structure*. Although 'texture' and 'structure' seem to be used interchangeably in popular usage, there are real differences in meaning. The small particles of clay and silt are not spread uniformly throughout the soil. The clay coats sand particles. Individual clay units join together into 'clay domains' rather like the leaves of a book. This is a kind of parallel orientation, with the clay domains forming coatings around sand particles or soil-structure units. A further distinction between texture and structure lies in the role of organic matter. Humus or humic colloids considerably influence the properties derived from texture. Organic matter can improve the water-holding capacity of sand and can improve the drainage properties of clay. This is achieved by promoting structure formation. Figure 18.3 illustrates how clay colloids and humic colloids provide various types of 'bridges' or linkages between the mineral particles in soil. This is the **Emerson model** of structure formation and illustrates that attraction between particles depends upon electrostatic forces on the surfaces of colloids. These electrostatic forces are discussed on pp. 329–31. The role of organic matter lies in providing a strong and stable structure due to the fact that the humic colloid is **hydrophobic**, i.e. water-repellent. This means that the structural unit is stable and is more likely to survive in the face of wetting and of raindrop impact. The **grade** or

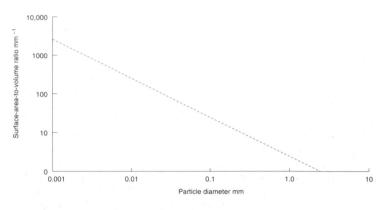

Figure 18.2 *The relationship between the diameters of mineral particles in soil and their reactive surface areas (surface area-to-volume ratio).*

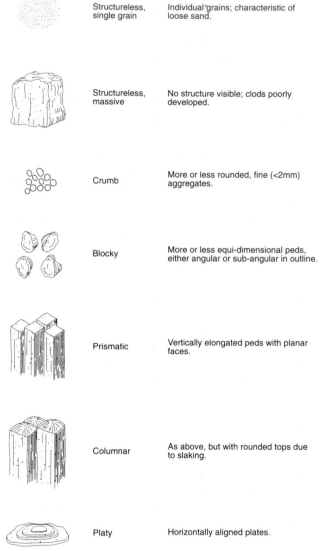

	Structureless, single grain	Individual grains; characteristic of loose sand.
	Structureless, massive	No structure visible; clods poorly developed.
	Crumb	More or less rounded, fine (<2mm) aggregates.
	Blocky	More or less equi-dimensional peds, either angular or sub-angular in outline.
	Prismatic	Vertically elongated peds with planar faces.
	Columnar	As above, but with rounded tops due to slaking.
	Platy	Horizontally aligned plates.

Figure 18.4 *The shapes of the common soil structural units.*

stability of structure is an important property in studies of soil erosion; usually three grades of structure are recognized – weak, moderate and strong (Plate 18.1).

The **kinds** or **shapes** of aggregates are summarized in Figure 18.4. *Structureless single grain* consists of loose, individual particles as found in raw sands. *Structureless massive* consists of a large mass of compacted soil with no recognizable aggregates. It may be found in wet and raw clays. *Crumb* is characteristic of soils with **mull** humus and a very active soil faunal population, especially earthworms. **Blocky**, either angular or sub-angular, is common in many arable topsoils and subsoils in medium-textured soils. **Prismatic** is characteristic of subsoils in clay soils affected by shrinking and swelling. **Columnar** shows some slaking (dispersion) and is associated with the subsoils of clay soils with a large sodium content. Both prismatic and columnar form very hard and dense horizons which are difficult to cultivate with farming implements and which greatly impede

water and root penetration. **Platy** structures can occur in a variety of soils. They are characteristic of compacted clay soils and also of silts. In the latter case drying and crusting after early spring rain gives platy aggregates which can cause serious problems with the emergence of young seedlings (Plate 18.2).

Water

The total volume of pore space in soil is its **porosity** or **air capacity**. It is calculated from the **bulk density**, the weight of soil per unit volume. Assuming that the mineral density of soil particles is 2.65 g cm^{-3}, then

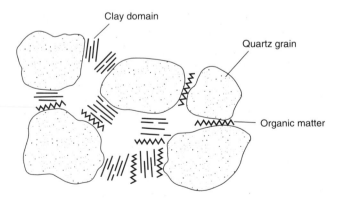

Figure 18.3 *The arrangement of quartz particles, clay domains and organic colloids in a soil aggregate. Source: After Emerson (1959).*

$$\text{Porosity} = \left(\frac{BD}{PD}\right) \times 100 = \left(\frac{BD}{2.65}\right) \times 100$$

where BD = bulk density and PD = particle density. Bulk density values range between about 0.8 g cm⁻³ and 2.4 g cm⁻³, which is equivalent to a range of soil porosity from 70 per cent to 10 per cent respectively. After a very wet period most of the porosity will become filled with water, and the soil would be described as *saturated*. Under the influence of gravity, water would drain out of the larger **transmission pores**. This water which quickly drains away is termed **gravitational water** and it drains out of pores larger than about 0.05 mm diameter. When it has all drained away the soil is said to be at **field capacity**; i.e. at the upper limit of wetness at which a soil can retain water without gravitational loss.

Smaller pores less than 0.05 mm diameter can hold water against gravitational removal owing to capillary forces. Such water is classed as **capillary water**. Capillary water forms the bulk of **available water** for plant roots. The plant root expends energy in absorbing water from the soil, but eventually there comes a point where, as the soil dries, the forces between solid phase and water exceed the energy available to the root for water absorption. This limit of wetness, below which plants can no longer extract water, is the **wilting point**. Between field capacity and wilting point water is available for plant growth and is therefore termed *available water*. Capillary water and available water are the same for many plants, though there is some debate whether all plants are able to utilize all available water. Some water is retained in the soil under the driest of natural conditions; such water is called **hygroscopic water**.

Soil water is thus best regarded in terms of the energy with which it is held by the solid phase in the soil. The smaller the water content, the more tightly the water is held by solid particles. This force is measured in units of **suction**, i.e. the force required to remove a certain proportion of the soil water. Suction is measured in pascals or **bars** or **atmospheres** (10^5 pascals = one atmosphere = one bar = 1000 mbar). Table 18.2 shows the suction with which different classes of water are held in soil. Wilting point is 15 bars suction, the limit between capillary water and hygroscopic water is over 31 bars, and the water in an oven-dried soil is held at over 10,000 bars. Field capacity has been defined at various suctions in the range 0.33–0.05 bars. Table 18.2 also indicates the pore diameter corresponding to the soil suction, the physical appearance of the soil, and the availability of the water to plants.

Plate 18.1 *In semi-arid lands, the lack of protection from vegetation means that raindrop impact breaks down surface structure, allowing an impermeable crust to form on drying (scale in cm). Photo: Ken Atkinson.*

The amount of available water which can be stored in any soil is influenced by soil texture, soil structure and the organic matter content. These factors have a marked effect on the size and distribution of the pore spaces. The influence of texture on **storage capacity** or *available water-holding capacity* is indicated in Table 18.3. In many parts of the world soil moisture is probably the major factor limiting crop production, so this property of soils is of enormous economic importance. The most accurate method of determining the water content of a soil sample is to measure the loss in weight when a moist soil is dried in an oven at 105° C overnight. The moisture content is expressed as the loss in weight as a percentage of the oven-dried soil.

Plate 18.2 *Vertical view down on to a cylinder measuring the infiltration rate of the soil. The added water has destroyed the coarse, blocky structure shown in the surrounding dry soil.*
Photo: Ken Atkinson.

Drainage and infiltration

The content of available water in soil correlates fairly closely with total pore space, i.e. *porosity*. Fine textures like clays and clay loams are able to hold considerably more available water than coarse textures such as sands and sandy loams (Table 18.3). Excess gravitational water will drain away from soil in **macropores** or **transmission pores** larger than about 0.05 mm diameter. The ability of a soil to

allow water to pass through in this way is its **permeability**. Again it is closely linked with soil texture and soil structure. It has no relationship to total porosity; clay soils with high porosity usually have low permeability, and vice versa with sandy soils. The important characteristic of soils with a high permeability is their high content of large pores, wide cracks or faunal burrows and channels (Plate 18.3). The soil physicist Darcy defined the ability of a porous medium to transmit a fluid as the **hydraulic conductivity**, K, in units of centimetres transmitted per hour. It is almost identical to permeability. Table 18.4 shows some typical values of K for representative textures and structures.

Soils with a horizon or horizons of low hydraulic conductivity will not allow gravitational water to drain away. This will promote waterlogging both in and above such horizons, to the exclusion of air and adverse effects on the respiration of plant roots, macro-organisms and micro-organisms. Waterlogging can also bring about chemical changes through the process of gleying under conditions where oxygen is excluded (**anaerobism**) (Chapter 19). Except for rice, agricultural crops are stunted or killed by anaerobic conditions. It is therefore often necessary to provide artificial drainage to remove excess water from the soil. A variety of methods are available, including open drains or ditches, tile or plastic drainpipes in the subsoil, mole drains or subsoil ploughing. Investment in artificial drainage has been the largest area of capital investment in British agriculture during the twentieth century. Figure 18.5 illustrates how a poorly drained soil may be improved by a combination of tile drainage and mole drainage, installed at 90° to each other.

In order that water can be stored in soil, it is first necessary for it to enter downwards from the surface. The rate at which a soil can absorb water, defined as the volume of water passing into a unit area of

Table 18.2 *Different types of soil water.*

Suction held (bars)	Water constant	Pore diameter (mm)	Type	Physical state	Availability
10 000					
1000			H	Dry	Unavailable
31	Hygroscopic coefficient	0.001			
15	Wilting point	0.002	C	Moist	Available
0.33	Field capacity	0.01		Wet	
0.05		0.06	G		
0.001				Saturated	Unavailable transient

Note: H hygroscopic, *C* capillary, *G*, gravitational.

soil per unit time, is the **infiltration rate**. Its units are velocity, cm hour^{-1}. Initially in dry soils infiltration rates can be high, especially in coarse-textured soils and in heavy-textured soils with surface cracking. The infiltration rate then falls as pores fill with water, as cracks close up owing to swelling clays, and as structure starts to collapse in the wet state. Infiltration rates can vary from over 50 cm of water per hour in coarse permeable sands to as low as 0.02 cm of water per hour in low-permeability clays.

Infiltration rates of soils can be measured in the field by means of a commercial infiltrometer or a home-made device. The commercial infiltrometers are often double-ring, with the ability to maintain standard moist conditions in the outer ring. In home-made infiltrometers the vessel can be plastic piping or a tin. Three broad techniques are available. The first is to note the time required for a volume of water, say 250 ml, to infiltrate completely. A second technique is to construct a scale on the inside of the pipe or tin, add 250 ml of water to the container, and note the time taken for unit amounts, say 50 ml, to infiltrate. The level is then topped up after each reading. A third set of methods involves an inverted bottle, with a suitable air intake, so that the level of water in the pipe or tin is maintained at a constant level. In this case the scale is on the bottle. The latter two methods are designed to maintain a more or less constant head of water. The rate of infiltration may initially be rapid but it generally decreases with time and approaches a constant value. The infiltration can be shown on a graph of cumulative infiltration versus time.

Colloidal properties

Clay minerals

The weathering of primary minerals in rocks and loose, transported deposits (e.g. glacial tills, loess, etc.) produces a range of weathering products. These products are transformed during the process of soil formation (see pp. 337–9). Of great importance in the soil are the new clay-sized minerals, or **clay minerals**, which are formed from the weathering products. 'Clay' has two different but related meanings. It refers to the size fraction of less than 0.002 mm diameter and also refers to secondary clay minerals which are synthesized from chemical weathering. These distinctive minerals have colloidal properties, i.e. the very small particles carry an electric charge. It is also possible in soils to have clay-sized particles consisting of disintegrated fragments of

Table 18.3 *Storage capacity of soils (cm water/30 cm soil depth).*

Soil texture	Field capacity	Wilting point	Available water
Sandy loam	5.6	2.8	2.8
Loam	8.4	4.3	4.1
Clay loam	9.9	5.3	4.6
Heavy clay	11.9	6.3	5.6

rock; such 'rock flour' does not have colloidal properties. Clay minerals are alumino-silicates, formed from the fusion of silica and alumina. The silica is in the form of a sheet of silica tetrahedra. Figure 18.6a shows the silicon (Si) atom at the centre of a tetrahedron bounded by four oxygen atoms (O). The

Plate 18.3 *Cracks in dry clay allow rapid infiltration of any precipitation or irrigation water. However, the cracks will close up as the clay expands on re-wetting.*
Photo: Ken Atkinson.

Table 18.4 *Hydraulic conductivities of soil.*

Texture	Structure	Hydraulic conductivity K (cm hour^{-1})
Coarse sand	Single grain	> 50
Sandy loam	Blocky, fine crumb	6–12
Loam, silt loam	Blocky	2–6
Clay, clay loam	Blocky, prismatic	0.5–2
Clay, clay loam	Blocky, prismatic, fine platy	0.25–0.5
Clay, heavy clay	Massive, fine columnar	< 0.25

alumina unit is shown in Figure 18.6b. It consists of an aluminium atom (Al) equidistant from six oxygens (O) or hydroxyls (OH). In the silica sheet the three oxygens at the base of the tetrahedron are shared by two silicons of adjacent units. The sheet can be visualized as two layers of oxygen atoms with silicon atoms fitting into the holes between. In the alumina unit each oxygen is shared by two aluminium ions, forming sheets of two layers of oxygen (or hydroxyl) in close packing, but only two-thirds of the possible octahedral centres are occupied by aluminium.

Clay minerals are formed by the silicon–oxygen and aluminium–oxygen structural units being bonded together so that sheets of each result. Clay minerals thus have a platy, crystalline structure. In the soil other ions, usually of similar size, can take the place of silicon and aluminium by a process of isomor-

phous substitution. The different types of clay minerals are determined by three features: the ways in which the silica and alumina sheets are stacked into layers, the bonding between the layers, and the substitution of other ions for Si and Al.

Figure 18.7 gives a schematic representation of the structure of five common clay minerals. **Kaolinite** is made of a silica sheet and an alumina sheet sharing a layer of oxygen atoms. The layers are held together by strong hydrogen bonding and the structure is non-expanding. **Illite** or clay mica has repeating layers consisting of one alumina sheet sandwiched between two silica sheets. The layers are firmly bonded together by potassium (K) ions, which are just the right size to fit into the hexagonal holes of the silica sheet. **Montmorillonite** has a similar structure to illite, except that there are no potassium ions to bond the layers together, and water enters easily between the layers. Thus the wet clay can expand to several times its dry volume. **Vermiculite** resembles montmorillonite except that absorption of water between layers is limited to two thicknesses of water molecules. **Chlorite** is made of mica layers held together by brucite sheets. Figure 18.8 illustrates how the alumina and silica sheets condense together to give the structures of kaolinite and montmorillonite.

The volume change caused by wetting is an important physical property of clays. Dry sand and silts can take up water when the air in pore spaces is replaced, but that gives no increase in volume. With clays, water can give forces of repulsion between particles

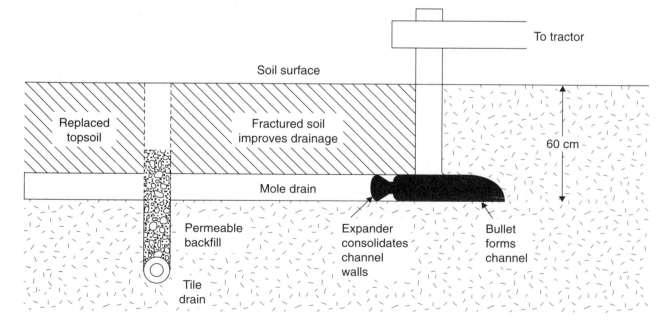

Figure 18.5 *Artificial drainage in poorly drained soils. Mole drainage runs perpendicularly to tile drains.*

(a)

(b)

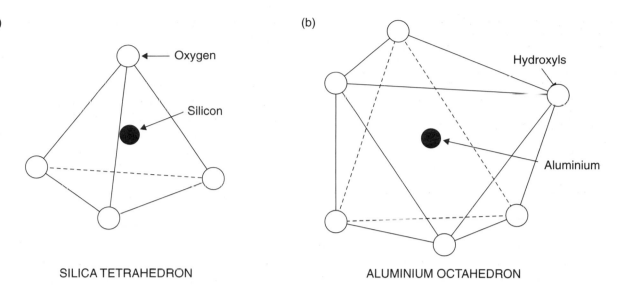

SILICA TETRAHEDRON

ALUMINIUM OCTAHEDRON

Figure 18.6 *The basic building blocks of clay minerals: (a) a silica tetrahedron, (b) an alumina octahedron.*

so that the volume increases as water content increases. Swelling increases with increasing surface area of the clay particles. In turn, surface area depends on the thickness of the crystalline particles. It increases from the thicker kaolinite particles to the thin particles of montmorillonite (Table 18.5, Plate 18.4).

Cation exchange

As mentioned in the previous section, the replacement of aluminium or silicon by an ion of similar size in the octahedral or tetrahedral sheets is known as **isomorphous substitution**. It is possible for aluminium (Al^{3+}) to replace some of the silicon (Si^{4+}) in the tetrahedral sheets. Similarly magnesium (Mg^{2+}), iron (Fe^{2+} or Fe^{3+}) and calcium (Ca^{2+}) may replace Al^{3+} in octahedral sheets. When the replacing ion has a lower positive charge than the ion it replaces, the clay mineral has a net negative charge. These substitutions account for most of the negative charge in the 2:1 and 2:1:1 minerals, but only a minor part in the 1:1 kaolinites. A second source of electric charge is unsatisfied charges at the edges of the particles, the broken bonds. The hydroxyl (OH^-) groups at the edges become ionized at high pH values and give an increasing negative-charge capacity as pH rises. This charge is thus pH-dependent. The ease with which the hydrogen ion (H^+) can be exchanged also increases as the pH increases and thus the total charge due to 'broken bonds' increases as pH increases. Conversely at low pH values many positively charged sites are found on the

clay colloids, though the net charge of the colloid is overall negative.

The overall net negative charge of clay minerals is the **cation exchange capacity** (CEC), the capacity of the negatively charged colloid surface to attract positively charged ions (cations). Cation exchange capacity is usually given as milliequivalents per 100 g soil. The equivalent weight is the weight, in grams, of that element needed to displace one gram of hydrogen. For monovalent cations (Na^+, K^+) the equivalent weight is the same as the atomic weight; for divalent cations it is half the atomic weight (Ca^{2+}, Mg^{2+}) and for trivalents one-third (Fe^{3+}, Al^{3+}). Since the amounts involved are very small, the term *milliequivalent* (EW/1000) is used. The value me 100 g^{-1} therefore represents the number of milligrams of particular elements which can be held by 100 g of a particular soil. The average electric charges (CEC) on the common clay minerals are given in Table 18.6.

The net negative charge on the clay colloids is balanced by **exchangeable cations** which are attracted to the surface of the clay particles. These are positively charged ions in the soil solution (H^+, Ca^{2+}, Mg^{2+}, K^+, Na^+). They are termed 'exchangeable' because one cation can be readily replaced by

Table 18.5 *Size and swelling of clays.*

Mineral	*Thickness* *nanometres*	*Surface area* *($m^2\ g^{-1}$)*	*Volume* *change*
Montmorillonite	2	800	High
Illite	20	80	Medium
Chlorite	20	80	Medium
Kaolinite	100	15	Low

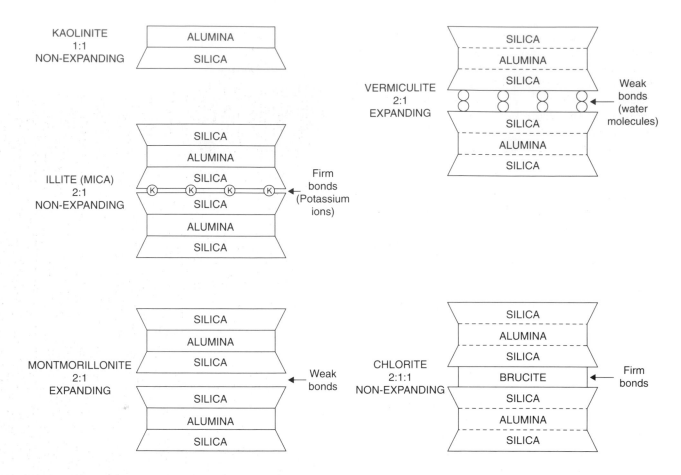

Figure 18.7 *The arrangement of silica and alumina sheets in common clay minerals.*

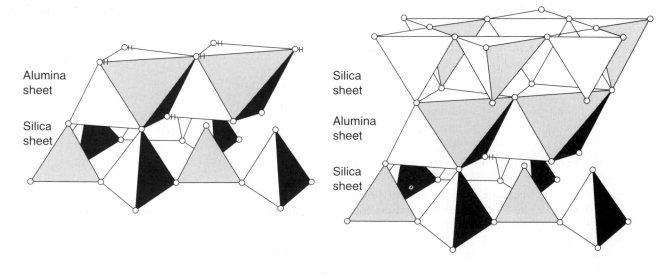

KAOLINITE STRUCTURE   MONTMORILLONITE STRUCTURE

Figure 18.8 *The arrangement of sheets of silica tetrahedra and alumina octahedra in forming the structures of kaolinite and montmorillonite.*

another of equal valence, or by two of half the valence of the original one. For example, if a clay containing sodium as the exchangeable cation is washed with a solution of calcium chloride, each calcium ion will replace two sodium ions, and the sodium will be washed out in solution. This process is called **cation exchange** or **base exchange**. It can be written as the chemical equation:

$$Na_2\ Clay + CaCl_2 = Ca\ Clay + 2NaCl$$

The total quantity of exchangeable cations held is the *cation exchange capacity*. The predominant exchangeable cations in soils are calcium and magnesium, with lesser amounts of potassium and sodium. Aluminium and hydrogen are common in acid soils. The proportions of these cations found on the colloids of any particular soil are governed by the parent rock and by the nature and intensity of weathering and leaching. Calcareous soils over limestone will contain mostly calcium. Clays deposited in sea water will have mostly magnesium and sodium. Leaching removes the cations which form bases (calcium, sodium, etc.), leaving a clay with the acidic cations, aluminium and hydrogen. The influence of hydrogen ions on the exchange sites was originally thought to give soils acidic properties, but it was later found that acid clays had aluminium rather than hydrogen as the exchangeable ion. In very acid soils the clay minerals themselves start to dissociate, releasing aluminium which can then move on to the soil complex. The process of cations fixing themselves on to exchange sites on colloids is termed **adsorption**. The cations are not all held in a layer right at the clay surface but are present as a **diffuse double layer**, as shown in Figure 18.9. The inner layer is the highest concentration of cations at the colloid surface, attracted by coulomb electrical forces; the outer layer is a diffuse 'cloud' of cations whose thermal energy makes them diffuse away from the colloid surface.

Table 18.7 illustrates the cation exchange characteristics of six contrasting soils. The values for the four commonest base cations (Ca, Mg, K, Na) are given, together with those for hydrogen (H). The total cation exchange capacity is the sum of these five ions, and the percentage base saturation (% BS) is the proportion of the CEC occupied by these four base cations. The pH values (pp. 334–5) are directly related to % BS.

Organic colloids

The values of the cation exchange capacities for clay minerals range from a low of about 5 me 100 g^{-1}

Plate 18.4 *Black tropical soil (FAO: Pellic Vertisol) composed of expanding 2:1 montmorillonite clay minerals. Repeated expansion and shrinkage due to wetting and drying cause thorough mixing and overturning of the soil. (Scale in 10 cm units.)*
Photo: Ken Atkinson.

to a high of about 150 me 100 g^{-1}, depending on the type of clay mineral (Table 18.6). Organic colloids or humic colloids have much higher activity values in the range 150–300 me 100 g^{-1} (Figure

Table 18.6 *Electrical charges on clay minerals.*

Clay mineral	Charge (me 100g^{-1})	Source of charge
Kaolinite	5–15	Broken bands Ionisation of OH
Illite, Chlorite	20–40	Ion substitution
Montmorillonite	80–100	Ion substitution
Vermiculite	100–150	Ion substitution

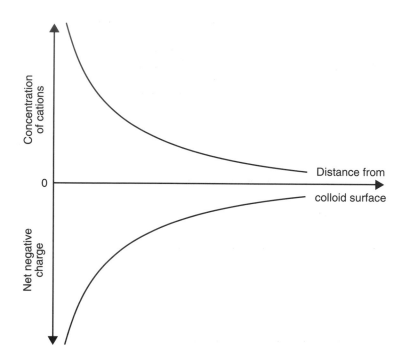

Figure 18.9 *How the distribution of adsorbed cations and the strength of the net negative charge varies with distance from the surface of a soil colloid.*

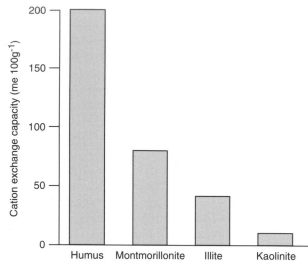

Figure 18.10 *The cation exchange capacities of humus and common soil clay colloids.*

18.10). The reasons for this high activity are not fully understood, but it appears to derive from the negative charges of phenolic (OH) and acid carboxyl groups (COOH) which occur in both humic and fulvic acids.

Organic matter is that fraction of the soil which is derived from plant and animal remains which are added to the soil surface and which are decomposed by soil organisms. The organisms range from the larger soil fauna (earthworms, ants) to soil micro-organisms (bacteria, fungi, actinomycetes). The decomposition of the organic material is the process of *humification*, and results in a dark-coloured amorphous material known as *humus* which gives the surface soil its dark colour. It is not an easy material to study, and its chemistry varies, depending upon the prevailing soil conditions and the nature of the original plant material. During the course of humi-

fication the original plant material quickly loses the most readily decomposed fractions (sugars, polysaccharides, amino acids), although it takes longer to break down the more resistant carbohydrates (celluloses). The most resistant fraction of lignin tends to accumulate in humus.

During humification, the many organisms and micro-organisms in soil are also synthesizing organic molecules (proteins, polysaccharides) which are added to the soil reserve of organic matter on their death. Thus humus is not simply a residual product, after the soil organisms have attacked and partially decomposed the dead plant structures. Much of it consists of freshly made microbial products which are very influential in giving humus its important properties. Figure 18.11 illustrates how an active organic cycle can rapidly break down large quantities of organic matter by the activities of bacteria and fungi. Larger organisms such as earthworms act to mix the organic matter throughout the topsoil. The end result is a relatively stable mild humus or **mull** (pH >5.5).

Table 18.7 *Cation exchange values for various soils.*

Soil	Exchangeable cations (me 100 g^{-1})						% BS	pH
	Ca	Mg	K	Na	H	CEC		
Cambisol (Scotland)	3.0	27.5	0.2	0.1	3.6	34	90	6.5
Chernozem (Russia)	30.5	1.8	0.5	0.2	0	33	100	7.3
Podzol (Scotland)	0.6	0.7	0.2	0.1	37	39	4	4.3
Ferralsol (Kenya)	4.8	1.2	0.3	0.2	3.5	10	65	5.5
Luvisol (Canada)	22.8	0.6	0.3	0.1	3.8	28	86	6.2

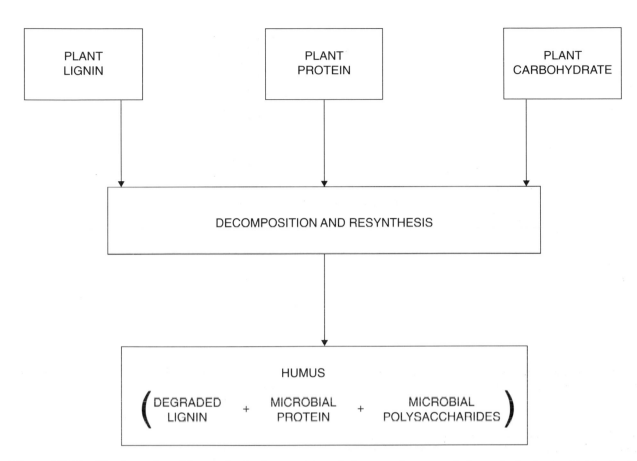

Figure 18.11   *The formation of humus by the decomposition of plant constituents and the synthesis of microbial products.*

If the organic cycle is less active, with slow decomposition and the build-up of only partially fermented litter, a very different surface organic horizon results. Acid soil conditions result from an acid parent material, or excessive leaching or acid-tolerant vegetation. The population of bacteria is reduced, and slow decomposition by fungi will produce a raw humus or **mor**. This is usually layered into litter (L), a fermentation (F) layer and a very thin humus (H) layer. Mor is not well decomposed, or mixed into the soil profile, in the absence of both bacteria and earthworms. An intermediate stage (pH 4.5–5.5) produces **moder** humus. Figure 18.12 shows the three types of organic material which are found in well drained natural soils. The nature of the soil, climate and vegetation will determine which of the three types is formed. In poorly drained, waterlogged situations the lack of oxygen excludes many fauna and soil micro-organisms. Decomposition is very slow, and the remains of plant build up into a **peat**.

In agricultural soils the addition of organic material to the soil surface is lower than under natural conditions owing to losses by harvesting, stubble burning and increased rates of erosion. Cultivation and disturbance of the soil surface also increase the rate of

decomposition and mineralization. Manuring and the rotation of grass leys are important as possible means of maintaining soil organic levels. Mineral soils commonly contain 1–10 per cent organic matter, with arable soils usually at the lower end of this range. A simple method of estimating the organic matter of the soil is to ignite a sample at a high temperature and determine the loss in weight.

Humus has many beneficial effects on soil. First, as already noted, the cation exchange capacity of humic colloids is very high, due partly to its high specific surface and partly to the density of phenolic (OH) and carboxyl (COOH) groups. The size of the negative charge is heavily pH-dependent, typically doubling from 120 me 100 g^{-1} at pH 5.0 to 240 me 100 g^{-1} at pH 8.0. This is a result of the ionization of the OH groups at higher pHs. The second role of organic matter is that it is the chief source in the soil of the plant nutrients nitrogen and sulphur, and an important source of phosphorus. Ninety-eight per cent of the nitrogen absorbed by plants comes from the mineralization of soil humus. The microbial protein in humus is mineralized to nitrate, in which chemical form it is absorbed by plants, whence it becomes a constituent of plant

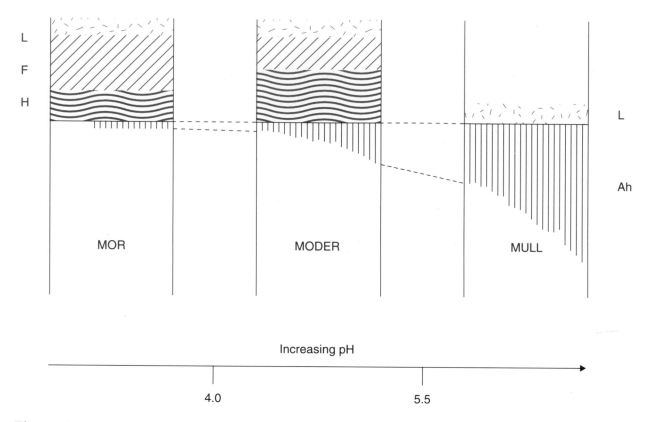

Figure 18.12 *A diagrammatic representation of the three main types of surface organic matter in soils. Source: After Avery (1990).*

proteins. Sulphur is an essential constituent of some plant proteins. Organic sulphur compounds in humus are mineralized to give sulphate, which is the form absorbed by plants. Phosphorus plays a fundamental role in a very large number of enzymic reactions in plants, and is a constituent of the cell nucleus and essential for all cell division. About 50 per cent of the phosphorus absorbed by plants comes from organic matter and about 50 per cent from phosphorus in rock minerals, e.g. apatite. The third role of humus is to act as a source of most of the micro-nutrients which plants need. Micro-nutrients are needed only in very small quantities, but they are absolutely essential. The main micro-nutrients are iron, manganese, copper, zinc, molybdenum and boron. Those nutrients which are also metals, e.g. iron, manganese, zinc and copper, can be held in organic molecules in the form of **chelates**, where the metal ion is held in the form of a 'chelate ring'. The chelates will be decomposed by micro-organisms to release the nutrient ion into the soil solution, whence it can be absorbed by plant roots. As well as providing nutrients to plants, it must not be forgotten that humus also supplies food for bacteria and other living organisms essential to a productive soil. The final role of humus is its influence on the soil's physical

properties. Thus it acts to improve the soil's water-holding capacity through its effects on soil structure (pp. 323–4). It is especially efficacious in improving aggregation in sands and sandy loams. Similarly, in heavy clays the humic colloids improve structure formation and aeration. The role of humus in darkening the soil surface influences the thermal absorption and radiation characteristics of a soil. A darker soil will heat up more rapidly than a lighter soil, owing to its lower albedo, though it will also cool faster at night.

pH or soil reaction

The reaction or pH of a soil greatly influences the growth of higher plants and of micro-organisms in the soil. pH is defined as the negative index of the logarithm of the hydrogen ion (H^+) concentration. For pure water, the amount of dissociation into H^+ and OH^- ions is very small. Thus:

$$H_2O = H^+ + OH^-$$

At 25° C the product of the ionic activities equals 10^{-14} g ions (moles) litre^{-1} i.e. the activity of each ion is 10^{-7}. Hence:

$$pH = -\log_a H^+ = 7$$

and

$$pOH = -\log_a OH^- = 7$$

At neutrality pH - pOH = 7. Thus each division below or above pH 7 represents a tenfold decrease or increase in acidity. Natural rainwater has a pH value of about 5.5, reflecting the pressure of carbon dioxide in the atmosphere with which the rainwater comes into equilibrium. There is a close connection between soil pH and the degree of **base saturation** of the soil colloid complex. The higher the proportion of the cation exchange sites which are satisfied by hydrogen (H^+) and aluminium (Al^{3+}) the lower will be the pH. Table 18.8 illustrates the terms used to describe the increasing acidity with lower pH and increasing alkalinity with higher pH.

Soil acidity is probably the most common and apparently simple test performed on soil. In detail, determining the accurate level of pH is affected by a range of analytical problems. However, field pH kits using indicators and electronic field pH meters are commonly used, as also are electronic meters in the laboratory. The reason for its widespread testing is twofold. First, pH reflects a range of important soil processes, including leaching, podzolization, calcification, salinization and humification. It is also much influenced by fertilizer use on agricultural soils, as the continual use of inorganic fertilizers leads to progressive soil acidification. Second, soil pH has important indirect effects on plant growth. Figure 18.13 illustrates the relative availabilities of major and minor nutrients according to pH values. Soils in the range pH 5.5 to 7.0 are more fertile than those higher or lower. The adverse effects of extreme acidity (low pH) or alkalinity (high pH) on plant growth are twofold, as illustrated in Figures 18.13 and 18.14. Extreme soil pHs lead to high solubility of particular metal elements; such high concentrations can easily give rise to toxicities which kill plants. Second, the influence of soil acidity and soil alkalinity on plant growth is due to indirect effects on the availability of plant nutrients.

Values of pH below 5.0 usually indicate a deficiency or unavailability of plant nutrients such as calcium, magnesium, phosphorus, molybdenum and boron. Such soils may also indicate toxic amounts of zinc, manganese, nickel and other elements due to increased solubility. Soil micro-organisms are also most active and beneficial at pH values in the range 6.0 to 8.0. These important facts explain the importance of the farming practice of liming as a means of raising pH to about 6.0 to 6.5, enhancing the

Table 18.8 *Soil acidity and alkalinity.*

Soil type	pH value
Very strongly alkaline	> 9.0
Strongly alkaline	8.5–9.0
Moderately alkaline	7.9–8.4
Mildly alkaline	7.1–7.8
Neutral	7.0
Slightly acid	6.1–6.9
Medium acid	5.6–6.0
Strongly acid	5.1–5.5
Very strongly acid	4.5–5.0
Extremely acid	< 4.5

availability of nutrients to plants. Values in the range 8.0 to 8.5 often indicate the presence of free calcium carbonate ($CaCO_3$) and the low availability of phosphorus, manganese, zinc and copper. Values higher than pH 8.5 indicate the presence of sodium carbonate (Na_2CO_3) and/or high exchangeable sodium.

Soil fertility

The fertility of a soil is its ability to support a desired crop at an adequate level of yield and quality. It must be capable of providing sufficient water, air and nutrients for satisfactory crop growth. Large areas of the world's soils still suffer from limitations on agricultural productivity because of their inability to provide one or more of these three in an optimal amount. The nutrient requirements of different plant species vary considerably, and therefore soil fertility varies for different species. In general, however, any form of agriculture removes a large amount of nutrients in the harvested crop, and therefore losses need to be replaced by the application of manures and fertilizers.

Plant roots absorb nutrients from the soil in the form of cations (ions carrying a positive charge, e.g. potassium, K^+) and anions (ions carrying a negative charge, e.g. nitrate, NO_3^-). The mechanisms of absorption must involve the expenditure of a large amount of energy, as the nutrients are concentrated up to 100 times more in the plant sap than in the soil solution. The source of energy is the plant carbohydrates which are oxidized and converted into carbon dioxide (CO_2) by respiration. The concentration of nutrients in the soil solution is very dilute, so, when a plant absorbs a nutrient, supplies in the soil solution need to be replenished quickly from a nutrient store. For cations the most important store is the exchangeable cation store on the clay and humic colloids. When the concentration of a

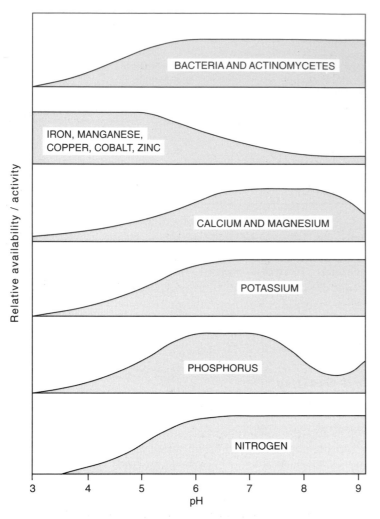

Figure 18.13 *The influence of pH on the availability of plant nutrients and on the activity of micro-organisms.*

in soils. The sum of A, B, C and D in Table 18.9 is the *total nutrients*. The total content of a nutrient can be measured by digesting the soil with a strong acid, and is a relatively straightforward value to determine. The content of available nutrients is more problematic to measure, as there are many possible dilute extractants; the value obtained is always a function of the chemical extractant which is used. From Table 18.9 categories C and D are available, plus these amounts of A and B which will be released by the weathering of minerals and the microbial decomposition of humus. Figure 18.15 illustrates the concept of nutrient availability with respect to cation nutrients (e.g. potassium, calcium).

A final important feature of nutrient availability in soils is that of **ion antagonism**. By this mechanism one nutrient element exercises an influence on another so that the latter may not be absorbed by the plant. For example, calcium and iron are antagonistic, as are calcium and potassium. There may be ample iron or potassium in a soil for normal plant growth, but if the content of calcium is high the iron and potassium will become *unavailable* for plant growth. This condition of iron or potassium deficiency shows itself by a yellowing of the plant leaves (**chlorosis**). The plant root is unable to absorb much iron and potassium physiologically in the presence of large numbers of calcium ions.

Conclusion

The properties of soils are important for determining soil fertility. The health and welfare of the human race depend upon the ability of soils to provide a sustainable yield of good-quality food. The capacity of soils to do so reflects their physical, chemical and biological properties.

Physical properties depend upon soil texture or particle size, and soil structure or aggregation. These two properties determine the ability of the soil to retain moisture for plant growth, to allow the drainage of excess water and to permit rainwater to infiltrate into the soil. Water in soil is a key element in soil fertility, and the ability of the soil to retain sufficient water for plants, but not enough to exclude oxygen from larger pores, is dependent on texture and structure.

The chemical aspects of fertility are greatly influenced by the colloidal properties of soil. The cation exchange capacity, percentage base saturation and soil pH are all interrelated. The ability of the colloids to hold cations against leaching losses is an important aspect of chemical fertility.

particular cation in the soil solution is lowered by plant absorption, cations immediately leave the exchange sites and enter the soil solution to maintain equilibrium. Exchangeable cations are thus available to the plant almost immediately. Another store of nutrients in the soil is the organic matter, but in this case the organic molecules need to be decomposed or **mineralized** first to release their nutrients. Nutrients may also be bound up in the crystalline structure of minerals in soil, which in that case have to be weathered in order to release the ion into the soil solution or on to the colloidal exchange sites.

An important concept in soil fertility is that of **available nutrients**, i.e. those nutrients in the soil solution and soil stores which the crop can reasonably be expected to absorb over the course of its growing season. Table 18.9 lists the forms in which a nutrient or indeed any chemical element may exist

Table 18.9 *Forms of nutrients in soils.*

A As part of mineral crystal structure
B As part of organic molecules
C Adsorbed on to clay and organic colloids
D As an ion in the soil solution

The organic matter in soils greatly modifies the physical properties and can mitigate the adverse properties of loose sands and heavy clays. It has colloidal properties, provides nutrients such as nitrogen and sulphur, and provides energy for soil micro-organisms.

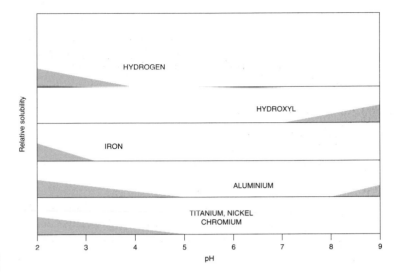

Figure 18.14 *The influence of pH on the solubility (toxicity) of chemicals in soils.*

Key points

1 The important physical properties of soil are texture, structure, water-holding capacity, permeability, infiltration capacity and aeration. These properties result from the mineral part of the soil (sand, silt, clay) and the organic fraction (humus, raw organic matter).

2 The colloidal fraction in soil consists of clay minerals and humic colloids. They have net negative charges with the ability to hold exchangeable cations by adsorption. The amount and nature of exchangeable cations govern many chemical properties, such as base status and soil reaction.

3 The fertility of a soil reflects its physical, chemical and biological properties. The yield of crops will reflect any adverse fertility factors affecting water, air, nutrients or physical support. Any limitation on the soil's ability to provide these four essential factors has serious consequences for soil fertility.

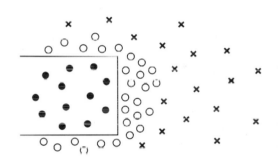

● Ions held in colloidal particles (unavailable)
○ Ions adsorbed on colloidal surfaces (exchangeable and available)
✕ Ions in solution (available)

Figure 18.15 *The distribution of cations in relation to soil colloids, and their relative availability to plants. Source: After Briggs (1977).*

Further reading

Brady, N.C., and Weil, R.R. (1995) *Nature and Properties of Soils*, eleventh edition, London: Macmillan.
The latest edition of a comprehensive and popular textbook. The examples are mostly American.

Rowell, D.L. (1994) *Soil Science: Methods and applications*, London: Longman.

A detailed discussion of soil physical and chemical properties. Many examples of practical work in the field and the laboratory are given.

Wilde, A., ed. (1987) *Russell's Soil Conditions and Plant Growth*, eleventh edition, London: Longman.
The latest edition of this classic work covers all aspects of soil science in an advanced manner.

CHAPTER 19
Soil formation

Soils are derived from the rocks and minerals which make up the surface of the Earth. They may be developed on parent materials which have not been involved in any erosion cycle; thus hard or soft bedrocks weather *in situ* to give residual soils. Such country rocks are **residual** parent materials and will consist of igneous, sedimentary or metamorphic rocks. Alternatively the soil-forming parent material may have already passed through one or more cycles of erosion and soil formation; these are **transported** parent materials and consist of sediments that have been moved by ice (moraines, till, fluvioglacial deposits), wind (aeolian sands, loess), water (alluvial, marine, lacustrine) and gravity (colluvium). In Britain these deposits form the majority of parent materials, many of which date from Pleistocene times.

Soil development, soil profiles and soil horizons

When considering soil formation it is important to distinguish two related but fundamentally different processes which are occurring simultaneously. The first is the **formation** *of soil parent materials* by the weathering of rocks, rock fragments and sediments. This set of processes is carried out in the *zone of rock decomposition* or **zone of weathering**. The end point is to produce parent material for the soil to develop in. This material is referred to as C horizon material. This applies essentially in the same way for glacial deposits as for rocks. The second set of processes is the *formation of the soil profile* or *solum* by **soil-forming processes** which change the C horizon material into A, E and B horizons. This is carried out near the surface in the **zone of soil formation**. Figure 19.1 illustrates two soil profiles, one on a hard country rock, e.g. granite, and one on a glacial deposit. In the latter case the C parent material has been much altered from the glacial deposit which was originally laid down by the ice. The zones of soil formation and weathering are not always close and

juxtaposed. In tropical regions the weathered material can be as deep as 60 m. In that case soil-forming processes will be going on in the soil profile at the surface, whilst rock breakdown and weathering will be operating at the junction of the weathered residue and fresh rock many metres below the solum.

Soil development is a complex of many processes acting over many years. Figure 19.2 shows the interconnections between the main processes. Soil development is viewed as two sets of processes, namely **weathering** and **morphogenesis**. Atmosphere and hydrosphere provide gases (oxygen, carbon dioxide, nitrogen) and water which support plants and organisms (soil fauna and soil organisms) which provide the soil with its organic matter and organisms. Parent material weathers under the influence of the atmosphere and hydrosphere (carbon dioxide, oxygen, water) to produce four components in soil: a relatively **resistant residue** consisting of quartz, feldspars and heavy minerals (e.g. zircon, iron minerals); secondary minerals or **alteration compounds** synthesized by weathering processes and consisting of clay minerals and hydrous oxides of iron and aluminium; a component or *organic matter* derived from plant and animal residues; and finally a **weathering solution** containing cations, anions and silica.

Of these four components, only the resistant residue is relatively stable and changes only slowly. The organic matter is decomposed by the soil's fauna and micro-organisms to produce humus. The clay minerals and hydrous oxides undergo further alteration, depending on the amount of leaching and the type of ions in the weathering solution; new minerals can crystallize and previous ones can be altered. Ions in the weathering solution can be thrown out of solution and precipitated in the solum if chemical conditions allow (calcium carbonate, gypsum, soluble salts, secondary quartz from silica).

The final stages of soil formation consist of the processes of morphogenesis, i.e. the production of a distinctive *soil profile* with its constituent layers or *horizons* (Plate 19.1). The soil profile is the vertical

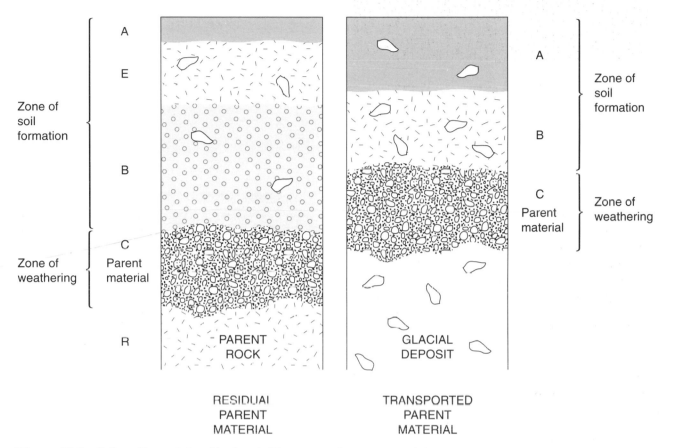

Figure 19.1 *Soil profiles on (a) residual and (b) transported parent materials.*

section through the soil; it is the fundamental unit for describing, sampling and mapping soils. The soil horizons are the distinct layers, roughly parallel to the surface, which differ in colour, texture, structure and content of organic matter. The clarity with which horizons can be recognized depends upon the relative balance of the migration, stratification, aggregation and mixing processes in Figure 19.2. Some soils tend to show striking horizonation (e.g. podzols) whereas in others the horizons are less distinct (e.g. vertisols). Table 19.1 lists the processes which create and destroy clear soil horizons. (See Plate 19.2.)

When horizons are studied they are each given a letter symbol to reflect the genesis of the horizon. There are many different schemes in use by the major soil survey organizations in the world. There are broad similarities, and full details can be found in their published soil memoirs (Soil Survey of England and Wales; Soil Survey of Scotland; National Soil Survey of Ireland; Soil Conservation Service of the United States). Internationally there are two commonly used soil classification schemes: the *Soil Taxonomy* of the USDA (United States Department of Agriculture) and the system used by FAO–UNESCO (Food and Agriculture Organization –

United Nations Educational Scientific and Cultural Organization). The FAO system is the one used in this book and details of it are given in the appendix.

The soil-forming environment

The nature of the soil profile at any particular place depends upon five main factors. These are: the past and present climate; the physical and chemical

Table 19.1 *The formation of soil horizons.*

Vertical redistribution of soil materials

- Leaching of ions in the soil solution
- Movement of clay-sized particles
- Upward movement of water by capillarity
- Surface deposition of dust and aerosols

Mixing processes

- Organisms (e.g. cambisols, chernozems)
- Cultivation of agricultural soils
- Creep processes on slopes
- Frost heave (cryoturbation)
- Swelling and shrinkage of clays (e.g. vertisols)

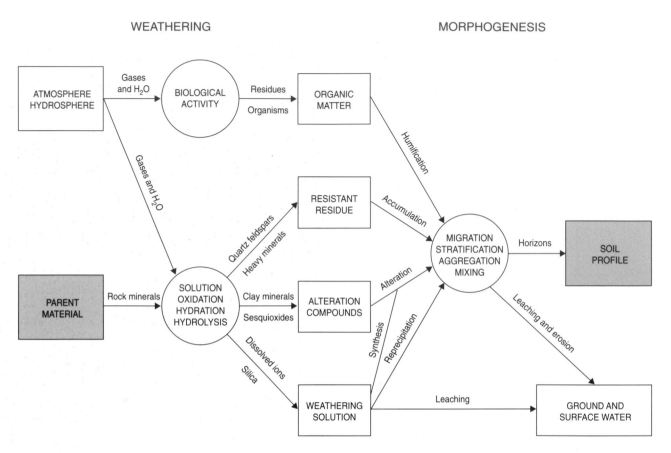

Figure 19.2   *A scheme of soil development.*

characteristics of the parent material; relief and hydrology; the length of time during which soil-forming processes have been active; and the ecosystem, including vegetation, fauna and the effects of human activities. Dokuchaiev, a famous Russian soil scientist of the nineteenth century, was the first to record the connection between the genesis of soil profiles and these five controlling factors (Figure 19.3). Later an American soil scientist, Hans Jenny (1941), expressed the relationship in his equation of soil formation:

$$s = f \, (cl, \, p, \, r, \, t, \, o)$$

where s = soil profile or property, f = a function of, cl = climate, p = parent material, r = relief, t = time and o = organisms, including humans. The importance of the work of Dokuchaiev, Jenny and other pedologists is that soils are recognized as 'independent natural bodies', each with a distinctive succession of horizons reflecting the combined effects of a particular combination of the five genetic factors. Over time, soils were considered to evolve towards a condition of equilibrium corresponding to a particular ecological climax.

The effects of climate and soil formation operate through precipitation and temperature. High rainfall produces intense leaching and strongly acid soils. Lower rainfall gives less marked leaching, with the possibility of calcium carbonate in soils over calcareous rocks and deposits. Temperature affects the speed of biochemical reactions in soil and the rate of evapotranspiration from the soil surface. Thus low summer temperatures retard the decomposition of organic matter and encourage its accumulation at the surface as peat.

The main properties of parent materials that influence soil formation are the permeability, base content, hardness, grain size and mineralogy of their weathering products. On drift deposits, poorly drained soils will form in fine-textured clays and silts. However, well drained brown earths (cambisols) occur where the deposits are more permeable. Hard igneous, metamorphic and sedimentary rocks disintegrate only slowly, to give shallow, stony, coarse-textured soils. Soft rocks give deep, less stony, loamy soils. Where rocks are base-deficient, it is common to have acid or podzolized soils. The clay mineralogy affects the potential for shrinking and swelling and the composition of the cation exchange complex.

Relief and the slope profile influence hydrology and soil water régime. On undulating ground with slowly permeable parent materials, surface water-logging causes gleying on flat ground, but soils on slopes are drier, as most rainwater runs off the surface or through upper horizons to lower ground. Gleying reappears in valleys and basins where run-off and through-flow concentrate. The distribution of soils is shown is Figure 19.4a. In permeable materials water penetrates to the subsoil, leaving higher ground well drained. On lower land soils are affected by groundwater, as shown in Figure 19.4b.

The length of time a soil remains undisturbed by erosion or deposition is important in its evolution. The glaciations of the Pleistocene era removed the old soils from much of Britain and other countries of mid to high latitudes and deposited thick drift in other regions. Soil formation began again on new surfaces after the final retreat of the ice some 10,000 years ago. Old soils are more common in low latitudes where the soil cover was not eroded or buried during the Pleistocene.

Human activities have many effects on soils. Indirectly, the native vegetation can be modified or removed. Directly, soils are changed by agricultural

Plate 19.1   Carboniferous limestone is weathered by physical and chemical processes to give a regolith of rock fragments and clay minerals. Morphogenesis produces the solum of a dark humic (Ah) horizon above a lighter-coloured weathered (Bw) horizon.
Photo: Ken Atkinson.

Figure 19.3   Factors affecting the formation of soils.

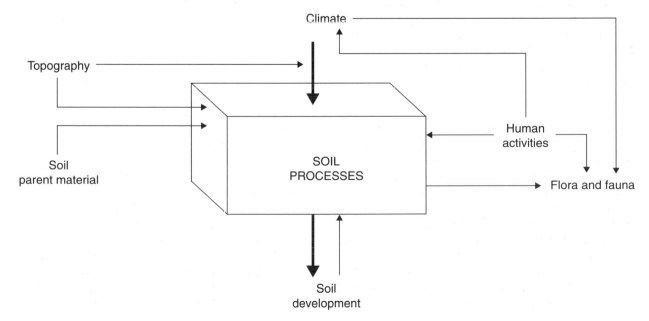

Table 19.2 *Processes which form soils.*

Permafrost soil processes (cryopedology)
Leaching
Clay translocation
Podzolization
Decalcification
Calcification
Gleying
Rubefaction
Salinization
Alkalization
Solodization
Laterization

Plate 19.2 *An arctic brown soil (FAO: Gelic Cambisol) with the top of the permafrost at the base of the soil pit. Mixing by frost action (cryoturbation) in the active layer moves black tongues of organic material into the subsoil. Profile depth 75 cm.*
Photo: Ken Atkinson.

influences vegetation and human activities and is itself affected by altitude and topography. Figure 19.5 shows the relationship of climate, altitude, slope and soils for northern England. Precipitation increases with altitude, giving intense leaching and podzolization on well drained sites, chiefly steep slopes. Waterlogging occurs on high ground owing to higher rainfall, lower evaporation and lower transpiration. Organic matter accumulates as peat on summits, or as peaty surface horizons (stagno-podzols and stagno-humic gleys) for the same reasons. On lower ground, where rainfall is lower and temperatures are higher, leaching is weaker, weathering is stronger and brown earths tend to form. Podzols are restricted to coarse-textured base-deficient parent materials.

The combined influence of the five factors of soil formation is to produce a set of **soil-forming processes** which produce the world's distinctive soil profiles and their constituent horizons. The processes are listed in Table 19.2 and will be discussed in the remainder of this chapter, except for **permafrost soil processes** (**cryopedology**), which are covered in Chapter 24, and *rubefaction*, which is analysed in Chapter 26.

Leaching, decalcification, calcification

The process of **leaching** is caused by the continual washing of the soil with rainwater. Rainwater has a natural pH of about 5.5, owing to dissolved carbon dioxide, making it a weak hydrocarbonic acid (H_2CO_3). In some regions atmospheric pollution by sulphur dioxide (SO_2) and nitrogen oxides produces an even more acid leachate, 'acid rain'. Also, as it passes through the surface organic horizon, it dissolves organic acids from decomposing plant residues. It is thus able to dissolve and decompose minerals and carry away cations and anions dissolved

practices. For example, pollen analysis (*palynology*) of upland Britain shows that the clearance of the upland deciduous forest by Neolithic people led ultimately to the development of heathland. Deforestation broke the nutrient cycles of the brown earths under deciduous trees and led to the acidification of soils and the invasion of heather. This caused acid humus and thin peat on wetter sites, both leading to podzolization. Soils used for arable agriculture have their relations with soil-forming factors changed by ploughing, draining and the use of lime and fertilizers.

The five factors of soil formation do not operate as single independent factors, of course. Climate

(a) Impermeable parent material

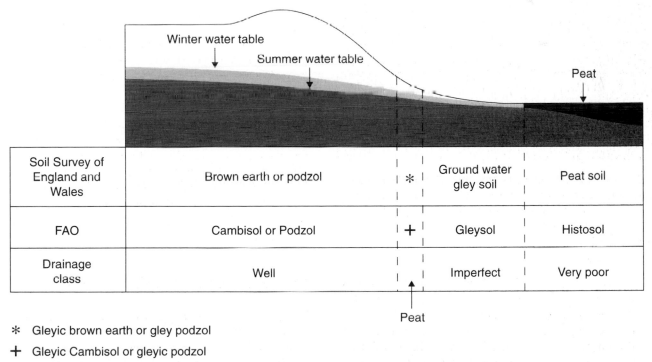

Soil Survey of England and Wales	Surface water gley soil	Gleyic brown earth	Brown earth	Gleyic brown earth	Surface water gley soil
FAO	Gleysol	Gleyic Cambisol	Cambisol	Gleyic Cambisol	Gleysol
Drainage class	Poor	Imperfect	Well	Imperfect	Poor

(b) Permeable parent material

Soil Survey of England and Wales	Brown earth or podzol		*	Ground water gley soil	Peat soil
FAO	Cambisol or Podzol		+	Gleysol	Histosol
Drainage class	Well			Imperfect	Very poor

Peat

* Gleyic brown earth or gley podzol

+ Gleyic Cambisol or gleyic podzol

☐ Never waterlogged; unmottled   ▨ Seasonally waterlogged; strong mottling

☐ Occasionally waterlogged; slight mottling   ■ Permanently waterlogged; typically grey or blue grey

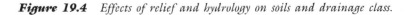

Figure 19.4 *Effects of relief and hydrology on soils and drainage class.*

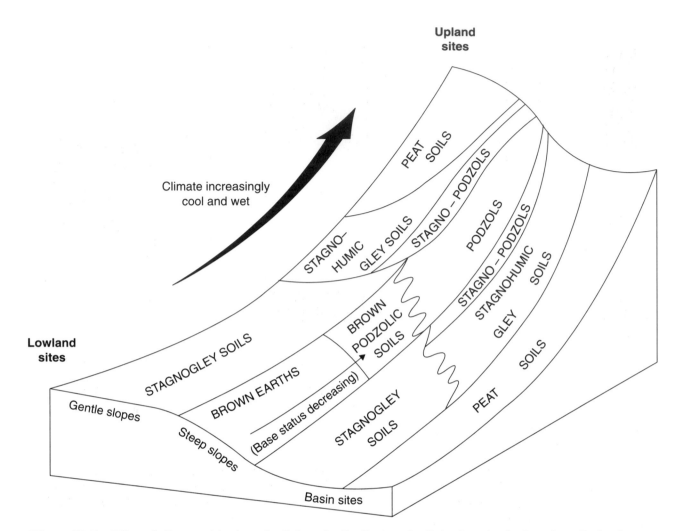

Figure 19.5 *Effects of climate, altitude and relief on the distribution of soils in the uplands of northern England.*

in the soil solution. The basic cations (bases) held on the soil colloids (calcium, magnesium, potassium, sodium) are released from the colloid surface and replaced by hydrogen or aluminium ions. This leads to a lower pH and percentage base saturation as leaching progresses. Thus soil pH is a good general indicator of the intensity of leaching, which is related to the amount of annual rainfall and the chemistry and texture of the soil parent material.

The leaching of bases is an important process ecologically and agriculturally, as the ions are moved downwards out of the rooting zone of plants. In natural vegetation it is important that deeply rooting species (e.g. grasses, deciduous trees in the temperate zone) are able to recycle nutrients from deep in the subsoil and thus act as a kind of 'nutrient pump'. Shallow-rooting species (e.g. coniferous trees and many heath plants) are at a disadvantage and, other things being equal, will not be able to counteract leaching losses as effectively. In farming practices,

lime and fertilizer are applied in order to balance both the leaching losses and the heavy withdrawal of nutrients by crop yields. The French pedologist Duchaufour has coined the term *lixiviation* to designate the process of chemical leaching.

The main soil profile formed by leaching is the brown earth soil (FAO Cambisol or Phaeozem). The sequence of horizons is Ah/Bw/C/R, with Ah replaced by Ap where ploughed (Figure 19.6). The colour, texture and pH of the soils vary with the type of parent material. Sandy textures can lead to a 'slight weathering-intense leaching' régime which produces low pH values in the range 4.0–5.5. In the past these brown earths were called 'low base status brown earths', in order to distinguish them from 'high base status brown earths' on less acid parent materials (e.g. basic igneous rocks, calcareous deposits). The acid variants have partially decomposed litter (F and/or H horizon) at the soil surface with a thin eluvial Ea or Ae to give a F/H/Ae/

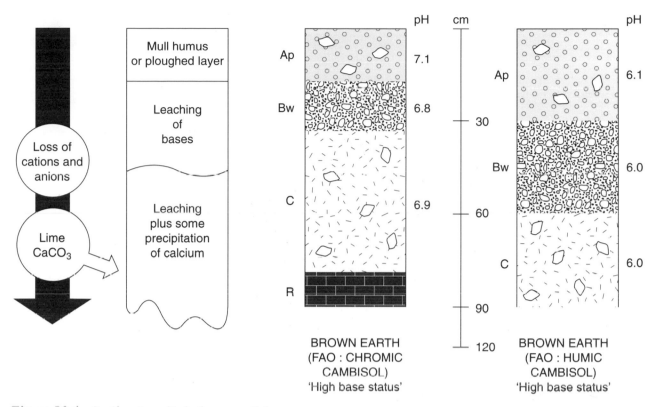

Figure 19.6 *Profiles of cambisols (brown earths).*

Bw/C/R sequence of horizons. On more basic parent materials the soils have a strongly formed structure, a relatively high percentage base saturation and a high cation exchange capacity, due to expanding clay minerals vermiculite and montmorillonite. A variant has formed on ultrabasic rocks (e.g. on serpentines in Scotland or the Lizard peninsula in England) in which magnesium is the dominant exchangeable cation, often exceeding calcium by a factor of five, and in which the contents of heavy metals such as nickel, chromium and cobalt are exceptionally high. Pastures on these soils can give rise to heavy-metal poisoning in cattle.

The removal of free calcium carbonate from soils by leaching is called **decalcification** and leads to a lowering of soil pH. In humid regions free calcium salts will be washed out of the soil profile but as the rainfall decreases in semi-arid and arid regions they may not be entirely removed but instead be deposited in the subsoil. This is called **calcification** and occurs with calcium carbonate ($CaCO_3$) and calcium sulphate (gypsum $CaSO_4.2H_2O$). Where these accumulate in the lower profiles, calcic (Bk and Ck) and gypsic (By and Cy) horizons are formed. As gypsum is more soluble than lime, the gypsic horizon is found below the calcic horizon. It also disappears first from the soil profile in a sequence of soils from arid to humid regions. The sequence in Figure 19.7 shows these changes taking place along a north–south transect in central Canada as one moves from a humid to a semi-arid climate.

Podzolization

Podzols were first named by peasants in the Russian *taiga*, or coniferous forest, who noticed a distinct white horizon below the surface litter and at depth a black layer. Believing the black layer to be charcoal from past forest fires, they called the white layer 'podzol' or literally 'ash soil'. This is not, of course, the way in which the soils are formed, but it describes well the pale surface horizon of **eluviation** (outwashing) overlying the blackish or orange horizon of illuviation (in-washing). (Plate 19.3.) Podzols are the result of intensive leaching and the translocation of sesquioxides. These processes occur typically under coniferous woodland, mixed forest vegetation and heath, especially in sub-arctic and temperate climates. These vegetation types produce an acid humus layer on the soil surface. This acid humus or **mor** decomposes only slowly and only partially, releasing organic acids (fulvic acids) and reactive organic chemicals (polyphenols, hydroquinones) which can form

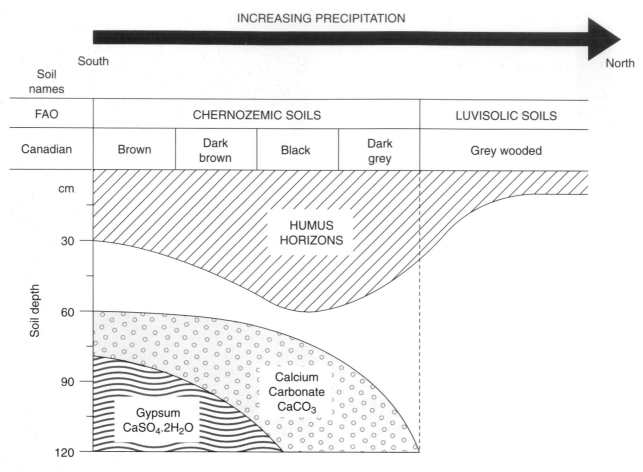

Figure 19.7   *The position of lime and gypsum horizons along a transect of increasing aridity from north to south in central Canada.*

complexes with the iron and aluminium cations in the soil. The chemical complex is often in the form of a **chelate**, as illustrated in Figure 19.8. In this form the normally immobile iron and aluminium can migrate easily downwards in percolating water. These sesquioxides are deposited in the B horizon together with any humic colloids. The resulting soil is strongly acid at the surface (pH 3.0), with a bleached sub-surface horizon (Ea) above B horizons where humus, iron and aluminium have been precipitated. As one might expect, the most distinctive podzol profiles are found where drainage is good and the parent material is acid (e.g. acid igneous and metamorphic rocks; sandstones; depositional acid sands).

Most soil classifications recognize different types of podzol, depending on local conditions. Three varieties are illustrated in Figure 19.9. The typical humo-ferric podzol (FAO: Orthic Podzol) has accumulations of organic material (Bh or Bhs) overlying the horizon of iron accumulation (Bs). (Plate 19.4.) This is the normal profile found under coniferous and mixed forest in Russia and North America, where

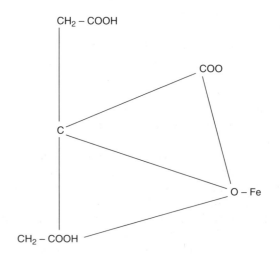

Figure 19.8   *A chelate structure holding an atom of iron.*

the parent materials have a reasonable content of iron. They are common on coarse, non-calcareous materials or on materials like glacial tills from which free lime has been removed. By contrast, humic podzols have strong Bh horizons but lack a horizon of iron accumulation. They form under cool, moist conditions and usually on parent materials with a low iron content.

The third type is the iron pan stagnopodzol (FAO: Placic Podzol), where 'placic' signifies the presence of a thin ironpan (Bf) above or within the podzolic Bs horizon. Frequently the Bf is so impermeable that it blocks the downward percolation of water, causing gleying in the bleached horizon (Eag). The reducing conditions so produced lead to the reduction of ferric iron (Fe^{3+}) to the more mobile ferrous iron (Fe^{2+}), which causes further iron removal from the Eag. In Britain placic podzols are also common in uplands, where high precipitation causes peaty surface horizons. Descending soil water reaches less weathered iron-rich parent material below seasonally waterlogged upper horizons. The soils thus show features of both gleying and podzolization, hence the name 'stagnopodzols' in the Soil Survey of England and Wales. The former name for the placic podzol is the 'peaty gley podzol with ironpan', under which appellation it occurs in older Soil Survey reports.

It is likely that several different development pathways can lead to this soil profile. The Scottish pedologist FitzPatrick has suggested an alternative evolutionary sequence after observing that the ironpan frequently occurs above a dense subsoil; in such cases the ironpan picks out a physical interface in the subsoil, the interface marking the upper limit of permafrost conditions in Pleistocene times. Such compact subsoils are termed **fragipans** and they have a high density and low porosity which cause precipitation of the sesquioxides. Another pathway for the development of soil with thin ironpans is dependent on vegetation change. From evidence of archaeology and radiocarbon dates, it is clear that there are sites in North Wales and Scotland where acid brown soils under forest preceded the placic podzols. Thicks Bs horizons below the pan suggest that the Eag and Bf horizons of the present soils resulted from the formation of surface peat between 2000 and 1000 years BP following the replacement of deciduous forest cover by moorland and heathland plants. Whether this vegetation change resulted from climatic deterioration or human influence is not always clear. In this case the placic podzol is a *poly-cyclical* soil, reflecting more than one cycle of soil formation.

Plate 19.3 *Leaching of fluvioglacial sands has produced a bleached eluvial (Ea) horizon. Precipitation of medium acidity is made more acid by organic acids from the surface litter. Profile depth 2 m. Photo: Ken Atkinson.*

Clay formation and translocation

The formation of clay-sized particles is a fundamental feature of soil formation. The colloids consist of clay minerals and hydrated oxides of iron and aluminium. In many soils clay content increases from the A horizon down to the B horizon and then decreases in the C horizon. The B horizon may acquire its higher clay content in two ways; either percolating waters with chemical elements in solution are precipitated in the B horizon to form new clay minerals or percolating waters carry clay minerals from the A horizon in suspension, which are deposited in the B.

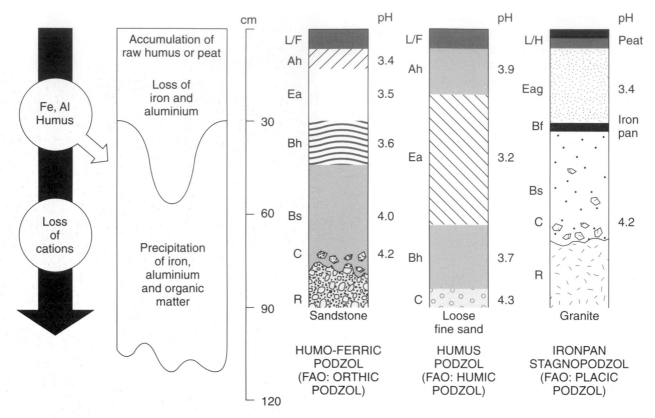

Figure 19.9 *Various types of podzol soil.*

The process of clay mineral formation *in situ* in the B horizon gives cambic B or weathered B horizons designated Bw. As the name suggests, the process is characteristic of cambisols (or brown earths). The type of clay mineral formed in this way depends upon the parent material and the degree of leaching in the soil profile. In freely drained soils with intense leaching, silica tends to be removed, thus tending to produce low-silica 1:1 clay minerals such as kaolinite. Soil age may also be a factor here, as kaolinite is the most resistant clay mineral and tends to accumulate over prolonged periods of weathering (e.g. tropical Ferralsol soils). In poorly drained soils, especially where calcium and magnesium occur, there is usually enough silica and bases to form montmorillonite and vermiculite (e.g. in vertisols and chernozems). Under moderate leaching and a reasonable supply of bases, illite and chlorite are often the dominant clay minerals. They are the commonest clay minerals in British soils. Illite, which has a mica-type structure, may also be formed directly from mica minerals in parent materials. In acid soils such as podzols the increase in clay-size material in the B horizon is due to the formation of hydrated oxides of iron and aluminium rather than clay minerals. It results from the total loss of silica from the soil profile under the influence of organic acids. In that case the B horizon would be termed Bs, the spodic (podzolic) B.

The second case, the movement in suspension of discrete clay particles by water percolating to lower levels in the soil, produces clay enrichment in the B to give the Bt, the luvic (argillic) B. This form of clay leaching is a dominant process in luvisols and acrisols (argillic brown earths) and is variously termed clay translocation, clay eluviation, clay illuviation (in-washing) or *lessivage* (the French term). However, it can also occur in other soil types. Various factors favour clay translocation. It operates best in slightly acid conditions when the clay particles are dispersed and not flocculated by the presence of calcium. The clay is usually precipitated in dry periods and moved in wet periods. Thus areas with climates with marked wet and dry seasons, e.g. in Mediterranean, savanna and continental regions, favour clay translocation. This may also explain why the process is common in southern and eastern England but not in western Britain or Scotland. The clays are deposited as oriented clay or coats (called cutans) on the surface of structural aggregates or in pores and around stones. These are sometimes visible to the naked eye but often they need to be identified as thin cutans by the study of thin sections under a petrological microscope (Figure 19.10). When the individual clay plates are deposited they become oriented parallel to each other; the entire cutan has the property of bi-refringence, hence the term 'birefringent clay'.

The typical horizon sequence for luvisols (argillic brown earths) is A/E/Bt. This is illustrated in Figure 19.11, an uncultivated soil from a beechwood in the Chilterns. The parent material is colluvium with many chalk fragments. The leaching of calcium carbonate from the upper horizons will have taken place before the clay minerals were translocated.

Gleying

Soils which are affected by temporary or permanent waterlogging have very distinctive profiles. When pore space is occupied by water rather than air, a series of reduction processes replace the oxidative processes in well aerated soils. One of the main reduction reactions is that involving iron oxides which reduce from ferric to ferrous compounds according to the equation:

$$Fe(OH)_3 + e^- + H^+ = Fe(OH)_2 + H_2O$$

ferric hydroxide · electron · from organic matter · ferrous hydroxide

The process of ferric iron reduction to more mobile and grey ferrous iron compounds is partially chemical, partially carried out by anaerobic micro-organisms and partially carried out by the products of decomposing organic matter. The process is known as **gleying**. When the soil or horizon is permanently gleyed it has a uniform grey or blue-grey colour. Where the soil or horizon is gleyed only temporarily or seasonally, and reoxidation can take place, the soil shows reddish-orange *mottles*, or segregations of ferric oxides. In tropical and subtropical climates, dehydration often produces hard, black iron–manganese concretions. Mottles and concretions are usually found in more porous spots within the horizon or along root and faunal channels, where air can enter.

Waterlogged soils form the basis of the paddy system of rice cultivation in many tropical and sub-tropical regions. The typical soil profile of a rice paddy is shown in Figure 19.12. The sheet of water at the soil surface supports algae, which keep the water oxygenated and some of which are able to fix atmospheric nitrogen. Below it is a brown aerobic horizon and below that a thick blue-grey gleyed layer.

High-yielding varieties of rice require a high level of nitrogen. Nitrates which form in the surface soil

Plate 19.4   *A humus–iron podzol (FAO: Orthic Podzol) beneath acid heath on quartzite gravels. The horizon sequence is L/F/Ea/Bh/Bf/C. Profile depth 1 m.*
Photo: Ken Atkinson.

layer are able to diffuse into the anaerobic reduced soil horizon below. Here they are reduced to the nitrogen (N_2) and nitrous oxide (N_2O) gaseous forms and are lost to the atmosphere. The two most commonly used fertilizers are ammonium sulphate and urea; these are placed in the anaerobic layer to prevent the oxidation of ammonium ions to nitrates, thereby reducing losses by denitrification.

In temperate regions gleyed soils are of two main types. Surface-water gleys result from drainage being restricted by a slowly permeable subsoil; they are also known as stagnogleys. In upland locations with high rainfall a peaty surface can form to give a stagno-humic gley soil (Figure 19.13). On permeable parent

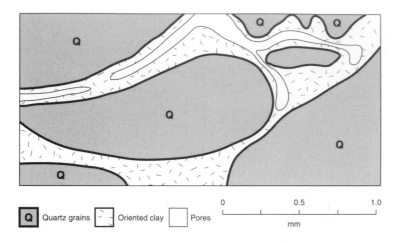

Figure 19.10 *Cutans of oriented clay formed by the translocation of clay particles into the Bt horizon.*

materials, gleys will form only in topographical hollows and at low points in the landscape under the influence of groundwater (Figure 19.14). These are classed as groundwater gleys. As a group, gley soils can have a wide range of pH values, being acid, alkaline or even calcareous. The pH and base status

is mainly determined by the acidity and basicity of the parent material. Groundwater gleys tend to have higher pH values, owing to the low-lying sites receiving bases washed in from surrounding slopes.

Salinization, alkalization, solodization

Under arid and semi-arid conditions, leaching of the soil profile is very weak. In normal desert soils it is still sufficient to remove soluble salts, though the less soluble gypsum ($CaSO_4.2H_2O$) and lime ($CaCO_3$) accumulate as distinct layers (By and Bk horizons). If, however, there is a higher input of soluble salts into the soil, leaching may not be sufficiently powerful to remove them and they accumulate, usually as a salt-enriched surface (Az horizon) or salt crust. The enrichment of salts is common wherever groundwater comes close to the soil surface, such as in river flood plains and low-lying depressions. Also, salts will accumulate where waters from inland drainage accumulate or where lakes exist or existed in the past. Coastal areas can also accumulate salt from aerial sea spray ('cyclical salt'), and from the

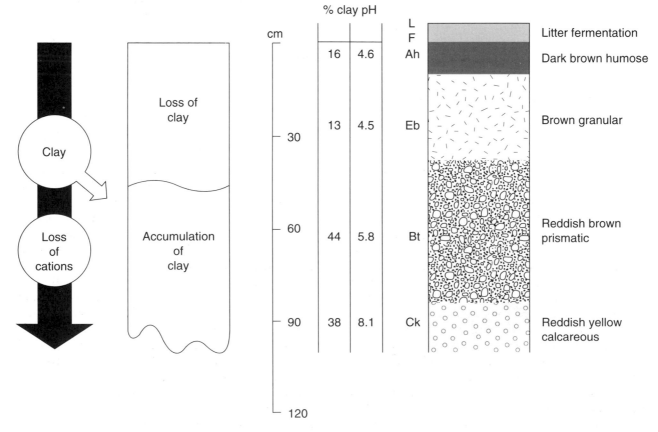

Figure 19.11 *A luvisol soil, showing the effects of clay translocation.*

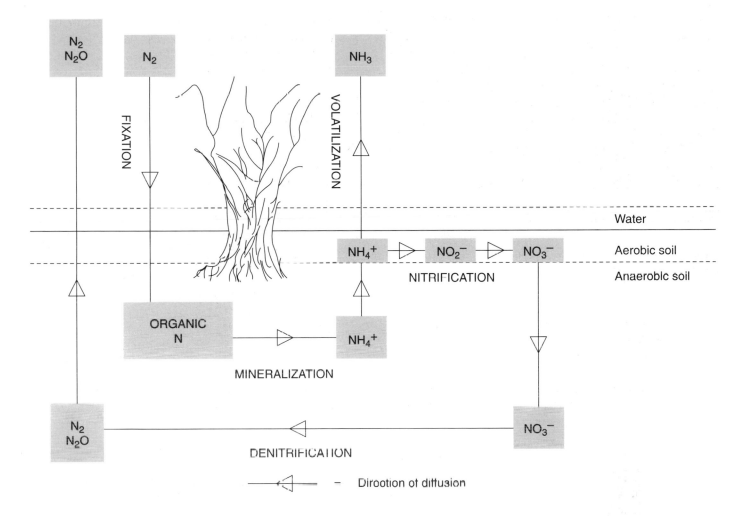

Figure 19.12 *The nitrogen cycle in a flooded rice paddy.*

incursion of salt water into the coastal aquifer. However, faulty irrigation schemes have been a major reason for the spread of salt-affected soils in the twentieth century. Misuse by agriculture has resulted from the over-application of irrigation water or failure to provide efficient drainage to remove surplus water from the soil. As the water table rises, capillary forces are able to move water containing ions, and water vapour, to the soil surface, where, under the intense prevailing evaporation, salts are deposited in pores and on the surface. This is the so-called 'wick' effect. Where salt crusts are formed, they usually consist of a finely comminuted salt dust which can be blown up into the atmosphere, eventually to come down by gravity or in rain to influence soil formation in surrounding areas. Salinity is also added to the soil surface by the addition of fertilizers to the irrigation water, much of which will not be taken up by plants and will increase the salt content of the surface horizon. Salty soils will also occur where saline

groundwater results from the presence of salt deposits in the geological column.

Saline soils described above are classified as *solonchaks*, though they have also been called **white alkali soils** in the United States. They contain sulphates and chlorides of sodium and potassium, though magnesium and nitrate ions may also occur. They show white salt efflorescence at the surface, but usually show no change in structure down the profile (Figure 19.15). They are usually low in humus, on account of the low productivity of natural vegetation on such soils, and the consequently low input of dead plant residues. The pH values are in the range 8.0 to 8.5 but go no higher because of the high concentration of neutral soluble salts. Soil-forming processes are inhibited and profile development is minimal.

If a situation arises whereby salts no longer accumulate at the surface, there will be far-reaching changes in the soil profile. For example, if the water table falls or rainfall increases, rainfall may wash the

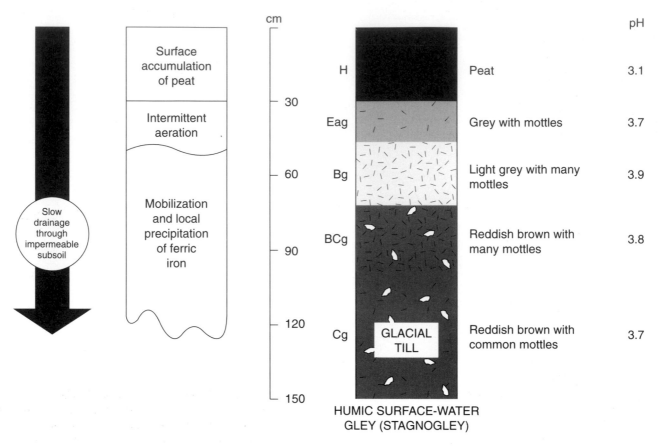

Figure 19.13 *Profile of a humic stagnogley (surface-water gley) soil.*

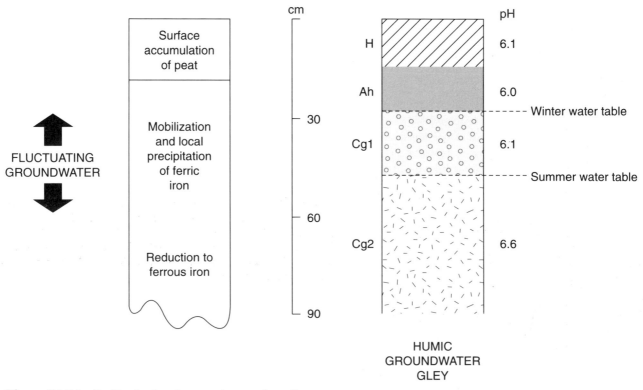

Figure 19.14 *Profile of a humic groundwater gley soil.*

Plate 10 *A brown earth soil (FAO: Eutric Cambisol) developed under oak woodland in glacial till at Loch Lomond, Scotland. The surface humus (H) overlies a thin, brown leached zone (Eb) above a thick, reddish brown horizon (Bw) formed by the weathering of rock minerals.*

Plate 11 *An iron podzol soil (FAO: Ferric Podzol) formed in fluvioglacial sands beneath spruce in the boreal coniferous forest zone of Canada. The litter and fermentation layers (L and F) at the surface overlie the white eluvial horizon (Ea) and the orange-brown illuvial iron horizon (Bs)*

Plate 12 *A humus-ironpan stagnopodzol soil (FAO: Placic Podzol) developed in acidic glacial till in Glen Fiddich, Scotland. A peaty surface (O) overlies a grey leached and gleyed subsurface horizon (Eag). Then comes a dark coloured horizon of illuvial humus (Bh), above the thin ironpan (Bf) which follows a wavy course through the soil.*

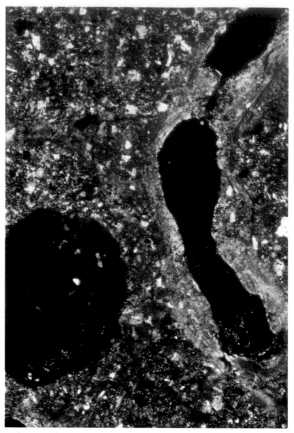

Plate 13 *Photomicrograph of an horizon formed by clay illuviation (Bt) in an argillic brown earth soil (FAO: Orthic Luvisol). Under cross-polarized light, the pores in the soil to the right appear black (isotropic), with the bright yellow colours showing the coatings of clay which line the pores (bire-fringence). The black round object to the left is organic material (frame width = 2 mm).*

Plate 14 *A groundwater gley soil (FAO: Eutric Gleysol) formed in estuarine clay in Stirlingshire, Scotland. Dull bluish colours at the base of the profile indicate anaerobic reducing conditions (Cg), whilst above are brown mottled horizons (Ag and Bg), which, though still gleyed, have brighter colours indicating better aerated, partially oxidizing conditions.*

Plate 15 *A black chernozem soil (FAO: Calcic Chernozem) in the prairie of Alberta, western Canada. This is one of the most fertile soils in the world. The humus-rich surface (Ah) overlies weathered (Bw) and calcic (Bk) horizons. Lime deposition is clearly seen, as is the water table at the base. The scale is in feet.*

Plate 16 *A salt crust has formed on the surface of an irrigated field on sands in North Africa. The salinity results from upward capillary movement of soil water under high rates of evaporation. This is worsened by the addition of chemical fertilisers to the irrigation water. The scale is in centimetres.*

Plate 17 *A lateritic soil (FAO: Rhodic Ferralsol) developed under tropical rain forest in Ghana, West Africa. A thin surface humus (Ah) overlies the dominant red-coloured (Box) horizon. The profile is 2 m deep and intensely leached. It consists of ferric oxides, kaolinite and quartz, with no weatherable minerals.*

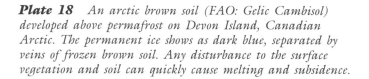

Plate 18 *An arctic brown soil (FAO: Gelic Cambisol) developed above permafrost on Devon Island, Canadian Arctic. The permanent ice shows as dark blue, separated by veins of frozen brown soil. Any disturbance to the surface vegetation and soil can quickly cause melting and subsidence.*

Plate 19 *A red Mediterranean soil or terra rossa (FAO: Chromic Luvisol) developed on limestone in northern Libya. The process of rubefaction produces hematite iron oxides which give the bright red colours masking other soil properties.*

INCREASING RAINFALL, LEACHING AND PROFILE DIFFERENTIATION

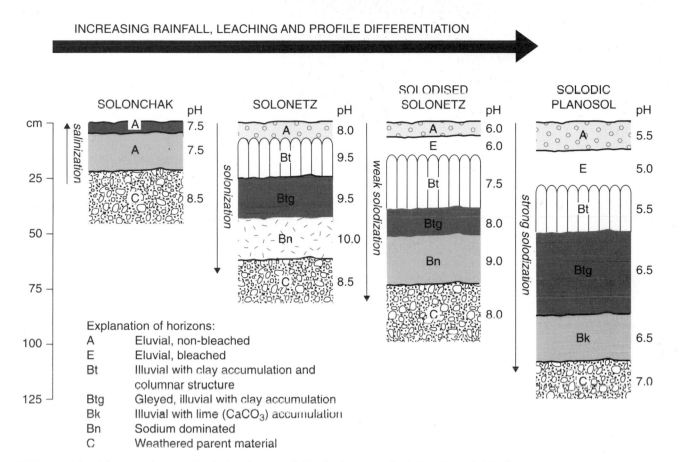

Figure 19.15   *Formation of solonchak, solonetz, solodized solonetz and solodic planosol (solod).*

salts through the profile. If the salts are mainly calcium salts, soil formation will go in the direction of xerosols or chernozems. However, if a significant proportion of the exchangeable cations are sodium (perhaps over 15 per cent) sodium carbonate will be formed, owing to the carbonate (CO_3^{2-}) and bicarbonate (HCO_3^-) anions continually being produced by plant roots and organisms. This gives the soil a pH of 9.0 or over, a structure which becomes unstable and deflocculated and a soil surface which is dark-coloured, often black, owing to dispersed humic particles. Dispersed clay particles are washed down in the profile and form a clay pan which dries into hard columnar units. These soils are called **solonetz** and are also termed *black alkali soils* in the United States.

As the leaching of salt proceeds, and more clay and organic matter moves into the clay pan, a distinct pale horizon (E) forms above a spectacular columnar structure, the top of the columns having a white amorphous silica coating. Soils with this striking profile are known as *solodized solonetz*. As even more leaching takes place, significant amounts of sodium are removed from the exchange complex and the B

horizon structure is lost. The resulting soil has a loose, coarse-textured acidic A horizon over a hard, compact B horizon with a pH neutral to acidic. This soil profile is a **solodic planosol**, previously called **solod**. The succession of soils from solonchak to solodic planosol is shown in Figure 19.15.

Soils affected by salinity and alkalinity have low fertility. High concentrations of soluble salts are harmful to plants. The salinity of the soil is measured by the **electrical conductivity** (EC) of a saturation paste or of water extracted from a soil which has just been saturated with water (the *saturation extract*). The units are mS cm^{-1}, millisiemens per centimetre. Table 19.3 shows typical values. The alkalinity of soil

Table 19.3   *Significance of EC values.*

EC mS cm^{-1}	Effect on crops
0–2	Negligible
2–4	Sensitive crops: reduced yield
4–8	Many crops: reduced yield
8–16	Only tolerant crops yield satisfactorily
> 16	Only a few very tolerant crops yield satisfactorily

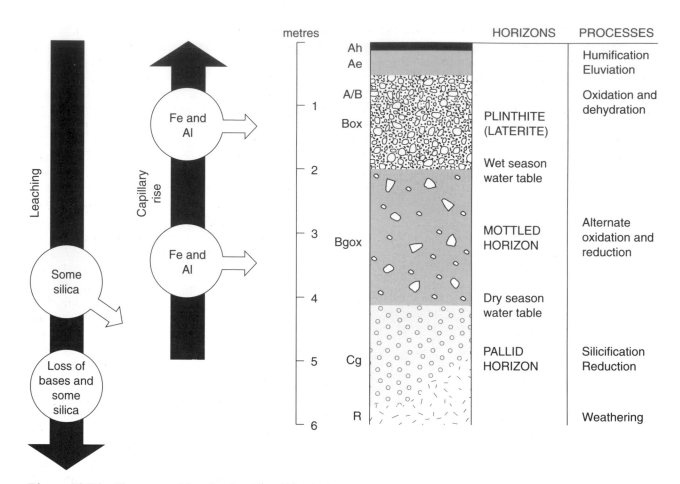

Figure 19.16   *The process of laterization (ferralitization).*

is measured by three properties: the pH, the **exchangeable sodium percentage** and the **sodium adsorption ratio** (the ratio of sodium ions to the square root of the calcium plus magnesium ions):

$$SAR = \frac{Na^+}{\sqrt{Ca^{2+} + Mg^{2+}}}$$

Laterization

Tropical regions are well known for the speed and intensity of processes of weathering and soil formation. Where abundant moisture is available to match the prevailing high temperatures, the soils are the product of intense tropical weathering which removes all the geochemically unstable elements (potassium, sodium, magnesium and calcium) from the soil and concentrates sesquioxides (oxides of iron, aluminium and manganese) and silicon. The rapid breakdown of rock minerals and the thorough leaching of base elements in tropical environments has been termed **katamorphism**.

The resulting soil profile is shown in Figure 19.16. At the surface, organic matter is rapidly decomposed, so that the humus Ah horizon is very thin. Some surface eluviation may produce a thin, eluvial layer (Ae) but it is usually masked by the red iron colours. The surface horizons are commonly high in concretionary iron particles about the size of a large pea. They thus appear gravelly but the gravels are pedologically formed, not alluvially formed. This is the horizon of concretionary pisolithic ironstones (A/B) and may be between 1.0 m and 1.5 m thick. Below it is the main laterite (**plinthite**) horizon with a strongly developed accumulation of iron and other sesquioxides. This horizon is again about 1.0 m to 1.5 m thick and is the horizon of cemented ironstone sesquioxides (Box). With depth the soil becomes paler, with distinctive red mottles. This is the mottled zone (Bgox), which undergoes alternate oxidation and reduction, owing to the seasonal changes in the depth of the water table. With depth the soil becomes paler and paler, often becoming quite white. This is the pallid horizon (Cg), which is soft, with rock structures often preserved. It is the zone of permanent reducing conditions and continues down to the weathering rock zone. Thus a well developed lateritic soil (FAO: Ferralsol) will have four distinctive zones: a surface zone, a lateritic zone, a mottled zone and a pallid zone. There is quite a bit of variety from place to place, with the lateritic and pallid zones varying greatly in thickness. The bright red colours of lateritic soils are due to the presence of haematite (α Fe_2O_3) and goethite (α $FeO.OH$) iron oxides; in the pallid zone iron occurs as the hydrated iron oxide lepidocrocite (γ $FeO.OH$) and as ferrous compounds.

Owing to the intense leaching, tropical lateritic soils are deficient in major and minor nutrient elements. Many elements are simply removed from the soil profile, or bound up in an unavailable form in the iron oxides (e.g. phosphate). There has been some success in growing commercial crops when fertilizer is applied (e.g. in parts of Australia and Africa). However, physical conditions are often difficult. In the 1940s the Groundnut Scheme of the British government unsuccessfully attempted to cultivate groundnuts on tropical soils in Tanzania (then Tanganyika). It was the induration of the soil structure which proved to be the most difficult physical factor of soil fertility to ameliorate. The gravelly, concretionary surface erodes mechanical implements very quickly and under the Groundnut Scheme it was found that a conventional set of discs would last little more than a month! Lateritic soils are still very much 'problem soils', as they occur in less developed countries which are keen to improve their agricultural productivity; shifting cultivation is still the normal practice in many of these areas.

The soil catena: the topographical factor in soil formation

We have seen earlier in this chapter, in the section on *the soil-forming environment*, that the Russian soil scientist Dokuchaiev in the late nineteenth century was the first to recognize that soils are independent natural bodies reflecting the effects of zonal and local soil-forming agents. By the end of the century he realized, however, that some soils are not only a factor of local zonation but may exist in several zones (gleys, alkaline soils, limestone soils); these he called **transitional soils**. There are also soils which do not appear to be controlled by zonal effects, which he called **abnormal soils** (alluvium, aeolian deposits). Sibirtsev, a follower of Dokuchaiev at the beginning of the twentieth century, called the three types of soils **zonal**, **intrazonal** and **azonal** soils. This concept of **soil zonality** classifies soils which primarily reflect climate and vegetation as normal or zonal. Those soils which reflect some local factor such as excess water or carbonates (which relate to relief and parent material) are intrazonal and may occur in several geographical zones. Similarly, azonal soils cross zonal boundaries and are essentially young, unweathered parent materials.

During the twentieth century one of the major additions to the concepts of soil science has been the innovative concept of the **catena**, introduced by Milne in the 1930s whilst studying the regular repetition of soils in East Africa. The term (Latin *catena*, 'chain') was originally used to designate a complex mapping unit, but later soil scientists came to recognize it as a fundamental unit in the landscape, reflecting the ways in which soils are influenced by slope or the topographical factor of soil formation. Milne recognized two types of catena. In the first type the parent material is uniform and differences between soils in the catena result from different processes at the surface and in the subsurface along the slope. In the second type one parent material is superimposed on another, and the slope thus cuts across them both. The upper part of the catena is on one parent material, and the lower slope on another. Hence a parent material or geological factor is added to slope effects. In Britain, as in most countries, both types of catena are well represented.

Figure 19.17 shows Milne's catena in East Africa on a uniform granite outcrop. The soil sequence is from grey loams on hill crests, changing into red soils on upper slopes and to yellow soils on lower slopes, and finally to black clays in depressions. Downslope there are changes in soil texture as well as changes in soil colour; the change is from sandy clay and sandy clay loam to loamy sand and sand. The processes operating on the slope are chiefly processes of erosion, transport and deposition. Granite is weathered on the upper slope, giving clays and sands as weathering products. Rainwash transports and sorts these particles differentially downslope. Coarse sand particles are moved to the lower slope as a yellow sand; fine clays are transported out of the system or are deposited in depressions. Milne thus recognized that hillslope erosion is an important part of soil formation on slopes in tropical and subtropical regions, with the slow but steady movement of solid material from one profile on the slope to another.

The soil profile of each member of the catena is related to every other member of that catena. Hence individual soils are like individual links in a chain. The effect of slope in giving rise to soils with very different drainage conditions and water content has

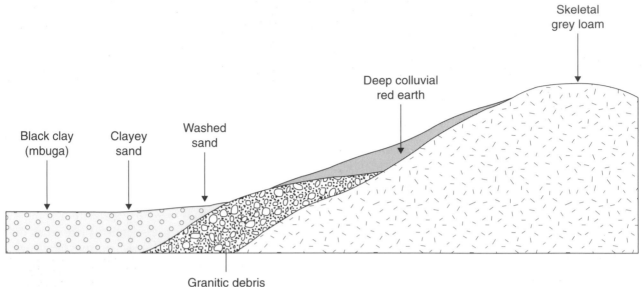

Figure 19.17 *Milne's soil catena in East Africa.*

been noted already in the sections in this chapter on the soil-forming environment and on gleying. The sequence of soils formed along a transect from a hill crest to an adjacent valley bottom is a **hydrological sequence**. An example of an acid parent material is shown in Figure 19.18. Freely draining soils in the upper part of the slope have bright, well oxidized colours. Progressively less well drained and increasingly gleyed soils occupy the lower positions of the catena. A greyish subsoil with ochreous mottling denotes imperfect drainage. A continuously gleyed horizon with much ochreous mottling shows poor drainage, whilst a dark peaty surface overlying a blue-grey horizon reflects very poor drainage. Such sequences are common on uniform parent materials, especially on glacial tills. In Scotland the Soil Survey maps hydrological sequences as *soil associations* when formed on uniform parent materials in the landscape.

The movement of water is the principal reason for the differences in soils downslope in humid temperate regions. Subsurface lateral flow is more important than overland flow. Chemical cations and anions will be carried in the water and will thus tend to accumulate relatively in downslope locations. The increased content of cations downslope causes a parallel increase in the pH of the mineral horizons below the surface peat. Often pH can reach 7.0 in the Humic Gleysol and the Eutric (i.e. base-rich) Histosol. There is also the increased weathering by hydrolysis because of the increased wetness. If

there are many ferromagnesian minerals in the parent material, weathering of these can cause exchangeable magnesium to exceed exchangeable calcium in the wetter soils of the catena. Thus a soil catena often shows a chemical sequence, with more mobile and weatherable chemicals (magnesium, manganese) reaching the lower slope while less mobile elements (iron, titanium) are retained at midslope. If rainfall is very high, and parent materials acid, many of the bases are lost from the catena entirely, and the soils may have similar low pHs on all parts of the slope.

The age factor in soil formation

Most British soils have been forming for only the short period of time since the retreat of the last Pleistocene ice sheets some 10,000 years ago. However, some soils show horizon features, particularly clay translocation and rubefaction, which can be attributed to interglacial periods in the Pleistocene or even to pre-Pleistocene times. Such older soils or horizons are termed **relic soils** if they occur at the surface and **buried** or **fossil soils** if they occur at depth. Periglacial features in soil, especially **fragipans** marking a former permafrost subsoil, are common relic features in British soils. Deeply weathered profiles on a range of igneous rock in Scotland are interpreted by many soil scientists as buried features formed in interglacial or pre-glacial times.

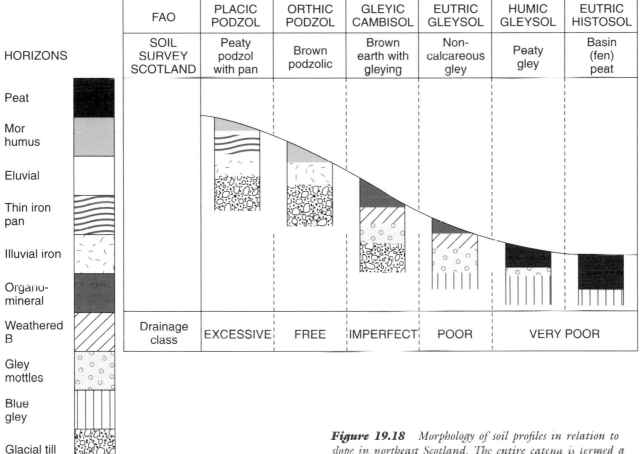

HORIZONS

Peat

Mor
humus

Eluvial

Thin iron
pan

Illuvial iron

Organo-
mineral

Weathered
B

Gley
mottles

Blue
gley

Glacial till

	FAO	PLACIC PODZOL	ORTHIC PODZOL	GLEYIC CAMBISOL	EUTRIC GLEYSOL	HUMIC GLEYSOL	EUTRIC HISTOSOL
	SOIL SURVEY SCOTLAND	Peaty podzol with pan	Brown podzolic	Brown earth with gleying	Non-calcareous gley	Peaty gley	Basin (fen) peat
	Drainage class	EXCESSIVE	FREE	IMPERFECT	POOR	VERY POOR	

Figure 19.18 *Morphology of soil profiles in relation to slope in northeast Scotland. The entire catena is termed a soil association or hydrologic sequence.*

Land in mid- and low latitudes which was not affected directly by glaciations in the Pleistocene often contains much older soils, some dating back to the Tertiary period. The Tertiary period was a time of plain formation in Australia, Africa and South America, when deeply weathered soil profiles up to 50 m deep were developed under hot and humid climates. The deeply weathered soil profile has traditionally been called **laterite** (FAO: **plinthite**), consisting of hard, reddish, sesquioxide-rich material. The Tertiary period had humid times or pluvials alternating with arid episodes. The laterite profiles were formed during humid and hot conditions (Figure 19.19) and during these times the Box horizons were soft, spongey and permeable to water. However, at the start of arid periods the laterite horizons irreversibly hardened, so that an impermeable hard layer was developed at about 1 m depth in the soil. As aridity increased, and the vegetation changed from tropical rain forest to sparse grassland, considerable erosion removed the topsoil horizons. The hard laterite acts as a cap rock and protects the landscape from backward erosion (Figure 19.20). It often forms

scarps which are locally called 'breakaways'. There were at least two periods of laterization in the Tertiary period: the Miocene and the early Pliocene. In the intervening arid times a landscape similar to the one outlined above would be shaped. A new pluvial climate would form a new laterite profile on newly exposed surfaces. Such landscapes in areas such as Africa, Australia and South America are often *poly-cyclical*, and when studying geomorphology and soils it is important to know which particular surface is in the immediate neighbourhood. Where the laterite profile is found in present-day desert landscapes it is a *relic profile*.

Laterites appear to be a product of the Tertiary period only, as relic laterite landscapes have not been found in any other geological period. In areas where there have been no arid episodes since the Tertiary, soft spongey laterite still exists. The material is used extensively as a building material in India and Thailand; it is dug out of the ground, shaped into bricks and allowed to dry irreversibly in the sun. This has given laterite its name (Latin *later*, 'brick'). The present climatic environment bears little relationship

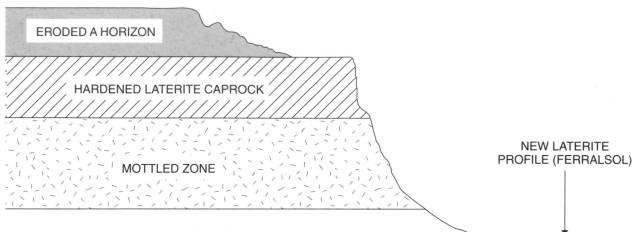

Figure 19.19 *Laterite (plinthite) acting as a caprock in a polycyclic landscape.*

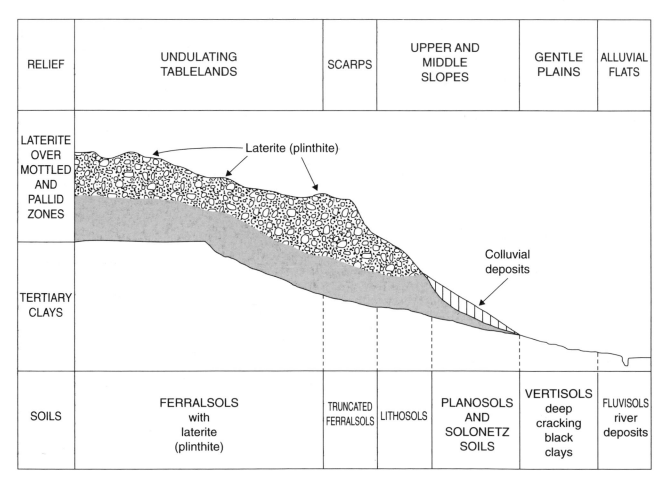

Figure 19.20 *Soil catena on denuded laterite (plinthite), Queensland, Australia.*

to the original climate at the time of laterization. In Australia laterite is found in areas ranging from a rainfall of 50 mm to 5000 mm. Laterite is found over a wide range of parent rocks, though some iron content is necessary. All igneous rocks and many sedimentary rocks will give laterite, though quartzites and pure limestone will not. Some limestones which are rich in iron will give laterites, as is the case in Cuba.

Conclusion

Soils are made from rocks and sediments by a two-stage sequence of development. Weathering produces the parent material, which is then acted upon to produce the soil profile. The particular profile formed at any point on the Earth depends upon the prevailing climate, vegetation, rock type and topography, together with the length of time of soil formation. Generally soils in mid- and low latitudes are deeper, more weathered, redder and less fertile than soils of higher latitudes.

The fundamental unit for studying soils is the soil profile, the vertical section from the soil surface down to underlying rock. The profile is formed by vertical movements of water and materials, both downwards and upwards. These processes are leaching, decalcification, clay translocation, podzolization and laterization in a predominantly downward direction. Processes of salinization, alkalization and calcification involve precipitation of chemicals *in situ*, or by upward movement. The processes of rubefaction and gleying give distinct colours to soil.

The main departure from the central concept of the soil profile comes with the concept of the soil catena. In temperate climates soil changes along slopes are largely conditioned by hydrology. In subtropical and tropical climates soil changes long a catena are also greatly influenced by the movement of mineral particles by rainwash erosion. In all natural regions it is necessary to be aware of the geomorphology and landscape history, as polycyclical soils are common features in polycyclical landscapes.

Key points

1 Soils are formed by a series of soil-forming processes occurring within the top 1–2 m of the soil surface. Many processes will be going on simultaneously, but the soil formed will mostly reflect the dominant one. High rainfall plus a permeable parent material favours leaching. Acid vegetation and parent material promote podzolization. Intense weathering and leaching give rise to laterization. Poor drainage gives gley soils, which, under arid and semi-arid climates, become saline and alkaline soils.

2 The soil formed under the prevailing climate and vegetation conditions is the zonal soil. This corresponds to the climax vegetation and will be found on flat, well drained sites. In the real world, however, the zonal profile will be considerably modified by local factors (drainage, limestone rocks), by topography (giving soil hydrological sequences or soil catenas) and by the age factor (the length of time particular processes have been operating).

3 The concept of soil zonality was developed in the late nineteenth century but is still relevant today, as it fits neatly into ecological concepts of biomes and ecosystems. An understanding of soil forming processes is vital to a thorough understanding of the potential of the world's soils for agriculture and forestry.

Further reading

Avery, B.W. (1990) *Soils of the British Isles*, Wallingford: CAB International.
The definitive account of the distribution, formation and properties of British soils. A good source of example and illustration.

FitzPatrick, E.A. (1980) *Soils: Formation, classification and distribution*, London: Longman.
A comprehensive discussion of world soil types, based on the FAO classification scheme.

Soil Survey of England and Wales (1983) *Soil Map of England and Wales*, Harpenden: Soil Survey of England and Wales.
Six bulletins (Nos 10–15) describe the soils of the six major regions of England and Wales, together with accompanying maps at a scale of 1:250 000. Many fine photos and diagrams explain soil distributions and soil processes.

CHAPTER 20

Vegetation systems

Vegetation clothes the Earth and provides the vital link between the sun and the Earth's ecosystems. That part of the incoming solar radiation which is fixed by the process of photosynthesis in the leaves and stems of plants provides the organic molecules which underlie all life on Earth. The golden rule is: 'No plants, no organisms, no human life'. Vegetation is not a haphazard material. Plants are governed in their distribution by a range of physical and biological factors. In living together in communities, plants are influenced by a range of mutual relationships, rather as human beings in human societies are ruled by sets of relationships. This chapter explores the nature of plant–environment relationships, and the key concepts which are used in the study of vegetation communities.

Units and scale of study

Vegetation is only one component of the world's landscape and of global ecosystems, but special importance is attached to it, as it is the basis of productivity. It fixes carbon through photosynthesis, builds up organic matter in soil, provides food and shelter for animals, stabilizes soils, and influences the hydrological cycle. In short, vegetation provides the life-supporting properties of the biosphere; it supports the food webs of herbivores, carnivores and decomposers which make up the world's fauna on land and in the sea.

The plants which make up natural vegetation do not exist just as individuals – they also live in communities. In the idea of **plant community** is the concept of all the relations between plants. These inter-relationships are mostly beneficial to all parties; they are called **obligate symbiotic relationships**. The driving force in developing these mutual bonds is *co-evolution*, the development over time of mutually beneficial connections between organisms in a particular environment. Plants and animals have evolved these obligate ties by living together, and together

they form an organically interdependent community. Thus an ecosystem such as an oak woodland can be understood and explained only by referring to the functions and processes involved in the interaction of its component parts, including climate in the air layer, the soil layer and the living organisms of both.

The second property of plant and animal communities is that they are collections of organisms whose common denominator is tolerance of the particular environment that they share. Some ecologists believe that this is the most important feature of communities, and any other relationships are optional rather than obligatory (**facultative relationships**). This is the '**individualistic community**' concept of the American ecologist Gleason, put forward in 1939. It is in contrast to the **organismic community** concept of Frank Clements of 1916 whereby the plant community is likened to a 'super-organism' which functions by means of the connections between all organisms in the community. Most ecologists would probably lie closer to Clements than to Gleason in their views, and would recognize that there are plant communities that repeat themselves over geographical space. A plant community can therefore be defined as : 'the collection of plant species growing together in a particular location that show a definite association'.

Vegetation can be studied at a range of different scales, starting at an individual plant at the lowest level, right up to the vegetation of the entire globe i.e. the **biosphere** or **ecosphere**. The Canadian ecologist Stan Rowe proposed a system of nested levels for both vegetation and ecosystems. Each level occupies a smaller and smaller area. Table 20.1 shows a modified version of Rowe's system. Biogeographers have historically studied vegetation at all scales; the larger-scale **biomes** and **formations** were popular in the nineteenth century, when the world's surface was first being explored and mapped. This generalized scale of working has enjoyed revived popularity in recent years as our techniques for studying global systems have improved. For much of the twentieth century, however, biogeographers have focused on the *plant community* level, as it is a very convenient scale for fieldwork. Biologists also work at various levels of study, and there is overlap between the subject matter of the biologist and the geographer. Table 20.2 shows the views of the American ecologist Odum on similar levels of study in biology.

The concept of the ecosystem

The previous section introduces a real distinction between 'vegetation' and 'ecosystem' which must now be explained. The term 'ecosystem' was first used by Arthur Tansley in 1935, writing about British vegetation. For him the ecosystem is: 'the whole system, including not only the organism-complex but also the whole complex of physical factors forming what we call the environment'. The terms 'landscape', 'environment', 'terrain' and 'ecosystem' are often used interchangeably by ecologists to mean a specific land system or land area whose interrelated parts are rocks, landforms, soils, topoclimate and organisms. What emerges is that it is impracticable to understand the dynamics of any vegetation communities without parallel examination of geology, topography, soils, hydrology and micro-climate which together make up the habitat of plant life. The value of the ecosystem concept is that, by focusing on living organisms and physical environ-ment together, it broadens understanding of what is functionally and structurally important in the land-scape. The ecosystem is a geographical unit of air, earth and water that encompasses all living organ-isms, whether flora or fauna; it forms a layered struc-ture of interactive parts at the earth's surface.

Since Tansley's first definition there have been many others, as understanding of ecosystems has grown. Many stress the role of energy in ecosystems (see pp. 384–6); thus Billings defines an ecosystem as: 'an energy-driven complex of community and environment'. However, the key properties are the relationships between organisms (both plants and animals), and between organisms and their physical environment. Thus a useful definition of an ecosystem would be: 'the biological and non-biological components of the landscape which exist as an adjusted system whose parts are interrelated'. Figure 20.1 portrays the special fields or disciplines through which scientists have traditionally studied the 'environment'. Scientific knowledge has advanced by reduction, by the analysis of systems into subsys-tems. The ecosystem approach is strikingly different. It tries to conceptualize and study 'environment' in an integrated, holistic manner. The study of special-ized fields or disciplines can be justified only by refer-ence to the wholeness of the environment and the hierarchy of its functional wholes.

Table 20.1 *Scales of vegetation and ecosystem study.*

Scale	Vegetation	Ecosystem
Large	All vegetation	Biosphere or ecosphere
	Vegetation formations	Biomes
	Vegetation types	Regional ecosystems
	Plant communities	Local ecosystems
Small	Species populations and individuals	Single organism -habitat system

Source: After Rowe.

Table 20.2 *Scales of organization in biology.*

Large-scale	Biosphere
	Ecosystem
	Community
	Population
	Organism
	Organ system
	Organ
	Tissue
	Cell
Small-scale	Protoplasm

Source: After Odum.

Environmental factors

The growth of plants imposes demands upon the plant's environmental resources. For successful growth the plant requires six essential factors: light, heat, moisture, air, nutrients and physical support. Plant carbohydrates are photosynthesized by short-wave solar radiation, atmospheric carbon dioxide and soil moisture. For growth and development this process must take place faster than the rate of break-down of carbohydrate by plant **respiration**. Second in importance to this fundamental process of **photo-synthesis** is the process of *transpiration*, whereby moisture is absorbed from the soil by plant roots, transported up the stem via the xylem tissue to the leaves, where it is evaporated into the atmosphere through leaf pores or stomata. Transpiration is not only a cooling mechanism for leaves exposed to solar radiation, but also the mechanism by which the majority of nutrients are absorbed into the plant and moved within it.

Figure 20.2 shows the position of plant commu-nities in relation to the different environmental

Figure 20.1   *The individual scientific disciplines which comprise the holist study of ecology and ecosystems.*

factors which control their structure, productivity and distribution. The environmental factors are not themselves independent variables but are typically influencing each other as well as the plant community. Thus increased radiation brings an increase in temperature which brings a decrease in soil moisture. An increase in slope angle brings a decrease in soil depth which brings a decrease in soil moisture. The directions of the controls are indicated by the arrows in Figure 20.2. It is a feature of vegetation that two or more factors can act together to produce a net effect which is larger than the sum of the separate effects when each factor operates alone; this is called a *'synergistic'* effect.

Figure 20.2 illustrates also that there are some environmental factors which directly and locally affect the plant community, whilst others seem more distant, with only indirect effects. Thus it is possible to distinguish between *direct* and *indirect* environmental factors. Soil conditions (pH, nutrients, soil oxygen, depth, absence of toxic chemicals) and soil moisture (sufficient available moisture for transpiration) directly affect plant growth, as also do human land-use practices. Other factors are important in explaining plant distributions, although they themselves have no direct effect on plant growth. Factors such as aspect and slope, for example, are indirect factors which are nevertheless important because they explain, through correlations with direct growth factors, a high proportion of plant distributions.

Limiting factors and range of tolerance

The relationships between plants and their environment form the cornerstone of modern ecology. In the past some biogeographers have studied a single species and the environmental conditions that control it; this is the science of **autecology**. Others have studied the plant communities as a whole, especially the links between organisms; this is the science of *synecology*. The modern ecosystem approach would contain elements of both: the entire community would be studied, in terms of the relations between the living components, and between the living components and their physical environment.

In the nineteenth century the German soil chemist Liebig formulated his famous **law of the minimum** to express the influence of an environmental factor on plant growth. He argued that, if growth depends on several factors, what is important is that factor which is in short supply. It is of little use, for example, if all factors are favourable but one (perhaps a soil nutrient) is missing or in low supply. Productivity in this situation will be zero or low, irrespective of the abundance of other factors. The law of the minimum states: 'growth is governed by the factor which operates at a minimum'.

In the real world environmental conditions exist as **gradients**. There may, for example, be a pH gradient from a basic igneous rock (e.g. basalt) to an acid

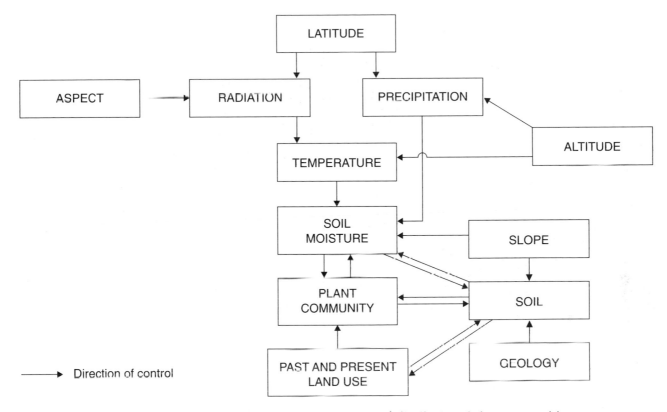

Figure 20.2 *The environmental factors which control the nature and distribution of plant communities.*

igneous rock (e.g. granite). A moisture gradient may go from a wet, valley bog to dry ridge crests in the same valley. The change in the performance of a plant species along such a trend is called an environmental gradient (Figure 20.3). There will be upper and lower threshold values on the gradient beyond which the species cannot survive. These points are the *upper limit of tolerance* and the **lower limit of tolerance**. In Figure 20.3 this tolerance range is shown as a broad-based normal curve, though in reality it may be much narrower for a particular species on a particular gradient. The **ecological optimum** for the species is that part of the tolerance range where the vigour of the plant is at a maximum.

In the real world there are two complications to the concept of tolerance range. First, a species has a separate range for each environmental factor. (Plate 20.1.) Each separate response includes a different range and optimum. When all ranges are added together we get the **ecological amplitude** of the species. This is a multidimensional 'hyperspace' which is not easy to define or represent. However, it is a useful concept for summarizing the sum total of the effects of all environmental factors. The second complication is that plants differ in their ability to utilize a resource which is in limited supply. (Plate

20.2.) The process of **competition** will eliminate the less efficient plant, or the less efficient species. Competition between individuals of the same species is **intraspecific competition**, and often occurs at the beginning of successions on fresh, bare surfaces where colonizers are competing for space. Competition between species is **interspecific competition** and is universal, and results in actual species ranges that are much narrower than their full tolerance ranges. Figure 20.4 shows the viability of four species with different competitive abilities along an environmental gradient (say, soil moisture). Species A is not competitive, and therefore remains a secondary component of the final communities. Species B has a wide range of tolerance but is only moderately competitive, and therefore dominates at the wet end of the gradient. Species C is highly aggressive and dominates in its narrow range, as does species D for most of its wider range. The resulting structure of the plant communities along the gradient, in terms of dominant and secondary species, is shown in Figure 20.4e. The pattern of plants which results is thus the result of two broad influences: first, the **range of tolerance** of the species to an environmental gradient and, second, the interspecific competition between the plants. (Plate 20.3.)

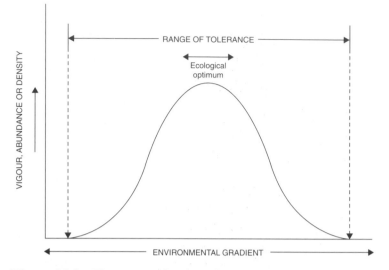

Figure 20.3 *The range and ecological optimum of a plant species.*

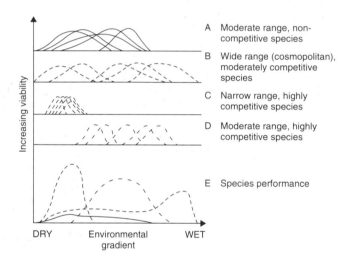

A Moderate range, non-competitive species

B Wide range (cosmopolitan), moderately competitive species

C Narrow range, highly competitive species

D Moderate range, highly competitive species

E Species performance

Figure 20.4 *The viability of four plant species of different competitive abilities (A–D) in relation to an environmental gradient, and the resulting effects on the distribution of species (E).*

The concept of the ecological niche

The way in which the distribution of organisms is influenced by the physical and biotic factors of the environment is the essence of the concept of the ecological niche. The **ecological niche** of a species is defined, first, by the functional role of the species in its community (i.e. its trophic position as described in Chapter 22), and, second, by the position of the species along environmental gradients such as temperature, moisture, soil pH, soil fertility and other factors. The American ecologist Hutchinson introduced in the 1940s the concept of the multidimensional 'hypervolume' where the limits of a species' existence are defined by a large number of variables. He also drew a distinction between the fundamental

niche and the realized niche. The **fundamental niche** is the maximum 'theoretically inhabited hypervolume' where a species, free from any sort of interference from another species, can occupy the full range of variables within the community to which it is adapted, to the outer limits of its tolerances. The **realized niche** is a smaller 'hypervolume' which is occupied by the species under interference from the competition of other species. A niche is thus defined by n variables, and can be defined as: 'the limits, for all important environmental features, within which individuals of a species can survive, grow and reproduce'. Figure 20.5 illustrates the results of fieldwork carried out around a saline lake. The densities of three plant species – wheatgrass, saltgrass and red samphire – were recorded in quadrats, and the salinity of the soil was measured for each quadrat. The graph thus shows density of plants in a salinity gradient. An inspection of the graph shows that wheatgrass is found where soil salinity values are lowest, i.e. it is not very tolerant of salinity. In this environmental gradient it has the narrowest realized niche, owing to its lack of adaptation to saline conditions. Red samphire has a wider realized niche; it is a halophyte (a plant tolerant of saline areas) and is found where salinity levels are the highest. Saltgrass has the widest realized niche, due to its tolerance of a wide range of salinity levels. It also reaches its maximum densities where competition from wheatgrass and red samphire is less intense, i.e. near or beyond their limits of tolerance.

Vegetation zones of Earth

In describing the terrestrial vegetation of the entire Earth, large units of vegetation known as vegetation **formations** or **biomes** are used (see Table 20.3). Different biogeographers have proposed different schemes, but all agree that formations and biomes are strongly correlated with regional climate; climate determines what kind of vegetation can be expected in a particular locality. Climatic diagrams often accompany the vegetation classifications. They show months of the year on the horizontal axis, and both °C and mm precipitation on the vertical axis, and can thus indicate the mean temperature and precipitation at a particular locality over the course of the year. The length and intensity of relatively humid and relatively arid seasons are shown, as well as the duration and severity of a cold winter, and the possibility of late or early season frosts. Figure 20.6 shows the climatic diagram for Douala, Cameroun, West Africa, a city close to the equator. The climate illustrated is typical of the equatorial zone.

The German biogeographer Walter in 1976 recognized nine climatic zones and corresponding vegetation zones. They were labelled I–IX. Transitional types were given a double label, e.g. I (III). Mountainous regions were designated X, with a second label, e.g. X (V), indicating the climatic zone from which the mountains rise (Figure 20.7). The equatorial zone (I) lies between about 10° N and 10° S. The daily variation in temperature is greater than the annual variation, which is in the range 25–7° C. Generally annual rainfall is high, with the maximum occurring at the times of the equinoxes. Vegetation is classed as the *tropical evergreen rain forest zone* (I). The *tropical zone (II)* is located between 10° and 30° N and S approximately. Some seasonal variation in mean daily temperature is noticeable. Rainfall is at a maximum during the summer rainy season, and in the cool season there is a dry season that increases in duration with increasing distance from the equator. *Tropical moist (2)* and *dry deciduous forests and savannas (2a)* occur here. The *subtropical dry zone (III)* is the hot desert zone and is located poleward of 30° N and S. Rainfall is very low, daytime temperatures are very high, and at night in the winter months the temperatures may drop to zero. The main tracts of these *subtropical deserts and semi-deserts (3)* occur in the Sahara and Libyan deserts of northern Africa, the Arabian desert in Asia and the Middle East, interior Australia, south-western parts of North America, south-western Africa, northern Chile and Pakistan.

The *transitional zone with winter rain* (IV) is located at latitudes approximately 40° N and S. In this Mediterranean climate there is rain in winter, a long summer drought and no cold season, although frosts do occasionally occur. The *sclerophyllous forests of the winter rain regions (4)* occur along the Mediterranean coasts, in central and southern

Plate 20.1   *Trees suffer stress at the limits of their range. Here alpine fir (Abies lasiocarpa) in the Canadian Rockies has been blasted on one side by wind and ice crystals to give a misshapen form ('krummholz'). Photo: Bill Archibold.*

Table 20.3   *Climate and vegetation zones of the Earth.*

Climatic zone		Vegetation zone	
I	Equatorial	1	Tropical evergreen rain forest
II	Tropical	2	Tropical moist forests
		2a	Dry deciduous forests and savannas
III	Subtropical dry	3	Subtropical deserts and semi-deserts
IV	Transitional and winter rain	4	Sclerophyllous forests of winter rain regions
V	Warm temperate	5	Temperate wet evergreen forests
VI	Typical temperate	6	Deciduous forests
VII	Arid temperate	7	Steppes
		7a	Semi-deserts and deserts with cold winters
VIII	Boreal cold temperate	8	Boreal coniferous
IX	Arctic	9	Tundra

Source: After Walter (1976).

Plate 20.2   *The effect of competition for water on vegetation. In southern Spain bunch grasses of esparto (Stipa tenacissima) are spaced widely in order to exploit low soil moisture reserves efficiently.*
Photo: Ken Atkinson.

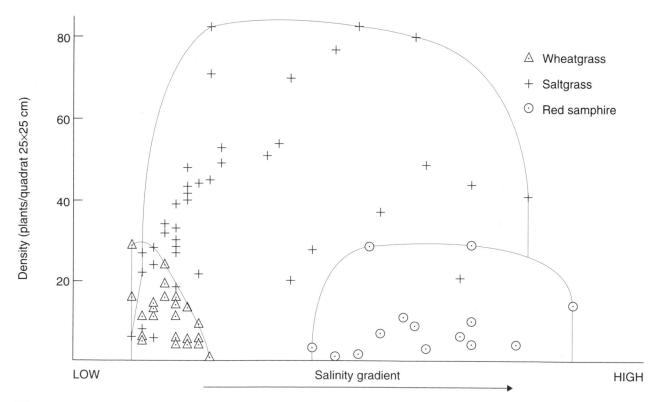

Figure 20.5   *The realized niches of three plant species along an environmental gradient of soil salinity.*

Plate 20.3 *The effect of grazing on vegetation. Sheep are excluded from grazing on this small island in upland Britain, thus allowing Scots pine* (Pinus sylvestris) *to regenerate.*
Photo: Ken Atkinson.

California, central Chile, the Cape of Good Hope in South Africa, and south-western and southern Australia. *The warm temperate climatic zone* (V) has scarcely any or no winter. It is extremely wet, especially in summer. *Warm temperate wet evergreen forests* (5) are most extensive in eastern Asia. They are also located on the south-eastern coast of Australia, the North Island of New Zealand, the east coast of South Africa, south-eastern Brazil, parts of southern Chile, the higher regions of Central America, and Florida. The *typical temperate climatic zone* (VI) has a cold but short winter in continental locations or a winter almost free of frost with cool summers in oceanic localities. The *deciduous forests of the temperate zone* (6) occur in large parts of

Figure 20.6 *Climatic diagram of Douala, Cameroon: a station, b height above sea level, c record length (years), d mean annual temperature (° C), e mean annual precipitation (mm), f mean daily minimum of coldest month, g lowest recorded temperature, h mean daily maximum of warmest month, i highest recorded temperature, j mean daily temperature range, k mean monthly temperature, l mean monthly precipitation, m humid season (light shading), n mean monthly rainfall over 100 mm. Source: After Walter (1976).*

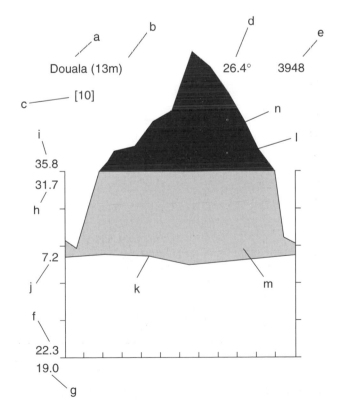

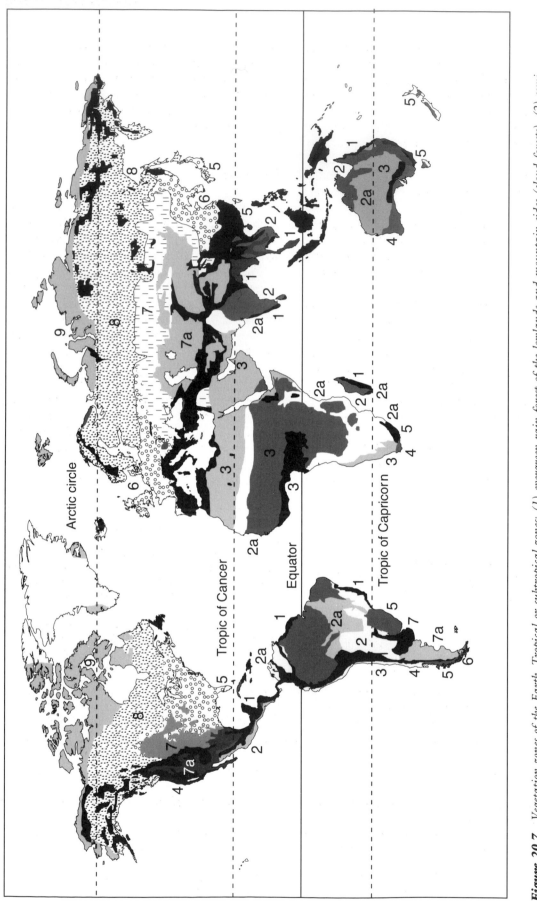

Figure 20.7 Vegetation zones of the Earth. Tropical or subtropical zones: (1) evergreen rain forest of the lowlands and mountain sides (cloud forests), (2) semi-evergreen and deciduous forest, (2a) dry woodland, natural savanna or grassland, (3) hot semi-deserts and deserts, polewards up to latitude 35°. Temperate and arctic zones: (4) sclerophyllous woodland with winter rain, (5) moist warm-temperature woodland, (6) deciduous (nemoral) forest, (7) steppes of the temperate zone, (7a) semi-deserts and deserts with cold winters, (8) boreal coniferous zone, (9) tundra, (10) mountains. Source: After Walter (1976).

western and central Europe, eastern North America and East Asia. In the southern hemisphere this zone is restricted to a small area of southern Chile. In the *arid temperate climatic zone* (VII) large temperature contrasts occur between summer and winter. Little precipitation is received. The *steppes of the temperate zone* (7) and the *deserts and semi-deserts with cold winters* (7a) occur across Eurasia from the Black Sea to the Himalayas, and in the grassland regions of Canada and the United States. In the southern hemisphere this zone occurs in the pampas of Argentina, the semi-desert of Patagonia and the tussock grasslands in the South Island of New Zealand.

Cool, wet summers and cold winters lasting more than six months occur in the *boreal or cold temperate climatic zone* (VIII). The boreal coniferous zone (8) occurs across northern parts of North America and Eurasia, but it is absent in the southern hemisphere. The *arctic climatic zone* (IX) is characterized by low precipitation distributed over the entire year and by low temperatures. Summers are short and wet, with twenty-four-hour days, while winters are very long and cold, with twenty-four-hour nights. The *tundra zone* (9) encircles the North Pole in the Arctic, and similar vegetation is found in the southern hemisphere on the southernmost tip of South America and on many small islands in the Southern Ocean. Walter's classification of climates and vegetation is a general zonal one. Variations will occur within zones caused by factors such as proximity to oceans, the influence of trade winds and monsoons, and the presence of major mountain ranges, as well as local micro-environmental differences caused by topography and soil types.

Conclusion

The British ecologist Tansley revolutionized the study of natural systems of vegetation and soil in the 1930s when he introduced the concept of the ecosystem. Since then the Canadian ecologist Rowe and the American ecologist Odum have both placed the ecosystem within a hierarchy of ecological units. The distribution of plants is governed by a range of environmental factors. There are also strongly competitive relations between species (interspecific competition) and between individual plants (intraspecific competition). The concept of the ecological niche was introduced by the American ecologist Hutchinson to summarize the sum total of physical and biological controls. The natural regions of the world are largely climate-determined. The biome is the major large-scale ecological unit. The scheme proposed by the German biogeographer Walter is a basis for dividing the Earth into eleven major vegetation zones.

Key points

1 The unit of the plant community was first studied by the American ecologist Clements at the beginning of the twentieth century. A group of plants with similar tolerances are able to live together in a definite association. Some ecologists dispute the existence of communities as major units, and stress the life of plants as individuals governed by their own limiting factors; in this view, albeit only held by a minority, any associations are coincidental.

2 The main concepts which govern our knowledge of the nature of vegetation, and which determine the methods by which it is studied, are six in number. Three of these relate to the units of study; these are the concepts of the plant community, the ecosystem and the biome. Three of these relate to the factors which govern the distribution of plants as individuals, species or growth-forms; these are the concepts of range, limiting factors and ecological niche.

3 Climate is the major determinant of the Earth's major ecosystems or biomes. Climate brings in its effects through temperature, precipitation, radiation and seasonality. Famous climatologists such as Köppen and Thornthwaite have always stressed the links between climate and vegetation. The scheme by the German biogeographer Walter provides a basis for dividing the Earth into eleven major biomes.

Further reading

Archibold, O.W. (1995) *The Ecology of World Vegetation*, London: Chapman and Hall.
The most up-to-date text on world vegetation regions and their flora. Written in an advanced but accessible manner.

Kent, M. and Coker, P. (1992) *Vegetation Description and Analysis: A practical approach*, London: Belhaven Press.
A detailed discussion of the principles of vegetation distributions, and the methods used to study vegetation communities through fieldwork and computer analysis.

Tivy, J. (1993) *Biogeography: A study of plants in the ecosphere*, third edition, London: Longman.
A popular course text for biogeography students which covers both principles and regional examples. Less advanced and less up-to-date than Archibold.

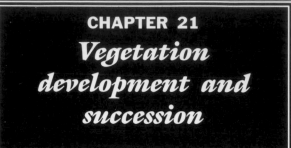

CHAPTER 21
Vegetation development and succession

As with all elements of the natural landscape – rocks, slopes, rivers, soils – vegetation communities have a history of development. The changes which occur in the characteristics of a particular species over many generations are called *evolution*. Evolutionary change takes place over millions of years and is achieved by the mechanism of natural selection, first proposed by Charles Darwin in 1859. In this way, the modern characteristics of plant and animal species result from a long history. The present distribution of species is greatly influenced by past events, too. The theory of continental drift explains why some species are widely distributed across continents now separated by thousands of miles. The changing patterns of distribution over time are preserved in the fossil record. The latter may take the form of hard-rock fossils or, for Quaternary peat and lake sediments, the plant pollen grains and spores which they contain. Vegetation communities are also subject to short-term changes over the order of, say, thirty to 100 years. Natural disturbances by floods, volcanic activity, hurricanes, disease and fire can completely alter the vegetation of an area. The habitats of the plants are changed, favouring a new set of communities. These short-term changes reflect the dynamic nature of vegetation; the processes occurring in the successional development of plant communities will be discussed in this chapter.

Succession and climax

Ecological succession is the term used to signify the changes in the composition of a community over time. It refers to the sequence of communities which replace one another in a given area. The entire sequence of stages is referred to as the **succession** or **sere**, each temporary stage in the succession being called a **seral stage**. Plant species invade the site when conditions are favourable, and are eliminated when the succession leads to unfavourable local conditions. Thus seral stages are defined by the changing dominance of plant species, together with associated soils and fauna. Each seral stage has a distinct ecology which is ephemeral in the sense that it prepares the ground for the succeeding ecosystem. The final endpoint of the succession is called the **climax community** (Figure 21.1). Considerable attention has been paid to successions in biogeography because they reflect the dynamic nature of ecological communities, and illustrate the importance of the time factor in the development of plant communities. Figure 21.2 has been adapted from the work of the British biogeographer Eyre to illustrate the detailed terminology which has built up around the theory of succession. If succession commences on a bare surface which has not previously been occupied by a plant community, it is called a **primary succession** or **prisere**. (Plate 21.1.) A new land surface of coastal sand or volcanic lava and ash is a

Figure 21.1 *The classical view of unidirectional plant succession, starting with an unvegetated surface and ending with the climax community.*

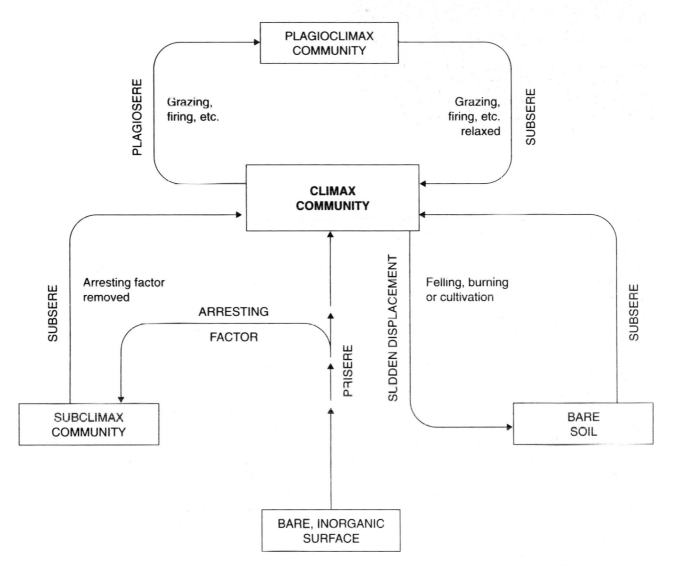

Figure 21.2 *The diagrammatic representation of priseres, subseres and plagioseres. Source: After Eyre (1968).*

typical site. Primary successions on new volcanic surfaces such as Krakatoa in the Pacific Ocean, Surtsey Island in the Atlantic or Mount St Helens in the US state of Washington, are all classic examples. However, where the community develops in an area which was previously vegetated, but from which the community was removed, the sequence is called a *secondary succession*. Examples would include areas cleared for farming but then abandoned to recolonization by plants and animals. Another example would be an area of clear-felled forest (Plate 21.2), or any community destroyed by fire.

Secondary successions are usually more rapid than primary ones, as seeds and seedlings are often present, and initial soil conditions are more favourable than on a bare sterile surface. The **primary stage**, when plants are first gaining a presence in the prisere, is much shorter in the secondary succession.

Figure 21.2 also illustrates the importance of human activities in determining the nature and composition of plant communities. Activities such as grazing by domestic animals or deliberate burning will maintain a vegetation community which would otherwise be a natural climax. Such communities are designated **plagioclimaxes** ('changed' climaxes) or, in the American literature, **disclimaxes** ('disturbance' climaxes). For example, overgrazing of natural prairie grassland can produce a degraded plagioclimax which could return to natural prairie if the grazing pressure was removed. The succession from plagioclimax to climax is a type of *subsere*. A subsere is a succession which proceeds from a subclimax community (say, a plagioclimax) to climax. Another example would be where the primary succession is halted by a dominant environmental property (for example, excessive wetness). This factor maintains the

Plate 21.1   *Primary succession or hydrosere along the banks of the river Cree, Scotland. The four vegetation zones visible are reeds (*Phragmites *spp), alder (*Alnus glutinosa*), mixed deciduous saplings and finally oak woodland (*Quercus robur*). Photo: Richard Smith.*

community in a condition which is subclimax and not yet climax. Over time the influence of the arresting factor will be lessened or removed, and the subclimax community will proceed to climax via a subsere. In order to understand fully the complexity of vegetation patterns which one meets in the real world, it is important to know whether a particular community is seral, subclimax, plagioclimax or climax. This is the **ecological status** of the ecological community, and is not always an easy attribute to discover. However, it is worth undertaking an analysis of it, as it indicates the relationship of a particular community to the hierarchy of plant communities in the ecological succession.

Classification of successions

Ecological processes occurring in successions have received considerable attention from biogeographers.

One classification of seres is based on the nature of the surface from which the primary succession starts. Those habitats where drought is the main limiting factor are referred to as **xeroseres**. Two common situations are where bare rock dominates (new volcanic lava or a recent scree) and where sand dominates (coastal sand or fluvioglacial sands or aeolian sand). The former cases are **lithoseres** and the successional processes are directed at the weathering of the bare, consolidated rock, and the production of a soil upon it (p. 374). In the latter cases successional processes strive to stabilize the unstable sandy environment so that stable plant and soil communities can develop; these are **psammoseres** (p. 375). Other seres start with almost the opposite type of conditions unfavourable to plant growth. These are habitats dominated by water (lakes or marshes) and hence are termed **hydroseres**. The freshwater hydrosere is called the *hydrosere* (p. 376), whereas the salt-water hydrosere is a **halosere** (p. 379). Figure 21.3 illus-

Plate 21.2 *Secondary succession of grasses, herbs, shrubs and tree seedlings invading a clear-cut made by forestry operations in British Columbia, Canada.*
Photo: Ken Atkinson.

trates the classification of primary successions. The general principle governing the sequential changes which occur during a succession typically follows the **principle of competitive replacement**, which states that: 'a plant community in a succession creates conditions which are more and more favourable for more complex and demanding communities which will out-compete and replace it'. The initial habitat conditions are very demanding, and only a small number of plant species can survive them. However, the net effect of those *pioneer* plants is to create less severe conditions which in turn can support a greater diversity of plants. Pioneers arrive and become established at a site, and with time alter the habitat so that other species can survive. The diversity in species and in the structure of these seral communities increases with time until environmental conditions become stabilized, and a self-perpetuating climax community is formed. Many of the early studies of succession were carried out by the US ecologist

Clements in 1916. He envisaged succession as an orderly and predictable evolution, following definite pathways to a predictable climax (p. 381). He envisaged succession as five basic processes: nudation, immigration, ecesis, reaction and stabilization. These terms are defined in Table 21.1.

Whilst the broad principles of Clements's theories of succession form the basis of our understanding, several amendments of his basic model are now accepted. Three alternative models are illustrated in Figure 21.4. The **facilitation model** mostly follows Clements and envisages the establishment of plant communities which modify the physical conditions so that they become favourable for late successional species. In the **tolerance model** successive stages in the succession depend upon the competitive abilities and life spans of the plants so that, for example, the longer-lived species associated with later stages will persist in the community. The third alternative, the **inhibition model**, envisages the initial plant cover modifying the physical habitat so that it is *less* favourable to colonization by other species, and succession can occur only when the inhibitory species are removed.

The classical views of succession are thus becoming modified as it is realized that it is a more complex process than was first envisaged by Clements. There are some elements of orderliness and predictability, but there are also elements of apparent disorderliness and unpredictability due to factors of habitat variability, propagule availability and replacement by pure chance (randomness). Successions appear to be less 'directional' than in the classical model. Initial sites are usually very variable, and the availability of propagules to colonize and replace can differ between sites. Also external environmental fluctuations can occur through the course of a succession. In the matter of competition between a wide range of species combinations it may be very difficult to

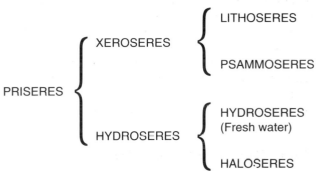

Figure 21.3 *The classification of priseres according to dry (xero-) and wet (hydro-) starting points. Source: After Eyre (1968).*

Table 21.1 *Fundamental successional processes.*

Term	Process
Nudation	Initial creation of bare surface
Immigration	Arrival of available propagules
Ecesis	Establishment of propagules
Reaction	Interaction of plants and of plants and habitat
Stabilization	Creation of equilibrium communities

Source: After Clements.

predict an outcome; the element of chance enters the process of species replacement. The importance of these plant factors gives rise to what are termed **allogenic successions** (externally induced successions), which contrast with **autogenic successions** (self-induced successions), where distinct seral changes result from autogenic habitat modification.

The question of differences between species in seed dispersal, germination rates, growth rates and longevity is important in studying successions. Pioneer species and early successional plants are usually short-lived 'r species' producing many easily dispersed seeds which have long viability and are capable of dispersal over long distances. The shade tolerance of these plants is low, and growth rates are rapid. By contrast late successional plants are long-lived 'k species' with slow growth rates, large size and high shade tolerance. Their seeds are also large, of short viability and capable of dispersal for only a short distance.

Lithoseres

The development of vegetation communities over time on a fresh rock surface is called a **lithosere**. In some instances the new unvegetated mineral surfaces are already unconsolidated. Many deposits resulting from the glaciations of the Pleistocene era give mineral landscapes which are already prepared for the invasion of land plants; they give media such as glacial tills, fluvioglacial sands and aeolian loess silts in which plants can readily take root. Similarly fresh volcanic ash and recently deposited river alluvium can quickly be covered by land plants. A lithosere on a hard-rock surface, however, presents very different and much more hostile conditions. Bare rock is classed as a xerosere, as the surface is extremely dry, owing to the rapid flow of any precipitation. The first plants to colonize such a surface must be able to withstand

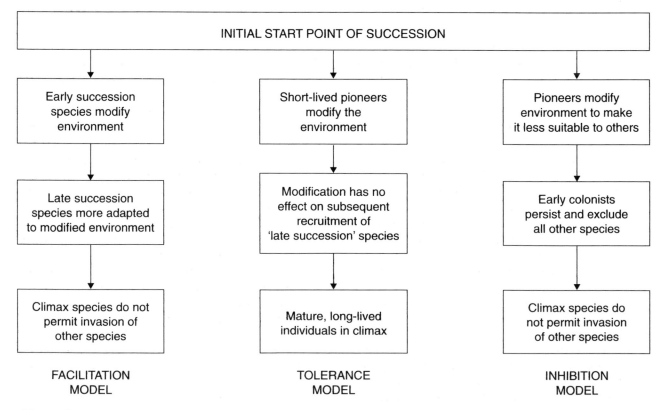

Figure 21.4 *Modern views of plant succession according to the facilitation model, the tolerance model and the inhibition model.*

complete drought and must be able to cling to a bare rock surface devoid of soil. This **pioneer community** typically consists of crustose lichens which form the white, black or orange growths on stone walls, gravestones or boulders. These lower plants gradually weather the crystals and cements in the rock by the process of **chelation**; chelation is the process whereby organic molecules, either from humification or from root secretions, can form a soluble complex with metals, especially iron and aluminium. They can also add organic material to the thin, raw soil, and in so doing increase its water-holding capacity and content of plant nutrients. A **second stage community** of mosses is able to establish itself, which in turn accelerates the weathering by chelation and also **hydrolysis**. The surface is becoming less susceptible to drought as the depth of water-holding soil increases. The **third stage community** consists of hardy grasses (e.g. sheep's fescue, *Festuca ovina*) and annual herbs. The *fourth stage community* will consist of shrubs such as bramble (*Rubus fruticosus*) and dog rose (*Rosa canina*). Soil is thickening continually, and the water content and nutrient content increase in step with the increase in soil colloids; clay minerals are synthesized from the products of rock weathering, and organic colloids are formed from humification. The *late successional community* would witness the arrival of the first tree seedlings such as birch (*Betula* spp.), rowan (*Sorbus aucuparia*) and ash (*Fraxinus excelsior*). Eventually deeper rooting trees of the *climax vegetation* (e.g. oak, *Quercus* spp.) will be able to colonize.

As the weathering of hard consolidated rock proceeds slowly, the lithosere takes hundreds of years to reach a climax condition. The key variables are, first, the susceptibility of the rock to weathering and, second, the weathering intensity as influenced by climate. Soil-forming processes in other primary successions involve either stabilization of sand (psammoseres) or siltation of water bodies (hydroseres and haloseres). These physical processes are generally much more rapid, and thus the corresponding succession to climax is much quicker. In Britain typical sites for lithoseres are landslides, scree slopes and some cliffs. Where bare rock surfaces have been formed by human activity, e.g. quarrying, the scars formed on quarry faces or on discarded rock waste remain visible for a long time. In such cases, if reclamation is required, it is usual for the slow natural processes to be speeded up by planting, stabilizing loose surfaces, and even importing topsoil. By these means the primary succession can be speeded up tenfold. Figure 21.5 illustrates the early stages of a lithosere on Carboniferous limestone, a hard rock prominent in the English Pennine uplands. The bare rock surface of this karst landscape is colonized by **crustose lichens** initially, which commence the processes of physical and biochemical weathering. Several generations of these organisms provide humic remains, which in turn can be invaded by **prostrate mosses**. Once deeper depressions have been hollowed out, thick **cushion mosses** can fill them and start the processes of soil formation. Fine mineral particles are released by weathering, and often also blown or washed into the site. Over time a granular Ah horizon is formed beneath a grass turf, and a lithosol is formed, which over limestone would be a rendzina soil (FAO: Calcaric Lithosol). (Plate 21.3.)

As well as vegetation development and soil deepening, the lithosere is characterized by invasions of animals and micro-organisms which increase in abundance and variety as the succession proceeds. Humus that accumulates consists mainly of the droppings of small arthropods (mites), collembola and insect larvae. These all play an important role in breaking residues. On base-rich limestones the chemical conditions are favourable for earthworms, which will invade the thin soil and thoroughly mix mineral and organic particles. In more acidic situations earthworms are inhibited, and the undecomposed organic matter accumulates as a peaty ranker.

Psammoseres

The **psammosere** is a succession which starts its development on bare, loose sand, either on sea or lake shores. The beach itself has no plant cover as the waves continually move the sand, whose abrasive action will destroy any rooted plants. (Plate 21.4.) Flora are restricted to microalgae and diatoms, often attached to sand grains. Some organic matter will be brought in on each tide, and decomposition will be carried out by organisms such as lugworms living in the sand. A considerable population of micro-organisms, nematodes, cocepods and worms provides food for large predatory worms, and many species of filter feeders are found within the sand.

In coastal regions with strong and consistent winds, sand is moved inland and deposited as dunes. The first deposition of *embryo dunes* is initiated by the deposition of sand around pioneering plants such as saltwort (*Salsola kali*) and sea rocket (*Cakile maritima*). Continued growth of the dune depends upon stabilization of the shifting sands by grasses, especially marram grass (*Ammophila arenaria*) in Europe. This species is salt-tolerant and drought-tolerant and spreads vegetatively by rhizomes, laterally growing underground stems. Rhizomes and extensive root

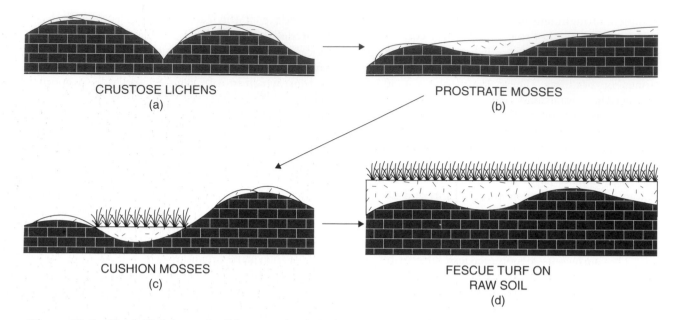

CRUSTOSE LICHENS
(a)

PROSTRATE MOSSES
(b)

CUSHION MOSSES
(c)

FESCUE TURF ON
RAW SOIL
(d)

Figure 21.5 *The initial stages of a lithosere on hard Carboniferous Limestone in the English Pennine uplands.*

systems help to bind the sand, and to convert the mobile dune into a fixed dune which is more favourable for a larger number of less xerophytic and less hardy plants. The biomass of marram decreases as one moves inland and the dunes become older. A larger range of grasses and annual herbs are able to colonize these *yellow dunes,* they are semi-fixed dunes, much of whose surface is still not covered by vegetation. Plants must be able to withstand stress from both heat and drought. Sands hold little moisture, and the water table will be several metres below dune surface. Spring annuals flower and set seed before the higher soil temperatures of summer.

Soils in embryo and yellow dunes consist mainly of quartz grains. They have a low cation exchange capacity, are deficient in nutrients, and have high calcium carbonate contents from shells. This, together with salt spray, makes them very alkaline, with a high pH. Nitrogen is considered to be critical for marram grass and sea buckthorn, and most is fixed from atmospheric nitrogen by nitrogen-fixing bacteria which live in the root zone of both these plants. Beyond the yellow dune zone, the **grey dune zone** is marked by the appearance of mosses and lichens. These older and more stable dunes have a larger cover of plants, with a more diverse flora of grasses, heaths and shrubs. Non-maritime species (e.g. Scots pine, *Pinus sylvestris*) start to make their appearance. Wherever depressions or 'slacks' occur in the dune zones (often caused by wind 'blow-outs'), marshy areas of rushes (*Juncus* spp.), sedges (*Carex* spp.) and willows (*Salix* spp.) show that the water table is near the surface. It is possible to get

local hydroseres or haloseres, depending on inundations by the sea, around the ponds of these 'slacks'.

As one moves inland the soil changes in psammoseres on sand dunes are striking. Acidity increases and pH declines as the influence of sea shells and salt spray lessens, and leaching remains strong through the coarse sand. Organic matter in the topsoil increases with distance inland, as the more varied and more abundant plant cover gives a larger litter input to the soils. Micro-organisms are more abundant and produce more humus, which in turn is able to hold more moisture, to provide more nutrients, and to give greater stability to soil surfaces. Eventually a deep, humus-rich soil with a thriving faunal population of earthworms, snails and insects will result. In this manner the original sterile area of sand has been transformed completely by the succession of communities (Figure 21.6).

Hydroseres

All lakes are ephemeral features of the landscape because they are gradually filled in with sediments. This silting process occurs independently of plant successions, but the process is considerably speeded up on lake margins by the development of **hydroseres**. The littoral zones of ponds and lakes are shallow sediment-receiving zones where the shallow water allows concentric zones of aquatic vegetation to develop, with one community replacing another as the depth of water changes in space and time. Figure 21.7 shows a transect of vegetation zones which one would meet in passing from the open water of the centre of the lake on to surrounding dry land.

The pioneer community of water lilies (*Nymphaea alba*) establishes itself when the lake is reduced to about a metre in depth. This plant has the effect of accelerating silting, both by reducing water velocity and by adding organic debris to the lake bottom. When the depth is further reduced, a second-stage community of bulrushes (*Scirpus lacustris*) and reeds (*Phragmites communis*) develops. Silting continues and the water shallows enough to allow a zone of pond sedges (*Carex* spp.). The emergence of a marsh surface above the water surface allows tussock sedges to develop, especially the cotton-grasses (*Eriophorum* spp.). Eventually higher ground supports a deeper soil, with a better drained surface which allows tree seedlings to survive. Alder (*Alnus glutinosa*) and willows (*Salix* spp.) can tolerate wetness and form *carr* woodland. In turn drier conditions allow Scots pine (*Pinus sylvestris*) to grow which eventually will be supplanted by the climax oak woodland (*Quercus robur*).

Each stage in the hydrosere is characterized by a vegetation zone where the plants have specific tolerances of waterlogging and the degree of soil wetness. Plants which can tolerate waterlogged soils are **hydrophytes**. Excess water excludes air (oxygen) from the pores of waterlogged soils, and this shortage of oxygen causes problems with root respiration. In well drained soils oxygen enters the root by diffusion from the soil atmosphere. However, the rate of oxygen diffusion in water is about 10,000 times slower than in air. The hydrophytes contain much more spongey tissue (**aerenchyma**) with thin walls and large air spaces that permit air to diffuse through roots and stems. This allows sufficient oxygen to be conducted from the atmosphere to the roots. Differences in the rate of oxygen diffusion affect the zonal distribution of hydrosere plants; moving from zone A to zone D in Figure 21.7 is to move through zones where oxygen moves less freely through the plant tissues.

Another problem with waterlogged soils in hydroseres is that the soil chemistry has undesirable characteristics. Elements such as iron and manganese are in a reduced state (ferrous and manganous salts). In this state they are very soluble and can be absorbed in toxic quantities by plants. Wetland species have the ability to transform the toxins into less harmful states (ferric and manganic) by the diffusion of oxygen outward from roots. Deposition of iron and manganese oxides around the roots of many hydrophytes is evidence of this.

Plate 21.3   *Early stages of a lithosere on limestone pavement in northern England. In the gryke there is more shelter, more soil and protection from grazing for the higher plants which are colonizing. Photo: Ken Atkinson.*

Zones E, F and G in Figure 21.7 represent the arboreal stages of the hydrosere. Alder (*Alnus glutinosa*) is more tolerant of marshy conditions than Scots pine (*Pinus sylvestris*) and oak (*Quercus robur*), both of which require a dry, well aerated soil for seedling establishment. Tree seedlings are also more demanding of nutrients, and this depends upon deeper soils with larger humus contents and more active nutrient cycles. The ability of alder to fix atmospheric nitrogen through micro-organisms living in nodules on its roots is a valuable input in raising soil fertility. Pine invades rapidly and forms a clear community before it is replaced by oak, which is able to out-compete pine for light.

Plate 21.4 *Stressful habitats at the start of successions support only a few well adapted species. Only a grass and a chenopod are able to tolerate the arid and saline environment in the pioneer stage of this psammosere. Photo: Ken Atkinson.*

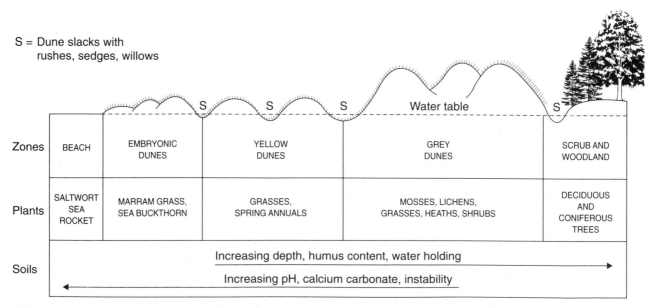

Figure 21.6 *The sequence of plants and soils found in a typical psammosere on coastal or lakeside sand dunes.*

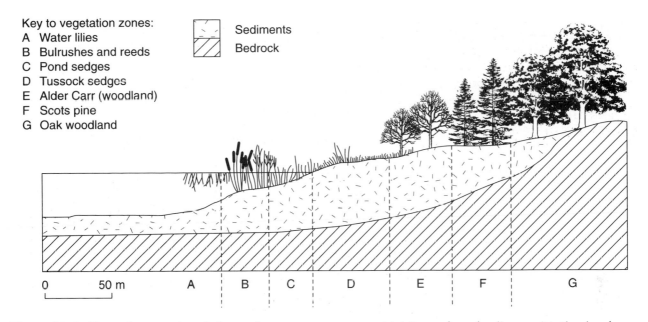

Key to vegetation zones:
A Water lilies
B Bulrushes and reeds
C Pond sedges
D Tussock sedges
E Alder Carr (woodland)
F Scots pine
G Oak woodland

☐ Sediments
☐ Bedrock

Figure 21.7 *Vegetation zones in a hydrosere from open water to terrestrial forest where the climax vegetation is oak woodland.*

Haloseres

Haloseres are found in salt marshes where silts can sediment in sheltered estuaries or where the coastline is protected by islands, bars and spits. Dominant plants have to be adapted to the stresses brought about by inundation by tides and additions of salt. Continual deposition of tidal silt raises the level of the land, and a change in vegetation is towards those species which are increasingly tolerant of prolonged exposure to the atmosphere. Colonization by flowering plants commences as soon as there is a stable soil surface. The rate of deposition increases as vegetation cover increases, together with the binding effects of their root systems. The pattern of vegetation communities reflects the frequency and duration of tidal flooding, and the effects which these two factors have on sediment accumulation. The **principle of competitive replacement** occurs in response to these environmental changes. As salt marsh plants have to survive periodic inundation by sea water, the vertical zonation reflects relative tolerance of salinity. Salinity causes problems to plants in water absorption, owing to the low external water potentials, and in nutrient absorption. Some species, e.g. *Spartina*, overcome this by excluding sodium from root uptake; other species, e.g. *Suaeda*, move potassium out of older leaves before they are shed in order to maintain internal ionic balance. **Halophytes**, plants tolerant of high salt content, also require an increase in tissue solute concentrations so that the osmotic potential of the plant exceeds that of the external soil solution.

The halosere shows a distinct vertical zonation, reflecting decreasing flooding as the level of the marsh is built up through accretion and the higher parts are flooded only by occasional high spring tides. The characteristic species of the lower marshes are summer annuals such as saltwort (*Salicornia herbacea*) and seablite (*Suaeda maritima*). Both are resistant to high concentrations of sodium, although many seedlings are destroyed by unstable substrates. When once the mudflat has achieved a certain height above the high-tide mark, rainfall will start to leach out the salt. Thus a less salty and more stable surface will allow a more diverse collection of plants to become established. These include the grass *Puccinella maritima*, sea lavender (*Limonium vulgare*) and sea aster (*Aster tripolium*). Marsh soils are typically waterlogged and anaerobic, with problems associated with oxygen deficiency and possible chemical toxicity. Some species, however, are restricted to the better drained banks of creeks or to areas of upper marsh; these inlcude the dwarf shrub sea purslane (*Halimione portulacoides*) and sea wormwood (*Artemisia maritima*).

The ecology of haloseres has been transformed in the past two centuries by the colonization of European marshes by *Spartina anglica*, cord grass, which is able to raise marsh levels by silt accretion much more rapidly than other marsh species. *Spartina anglica* (previously known as *Spartina x townsendii*) is a recently evolved hybrid of *Spartina maritima* (a West European species) and *Spartina alternifolia* (an introduction from North America in the 1820s). *Spartina anglica* has spread rapidly, both naturally and also through planting programmes, and has become the dominant species in many salt marshes. It is a large, vigorous grass which has quickly colonized low mudflats and open salt pans. These areas were previously unvegetated. Thus the accidental transhipment of a species from America to Europe has resulted in a new hybrid with distinctive properties. The processes of succession have been dramatically speeded up by this hybrid. In turn *Spartina anglica* is replaced by sea poa (*Puccinella maritima*), especially where the grass is intensively grazed. *Puccinella* withstands the grazing and trampling of sheep and cattle much better. Extensive die-back of *Spartina* has also been reported because of accumulated plant litter and toxic conditions in very fine-textured sediment.

Ecosystem changes through succession

A number of important changes occur in ecological communities during the course of succession. Through succession there is a progressive increase in the height of the dominant plants, increasing diversity of growth form, and increasing differentiation of communities into strata. Species diversity increases during the development stages of the succession, but this trend is reversed in the later stages, and the climax does not have the highest species diversity. Progressive soil development shows itself in increasing depth of soil, a larger organic matter content and sharper differentiation into soil horizons. The nutrient content of the soil also increases, as colloidal content is increased by mineral weathering and by humification. The stock of nutrients held in the standing crop (i.e. the biomass) also increases as the biomass itself gets larger towards climax. The US ecologist E.P. Odum studied the overall changes in community attributes during the course of succession. His findings are summarized in Table 21.2. Pioneer communities are characterized by extreme and variable physical conditions. Pioneer communities are composed of few species, but their densities and primary productivities are high, owing to their efficient seed dispersal and wide ecological tolerance. The pioneer community is fast-growing but of low stability. Mature communities have greater complexity of structure and species show increased specialization. The intricate organization gives a higher level of homeostatic regulation in mature

Table 21.2 *Community changes through succession.*

Attribute	Early	Late
Organic matter	Small	Large
Nutrients	External	Internal
Nutrient cycles	Open	Closed
Role of detritus	Small	Large
Diversity	Low	High
Nutrient conservation	Poor	Good
Niches	Wide	Narrow
Size of organisms	Small	Large
Life cycles	Simple	Complex
Growth form	r species	k species
Stability	Poor	Good

Source: After Odum.

Table 21.3 *Energy changes through succession.*

Attribute	Early	Late
P/R ratio	> < 1	≃ 1
P/B ratio	High	Low
B/E ratio	Low	High
NPP	High	Low
Food chains	Linear	Webs

communities. This means that internal regulation and symbiosis are well developed, so that external effects have declining importance and impact.

Energy exchanges are a vital part of ecosystems (see Chapter 22). During succession there are some important changes in energy attributes, as listed in Table 21.3. The P/R ratio (the ratio of gross production to respiration) indicates that the communities tend to achieve a balance of these in the later stages of succession; thus all solar energy fixed by the community is used in respiration. The P/B ratio (the ratio of gross production to biomass) decreases during succession. The B/E ratio (the ratio of biomass to a unit of energy flow) is the inverse of the P/B ratio, and is called the biomass accumulation ratio. It increases from close to 1 in the annual herb stage to 2–5 in the perennial herb–low shrub stage, to 5–10 in the high shrub–young forest stages, to 30–50 in climax forests. Thus biomass in relation to productivity reaches a maximum in the climax. Net primary production (NPP) does not reach a maximum in the climax, however. It is highest in the early stages of succession, when the vegetation consists of fast-growing species. In the climax vegetation the biomass shows a stable, steady state.

Theories of climax vegetation

The concept of climax vegetation arose at the beginning of the twentieth century. At a time when soil scientists were working with the idea of 'zonal soils', and climatologists were defining 'climate regions', ecologists started to conceptualize a stable type of natural vegetation which would be in complete equilibrium with climatic and soil conditions. The theory was put forward that, with no human interference, the end point of succession will be a self-sustaining and self-perpetuating community. This community will be the one which can compete most successfully in the prevailing soil and climatic conditions. The US ecologist Frederick Clements proposed the term *climatic climax vegetation*, defined as 'vegetation in stable equilibrium with climate and soil, given undisturbed conditions and free soil drainage'. Clements is credited with advancing this **monoclimax theory**, or *climatic climax theory*. He developed the concept of the plant community as an 'organism' which followed a sequence of stages as it developed into a mature state. The mature state or 'climatic climax' would be in equilibrium with the regional climate and the zonal soil, provided there was relatively long-term stability. All communities would reach this end point through plant succession, no matter what had been the initial starting point. Thus a psammosere and a hydrosere in a given region would ultimately reach the same steady-state vegetation. In southern Britain both would finish up as climax oak woodland; in Scandinavia both would finish up as coniferous forest.

The British ecologist Arther Tansley, who coined the term 'ecosystem' (see Chapter 20), was engaged in studying the interaction of plants and soils, and also the influence of grazing and other activities of animals on plants. He was, of course, interested in the ways in which at any location plants, fauna, soil and climate form an interacting ecosystem or equilibrium. Although he agreed with Clements in theory about equilibrium and climax vegetation, he disagreed that there would be *one* climatic climax and that plant communities behaved as organisms. He believed that the term 'organism' was best reserved for individual plants and animals. He argued that ecological communities are essentially complex physical-biological systems. He regarded the biosphere as a vast number of such systems, each one tending towards its own state of maturity and equilibrium. The time required to reach Clements's climatic climax was in practice too long to make the concept realistic. Other environmental factors would be powerful enough to hold a community relatively stable for considerable periods of time. Thus soil factors of drainage or chemistry (**edaphic climax**), or topographical factors (**topographical climax**), or human and animal activities (**biotic climax**) would prevent a true climatic climax from forming. Within any climatic region, different plant communities could be in relatively stable equilibrium with any one or combination of the above factors. This is the **polyclimax theory** of Tansley.

A third theory of climax vegetation is that of the American ecologist Whittaker, whose ideas are similar to Tansley's. Whilst studying the vegetation patterns of the Great Smoky Mountains, in Tennessee and North Carolina, he developed his **mosaic theory** to describe what he called **climax pattern**. He noted that similar patterns of environmental and biotic pressures do repeat themselves, and the vegetation is repeated like similar patterns within a mosaic. He noted also that only 60 per cent of the vegetation could be placed in these types, and that there was considerable gradation across community boundaries. These 'transitional' communities are termed **ecotones**.

The fourth concept of climax was first suggested by the French ecologist Aubréville whilst studying the tropical rain forests of the then French West Africa in 1938. The theory was revived in the 1980s, when there was renewed interest in tropical vegetation. It is the **cyclical climax theory**, and its modern supporters argue that it is valid for many ecosystems outside the tropics too. According to this theory, forests are regarded as areas of cyclical succession of growth and decay. As the cycles are out of step, the forest has the appearance of a mosaic, owing to different cycles operating side by side. There are three key elements in the cycle. First, an *optional phase* of trees of roughly equal age is established. Second, the optional phase deteriorates into a *decay phase* caused by the collapse of the forest over the greater part of a particular area. Young plants are now able to become established, but these young plants are often not the original tree species. Thus the collapsed primary forest is succeeded by a different tree community which, when it in turn collapses, is replaced in the third stage, the *mature* phase, by another even-aged forest community. According to the cyclical theory, what one encounters in a primary forest is not a constant steady state but a regularly recurring cycle. As different parts of the landscape are at different stages of the cycle, a patchwork mosaic results. Figure 21.8 illustrates the cycle in a tropical forest when an opening in the tree canopy is caused by a natural tree fall through disease or by ageing. After the gap has opened, light and

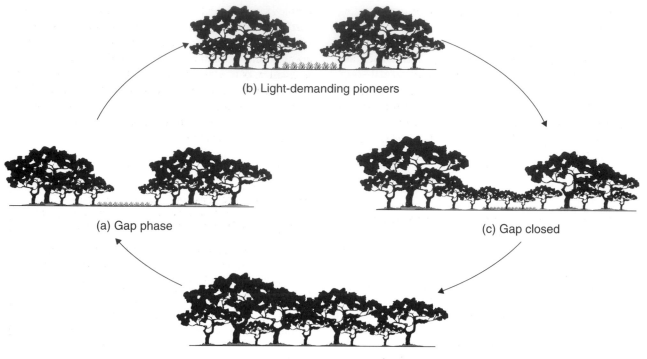

Figure 21.8 *The theory of cyclical climax as illustrated from an area of tropical rain forest. A clearing caused by a natural catastrophe, old age or human activity is colonized by species different from those originally growing there.*

temperature increase on the forest floor, killing the shade-tolerant understorey. Young seedlings of light-demanding species become established, and the gap becomes closed by the upward growth of these pioneers and by the crowns of trees surrounding the gap growing into the open. When the gap is closed again, no more light-demanding seedlings become established, but adults may persist. Seedlings of forest species will become established, and understorey plants adapted to low light and temperatures will thrive again.

In addition to the tropics, cyclical development has also been reported from the species-rich forests of eastern Europe. Here, as in the tropics, it is noted that trees which die are replaced not by the same species but by different species from the forest. There may be several causes. One is competition for light. Beneath trees that give heavy shade (e.g. common beech, *Fagus sylvatica*) the young plants of light-requiring trees like beech are unable to mature in the shade. Another reason to explain the failure of a plant to thrive on the same site as its parents may be the fact that different species do not have identical nutrient requirements. If a particular species has removed nutrients from a site for a long time, the same species may not be able to thrive on the same site. A biological mechanism involving animals has

recently been reported by the US ecologist Janzen from his studies of tropical forests in Costa Rica. In explaining why individuals of a species are scattered in the forest, with no near neighbours, he came to the conclusion that intense predation of tree fruits when they fall to the ground by insects, especially ants, means that a tree can reproduce only when its seed is carried far from the parent by birds or monkeys. Seed mortality is always 100 per cent, so that regeneration *in situ* is never possible.

Conclusion

The study of ecological successions has formed an important part of twentieth-century biogeography. From the stimulus provided by the early work of Clements have come classic studies by Tansley, Hutchinson, Whittaker and Aubréville, among others. The processes operating in primary and secondary successions are quite well known, and are reflected in the changing characteristics of soils and plant communities with time. The overall strategy of successional development, and the nature of the climax community in it, is less well understood. There are four main theories of climax vegetation, and each may have some relevance in a particular situation. Owing to clear differences in growing conditions we must not expect successions to have the same overall strategies in tropical, temperate and polar climates. Overall, however, the polyclimax theory of Tansley has many supporters.

Changes in soils and plant communities during the course of a succession not only involve changing soils and plant species but also are accompanied by changes in ecological processes relating to energy flow, diversity creation, nutrient cycling and ecological stability. Later stages of successions show increases in species biomass, in the structural complexity of the community and in the efficient utilization of energy and nutrient resources. Structure and function become increasingly self-regulated in climax communities, though species diversity usually declines in climax communities as poorly adapted species are eliminated.

Key points

1 Primary successions or priseres start when new habitats on land and water become available for colonization. Successive plant communities occupy the sites, starting with pioneer communities and finishing with climax vegetation. The principle of competitive exclusion operates, whereby each community creates conditions favourable to a succeeding community, which eventually out-competes and replaces it.

2 Secondary successions occur when a land surface utilized by human activity is made available to recolonization when the land use is abandoned. Thus old farmland becomes reinvaded by a pioneer community, or a forest clearing used for farming, as in tropical shifting cultivation, is abandoned.

3 Climax vegetation marks the end point of succession. It is the most stable, conservational and massive ecological community. Over the years of studying climax vegetation, four views of climax have arisen, namely monoclimax, polyclimax, mosaic climax and cyclical climax. Mosaic and cyclical climaxes have been reported from particular biomes, but the polycyclical theory appears to be the most widely relevant on a global scale.

Further reading

Colinvaux, P. (1980) *Why big fierce Animals are Rare*, London: Penguin.
An extremely readable and perceptive paperback, which covers many principles of ecology. Chapter 12 gives an interesting, modern account of succession.

Eyre, S.R. (1968) *Vegetation and Soils: A world picture*, second edition, London: Arnold.
A classic text with a detailed discussion of the principles of succession (especially chapter 2). Full of examples from all world biomes.

Miles, J. (1979) *Vegetation Dynamics*, London: Chapman and Hall.
A slim volume which covers key concepts in a lucid and simple manner.

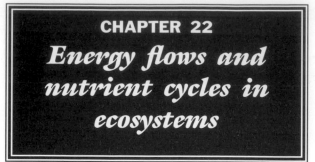

CHAPTER 22
Energy flows and nutrient cycles in ecosystems

As well as being distinctive assemblages of plants and animals, ecosystems also carry out work. Solar energy is captured by plants and passed on to other organisms. All life ultimately depends upon the sun's energy through the process of photosynthesis. Together with the flow of energy, ecosystems also depend upon the circulation of sufficient nutrients to maintain the healthy vegetative growth. Nutrients enter ecosystems from the atmosphere and from rock weathering, and can exit the ecosystem into the atmosphere and into drainage water. Nutrient cycles thus link the air, the rocks and the soils of the abiotic environment with the organisms of the biotic component of the ecosystem. Although each nutrient element has its own unique cycle, it is possible to generalize about different types of nutrient cycle. The processes involved in energy flow and nutrient cycling are analysed in the present chapter.

General principles of energy flow

The growth of interest during the present century in the study of the world's ecosystems has brought new ideas, concepts and models into the ways that humans study nature. Central to investigations of ecological systems is the need to understand their structure and function at various size levels, ranging from simple, local communities to the global biosphere as a whole. In the 1960s the US ecologist E.P. Odum suggested that ecology could best be defined as 'the study of the relationships between structure and function in nature'. Table 22.1 simplifies the major items which are studied under the two fundamental headings of structure and function.

The behaviour of energy in ecosystems is referred to as 'energy flow' because energy transformations are directional, in contrast to the cyclical behaviour of nutrients. Green plants photosynthesize organic compounds from water and carbon dioxide, using incident solar radiation as the energy source. Solar energy is thereafter fixed in a chemical form in the photosynthates until released as thermal energy during the respiration of plants and animals and the decomposition of organic matter. Total organic material fixed by photosynthesis over a unit time is called the *gross production* or **gross primary productivity** (GPP). The proportion which remains after respiration losses in the plant is termed *net production* or **net primary productivity** (NPP). The organic matter comprising plants or vegetation at any one time is the **biomass** or **standing crop**. Thus the energy flow of an ecosystem (E) can broadly be defined as :

$$E = GPP + R$$

where E = energy flow, GPP = gross primary productivity and R = plant respiration.

Figure 22.1 shows an energy flow model for any terrestrial ecosystem. Incoming solar radiation is fixed by plants and passed on in turn to herbivores, carnivores and decomposers (or detrivores). The transfer of energy between each component (box) in the ecosystem to the next highest involves a loss of energy as heat, following the second law of thermodynamics ('the change of state of energy involves degradation of some of it into a lower state (heat)'). Thus *radiant energy* from the sun is converted by plant photosynthesis into *potential energy* which, when utilized by plants and animals, is dissipated as *heat energy*. (Plate 22.1.)

Table 22.1 *The subject matter of ecology.*

Structure	1	Composition of the biological community	Species, numbers, biomass, etc.
	2	Quantity of abiotic materials	Nutrients, water, etc.
	3	Environmental gradients	Temperature, light, etc.
Function	1	Rate of energy flow	
	2	Rate of nutrient cycling	
	3	Regulation by the physical environment	

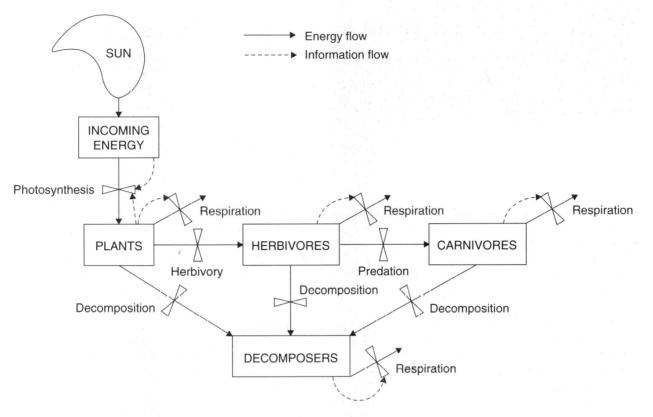

Figure 22.1   *The general model of the flow of energy through different trophic levels in ecosystems.*

Figure 22.1 is a universal model of energy flow. In real ecosystems living organisms are linked together by 'feeding' relationships or **trophic relationships** or *food chains and webs*. Each 'step' in the flow of energy is a **trophic level**, and organisms can be classified according to their functional trophic level. '**Producers**' (or **autotrophs**) have the ability to fix carbon through photosynthesis via green chloroplasts in their leaves. **Herbivores** are the primary consumers of organic molecules fixed by the producers. **Carnivores** are secondary consumers, living off the organic molecules of the herbivores. There may be several levels of carnivores in any one ecosystem; in such cases the ultimate level will be occupied by the '**top carnivore**'. The final group of organisms in an ecosystem are '**decomposers**' or **detrivores**, small animals, bacteria and fungi which can break down the complex organic chemicals of dead material and waste products. It should be recognized, however, that a food chain model is a simplified attempt to structure energy flow; in reality a particular species can occupy a position at several trophic levels. Thus some species of algae and bacteria can act both as photosynthesizers (autotrophs) and as grazers (heterotrophs). Foxes also can obtain part of their feed from eating the fruits and

leaves of plants (herbivore) and part by eating herbivores such as rabbits, mice and voles (carnivore). The fact that energy flow occupies such a central position in modern ecology owes much to the classic work of the US ecologist Raymond Lindeman in the 1940s. His 'trophic-dynamic' concept of ecology brought a focus at the level of entire ecosystems. In the 1920s the English ecologist Charles Elton had written about 'the pyramid of numbers' in terms of how food webs are organized in ecosystems. He noticed that a large number of green plants support a smaller number of herbivores which support a smaller number of top carnivores (Figure 22.2a). Such a pyramid is common for both populations and for species. Other ecologists noticed that a similar trend was evident in the total weight of living organisms (biomass) at the different trophic levels (Figure 22.2b). Whilst such pyramids of numbers of individuals, numbers of species and biomass are common, they are not universal. There are exceptions both on land and in aquatic ecosystems. One complication may be that one component of the ecosystem may have a rapid **turnover rate**, so that biomass figures would underestimate its importance. Lindeman realized that if organic material is looked on as a fuel or food energy (calories or joules), then an **energy**

Plate 22.1 *Massive western red cedar* (Thuja plicata) *and saplings of western hemlock* (Tsuga heterophylla) *in the Pacific rain forest of Canada. This climax ecosystem has evolved to regulate the flow of energy to maximum efficiency. Photo: Ken Atkinson.*

pyramid will always be found in nature (Figure 22.2c). Whenever an ecosystem is described in terms of a rate of energy flow (calories per square metre per day, cal m^{-2} d^{-1}, or thousands of joules per square metre per year, kJ m^{-2} yr^{-1}) through the different trophic levels, a pyramid shape will always result, following the second law of thermodynamics.

Biomass and productivity

Biomass is the mass of living organic material in a specific area or ecosystem. The units are weight per unit area (g m^{-2} or kg ha^{-1} or t ha^{-1} or t km^{-2}). Changes in biomass from year to year indicate the amount of energy or carbon fixed by photosynthesis and incorporated into an ecosystem. Biomass can be measured by harvesting the above-ground plant parts in a sampling plot (i.e. the shoots) by clipping at ground level. Large shrubs and trees are difficult to harvest in this way and usually some parameters of the trees (e.g. DBH, diameter of the trunk at breast

height) is measured and biomass calculated using yield tables or appropriate formulas. Underground biomass (i.e. root biomass) is more difficult to measure, but an estimate can be arrived at by washing the plant material from a volume of soil taken beneath the sampling plot.

Harvested material is dried in the laboratory at 80° C until it reaches a constant dry weight, which it usually does in about twenty-four hours. *Dry weight* provides the best estimate of biomass because fresh weight (or *wet weight*) includes the water content, which varies widely among plant species and even between fresh samples from the same species.

The energy content of different plant species or plant parts also varies. Dry weight can be converted to energy content if the calorific value of the material is known. The units are calories per gram (cal g^{-1}) or joules per gram (J g^{-1}). This value can be determined in the laboratory, using a calorimeter. Energy contents have been determined for many species and foods; Table 22.2 gives examples of some of them.

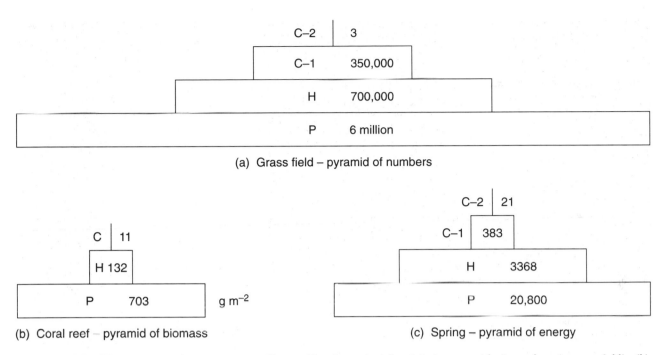

Figure 22.2 *The structure of ecosystems according to Eltonian principles: (a) the pyramid of numbers (a grass field), (b) the pyramid of biomass (a coral reef), (c) the pyramid of energy (a spring).*

Table 22.2 *Energy content of various plants and foods (kJ 100 g⁻¹).*

Plants and crops	
Potato	1646
Oats (seed)	1774
Mustard (seed)	2532
Green algae	2280
Sunflower (seed)	2829
Pine (seed)	2507
Spartina grass	1707
Wheat (seed)	1793
Foodstuffs	
Brown wheat flour	1377
Oatmeal	1587
Cornflour	1508
Whole milk	275
Double cream	1850
Cheddar cheese	1708
Brie	1323
Vanilla ice cream	814
Fats and oils	
Butter	3031
Olive oil	3696
Rapeseed oil	3696
Lard	3663

The increase in biomass (ΔB, delta B) with time (Δt, delta t) is a measure of **net ecosystem production** (NEP). In order to use data collected by the harvest of dry matter for calculating **net primary productivity** (NPP), the losses of biomass to herbivores (grazers) and detritus must be taken into account (Figure 22.3). In equation form:

Net ecosystem production = ΔB

Net primary productivity = Gross primary productivity – Respiration of plants

$$NPP = GPP - Rp$$
$$= \Delta B + G + D$$

where G = grazing and D = decomposition. In controlled experimental conditions, say in a laboratory or greenhouse, the flows D and G can usually be made very small so that

$$NPP = \Delta B \, / \, \Delta t$$

However, under field conditions, especially in natural ecosystems, the measurement of NPP is much more difficult because flows D and G must also be measured.

Energy flows in deciduous woodland

The study of energy flows in ecosystems is a long-term investigation. Populations of all plants and all animals (both vertebrate and invertebrate) are studied by sampling methods for many years. This involves counting their numbers, measuring their weights and sizes, and determining their energy contents in the laboratory using a calorimeter. All data for plants and

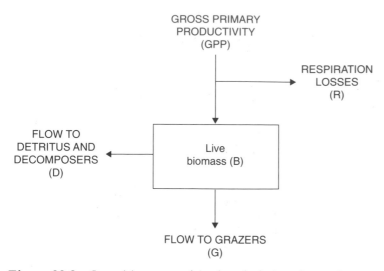

Figure 22.3 *Quantities measured in the calculation of net primary productivity and net ecosystem production.*

animals can be expressed as *energy equivalents*. Note that one calorie per gram is equivalent to 4.186 joules per gram (1 cal g^{-1} = 4.186 J g^{-1}), although usually energy content is expressed as kilocalories, where 1 kcal = 1000 cal or 4186 J. Figure 22.4 is an energy flow diagram for Wytham Wood, Oxfordshire. This famous deciduous woodland consists of oak and other species, with an understorey of shrubs and a ground flora of herbs. All figures in the diagram are thousands of joules per square metre per year (kJ m^{-2} yr^{-1}). The values within the boxes are for primary and secondary production (P); the values in circles are for consumption (C) or are energy inputs. These values are also shown in Table 22.3, together with values for **biomass** (kilojoules per square metre).

Table 22.3 *Productivity and energy values for Wytham Wood.*

| Units | *Ecological process* | | |
	Biomass (kJ m^{-2})	Consumption (kJ m^{-2} year^{-1})	Primary and secondary production (kJ m^{-2} year^{-1})
All trees and shrubs			26 × 10³
Oak trees	1 × 10⁵		5 × 10³
Total litter			13 × 10³
Caterpillars	41	356	40
Predatory beetles	38	380	38
Spiders	0.5	12	3
Great and blue tits	0.02	23	0.17
Shrews	0.00075	17	0.15
Voles and mice	0.16	105	1.2
Tawny owls	0.01	2.1	0.01

Note: Where no values are given no information is available.

There are many reasons for studying energy flow in both natural and cultivated ecosystems. One reason is that the *ecological efficiency* of different ecosystems can be compared. From the data for Wytham Wood the efficiency of the trees and shrubs can be calculated as follows:

$$\text{Efficiency} = \text{Output per unit input}$$

$$\frac{\text{Ecological}}{\text{efficiency}} = \frac{\text{Energy produced}}{\text{Energy received}}$$

$$= \frac{(5 \times 10^3) + (21 \times 10^3)}{1 \times 10^6}$$

$$= \frac{(26 \times 10^3)}{(1 \times 10^6)}$$

$$= \left(\frac{26 \times 10^3}{1 \times 10^6} \times 100 \right) \%$$

$$= 2.6\%$$

This low figure is characteristic of terrestrial ecosystems. Many authorities estimate that about 2 per cent of the usable light energy is fixed by plants. As only about 50 per cent of incident solar radiation is in the visible range of wavelengths absorbed by chlorophyll, the efficiency of total radiation is only about 1 per cent. In water bodies much radiation is absorbed by the water and its impurities, so efficiencies are much lower, probably in the range 0.1–0.2 per cent on average.

In attempting to understand ecosystems, biogeographers frequently subdivide organisms into *trophic levels*, where trophic level is defined as : 'the level at which an organism feeds in food chains or food webs. It gives the level or stage at which food energy passes from one organism to another.' In Wytham Wood insects, caterpillars, earthworms and mites are primary consumers, whilst owls, weasels and titmice are tertiary consumers. Note that titmice can also be primary consumers and secondary consumers, and weasels can also be quaternary consumers. A species thus often occupies different trophic levels within the same ecosystem.

One limitation of such energy flow diagrams is that they display average conditions only; they do not show variations over time. There are great *seasonal* changes in the feed value of deciduous trees and shrubs throughout the year. Leaf and bud growth leads to large seasonal populations of caterpillars and insects. Earthworms are most active in spring. Populations of birds and mammals also vary during their breeding seasons. Very large **annual** variations occur between years, largely dependent on climatic conditions. There are large *interannual*

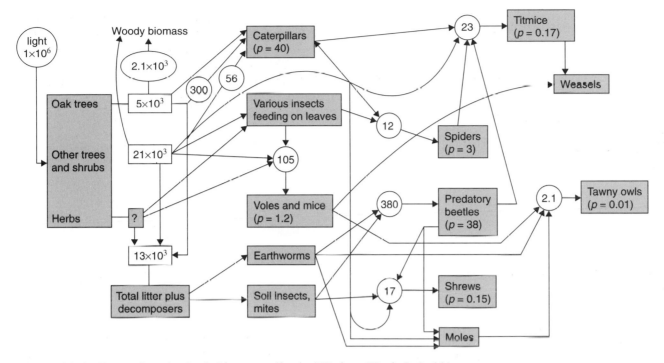

Figure 22.4 *Energy flows in the deciduous woodland of Wytham Wood, Oxfordshire.*

variations in the populations of insects, small animals, small birds and greenery. *Human influence* could also be important, as timber is harvested from the wood each year. This could have large effects on the ecosystem, especially if the rate of harvesting exceeds the rate of primary production.

Energy flows in prairie grassland

Figure 22.5 shows the results of an intensive study of a prairie grassland ecosystem in Saskatchewan, Canada, as part of the International Biological Programme (IBP). The prairie vegetation consists of grasses, sedges, sage and herbs. Biomass above ground was measured using a harvesting technique with 0.25 m² quadrats (50 cm × 50 cm). Shoots were clipped to ground level and litter was swept from the same plots. Green leaves and stems were separated from dead shoots. The distribution of biomass in the various components is given in the pie chart in Figure 22.5. These biomass values by themselves do not indicate the annual net primary productivity (NPP), as productivity is a rate measure of biomass accumulation over time. Also studies of populations and consumption rates of natural herbivores indicated that 26 g m⁻² of green plant material was consumed during each growing season in the proportions indicated in the second pie chart in Figure 22.5. Underground parts were studied by means of soil

cores taken at monthly intervals from the same quadrats. They were divided into 10 cm segments and the soil was removed by washing. An attempt was made to separate dead material from live material, but it is not easy. It is difficult to distinguish dead roots from live roots, and to decide what is current and what is old root. Consequently a conservative estimate of annual root production was arrived at by taking the difference between maximum and minimum dry weights in each layer, to a maximum depth of 1.5 metres. From Figure 22.5 it can be seen that net radiation for the growing season of 240 days averaged 17,130 kJ m⁻² day⁻¹. The percentage figures on arrows are percentage net radiation if above the NPP compartment and percentage NPP if below the NPP compartment. A useful statistic in describing ecosystems is the annual turnover of above-ground biomass. In the prairie ecosystem this is the annual turnover of shoot biomass, where

$$\% \text{ annual turnover of shoots} = 100 \times \frac{\text{NPP of shoots in g m}^{-2}\text{ yr}^{-1}}{\text{Total above-ground biomass in g m}^{-2}}$$

In this calculation total above-ground biomass includes the mean standing crops of green shoots, dead shoots plus litter

$$= 100 \times \frac{495}{2767 + 411 + 75}$$

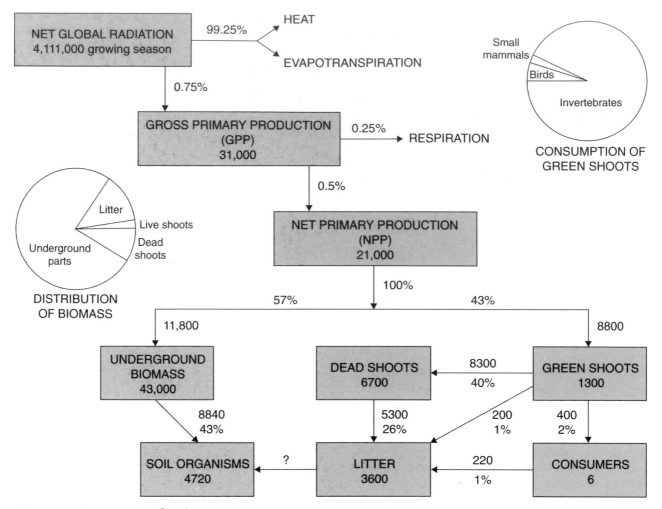

All energy values are kJ m^{-2} yr^{-1}

Figure 22.5   *Energy flows in natural grassland on the Canadian prairies.*

$$= 100 \times \frac{495}{3253}$$

$$= 15.2\%$$

Another useful descriptor of rates of turnover is *mean residence time*, which is simply the reciprocal of the annual turnover rate, when the turnover rate is expressed as a decimal fraction. Thus:

$$\text{Mean residence time for shoots} = 1/0.152$$
$$= 6.6 \text{ years}$$

Ecosystem production

Our knowledge of the production ecology of ecosystems has increased enormously since the start of the International Biological Programme (IBP) in the 1960s. This world-wide programme to learn more about the world's biomass was based on the theme

'the biological basis of productivity and human welfare'. The aim was to solve the fundamental equations of production ecology:

$$GPP = NPP + R(A)$$
$$NPP = GPP - R(A)$$
$$NEP = NPP - R(H)$$
$$= GPP - (R(A) + R(H))$$
$$= \Delta B$$

where GPP is gross primary production, NPP is net primary production, NEP is net ecosystem production, R (A) is autotroph respiration, R (H) is heterotroph respiration and B is biomass. Table 22.4 shows data on the primary biological productivity of the world's main ecosystem types. This table shows the area of each of the named biomes, the mean rate of NPP for each, the mean biomass of each, and the total biomass of each (mean biomass × area).

In terms of the degree of dominance at the global scale, Table 22.5 shows the proportion of the world's biomass which is found within each biome. Tropical forests contain two-thirds of the world's biomass, and a further quarter is held within temperate and boreal forests. Grasslands contain only 5 per cent, whilst aquatic ecosystems are veritable deserts in comparison. Table 22.6 shows additional information. LAI (leaf area index) is the area of leaves per unit area of ground. It ranges from highs of about 23 in some swamps and marshes down to about 0.5 in polar deserts. LAI has become an important factor in explaining differences in productivity both within and between biomes, as the leaf is the organ of photosynthesis. The length of time a plant can photosynthesize is equally important (evergreen v. deciduous). This is the index LAD (leaf area duration), which is not shown here. The percentage of NPP consumed by herbivores is also shown in Table 22.6. There are very real differences between forest biomes, grassland biomes and aquatic ecosystems, with increasing turnover rates in these groups. Similarly the ratio NPP/B contrasts the rapid turnover rates of lakes and oceans with the very low figures for terrestrial ecosystems, especially forests.

Global carbon budgets

Studies of the production ecology of the world's biomes are not just of theoretical interest. In recent years concern has arisen over the effects of increasing carbon dioxide in the earth's atmosphere due to the burning of fossil fuels. Most of the CO_2 released is absorbed by the sea or remains in the air, which are the major 'stores' or 'sinks' of carbon. However, a further 'sink' could be the world's vegetation, through increased photosynthesis. In the 1980s the burning of fossil fuels was releasing 5.4 billion tonnes of carbon per year, although the atmosphere content rose only by 3.4 billion tonnes per year. Scientists have been keen to discover what amount of this 'missing' carbon has been absorbed by the ocean and vegetation respectively. It is not easy to measure either the ocean or the vegetation system for small changes against the background of the huge scale of natural carbon exchanges. However, the study of tree rings in the boreal coniferous forests of the northern hemisphere has shown variations in growth rates over the past century. The width of the annual tree ring reflects the growth rate of the tree. When combined with data on tree area and tree density, tree ring data indicate whether the trees are growing faster. Any gain in overall wood volume shows that the

Table 22.4 *World productivity values.*

Ecosystem type	Area (10⁶ km²)	Mean NPP (kg m⁻² yr⁻¹)	Mean biomass (kg m⁻²)	World biomass (10⁹ t)
Tropical rain forest	17	2.20	45	765
Temperate deciduous	7	1.20	30	210
Boreal forest	12	0.80	20	240
Savanna	15	0.90	4	64
Temperate grassland	9	0.60	1.6	14
Tundra and alpine	8	0.14	0.6	5
Desert	18	0.10	0.7	13
Open ocean	332	0.13	0.003	1.0
Reefs	1	2.50	2.0	1.2
Estuaries	2	1.50	1.0	1.4

Table 22.5 *Global biomass within individual ecosystems (%).*

Forests	
Tropical rain forest	42
Tropical seasonal	14
Tropical evergreen	10
Temperate deciduous	11
Boreal	13
Grassland and desert	
Savanna	4
Temperate grassland	1
Tundra and alpine	0.3
Desert and semi-desert	1
Aquatic	
Open ocean	0.5
Reefs	0.6
Estuaries	0.7

Table 22.6 *Production characteristics.*

Ecosystem type	LAI (m² m⁻²)	% NPP consumed by herbivores	Ratio NPP/B
Tropical rain forest	6–16	7	0.04
Tropical evergreen	5–14	4	0.04
Temperate deciduous	3–12	5	0.04
Boreal forests	7–15	4	0.04
Savanna	1–5	15	0.23
Temperate grassland	5–16	10	0.33
Lakes		20	25
Open ocean		40	42
Reefs		15	2
Estuaries		15	2

forest is acting as a carbon sink, turning CO_2 into wood. Any decrease in overall volume shows that the forest is acting as a carbon source, releasing CO_2 into the atmosphere. The US ecologist Auclair has studied the boreal coniferous forest of the northern

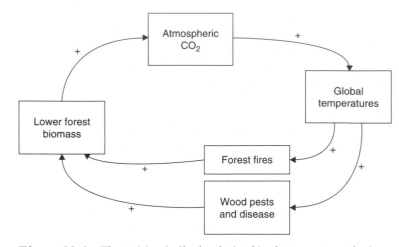

Figure 22.6 *The positive feedback relationships between atmospheric carbon dioxide (CO_2), global temperatures and the incidence of forest fires.*

hemisphere and concludes that the biome has been changing from a carbon source to a carbon sink, and recently back to a source again. Before 1890 boreal forests were a source of CO_2 because of logging and forest fires. After 1920 a rapid increase in the rates of tree growth exceeded the losses due to felling and fire, turning boreal forest from a carbon source to a carbon sink. However, in the late 1970s the boreal forests began to lose wood rapidly and once again became a source of CO_2. One estimate is that, if previous trends had continued, the boreal forest could have absorbed another 15 billion tonnes of carbon between 1976 and 1996. Instead, increased harvesting, fires and pests removed more timber than was assimilated. One hypothesis is that increased loss of trees from fire and pests (by insects and by micro-organisms) result from the increasing temperatures through the twentieth century. Thus a positive feedback may be in operation. Increasing levels of atmospheric CO_2 could be causing increases in temperature that in turn help to destroy trees, which releases more CO_2, causing even higher temperatures, and so on. These positive feedback relationships are illustrated in Figure 22.6.

Because tropical forests of different kinds are so dominant in the carbon budget of the globe (see previous section), it is vital to know what is happening to these tropical carbon stores. Until recently it was believed that they were acting as sources of CO_2 in the atmosphere. It was argued that tropical deforestation was releasing billions of tonnes of CO_2 into the atmosphere from the decomposing plant material on the cleared floor. However, the source may be smaller than was once thought. Farmland is often abandoned after a few years and

forests may then grow back fast enough to reabsorb as much CO_2 as is released by deforestation elsewhere. A current view is that the release and absorption of carbon are in balance and the tropics may not be as significant a source of CO_2 as was first thought. Tropical deforestation still has adverse effects on biodiversity, ecosystem stability and wilderness, of course.

Plant nutrients

The essential nutrients needed for plant growth are eighteen in number. Three of these – carbon, hydrogen and oxygen – comprise over 90 per cent of plant tissue and come from water and atmospheric carbon dioxide and oxygen. The remaining fifteen nutrient elements come largely from the soil, though there is the possibility of some absorption through the stomata on leaves. The fifteen soil-derived nutrients can be classified into **major nutrients** and **minor nutrients** on the basis of the amounts needed by plants. Thus nitrogen, phosphorus, potassium, calcium, magnesium and sulphur are required in large amounts. The minor nutrients are iron, manganese, copper, zinc, molybdenum, boron, chlorine, cobalt and selenium. The minor nutrients are also known as **trace elements**. There are two important features of nutrients which govern their cycling in ecosystems and their behaviour in soils. These are, first, whether or not the element participates in a cycle involving gaseous atmospheric components, and, second, the chemical form by which the nutrient is absorbed by the plant. In addition to carbon (C), hydrogen (H) and oxygen (O), cycles which involve gaseous components are nitrogen (N), sulphur (S), chlorine (Cl) and selenium (Se). The major store of these nutrients is the atmosphere, though only nitrogen and sulphur are of major importance. The remaining eleven nutrient elements have no gaseous form, though of course they can and do exist in the atmosphere as dust. In this group it is possible to distinguish between the **base cations** which are absorbed by plants as the positively charged ion (cation), and those absorbed as the negatively charged ion (anion). As we shall see, the distinction is vital for the nature of the respective nutrient cycles. In the cation group are potassium (K^+), calcium (Ca^{2+}), magnesium (Mg^{2+}), iron (Fe^{2+} or Fe^{3+}), manganese (Mn^{2+}), copper (Cu^{2+}) and zinc (Zn^{2+}) and cobalt (Co^{2+}). Nutrient elements which cycle and are absorbed primarily in the anion form are nitrogen (NO_3^-), phosphorus (PO_4^{3-}), molybdenum (MoO_4^-), boron ($B(OH)_4^-$), chlorine (Cl^-) and selenium (SeO_4^{2-}). Table 22.7 gives a classification of the

Table 22.7 *A classification of nutrient cycles.*

Store	Cationic	Anionic
Atmosphere	–	N, S
Lithosphere	K, Ca, Mg	P

major nutrient cycles on the basis of the main store (atmosphere or lithosphere) and the main chemical ion in the cycle (cation or anion). From the table it can be seen that the metallic cations form a group, nitrogen and sulphur have some general similarities (with some contrasts in detail), and phosphorus has a somewhat unique cycle.

The nitrogen cycle

Nitrogen (N) is the plant nutrient needed in greatest quantities (after carbon, oxygen and hydrogen), and forms an essential part in the structure of plant proteins. Its cycle is complex, involving atmosphere, soil and organic material, and depends upon the activities

of a range of specialized micro-organisms. The main features of the cycle are shown in Figure 22.7.

Organic materials are added to the soil surface upon the death of a plant or its organs. Waste products are also added which contain significant quantities of nitrogen. In whatever form, the conversion of this organically bound nitrogen into a form in which it can again be absorbed by plants (e.g. nitrate, NO_3^-) is referred to as **mineralization**. In detail, mineralization comprises several distinct and separate steps which have their own particular chemistry and microbiology. The first step is the breakdown of the organic nitrogen molecules (largely proteins) into ammonia (NH_3) or ammonium ion (NH_4^+). Under well drained, slightly acid conditions NH_3 is produced in large quantities; at neutral or alkaline pH, NH_4^+ predominates. This stage is known as **ammonification** and is carried out by a wide range of heterotrophic soil bacteria which gain their energy from organic carbon. The NH_4^+ ion can be readily absorbed by plants and micro-organisms in theory, but in reality most is used by a specialized group of

Plate 22.2 *Natural forest fires help to cycle nutrients. There is some loss of carbon, nitrogen and sulphur in smoke, but many bases like potassium, calcium and magnesium are concentrated in the ash.*
Photo: Alberta Forest Service.

THE NITROGEN CYCLE

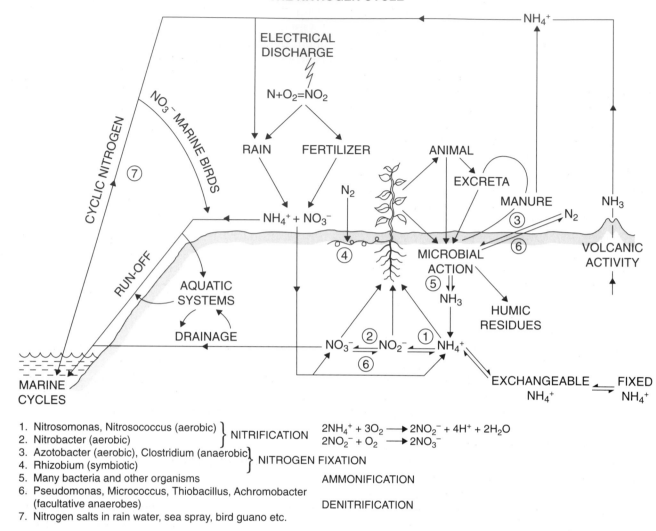

1. Nitrosomonas, Nitrosococcus (aerobic) } NITRIFICATION $2NH_4^+ + 3O_2 \longrightarrow 2NO_2^- + 4H^+ + 2H_2O$
2. Nitrobacter (aerobic) $2NO_2^- + O_2 \longrightarrow 2NO_3^-$
3. Azotobacter (aerobic), Clostridium (anaerobic) } NITROGEN FIXATION
4. Rhizobium (symbiotic)
5. Many bacteria and other organisms   AMMONIFICATION
6. Pseudomonas, Micrococcus, Thiobacillus, Achromobacter
   (facultative anaerobes)   DENITRIFICATION
7. Nitrogen salts in rain water, sea spray, bird guano etc.

Figure 22.7 *The nitrogen cycle.*

nitrifying bacteria which obtain their energy by oxidizing NH_4^+ or NH_3. These chem-autotrophic bacteria obtain energy by carrying out a chemical reaction rather than from organic carbon already assimilated by a plant or animal. The processes which convert NH_3 and NH_4^+ to NO_3^- are known collectively as **nitrification**.

Nitrification is a vital conversion for ecosystems and agricultural crops and has been studied in considerable detail. Two separate groups of chem-autotrophic bacteria are involved. The first group converts NH_4^+ to nitrite (NO_2^-) and consists of the aerobic bacteria *Nitrosomonas* and *Nitrosococcus*, which live in soil, freshwater and the sea. The second group oxidizes NO_2^- to NO_3^- and consists of the aerobic bacteria *Nitrobacter*. In addition to the need for oxygen, the processes also require a favourable pH (usually

between 5 and 8) and a suitable temperature. It follows that nitrification is much reduced in waterlogged, acid, alkaline or cold soils. Many micro-organisms have the ability to chemically reduce nitrous oxides (NO_3^-, NO_2^-, nitric oxide NO, nitrous oxide N_2O) under anaerobic conditions, when the compound is used as a substitute for oxygen. This process is known as **nitrate reduction**. When the reduction proceeds as far as the gaseous products of nitrogen N_2 and nitrous oxide N_2O the process is called **denitrification**. This extreme step is restricted to only a few genera of bacteria, namely *Bacillus*, *Micrococcus* and *Pseudomonas*. In waterlogged soils as much as 15 per cent of inorganic nitrogen may be lost to the atmosphere in this way. Even in well drained soils denitrification occurs because there will be anaerobic micro-environments where the diffusion of O_2 is slow.

The loss of gaseous nitrogen from ecosystems by denitrification is balanced by an approximately equal process of **nitrogen fixation** which brings organic nitrogen into plants and micro-organisms in the soil from gaseous N_2 in the atmosphere. The list of organisms that are capable of N_2 fixation has expanded enormously in recent years. The basic classification is into **free-living fixation**, carried out by aerobic bacteria, blue-green algae and anaerobic bacteria, and **symbiotic fixation**, carried out by root-nodule bacteria, root-nodule actinomycetes, and symbiotic associations with blue-green algae. Unlike nitrification, nitrogen fixation can readily occur in anaerobic soil conditions by either free-living anaerobes (e.g. *Clostridium*) or symbiotic blue-green algae (e.g. *Anabaena*). Table 22.8 shows the relative efficiencies of some N_2-fixing systems. The relative inefficiency of free-living N_2 fixers is clear, owing to their inability to obtain sufficient energy for the fixation process. On the other hand, blue-green algae are phototrophic and can get energy from photosynthesis, and are of great value in fixing N_2. In rice cultivation up to 50 per cent of the nitrogen requirement of the plant is met by N_2-fixing blue-green algae such as *Anabaena*, *Calothrix* and *Nostoc*.

The most important N_2 fixation occurs through the legume–*Rhizobium* symbiosis. It is estimated that legumes in agriculture fix 35 million tonnes of N_2 every year, 4 million tonnes are fixed in the rice crop, and 100 million tonnes are fixed in remaining terrestrial ecosystems. The importance of fixation by root nodule associations between actinomycetes (especially *Frankia*) and a variety of perennial non-leguminous plants is now being recognized. The plant genera which are known to form such nodules are *Casuarina*, *Hippophae*, *Myrica*, *Alnus*, *Dryas* and *Ceanothus*.

Other branches of the nitrogen cycle seem subsidiary, but can have important effects at the local scale. Lightning can produce N oxides in the atmosphere which are brought to the soil surface by precipitation. Significant quantities of N oxides are also produced by the internal combustion engine, and such pollution increases nitrogen inputs to local ecosystems. Human fixation of nitrogen is quantitatively much more important. Perhaps a quarter of all nitrogen fixation is by chemical industrial fixation for the nitrogenous fertilizer industry. The fate of these nitrogenous fertilizers is a cause of concern, as the NO_3^- anion is readily leached from soils and has the capacity to cause eutrophication of streams and lakes. It is also potentially toxic to humans; the disease methaemoglobinaemia ('blue baby syndrome') is due to high NO_3^- levels in drinking water

Table 22.8 *Relative efficiencies of N_2 fixation.*

Type of association	Organism	N fixed (kg ha^{-1} yr^{-1})
Nodules	Legumes	
	Tropical clover	900
	Temperate lucerne	45–675
	Non-legumes	
	Temperate alder	140
	Tropical Casuarina	50
Root		
	Temperate rye grass	60
	Temperate grassland	40
Blue-green algae associations		
	Tropical lichens	10–100
	Tropical Azolla	80–125
Free-living		
	Blue-green crusts	15–50
	Rice paddies blue-green	10–80
	Azotobacter	<1
	Clostridium	<1

which becomes reduced to NO_2^- in the human body, causing problems with oxygen uptake, particularly in infants. This has led to the designation of **nitrate vulnerable zones** (NVZs) in the United Kingdom to limit the use of nitrogen fertilizers.

The phosphorus cycle

Phosphorus (P) is a major plant nutrient which is absorbed as the anion orthophosphate (PO_4^{3-}). This is the chemical form which dominates its environmental cycle. As an anion it is denied an easily available and exchangeable reservoir on soil clays; being a solid, it is denied a large reservoir in the atmosphere like nitrogen and sulphur (Table 22.7). Phosphorus thus faces some unique problems. Most soils contain much phosphorus, but it is often a limiting nutrient because most is unavailable to plants. The main features of the phosphorus cycle are shown in Figure 22.8. One phosphorus store is in phosphorus-bearing minerals in rocks such as apatite. Phosphorus is released from these minerals by chemical and microbiological weathering. However, the major part of the phosphorus in soils is in the organic matter, largely as inositol phosphates, and the PO_4^{3-} will be released as decomposition and mineralization processes take place. The fate of PO_4^{3-} released by both mineral weathering and organic decomposition is crucial for the uptake and recycling of this nutrient. Phosphorus availability depends mainly on the pH of the soil. Under acid,

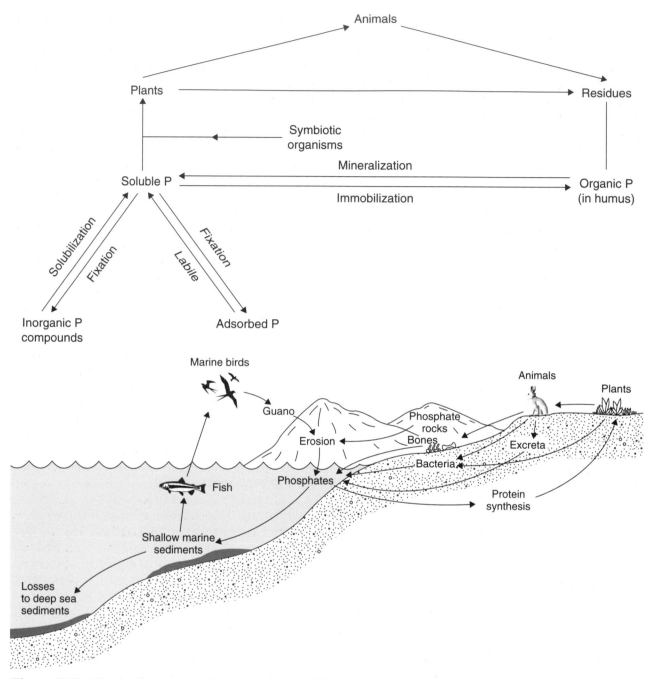

Figure 22.8   *The phosphorus cycle and its relative availability.*

low pH conditions, phosphorus is quickly precipitated as iron and aluminium phosphates; both of these cations are more soluble at low pH. Under alkaline, high pH conditions, phosphorus is precipitated as calcium phosphates in the presence of the high calcium concentrations usually present at high pHs. In all three precipitates – iron, aluminium, calcium – phosphorus is held tightly in a chemical form which is not available to plants. It is around

neutrality (approximately pH 7) that phosphorus is most available. Some adsorption of PO_4^{3-} will occur on to clay surfaces at neutral pH. Also microbial activity will usually be at a maximum at neutrality, which results in increased microbial mineralization, and the conversion of phosphorus from an organic to an inorganic form. The 'phosphorus problem' is that soluble and plant-available PO_4^{3-} is present only in low concentrations in most soils, and is quickly converted to unavailable forms. Micro-organisms are known to play an important role in solubilizing PO_4^{3-} from unavailable organic and inorganic stores. Large numbers of soil and marine micro-organisms are able to solubilize apatites and possess the enzyme phosphatase which will release PO_4^{3-} from organic phosphorus. Fungi can also form mycorrhizae which help phosphorus uptake by plants (see section on tropical forests in this chapter). As PO_4^{3-} is of limited solubility, very little phosphorus is deposited in oceanic sediments, when compared with the total biomass phosphorus. The amount deposited in oceanic sediments roughly balances the run-off from the terrestrial environment, which in turn equals the global input from rock phosphorus via weathering and mining. The oceans have a huge capacity for the immobilization of phosphorus in sediment, and act as an enormous pool of soluble phosphorus. The action of fish-eating sea birds in transferring marine phosphorus from the sea to the land is brought about by the large 'guano' deposits off the coast of Peru. The birds eat the fish, whose bodies are phosphorus-rich, and much phosphate is contained in the birds' droppings. A large tonnage of phosphorus is returned from the sea to the land in this way, and presumably similar if less spectacular returns are made in all coastal areas. However, terrestrial phosphorus is still lost to the marine environment; one calculation is that, whereas the world's rivers discharge 14 million tonnes of phosphorus into the oceans annually, sea birds can return only about 70,000 tonnes (0.5 per cent). This 'leak' of the nutrient is a further aspect of the 'phosphorus problem'.

Biogeochemical cycling of base cations

Base cations are those essential plant nutrients which are absorbed as the positively charged ion (cation) and which have no gaseous phase. From Table 22.7 we see that the main store is in rocks and minerals, and that the group contains the elements potassium (K^+), calcium (Ca^{2+}) and magnesium (Mg^{2+}). Other minor nutrients or trace elements which cycle in a similar way are iron (Fe^{2+} or Fe^{3+}), manganese (Mn^{2+}), copper (Cu^{2+}), zinc (Zn^{2+}) and cobalt (Co^{2+}). The term 'biogeochemical' reminds us that vegetation, soil, rocks, atmosphere and wildlife must never be considered separately in isolation; each is part of a continuously interacting ecosystem. In some respects the biogeochemical cycles of the base cations are the simplest cycles, as they do not have a gaseous component, although there are atmospheric inputs of precipitation, dust and aerosols. Considerable attention

was paid to the study of biogeochemical cycles by the Hubbard Brook ecosystem study in New Hampshire, which started in 1963 as a major experiment studying the biogeochemistry of a forest ecosystem, under the direction of the US ecologists F.H. Bormann and G. Likens. The ability of the watershed to retain nutrients was monitored, and entire watersheds were deforested in order to measure the effects on the export of nutrients. Figure 22.9 shows the general form of the cycle for all base cation elements. No fewer than nine fluxes can be identified in the cycle. **Weathering** releases the element from rock minerals and the ion becomes adsorbed by **cation exchange** on to clay minerals or humic colloids in the soil. **Plant uptake** is from soil water into biomass via plant roots. As a nutrient ion is absorbed by the plant, cation exchange releases an ion from colloid exchange sites, to maintain the concentration. Nutrients in plants are returned to the soil via litter into the soil organic matter, to be released again into the soil solution by **mineralization**. **Leaching** causes a loss of nutrients from the ecosystem into streams. *Precipitation input* from the atmosphere provides an import from outside the ecosystem, and dry deposition from dust can also take place. Some of the precipitation input can be absorbed by plants through **leaf uptake**; rainwater running across leaf surfaces can also leach ions back to the soil solution by **leaf leaching**. If the nutrient element forms insoluble and unavailable chemical compounds, it is being removed, even if only temporarily, from the cycle by **fixation**. The base cations held on the cation exchange sites of the soil's clay mineral and humic colloids are available to plants, and whilst held as adsorbed ions are not subject to leaching. Thus the soil colloids assume a pivotal importance in biogeochemical cycles. In turn the ability of the colloids is determined by soil pH. Acid soils have colloids with hydrogen ions occupying many of the exchange sites on the soil colloids. These soils are associated with low base saturation, and the availability of calcium, magnesium and potassium is much lower in acid soils than in near-neutral or alkaline soils.

In their studies of the calcium cycle in the natural forest of Hubbard Brook, Bormann and Likens discovered that the forest ecosystem is extremely conservative in its nutrient cycling. The precipitation input of calcium of 2.6 kg ha^{-1} yr^{-1} is matched by a loss in stream output of only 12 kg ha^{-1} yr^{-1} of calcium. This is a small rate of loss considering the large amounts of calcium in the calcium stores of the watershed, and is probably balanced by 9.1 kg ha^{-1} yr^{-1} released by weathering. The main stores and flows of calcium in Hubbard Brook are illustrated in

A BIOGEOCHEMICAL CYCLE

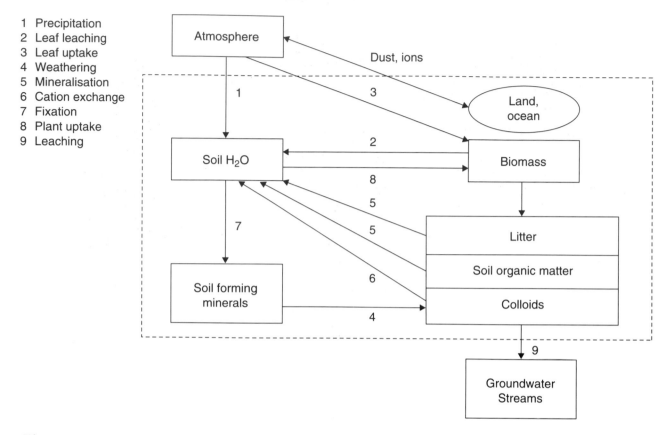

1 Precipitation
2 Leaf leaching
3 Leaf uptake
4 Weathering
5 Mineralisation
6 Cation exchange
7 Fixation
8 Plant uptake
9 Leaching

Figure 22.9 *General model of cycles of nutrient cations (base elements).*

Figure 22.10. The conclusion is that natural ecosystems have many nutrient-conserving mechanisms – in the living biomass, in the soil and in the micro-organism population. They result in the nutrients being recycled in a very efficient and tight manner. One effect of human interference is to break such conservational cycles and to cause serious depletion of nutrients from the ecosystem. As part of their experimental work in the Hubbard Brook catchment Bormann and Likens experimentally clear-cut several small watersheds and monitored dissolved nutrients in the stream water. The results are shown in Figure 22.11. The low figures in the sixty-year-old forest are increased enormously upon deforestation. This is due to the increased mineralization of litter and plant debris, the elimination of plant uptake, and the destruction of the buffering power of soil humus colloids. With time recovery will be brought about by the reinvasion of the cleared sites by ground vegetation, shrubs, seedlings and ultimately trees. The vegetation will eventually re-establish the nutrient cycles and lead once more to nutrient conservation. Temperate forests such as those of Hubbard Brook have the main store of nutrients in the litter layer,

slightly less in the living biomass of trees, shrubs and ground vegetation, and even less on the soil colloids. This is particularly so when considering nutrients such as nitrogen, sulphur and phosphorus, which occur mainly in organic forms. The situation is in marked contrast with the situation in tropical forests, where the bulk of the nutrients are in the living biomass, with only small proportions in the soil or the litter. This is illustrated in Figure 22.12, which contrasts the sizes of stores and flows in a tropical rain forest with those in a temperate coniferous forest. The mechanisms by which tropical rain forest is able to be sustained, despite the low content of soil and litter, are discussed in the next section.

Nutrient cycling in tropical rain forests

The relative proximity of tropical rain forests to the sun, when compared with higher latitudes, leads to faster, more dynamic systems, owing to the greater input of solar energy. Figure 22.13a shows the stores and flows (or fluxes) of nutrients within the rain

Calcium cycle
kg ha^{-1} yr^{-1}

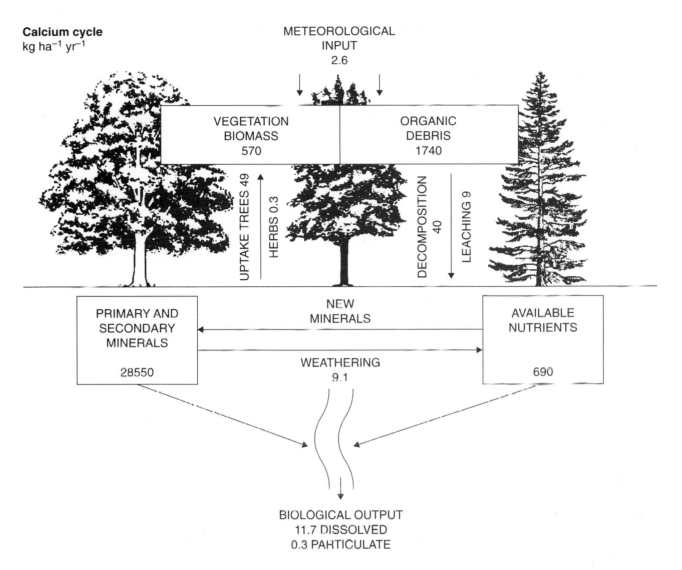

METEOROLOGICAL
INPUT
2.6

VEGETATION BIOMASS 570 | ORGANIC DEBRIS 1740

UPTAKE TREES 49

HERBS 0.3

DECOMPOSITION 40

LEACHING 9

NEW MINERALS

PRIMARY AND SECONDARY MINERALS 28550

AVAILABLE NUTRIENTS 690

WEATHERING 9.1

BIOLOGICAL OUTPUT
11.7 DISSOLVED
0.3 PARTICULATE

Figure 22.10 *The calcium cycle in Hubbard Brook, New Hampshire.*

forest. It is immediately clear that the vast bulk of nutrients are stored within the living biomass (the biota), and there is an absence of nutrient reserves outside the biota. However, there are several exceptions to this, as in forests on young volcanic soils (e.g. in Zaire or in the Pacific), where the nutrient input from weathering can be large. Also the flood plains of tropical rivers are similar, where annual floods supply large volumes of nutrient-rich sediments to the system. Generally, however, nutrient reserves in the soil component of the ecosystem are low. There are five main reasons for this.

1 The cation exchange capacity of the soil is small, owing to the presence of less reactive kaolinite clay minerals and oxides of iron and aluminium ('sesquioxides'). These colloids are formed under the influence of high temperatures and high leaching rates. Unlike large lattice clay minerals, they can hold few nutrients by ionic bonding.

2 Decomposers like termites and ants flourish within the continually maintained organic debris of the forest; they quickly decompose litter on and in the soil (see Plate 28.2).

3 Micro-organisms (fungi and bacteria) thrive in the hot and humid conditions at the soil surface, and are capable of completely removing nutrients from the soil surface.

4 Trees have a high capacity for the uptake of nutrients through their symbiotic relationships with a root fungus. This relationship gives mycorrhizae (particularly the type VAM, vesicular arbuscular mycorrhizae), and is an association of a fungus with the root of a higher plant. They are present

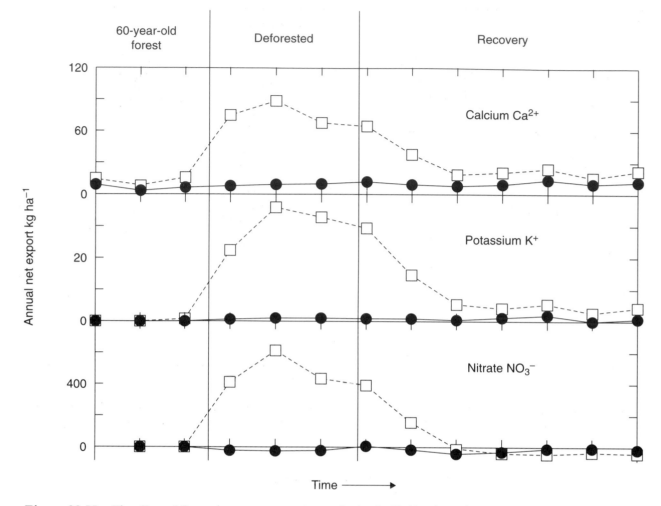

Figure 22.11   *The effects of forest clearance on nutrient cycles in the Hubbard Brook study.*

in most latitudes, but are particularly ubiquitous in the tropical zone. VAM are fungi which penetrate the root in order to feed on the cell contents. The benefit to the tree is that it is able to absorb nutrients from the fungi, which in turn are able to extract ions from dilute solutions such as soil water. The fungus thus acts as a nutrient pump for the tree, and is especially important to tropical trees because they are on old ferralitic soils (i.e. heavily leached soils with low cation exchange capacities). Indeed, so integrated and refined is this mycorrhizal relationship that the fungi are often in direct contact with organic litter and can transfer nutrients from it direct to the roots.

5 The physiology of tropical trees, especially their root systems, is such that they have the ability to pump large volumes of water from the soil, 'filtering' it for nutrients as they do.

Nutrient cycles within the rain forest are radically altered by human clearance for agriculture. The traditional peasant system is that of shifting cultivation, whereby a patch of forest is burned on a rotation basis (Plate 22.3). Crops are cultivated for several years in the burned area, until it is abandoned, thus allowing the forest to reinvade. The patch may be reused on a twenty to thirty-year rotation. The essence of this 'slash and burn' system is that the nutrients in the biomass are quickly released into the soil and litter compartments of the cycle. Harvesting of the crops then takes nutrients, as well as energy, out of the system (Figure 22.13b). Even if fertilizers are added, the effects are temporary, and cropping of the patch becomes unsustainable after a few years. These traditional indigenous technologies are well adapted to their local environment unless disturbed by rapid population growth, economic exploitation from outside, or imposed land tenure

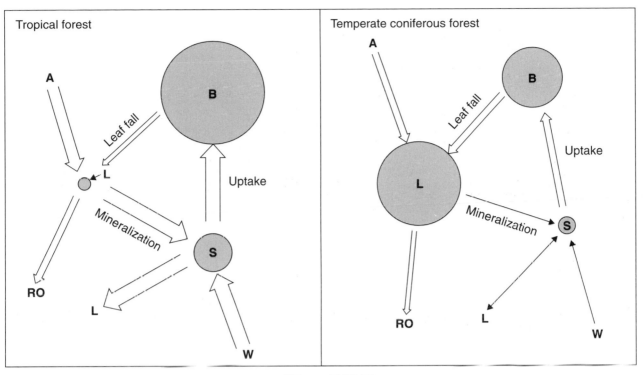

NUTRIENT CYCLES

Figure 22.12 *Diagrammatic comparison of nutrient stores and flows (a) in the tropical rain forest with those (b) in the temperate coniferous forest. B = Biomass store; S = Soil store; L = Litter store; A = Aerial input; RO = Run off; L = Leaching; W = Weathering input.*

changes. However, pressures from one or more of these three factors render a complex, sophisticated and elegant system unsustainable as inappropriate alien technologies replace traditional practices. Tropical rain forests are frequently replaced by plantation crops (oil palms, rubber, cacao, coffee, sisal). Such crops normally reduce the total amount of nutrients within the cycle. Stem flow is accelerated, and management techniques like weeding increase surface run-off. This, in conjunction with harvesting, leads to large losses of nutrients (Figure 22.13c). The biomass and soil stores are likely to hold the bulk of the nutrients. The net result of both shifting cultivation and plantation agriculture is to drastically alter nutrient cycles from a fundamentally stable natural system, comprising a diverse complexity of components where every niche is filled in order to maximize efficiency (Plate 22.4), to an unstable condition where there are few components or, in the case of the plantation, just one. The human objective is to harvest nutrients with a minimum of inputs. Inevitably, where such land use systems expand in their areal coverage within the tropical rain forest zone, the entire ecosystem becomes unstable (see Chapter 28).

Conclusion

The International Biological Programme, which commenced in 1964, has stimulated much data collection and ecological stocktaking in all the world's biomes. The aim is to solve the fundamental ecological equations dealing with productivity, biomass, nutrient status and energy assimilation. Rates of photosynthesis in different biomes vary with light intensity, temperature, moisture and soil nutrient content. Thus latitude is a great determinant of productivity on land through its effects on radiation, temperature, moisture and the length of the growing season. By contrast productivity in oceans is much more closely linked with the availability of nutrients. The productive zones in the oceans occur where the mixing of ocean currents brings sedimentary particles to the surface to feed the phytoplankton. This occurs more readily at mid to high latitudes.

Soil micro-organisms play a key role in nutrient cycling. The largest store of nutrients in many ecosystems is the organic matter, whether living in biomass or dead in litter and humus. The nutrient elements contained in those stores are mineralized

Plate 22.3   *A 'swidden' or burnt patch made by shifting cultivators in tropical rain forest in Sarawak. Crops will benefit from the nutrients in the ash 'fertilizer'.*
Photo: Ken Atkinson.

by microbial pathways of decomposition, releasing cations and anions which can again be absorbed by plants. These microbial processes are mostly carried out by a wide range of general-purpose soil microorganisms, but in the case of nitrogen and sulphur many reactions are carried out by highly specialized autotrophic bacteria.

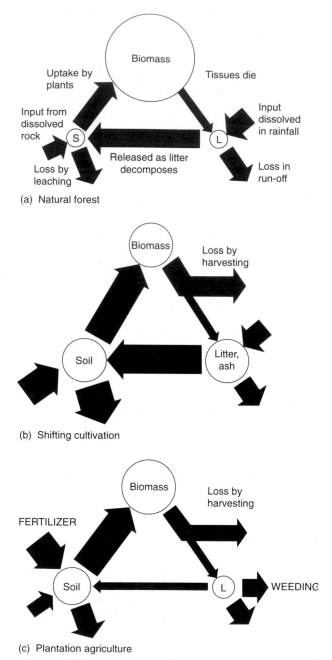

Figure 22.13   *Nutrient flows in the tropical rain forest and the impacts of agriculture: (a) nutrient cycling in undisturbed forest, (b) nutrient cycling under shifting cultivation, (c) nutrient cycling under tropical plantation crops.*

Key points

1 Gross primary productivity (GPP) is the sum total of energy fixed by autotrophic organisms through photosynthesis. That energy which is not used by the autotrophs themselves for respiration is termed net primary productivity (NPP). Some of the NPP will be grazed each year by herbivores, and some organisms will die and become decomposed. The remainder will cause an increase in biomass.

2 Production in ecosystems depends on an assured supply of nutrients, including water, in addition to light and heat energy from the sun. Two types of nutrient cycle provide the many nutrients necessary to plants. Gaseous cycles provide carbon, oxygen, nitrogen and sulphur to the biosphere through fixation. Sedimentary cycles provide elements such as potassium, phosphorus and calcium through the weathering of rock minerals.

3 Living plant material (biomass) and dead organic matter (litter and humus) contain a great reservoir of nutrients. This reservoir is released through decomposition by micro-organisms. Individual species (e.g. coniferous trees) and ecosystems (e.g. tropical rain forests) have evolved many mechanisms for cycling nutrients efficiently, with a minimum of loss from the system. The efficiency of nutrient cycling is perhaps the hallmark of a climax vegetation

Further reading

Bormann, F.H. and Likens, G.E. (1979) *Pattern and Process in a Forested Ecosystem*, New York: Springer.
The detailed account of the Hubbard Brook experimental catchment. Combines useful illustrative data with a clear enunciation of principles.

Grant, W.D. and Long, P.E. (1981) *Environmental Microbiology*, Glasgow: Blackie.
The principles of nutrient cycling are clearly presented with, as the title suggests, particular detail on the role of micro-organisms.

Trudgill, S.T. (1988) *Soil and Vegetation Systems*, second edition, Oxford: Clarendon Press.
A detailed treatment of productivity and nutrients in ecosystems, using a systems approach in the text and diagrams.

Plate 22.4 *A lateritic soil (FAO: Rhodic Ferralsol) with thin humus (Ah) over infertile red earth (Box). Roots are concentrated at the surface in order to 'catch' nutrients as they are released by mineralization. The pen shows the scale.*
Photo: Ken Atkinson.

CHAPTER 23
Diversity and stability in ecosystems

A natural ecosytem is a self-regulating community of organisms in equilibrium with their physical environment. Species become adapted through processes of natural selection to the conditions – biological and non-biological – which exist in the ecosystem. These processes of natural selection have worked through evolution to give us a world which can support a rich variety of species. Precisely how many species are known has been estimated by E.O. Wilson to be 1.4 million, including all plants, animals and micro-organisms. However, the accuracy of this figure is qualified by the fact that biologists agree that as an estimate it is probably less than a tenth of the number that actually live on earth! In fact the true number probably lies somewhere between 10 million and 100 million species.

Table 23.1 shows the number of species of living organisms known at present. Each group of organisms shows immense variety, which gives our first definition of **diversity**, namely the genetic diversity of organisms. Implicit in this definition is the idea of 'genetic banks' and 'genetic resources'. An alternative definition of diversity is simply the number of separate species in a defined geographical area.

Why are there more species in some areas than in others? The question has intrigued ecologists for some time, as it raises fundamental questions concerning *speciation* (the rate at which new species are formed), *adaptation* (the process of acquiring

structural, physiological and behavioural characteristics which improve on organism's chances of survival in a particular habitat) and *evolution* (changes in an organism's genetic make-up through time). **Biodiversity** has recently become a popular term for the variety of organisms at all levels (varieties within species, species, genera and families). It is also used to include the variety of ecosystems at different spatial scales. As biodiversity has become more prominent in political and economic debates at an international level it has given rise to the term **biocomplexity**, which is biodiversity within both a natural systems and a social systems context. Thus biocomplexity would include considerations of native rights, trade, aid and political–economic issues beyond purely biological features.

The present chapter examines the nature of diversity and the ways it can be measured. This enables us to compare different ecosystems at the local, field scale or at larger national and international scales (pp. 405–6). Then the reasons for such large contrasts in diversity in the world are discussed. Are the contrasts related to the fact that some places have had more time to evolve new species, or to the larger flows of energy in some ecosystems (pp. 406–8)? Do the factors of environmental stress and its converse, environmental stability, have any influence on diversity (pp. 408–9)? Or is the size of a particular habitat the important influence?

Ecosystem stability is a property of ecosystems which has aroused much interest in recent years. The reasons for the interest have much to do with concern about human impacts on the world's natural ecosystems, and uncertainty about how these ecosystems may respond to disturbances. However, the study of stability has proved to be difficult, partly owing to different meanings which can be attached to the term 'stability' and also partly from lack of basic information on many ecosystems. Definitions of stability are discussed (pp. 411–12) and an example taken from data about British birds (pp. 412–14). Many attempts have been made by ecologists to construct

Table 23.1 *Number of living species.*

Insects	751,000
Other animals	281,000
Higher plants	248,400
Fungi	69,000
Protozoa	30,800
Algae	26,900
Bacteria	4,800
Viruses	1,000
Total	1,412,900

Source: After Wilson (1992).

a general ecological theory linking diversity and stability (pp. 414–16). Finally there is a discussion of current interest in biodiversity from a variety of viewpoints (pp. 416–20).

Definitions of diversity

Diversity of ecosystems is not an easy property to define or measure. First it is necessary to define precisely the limits of the community being described in time and space (the temporal and spatial bounds). Thus one can talk of the diversity of seabirds on an island in spring, the diversity of plants in an oak woodland or the diversity of insects in the whole of the Arctic Tundra biome. Very often the boundaries in space, time and community are set by the logistics of the sampling programme of a particular field study. Two major characteristics make up diversity – the number of species in the system (**species richness**) and the evenness of species within the system (**equitability** *of species abundance*). Consider the data in Table 23.2, which show the relative dominance of five tree species in three woodlands. Woodland A has perfect evenness and the largest number of species. Woodland B has fewer species, but is still relatively even. Woodland C (a pine plantation) has few species and very uneven equitability.

One index of diversity is the total number of species, which gives species richness. Thus for the three woodlands the values of 5, 3 and 2 respectively would reflect the different diversities. However, that would be a crude, unweighted measure, taking no account of the relative proportions. A better index of diversity would take into account both species

Table 23.2 *Relative dominance of tree species in three woodlands (%).*

Woodland	Oak	Ash	Birch	Alder	Pine
A	20	20	20	20	20
B	40	30	30	0	0
C	0	0	10	0	90

richness and relative abundance. Several measures do so, but the most widely used is the *Shannon Diversity Index*, which is calculated by:

$$H^1 = - \sum_{i=1}^{n} P_i \log P_i$$

where S = number of species, P_i = proportion of ith species as a proportion of total cover and log = log base$_n$ (usually $\log_{10}$). The Shannon index is also known (correctly) as the *Shannon–Wiener Index*, and (incorrectly) as the *Shannon–Weaver Index*. It is derived from the complex mathematical field of information theory and hence its alternate name of **information theoretic index**. An example of the calculation of the Shannon index for the data in Table 23.2 is given in Table 23.3, using $\log_{10}$. The most diverse woodland is community A, with five species of equal dominance. The second most diverse is woodland B, which has fewer species than woodland A and also has a more uneven distribution, with oak being dominant. Woodland C, a pine plantation, has the lowest diversity, being almost a monoculture. The Shannon–Wiener values which reflect this trend are 0.70, 0.48 and 0.15 respectively. Despite the relative ease with which the index can be calculated, much discussion of the diversity and complexity of ecosystems is still based

Table 23.3 *Shannon indices for three woodlands.*

Species	Cover (%)	Proportion (Pi)	Log Pi	Pi log Pi
Woodland A				
Oak	20	0.2	−0.70	−0.14
Ash	20	0.2	−0.70	−0.14
Birch	20	0.2	−0.70	−0.14
Alder	20	0.2	−0.70	−0.14
Pine	20	0.2	−0.70	−0.14
			−Σ Pi log Pi =	0.70
Woodland B				
Oak	40	0.4	−0.40	−0.16
Ash	30	0.3	−0.52	−0.16
Birch	30	0.3	−0.52	−0.16
			−Σ Pi log Pi =	0.48
Woodland C				
Pine	90	0.9	−0.05	−0.05
Birch	10	0.1	−1.00	−0.10
			−Σ Pi log Pi =	0.15

(a) Four species

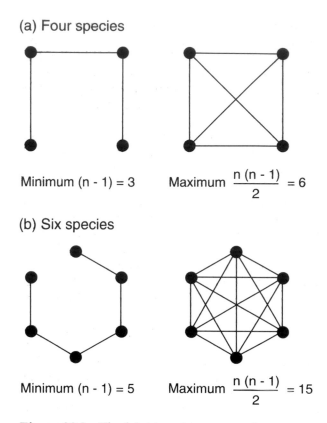

Minimum (n - 1) = 3   Maximum $\dfrac{n\,(n-1)}{2}$ = 6

(b) Six species

Minimum (n - 1) = 5   Maximum $\dfrac{n\,(n-1)}{2}$ = 15

Figure 23.1 *The definition of connectance in ecosystems; (a) a system with four species in the community; (b) a system with six species in the community.*

on species richness rather than on both richness and evenness. Other indices of complexity are more difficult to handle, because of the sophistication of the data required. Thus **connectance** in an ecosystem describes the actual number of interactions between species divided by the number of possible interactions between species. For example, a community of *n* species can have a minimum connectance of $(n-1)$ and a maximum connectance of

$$\left(\frac{n\,(n-1)}{2} \right)$$

Thus a community of four species can have a minimum of three interactions and a maximum of six. Connectance is important as an index of how strongly all the species in the system interact; if it were possible, it would be very useful to distinguish pairs of species which interact from those which do not (Figure 23.1).

The ecologist Whittaker considered that the diversity of any geographical area is made up of two components – the *within-habitat* diversity of a particular habitat (e.g. a field or a woodland) and the *between-habitat* diversity caused by micro-variations in the habitat or environment. The within-habitat diversity is called the **alpha diversity** (α) and the between-diversity the **beta diversity** (β). Beta diversity represents more species being supported when an area is broken down into smaller habitats, in which animals and plants become spatially isolated and avoid competition. Beta diversity can be measured by setting up a transect along an environmental gradient (e.g. of slope or wetness or soil type) and measuring species at equidistant sample points. A suitable community coefficient (see Chapter 20) is used to measure similarity in species composition between any two sample points. A graph of similarity against distance is constructed, so that the distance necessary to reduce similarity by 50 per cent can be determined. The beta diversity is the reciprocal of this distance:

$$\beta = \frac{1}{D}$$

where D = distance required to reduce similarity by 50 per cent.

Factors influencing diversity

Time and energy

One of the basic patterns of species diversity on the globe is the inverse relationship with latitude, i.e. species diversity increases as one travels from the poles to the equator. This is a dramatic relationship, being shown in all groups of organisms (plants, animals, insects, marine organisms) and from the geological record can be seen too in past ages. Figure 23.2 shows this trend for Permian molluscs, Cretaceous plankton, modern breeding birds and higher plants in Canada. The relationship is all-pervasive; twenty species of tree in northern Canada rise to 600 species at the equator, and ten species of marine crustaceans in the Arctic Ocean increase to 100 species in the Pacific. There are areas of low species diversity in the Tropics, as for example in arid tropical deserts, where diversity is lower than in temperate forests. However, this is due to dissimilar environments; when similar habitats are compared, the latitudinal trend is always strong.

There have been many attempts to explain this trend. Many hypotheses have been put forward, covering almost every environmental factor which changes with latitude. However, explanations concentrate on four main possibilities – time, energy, stability and geographical separation.

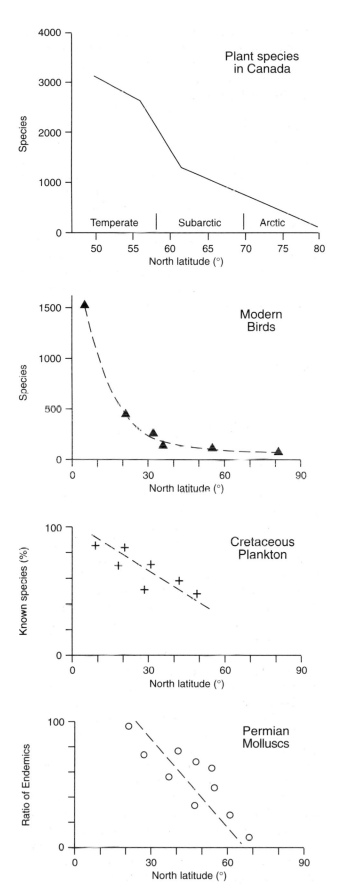

The Time Hypothesis is one of the oldest suggestions, and is based on the length of time which has been available in tropical forests for speciation mechanisms to evolve. It is argued that the absence of glaciation in tropical regions during the Quaternary era means that they contain a larger legacy of species from pre-glacial times; the implication is that the number of species in mid and high latitudes will catch up in time, hence their lower diversities are temporary. This is a doubtful hypothesis, for several reasons. The geological fossil record shows that the diversity gradient existed before the Quaternary, and seems to have been evident throughout most geological time. Also, as research in the tropical forests continues to increase our understanding of them, it becomes clear that they did suffer major climatic changes in the Quaternary; in fact it is difficult to find habitats on the earth which did not suffer disruption and disturbance during that time.

One aspect of the Time Hypothesis which is ecologically significant is the variation in diversity during the course of an ecological succession. The early stages of successions have low diversity, whether we are considering sand dunes, salt marshes, bare rock surfaces or wet lakeside habitats. These are rigorous environments where only a few opportunist species can survive. As succession progresses through time, and growing conditions ameliorate, the successional communities become richer and richer in plant, animal and insect species as more and more newcomers arrive by processes of replacement and immigration. Low diversity at the pioneer stage of the succession is replaced by higher diversity in the middle stages. In this sense, time is clearly an important local factor for such local situations. In the final stages of the succession, however, in the climax community, diversity usually declines; a small number of dominant species become established and shade out the larger number of successional species. Figure 23.3 shows changing diversities over the time of a succession in Brookhaven experimental forest, New York State. The number of species has been counted in plots of 0.5 ha size; diversity reaches a maximum in the herb stage, decreases in the shrub stage, increases again in the young forest before falling to a lower level in the climax community. In this sense the 'time factor' is important locally, but this situation is not comparable with the major latitudinal gradient of diversity.

Figure 23.2 *The inverse relationship between latitude and species diversity: (a) plant species in Canada, (b) breeding birds, (c) plankton in Cretaceous times, (d) molluscs in Permian times. Source: After Colinvaux (1973).*

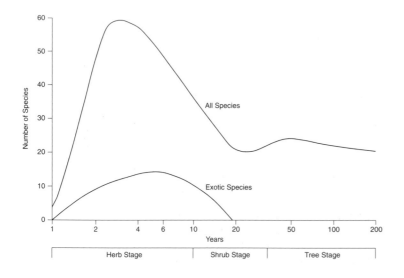

Figure 23.3 *The changes in plant diversity during the course of a succession.*

The second explanation concentrates on energy available for speciation and ecological processes, and is referred to as the Species–Energy Hypothesis. Briefly, more energy leads to more species. Variations in diversity are caused by variations in solar energy, which lead to variations in ecological productivity. With their high inputs of solar energy, tropical regions have more biomass and hence more structural niches. With many niches, many species are able to coexist even under conditions of intense competition and intense predation. MacArthur studied the relationship between bird diversity and foliage-height diversity (a measure of structural diversity) over a range of different habitats in the United States and found that the more complex the physical structure the more specialized the niche, and hence the more species (Figure 23.4).

However, there are several difficulties with this hypothesis. The temperate west coast forest of North America has a large biomass but much lower species diversity than the tropical forest, which has only a slightly larger biomass. Equally there are high-energy (high-productivity) ecosystems which have very few species; estuaries and salt marshes are examples, as also are the human-affected, high-energy communities of farming systems (high biomass, few species) and polluted lakes like Lake Erie (affected by eutrophication and having high productivity yet few species). On limestones in Britain one of the most species-rich habitats is limestone grassland, yet it has a low biomass. The connection between energy and diversity is thus not a consistent one. A more recent suggestion is that population size may be important. A larger energy flux could be expected to lead to

larger populations, and larger populations will have lower probabilities of extinction, given some environmental or ecological stress. This is a new way of looking at the Species–Energy Hypothesis and remains to be tested.

Stability and speciation

Intuitively it seems logical that places which are older and more productive should have more species. However, there are so many exceptions that neither the time hypothesis nor the energy hypothesis can be said to provide a general theory. On energy grounds many more species could exist almost anywhere. The vital question is thus: why have more species accumulated and survived in tropical regions than in high latitudes? A plausible hypothesis has been put forward by Sanders from a study of species diversity in oceans, seas and estuaries. The Stability Hypothesis suggests that areas of environmental risk, instability and stress are places of low diversity where only opportunistic species can survive. Thus estuaries, salt marshes, deserts and polar regions are areas of frequent stress and unpredictable hazards; they all have low diversity because of the high probability that species can be eliminated by the hazard. This is despite the high productivity of both estuaries and salt marshes. Conversely, regions of stable environmental conditions, such as the tropical forests and

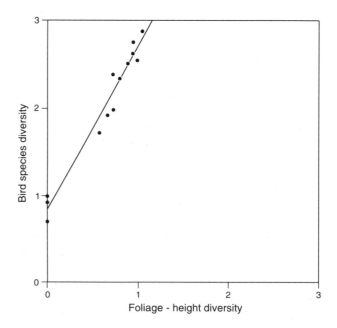

Figure 23.4 *The relationship between the structural complexity of vegetation and the diversity of bird species. Source: After MacArthur (1965).*

the ocean depths, where climate changes little from season to season and from year to year, are areas of high diversity. This is so even though oceans are ecosystems of low productivity and low energy fluxes. Sanders has summed it up thus: 'all places of high diversity have stable or predictable environments; all places of low diversity are either places of unpredictable hazard or are ephemeral'. Environmental stability leads to community diversity by providing reliable food resources, so that species can be more specialized in their feeding habits and thus are able to occupy narrower niches. It has also been shown by the ecologists MacArthur and May that groups of competing species can tolerate a greater degree of overlap between their respective niches when environmental conditions are more constant. Both these factors enable more species to exist on the same food resource and give a higher overall diversity. Bird and insect diversity in the tropics appears to result from the 'narrow niches' factor, allowing specialized diets to evolve. Tropical trees, however, are more problematical in that many species occupy apparently uniform habitats. In such circumstances, why are some trees not eliminated by interspecific competition? The question is not resolved yet, but two factors are being investigated. The first is that the tropical environment (e.g. soil cover) is less uniform than may at first appear, and micro-differences in soil properties may support specialized species. There is likely, therefore, to be more micro-variation in the physical environment than has been recognized so far. Second, tropical forests have rapid turnover rates, with trees and branches falling all the time and making gaps and clearings which can be colonized by equivalent but non-competing species. This is a continuous process and gives rise to the large spatial variability in what superficially may appear to be a uniform community. Figure 23.5 summarizes the stability hypothesis. As environmental stress increases, the diversity of the community decreases. Also, communities become increasingly under the control of the physical environment (climatic hazards, soil limitations, tides). Conversely, communities in regions of little stress are controlled by essentially biological processes (physiology and ecology).

In addition to differences in environmental stability, some ecologists have pointed out, real regional differences in gene pools must exist, so that speciation rates in the tropics are higher than in non-tropical latitudes. Thus in theory there would be a latitudinal gradient even if extinction rates were uniform all over the globe. One way to achieve high speciation rates in the tropics would be for the distribution of species to be more local, fragmentary and

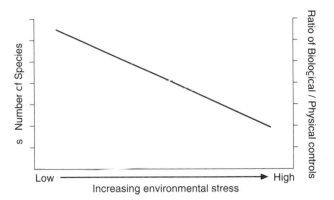

Figure 23.5 *The stability hypothesis to explain diversity. As the number of community-biological relationships in the ecosystems increases, they act as a buffer against physical controls, and the number of species increases.*

patchy. Geographical barriers permit species to become spatially isolated, to evolve in different directions (divergent evolution) and eventually to prevent interbreeding. In this way tropical 'fringe' species can survive long enough to adapt to local conditions and become a new species ultimately. In harsh environments (e.g. the Arctic) 'fringe' species are totally eliminated. Speciation in this manner is called *allopatric speciation* and the 'speciation hypothesis' explains a mechanism enabling greater diversity to arise in the tropics. The reasons why this greater diversity is able to survive is a more complex question but is clearly related to a combination of two important factors – environmental stability (which allows more niche specialization and niche overlap) and a larger energy flux (which speeds up mechanisms of speciation and divergent evolution).

Area

Biological diversity at any particular moment in time for any particular geographical location is the balance of immigration and extinction. In successions on new land surfaces (volcanic islands, salt marshes), new species arrive and colonize; initially the rate of immigration exceeds the rate of extinction, but as more species compete for space the rate of extinction increases until it equals the rate of immigration, and a state of dynamic equilibrium exists. New species arrive, old species disappear and the composition is always changing, but the number of species at any particular moment is constant.

MacArthur and Wilson developed a theory of island biogeography to explain why some islands have many species, others few. Their theory uses the **area**

effect and states that, other things being constant, there is a direct relationship between the number of species on oceanic islands and their size; larger islands have more species. This species–area relationship can be expressed as a mathematical equation:

$$S = CA^Z$$

where A = area, S = number of species, and C and Z are constants. The value of Z depends on the group or organizations being considered (e.g. trees, birds), and takes on values between 0.1 and 0.4. Thus one can say, approximately, that if area increases by a factor of 10 the number of species doubles. The larger the area the larger the diversity, mainly owing to the lower rates of extinction which result from more space and larger populations of each species. Immigration rates and extinction rates on large islands reach equilibrium only after many more species have colonized than on small islands. The second effect in the theory of island biogeography is the *distance effect*, whereby, the farther away from continents the islands are, the lower is the species richness. This is explained by a lower immigration rate the greater the distance which new colonists have to travel. Extinction rates will be the same for islands of similar size, but the extinction rate will reach equilibrium with the immigration rate at a level of fewer species on more distant islands.

An important corollary of the theory of island biogeography is to be able to determine the minimum critical size of populations and habitat in order to conserve and sustain plant and animal species in a particular habitat. What size of oak woodland is necessary to preserve its plant and animal population? One argument from $S = CA^Z$ may be that for 100 ha of woodland it is better to have ten patches of 10 ha (each with two species) than 100 ha (with its four species). However, if the two species are the same in each case, there would clearly be no increase in diversity. The discussion is complicated by the fact that different species have different ranges. Sparrow hawks require larger territories than warblers, and thus the conservation of sparrow hawks requires large nature reserves. Unfortunately too little is known about the territories of animal species, expecially in relation to emigration and immigration. For example, if one is keen to conserve hedgehogs, it is important to know whether the population in a particular habitat is an isolated entity or whether it reflects immigration of individuals from outside which is balanced by emigration to areas outside. This is where the DNA fingerprinting of individuals and the use of radio-tracking can establish the minimal critical size needed to conserve the population.

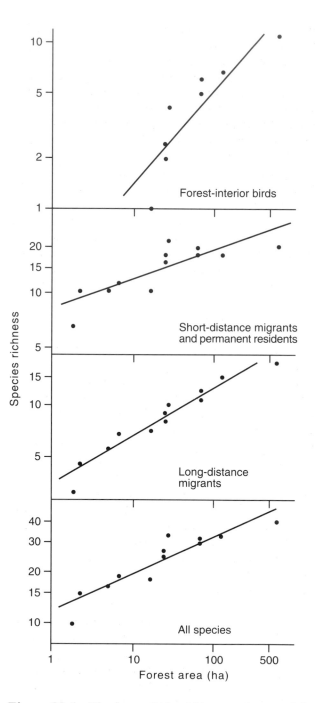

Figure 23.6 *The theory of island biogeography to explain species diversity. The species richness is plotted against the size of forest 'islands' for different bird populations in Illinois.*

Studies of birds in Britain have shown that the number of species found in woodlands does indeed reflect area. However, area seems to be an indicator of species diversity rather than a cause; a larger area is usually associated with greater habitat diversity in the form of floristic diversity and canopy height. Smaller woodlands possess fewer kinds of species than larger areas, and also contain smaller populations of particular species. This in turn leads to genetic drift, inbreeding and loss of genetic diversity, especially where habitat islands are physically isolated from each other. The net result is that population numbers may fall below a critical threshold, become vulnerable to a physical disturbance and may become locally extinct.

Figures 23.6 shows the relationships between forest area and species richness for twelve forests in Illinois. The relationships reflect the different ecologies of the four groups. Forest-interior species, most of which were long-distance migrants, were strongly dependent on forest area. Many long-distance migrants were not present in small forests, and thus would not be preserved if forest was present only in small patches. Species that do occur over the full range of forest areas (e.g. flycatchers, buntings) were birds of edge habitats.

On a global scale the tropical rain forests are being destroyed at a rate of 2 per cent per year, and it is estimated that the present area of 8 million km² is about half that in immediate post-glacial times. The rate of loss is increasing and will reduce the cover to 4 million km² by AD 2020. The question arises: what proportion of species will disappear? The answer will lie between 10 per cent (Z value 0.15) and 23 per cent (Z value 0.35). This elimination of 10–23 per cent represents 5–10 per cent of all species on Earth, at the most conservative estimate. The species–area equation accounts for most, though not all, of this loss. Hence many tropical countries try to preserve 'islands' of forest, as in Brazil, where a government law requires landowners to leave at least 50 per cent of their land under forest. Analysis of such 'islands' by ecologists shows that diversity decreases more rapidly the smaller the island. Winds and desiccation reduce shade-loving insects (ants, butterflies) in plots less than 10 ha in size, as well as amphibians, mammals and birds which depend on them. Large ground-dwelling mammals migrate quickly but some species of birds and monkeys flourish around the forest edges.

Definitions of stability

Stability of ecosystems is not an easy property to define. Indeed, the ecological literature suffers from confusion; in some cases the same term is used with different meanings, and in others different terms are used to convey the same meaning. Table 23.4 presents the most acceptable definitions of stability.

Table 23.4 *Definitions of stability.*

Term	Definition	Units
Stable	Returns to initial equilibrium after a perturbation	n.d.
Resilience	Speed of return to equilibrium after a perturbation	Time
Persistence	Time before variable changed to new value	Time
Resistance	Degree of change after a perturbation	n.d.
Variability	Variance of population densities over time	s.d or c.v.

Note: n.d., non-dimensional; s.d., standard deviation; c.v., coefficient of variation.

An ecosystem is *stable* if all variables return to the initial equilibrium position (defined as K) after suffering a perturbation or shock which has displaced the variables from their equilibrium position. How fast the variables return to their equilibrium is the *resilience*. If a system is unable to return to equilibrium it is *unstable* and therefore has no resilience. A special case is where biological populations do not return to equilibrium but cycle indefinitely (lemming in the Arctic, lynx in the Subarctic, red grouse in Britain). Figure 23.7 shows the reaction of four ecosystems to a perturbation. System A (a tropical rain forest) is stable; it does not depart far from equilibrium, and returns rapidly to it. System B is unstable; it passes beyond the stability domain and collapses. System C (a boreal coniferous forest) is stable, but less stable than A, owing to its larger displacement from K and the longer time needed to return to it (lower resilience). System D is the continual cycle referred to previously (Arctic populations); it is dynamically stable. All the above trends assume that the perturbations are equally strong. In the real world the behaviour of ecosystems will depend on the precise nature of the perturbation and on its magnitude. Table 23.5 lists the major perturbations (disturbances) which affect ecosystems.

Some perturbations cause very large changes in the abundance of species, like the severe British winter of 1962–3, which lowered bird populations. Others

Table 23.5 *Perturbations affecting ecosystem stability.*

Natural	Human-made
Drought	Deforestation
Freezing	Overgrazing
Fire	Agrochemicals
Insect pests	Acid precipitation
Disease	Oil pollution

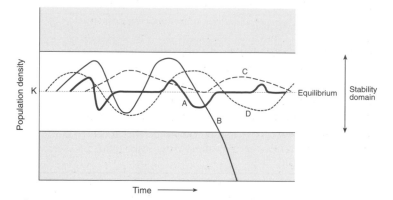

Figure 23.7 *The definition of stability. The behaviour of four different systems: A a stable tropical rain forest; B an unstable system; C a boreal coniferous forest, stable but less so than A; D a dynamically stable population which cycles.*

may involve the removal of some species, as in the outbreak of Dutch elm disease in 1975–85, which involved a long-term recovery. Perturbations need to be defined in terms of area of impact and time of impact. For that reason it is difficult to compare widely different ecosystems, although comparisons should be possible for similar ecosystems.

Two approaches to defining stability quantitatively are possible. The first, favoured by the ecologist MacArthur, uses information theory, in a similar manner to its use in the definition of diversity (pp. 405–6). Arguing that more ecosystem linkages and a more even flow of energy along them will give greater stability, one arrives at:

Measuring Stability

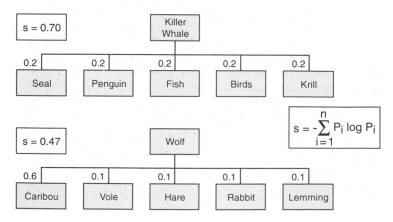

Figure 23.8 *Two food webs with contrasting stabilities. The killer whale receives food from a range of equally energetic sources and is potentially stable, whereas the wolf is overdependent on one source and is potentially unstable.*

$$S = - \sum_{i=1}^{n} Pi \log Pi$$

where S = stability, Pi = proportions of energy passing through the *i*th species. Two different stability situations are shown in Figure 23.8. The killer whale receives energy equally from five separate sources. This is a system with maximum choice, low information content and maximum uncertainty. In contrast the wolf subsists mainly on caribou, whose migration paths it follows, with lesser amounts of energy from a range of small mammals. This is a system of little choice, high information content and little uncertainty. The differences are reflected in the stability values of 0.70 and 0.47. MacArthur hypothesized that any failure of one energy pathway would be less severe the greater the number of pathways and the more even the distribution of energy between the pathways.

A second index of stability is the degree to which a biological population fluctuates, i.e. the variability of population densities over time. This can be measured by the standard statistical measures of variance (σ^2), standard deviation (σ) or coefficient of variation (cv) where

$$cv = \frac{\sigma}{\bar{x}}$$

where cv = coefficient of variation, σ = standard deviation and $\bar{x}$ = mean density. The variability of biological populations is important because it depends not only on internal properties of the ecosystems (intrinsic factors) but also on the nature and frequency of the perturbations (extrinsic factors). Populations vary more in climatically unpredictable ecosystems (arctic, subarctic, arid, semi-arid) than in predictable ones (tropical rain forests), suggesting that extrinsic factors may govern variability more than intrinsic ones. (Plate 23.1.)

British birds: a stability example

Information concerning the stability of biological populations (mammals, birds, insects) is notoriously difficult to obtain. It requires long-term studies to monitor long-term effects; long runs of population data are needed because short-term studies can be poor indicators of long-term trends. Charles Elton collected data from the Hudson's Bay Company in Canada to record population trends for the chief fur-bearing animals (see Chapter 24). He argued that, as hunting and trapping effort is not likely to change much from year to year, company records of furs and skins bought would be good indices of population numbers.

Another good data set is that provided by the monitoring programme of the British Trust for Ornithology (BTO), whose Common Birds Census (CBC) has recorded bird populations in sample plots since 1962. Figure 23.9 shows trends in the breeding population densities of five British birds since the 1960s. Numbers are relative to an arbitrary value for a base year of 100, except for the common crossbill, where the data show percentages of sightings. The data should be interpreted both in terms of the linear trend shown and also by its variability. The trends show 'increasers' (e.g. the jackdaw), 'decreasers' (e.g. the woodcock and the grey partridge) and 'marginal increasers' (e.g. the crossbill and the wren).

However, the crossbill and the wren show populations which are fluctuating, and both show large coefficients of variability. The wren suffered a collapse in numbers during the severe winter of 1962–3 but recovered quickly, i.e. it is resilient. The fluctuations since the recovery seem to correlate with the severity of individual winters. The common crossbill also fluctuates widely in numbers, for reasons which are not altogether clear. However, it is slowly increasing, owing to the expansion of its habitat by coniferous afforestation The consistent decline in the population density of the grey partridge is associated with two factors. First, there are declining numbers of nesting sites, especially as hedgerows are removed and field headlands ploughed; second, the increased use of agrochemicals leads to a decline in the availability of insect food. The decline in the woodcock population has still not been adequately explained. The increasing population of the jackdaw is probably related to its less specialized feeding habits, compared with many birds; it is therefore adaptable, flexible and less influenced by changes in farming practices. One thing which is clear from the forty-two farmland birds and the thirty-two woodland birds which the BTO censuses annually is that there is no common trend in the abundance and therefore stability of species; some species increase, others decrease. The majority of populations show long-term trends in abundance, but for different reasons and hence in different directions.

These overall trends have been of interest to ecologists who wonder whether they reflect long-term environmental changes or human impact. Steele has argued that there are two basic trends – 'red noise' and 'white noise'. 'Red noise' is the stability situation where the variability of populations increases with time; 'white-noise' is the stability situation where variability does not increase with time. In the former case the amplitudes of variability increase with time; in the latter case, the amplitudes are constant. Figure

23.10 plots the standard deviations of the logarithms (SDL) against the period over which the calculation was made. SDL increases with time for the 'red' but not for the 'white' population. Steele put forward the hypothesis that the oceans exhibit 'red' noise, whereas land ecosystems show 'white' noise, i.e. truly random effects and impacts. However, recent work on the BTO census data for British birds by the ecologists Pimm and Redfearn suggests that the variability of land 'noise' is coloured too. Figure 23.11 shows the example of the skylark. Figure 23.11a shows the densities on farmland for the years 1962–86, with the scale set at 100 for 1966. Figure 23.11b shows the data for the same population plotted as standard deviations of the logarithms

Plate 23.1 *A young snowy owl on its nest on Devon Island, Canadian Arctic. Snowy owls live almost entirely on lemmings and occasional hares. Populations are unstable, as they fluctuate greatly in response to lemming cycles.*
Photo: Ken Atkinson.

(SDL) of density against the period over which the calculation is made. Pre-1970 densities are ignored because before 1970 many bird populations were recovering from the crash in the hard winter of 1962–3. Had those data been included, the increase in SDL would have been more marked. For both nested years (2, 4, 6, 16) and non-nested years (2, 4, 8) SDL increases with period. Pimm and Redfearn found the same result for a range of birds, mammals and insects from various countries. The implication is that land 'noise' is also red, and populations show larger fluctuations as time goes by. The implication is that biological populations have no 'equilibrium level' but can build up to levels which make them susceptible to random crashes and possible extinction. In the case of British birds, widespread habitat changes have occurred during the second half of the twentieth century, perhaps at a faster rate than ever before, and they continue to affect the number and distribution of birds.

Relationship between diversity and stability

The debate on the relationship between diversity and stability has been going on for some time. For many years it was assumed, intuitively, that there was a causal connection between complexity on the one hand and community and ecosystem stability on the other. Perhaps the most compelling case for this view was expressed by Charles Elton in his book *The Ecology of Invasions by Animals and Plants* (1958). Elton made six main points.

1 Simple mathematical population models of ecosystems show large fluctuations in species numbers, with frequent extinctions.
2 Simple laboratory communities are also unstable, extinction being the norm.
3 According to the theory of island biogeography (p. 410), small oceanic islands with few species are more easily invaded by alien species than are large islands or continental communities.
4 Ecosystems simplified by humans (agricultural land, forestry plantations) are more susceptible to pest and disease outbreaks than are natural communities.
5 Species-rich tropical rain forests have fewer pest outbreaks than less diverse temperate forests.

Figure 23.9 *The population histories of selected British birds, 1962–88, illustrating different stabilities.*
Source: After British Trust for Ornithology.

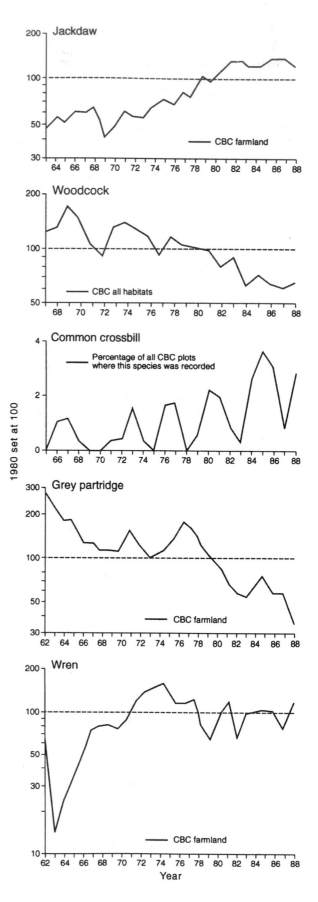

6 Coming from Elton's extensive fieldwork in the Arctic was the observation that extreme population cycles are found in areas of low diversity.

These observations have been supported by many others. Prominent in support of Elton has been MacArthur, who uses information theory to define stability (p. 412). The argument is that complex food webs are more likely to occur in species-rich communities. In a complex food web most of the consumers feed on several different organisms, and most prey organisms are attacked by more than one predator. In theory, the more cross-connecting links there are, the more chances the ecosystem has of compensating for a perturbation imposed upon it. In the 1960s MacArthur put forward the view strongly that the stability of an ecosystem is a function of the number of links in the web of its food chains. Figure 23.12 illustrates two ecosystems of very different complexity. In the simple case, the loss of any one component would cause the system to collapse (Plate 23.2). However, the loss of any one component of the complex system (say, the elimination of herons) would lead to an enlargement of other populations, which would fill the gap.

In the 1970s, however, a radical change took place in the diversity–stability debate. It became clear that any direct relationship between diversity and stability must be more complicated than was previously thought and was certainly not automatic. May has studied the mathematics of diversity–stability theory and sees no reason to suppose that stability increases with diversity. Others quote examples of species poor communities which are stable and species-rich communities which are unstable. Many species-poor communities can be very resilient, with the plants recovering quickly from an unusual drought, for example. Such are the species-poor heather moorlands of upland Britain. In contrast nearby species-rich limestone grassland communities can show very low resilience.

Other ecologists emphasize the value of links which have evolved over time (*co-evolutionary links*); human-modified systems (agro-ecosystems) have no co-evolutionary links between the interacting species. Farms are really ecosystems with a haphazard collection of species selected by the farmer. It should also be borne in mind that, among natural communities, stable complex systems have survived whilst unstable complex systems have disappeared. (Plate 23.3.) Other workers have discovered that in some ecosystems the more connected the ecosystem (the higher the connectance) the more amplified small variations become as they propagate through the well connected system. This of course is counter to the hypothesis of MacArthur.

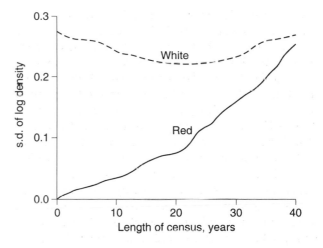

Figure 23.10 *The definition of 'white noise' and 'red noise' in census data for biological populations. Source: After Pimm and Redfearn (1988).*

The picture regarding diversity and stability is complicated and unclear. As we have seen (pp. 405–6, 411–12), there are various definitions of both diversity and stability. In the past different relations between different indices have been investigated, and so perhaps it is not surprising that there are different answers. What is clear is that ecosystem and community stability are dependent on environmental stability. A stable environment will promote community stability, and species-richness is then likely to increase over time.

The main conclusions seem to be as follows. First, the more species that are present in a community:

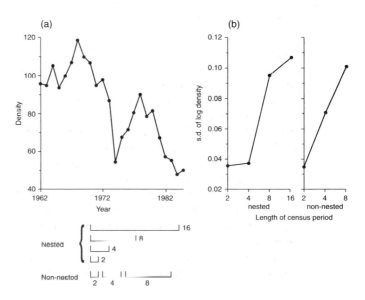

Figure 23.11 *The 'red noise' stability of the skylark: (a) population densities 1962–88, with 1966 set at 100; (b) same data plotted as standard deviation of logarithms of density.*

Plate 23.2 *A beluga whale in Arctic waters. These mammals concentrate together in large numbers at certain times of the year. At such times they are very vulnerable to natural or human hazards.*
Photo: Ken Atkinson.

1 The less resilient it will be.
2 The greater the change in composition and biomass when a species is removed.
3 The longer the persistence.
4 The less connected it should be, to be stable.

Second, the more connected a community:

1 The more resilient it will be.
2 The more resistant will be its biomass if a species is removed.
3 The longer its persistence.
4 The more likely it is to lose other species if one is removed.
5 The fewer species it must have if it is to be stable.

The fact that different diversity–stability questions elicit conflicting answers goes a long way to explaining why the debate has been so controversial in the past.

Threats to global biodiversity

The destruction of the world's vegetation is recognized as one of the most serious of human impacts. An immediate result is the extinction of plants and animals and the loss of habitat. It is estimated that between 1990 and 2015 between 2 per cent and 10 per cent of the flora of tropical forests will become extinct. Another threatened habitat is oceanic islands, very often with their own endemic species which are jeopardized by the introduction of competitive foreign species. It is estimated that 30 per cent of plants under threat of extinction are island endemics. About 1000 plant species are known to have become extinct in the past 2000 years, and about 25,000 are currently threatened. However these estimates are likely to be on the low side, given the lack of data for many regions.

The International Union for the Conservation of Nature (IUCN) issues Red Data Books of threatened species. In 1990 4500 animal species were listed

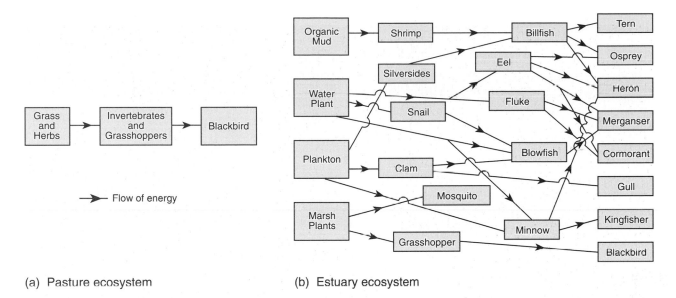

(a) Pasture ecosystem   (b) Estuary ecosystem

Figure 23.12 *Two ecosystems of strikingly different complexity. The simple blackbird case (a) is vulnerable to collapse if one component is adversely affected; the estuary ecosystem (b) is likely to survive the loss of one component. Source: After Woodwell (1967).*

as threatened; this figure represents 12 per cent of mammals, 11 per cent of birds, 4 per cent of fish, but only 0.1 per cent of insects. Table 23.6 lists the major threats to both plants and animals. Most of the threatened mammal and bird species live in tropical countries or on oceanic islands. In the latter case, flightless birds have suffered very badly from introduced vermin (especially rats). By contrast, many threatened reptiles, amphibians and fish live in temperate latitudes.

A special concern has been to protect entire ecosystems rather than to concentrate on one or two species within those ecosystems. About 20 per cent of all plants are classed as *endemics*, i.e. plants with a very restricted range and confined to a specific region. For example, the island of Madagascar, one of the most isolated of large islands, has 10,000 plant species, 80 per cent of which are endemic, i.e. found nowhere else. It also has thirty species of lemurs which are endemic too. The Mediterranean biome in California contains 25 per cent of all plant species found in North America, of which about 50 per cent are found nowhere else in the world. On the basis of regions which have unique species and which are threatened by extinction, Myers listed eighteen hot-spots (Figure 23.13), considered to have the highest conservation priority. Of the eighteen habitats, fourteen are tropical forests and four are Mediterranean ecosystems. The list is likely to be the very minimum, representing only those areas which are well documented. There are likely to be many more 'hot-spots', especially in oceans, lakes and rivers, which

will be added in the near future. The IUCN also recognizes 250 *Centres of Plant Diversity* (CPDs) which are particularly rich in plant species and which would safeguard a high proportion of the world's flora if they were to be protected. In contrast to 'hot-spots', though, CPDs are not classed on the basis of threat of extinction.

Part of the awakened interest in global biodiversity is undoubtedly economic as much as ethical in nature. The world's food supplies depend upon about 200 plants which have been domesticated, of which perhaps twenty are of major economic importance. The development of high-yielding varieties depends upon wild plants to donate genetic material to the cultivars for needed improvements, e.g. to improve resistance to pests and diseases (see page 513). Future breeding programmes will depend upon the availability of wild plants. Similarly, there is enormous potential among the plants of the world for medicinal use and the extraction of new drugs. The World Health Organization (WHO) lists 20,000 plants with medicinal uses, of which only 25 per cent

Table 23.6 *Threats to plant and animal species.*

Loss or fragmentation of habitat by cultivation, forestry, grazing, settlement
Over-exploitation for commercial gain
Deliberate or accidental introduction of competitive species
Deliberate eradication of pest species
Disease

Plate 23.3   *A climax ecosystem of Douglas fir* (Pseudotsuga taxifolia) *in the Pacific rain forest of western Canada. Species diversity is only moderate, as the fir excludes many other tree species by shading.*

have been studied as sources of new drugs. However, anthropologists and ethnobotanists estimate that among the world's indigenous peoples perhaps 50,000 to 70,000 plant species are used for medicines; again, only a few have been studied in detail and there is an urgent need to investigate them before they become lost for ever through extinction. Given the importance of biodiversity, it is not surprising that it has been the subject of several important conservation programmes. The *World Conservation Strategy* published by the IUCN in 1980 brought the issue to centre-stage. It proposed that countries should develop national conservation strategies, with biodiversity as one of several goals. More recently in 1992 in Rio de Janeiro UNCED (United Nations Conference on Environment and Development) proposed a Biodiversity Convention which amounted to a global strategy for maintaining biodiversity. In November 1995, at a conference of interested parties in Jakarta, Indonesia, it was decided that Montreal, Canada, would be the home of the United Nations Convention on Biological Diversity.

Conclusion

The diversity and stability of ecosystems have become an important field of study in biogeography as human society tries to measure and mitigate its adverse impacts on the natural world. The assessment of such impacts is made difficult by the lack of many long-term data sets which would enable conclusions to be drawn. Diversity can be measured relatively accurately from fieldwork, but stability is a more elusive property to assess. There are several different aspects of stability, and they can give contrasting indications of the stability of a particular ecosystem.

Long-term data on bird and animal populations appear to offer the best bet for assessing stability, but such data are not common. The diversity–stability debate which has occupied so much attention in the past few decades urgently needs such field data; it has relied too heavily on the results of laboratory studies and computer modelling for its predictions.

Future conservation at an international level needs to concentrate on the eighteen 'hot-spots' which

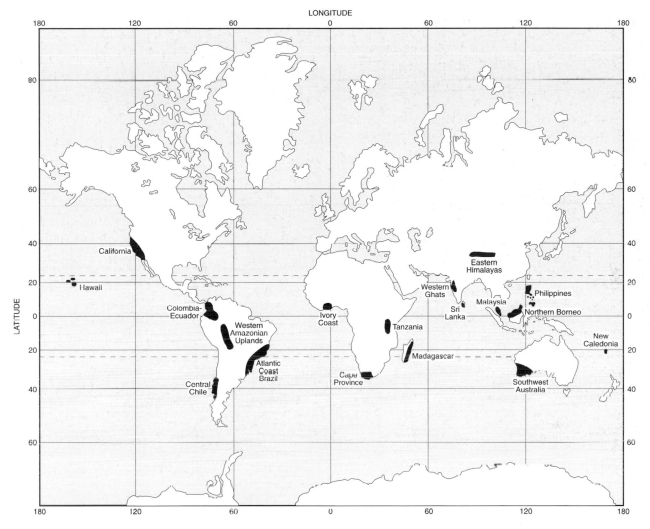

Figure 23.13 *The eighteen hot-spots of the world – areas of exceptional ecological diversity. Source: After Wilson (1992).*

contain a high proportion of the world's species and which are particularly vulnerable. These all occur in tropical and Mediterranean biomes, in areas where the pressures of economic development increase year by year.

Key points

1 Diversity is measured by the Shannon index, which takes into account the number of species and the evenness of distribution of species. Several indices are available to measure stability, depending on the aims and nature of the investigation.

2 The relationship between diversity and stability has come under close scrutiny recently, as serious doubts have arisen concerning the old adage 'more diversity means more stability'. Hence it is important to define carefully *a priori* what particular property of stability is being studied. Diversity and stability are both relative concepts; they need to be defined in terms of geographical space, community and time.

3 The upsurge of interest in biodiversity at an international level has been brought about by the increasing concern for the diversity of species, especially in a number of tropical locations. What is certain is that many more species on Earth have yet to be discovered than are at present known. Only by the scientific study of diversity and stability, according to the principles and methods discussed in this chapter, will it be possible for effective policies to be formulated at international and national levels, and for those policies to be translated into action programmes at a more local level.

Further reading

Groombridge, B., ed. (1992) *Global Biodiversity: Status of the earth's living resources*, London: Chapman and Hall.
A detailed study of the current situation regarding threats to species and to ecosystems of high diversity.

United Nations Environment Programme (1995) *Global Biodiversity Assessment*, Cambridge: Cambridge University Press.
The official and detailed study coming out of the Rio conference on global biodiversity. Packed with useful discussion and example.

Wilson, E.O. (1992) *The Diversity of Life*, London: Penguin.
This is perhaps the most scholarly and up-to-date account of the nature and causes of diversity. Very readable and full of examples.

Public and scientific interest in polar environments has never been greater. The twentieth century has witnessed expanding concern for the polar regions which has built on earlier limited contacts. First interest in the eighteenth and nineteenth centuries was commercial, with the entry of whaling and fishing fleets from Europe and the United States. The nineteenth century witnessed the dramatic attempts in the Canadian Arctic by Royal Navy expeditions to find the North West Passage. Similarly in the Russian Arctic hectic explorations brought increasing geographical knowledge. The pace of contacts and 'map-making' has quickened relentlessly in the twentieth century. Public interest was ignited by the heroic exploits of Peary, Cook and Stefansson in the Arctic, and of Scott, Amundsen and Shackleton in the Antarctic, together with the first whaling activities in the Southern Ocean in the early 1890s. Interest between the First and Second World Wars was based on questions of sovereignty over Arctic lands, with the United States, Canada, Denmark, Norway and Russia, in particular, wishing to stake their territorial claims. Increased whaling, and political agreement over the Antarctic continent, typified a more collaborative approach in the southern polar regions.

Since 1945 renewed interest has come from several new directions. The 'Cold War' between the West and the former USSR brought great defence interest in the Arctic regions with the advent of intercontinental ballistic missile systems (ICBMs), military early warning systems (e.g. the DEW line – Direct Early Warning) and the reality of nuclear submarines operating beneath polar ice. Also, the period from the 1950s to the 1970s saw a great expansion in the search for non-renewable mineral resources (especially petroleum and non-ferrous minerals) by a commercial sector worried about the future availability of such materials. During the 1980s and 1990s have come three further stimuli. First, there is concern about protecting and conserving 'wilderness' which has brought steps in setting up wildlife

sanctuaries and national parks, on both a national and an international scale. Second, there has grown up an increasing concern for the welfare of aboriginal peoples in polar areas and a desire to give such peoples more 'rights' and a growing voice in how their environment should be used. Third, there has been growing concern about global environmental changes – climate warming, ozone depletion, pollution of ecosystems, biodiversity; such concern is rightly the province of the general public, as well as of the international science community. It is very clear that polar landscapes play a vital role in all these world-wide systems; polar landscapes act as an early-warning device, a kind of environmental quality barometer, and also play a pivotal role in regulating the direction of these global problems. It can confidently be predicted that interest in polar landscapes is set to soar to new levels in the twenty-first century.

Distribution of polar landscapes

The harsh polar environment is typified by the treeless tundra biome. In high latitudes the lack of summer warmth does not allow trees and large woody shrubs to grow, and the vegetation consists of grasses, sedges, low shrubs, mosses, lichens, and highly specialized flowering plants. About 10 million km² of tundra are located in the Arctic regions of Eurasia and North America, but only about 50,000 km² around the southern Antarctic latitudes. A further 10 million km² of mountain tundra is found at high elevations throughout the world, though it is not discussed in this chapter.

Polar climates are dominated throughout the year by cold, dry Arctic air masses (cA), with some incursions of humid, milder polar maritime (mP) air masses in summer; this effect is especially noticeable in north-western North America (Alaska and north-west Canada), where summer warmth is advected inland from the warm waters of the Japanese current in the Pacific, and likewise in north-west Europe

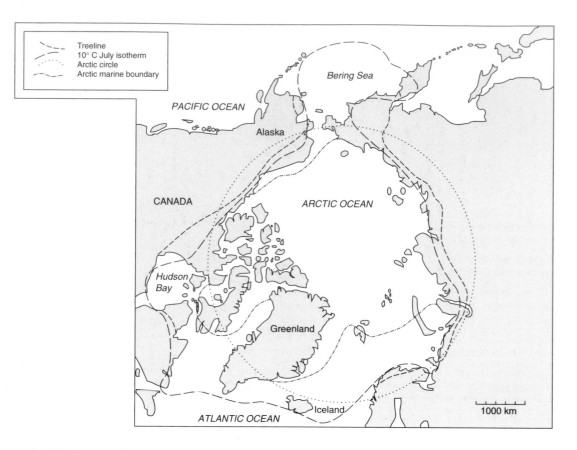

Figure 24.1 *The limits of the Arctic environment: the 10° C July isotherm, the treeline and the Arctic marine boundary.*

(Iceland and Scandinavia) under the influence of the North Atlantic Drift in the Atlantic. Temperatures rise above freezing for only two to four months per annum, and the average temperature is below 10° C in the warmest month. Incoming solar radiation and length of day are the two elements which give greatest contrast between winter and summer. Although the summer is brief, net radiation can be high; in the Canadian Arctic it is about 109 Wm^{-2} day^{-1} in July, compared with 133 Wm^{-2} day^{-1} at 49° N on the Canada–United States border. The contrast between eight to eleven months with a large negative radiation balance, and one to four months with a large positive radiation balance is an important environmental control.

Annual precipitation in polar regions is low, hence the label 'cold desert'. Most polar regions will receive less than 250 mm of precipitation annually, as the cold air is able to hold little moisture, and though relative humidity may be high, absolute humidity is always low. Sixty per cent of precipitation occurs as snow. Throughout polar regions lack of available water may be as limiting an ecological factor as extreme cold, exacerbated by soil water being frozen

for much of the year. However, precipitation figures for polar stations are notoriously unreliable, owing to the difficulty of measuring snowfall accurately. Generally precipitation declines at higher latitudes, where temperatures are colder, and where air masses from temperate latitudes have more difficulty in penetrating.

Under Köppen's system of climate classification, polar climates are designated E, or treeless polar climates. The limit of E climates is the isotherm for 10° C for the warmest summer month, normally July in the northern hemisphere and January in the southern hemisphere. For the Arctic this isotherm reflects latitude, ocean currents and continentality (Figure 24.1). It includes most of the northern coasts and islands of Alaska, Canada, Scandinavia and Russia, all of Greenland and Svalbard, and the northern two-thirds of Iceland. Anomalies occur where it is pushed north by the air masses associated with the North Atlantic Drift (Scandinavia) and the Japanese current (north-west North America), and where it is pushed south by the cold Labrador current (Labrador, Quebec) and the Bering current (Bering Sea, Kamchatka).

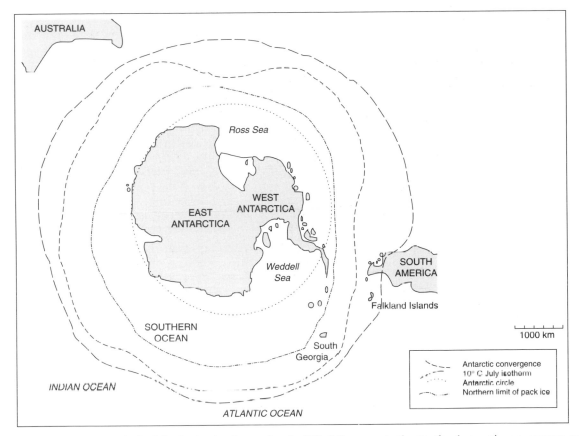

Figure 24.2 *Important physical limits in the Antarctic: the 10° C January isotherm, the Antarctic convergence and the northern limit of pack ice in winter.*

The 10° C summer isotherm boundary has been popular with ecologists, as it corresponds reasonably well with *treeline*. Treeline is the limit beyond which trees do not grow, and is one of the world's major ecological boundaries. In some areas the transition from boreal coniferous forest to tundra is abrupt, whilst in other places the boundary is more gradual over hundreds of kilometres. In the latter case the *forest–tundra* zone consists of open ground with sparse stunted trees, an **ecotone** which has plants from both the forest and the tundra. Isolated stands of trees can grow north of treeline where local conditions permit; topography providing shelter is the commonest reason, but better soils and more available water may be locally important. Although generally a good fit, the 10° C summer isotherm lies 100–200 km north of treeline over much of North America and Eurasia. This indicates that the precise position of treeline reflects mean annual net radiation rather than temperature. F.K. Hare has studied the climatology of treeline in Canada and points out that it corresponds to the average summer position of the Arctic Front which separates maritime Polar (mP) air masses to the south from cold maritime and

continental Arctic (mA and cA) air masses to the north. There is a sharp decrease in net radiation between boreal forest and tundra, and part of it is also due to the albedo of the surface, more incoming solar radiation being reflected from tundra vegetation than from coniferous forest vegetation. Thus treeline is a polygenetic ecological feature; its primary control is macroclimatic, based on the Arctic Front, but its course in detail is modified considerably by vegetation, soil surface and water bodies. Because of the anomalies, some ecologists feel that a closer fit is given by the **Nordenskjöld line**, an isopleth joining places where the mean temperature of the warmest month equals (9–0.1 K), K being the mean temperature of the coldest month in °C.

For the Antarctic the 10° C summer isotherm encircles the whole of continental Antarctica and also includes the tip of South America, Tierra del Fuego, and many islands in the Southern Ocean, including the Falkland Islands and South Georgia. It seems to separate quite well those islands which are treeless from those which are not. However, the Antarctic is so much dominated by the Southern Ocean, and is so obviously a marine environment rather than a

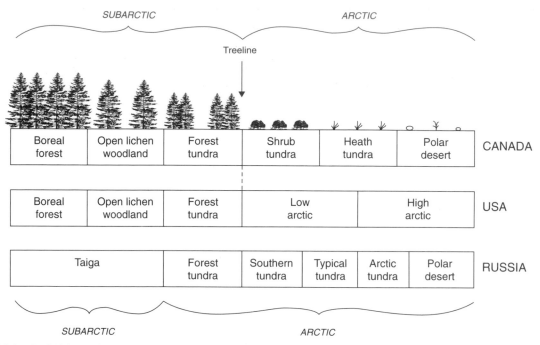

Figure 24.3 *Definitions of Arctic ecosystems according to US, Canadian and Russian usage.*

terrestrial one, that many authorities prefer to use a maritime boundary. The most generally accepted is the Antarctic convergence, a sharp boundary between Antarctic surface water and slightly warmer and more saline subantarctic surface water. It marks the point where colder Antarctic water moving north and east sinks below subantarctic water moving east and slightly southwards. It varies between latitudes 45° S and 62° S, being farther south in the Pacific than the Atlantic, and in the summer than the winter (Figure 24.2). The equivalent Arctic maritime boundary is between cold, less dense surface water from the Arctic Ocean and warmer, more saline water from the south. Again the contrasts in temperature, density and salinity are reflected in very real ecological differences between the water masses, with distinct communities of plants and animals. Figure 24.1 shows how the Arctic maritime boundary is pushed to above 80° N by the North Atlantic Drift and to the east of Novaya Zemlya. It also extends up the west coast of Greenland and penetrates the Bering Strait to go from Wrangel Island almost to Prince Patrick Island in the east.

Vegetation and soils in high latitudes

Vegetation is only one component of ecosystems, but special importance is attached to it, as it provides the basis for natural productivity, fixes carbon by photosynthesis, builds up organic biomass, stabilizes and influences soils, provides food and shelter for animals and influences the hydrological cycle. There is no commonly agreed system for naming Arctic ecosystems, as is illustrated in Figure 24.3, where the American, Canadian and Russian views are represented. The Canadian view of three zones north of treeline (shrub tundra, heath tundra, polar desert) contrasts with the US division into Low Arctic and High Arctic only. The Russian view is more complicated, recognizing five zones, though other polar scientists would include forest–tundra in the Subarctic rather than the Arctic. In addition to a recognizable zoning of vegetation with latitude, there is a strong patterning of plant cover in relation to landforms and soil. Whilst a superficial glance may suggest uniform plant cover, with little variation from place to place, in fact there is a high degree of spatial variety which is conditioned by local variations in microclimates, drainage conditions and particularly soil conditions. The **soil catena**, the sequence of soils and vegetation communities according to landforms, is a major unit in polar landscapes. Figure 24.4 shows an example of a soil catena at Rankin Inlet, Canada, on the western coast of Hudson Bay in the heath tundra zone (Low Arctic). Topography and depth-to-permafrost are crucial in influencing vegetation communities and soil profiles. The ridge crests are freely draining and therefore the permafrost table is relatively deep at 1–2 m. If the ridge is composed of hard igneous and metamorphic shield rocks, soil formation will have been minimal and the surface is likely to be bare rock

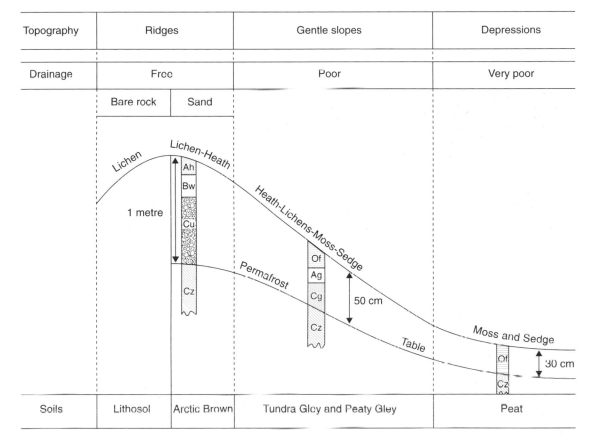

Figure 24.4 *The sequence of soils, vegetation and permafrost along a topographic catena in the Low Arctic (Rankin Inlet, North West Territories of Canada).*

(*Lithosol* soil) with a sparse cover of lichens and *Rhacomitrium* moss. Ridges of glacial till, sand or gravel will have a well drained Arctic brown soil above permafrost, with a surface vegetation of lichens and dwarf heath (crowberry *Empetrum nigrum*; Labrador tea *Ledum decumbens*). (Plate 24.1.) The Arctic brown profile is slightly acidic and leached, and has a weakly developed surface humus horizon (Ah) and weathered horizon (Bw).

On smooth slopes soil drainage deteriorates and the permafrost is shallower. The summer melt leads to waterlogging and gleying, hence giving tundra gley and tundra peaty gley soils. Plant communities reflect the change in moisture conditions, with more wet-tolerant species (mosses and sedges) forming a mixed community. The low points of the catena in topographical hollows consist of flat, wet meadows where permafrost occurs at a shallow depth. Typically peaty organic matter, frozen in winter but a black ooze in summer, rests on top of permafrost at 30 cm. The wetland vegetation is dominated by mosses and sedges of the *Sphagnum, Carex* and *Eriophorum* genera. The equivalent catena in the polar desert (High Arctic) is shown in Figure 24.5.

Because of the lower temperatures, lower precipitation and lower biomass of vegetation, the soils are thinner and the permafrost table is much shallower. Soils are immature and weakly developed, with no B horizons. A horizons are thin and the gleying in depressions replaces the peat of the heath tundra. Owing to the aridity, soils become more calcareous and more saline as one moves into the polar desert. Catenary relationships are less clear than in the Low Arctic, owing to the patchiness of the plant cover and the lack of soil moisture. Arctic soils have a low nutrient content, owing to the slow rates of organic matter mineralization, soil weathering and soil chemical reactions generally. Plant growth is severely limited by the low nutrient-supplying power of soils, as witnessed by the lush growth when nutrients are added. This is evident from reseeding and fertilization experiments carried out near development sites in the Arctic, and also from the response of plant growth to nitrate and phosphate provided by animal and bird droppings. Thus thick patches of vegetation are found beneath bird cliffs, around animal burrows, in musk-ox meadows and even around rotting skeletons.

Plate 24.1 *Although superficially featureless, small changes in the Arctic landscape produce significant changes in the spatial patterning of ecosystems. Along the coast of the Arctic Ocean the ridges will have dry habitats, whilst low-lying areas close by will be saturated, boggy terrain in summer. This gives the area a high spatial variability in soils and plants. Photo: Ken Atkinson.*

Plant adaptations in polar regions

Polar landscapes present many problems for the growth of plants. Among them are a negative radiation balance over the year, low temperatures, a paucity of plant nutrients in the soil, desiccating winds, with foliage blasting from ice crystals and dust, and, last but not least, an active layer above the permafrost which suffers seasonally from both frost heave and waterlogging. Low temperatures seem to cause most physiological adaptations but the other environmental controls are important. Heat exchange reactions are vital in many of these low-temperature adaptations. Many plants are low-growing so that they can remain in the warmer layer of air just above the ground surface known as the *boundary layer*. Plants growing with a rosette habitat are also common; here it has been shown that much energy is lost, owing to edge effects; therefore, with a low edge-to-surface-area ratio, rosette plants will lose less energy. This adaptation is shown by the purple mountain saxifrage (*Saxifraga oppositifolia*), the most

northerly plant species in North America, and Lapland rosebay (*Rhododendron lapponicum*). Other adaptations include the formation of a small cushion of air between the base of the plant and the ground, so that the plant has very little contact with it when the ground is frozen. There are many mechanisms for raising the temperature of the flower head, a common one being *heliotropism*, when the flower head tracks the solar path (arctic poppy, *Papaver radicatum*). Other adaptations include reflectance of solar radiation on to the centre of the flower by light-coloured petals (mountain avens, *Dryas octopetala*, and arctic buttercup, *Ranunculus oxytropis*). The absorption of solar radiation by the brown bract of the sedge *Eriophorum vaginatum* is given out in turn as long-wave radiation which is trapped by the fluffy head; this mechanism can raise the temperature of the seed head by up to 15° C.

As low temperatures also mean slow rates of mineralization of organic matter and low availability of nutrients in consequence, plants have adapted by having slower growth patterns and by being able to

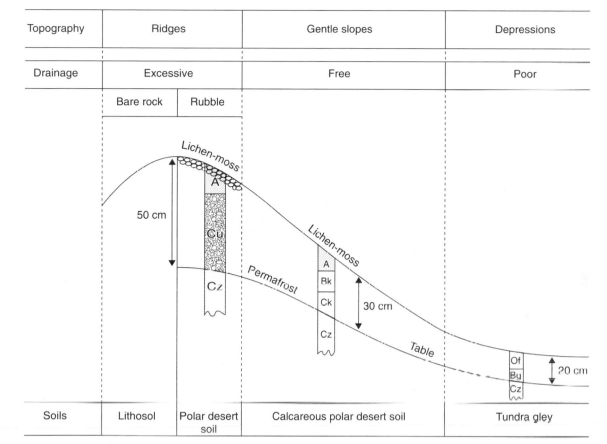

Topography	Ridges		Gentle slopes	Depressions
Drainage	Excessive		Free	Poor
	Bare rock	Rubble		
Soils	Lithosol	Polar desert soil	Calcareous polar desert soil	Tundra gley

Lichen-moss

A
Cu
Cz

50 cm

Permafrost

Lichen-moss
A
Bk
Ck
Cz

30 cm

Table

Of
By
Cz

20 cm

Figure 24.5 *The sequence of soils, vegetation and permafrost along a topographic catena in the High Arctic (Prince Patrick island, North West Territories of Canada)*

store nutrients. Many plants can reabsorb nutrients from dead leaves, roots and stems. More efficient methods of nutrient acquisition have been evolved, e.g. a more branching root network with greater surface area and the ability to release ecto-enzymes such as phosphatases into the soil to release bound nutrients. Adaptations to low precipitation and frozen soil water are shown in a range of xerophytic characteristics of the plants. These include sclerophyllous leaves with thick waxy cuticles, stomata in pits in leaves to reduce water loss, hairy leaves and stems (pubescence) to maintain the boundary layer around the plant together with slow-moving air, and a reduction in the surface area of leaves, by having smaller leaves (microphylly). The fact that the growing season is so short means that many species complete their life cycle over two or three years, building up nutrients for flowering one year and then building up nutrients for setting seeds the next. Under a three-year cycle, the sequence would be buds–flowers–seeds. Vegetative reproduction is common, too, among Arctic plants. By using creeping root stocks or adventitious roots and buds, plants can propagate even in years of exceptionally

severe climate. **Vivipary**, the germinating of young plants on the parent plant rather than in the soil, is common in the grass genera *Festuca* and *Poa*, and in plants such as nodding saxifrage (*Saxifraga cernua*) and alpine bistort (*Polygonum viviparum*). These plants have most of their flowers replaced by bulbils that fall off and take root.

Another distinctive adaptation is for different parts of plants to have to be able to carry out their functions at different times of the year. Owing to the waterlogged nature of the active layer in spring and summer, soil heat capacity is high, soils are slow to warm up and the air is warmer than the ground. The reverse is true in the autumn, when the ground retains heat whilst air temperatures drop below freezing. This means that plant roots must continue to absorb nutrients after the aerial part of the plant has died. The ecologist Raunkaier discovered in the 1930s that there is a higher concentration of plants with meristematic buds just above the ground surface (the **chamaephytes**), and at or just below the surface (the **hemicryptophytes**). This is due to the plants seeking shelter by positioning their reproductive organs beneath an insulating cover of snow. However, whilst the snow provides a buffer, a protecting cover in winter, there is the disadvantage that the plants have to wait for it to melt before they can start growing in spring. This is why south-facing slopes are favoured in the northern hemisphere. Where snow patches persist late into the summer they usually give rise to distinctive snow-patch plant communities (or **chionophilous**, snow-loving vegetation). These may be topographical hollows or north-facing slopes where the snow lingers until July. Such areas are easily recognized as dark patches on air photos and are frequently associated with dwarf birch (*Betula nana*), Arctic bell heather (*Cassiope tetragona*), bilberry (*Vaccinium uliginosum*) and cranberry (*V. vitis-idaea*).

The waterlogging of 'active layer' soils in summer can also cause problems with oxygen deficiency and the production of respiratory products which may accumulate to toxic levels, given the fact that the diffusion of gases in water is roughly 10,000 times slower than in air. This is particularly a problem after the spring melt, and therefore at a time when growth processes should be at their maximum rates. Plants adapt physiologically by having greater porosity in the aerenchyma cells of their roots so that air can diffuse more easily from shoots to roots and into the surrounding soil, i.e. the rhizosphere. The higher incidence of sub-zero temperatures in polar regions means that plants must be able to recover from complete freezing, thus extending their growing season into the autumn, when the first frosts of winter would kill or make dormant plants adapted to higher temperatures. One biological mechanism is to have a higher than normal concentration of water-soluble proteins in the cell cytoplasm. This accumulates under the influence of abscisic acid (ABA) and lowers the risk of ice crystal formation. Plants also synthesize 'cryopreservatives' for the winter months and decrease the amount of lipid saturation in membranes. If ice crystals do form plasmolysis leads to an accumulation of cryotic salts (calcium Ca^{2+}, potassium K^+) around membranes, which may lead to their denaturation. The 'cryopreservatives', which are usually organic acids, stop this increase in ionic strength by relative dilation, and prevent membrane damage. The extent to which plants can adapt to cold are illustrated by the moss campion, *Silene acaulis*, which under experimental laboratory conditions has recovered after chilling to $-198°$ C!

Energy and ecological productivity

The first ecologists who visited polar regions invariably commented on the simplicity, fragility and low productivity of both the terrestrial and the marine ecosystems. In general there are fewer species, fewer ecosystem linkages, and much less primary and secondary production than in temperate and tropical latitudes. Whilst these generalizations are true, more recent work has highlighted some specialized and sophisticated features of the ecosystems too. For example, it is now clear that Arctic food chains which support the top predators are very long, with polar bears, for example, occupying a clearly defined fifth trophic (feeding) level. Charles Elton, in his early expeditions to Svalbard in the 1920s, also recorded how integrated were the terrestrial and marine ecosystems, and how readily energy and materials are able to move from one to the other, via birds and sea mammals.

The environmental factors which favour ecological productivity become less favourable as one moves polewards. Solar radiation, moisture, air/soil temperatures and plant nutrients become limiting in high latitudes. Many ecologists rank shortages of nutrients, especially nitrates and phosphates, as the severest limit on productivity, given the successful way that plants have adapted to cold. In Arctic terrestrial ecosystems, plant production accounts for about 90 per cent of the energy flow, and decomposition in the soil by invertebrates and micro-organisms accounts for about 8 per cent, with the activity of herbivores and carnivores contributing 2 per cent. In the Low Arctic in Alaska the most productive ecosystems are wet sedge moss, cotton grass, dwarf shrub communities, which yield 1.0–2.5 t ha^{-1} yr^{-1} above-ground net plant production. In these communities nitrogen fixation by alder (*Alnus crispa*) and avens (*Dryas* spp.) is an important input, with nitrates also being added by blue-green algae on wetter sites. Daily rates of production during the short growing season are comparable with those of similar communities in the temperate zone, ranging from 3 g dry matter per square metre in the polar desert of the High Arctic to 225 g in wet meadows in Alaska. However, production values in a small area may halve between the wetter and drier parts of the arctic meadow, and values of biomass may vary a hundredfold between hummocks and hollows 1 m apart.

Data on energy flow in polar marine ecosystems are even scarcer than those for terrestrial ecosystems. (Plate 24.2.) Figure 24.6 shows the pathways of energy flowing through a marine food web in Lancaster Sound, Arctic Canada (75° N). The trophic-dynamic model has been constructed according to the principles discussed in Chapter 22. The model simplifies the complex reality of food chains, but it highlights the main components

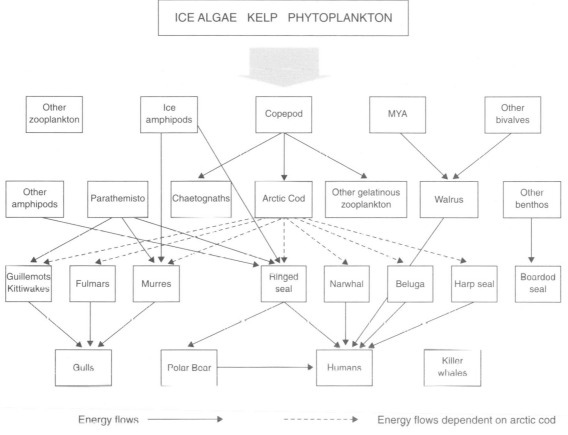

Figure 24.6 *Pathways of energy flow through the marine ecosystem of Lancaster Sound, Arctic Canada. Source: Welch et al. (1992).*

(compartments) of the ecosystem and the principal energy pathways. Phytoplankton, ice algae and kelp fix 89 per cent, 10 per cent and 1 per cent of the gross primary production respectively. Primary production peaks sharply in June, July and August, when climate conditions (light, temperature) are favourable, and when the retreat of sea ice allows more primary production in open water. The average primary production is 60 g carbon m^{-2} yr^{-1}, but there are large variations, depending on ocean currents and nutrients. A restricted number of amphipods, copepods and bivalves make up the herbivore compartments of the food web. A critical role in the ecosystem is played by Arctic cod, with 125,000 tonnes being consumed by marine mammals and 23,000 tonnes by seabirds annually. The fish consume micro-sized animals and concentrate the energy into larger 'packets' which can be eaten efficiently by seal, whales and birds. In Figure 24.6 the links in the food web which relate to Arctic cod have been highlighted. It is clear that the stability of the entire functioning of the ecosystem is dependent upon this one species. The fragility of the Arctic relates to the concentration of energy flow through a very restricted number of species. Thus variations in the populations of seals, whales and polar bears from year to year are due to changes in the abundance of Arctic cod, which in turn will reflect fluctuations in the physical environment (temperature, ice extent, nutrients).

The trophic-dynamic model shown in Figure 24.6 is incomplete, owing to the difficulty of measuring the activities of bottom-dwelling organisms, the migration of fish and marine mammals into and out of Lancaster Sound, and the size of winter populations. For these reasons energy-flow studies of oceans and seas are approximate at best.

A major control of polar marine ecosystems is the extent of sea ice. Ice-covered water has low light penetration, a low biomass of primary producers and zooplankton, low productivity and low rates of sedimentation. The cover of sea ice varies from 14 million km^2 to 7 million km^2 in the Arctic, and from 20 million km^2 to 4 million km^2 in the Antarctic. Even in the depths of winter, currents, winds and upwellings keep some areas ice-free either as leads or as large patches of open water (*polynyas*). These areas are important ecologically, as they allow light to support phytoplankton and hence maintain a more

Plate 24.2 *Conservation of polar wildlife depends upon good scientific data but collecting the data is expensive. Walrus suffered from overhunting in the past for their ivory tusks. Here animals are being tagged for population studies. Photo: Ken Atkinson.*

productive food web of zooplankton, fish, sea mammals and birds. Such marine oases show productivity rates comparable with those of the North Sea.

Arctic landscapes can also be regarded as unstable from two other viewpoints, in addition to the concentration of energy along few food chains. They have low resistance to change. The removal of vegetation, for example, destroys an insulating cover at the surface which allows heat to penetrate the soil and melt the permafrost. Once started this is a difficult process to stop, and usually leads to large thaw lakes. Also the impact of the disturbance will take a long time to fade. Recovery times are long in the Arctic and hence the ecosystem's resilience is low.

The large fluctuations in the population size of Arctic land and marine mammals are a clear sign of young and unstable ecosystems. Charles Elton studied the records of the Hudson Bay Company and recognized the eleven-year cycle of the lynx and the snowshoe hare (Figure 24.7). Arguing that hunting and trapping pressure would not vary much from year to year, he noted the cyclical nature of

the population harvests, with the predator following the abundance of prey. The four-year population cycle of the lemming is well known, and it in turn influences the size of the population of arctic foxes, snowy owls and gyrfalcons, whose numbers fluctuate in sympathy. Population cycles have also been noted in caribou and walrus, and are likely to be more common than is realized, given that there are not many long-term data on most Arctic animals. Variability is one indicator of ecosystem stability (see Chapter 23), and with their high variability Arctic populations appear unstable. An adverse impact, either through a natural cause (climate, disease, for example) or because of human action (overhunting or pollution, for example), entails a high probability of extinction if the impact occurs at the trough of the cycle. Thus musk oxen have been eliminated from Russia (though now reintroduced from Canada), and the musk-ox population of Canada was on the verge of extinction in 1917, when a total ban on hunting was introduced. Bowhead whales were also on the verge of extinction before protection.

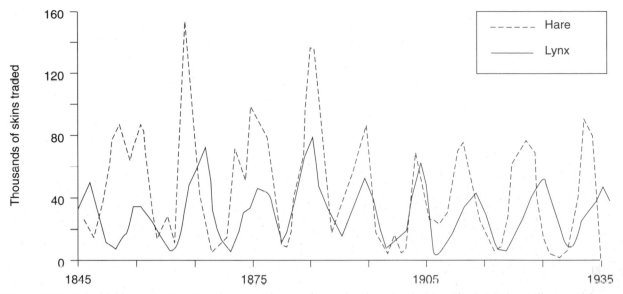

Figure 24.7 *The highly fluctuating populations of lynx and snowshoe hare in the Canadian Arctic as illustrated by figures for skins traded by the Hudson Bay Company, 1845–1935. Source: Hudson Bay Company records.*

Construction problems

One definition of cold regions is that they are places where engineering is complicated by the natural freezing and thawing of the ground, which significantly increase construction costs. Permafrost covers more than 50 per cent of the land surface in Canada and Alaska, and significant parts of Russia and northern Scandinavia. It greatly complicates work on any engineering construction associated with economic development, e.g. housing, industrial plant, roads, railways, airstrips or pipelines for oil and natural gas. The term 'permafrost' is misleading, as over a period of years the frozen soil may slowly thaw or slowly increase in extent owing to changes at the ground surface which modify the exchange of heat energy through it. The surface layer thaws and refreezes every year (the *active layer*) with the change of the seasons. The thickness of the permafrost in the subsoil is often changed by the removal of vegetation and the building of human-made structures. If the insulating cover of vegetation is removed, and perhaps a heated building placed directly upon the ground surface, it is hardly surprising if the permafrost thaws, causing subsidence and distortion of the building. Subsidence is a more serious problem with ice-rich permafrost, which is found in soils which would otherwise be very wet, such as gleys and peats. Coarse sand and gravel soils are well draining and present less of a problem.

Many historic buildings show the effects of subsidence, as in the gold-rush town of Dawson in the Yukon. (Plate 24.3.) What is surprising is that even modern buildings can exhibit subsidence, as for example in Anchorage, Alaska. It can be overcome by supporting buildings on 'piles' drilled into the permafrost, allowing cold air to circulate between the ground surface and the floor of the building. In that way the heated building does not come into contact with the ground. (Plate 24.4.) For large buildings (industrial plant, power stations, oil storage facilities, for example), it is necessary to ensure that the building is on a thick pad of stable but well draining aggregate. In some instances the insulating power of the pad is enhanced with fibreglass insulation, and by metal pipes running through it which permit air to circulate in winter. Pads are also used to prevent permafrost melting and land subsidence when roads, airport runways and railway tracks are constructed. However, unsuitable construction techniques still lead to degradation of ground ice, and it has been reported that a recently constructed railway line has shown considerable distortion and buckling at Yamburg, western Siberia, only a year after its completion.

The town of Inuvik, North West Territories, Canada, illustrates the problems of construction in the tundra. A modern town of 3500 population was completed in 1961 as a regional administrative centre. A hospital, a large secondary school, a diesel generating station, oil storage tanks and the usual collection of shops, churches, hotels and residences were built on terrain underlain by permafrost. An airport, road system and river docks were also

Plate 24.3 *Building directly on to ground underlain by permafrost is hazardous. This church in Dawson, Yukon, was built during Klondike gold rush days. Melting of the permafrost has caused the subsidence and tilting.*
Photo: Ken Atkinson.

station, a low-tech and low-cost solution is necessary. Typically the buildings are sited on pads of sand and gravel 1–2 m thick, placed on the original ground surface with as little disturbance as possible. Upon the pads wooden blocks are set to act as bearers for short columns which support the main joists of the building. In some cases the joists may rest directly upon a number of bearers consisting of two large opposed wooden wedges which can slide over each other when force is applied to one of them with a sledgehammer. In this simple way subsidence or heaving movements in the pad or ground beneath can be corrected by adjusting the wedges. Hence the building can be maintained in a horizontal position if settlement occurs. Both the columnar and the wedge arrangements ensure that an air gap, ideally about 1 m high, is maintained beneath the underside of the buildings. As conventional piped systems of water supply and sewage disposal are not practical, and utilidor systems are too costly, smaller settlements usually have these utilities serviced by truck.

Impacts of oil and gas fields

Political crises in the Middle East in the 1970s persuaded many international oil companies to look for oil reserves in Arctic regions. Significant supplies of oil and natural gas have been located in the Arctic regions of Alaska, Canada and Russia. Much development has already taken place, with the certainty that the impact of hydrocarbon extraction will be far greater in the future. The operation and drilling of wells for oil and natural gas carries great environmental risks. These are compounded by the transport of these hydrocarbons by tankers and pipelines across the tundra or through polar seas to 'southern' markets; there is also the necessity to service the needs of remote polar communities with fuels and supplies. There have been hundreds of accidental spills of crude oil, diesel oil, petrol and jet fuel kerosene as oil and gas deposits have been developed in Alaska, Arctic Canada and Arctic Russia. The spills have come into marine ecosystems from offshore well sites, ocean-going vessels and shore-based facilities, and into terrestrial ecosystems from exploration wells, production wells, storage tanks and oil pipelines. Contamination by petroleum hydrocarbons has reached very high levels on land and in marine sediments. These chemicals enter the food chain and are suspected of causing cancers in Arctic fish and mammals.

constructed. An innovative feature was the municipal *utilidor* system which carries the water supply, the heating pipes and the sewers in large aluminium-clad units on wooden piles. In this way the pipes for the utilities do not have to be buried in the sensitive permafrost. Figure 24.8 illustrates a variety of construction techniques necessary in the tundra.

Over the years, considerable experience has accumulated on how to avoid the dangers of abusing the sensitivity of permafrost terrain. However, many of the remedial measures are high-tech and high-cost. In smaller villages with perhaps only single-storey prefabricated buildings for housing, along with a school, a shop, a police station, a mission and a nursing

At present the technology for clearing up oil spills is not adequate to remove oil from ice-covered

Plate 24.4   *Modern service facilities in the Alaska oilfield at Prudhoe Bay minimize ground disturbance as much as possible by building on stilts to avoid melting the permafrost.*
Photo: Bill Barr.

waters, and there must be a high probability of a major environmental disaster in Arctic waters. An oil spill is much more persistent in polar regions, as oil degrades ten to twenty-five times more slowly at 5° C than at 25° C. Ice cover reduces the spreading and evaporation of the oil (spreading is two-thirds faster in the absence of ice), and thus organisms are exposed to oil for much longer periods than in temperate waters. The direct effects of exposure to oil on marine mammals and birds are lethal; very serious population crashes can occur if oil hits at breeding times. The colonial nesting habit of birds and the colonial gatherings of marine mammals make them particularly vulnerable if a spill occurs at the wrong time in the wrong place.

Table 24.1 lists the range of disturbance of Arctic ecosystems which comes from oil and gas development. Although the areas of direct impact are mostly small, the cumulative impact of large developments can affect larger areas. The indirect impact can lag behind construction by many years, and the total area eventually disturbed can greatly exceed the original site. Spills, thermokarst and flooding are important impacts of all activities, and result from permafrost melting and the disruption of drainage lines. The impacts on aesthetics and 'wilderness' are universal. The other impacts listed are mostly self-explanatory. Perhaps the area where least is known is the cumulative impact on wildlife, particularly land mammals and birds, both wildfowl and land birds. The effect of human activities and noise on breeding, calving and migration routes is not well researched. It is certain, however, that there will be a negative impact on wildlife populations. The passage of tankers or supply ships in Arctic pack ice, for example, leaves a jumble of broken ice ('freeboard') which can also disrupt the movements of native people as they travel over the polar ice.

The trans-Alaska pipeline (TAP) is perhaps the most famous oil pipeline in the Arctic (Plate 24.5), though there is also a smaller 27 cm diameter pipeline from Norman Wells, North West Territories,

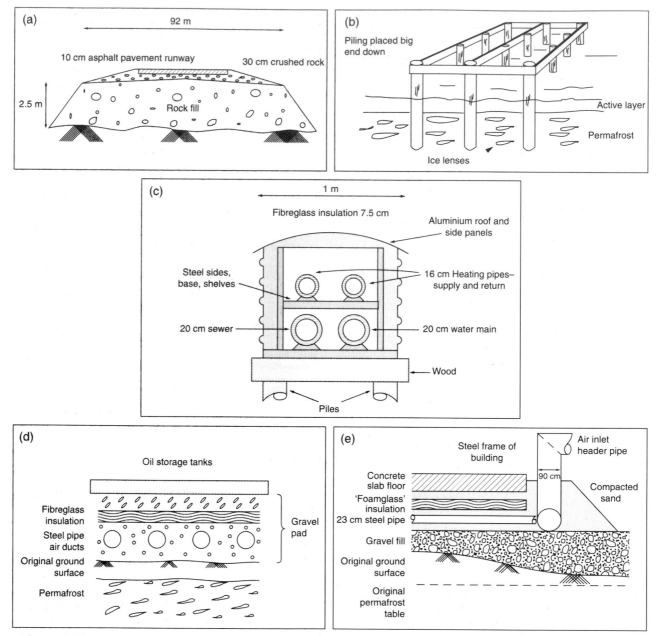

Figure 24.8 *Remedial measures for the construction problems posed by the Arctic climate and permafrost: (a) thick gravel pads needed for aircraft runways, (b) foundations for houses and wooden buildings, (c) the high-tech 'Utilidor' system for domestic services, (d) gravel pads, insulation and special cooling measures for oil storage tanks, (e) gravel pads, insulation and special cooling measures for heavy industrial plant.*

to Zama, Alberta, as well as several oil pipelines in European Russia and north-west Siberia. TAP runs from the Prudhoe Bay oilfields to the ice-free port of Valdez for a distance of 1280 km. Over half the distance is underlain by permafrost, much of which was unforeseen, with the result that despite the original estimate of $900 million, by completion in 1977 the pipeline had cost seven times as much. Although delays and cost increases were ascribed to the activities of native peoples and environmental

groups, much was due to failure to recognize the problems of the ice-rich permafrost.

TAP is a remarkable engineering achievement. The temperature of the oil in the pipe is 65° C, and for more than half its length it is above ground. The supporting members have been drilled into the permafrost, and each has an automatic refrigeration system which maintains the permafrost around the footing. The beam supporting the pipe is wide enough to allow the pipe to move laterally with temperature changes. For those sections of the course where the pipeline is buried, special insulation coatings 10 cm thick surround the pipe, and in particularly sensitive areas refrigeration pipes are installed in the trench below the pipeline to ensure minimal damage to the permafrost.

The situation with gas pipelines is different from the conditions of oil pipelines. Most tundra soils exhibit the process of 'frost heave' whereby water moves to the point of freezing ('the freezing front') and becomes incorporated into the freezing material. This process can cause the soil to double in volume, and is additional to the well known expansion of water on freezing of 9 per cent. Natural soils vary greatly in their susceptibility to heave, so the effects may induce bending in any buried pipeline. In permafrost areas the temperature of gas in a buried pipeline must be below 0° C. If it were not, the pipe would cause thawing and subsidence. However, where the pipe with its chilled gas passes through patches of unfrozen ground, freezing occurs around the cold pipe, with the likelihood of heave.

North-west Siberia is underlain by sedimentary rocks which contain numerous hydrocarbon resources. Oil reserves occur in the central region, with natural gas reserves in the Arctic landscapes along the coast. Development of the natural gas resources has only just begun, although there are no fewer than six large diameter pipelines already in operation. The arrangement is for hundreds of gas wells to produce gas for two processing plants, which will pipe the gas to Europe for sale. Early indicators are that development activity has had adverse effects on the tundra. Drilling sites were not initially on pads, resulting in thermal degradation, subsidence and the formation of ponds due to heat loss from buildings ('thermokarst'). On a much larger scale is the widespread degradation caused by tracked vehicles disrupting the tundra vegetation cover. Even a single pass of a tracked vehicle over sensitive terrain causes sufficient disturbance to initiate thermal degradation. The development planned for the three northern peninsulas of north-west Siberia is far larger than any development planned in North America. In the 1970s there were proposals in Canada for a gas pipeline to run from the Mackenzie valley to link up with the natural gas network in Alberta. However, owing to an adverse environmental impact study, combined with a falling price for gas, these developments have been postponed until the next century. However, the environmental damage which has already occurred in Siberia is enormous, and it is to be hoped that North America does not repeat it.

Table 24.1 *Environmental disturbances by oil and gas.*

	Disturbances	Impacts
All activities		Spills Thermokarst Aesthetics Wilderness Flooding
Oilfields	Prospecting wells Production wells Distribution pipes Gravel pads Storage tanks	Air pollution Rubbish Drilling wastes Gravel pads Wildlife impact
Transport corridors	Roads Pipelines Tanker routes	Roadside dust Drainage disruption Noise Wildlife impact 'Freeboard'
Seismic trails	Off-road vehicles	Vegetation Soils
Materials sites	Gravel quarries	Vegetation Soils
Camps	Rubbish Sewage Services Pads	Vegetation Soils

Arctic pollution

Although polar regions are still widely regarded as remote and pristine environments, in reality they have been subject to pollution from distant sources in Europe, North America and Eurasia. Indeed, in the 1950s the discovery of DDT residues in Antarctic penguins was one of the first indicators that pollution is a global problem. Although the concentrations of contaminants are generally lower than in temperate regions, their presence is serious because of their persistence, due to slow turnover rates in polar ecosystems. Cold temperatures slow down degradation processes and tend to condense volatile organic pollutants. The cold slows evaporation rates also, and this may lead to a continuous transfer of organic chemicals from warmer parts of the world. Mammals and birds in polar regions are long-lived organisms, at the top of long food chains (e.g. whales, seals and polar bears) and they have high levels of body fat, which stores contaminants in the body. Many native Arctic peoples eat large amounts of wild game or 'country food'; fat, liver, kidneys and hearts (often regarded as the 'choicest' parts) are organs where the pollutants are most liable to accumulate. Because of their great chemical persistence, stability in biological

Plate 24.5   *For much of its course the trans-Alaska oil pipeline is above ground. The units on top of the vertical members are automatic refrigeration systems which maintain the permafrost around each footing. The design allows the pipe to move laterally on the supporting beam under the influence of temperature changes.*
Photo: Martin Duddin.

systems and solubility in fat, polychlorinated biphenyls (PCBs) illustrate how pollution levels can be biomagnified by each link in the food chain, so that levels in the blubber of seals and whales are about 400 million times those in the Arctic Ocean, and magnified about 3200 million times in polar bears and

Table 24.2   *Contaminants from distant sources in polar regions.*

Chlorinated organics	Industrial chemicals PCB, HCB, dioxins, furans Agricultural pesticides HCH, DDT, DDE, chlordane, toxaphene organophosphorus
Heavy metals	Mercury, cadmium, lead, arsenic, selenium
Radionuclides	Strontium-90, caesium-137, plutonium-239
Acid precipitation	Oxides of sulphur and nitrogen

humans (Figure 24.9). Table 24.2 lists the main contaminants from urban and industrial areas which have been detected in polar regions. Owing to proximity, the levels are higher in the Arctic than in the Antarctic. Chlorinated organics have been in widespread use in agriculture since the first insecticide, dichlorodiphenyltrichloroethane (DDT), was introduced in the 1940s. Polychlorinated biphenyls have been used in paints, plastics and electrical and mechanical equipment since the early 1930s. The commonest pollutant of this group detected in the Arctic is hexachlorocyclohexane (HCH) in terrestrial ecosystems, and toxaphene, chlordane and PCBs in marine ecosystems. The effects on wildlife can be very serious, with reproductive failure in mammals, eggshell thinning in birds, and reduced egg hatching in fish, in addition to increased cancers in animals. Heavy metal concentrations are more difficult to interpret, as there is a background concentration from

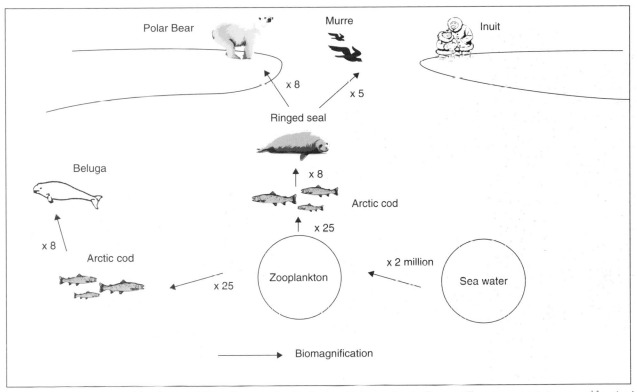

Figure 24.9 *Biomagnification of PCB pollutants in the food web of the Arctic Ocean. The amount of 'biomagnification' ('multiplication) at each link in the web is indicated.*

local rock sources. However, there are more data on mercury and lead, whose levels are extremely high in the kidney and liver of marine mammals.

Radioactive pollution by radionuclides (long-lived fission products) has come from inefficient waste disposal, from nuclear weapon testing in the atmosphere between 1952 and 1978, and from accidents like that at the Chernobyl nuclear power station, Ukraine, in 1986. There has also been pollution from nuclear-powered vessels (mostly submarines and ice-breakers), and possibly from the secret dumping of nuclear waste in Arctic waters. Average levels of radioactivity and annual fall-out have decreased steadily since the ban on weapon testing, but there was a major input of caesium-137 in Scandinavia after the Chernobyl nuclear accident. Radionuclides are absorbed by lichens, which are eaten by reindeer (or caribou in North America). Again, native people living off the meat and milk of the herds are very vulnerable. Acid precipitation, resulting from the reaction of sulphur and nitrogen oxides with water vapour, is the best known long-range pollution problem (see Chapter 10). Although sulphate deposition from acid deposition is less than 3 kg ha^{-1} and acid levels are ten times lower than in industrial temperate areas, a continuous acid load can lead to lower biological productivity in freshwater lakes and the gradual release of heavy metals, which are more soluble in acid conditions. About 95 per cent of the input of sulphur into Arctic regions arrives during the winter months when air currents are more favourable. During the winter the acids accumulate in the snow; when released by spring melt they can leach soils and acidify lakes. Most pollutants enter the Arctic Ocean and the Antarctic landmass by air currents. However, ocean currents, especially the North Atlantic Drift, and the northwards-flowing rivers Ob, Yenisey and Lena in Russia, like the Mackenzie in Canada, also connect the Arctic eco-systems with industrial regions (Figure 24.10).

Environmental impacts of polar tourism

Polar regions have become an important destination of eco-tourism as visitors are attracted by the wildlife populations and wilderness values. As with all tourism, however, there is a danger of the tourists destroying the very thing which attracts them to the area. The three types of tourism are overflights, cruise visits and land-based visits.

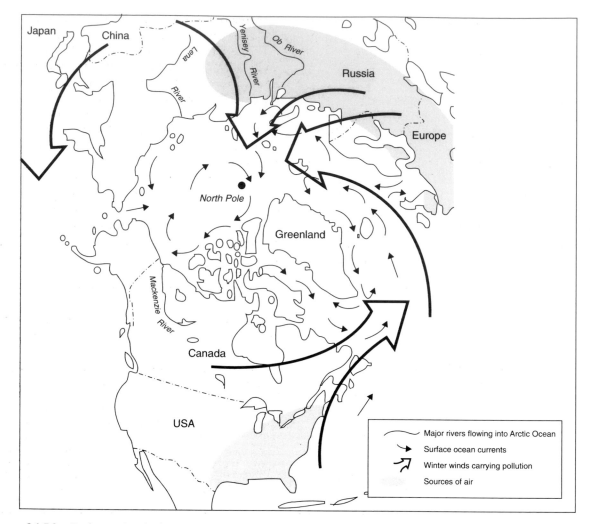

Figure 24.10 *Pathways for the long-range transport of pollutants into the Arctic from North America, Europe and Russia by wind, ocean currents and rivers. Source: After Twitchell (1991).*

Overflights are used in both Antarctica and the Arctic. Impacts will be greater from low-flying aircraft. Animals and birds can be greatly disturbed, with heavy loss of breeding success. In the north the main wildlife concerns relate to ungulates such as caribou, as well as birds. In Antarctica low overflights of penguin colonies have brought considerable destruction by causing panic, desertion and predator attack. Hydrocarbon residues from aircraft fuel can be scattered over a wide area by wind.

Cruise tourism has many more impacts. There are real risks of water pollution from diesel spills and waste and sewage disposal. Any shipwrecks are likely to cause considerable disturbance to wildlife, whether directly from fuel spills or indirectly. Although no permanent land-based construction is required, repeated visits can create pressure on vegetation and raise problems of waste disposal. There is also the possibility that bird and plant diseases may be intro-

duced, not to mention the introduction of exotic plant species into the region.

Land-based tourism will have the greatest impact on the polar environment, owing to the need for a full range of support facilities for transport (airstrips, roads, harbours), accommodation (hotels, lodges, camp sites) and the usual range of tourist attractions (shops, trails). There is increased competition with native flora and fauna for ice-free land and fresh water. Water pollution, the disposal of rubbish and sewage, and disturbance of the breeding and feeding patterns of wildlife are all direct negative impacts. Unsuitable travel through sensitive areas or uncontrolled souvenir-hunting and trampling can destroy sensitive ecosystems. Disruption of permafrost could occur in extreme cases. On the positive side, tourism can build up support for conservation and help to raise finance for it. However, it may be as harmful to the physical environment as are mining and mineral activity. It is necessary, therefore, to keep it under reasonable controls if it is not going to destroy the landscape which attracted the tourists in the first place.

Conclusion

Polar environments have become better understood since the great strides in exploration and discovery of the final years of the nineteenth century and the early years of the twentieth. The Antarctic region is unique on Earth in being the only entire ecosystem to be managed under the Convention on the Conservation of Antarctic Marine Living Resources (CCAMLR), which came into effect in 1982 under the Antarctic Treaty System. The convention confers a degree of protection unparalleled elsewhere. In the Arctic the eight Arctic countries (Canada, Denmark, Finland, Iceland, Norway, Russia, Sweden and the United States) have adopted the Arctic Environment Protection Strategy (AEPS), which aims to protect the fragile polar ecosystems. However, so far there is no unanimous view on how the Arctic should be protected, as conservation is defined as 'rational use'. There is little agreement on how 'rational use' should be interpreted; some countries define it as 'no use' whilst others clearly intend to use the area for non-renewable resources (metals and energy) and even for the harvesting of marine renewable resources (fish, seals, whales), The maintenance of healthy ecosystems remains an important responsibility which will not be easy to fulfil in the Arctic.

Key points

1 Polar ecosystems have low productivity and ecological diversities. Animal species have to hibernate or out-migrate during the harsh winters, and all biological activity is concentrated in a brief summer period. Soil processes and ecological mechanisms act at a low intensity.

2. The geomorphology and soils are dominated by the presence of permafrost in the subsoil. As permafrost is sensitive to any change in the surface vegetation it is easily disturbed by human activity. Any interference with the insulating properties of soil and vegetation, and any addition of heat to the surface (perhaps through industry and building), will inevitably cause permafrost melting and ground subsidence.

3 Polar ecosystems have a low resistance to outside impacts, and low resilience means that recovery is a long-term process. Polar ecosystems are also very variable in time and space. Soil and vegetation conditions change quite rapidly over short distances, under the influence of the catena relations of soils, plants, permafrost and slope. Time variability causes big contrasts in weather and biological activities from one year to the next. It is another factor which makes polar environments so fragile and so unpredictable.

Further reading

French, H.M., and Slaymaker, O., eds (1993) *Canada's Cold Environments*, Montreal: McGill-Queens University Press.
A collection of studies on Canada's alpine and arctic ecozones and their physical and biological features. It deals comprehensively with impacts on northern development, and is especially strong on the role of climate and its impact on the lives of Canadians.

Stonehouse, B. (1989) *Polar Ecology*, Glasgow: Blackie.
A scholarly yet readable analysis of the ecosystems of both Arctic and Antarctic regions. Covers both the physical and the biological aspects.

Williams, P.J. (1986) *Pipelines and Permafrost: Science in a cold climate*, second edition, Ottawa: Carleton University Press.
The second edition of a classic study which considers the opportunities and limitations of the polar environment for development. It deals with much more than pipelines.

CHAPTER 25

Mountain environments

No landsystem inspires emotions quite like mountains, which are hostile, rugged and remote in their physical character and inspire awe, as well as having strong spiritual associations. They were often thought to be the abode of the gods, peripheral to most people's lives and marking ethnic frontiers and political boundaries. Yet they have been home to others – the Incas and Tihuanacos of South America and the Kurds, Tibetans and Nepalese of Asia, often persecuted by intolerant societies – and a refuge against political storms throughout history. The mountains of Afghanistan, Kashmir, Vietnam, Iraq and the former Yugoslavia are among more recent intractable battlegrounds. Mountains cover 20 per cent of the continents but understandably house only 10 per cent of world population. They influence a further 40–50 per cent indirectly through their resources and their role as global 'weather makers' and 'water towers' – denoting their impact on planetary and synoptic meteorology and hydrology – and sustain 24 per cent of global tourism.

In *topographic* terms mountains are landscapes of steep slopes over 600 m in altitude or high altitudes above 2 km regardless of slope, thereby including the Tibetan plateau (over 5 km) and the Bolivian Altiplano (over 4 km), for example, which have other **montane** characteristics. This definition does not meet with universal acceptance, and high *relative* relief may be preferred. The Hispanic term **sierra** is reserved for saw-toothed peaks over 2 km high. Mountains usually occur in linear chains, ranges or **cordillera** but also embrace single eminences worthy of a distinct name. These latter may be constructional in the case of strato-volcanoes or erosional relicts in the case of **monadnocks**. In *structural* terms mountains are large-scale, elevated crustal disturbances characterized by intense folding, metamorphism and granitic intrusion – the very essence of morphotectonics. *Climatic* character shares extremes of temperature and precipitation of the mountain climate with hydrologic and geomorphic processes on steep slopes. *Ecological* definition emphasizes the presence of one or more **montane forest–timberline–alpine** elements or their *ecotones*. A proviso that these may be contemporary or *Pleistocene* acknowledges *Holocene* climate change, biotic **refugia** (survival habitats) and relict landforms. The terms 'mountain' and 'alpine' environments often become entangled. Although they are not synonymous, confusion is excused by the number of related terms derived from the Latin name *alpes* for the snow-covered mountains bordering northern Italy – the type Alps.

Physical geology of the principal mountain systems

Mountains are high-energy, high-stress and high-sensitivity landscapes reflecting orogenic periods of Earth history, three of which are still clearly recognizable. Spatially they are formed and relocated by plate motion and defy conventional geographic distributions and concepts of morphogenetic regions based on latitudinal zones (Figure 3.14). Mountains extend from the poles through every latitude to the equator and almost continuously for 29,000 km through the Americas–Transantarctic mountains. The Tethyan orogens extend through 150° longitude from Morocco to Indonesia. *Altitude* simulates *latitude*, replicating the eco-climatic gradient of 90° of latitude in just seven vertical kilometres in the equatorial Andes. Their spatial pattern is, of course, set by plate tectonics and its imprint on representative mountain chains is now examined.

The Americas: Andes, Rocky Mountains and coast ranges

Differences between the American 'cordilleran' systems are explained by asymmetric motion of the American and Pacific basin plates noted earlier (Chapters 3, 4). Both continental orogens are of late

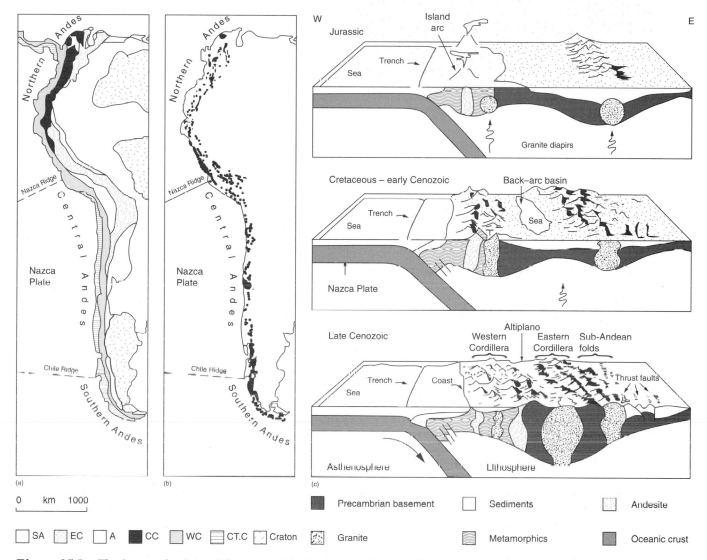

Figure 25.1 *The form and origin of the Andes. (a) Principal Andean cordillera, etc.; SA Sub-Andes, EC Eastern Cordillera, A Altiplano, CC Central Cordillera, WC Western Cordillera, CTC Coastal Cordillera. (b) Granite batholiths. (c) Formation of the Peruvian–Bolivian sector from island arc to arc–continent collision. Source: In part after James (1973).*

Mesozoic–Quaternary age but include exposed or rejuvenated elements of Proterozoic systems, up to 2 Ba old in the central Andes. The type-cordilleran Andes were formed by 'head-on' motion of the Nazca plate after *c.* 190 Ma. Older, late Panthalassic Ocean marine sediments and associated rhyolite-granite intrusions were compressed and thrust-faulted with older terraines to form the eastern Andes. Active subduction, as the Atlantic opened after 135 Ma, formed the western Andes by progressive arc–continent collision with the South American plate and associated landward intrusive and volcanic activity (Figure 25.1). **Molasse** sediments from vigorous syn-formational erosion, and ignimbrite deposits from eruptive volcanoes, collected in inter-cordilleran and marginal basins. These are best developed in the Altiplano (sediment) of southern Peru and Bolivia, and Puna plateaux (ignimbrite) of northern Chile and Argentina.

The Andes rise steeply from a narrow Pacific coastal plain, nowhere more than 125 km wide, and extend the entire length of South America. Four divergent lower cordillera, emerging from the Central American–Caribbean orogens, fuse into the Cordillera *oriental* (east) and *occidental* (west) of southern Colombia and Ecuador, over 200 km wide but rising to peaks above 5 km. The Andes widen steadily through Peru to a maximum 750 km in northern Chile–Bolivia, where the main east and west chains (5.5–7.0 km high) diverge around the Altiplano. The system narrows again to less than 250 km and falls to 2.5–4.0 km in southern Chile, south of the highest

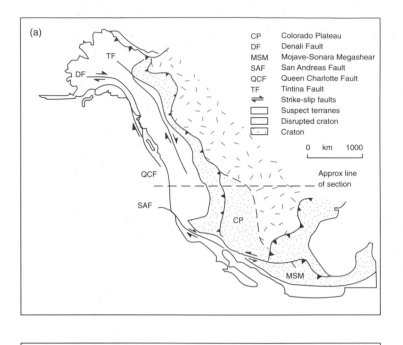

CP	Colorado Plateau
DF	Denali Fault
MSM	Mojave-Sonara Megashear
SAF	San Andreas Fault
QCF	Queen Charlotte Fault
TF	Tintina Fault

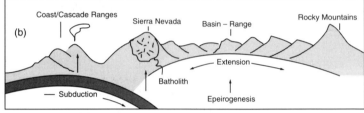

Figure 25.2   *General structure (a) of the North American cordillera, with (b) a representative West–East cross-section. Source: In part after Howell (1995).*

peak, Nevos Ojos del Salado (7084 m) on the Chile–Argentina border, before running out in Tierra del Fuego.

North American orogens are more complex. Head-on subduction has combined with the split and lateral movement of an older (Farallon) plate *and* oblique continental override of the East Pacific Rise mid-ocean ridge. The latter induced transform faulting, including the San Andreas fault, and crustal extension forming basin–range systems (Figure 25.2). The Rocky Mountains are of Mesozoic origin, with older sediments thrust eastward over continental cratons. They rise steeply above the interior Canadian prairies and High Plains of the United States, 750 km east of the Pacific in Canada and 1500 km inland in Colorado, and widen from 250 km in northern British Columbia to 600 km in Wyoming. Peaks rise steadily in the same direction from 2.3–3.9 km in Canada (highest peak, Mount Robson, 3954 m) to 4.2 km in the Wind River range (Wyoming) and 4.4 km in the Park and Sangre de Cristo ranges (Colorado). The Brooks range in Alaska is of similar

age and both systems may have A-subduction elements due to Atlantic spreading. They formed a buttress, against which Pacific plate subduction established younger and still active coastal cordillera comprising accret*ed*, accret*ing* and oceanic arcs from Mexico to the Aleutian Islands.

This landsystem is complicated by massive **dispersion tectonics**. Slivers of craton, accretion prism and arc were displaced northwards along strike-slip faults 10^{2-3} km long, owing to oblique subduction and transform plate motion. This consumed the former Farallon plate and is now working on the Cocos plate, creating a further complication. By the Oligocene (late Cenozoic), westward motion of the North American plate had overridden the East Pacific Rise, leading to thermal epeirogenesis and basalt effusion above hot-spots 1000 km inland. This crustal extension caused large-scale asymmetric rifting of the basin and range structures, centred on Nevada–Utah, and the San Andreas transform fault. Subduction tectonics created the Alaska range and coastal mountains of Alaska and British Columbia, up to 250 km wide, and the highest coastal cordillera (4.0–6.2 km), including the highest peak in North America, Mount Denali (6194 m). Lower cordilleran systems lie inland, separated by narrow coast-parallel intermontane basins. Farther south, the coastal ranges from Washington to California rarely reach 3 km but the more extensive Cascade and Sierra Nevada ranges rise to 3.5–4.4 km across the Willamette and Sacramento–San Joaquin basins. They include the ice-capped strato-volcanoes of Mount St Helens, Mount Hood, Mount Rainier, etc., and the highest mountain in the contiguous states, Mount Whitney (4418 m).

The Andes, coast ranges and Rocky Mountain systems are high enough to support mountain icefields intermittently throughout their length, including equatorial Ecuador, with a snowline at approximately 4.6 km. Substantial valley glaciation with tidewater glaciers occurs polewards of 50° in southern Chile and Alaska, despite lower altitudes. The St Elias and Juneau icefields of Alaska are among the largest outside polar circles (Plate 25.1). Rapid sea-floor spreading and subduction (Andes, 5–10 cm a^{-1}; North America, 5–7 cm a^{-1}) promote continuing active uplift (Andes, 10–70 cm ka^{-1}; North America, 30–60 cm ka^{-1}), strato-volcanic eruption and highly active rivers and slopes outside the intermontane plateaux (see Colour Plate 3.3). Both cordilleran systems inevitably have a major impact on global and continental climate, and act as the principal continental watersheds in both orographic (enhanced rainfall) and topographic senses. Watershed asymmetry divides

Plate 25.1 *Cirque and outlet glaciers of part of the Juneau icefield, Alaska.*
Photo. Kenneth Addison.

many short, swift Pacific coast rivers from a few massive basins draining to the Arctic and Atlantic Oceans and the Gulf of Mexico. Intermontane basins in both continents typically channel inland, and sometimes isolated, drainage networks into large and usually saline lakes (Andes: Titicaca, Poopo and Salar de Atacama; North America: Great Salt Lake, former Lake Bonneville).

Eurasia: Pyrenees, Alps and Himalayas

Alpine and Himalayan mountain systems represent continental collision, where indentation of one plate into another generates widespread thrusting, plate fragmentation and A-subduction epeirogenesis. 'Afro-European' collision took place in stages, with the direction of thrusting and subduction changing as microplates were detached from, and jostled with, each other as contact was made elsewhere. The African plate moved first east, then west, relative to Europe during the Mesozoic before the principal northward drive in the Cenozoic. This complicated pattern of Tethys Ocean closure is imprinted on the contorted shape of individual European Alpine ranges and associated Mediterranean peninsulas. In the west, the Pyrenees and associated Ebro (northern Spain) and Aquitaine (southern France) basins were formed by thickening and thrusting of European

continental plate at the edge of the collision. The Pyrenees are 400 km long and 30–90 km wide, with peaks between 2.0 km and 3.4 km high, culminating in the Pic d'Aneto (3414 m).

The main Alpine ranges sweep from south-east France (Alpes Maritimes) to eastern Austria (Hochschwab) in a 1000 km-long arc. They were formed by southward subduction of European plate and northward thrusting of nappes off the same plate boundary, with Tethys ophiolite, shelf carbonate, flysch sediments and African plate (Figure 25.3). Each element forms distinct ranges. Carbonates dominate the northern Helvetic Alps and Tethyan mélange forms the more southerly Pennine Alps. The former rise to 4275 m (Finsteraarhorn) and 4158 m (Jungfrau) and the latter to 4634 m (Monte Rosa) and 4478 m (Matterhorn). Between these ranges, flakes of the crystalline European continental basement form individual massifs, including the highest European mountain, Mont Blanc (4807 m). Basement flakes also appear in the Jura (1.0–1.6 km), the youngest (late Tertiary/Quaternary) element of the system. They lie north of basins containing lakes Geneva, Neuchatel and Constance. African **klippes** (nappe fragments) form some of the highest summits (Figure 25.4).

Other Alpine orogens form lower relief to the east, where the Carpathian Alps sweep through Slovakia and Bulgaria (1.0–2.6 km high) and peninsula and

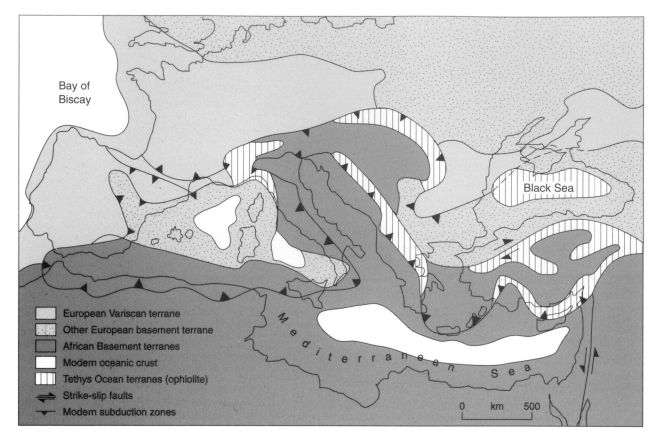

Figure 25.3 *The principal terranes and plates involved in the formation and structure of the Alps. As the African plate drives into Europe, minor platelets pirouette in a complex mosaic of thrusts and subduction. The Arabian plate shears past the African plate along the Dead Sea strike-slip zone. Source: After Howell (1995).*

island systems flank Mediterranean marine basins – the Apennines (2914 m), Dinaric Alps (2522 m) and ranges running through Greece (2917 m) and Bulgaria (2952 m) into Turkey – which demonstrate rotation and reverse motion in microplates. There is

now little measurable uplift in the Alps but volcanic arc activity in the Mediterranean is indicative of continuing tectonic activity.

The Himalayas, like the Alps, are specific ranges but also define a larger system extending for approxi-

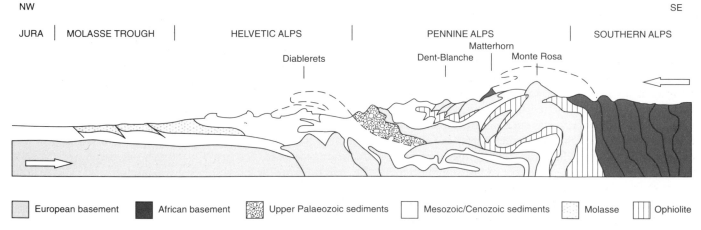

Figure 25.4 *Cross-section through the principal structures of the Alps. Large nappes of African plate and young European sediments were thrust north-westwards during the continental collision of their respective plates and incorporated Tethys ophiolite. Source: After Park (1982).*

mately 3000 km between Afghanistan and South East Asia (Figure 25.5).This is the younger and less complex Asian extension of Alpine–Middle Eastern Tethyan orogens of Turkey, Iraq and Iran. Cenozoic and continuing collision and indentation of the Indian plate are creating a narrow, 250–350 km-wide orogen of ocean ophiolite, accretionary prism, continental crust and A-subduction intrusion. This represents progressive Tethys Ocean closure and, *c.* 40 Ma ago, eventual continental collision. Terranes are stacked as four parallel units forming the Himalaya ranges along a 2000 km front, arc-on to the continent, from Kashmir to northern Burma. The Karakoram batholith to the north-west is a related orogen currently experiencing the highest global uplift rates. Stretches of the rivers Tsangbo (Brahmaputra), Sutlej and Indus, like the Stikine, Columbia and Colorado in North America and the upper Rhine, Inn and Rhône in the Alps, are antecedent where they maintain incision through emergent structures.

Oceanic crust and sediments mark the collision boundary along the Indus–Tsangbo suture, with the Tibetan plateau rising to the north (Figure 25.6). Indentation accommodates crustal shortening by expelling continental crust laterally, raising other ranges sub-parallel to its path like a bow wave. The Hindu Kush to the west and Hengduan Shan to the east may therefore be regarded as extended parts of the Himalayan system, whereas interior thrust and strike-slip Kunlun Shan and Tien Shan ranges, north of Tibet, are not. The Himalayas and Karakoram ranges support a large number of peaks over 7 km, including Everest (8848 m), K2 (8611 m), Makalu (8475 m), Dhaulagiri (8172 m) and Annapurna (8078 m). Indentation at 2–5 cm a^{-1} and tectonic uplift of some 4 m ka^{-1} ensure that the Himalayas are amongst Earth's most geomorphically active areas. Evergreen forest grows at 2500 m on southern slopes in annual average temperatures of 10° C but appears as fossils in Pliocene sediments 5900 m high and at –9° C in the northern Himalayas. This, and the occurrence of only the last Pleistocene glaciation in the Nan Shan range, is powerful evidence of the recent rapid formation of these mountains.

Himalayan ranges and the Alps sit astride, and accentuate, meridional climatic divisions of Asia and Europe. Their zonal distributions sharpen north–south thermal contrasts and inhibit moisture transfer, especially in Asia, where the Himalayas confine the monsoon mostly to the Indian subcontinent. They exert less emphasis on global atmospheric circulation than American cordilleras but the elevation and size of the Tibetan plateau profoundly disturbs the Asian subtropical jet stream and impacts on hemispherical

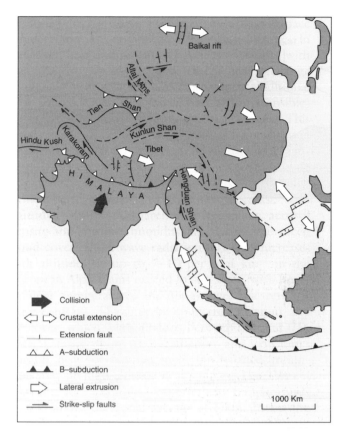

Figure 25.5 *Tectonics and related landsystems in the Himalayas and East Asia region. Collision, indentation and lateral extension continue to elevate orogens and pull rifts and ocean basins apart. Source: Modified from Windley (1995).*

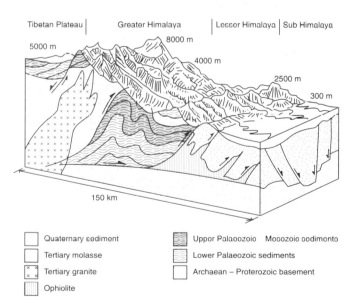

Figure 25.6 *Block section through the Himalayas. Source: Modified from Vuichard, Institute of Mineralogy, University of Berne.*

Environmental problems in mountain areas

Mountain environments are naturally prone to catastrophic geophysical processes – landslides, debris flows, earthquakes, volcanic eruptions, glacier lake bursts and flash floods. Quite simply, that is how their landsystems evolve, and any one may trigger others. There are also biometeorological hazards in high-altitude living. These become apparent above 3 km altitude and include anoxia, mountain sickness and pulmonary oedema due to the rarified atmosphere, and windchill, frostbite and snowblindness due to exposure to the cold and snow. Yet all these are tolerated perennially by 10 per cent of Earth's human population and by a further 20 per cent of us seasonally in pursuit of recreation.

'Traditional landsystem'	Climatic and economic hazards
Mountain snowfield, cirque and valley glaciers; permafrost. Alpine tundra. Alpine rockwalls, moraine and colluvial debris.	Global warming – enhanced glacier melt and retreat. Increased rockwall and debris exposure.
Krummholz Forest-tundra ecotone/high pasture.	Landslides and debris flows. Accelerated erosion.
Timberline Natural/planted coniferous forest.	Tourist development – replacement of traditional farming, heavy construction, slope and vegetation damage; accelerated erosion.
Valley floor slope-channel system. Intensive pasture/arable farming	Fluvial erosion, soil loss and loss of farmland

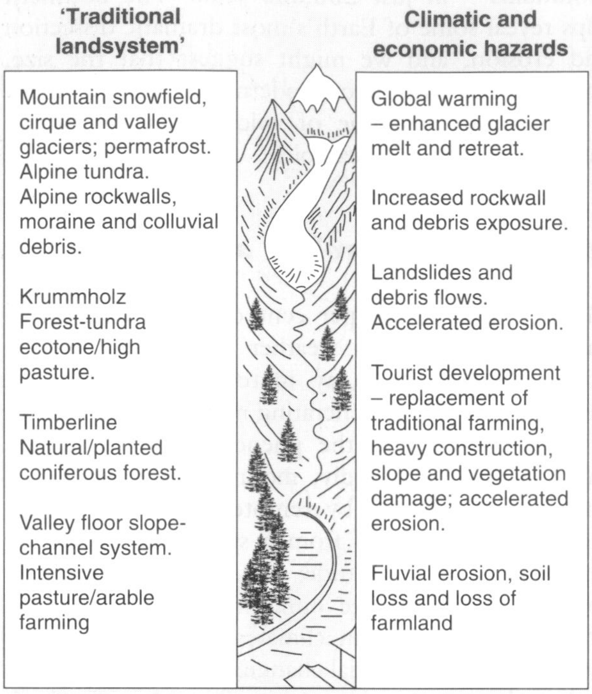

Figure 1 *The character and contemporary hazards of a traditional alpine landsystem.*

Plate 1 *The principal elements of alpine landsystems, Arolla, Switzerland. Dwindling glaciers, active and fossil talus slopes and frost-weathered pinnacles support a sensitive ecosystem above the forest belt. Traditional farms and tourist hotels are uneasy neighbours and every component has an uncertain future in a warming and economically insecure world.*
Photo: Kenneth Addison.

Mountains themselves are under stress. The indigenous populations are mostly citizens of less developed countries, often pressed for cultivable land and natural resources and looking increasingly towards the more marginal mountains. In the developed world, sustained growth exploits the tourist and hydroelectric potential of the mountains and their water, mineral and timber resources. Demand for open spaces, scenic quality and solitude conflicts with the impact of atmospheric pollution, the spread of economic infrastructure and tourism overdevelopment.

These pressures threaten the fragile stability of the mountain geoecosystem, leading to serious landscape degradation and the diminution or ultimate loss of resources. We are only just becoming aware of the harm already done or 'in the pipeline' by irreversible reactions and as yet have no coherent strategy of response. We must also beware of overreaction or mistaking the degree of vulnerability to human, rather than natural, changes. Apparent linkages between Himalayan deforestation and flooding in Bangladesh may be one example where geographers' thirst for 'issues' may distort reality. However, as we strive to resolve existing pressures we are aware that global climatic change poses new threats to the mountains. The United Nations Conference on Environment and Development (UNCED), Mountain Agenda 1992, is a timely focus of attention. Figure 1 and Plate 1 identify a range of threats and sensitivities.

Plate 25.2 *Rime ice coating the windward side of a meteorological screen and buildings, deposited by sublimation in a cold moist airstream.*
Photo: Kenneth Addison.

by advection and sublimation (Plate 25.2). It was thought that orographic *enhancement* was exaggerated and that precipitation from onshore winds would occur anyway at the coast. Windward coast mountains undoubtedly experience some of Earth's highest precipitation levels and intensities, with distinct patterns reflecting an orographic role. Precipitation may increase up to 3–4 km in mid-latitudes at frontal disturbances but only to 2–3 km in the tropics, marking the more limited vertical extent of warm tropical clouds. Mean precipitation maxima are found between 0.5 km and 1.0 km, between 0.7 km and 1.5 km, and above 3 km in equatorial, tropical maritime and high-latitude mountains respectively.

Orographic effects may be greater in winter, and individual cells within frontal systems are capable of 'dumping' rain at rates of 5–20 cm hr⁻¹. The effect is confirmed by concentrations in, or just to the leeward of, ranges and marked downwind rain shadows. Tropical coast mountains such as Kauai (Hawaii) and the Dorsale range (Cameroon) experience amongst the highest global precipitation rates at 11,000–12,000 mm a⁻¹. British Columbia and Alaska coast ranges receive 2500–5000 mm a⁻¹, falling to 500 mm and 1250 mm respectively within 100 km downwind. Much of this precipitation falls as snow by virtue of their altitude and latitude and regional snow-lines reflect rain shadows, rising from 1.6 km on the west side of the southern British Columbia coast ranges to 2.9 km inland and 3.1 km in the eastern Canadian Rockies. In Europe it rises from approximately 1.7 km in Scandinavia to 3.3 km in central Europe and cannot be far above the semi-permanent snowbeds at some 1.3 km on Ben Nevis and the Cairngorms in Britain. Global values range from sea level in polar areas to about 4.5 km in moist equat-

orial régimes such as the Ecuadoran Andes, rising to 6.5 in the dry Andes and Tibet (Figure 25.8).

Mountain climate

So far we have looked at systematic effects with important consequences for regional snowline elevations and plant growth. Substantial local variations are imparted by mountain aspect, slope and topography, when the orientation and angle of slopes in relation to the solar path and local winds have their effect. Mountains do not simply experience broad meteorological trends influenced by altitude and latitude but play an active part in creating their own weather and climate. By acting as heat and moisture sources they emphasize the contrast between the *mountain* and *free* atmosphere at any particular altitude and create mountain atmosphere zones (Figure 25.9). Reference was made earlier to the impact of mountains which penetrate the mid-upper troposphere planetary wind belts. They, in turn, steer weather systems which influence the latitude of maximum orographic effect (polar front jet stream belts) and mountain arid zones (trade wind belts). We are concerned now with progressively more local impacts.

Mountain macroclimate

Mountain barriers disturb airflow in a zone less than 2 km thick and modify the regional climatic character, especially cloudiness, precipitation and regional winds, and often exhibit a downwind plume. The general pattern of orographic cloud and precipitation, associated with instability, may be varied by seasonal or transient effects such as the forced uplift of conditionally stable air, warm air downflow, mountain circulation systems and cold air drainage in more generally stable conditions. **Lee wave** disturbance occurs where there is surplus energy in airflow crossing a barrier and flow wavelength approximates that of the barrier. Clouds form where air ascends through a condensation level, either in the waves or in isolated **rotors**, and – in stable air – evaporates on the descending limb, creating stratiform lee wave or rotor clouds (Plate 25.3).

Fall winds are currents descending leeward slopes, distinguishable by their thermal character and development according to diurnal, seasonal or synoptic conditions. Warm air downflow is generated under particular lapse rate conditions which lead to warming on descent known as the **föhn** (Alps) or **chinook** (Canadian Rockies) effect. This occurs

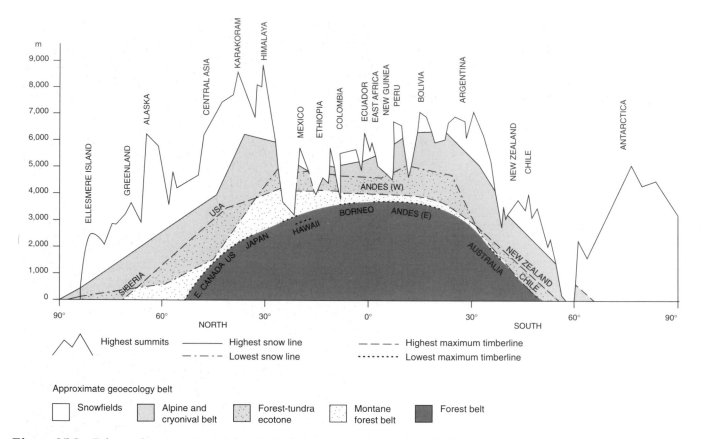

Figure 25.8   *Pole-to-pole cross-section of the principal elevations and geoecological belts of alpine mountains. Source: Modified from Ives and Barry (1974).*

strictly when stable air loses moisture over the barrier and descends at a dry adiabatic lapse rate with an absolute rise in temperature (Figure 25.10). Most currents warm on descent but are less powerful mechanically and thermally. The föhn or chinook occurs seasonally in most mountain systems; the native Canadian term means 'snow-eater', under-

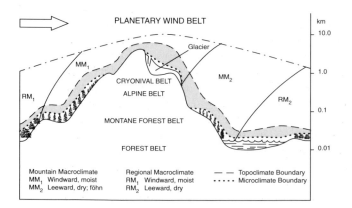

Figure 25.9   *The scale and variety of mountain climates. Mountain topoclimate (shaded) separates myriad surface microclimates from the broader mountain macroclimate. The logarithmic scale refers to the thickness of each layer, not to the absolute altitude. Source: Modified from Barry (1992).*

lining their important environmental influences as warm, desiccating winds.

Contrasting **cold air drainage** occurs as gravity flows of denser air. In their simplest form they are the **bora** (Adriatic) or **oroshi** (Japan) effects of cold air ponded up on windward slopes and draining through passes or other topographic lows. Cold air is also trapped beneath inversion layers in snow-bound mountains and, in winter, may trigger violent downslope **wind storms**. They are common east of the Rocky Mountains. Cold outflows are also widespread outside their type areas and generally relate to synoptic pressure systems. They are a major source of polar air outbreaks south of the European Alps. Cold outflows or **katabats** also form one element of diurnal **mountain circulation winds** (Figure 25.11). Daytime heating of confined valley air, especially on sunlit slopes, induces convective **anabatic** upflow and inflow, coupling valley and surrounding lowlands with a corresponding upper outflow. Evening cooling commences on upper slopes and reverses the circulation with surface katabatic outflow. This is a widespread small-scale phenomenon but is·also, in effect, the system developed over 2 M km² in the Tibetan plateau.

Mountain topoclimate and microclimate

Mountain winds connect the macroclimate with the immediate slope boundary layers where, in some cases, it also has its origins. The latter may be divided into a **topoclimate** zone up to 250 m thick, determined chiefly by slope geometry, and, at its base, a **microclimate** zone up to 15 m thick, modified by vegetation and slope material properties. Topoclimate embraces the effect of rugged topography in stimulating a mosaic of radiation, temperature, moisture, cloud and wind variations across the mountains. Its key components are slope angle and aspect, which allow us to calculate the short-wave radiation flux for any particular latitude and time. Their impact is illustrated by the receipt of 1.2, 2.3 and 4.3 times the incident radiation on south-facing (**adret**) compared with north-facing (**ubac**) slopes at 5°, 25°

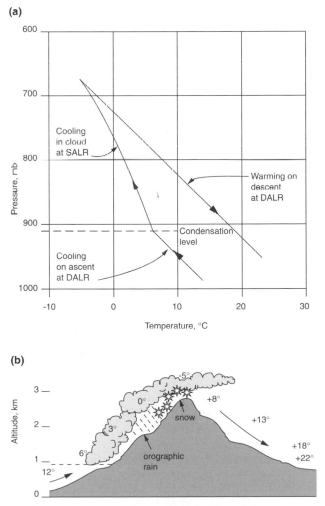

Figure 25.10 *The föhn or chinook effect, illustrated (a) by lapse rates and (b) in cross-section, showing comparative temperatures at the same altitude on windward and leeward slopes.*

Plate 25.3 *Lee wave clouds at several levels above mountains in the English Lake District. Photo: Kenneth Addison.*

and 45° latitude. Differences between east–west facing slopes are more subtle and their impact depends on other conditions and functions. Thus, afternoon direct sunlight may be more effective in melting snow than a similar morning flux because ambient temperature increase during the day has already raised its temperature. Insolation asymmetry accentuates other differences. Anabatic flow, for example, may be accompanied and enhanced by progressive development of mountain cumulus on sunlit slopes (Figure 25.11c).

Slope variability also generates random changes in parameters such as shade, wind exposure, air and water drainage. We move imperceptibly into the microclimate zone, with its most subtle interactions between atmosphere and surface. Here energy and moisture transfers are influenced at successively

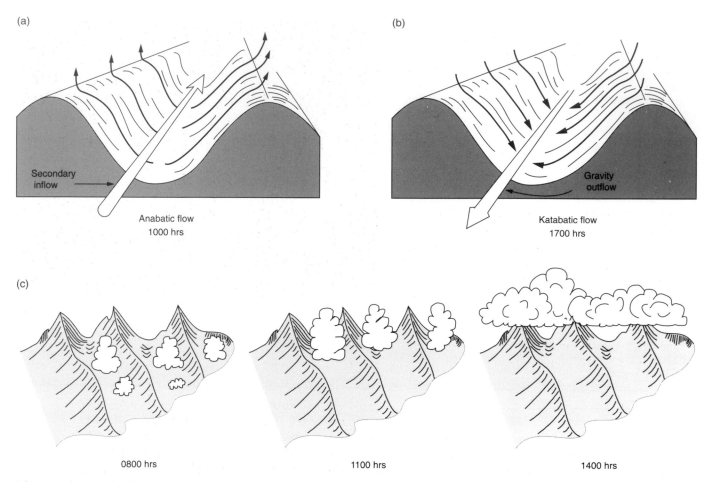

Figure 25.11 *The development of (a) anabatic and (b) katabatic mountain winds and (c) the associated development of rising, mountain cumuliform cloud. Source: Adapted, in part, from Barry (1992). See also Figure 10.18.*

micro-scales by the presence or absence of vegetation, the juxtaposition of snow, soil and vegetation surfaces, the albedo of snow and specific heat capacity and the void ratios of soils. An example of the extremely small spatial and temporal scale of micro-climate is where the influence of albedo and diurnal shifts in radiation and air temperature on melt potential at a glacier surface is evident right down to the evolution of suncups of up to 10 cm vertical amplitude (Plate 25.4).

Mountain ecosystems

Mountain climate, geomorphology, pedology and ecology are more integrated than in most terrestrial systems through their high degree of variability over short distances and mutual sensitivity to instability and environmental change. Emerging from lowland geomorphic and 'zonal' vegetation systems, integration focuses on the high mountain environment and its montane forest–alpine–cryonival zones (Figure 25.12).

North America		Eurasia	
Zone	Vegetation	Zone	Vegetation
Cryonival zone of permanent snow, ice + permafrost			
Alpine	Treeless alpine tundra	Alpine Upper	Snow, rocks, polsters
		Middle	Mats, polsters
		Lower	Grass Heath Tundra
Forest-tundra ecotone	Alpine treeline — Krummholz Meadow Heath Forest line	Subalpine	Alpine treeline — Krummholz Meadow Heath Forest line
Subalpine	Closed conifer forest (mainly spruce and fir)	Montane Upper	Closed conifer forest (mainly spruce, fir, larch-silviculture)
Montane	Closed, mainly mixed coniferous and deciduous forest	Middle	
		Lower	Closed, mixed forest (spruce, fir, pine, oak, beech)

Figure 25.12 *The geoecological zones of mid- to high-latitude mountains in the northern hemisphere. Source: Modified from Ives and Barry (1974).*

It is not prescribed by fixed elevation so much as by the distribution of accordant **geoecological** features and allows us to resolve the term **alpine**. From Latin origins, its use has both narrowed – to identify the zone between permanent snowline and treeline and, in the term **alp**, a bench overlooking a glacial trough or the high meadow often developed there – and widened, to refer descriptively to high mountains. Ecologists may settle for the treeline–snowline belt in locating an **arctic–alpine** flora, which has similarities with **arctic tundra** ecosystems, although diversified by the greater range of high mountain climates and their smaller individual areas. Net arctic–alpine primary production is low, averaging 140 g m^{-2} yr^{-1} dry weight with a range between 400 g m^{-2} yr^{-1} (alpine meadow) and 40 g m^{-2} yr^{-1} (nival belt), compared with montane/boreal forest ranging between 800 g m^{-2} yr^{-1} and 1800 g m^{-2} yr^{-1}. Geomorphologists, however, incorporate this belt with the **cryonival** zone of permanent snow cover, permafrost and associated alpine glaciers and mountain icefields, into an **alpine landsystem** dominated by cryogenic processes.

Plate 25.4 *Sun cups formed on the sunlit face of snowpack micro-relief. Their shape and orientation will change as the sun moves around during the day.*
Photo: Kenneth Addison.

Montane forest

Altitudinal forest zones reflect temperature and moisture limitations. Summer temperatures generally determine the timberline, especially in extra-tropical mountains. Thermal effects are also implicit in tropical *tierra* zonation, with its *t. caliente* (hot/*tropical forest*), *t. templada* (temperate/*submontane–montane forest*), *t. fria* (cool–temperate/*montane–sub-alpine*) and *t. helada* (frozen/*alpine*) (Figure 25.13). Montane forest on cool mountain slopes mirrors the latitudinal temperate–boreal zone transition by the replacement of more thermophilous species (oak, beech, pine) with boreal communities (spruce, larch, fir, birch), or with rhododendron and giant ericaceae in Asian and African tropical mountains. A parallel transition from closed to open forest, with a strong incursion of alpine tundra vegetation, marks the approaching timberline. Moisture effects create interesting regional or topoclimate variations. Low moisture indices are reflected by more xeric communities in rain-shadow areas, replacing larch with pine, for example, whereas the moisture-source effect of high mountains permits a forest girdle to exist 0.5–1.0 km above the zonal timberline in arid mountains, especially in the Himalayas. Topoclimate may increase the frost-free or accumulated degree days substantially through shelter and raise the local timberline, or suppress it by acting as inversion frost traps and cold air drainage channels. This may lead to higher timberlines on exposed ridges, especially in mid-latitudes, unless checked by wind stress or physiological drought. Conversely, moist updraughts in tropical mountains elevate timberlines through the development of *cloud forest* around the condensation level. Human impacts on timberlines should not be overlooked. In general, forest clearance or environmental disturbance, including atmospheric pollution,

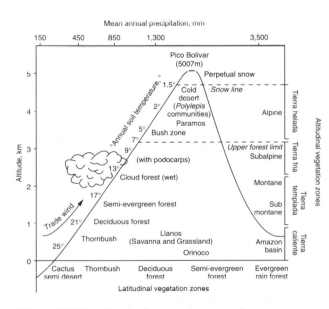

Figure 25.13 *Tropical mountain vegetation zones, Venezuelan Andes. Source: Huggett (1995).*

serves to depress them. Silviculture drives them back up artificially. In Britain, for example, the natural treeline would lie between 650 m and 800 m but most of that zone now supports remnant arctic–alpine flora and montane grassland.

Forest–tundra ecotone

Modern timberlines and Pleistocene snowlines are separated vertically by up to 1 km in tropical and arid mountains but are almost identical outside the tropics. Since modern timberlines and snowlines constrain the principal alpine vegetation zone, it is no older than the Holocene, and slopes which it has colonized are still adjusting to Holocene environmental change. The general location of this geoecological zone is shown in Figure 25.8. Edaphic, micro-relief and microclimatic factors, including snow-cover duration, assume increasing importance at the timberline, and the ecotone is usually marked by a lower forest limit and an upper, absolute alpine treeline. Sandwiched between them is the **alpine meadow** and/or **alpine heath**, dominated by alpine tundra flora interspersed with clusters of severely dwarfed conifers or **krummholz**. Wind pruning creates *flagged* krummholz but, in extreme cases, the conifers develop a creeping or *cushion* habit as basal branches root and no leading shoots survive above the winter snowpack (see Plate 20.1).

Alpine tundra

The term 'alpine tundra' for the dwarf shrubs, herbs, mosses and lichens of the alpine flora reflects strong taxonomic, life form and structural similarities with arctic tundra *sensu stricto*. **Cryophytes** or cold-tolerant alpine plants are almost all perennial and represent the three lower orders of Raunkaier's classification, emphasizing their dwarf ground-hugging habit. They must complete their annual cycles within a relatively short growing season with low summer temperatures of under 10° C for the warmest month; survive winter wind, snowblast and snow burial, physiological drought in exposed airstreams and on free-draining substrates, and frequent ground disturbance – especially as the nival zone is approached. Complex plant mosaics, gradients and adaptations develop, including the protection of pre-formed buds below ground in *geophytes* or within 50 cm in *chamaephytes*. It is seen best in rosette or cushion-shaped *hemicryptophytes*. Extreme moisture gradients over short distances juxtapose xerophytic with hydrophytic taxa.

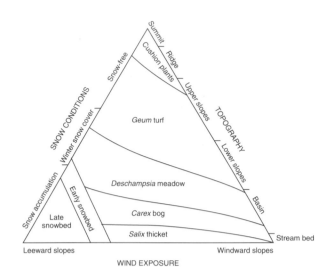

Figure 25.14 *Alpine vegetation pattern in the Beartooth mountains, Wyoming. Source: Billings (1974).*

Plants also have to contend with a variable extent of bare rock or debris surfaces. Figure 25.14 illustrates a typical segregation of sub-communities in response to the principal environmental gradients. Mountain fauna are less strongly zoned than the flora, although there are still substantial global variations in assemblages. A typical Rocky Mountain vertebrate foodweb is shown in Figure 25.15.

More specialized adaptations include high rhizome/root:stem/leaf mass, **vivipary** (production of bulblets instead of seeds), **autogamy** (self-pollination, removing dependence on insects) and a high incidence of **polyploidy** or genetic pre-adaptation facilitating the colonization of new substrates. These become essential to plants on exposed summits and in the cryonival zone, where late-surviving snow severely attenuates the growing season. Mosses and crustose lichens are often all that survive. Thoughts of survival also evoke the role of Quaternary plant **refugia** in alpine environments, especially in mid-latitudes. Arctic–alpine flora retreat polewards and upwards during temperate stages, and **nunataks** (alpine areas above glacier limits) remain their last refuge over wide areas during mid to high-latitude glaciation. This issue is complex but it seems likely that some colonization spreads from alpine areas during non-glacial cold stages and it has certainly sourced plagioclimax alpine communities after deforestation. Arctic–alpine flora may have originated in Cretaceous low-latitude Laurasia below 40° N and developed its modern character as less tolerant plants were 'sifted out' during polewards drift and by Plio-Pleistocene glaciation.

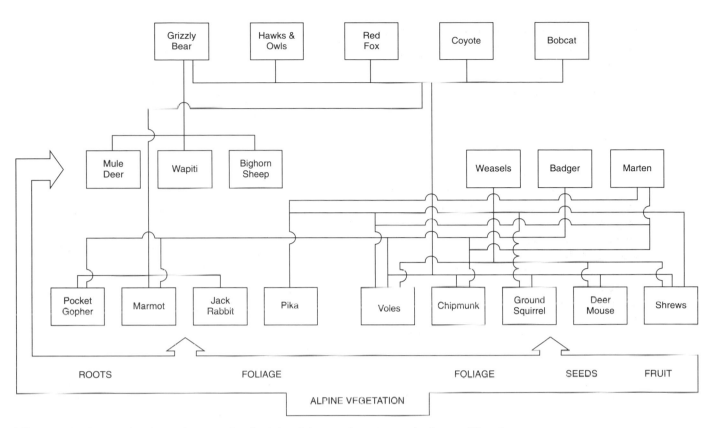

Figure 25.15 *Simplified vertebrate food web of the alpine tundra, Beartooth plateau, Wyoming. Source. After Hoffman (1974).*

Alpine landsystem

The Alpine landsystem develops its distinctive character through the integration of glacial, cryonival, slope and fluvial elements within the higher and spatially more restricted parts of the mountain systems. This is enhanced by close interaction with mountain climates and ecosystems and by partial geomorphic isolation by the timberline and glacially excavated lake basins. The latter both act as a major buffer, especially to sediment transfer, to predominantly fluvial systems at lower altitude. However, the landsystem is not isolated from endogenetic influences. Altitude, high gravitational potential and recent or ongoing tectonic activity combine to make it one of the most geomorphically active environments, in which catastrophic events may disguise continuous but less dramatic denudation.

Cryonival (snow and ice) belt

The alpine zone is a glacial–cryonival–slope landsystem of generally high energy and sediment transfer (Figure 25.16). This is best developed in areas of intense Pleistocene alpine glaciation which excavated deep glacial troughs and tributary cirques, oversteepened rockwalls and deposited considerable volumes of debris. During deglaciation, most of it was plastered indiscriminately or as retreat moraines on lower slopes. Rockwall and moraine-covered slopes are integrated fully into the alpine landsystem, which reworks this glacial 'inheritance' (Plate 25.5). The continuing role of modern alpine glaciers or mountain ice caps is covered in Chapter 15. Glaciers promote active denudation but their meltwater and sediment transfers may partially bypass the alpine slope system where glaciers terminate below the timberline. Their presence currently 'insulates' subglacial surfaces from other alpine processes but actively promotes them at their perimeter (especially in supraglacial rockwalls) by influencing radiation, moisture and wind aspects of the topoclimate.

The cryonival belt occurs in all alpine mountains, irrespective of whether they have a permanent snowline and/or glaciers, with geomorphic processes driven by short-term (diurnal/seasonal) mass and energy budgets of the snowpack and ground ice. **Nivation** concentrates frost and cold-chemical weathering and associated debris transport, under and

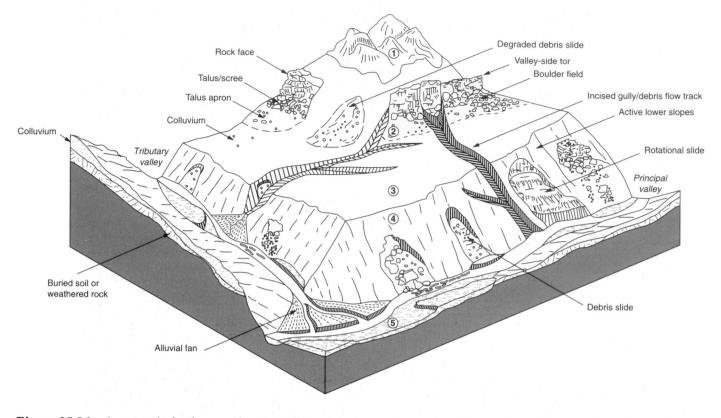

Figure 25.16   *A mountain landsystem, identifying (1) cryonival zone of snow, ice and permafrost, (2) rockwalls and rockwall processes, (3) degraded slopes and former high-level valley floors and benches, (4) active lower slopes and (5) valley floor system. Source: Fookes et al. (1985).*

around permanent and semi-permanent snow beds. These processes are especially active around their lower margins, in meltwater percolation and wet snow zones and in the ablation season. On low–moderate angle slopes they may erode shallow basins which feed debris fans or terraces through melt chutes or by solifluction. Debris, shed by frost weathering in rock-walls overlooking snow beds, slides over their surface to form a **protalus rampart** at their foot (Plate 25.6). The snowpack itself may move as slush avalanches during late-season melting, and more general avalanching from snowfields is a significant geomorphic agent, especially where avalanches incorporate rock debris and are focused in avalanche chutes.

Permafrost activity in the cryonival belt is *continuous* only on the highest summits and should be regarded as *sporadic* elsewhere, dependent on and contributing to local factors in the geoecological mosaic. Frost weathering plays a dominant role in eroding summit crests, producing sharp **aiguilles**, **tors** and residual **blockfields** as intermediate stages in the gradual **altiplanation** of alpine summits (Plate 25.7), and also on rockwalls when seasonal supplies of frost-weathered products form **stratified screes**. Cryoturbation reworks mountain-top debris on broad summits into patterned ground and contributes to downslope movement in both areas. It is often more effective on rockwall-foot talus slopes and reaches its climax in **rock glaciers** which are distinctive of arid alpine permafrost regions. Wind deflation of fine debris is also a significant process in arid alpine zones.

Alpine slope development

The evolution of high mountain slopes is subject, like all slope processes, to variation and much argument and modelling but follows the progressive removal of the rockwall at the expense of a developing talus/ colluvial foot slope (Figure 25.17). Rockwalls may be crowned by peaks or broader crests, dominated by frost weathering and shedding debris by solifluction and deflation. The rockwall is eroded at mean rates of 1–10 cm yr^{-1}, two to four orders of magnitude higher than the crest, which it steadily consumes. Initial rockwall height and steepness depend on the extent of glacial erosion and rockmass strength. Angles in general exceed 45°–50°, with major cliff elements over 65° and main valleys in the range 0.5–2 km deep.

Plate 25.5 *A typical alpine landsystem in the Zermatt region of Switzerland, with the Matterhorn (4478 m) to the left. Photo: Kenneth Addison.*

Rockwalls rarely extend continuously over this range and may be stepped in response to structural features or episodes of valley deepening.

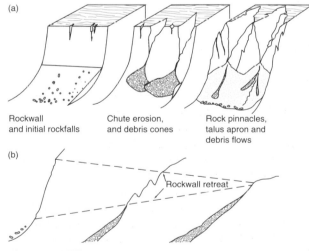

Figure 25.17 *Evolution of an alpine mountain slope after glacier retreat: (a) block section and landforms, (b) the related slope profile development.*

Rock*falls* and rock*slides* lead to their eventual destruction, exploiting the failure criteria of inherently oversteepened and therefore mechanically unstable rock (see Chapter 13). Major rockslides usually occur as a result of catastrophic destabilization along deep-seated failure surfaces involving very large rock volumes of 10^{5-10} m³. They depend on post-glacial unloading, as in the 1963 Vaiont slide in northern Italy (2.5×10^8 m³), or seismo-tectonic activity such as the 1964 Sherman Glacier slide, Alaska (2.3×10^7 m³). Frost weathering works directly, sending small fragments into free fall, and indirectly by widening discontinuities and weakening joint fill, which increases susceptibility to unloading and intense run-off. Unloading generates tension cracking in the upper rockwall, whereas intense run-off can operate across the entire face. Stimulated by rainfall or melt episodes, with power enhanced by high elevation and steep slopes, it flushes out loose debris and may promote rockslides. A combination of all three processes tends to work the rockwall into a series of chutes and intervening rock pinnacles which can locally intensify any one of them.

Plate 25.6 *A protalus rampart accumulating below active perennial snowbeds above Val d'Herens, Switzerland.*
Photo: Kenneth Addison.

Debris delivered to the rockwall foot accumulates initially as an extensive apron or series of **talus cones**. Progressive dissection of the rockwall leads to the emergence of an 'hourglass' shape with chutes feeding cones, which coalesce as the pinnacles are finally eroded (Plate 25.8). Debris supply to developing talus slopes is both relatively steady and markedly episodic, influenced strongly by seasonal melt, which in turn affects their form and onward transfer processes. Most rockfall and shallow slide debris enters the slope near the rockwall foot and is then reworked by a combination of slow, talus-wide processes involving creep, solifluction and slope wash and fast, more concentrated debris slide, mudflows and snow/slush avalanches. Rapid movement is common during spring melt, especially in the active layer when permafrost is present, and intense summer rainstorms after dry spells. Debris flows are the most rapid, channelled means of reworking talus and usually originate at the sharp contrast in slope angle and permeability

at the rockwall–talus boundary below chutes (see box, Chapter 13). Flowing at 5–12 m sec⁻¹, they form the clearest contrast with debris creep at 10–100 cm yr⁻¹. Progressive incorporation of glacial debris towards valley floors creates a geotechnically complex **colluvium**. The profile of the mature debris slope is usually concave, reducing from 35°–43° at the rockwall foot to 25°–35° in mid-slope (the approximate internal friction angle of granular debris) to 0°–5° at the foot. This reflects reworking according to grain/block size, shape and degree of disturbance.

Alpine soils

Pedological development reflects the inherent instability and sensitivity of alpine environments, with frequent occurrence of buried soils and the spatial mosaic of other geoecological processes. The shallow nature and short-term (seasonal) disturbance of

Plate 25.7 *Remnant boulder field, tors and summit altiplanation terrace (background) at about 1000 m in Y Glyderau, North Wales, frost-weathered in steeply dipping tuffs and siltstones.*
Photo: Kenneth Addison.

regolith discourages the development of anything more than raw, lithomorphic soils under a humic layer on crests and the steeper parts of talus slopes, corresponding to *rankers* (FAO classification). They are A/C soils with a medium–coarse texture, with *gelic* rankers on better drained slopes and gelic *gleyed* rankers where drainage is poor. *Cambisols* are found on gentler and more stable slopes, with a strongly humic Ao horizon due to slower decomposition rates at lower temperatures. Acid ericaceous and coniferous litter creates a **mor** humus. Cambisols show at least a weak development of continuous B/C horizons and appear most commonly as *gelic dystric* cambisols; they include *alpine histic* rankers and *alpine gleyic* cambisol soils. *Mollic* cambisols represent the most mature soil development towards valley floors and in association with the forest belt.

Alpine hydrology

Fluvial processes receive less attention in the alpine zone, although rivers eventually shift most slope-derived sediment, and yields from glaciated catchments are amongst Earth's highest. They are also responsible for deep incisions in areas of rapid uplift beyond the glacial environment; the erosive effect of mountain rivers is an order of magnitude higher than that of rivers in lowlands. Their régimes are markedly seasonal, with 'flashy' hydrographs, responding to seasonal and diurnal melt episodes and the high moisture fluxes, high relief, fast run-off generating character of the soil–vegetation–slope system. Snow and ice melt typically contributes 50–70 per cent of annual discharge in alpine mountain rivers, with a similar percentage occurring in just two or three summer months. **Jökulhlaups**, or glacier lake bursts, promote occasional but exceptional flood events.

Mountain river channels showed marked disequilibria in both sediment movement and channel form. Sediment delivery to upstream reaches is highly episodic, dependent on sediment transfers through the talus–colluvial slope and processes such as major rockslides and debris flows which override it. This and the immature development of rock slopes tend to create irregular beds which are rock-bound and

Plate 25.8 *Talus cones, talus sheets and a protalus rampart (left) below frost-weathered rock pinnacles in Nant Ffrancon, North Wales. Although they are largely inactive fossil forms their surfaces are scarred by recent debris flow tracks lines by levées. Photo: Kenneth Addison.*

stepped in places and armoured by large blocks in others. As a result, the suspended sediment load tends to be significantly higher than elsewhere. Timberlines and lakes buffer downstream reaches from slope sediment yields. Braiding is frequently found in mountain rivers where they enter flat valley-floor reaches or, especially, as they enter lowlands across sharp boundaries from regions of high orogenic uplift – as with the braided rivers of south-eastern South Island, New Zealand.

Conclusion

Mountains pack a wide range of environmental conditions into relatively restricted geographic areas, leading to paradox and conflicts of interest in terms of their human occupation and use. High altitude, steep slopes and more extreme weather conspire to make them one of the less inhabited and productive areas on Earth yet draw us in disproportionate numbers as tourists. Indigenous and inward-migrant populations grow as crowded lowland regions exceed human resource demands. Industrial regions exploit mountain water, hydroelectric potential, forest and mineral resources at a distance. Tourism exploits dramatic mountain scenery and alpine snowfields, bringing welcome income to the poorer indigenous communities. Yet it conflicts with their traditional life style and is driven by short-term economic interests inimical to the sensitivity and stability of their physical environment. Sustainable management of global mountains requires that we understand first the character and operation of their physical systems.

Key points

1 Mountains are areas of high absolute elevation or relative relief which develop their own climate, weather and ecosystems, distinct from surrounding landscapes of low to moderate altitude. They create an altitudinal zonation which extends the environmental attributes and processes of higher polar latitudes into equatorial mountains. Their high-energy, hostile alpine landsystems combine with environmental sensitivity to limit human population levels yet, paradoxically, stimulate a substantial tourist industry.

2 Mountains occur either as narrow orogens, the product of crustal shortening and isostatic uplift in response to tectonic compression and subduction, or more general areas of epeirogenic uplift. Three Phanerozoic orogens representing two supercontinental cycles form Earth's principal mountain systems today. Continuing uplift in American, Eurasian and New Zealand Cenozoic orogens coincides with Quaternary glaciation to provide spectacular alpine landsystems.

3 Mountains intrude into the mid-troposphere and create major weather disturbances, especially in Earth's jet streams. Global Rossby wave disturbance and its associated mid-latitude and monsoon climate impact are attributed to the Rocky Mountains, Andes and Tibetan plateau. Locally, the extension of topographic surfaces into the mid-troposphere alters energy and moisture balances compared with the free atmosphere and channels topographic winds.

4 Steep topographic, meteorological and edaphic gradients in mountains stimulate steep ecologic gradients and local diversity. Arctic–alpine tundra is sandwiched between a forest–montane forest belt on lower slopes and a cryonival belt around high mountain summits. The forest–tundra ecotone is primarily of Holocene age, as alpine glaciers receded from the timberline.

5 Alpine landsystems represent the integration of glacier, cryonival and slope processes above the timberline, which buffers them from lower-lying fluvial landsystems. High potential energy, continuing uplift in many orogens and Pleistocene glaciation ensure that the modern glacier–cryonival–slope landsystem is a high-energy, unstable slope and high sediment transfer system. Surviving alpine glaciers and icefields are threatened by global warming.

Further reading

Barry, R.G. (1992) *Mountain Weather and Climate*, second edition, London and New York: Routledge.
This detailed text may be too advanced for many but there are few books which can match the specialist attention paid to its subject. It is still possible to derive a greater understanding of mountain weather and climate without being drawn into its mathematical explanations.

Gerrard, A.J. (1990) *Mountain Environments: An examination of the physical geography of mountains*, London: Belhaven Press.
A comprehensive review of the ecology and geomorphology of mountains and Earth's principal mountain systems. Specific chapters on mountain climate and weather are omitted but relevant material appears as a component of chapters on geoecology, hydrology and glaciers. The geological origins of orogens are outlined and a whole chapter is dedicated to volcanoes before the book concludes by looking at mountains under pressure.

Ives, J.D. and Barry, R.G. eds (1974) *Arctic and Alpine Environments*, London: Methuen.
Although over twenty years old, this book of almost 1000 pages still sets standards for the breadth of cover of environments described here and in Chapter 24. It is a valuable reference work alongside other books which update individual subjects.

Stone, P. ed. (1992) *The State of the World's Mountains: A global report*, London and Atlantic Highlands, N.J.: Zed Books.
This book stems from popular concern for mountain environments and the cultural and social health of its inhabitants. Variable coverage and quality of illustration of the principal physical and human components of mountain environments are compensated for by its assessment of virtually all Earth's mountain systems.

CHAPTER 26
Mediterranean environments

Mediterranean environments are controlled by a distinctive climatic régime of hot, dry summers and cool, moist winters. This special and unique climate has great influence on natural processes (erosion, hydrology, soil formation, ecological processes) and on human activities (agriculture, forestry, conservation, water abstraction). Under the Köppen system of climatic classification, Mediterranean climates are designated Cs, i.e., temperate, with warm dry summers. A third letter indicates the summer temperature; thus a designates the warmest month above 22° C and b the four months above 10° C. Mediterranean climates are thus Csa. Köppen also defined the Mediterranean climate by the equation:

$$R_w \geq 3R_s$$

where winter precipitation (R_w) is at least three times the total amount of summer precipitation (R_s).

Climates defined in this way are found in five widely separated regions of the world, all of which occur on the western subtropical coasts of continents between latitudes 30° and 40°. These are shown in Figure 26.1 and are in the Mediterranean region proper, in California, Chile, South Africa and south-western and southern Australia. The total area of the world occupied by Mediterranean environments is only about 2 million km², about half of which occurs in the Mediterranean itself: southern Europe, North Africa, the Levant and the Mediterranean islands. Although plant species differ between each of the five main regions, evolutionary convergence has led to a vegetation in each which is dominated by ever-green woodland with sclerophyllous trees and ever-green shrubs. In all Mediterranean regions much of this woodland has been replaced by agricultural land, originally for the traditional dry-farmed crops of cereals and tree crops (e.g. the vine, olive, carob, almond), but increasingly for high-value irrigated land use (e.g. vegetables, citrus fruits, rice). Outside the limits of farmland, human impacts on the natural vegetation have been severe, mainly through grazing, ranching, wood collection and deliberate firing. The native woodland has therefore been replaced by dense scrub (*maquis* in France; *monte bajo* in Spain) or aromatic heath (*garrigue* in France; *matorral* in Spain). In California scrub known as *chaparral* is common, whilst the term *matorral* is used in central Chile. In South Africa the shrubby *veld* contrasts with the heathy *fynbos*. In south-western and southern Australia the term *mallee* is used for similar vegetation formations. Because of the difficulty of knowing how far vegetation has been influenced by human activities, it has become customary to use domesticated crop plants to indicate the limits of the Mediterranean climate. Figure 26.2 shows the range of the domesticated olive (*Olea europea*) in the Mediterranean region proper; the olive is a commonly used biological indicator of the distribution of the Mediterranean environment.

Climate, present and past

The typical climate which is found on the western coasts of continents in subtropical latitudes is ·' Mediterranean type with its hot, dry summers aı mild, wet winters. This seasonal contrast in temper-ature and precipitation is driven by seasonal changes in the position of subtropical high-pressure cells, and associated westerly jet streams in the upper tropos-phere. In the Mediterranean region proper the summer months are dominated by the eastward extension of the Azores high-pressure cell. The anti-cyclonic nature of this large-scale circulation gives rise to atmospheric subsidence and hence stability. Low-pressure weather systems can occur but are usually very local and weak. The summer heat is rein-forced by regional winds from continental tropical (cT) source regions which can cause a sudden decline in relative humidity to 20 per cent, and temperatures rise to above 40° C. These winds are known by a variety of local names – *scirocco* (Algeria and the Levant), *ghibli* (Libya), *khamsin* (Egypt) and *lebeche* (Spain). Other hot local winds can be very humid

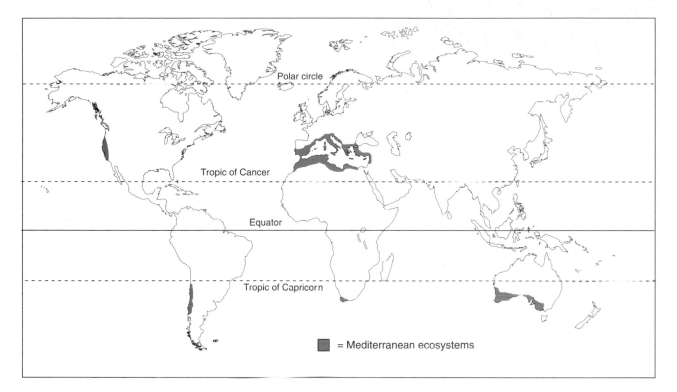

Figure 26.1   *The world distribution of Mediterranean environments.*

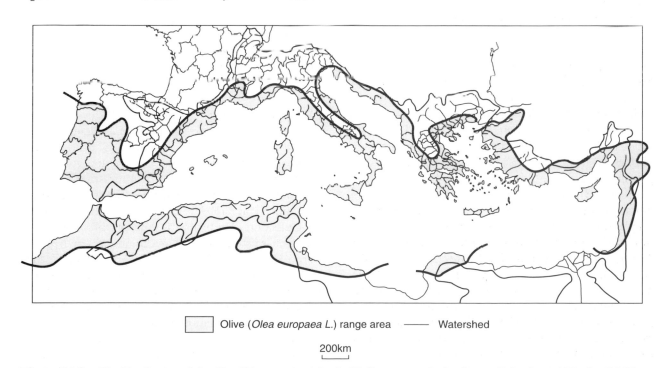

Olive (*Olea europaea L.*) range area —— Watershed

200km

Figure 26.2   *The distribution of the olive (*Olea europea*) in the Mediterranean basin. Source: Polunin and Huxley (1965).*

where there is a long fetch over the Mediterranean Sea (e.g. the *Levante* in southern Spain).

The high-pressure cell collapses quite suddenly in late October and early November. The subtropical high and its associated westerly jet stream move south to a position over the Sahara, allowing Mediterranean depressions to form and bring winter precipitation. These depressions are formed by incursions of air

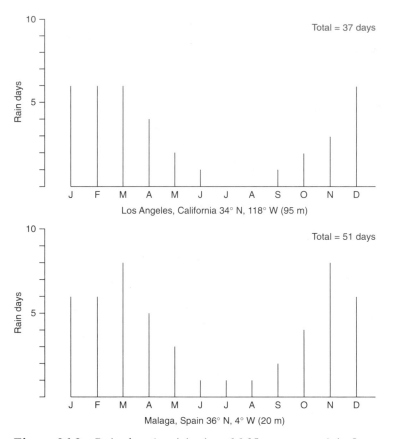

Total = 37 days

Los Angeles, California 34° N, 118° W (95 m)

Total = 51 days

Malaga, Spain 36° N, 4° W (20 m)

Figure 26.3 *Rain days (precipitation of 0.25 mm or more) in Los Angeles, California and Malaga, Spain, during the year.*

masses from many directions – mT air from the Atlantic, mP from the North Atlantic and north-west Europe, mA and cA from the Arctic and northern Russia, cP from Asia and cT from the Sahara. The formation of depressions (cyclogenesis) is stimulated by the relatively high sea-surface temperatures, which are approximately 2° C above mean air temperatures. Some 10 per cent of depressions enter the western Mediterranean from the Atlantic, and 20 per cent originate from the Sahara. The remainder, however, form as Mediterranean depressions in the lee of the Alps and Pyrenees from northerly cold and conditionally unstable air streams. The warming of this mP or mA air gives intense instability, with high precipitation along the warm front, and heavy showers and thunderstorms along the cold front. The boundary between these Mediterranean depressions and cT air flowing north from the Sahara is referred to as the Mediterranean Front. Depression tracks are complicated by relief effects, and by the influences of other airstreams entering the basin from outside. Winter weather in the Mediterranean is variable, also, owing to the mobility of the Subtropical Westerly Jet Stream, which can move northwards for long

periods. When this happens anticyclonic circulation is dominant, giving fine and settled weather.

The fact that anticyclonic circulation can re-establish itself during the winter – for 25 per cent of the time over the whole Mediterranean, but for 50 per cent of the time in the western basin – yields the important result that, although winter is a rainy period, there are relatively few rain days. Figure 26.3 shows the rain days for Malaga, Spain (precipitation 447 mm) and, for comparison, Los Angeles, California (precipitation 386 mm). Mean annual rainfall varies between 300 and 750 mm in Mediterranean regions, falling on forty to eighty rain days. Variations in precipitation totals result from altitude, with orographic rainfall being added to frontal, from rain shadow effects, and from the exposure of coastal areas to onshore winds from areas of cyclogenesis in the Mediterranean Sea.

Figure 26.4 shows the moisture balance for Malaga, Spain. Precipitation shows a bimodal distribution, with two peaks, in November and March. This annual pattern (régime) is characteristic of Spain, southern France, Italy and the Balkans. The double maxima are interpreted as transitional between the Continental Interior type, with a summer maximum, and the Mediterranean type, with the winter maximum. North Africa and the eastern and central Mediterranean show a simpler régime with a single winter maximum. The high summer temperatures lead to the high levels of potential evapotranspiration (PE) shown in Figure 26.4. The monthly potential evapotranspiration totals are calculated by a method developed by C.W. Thornthwaite on the basis of air temperatures (see Chapter 6). It gives a general guide only, but illustrates how the higher soil moisture levels of winter ('soil moisture recharge' and 'soil moisture surplus') fall rapidly in April to remain very low ('soil moisture utilization' and 'soil moisture deficit') until the rains of October and November. Available soil moisture is a dominant influence on the development and productivity of the region's vegetation, and hence on the entire ecosystem. It is an additional limitation on ecosystems that rainfall totals can vary greatly from year to year (interannual variations, measured by the 'interannual coefficient of variation', can reach 25–35 per cent), and that available records of rainfall show significant periods of wetter and drier rainfall.

In addition to the three unfavourable factors of the summer drought, the variability of rain from year to year and the general unpredictability of rainfall, a further important characteristic is the intensity of rainfall from Mediterranean fronts, falling on bare and dry soils in autumn. A raindrop may reach 6 mm diameter in size, giving a terminal velocity (maximum

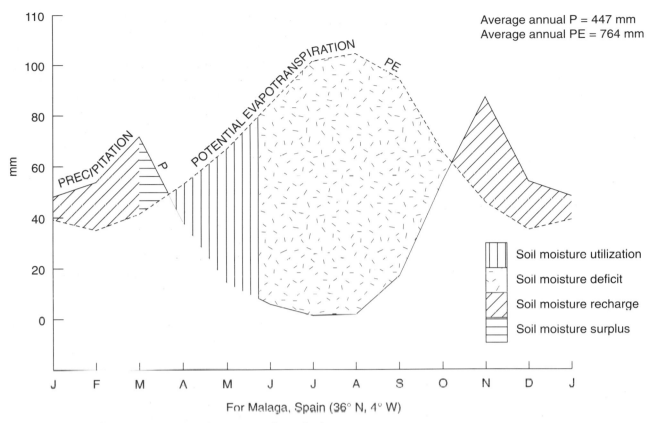

Average annual P = 447 mm
Average annual PE = 764 mm

Soil moisture utilization
Soil moisture deficit
Soil moisture recharge
Soil moisture surplus

For Malaga, Spain (36° N, 4° W)

Figure 26.4 *The seasonal water balance for Malaga, Spain.*

sustained speed) of 10 m s⁻¹. The amount of work done and the erosion caused by such storms, infrequent though they are, is out of all proportion to the relatively small amounts of precipitation involved. The question of climate change is a vital one, given the present interest in the two environmental processes of 'desertification' and 'global warming'. During the glaciations of the Pleistocene, the higher mountain ranges experienced ice caps (Alps, Pyrenees, Sierra Nevada, Tauros mountains), although the rest of the Mediterranean was outside the area of direct glaciation. The Pleistocene seems to have been dry in the Mediterranean (steppe vegetation), becoming wetter by 8000 BP in the early Holocene (postglacial). Since then it has become difficult to identify climatic cycles because of changes brought about by cultural and land-use events. It is difficult to distinguish natural climatic trends from anthropogenic (human) effects. However, it is possible to analyse climate records in the historical period, and these show persistent trends over long periods of time. For example, the mean rainfall in south-east Spain halved between 1890 and 1940. There have also been significant wetter and drier periods on a regional scale. Although the evidence is still debatable, the so-called Mediterranean Oscillation gives periods of lower rainfall in the western Mediterranean associated with higher rainfall in the eastern Mediterranean, and vice versa. Some see these trends as entirely random, whilst others detect clear cycles, even though the causes are as yet unknown.

Soil formation and distribution

Soils form an integral part of all ecosystems. They reflect the influences of their specific site (the factors of soil formation: climate, vegetation, topography, geology, time) and they strongly influence many surface processes (infiltration, overland flow, surface erosion). Also, they have important effects on land use through the supply of a rooting medium, water and plant nutrients for any cultivated crops.

The Mediterranean climate exerts a powerful influence on soil-forming processes. In the moist winter season, rates of weathering and leaching are at a maximum. Minerals in rocks and unconsolidated parent materials are subjected to chemical weathering along cracks and fissures in the subsoil. The weathering processes of hydrolysis and hydration are carried out by rainwater charged by carbon dioxide (CO_2) both from the atmosphere and from soil air, whose higher content of CO_2 comes from the activities of soil fauna and soil micro-organisms. pH values for rainwater of 5.5 readily attack soil minerals, and, where the parent rock is limestone, cause rapid dissolution by carbonation. Simultaneous with weathering during the winter months will be leaching, the removal of weathered products (cations and anions) and any free calcium carbonate (decalcification) from the soil profile. The rates of soil formation and thickening vary considerably between different rock types. Hard igneous and metamorphic rocks weather slowly, owing to the restricted length of the moist season and the generally low precipitation totals; in such situations there is usually a sharp and clear interface between solum and rock (i.e. the profiles have no C horizon). On softer rocks (e.g. chalks and marls) weathering proceeds fast enough to give a deeper profile, with fragments of rock and stones in a C horizon.

In addition to the leaching of ions from the soil, winter precipitation causes the leaching of clay and silt particles from the A into the B horizon to give a clay-enriched or textural Bt horizon. This is very evident in the field because of the clay and silt coatings (cutans) on stones and soil structural units, giving typically prismatic structures in the subsoil. The process is referred to as *argillation*.

The results of weathering and leaching in the winter months are thus the dissolution of the parent rock, the formation of a Bt horizon, and the production of secondary weathering products of clay minerals, oxides and hydroxides of iron and aluminium ('sesquioxides'), and silica. In the ensuing hot and dry season the non-crystalline (amorphous) iron and aluminium oxides become dehydrated and crystallize to form crystalline oxides; where the soil retains some moisture only partial dehydration takes place, and the browner hydrated oxides of iron are formed (*goethite* α FeO.OH and *lepidocrocite* γ FeO.OH). Where dehydration is complete (drier soil climate, well drained profile, porous parent material) the iron oxides take the form of anhydrous haematite (Fe_2O_3) which imparts a strong red colour to the soil. As this chemical reaction is irreversible, the development of a red hue will increase with time, and thus the degree of reddening can be used as an indicator of the age of a soil. In 1853 in Italy these red soils were first called *terra rossa* and the designation has remained ever since. The whole set of processes producing them is termed *rubefaction* (reddening). Figure 26.5 shows three profiles which are widespread in the Mediterranean region and

where the relative imprint of weathering, leaching, argillation and rubefaction varies. The Brown Mediterranean soil (FAO: Calcic Luvisol; *terra fusca*) is characteristic of more humid sites (higher rainfall, cooler summers, higher elevation, impervious parent material). By contrast the Red Mediterranean soil (FAO: Chromic Luvisol; *terra rossa*) is favoured on drier situations (lower rainfall, hotter and longer summer drought, low elevation, permeable parent material, greater age). The alluvial soil is characteristic of many of the alluvial plains and deltas around the Mediterranean. These *vega* soils have built up by the fluvial accretion of a mixture of clays, silts and sands. The profiles are typically black/dark brown in colour, with high organic matter content. Like the other two soils, they display good water-holding properties, owing to their heavy texture. Not surprisingly, they are mostly irrigated for high-value crops such as rice, vegetables and citrus orchards. The deep profiles of the alluvial soils are clearly of depositional origin, in contrast to the 'sedentary' nature of Brown and Red Mediterranean soils. However, even with the latter two types it is recognized that airborne salts, lime and dust can make an important addition to the profiles. Where leaching and accumulation processes are finely balanced, the degree to which soils are non-calcareous or calcareous, or non-saline or saline, can be due to the amount of atmospheric loading. Aerosol deposition of salts can be sufficient to yield saline soils, especially in coastal locations. Aerosol input of calcium and magnesium carbonates ('lime') can be sufficient to make the profiles calcareous, especially near a calcareous source rock (limestone, marl). Normally Mediterranean soils are non-calcareous, and in fact the absence of lime is a necessary condition for both argillation and iron oxide production. Calcareous soils must therefore indicate a secondary impregnation, either aerially, following a change in climate or surface cover, or by the burrowing and mixing of soil fauna.

The fertility status of Mediterranean soils shows both positive and negative features. Apart from the saline cases noted above, salt levels are low, and exchangeable sodium forms a minor proportion of cations on the exchange complex. pH values are close to neutrality and at the optimum range for many plant nutrients. The soils have high contents of calcium and magnesium, and a cation exchange complex that is base-saturated. However, the low organic matter content (typically between 1 per cent and 2 per cent at the soil surface) is a result of the high rates of mineralization and decomposition during the summer months. Unfortunately this leads to deficiencies of nitrogen and phosphorus, both major plant nutrients.

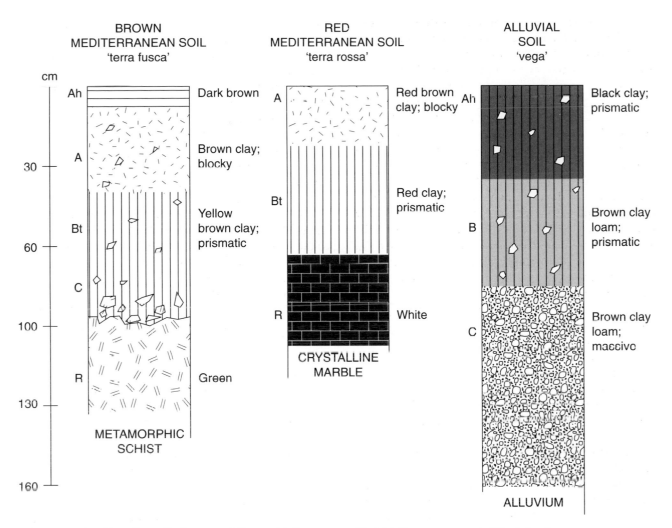

Figure 26.5 *The soil profile features of Brown Mediterranean, Red Mediterranean and Alluvial soils.*

Native vegetation has to adapt to these deficiencies, whilst agricultural management depends upon the application of suitable chemical fertilizers or the use of legumes in the crop rotation. From the point of view of physical properties, the soils show high water-holding capacities, owing to their clayey textures, if rain is able to infiltrate. Summer desiccation of the soil surface usually forms a cracking pattern, which can be especially deep in the case of alluvial soils. Such soils have high content of expanding lattice, montmorillonitic clay minerals. However, rehydration after the first autumn rain causes the clays to swell, thus closing the cracks and shutting off infiltration.

Development and adaptation of vegetation

The vegetation of the Mediterranean region has experienced many changes during the past 15,000

years, i.e. roughly during the time since ice sheets in Europe were at their Würm (Devensian) maximum. Vegetation is rarely static, and plant communities change rapidly in response to factors which control their structure and productivity (see Figure 20.2). In the Mediterranean region there have been significant prehistoric and historical changes in the two most powerful controlling factors, namely climate and human land use, and it is not easy to disentangle the relative importance of each. During the Glacial Epoch (20,000–15,000 years BP) the climate was distinctly cooler and drier. Mountains surrounding the Mediterranean became refugia for trees of both southern and northern European species which were unable to cope with the harsh glacial conditions farther north. Lowlands were mostly dry steppes, with or without groves of trees, depending on location.

Since glacial times (i.e. during the Holocene) trees have spread to lower altitudes as the climate has

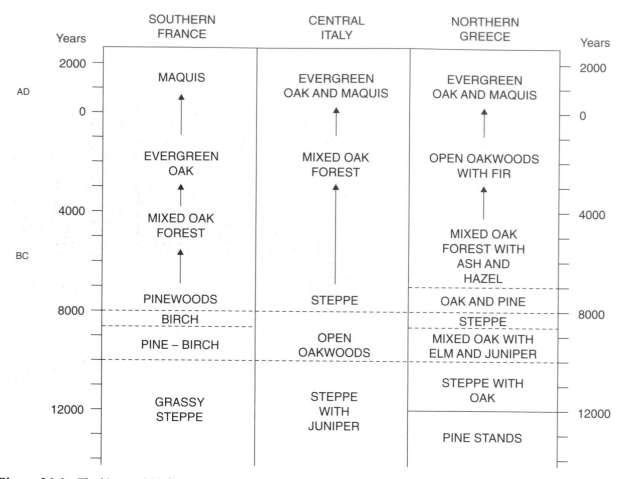

Figure 26.6   *The history of Mediterranean vegetation since the last glacial epoch.*

become moister and warmer. The details of the succession vary from location to location, and Figure 26.6 shows the slightly different sequence in France, Italy and Greece. The differences are due to regional conditions, but the overall pattern is first an invasion of northern types of coniferous and deciduous tree (pine, birch, elm, oak), with evergreen oak and chestnut becoming more widespread as temperatures increased to reach their Holocene maximum levels about 5000–6000 years BP. Lower soil moisture contents caused northern species to retreat to suitable altitudes and aspects, whilst at the lowest, hottest elevations scrub and open steppe vegetation developed.

This simple model of climatically controlled vegetation succession is greatly complicated by human occupation. The great antiquity of archaeological remains in the Mediterranean basin points to long and extensive 'attack' by human societies on a changing and emerging Mediterranean forest. (Plate 26.1.) Again, the details and dates of prehistoric and historical societies differ from region to region. Figure 26.7 shows the chronology of the southern

and south-eastern coastal areas of Spain. Here there is a particularly rich history of 'cultural waves' from the Palaeolithic through the Neolithic, Copper, Bronze, Phoenician, Greek, Roman, Moorish, and modern Spanish eras. Different types of agriculture have superimposed their imprint on the coastal landscape – pastoral farming, cereals, vineyards, orchards, vegetables; inland there is a long history of local pastoral transhumance following traditional sheep trails (*canadas*), and of dry farming for cereals, vines and olives. The net result of this prolonged exploitation of the wild vegetation has been, first, to reduce the natural resource base on which subsequent peoples could depend; second, to introduce, whether deliberately or inadvertently, species foreign to the area (olive, orange, cotton, sugar cane and many other species), and thirdly to accelerate natural rates of erosion by deforestation, grazing, burning and cultivation.

The changes brought about by the clearance and excessive exploitation of Mediterranean woodland are shown in Figure 26.8. The removal of trees affects many microclimate elements. Figure 26.8a shows the

Plate 26.1 *This remnant of Mediterranean evergreen oak woodland in southern Spain occupies rocky ground, and hence has avoided the clearance for wheat farming seen in the foreground.*
Photo: Kenneth Atkinson.

effects on soil temperature. Progressively higher maxima and lower minima occur under scrub oak and grassland. This leads to more rapid decomposition of soil organic matter, higher evaporation and drier soils. The erodibility of the soils also increases with clearance (lower soil organic contents lead to lower infiltration rates) as also does the erosivity of the rainfall; this leads to thinner, eroded soils over time (Figure 26.8b). The nature of the soils also changes in line with the changes in soil organic matter, dehydration and erosion. As illustrated in Figure 26.8c, Brown Mediterranean soils tend to evolve into Red Mediterranean soils as organic matter declines and as hydrated iron oxides convert to anhydrous haematite iron oxide.

The shrubby and steppe-like vegetation which is so characteristic of today's wild landscapes of the Mediterranean region can be viewed as a result of human pressures superimposed upon climatic trends. The effects of the human pressures, and their relationships with both progressive and retrogressive trends in vegetation, are shown in Figure 26.9. The climax vegetation is evergreen oak woodland (holm oak, *Quercus ilex*; cork oak, *Q. suber*) which will degenerate into *maquis* scrub under light exploita-

tion. *Maquis* is typically 1–3 m high, and today is more widespread than 'relict' evergreen forests. Many plants that are present in the forest but which prefer more open habitats grow abundantly in *maquis* (tree heath, buckthorn, holly oak, strawberry tree, myrtle, juniper). The net result is a dense, almost impenetrable shrub community, with plant species varying in different parts of the Mediterranean. Excessive exploitation leads to the formation of a low mixed heath, *garrigue*, which is a very diverse community of low shrubs and flowers, typically less than 1 m high. The community is colourful and aromatic, with species varying according to local conditions. However, common plants are rosemary, thyme, lavender, sage, broom and rock rose. The common feature of *garrigue* plants is their resistance to grazing by sheep and goats on account of their poisonous, thorny or 'oily' nature. Further degeneration can lead to the almost complete disappearance of shrubs and the formation of steppe grassland and stony pasture. (Plate 26.2.) Such eroded, rocky terrain supports only grasses (esparto), annuals (clovers) and bulbs (asphodels, tulips). Figure 26.9 also indicates the pathways of regeneration should the human impact cease, for example through the abandonment of

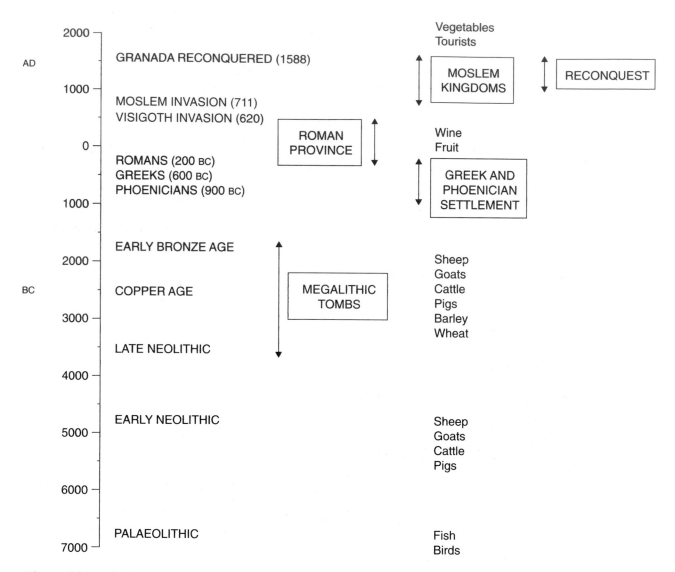

Figure 26.7 *The chronology of prehistoric and historical societies in southern Spain and their main food-gathering activities.*

agricultural land. However, regeneration of *garrigue* to *maquis* and then forest is clearly a much slower process than degeneration, as soil erosion will have reduced soil depth, water-holding capacity and nutrient content. Extreme degeneration can make regeneration impossible.

The long drought during the summer months means that plants must adapt to a severe period of water stress at a time when air temperatures are at a maximum. The xeromorphic (adapted to drought) strategies which allow plants to survive these adverse conditions are many and varied. **Annuals** are

ephemeral plants which grow only when conditions are favourable, i.e. in the cooler and moister Mediterranean winter. By germinating, growing, flowering, setting seed and dying within one growing season these plants exhibit the strategy of drought avoidance. **Geophytes** adopt a similar pattern but grow from a bulb or corm which is the vegetative resting stage after flowering (e.g. tulip, scilla, asphodel). **Succulents** are able to store water in swollen cells (e.g. cacti); a special group are **halophytes** (salt tolerators) such as salt-marsh plants which can survive saline soil conditions. Halophytes

Figure 26.8 (opposite) *The effects of woodland clearance in the Mediterranean environment: (a) temperatures of the soil surface under grassland, scrub oak and evergreen oak woodland, (b) degradation of the vegetation and reduction in soil depth, (c) degradation of the vegetation and the formation of different soil types.*

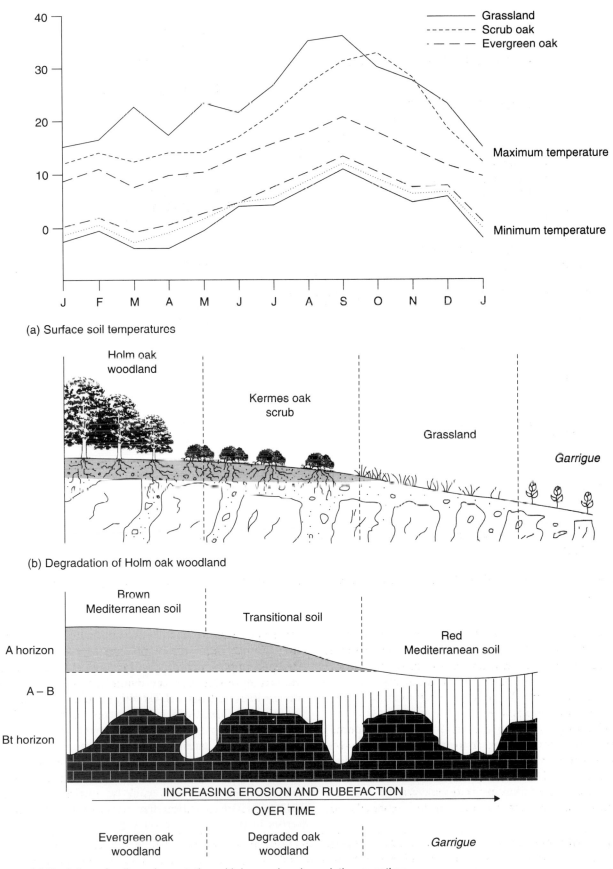

(a) Surface soil temperatures

(b) Degradation of Holm oak woodland

(c) Evolution of soils and vegetation with increasing degradation over time

Plate 26.2 *Overexploitation for millennia has produced this Mediterranean heath* (garrigue) *in northern Libya. Vegetation is being removed by overgrazing and collection for firewood.*
Photo: Ken Atkinson.

cope with salinity by two mechanisms: filtering at the root surface and expelling salt at the leaf surface. The succulence is able to dilute the salt within the plant. **Phreataphytes** are plants with deep tap roots allowing them to reach groundwater at depth (e.g. the carob tree).

A range of structural modifications in plants favour drought tolerance: needle-leaf form; the elimination of all leaves to give a photosynthesizing stem; sticky, waxy or hairy leaf cuticles (surfaces); leaf stomata sunk in surface depressions; loss of transpiring leaves in summer (drought-deciduousness); pale leaves and stems to increase reflectivity (higher albedo). A high ratio of below-ground roots to above-ground shoots favours moisture absorption by plants. Many Mediterranean plants are also aromatic, giving off oils and scents. This characteristic may serve several functions, including the lowering of surface temperatures through evaporation and higher suction ability of the plant for soil water. A widespread adaptation in Mediterranean regions is the sclerophyllous leaf type. This term refers to the small, thick-walled, rigid leaf cells which result from a build-up of sclerenchyma tissue around cell walls. The leaves do not bend easily or flutter, which could lead to water loss, and they are usually leathery and shiny. Many sclerophyllous plants (e.g. evergreen oaks, rosemary, thyme, erica) will transpire actively when water is available but will close their stomata during water stress to prevent transpiration. However, if stomata are closed for long periods, photosynthesis will be reduced and the plants be slow growers.

Distribution of the plant communities

The overriding characteristic of the Mediterranean is sparse, scattered remnants of the natural oak forest surviving amid widespread areas of shrub (*maquis*) and heath (*garrigue*) communities. The forest has the ecological status of climatic climax; the lower shrub and heath are plagio-climaxes after centuries of human impact (Plate 26.3). However, the visitor to the region cannot fail to be impressed by the many variations in plant types and cover which will be observed. For an explanation of such spatial variability reference should again be made to Figure 20.2, which shows the effects of regional and local factors in their influence on vegetation.

Plant cover responds most clearly to total rainfall, which will determine primary productivity. In coastal areas there are variations in rainfall due to regional

Plate 26.3 *Clearance of natural vegetation continues apace. Land similar to that in the background in being terraced in preparation for the planting of avocado orchards.*
Photo: Ken Atkinson.

position and local 'rain shadow' effects. Thus along the southern and eastern coast of Spain rainfall declines eastwards from Malaga (447 mm) to Cabo de Gata (122 mm) and increases again northwards (Valencia 472 mm). This trend reflects both the increasing protection from onshore cyclonic storms in winter, and the 'rain shadow' of the Sierra Nevada and the Spanish Meseta. Therefore the distribution of shrubland, heath, steppe and semi-desert is climatically controlled in a regional sense. As mountainous terrain is common in Mediterranean countries and islands, these trends are significant.

Increased elevation as one travels inland from the coast leads also to lower temperatures, reduced evapotranspiration rates and increased moisture effectiveness. The zonation of Mediterranean vegetation with altitude, because of these changes in climate, has been one of the most studied features of the ecosystem. Figure 26.10 shows the altitudinal gradients which are common to many upland massifs. Below 750 m the typical Mediterranean tree species will be found on land not cleared for cultivation.

Also in this zone the commonest Mediterranean conifer (Aleppo pine) will be planted extensively in afforestation schemes. Between 750 m and 1500 m deciduous trees replace evergreen, with chestnut occurring in the lowest sub-zone, and being replaced at higher elevations by deciduous oak, beech and elm. Between 1500 and 2000 m evergreen conifers become dominant, chiefly pine, silver fir, cypress and cedar. Above treeline at about 2000 m the sub-alpine and alpine zones consist of heaths and cushion vegetation. One further important influence which mountains have on the flora is to provide refugia which plants have been able to retreat to during glacial periods and have there evolved in isolation, giving rise to many endemic species. Shrub and ground flora are much richer on Mediterranean mountains than on the coasts because of the higher soil moisture contents. It is among these plants that endemism reaches high levels.

In addition to elevation, but equally obvious to the local observer, is the influence of aspect, which has an important effect on surface microclimates and

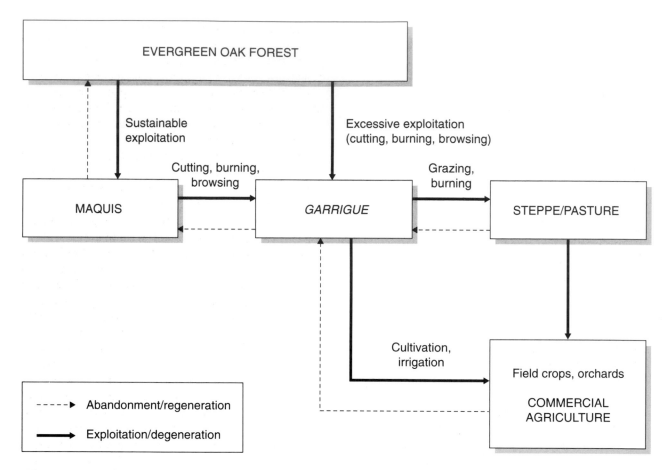

Figure 26.9   *The ecological dynamics of Mediterranean plant communities.*

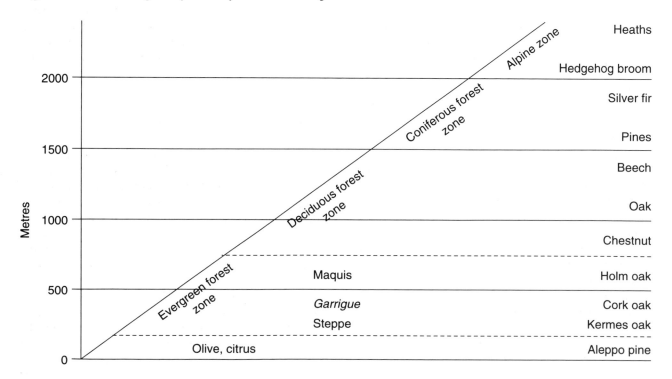

Figure 26.10   *The zonation of Mediterranean plant communities with altitude.*

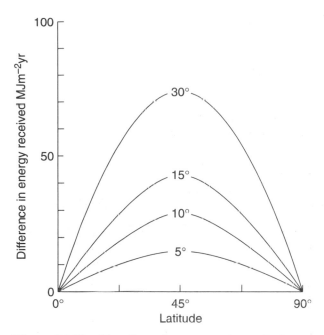

Figure 26.11 *The effects of aspect on solar radiation received at the ground surface, and the secondary effects of slope angle.*

vegetation growth, owing to the annual radiation budgets. It has long been recognized that aspect (compass bearing with regard to north and south) and topographical position (location on the slope profile) interact to influence air temperature, soil temperature and soil moisture. In mid latitudes the north and south-facing slopes show strong asymmetry in total annual incoming radiation. Figure 26.11 shows the effects of latitude and slope on the total amounts of short-wave radiation received by north and south-facing slopes. Assuming clear skies and no shading effects, the differences are greatest in middle latitudes and least in tropical and polar regions.

In reality, of course, a complication can be introduced at the bottom of deep valleys, where there may also be an effect due to shading by surrounding hills. All these influences can be studied by recording in the field the three parameters of aspect, gradient and angles to the surrounding horizon. These parameters can be used either in nomograms or in computer programs to calculate total annual and daily incoming radiation, including shading effects. This calculation allows a long list of ecologically important variables to be estimated: potential evapotranspiration, actual evapotranspiration, gross productivity, soil moisture content. The control of incoming radiation on plant communities is broadly viewed as the dual effect of temperature and moisture: a direct temperature effect on the plant via ambient temperature and biochemical reaction rates, and an indirect effect on the soil water

content via evaporation and transpiration rates. The two effects are well integrated in the real world and it is often difficult to disentangle them. However, research indicates that temperature has a greater influence on floristics (i.e., the plant species on different aspects) whilst soil moisture governs biomass of vegetation and percentage cover (i.e., the productivity of the plant community).

Other local factors which affect the distribution of plant species in the Mediterranean region are slopes and soils. Well drained, arid valley sides contrast with valley floors, where perennial or ephemeral streams allow denser riparian vegetation to grow. Valley bottoms are potentially hazardous habitats for plants because of flooding, sedimentation and abrasion but there are robust plants which can tolerate such conditions (tamarisk, retama broom, oleander). The effect of soil on vegetation distribution is largely a matter of the influence of soil on the amount of available moisture stored in the profile (AWHC: the available water-holding capacity). This brings out the expected contrasts between thin, rocky soils and deeper profiles. At the species level, though, there can be contrasts in the plants of basic soils (e.g. calcicole plants on calcareous soils over limestone) and acid soils (e.g. calcifuge plants on non calcareous soils over sandstones). Thus particular species of lavender, oak and rock rose, for example, are indicators of basic and acid soil conditions. Saline soils are characteristic of coastal salt marshes, where halophytic vegetation (saltwort, glasswort) will be found. As soil leaching is at low levels, geological substrates which contain salt and/or gypsum give rise to distinctive tolerant plants.

Fire in the landscape

Every summer newspaper headlines proclaim the ravages caused by fires in the Mediterranean regions, whether in Europe, California or Australia. Losses of property, life and vegetation seem to mount year by year; large conflagrations in the autumn of 1993 in California were followed by great mudflows in the spring of 1994, a sign that burning had accelerated erosion. Similarly in the early 1990s a series of extensive forest fires raged throughout Spain, Italy and Greece. The public and politicians link fires with land degradation, desertification and increased drought due to global warming. Whatever the scientific reality, public and politicians certainly rate fires as a major natural hazard, and a hazard which costs billions a year. Although the negative impacts of forest fires have probably been exaggerated, it

Plate 26.4 *Fire devastates thousands of hectares of countryside in Mediterranean regions each year. This fireburn has removed vegetation from a hillside on the French Riviera.*
Photo: Peter Smithson.

remains a significant environmental problem in Mediterranean regions. (Plate 26.4.)

Fire has been a common experience throughout the Quaternary era (i.e. since 2 million years BP) in the Mediterranean, and indeed many plants have had time to adapt to it. Adaptation can take the form of thick, protective bark (e.g. the cork oak, *Quercus suber*), or rapid regeneration after fire (e.g. esparto grass, *Stipa tenacissima*, dwarf fan palm, *Chamaerops humilis*, and many others), or by plants killed by fire but which regenerate quickly from seed (e.g. Aleppo pine, *Pinus halepensis*, and rock roses, *Cistus*). Owing to high resin and oil contents, many plants are very flammable, and after the hot, dry summer the plant material and dry litter can burn fiercely to reach 800° C. The hazard increases with the age of the vegetation, and *maquis* over thirty years old is very combustible. This gives rise to the paradox that the suppression of fire leads to greater hazard, as large fires every thirty years cause more damage and are more dangerous than smaller fires every ten years.

It is difficult to get a statistical picture of fires in the Mediterranean, as there is probably much under-reporting and also probably some confusion with practices such as the stubble burning of crops. There are some figures from Sardinia which suggest that 3 per cent of the *maquis* and 1 per cent of the woods are burned every year. Other figures from France suggest there may be 7000 fires per year, the majority of them are 'cause unknown'. Fires can be classified into **natural** (mostly due to lightning), **occupational** (caused by grazers and farmers in order to clear and control vegetation) and **wildfires** (resulting from accidents or deliberate firing). Statistics in the past fifty years point to increased frequency of fire but could easily reflect better reporting. In the early years only large fires were worth recording. If there has been a real change, it is probably the result of increasing tourism and rural recreation, with obvious increased risks of accidental firing. Suppression of small fires could also allow the build-up of biomass and litter which would be more susceptible to lightning strikes.

Land-use changes which cause the build-up of fuel could also contribute to increasing fire hazard. In those areas where agricultural land has been abandoned, reinvasion by scrub will provide more flammable biomass, as will any lessening of woodcutting, animal grazing or animal browsing. Modern forestry practices in Mediterranean regions also add to the risk. Foresters have mostly planted fire-promoting trees of the pine and eucalyptus genera, for example Monterey pine (*Pinus radiata*) in California, and Aleppo pine (*P. halepensis*) and maritime pine (*P. pinaster*) around the Mediterranean itself. In Spain, for example, 85 per cent of planting since 1950 has been of coniferous species, 13 per cent of eucalyptus, with only 2 per cent native evergreen oak (mostly cork oak). Although foresters deplore burning by grazers trying to increase grazing potential by stimulating grasses and herbs, one estimate is that a third of all fires occur in Aleppo pine plantations around the Mediterranean. It is also true that the expansion of settlement into the countryside (holiday complexes, camping facilities, suburban dwellings) greatly increases the cost and danger of fire damage.

Fire has important effects on the ecology. Underground plants emerge and bloom after a fire (e.g. squill), and the following moist season usually sees a flourishing of annual and perennial grasses and herbs which thrive on the injection of light, moisture and nutrients. Bulbs and tuberous herbs are usually prominent. *Maquis* recovers quickly, and kermes oak and strawberry tree can reach 1 m in height after two years. The conifers (e.g. pines and junipers) are killed by fire, but the pines recover quickly through seeds released from cones on burnt trees. Oaks are usually burnt back but not killed. The effects of fire are thus to maintain mixed communities (e.g. mixed pine–oak woodland rather than a monoculture of one species) of greater richness and diversity. In that sense, fire stimulates ecological processes, nutrient cycling and that vigorous regrowth of vegetation. It causes temporary bare land and therefore probably more erosion, but the adverse effects are short-lived. Overall, fire is a natural part of many Mediterranean ecosystems, stimulating productivity and diversity. It is doubtful whether fire causes serious land degradation or erosion, as its impacts are mostly temporary rather than permanent.

Desertification and soil erosion

The processes which lead to the erosion and degradation of the land surfaces of arid, semi-arid and sub-humid areas are collectively known as **desertification**. The term refers to adverse impacts on all ecosystem components (soil, water, vegetation, wildlife) and the process is considered to be due to human impacts on land use. Although normally associated with desert fringes (e.g. the Sahel on the southern margins of the Sahara), Mediterranean regions are highly susceptible to desertification processes because of the properties of their climate, soils and vegetation. Against the backcloth of an inherently sensitive physical environment, there has been a long history of often quite drastic human modification for agriculture, water supplies, urban and residential development, and tourism. These impacts have gone on for millennia in the Mediterranean region itself, and for several centuries in other Mediterranean areas.

The slopes of Mediterranean landscapes are rarely covered 100 per cent by plants; thus bare soil is open to erosion processes and the production of loose sediment, which is easily carried downslope. Plant cover is the major control of run-off and sediment yield, although another important variable is the intensity of rainfall, with low-intensity rain producing much smaller amounts of sediment and run-off. A vegetation cover of about 50 per cent seems to mark an approximate threshold between extreme and low erosion rates (Figure 26.12). Cover seems to be more important than biomass (size of vegetation), so that bushes and low clumps of plants are just as effective

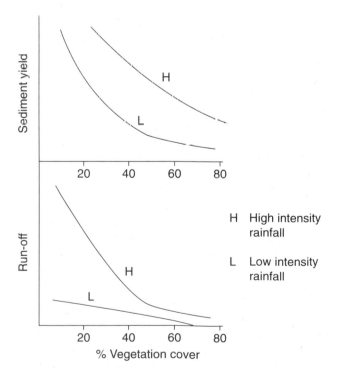

Figure 26.12 *The effects of vegetation cover on soil erosion (sediment yield and run-off) under low and high-intensity rainfall in the Mediterranean environment.*

Plate 26.5 *One end point of desertification is severe soil erosion. These 'organ pipe' badlands are formed by soil wash and gullying. The calcareous silty clays (marls) are coherent when dry, thus giving vertical faces. Photo: Ken Atkinson.*

as trees. The main effect of the vegetation, of course, is to break the fall of raindrops, thus preventing both rain splash and surface sealing. Surface crusts and seals are a common feature of soils in both arid and semi-arid regions. The force of the raindrop breaks the aggregates at the soil surface into loose mineral particles of sand, silt and clay, which on drying become repacked into a dense and impermeable surface 'skin'. If there are free chemicals such as salts, lime or gypsum in the soil, they provide a chemical cement which is extremely tough.

There are several additional influences of the Mediterranean vegetation on erosion rates. It is common to see shrubs growing on low mounds of soil. It was previously argued that, in addition to protection from rain splash, the roots have a binding effect, holding the soil against erosive forces. Whilst this will happen, there are additional processes at work. Any soil grains detached by splash may become 'caught' by the leaves of the plant, and will thus contribute to the mound. Plant litter also adds to the soil organic matter around plants, improving infil-

tration by producing a better formed and more stable soil structure. More infiltration leads to less overland flow and less erosion; there also comes a time when the mound can divert any flowing water around it. A clear positive feedback system thus exists between the plants and the mounds. Litter production and lower erosion rates create better soil structures and soil moisture conditions (more infiltration) as well as providing more plant nutrients (especially nitrogen). They also provide more attractive, shady habitats for soil fauna (worms, isopods), whose burrowing causes further mounding.

The pattern of erosion at the micro-scale is one of low-energy, minimum erosion beneath plants, and high-energy, maximum erosion on bare ground between vegetation. This simple model of the inter-actions between vegetation cover and erosion is complicated in the real world by other factors such as rock type, aspect, the nature of grazing by animals and the extent of burnt areas. The importance of rock type is illustrated by the 'badland' scenery which, though spectacular and seemingly important,

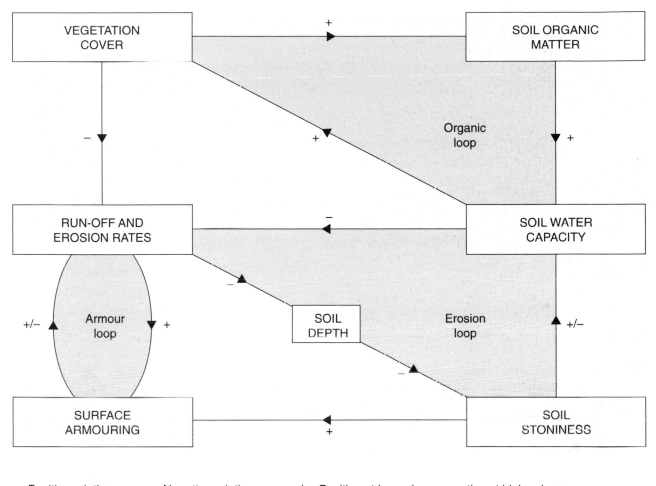

+ Positive relation   − Negative relation   +/− Positive at low values, negative at high values

Figure 26.13 *The Medalus desertification model. Source: After Kirkby et al. (1997).*

is localized and restricted to soft rocks such as marls and clays. Low infiltration rates lead to a dense system of gullies which forms an intensive network of channels and valleys with bare and steep slopes. The gullies are typically V-shaped, and surface-water erosion is obviously important. Additionally, subsurface erosion is caused by hydraulic gradients on steep slopes and the chemical solution of salts and gypsum in the marls, leading to underground pipe erosion. This piping produces tunnels which collapse in time and further concentrate gully erosion. Mass movement processes also contribute to the chaotic landscape of badlands. If the marls and clays are able to develop vertical cracks, the channels will take on a rectangular cross-section, with vertical walls and a flat valley floor, which get larger through the collapse along the cracks of walls of marl. Badlands give the appearance of extremely rapid rates of erosion, with fresh gullies and actively eroding slopes. (Plate 26.5.) However, whilst this may be true of some sites – for

example, where the badlands are eroding abandoned agricultural terraces – in other areas erosion seems to be very slow. There are badland sites in southern Spain where archaeological structures 4000 years old appear to have suffered little erosion.

Soils in Mediterranean areas often have a high stone content, whether due to stony colluvial and alluvial deposits or to their shallowness over bedrock. Whilst the bare surfaces of fine-grained soils can develop crusts with low infiltration rates, a layer of stones can protect the surface of stony soils from direct raindrop impact. The process of stones being concentrated at the surface by the erosion of finer particles is called **armouring**. Once armouring has developed, it reduces soil erosion by increasing infiltration and making the surface material harder to transport. The process of armouring is an important subsystem in the complex system which governs Mediterranean desertification. Figure 26.13 shows the Medalus desertification model, recently developed as a process-based model to simulate erosion in Mediterranean regions. Three loops provide feedback in the system. The organic loop controls vegetation cover; reduced vegetation cover leads to reduced soil organic matter, which leads to reduced soil water storage, which leads to reduced productivity. In the erosion loop an increase in stoniness due to an increase in erosion leads to increased soil moisture and increased vegetation. However, above a threshold of 50–70 per cent stones, soils have reduced water storage owing to the decrease in fines, and an increase in erosion thereafter completely erodes the soil, entailing irreversible degradation. The armour loop has positive feedback at low armour levels, owing to increased overload flow and roughness, but above 30–50 per cent armouring the armour reduces erosion rates.

Water supply problems: quantity and quality

During the past five decades three separate economic revolutions have impacted upon the Mediterranean region. First, mass tourism now increases the resident population by several factors during the critical summer months. Second, urbanization has caused an increase in the size of urban centres around the entire basin. Third, an agricultural revolution has caused a drastic transformation of the rural landscape, with greatly increased use of irrigation for high-value vegetable and orchard crops. These trends have been most striking in California and the Mediterranean region itself, but they are equally evident in the three Mediterranean areas of the southern hemisphere.

All these activities can be viable only if sufficient water is available to meet the increasing domestic, industrial and agricultural demand. These demands vary enormously between the different activities, from the low demands for drinking (potable) water, through the medium demands of industry, to the very high demands of agricultural irrigation. Table 26.1 shows the water requirements for different purposes; as a rough rule of thumb, it is said that an inhabitant of a Mediterranean area consumes ten times as much water through products and services as in food and drink. Figure 26.14 compares the water consumption of Spain, France, Greece and Italy.

With precipitation so seasonal, the value of groundwater supplies cannot be overestimated. The volume of water stored in rocks depends on their percentage of empty spaces, i.e. the **porosity**. Not all groundwater is available, however, and the proportion that can drain under gravity is called the *specific yield*. Groundwater is not stationary but flows through the rock. The rate of flow depends upon the porosity and the degree to which the parts are interconnected, i.e. the **permeability**. Rocks which are both porous and permeable, i.e. can both store water and allow water to flow through them, are called **aquifers**. Alluvial aquifers consisting of geologically young alluvial, terrace or fan deposits are important aquifers and follow all river valleys and deltas. Sedimentary rocks such as sandstone and limestone tend to have smaller pores, but fracturing and fissuring can contribute greatly to specific yield. Permeable basalt can be an important aquifer locally, but other igneous and metamorphic rocks such as granites, schists and gneisses may have no primary porosity but depend upon weathering to enlarge joints to provide some secondary porosity and permeability. Table 26.2 summarizes the porosity and permeability values of typical Mediterranean aquifers.

In cool temperate regions, groundwater and surface water sources both contribute to society's demand for water. Although the precise balance will vary, on average each is contributing about 50 per cent to supplies for drinking, for industry and for agriculture. Groundwater contributes also to the

Table 26.1 *Water demands.*

Source of demand	Tons of water needed per ton of produce/tissue
Domesticated animals/ humans	1
Industrial	
Paper	250
Nitrogen	600
Agricultural	
Sugar cane	1000
Wheat	1500
Rice	4000
Cotton	10000

Table 26.2 *Principal Mediterranean aquifers.*

Type	Porosity (%)	Permeability (m day^{-1})
Shallow alluvium	30–40	10–1000
Sandstone	10–30	0.1–10
Limestone	5–30	0.1–50
Karstic limestone	5–25	100–10000
Basalt	2–15	0.1–1000
Fresh igneous/metamorphic	2	0.000001
Weathered igneous/ metamorphic	10–20	0.1–2

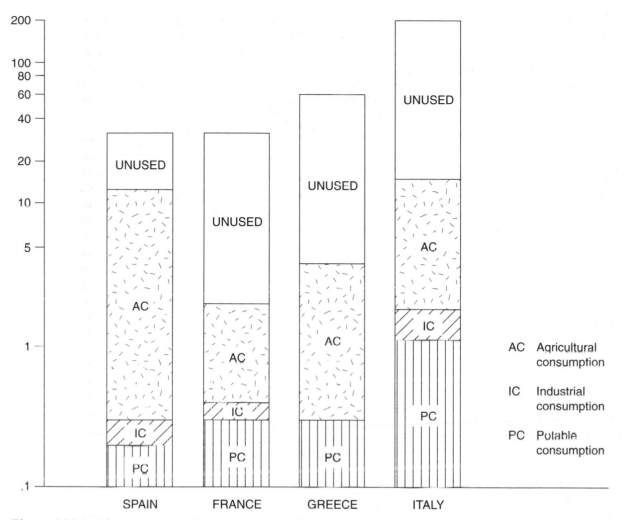

Figure 26.14 *The consumption of water in some Mediterranean countries.*

surface flow of streams, because they are partially fed by drainage from aquifers. Rivers also increase in discharge from source to mouth as they receive more surface run-off downstream.

In sub-humid Mediterranean environments, however, the hydrological cycle is less continuous. Higher rates of surface evaporation mean that smaller amounts of water are available for underground storage. As illustrated in Figure 26.15a, low rainfall is subject to loss by high rates of evaporation and evapotranspiration. The remainder either runs off the surface, evaporates at the surface or infiltrates locally. There is some groundwater flow, and in coastal areas fresh groundwater is in equilibrium pressure with saline groundwater.

Pumping of the groundwater resource has drastic consequences (Figure 26.15b). The groundwater level is lowered as surface run-off is reduced, giving reduced infiltration and recharge. Under natural conditions the fresh water–saline water interface is at

an equilibrium position; sea water extends under the land not at sea level but at a depth below sea level equal to about forty times the height of the fresh-water table above sea level. Extracting groundwater from coastal aquifers therefore not only lowers water tables but also causes an intrusion of sea water inland at depth. This extension of saline groundwater further into the aquifer creates a serious problem, and causes salt water to appear in wells, degrading crops and soil over large areas. Furthermore the recycling of irrigation water through salty soils continually increases the amount of dissolved solids in the groundwater.

There are several important aquifers around the Mediterranean, and many regions are dependent upon them for their supplies of drinking water, and supplies for industry and irrigation. Table 26.3 shows the importance of groundwater as part of the total water resources of the Mediterranean part of four countries. Spain stands out as a country where the

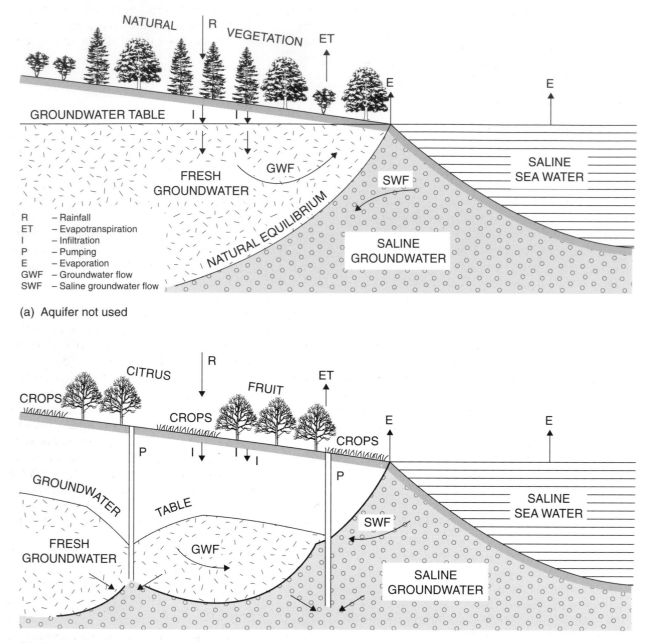

Figure 26.15 *The hydrological cycle in a coastal aquifer in the Mediterranean zone: (a) a non-exploited aquifer, (b) an aquifer exploited for irrigation, industrial and domestic uses.*

annual consumption of water is higher than the total groundwater resource. Increased pumping from an aquifer will cause a lowering of the groundwater level around the point of abstraction, to cause a 'cone of dejection' in the water table. Upconing of saline water occurs below bore-holes, owing to pumping. In the worst cases of massive overpumping from a large number of wells a regional reversal of hydraulic gradients takes place. The process is slow but virtually irreversible, and once an aquifer has been invaded

by saline water it is extremely difficult to restore the quality of fresh water. A salt-water aquifer has no value for irrigation or potable supplies. Serious intrusion has been reported from all along the Spanish Mediterranean coastline (especially the Campo de Dalias in Almeria), from Italy (Sardinia), from Greece (Argolides), from France (Roussillon and Var), from Lebanon and from Libya (Tripolitania). Control of sea water intrusion can be achieved by a modification of pumping (moving wells inland and/or reducing

Table 26.3 *Water resources and consumption (10^9 m^3 yr^{-1}).*

Country	Total resource	Groundwater resource	Total consumption
Spain	31.0	9.1	12.2
France	74.0	31.0	2.0
Greece	58.6	12.0	3.7
Italy	187.0	30.0	14.8

abstraction) to re-establish a stronger seaward hydraulic gradient. Artificial recharge of the aquifer is possible by constructing recharge barriers to cause surface spreading or by digging recharge wells. These control measures are very costly, however, and in many cases the only option is to cease using the aquifer and to develop entirely new sources. As the lower reaches of rivers became controlled and diked, natural flooding is prevented, and thus a pathway of natural recharge is removed.

The dominant factor behind water problems, however, has been the continual intensification of agriculture in the past four decades. Agriculture has changed from the traditional dry farming of cereals and olives, with some vegetable growing and sheep and goat grazing, to dominance by vegetable and orchard fruit production, particularly citrus. These new crops require irrigation. Agricultural technology has responded by using more and more powerful water pumps – from hand, through animal to diesel and electrically driven pumps; this technology has enabled water to be obtained from depths in excess of 400 m. Deeper and deeper wells are required to keep pace with declining water levels, and there is increasingly a problem with the quality of this deeper water.

Conclusion

The present ecosystems of Mediterranean regions can be regarded as the degraded remnants of a biome which was once dominated by mixed evergreen and deciduous forest. The two reasons for this situation are climate and anthropogenic influence. Mediterranean regions were cool and dry in glacial times (*c.* 15,000 years BP), treeless steppe being characteristic. Temperatures ameliorated in the Holocene to reach a maximum some 5000 years BP. A continuous forest cover of evergreen and deciduous trees had established itself by then.

The Mediterranean landscape is like a palimpsest; it shows the remains of successive human societies superimposed on the landscape and on one another. Those societies, in the Mediterranean region itself, have had successive impacts on the natural landscape from Neolithic settlement, Bronze Age people, Phoenicians, Greeks, Romans and Moors. Modern pressures have continued with agricultural intensification, irrigation and all the trappings of tourism. Tourist villages and golf courses now exist side by side with irrigated citrus plantations and horticulture under polythene sheeting. Increasingly the natural vegetation and soils are restricted, rather as in Britain, to inaccessible 'islands' and unwanted rocky hillsides.

Key points

1 Mediterranean ecosystems are dominated by the seasonality of the Mediterranean climate. Plants and animals have devised many strategies to adapt to, and survive in, the hot, dry summer. Production and reproduction take place during the more humid and cooler period from autumn to spring. This is the time when most weathering, geomorphological processes and soil-forming processes are operative too.

2 Despite the overriding importance of climate, there are variations in vegetation and soil according to regional and local factors. A zonation of ecosystems is found with increasing altitude. There is also a very powerful effect of aspect due to different annual radiation budgets on north and south-facing slopes. There is also a change in vegetation from valley side to valley floor, related to moisture availability.

3 The final set of controls on Mediterranean ecosystems are the many human pressures – hunting, grazing, deforestation, fire and cultivation. The net result of these pressures over millennia has been to reduce the Mediterranean forest to the shrubby and steppe-like vegetation we see today. Soil erosion has been accelerated too, making the depth of soil another key factor influencing vegetation. There are often big contrasts between the rocky outcrops and thin soils of ridges and upper slopes and the deeper soils of valleys and depressions.

Further reading

Fantechi, R., Peter, D., Balabanis, P. and Rubio, J.L. eds (1995) *Desertification in a European Context: Physical and socio-economic aspects*, Luxembourg: Official Publications of the European Union.
A volume containing a series of relevant studies of the European Mediterranean region. Very good accounts of research methods and recent research findings.

Levitt, J. (1980) *Responses of Plants to Environmental Stresses*, second edition, two volumes, London: Academic Press.
An analysis of how plants and ecosystems survive under stressful environments. The section on response to aridity is pertinent to Mediterranean and arid regions.

Polunin, O., and Huxley, A. (1965) *Flowers of the Mediterranean*, London: Chatto and Windus.
The classic guide to the Mediterranean flora. Excellent illustrations, and a useful discussion on the general physical geography of the region.

The dry lands of the world cover a large area of Earth's surface where moisture levels are limiting. In the tropics they extend from the savanna or seasonal forest zone to areas of extreme aridity in the desert cores. In polar regions and parts of continental interiors there are areas that qualify as dry in terms of their mean annual precipitation and the seasonality of water availability. Low temperatures reduce evaporation and so water levels are usually sufficient for some plant growth during the growing season. Clearly there are degrees of dryness that can be used to subdivide this large area. Climatologists have devised indices based on the inputs of precipitation relative to evaporation outputs in order to quantify the degree of dryness. For the purposes of this chapter we will retain the broad definition and consider dry lands as those areas of the world where there is a significant moisture deficit (Figure 27.1)

Climate

The core areas of the dry environments are the subtropical high-pressure zones that also act as the meteorological boundary between the tropical and temperate latitudes. The dominant surface air movement is away from the highs, with the flow being sustained by sinking air from higher levels as part of the Hadley cell of the tropics. Because the air is subsiding it tends to be warm and dry. An inversion of temperature develops near the surface (Figure 9.12) and so the core areas of the highs are generally cloud-free and deficient in rain. Where the highs

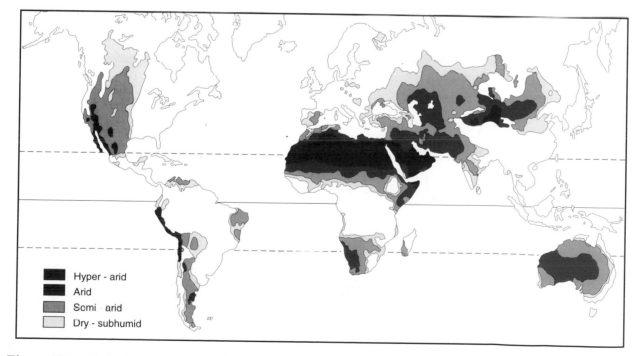

- ■ Hyper - arid
- ■ Arid
- ▨ Semi - arid
- ▫ Dry - subhumid

Figure 27.1 *Dryland areas of the world.*

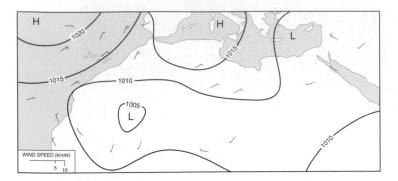

Figure 27.2 *Mean surface pressure over the Sahara in summer.*

remain fairly constant in position we find the main desert areas of the world – the Sahara, the Kalahari and the Great Australian Desert.

If we look at a map of surface pressure, the high-pressure centres that we would expect over the desert areas may be absent, especially in summer (Figure 27.2). On the contrary, there is often a weak low-pressure area. The lows are the result of intense heating of the ground surface during the cloudless days, taking temperatures above 40° C in summer. As they are a product of surface heating, they tend

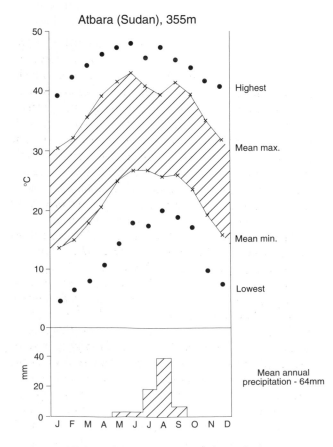

Atbara (Sudan), 355m

Figure 27.3 *Climatic data for Atbara, Sudan.*

to be fairly shallow and are replaced by relatively high pressure at higher levels of the atmosphere.

From what has been said, we would expect the climate of these zones to be characterized by little rain and extremes of temperature. The data for Atbara, Sudan (Figure 27.3), confirm this view. In midsummer, the mean maximum temperature is 42° C but in winter the mean minimum temperature is only 8° C and ground frost can occasionally occur. The very dry atmosphere helps by allowing long-wave radiation from the ground to escape to space with little counter-radiation from water vapour or clouds. Even on the coast at Bahrain (26° N), in the Arabian Gulf, night-time temperatures are cool in winter, though frost is very rare. Precipitation is very low. Rain falls on an average of ten days per year, with a mean annual total of 75 mm. Most of it falls in winter and spring, when temperate-latitude depressions extend their effects far south and do give occasional rain. Farther south, at Atbara (17° N) (Figure 27.3), the rainy season is associated in summer with the northward movement of the equatorial trough. Thus we find different rainfall régimes on opposite sides of the subtropical anticyclones, cool-season rains on the northern limb and hot-season rains on the southern limb. Where amounts are similar, the cool-season rain is more effective, as evaporation will be less at that time of year. As we move away from the central core of high pressure, rainfall totals generally increase; towards the Mediterranean climatic régime in a poleward direction (see Chapter 26) and into the seasonal or monsoonal régimes towards the equator.

Desert rainfall is notoriously unreliable. Several years without rain may be followed by heavy showers giving tens of millimetres. It is this variability that makes the average rainfall figures for desert areas almost meaningless. Annual rainfall totals at Al Wejd in north-west Saudi Arabia, for example, show a mean over a twenty-three-year period of 25.3 mm. Only eight years received more than the mean figure. It has been said that, for deserts, average rainfall is the total that never falls, though that is an exaggeration.

In some of the subtropical high-pressure belts additional factors reduce the likelihood of rain. On the west coast of the Sahara, the Kalahari and the Atacama deserts, cold ocean currents flow offshore. They cool the air and make it even more stable. Mist and fog may be frequent but rain is rare. One of the driest places in the world is Arica in the Atacama desert of Chile. Years have elapsed between rainstorms and even then only a few millimetres may fall. Conditions here are similar to those in other coastal deserts near a cold ocean current but in

Plate 27.1 *Small linear dunes in the Peruvian desert near Ica.*
Photo: Peter Smithson.

addition the prevailing winds blow from the southeast. To reach the Chilean coast, they must descend the main mountain barrier of the Andes, some 5000 m, which further emphasizes stability and dryness. The result of all these factors acting against the mechanisms of rainfall generation is to produce one of the driest parts of the planet (Plate 27.1).

Moving away from the core areas of the drylands, we encounter increased moisture availability. The areas polewards have been included in the chapter on Mediterranean environments (Chapter 26), but equatorwards we gradually change from true desert, through semi-arid environments to the savanna zone with deciduous woodlands and eventually into the monsoonal and tropical rain forests. The tropical rain forest will be discussed in Chapter 28 but the other areas are included here because of the importance of the seasonal dryness.

Desert

Vegetation

Arid and semi-arid land covers almost one-third of the land surface of the globe. Almost 60 per cent of

it is true desert. The remainder varies from steppe grassland to thorny scrub. In all cases, however, potential evapotranspiration greatly exceeds rainfall. In these areas of low and erratic mean annual precipitation, vegetation is sparse and the growing season short. The popular idea of desert areas as vast expanses of barren, shifting sand is false for all but a small part of this biome. Most deserts and semideserts support widespread, relatively sparse vegetation with a distinctive array of wildlife. Over time nature has evolved a great variety of ways of coping with extreme conditions of dryness and heat.

Desert vegetation consists mainly of short perennial grasses and thorny scrub (Plate 27.2). Only in extreme cases, such as rocky hamadas and regs, and the shifting sand dunes and sand seas of the Sahara, is vegetation absent. Even in those areas, locally developed lines of vegetation occur along wadis, with lusher growth around oases. In all cases, plants must be able to survive periods of drought, and thus xerophytic plants predominate. The adaptation of plants to desert conditions varies. For example, the saguaro cactus develops a widely spreading root system; the mesquite has roots that may reach depths of over 50 m; many cacti and agaves store water in their roots, stems and leaves. Some plants reduce water

Plate 27.2   *Desert vegetation in the southern Kalahari. Small amounts of vegetation survive on moisture provided by the alluvial fan emerging from the valley in the background.*
Photo: Peter Smithson.

loss through evaporation by controlling their stomata, while other have long dormant periods, growing and flowering briefly and irregularly when moisture is available (Plate 27.3). Some species of plants avoid excessive exposure to the sun and drying winds by growing largely underground.

Desert soils

The soils associated with desert conditions are typically little weathered, and lacking in humus. In the most extreme cases, no true soil exists, but, even where sufficient plant growth does occur to provide a surface accumulation of plant debris and a food base for soil fauna, the lack of leaching and chemical weathering leaves soils relatively infertile. Salinity may be a problem where the rock type produces saline groundwater, or where salty sea water seeps into aquifers as it does in many coastal areas. Winds blowing from the sea also may introduce salt. Constant evaporation from the surface draws water from the lower layers of the soil and leads to the accumulation of salts in the upper horizons. If the parent material is rich in sodium salts, solonetzic soils may develop. Practically no leaching occurs, so, even though the salt is soluble, it accumulates in the soil.

Nutrient cycles in deserts

Little information is available on which to base a discussion of general nutrient cycles in deserts. Total biomass is small and it is apparent that cycling is slow and involves very small quantities of nutrients. Net annual primary productivity is closely related to rainfall, and probably ranges from about 0.05 t ha⁻¹ to 0.2 t ha⁻¹. About 80 per cent of the organic material in a desert is underground, and cycling mainly occurs through the decay of root material. Leaching losses associated with the chemical oxidation of plant materials can be significant. Soil erosion may also represent a significant loss. Owing to the low rainfall and the limited weathering, the more insoluble compounds are present only in small quantities. Sodium and calcium tend to dominate in desert nutrient budgets (Figure 27.4).

Plate 27.3 *The desert in bloom. Winter rains in Namaqualand, northern Cape Province, South Africa, have helped to germinate the profusion of annual seeds and now carpet the desert with colour.*
Photo: Peter Smithson.

Savanna

Vegetation

As we move away equatorwards from the arid parts of deserts, vegetation becomes more abundant. Initially the grasses become dominant to provide a suitable habitat for grazing animals. Individual, and then small clumps of, trees gradually appear as rainfall levels increase. The trees are usually smaller than similar species in damper conditions and often appear gnarled. Many have developed protective mechanisms to stop grazing animals denuding them. For example, some *Acacia* species possess sharp thorns (Plate 27.4).

Figure 27.4 *Nutrient cycle of a hot desert. Data on such biomes are rare and only very general estimates can be given. Figures in boxes represent the amounts of stored phosphorus, potassium and calcium in kg ha^{-1} yr^{-1}; figures in circles represent annual flows of P, K and Ca in kg ha^{-1} yr^{-1} (T means trace). Dashed lines show minor nutrient flows.*

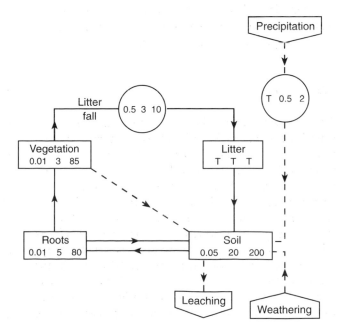

Plate 27.4   *Sharp thorns on the Acacia scrub provide some protection from grazing animals, Transkei, South Africa.*
Photo: Peter Smithson.

Savanna biomes cover approximately 11.6 per cent of the land surface of Earth, including many of the subtropical regions fringing the rain forests of Africa and South America. Although it is often referred to as a grassland biome, the savanna is an open woodland in many cases, with widely spaced and rather scrubby trees.

The formation of savanna is of considerable interest and dispute. Climatic factors alone cannot account for the character of these areas except in the drier parts. Although they experience a distinct dry season during which many plants become dormant, and although precipitation is variable, towards the ecotone with the tropical rain forest it appears that the climate could support a much more luxuriant and diverse flora. One possible reason for the disparity is that the savanna represents a form of *plagioclimax*, one which has been severely curtailed by human activities. Human-induced wildfires, in particular, have played a major part in the development of savanna, and many of the trees are fire-resistant. With the action of fire, and the voracious appetite of termites,

seeds rarely survive. In response, trees produce enormous numbers of seeds each year. *Acacia karoo*, for example, releases as many as 20,000, of which about 90 per cent are typically fertile. Few survive to grow into trees, however, as the scattered nature of the arboreal vegetation shows (Plate 27.5).

Many savanna trees are xerophytes. Their morphological and physiological resistance to water loss, and their ability to maximize the uptake of water, allow them to survive dry periods. They also have deep roots and flattened crowns. Some shed their leaves during the dry season in order to reduce transpiration. Savanna trees are often stunted, and may be overtopped by the tall grasses of this biome. Browsing by the savanna animals is a major constraint on tree growth and survival, and overgrazing is one of the main causes of savanna degradation.

Herbaceous savanna plants are dominated by a few species. African elephant grass is sometimes abundant, and may reach heights of several metres. The density of trees relative to grass is, to some extent, climatically controlled, and trees become

Plate 27.5 *Savanna vegetation and grazing animals, Natal, South Africa.*
Photo: Peter Smithson.

scarcer in the drier margins of the savanna. However, there are often subtle local variations in vegetation related to topography and drainage (Figure 27.5), particularly in the drier areas, where vegetation is attuned to the short growing season, and herb-layer plants grow rapidly once the rains come. Although the soils are relatively dry before the beginning of the wet season, there is no need for the rain to replenish soil moisture before plants can extract the water efficiently. Instead, the plants transpire at their full rate immediately, as much of the rainfall seems to be absorbed by the plants before it can move into the finer pore spaces in the soil.

Soils

The soils of the savanna are variable. They include ferralsols, acrisols, vertisols and luvisols. Their distribution is related to climatic, geological and geomorphological conditions. Slope processes are active, for the vegetation is often insufficient to prevent erosion and downwashing of nutrients. Consequently,

marked catena sequences develop on the hill slopes, grading from shallow stony soils to deeper, less well drained, base-rich alluvial soils (Figure 27.6). The ridge crests (or 'breakaways') are usually formed of

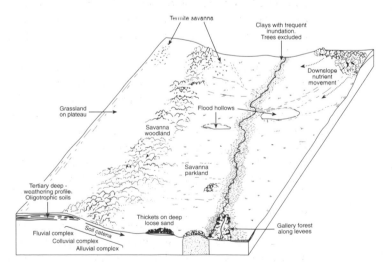

Figure 27.5 *Environmental relationships in a savanna area.*
Source: After Collinson (1977).

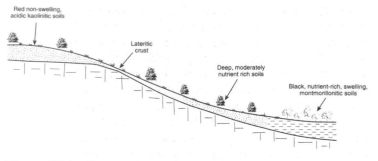

Figure 27.6   *Catena sequence in a savanna area.*

hardened iron oxides (laterite) and indicate old erosion surfaces.

Wildlife

The number of animal species is relatively low in the savannas, though their populations are large. Surprisingly, interspecific competition seems to be limited, and the food chains are short, with few secondary consumers. Most carnivores prey directly on herbivores. For example, lions attack mainly zebra, wildebeest, antelope and giraffe. However, many scavengers and decomposers, including mammals and insects, also feed on the lions' kills. Termites are very abundant, and their mounds are a major feature of the savanna landscape (Plate 27.6). These insects attack and macerate plant debris, making it more readily available for decomposition by other organisms. They also eat growing plants, especially during periods of drought.

Biogeochemical cycles in savanna areas

The nutrient cycles of savanna areas have been much less studied than those of the tropical rain forests which have been discussed in Chapter 22. As expected, net primary productivity is much less than that of the humid tropical forests. An annual total of 2–20 t ha^{-1} is typical, but quantities vary considerably with climate and tree cover.

Nutrient cycling is rapid, owing to the speedy breakdown of organic materials by soil organisms. High temperatures encourage chemical activity, and silica appears to be particularly soluble in these areas, often making up a major part of the total nutrient budget. Indeed, plant uptake of silica is so great that clots of amorphous silica may form in some leaves.

It is also apparent that the nutrients which are returned to the soil are removed less readily by leaching (Figure 27.7). This occurs for two related reasons.

Lower rainfall leads to less intense weathering and leaching, and the soils with clays that have a higher cation exchange capacity form stronger bonds with nutrients. Most nutrient loss in many of these areas results from soil erosion, a major problem facing agriculture in the savanna, where overgrazing and tillage leave the soil unprotected. The sudden, intense rainstorms characteristic of the savanna can result in rain splash and the erosion of unprotected surfaces.

Environmental problems of dry land

Water resources

The dry lands of the world present many problems of development, most stemming from the constraints determined by a climate with a lack of water. Depending upon the degree of aridity, water may be available from:

1 Perennial rivers whose headwaters are in wetter areas, such as the Nile, the Niger or the Tigris.

Plate 27.6   *A termite mound built around a thorn bush, South Africa.*
Photo: D.J. Briggs.

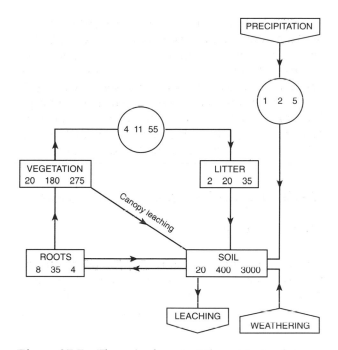

Figure 27.7 *The main elements of the nutrient cycle in a savanna area. Figures in boxes represent amounts of phosphorus, potassium and calcium in kg ha⁻¹; figures in circles represent annual flows of P, K and Ca in kg ha⁻¹ yr⁻¹.*

2 Seasonal rivers which are dammed to provide adequate storage to survive the dry season (and perhaps several drier-than-average years if the dam is large).

3 The diversion of water in aqueducts from wetter areas, such as the Los Angeles aqueduct.

4 Ground water, pumped to the surface artificially, by artesian pressure or through underground tunnels from the water table.

Let us have a look at these elements in turn to examine how the varying availability of water can affect the nature of water resources and their exploitation in an area.

Rivers

Many rivers begin their course in relatively wet mountain areas which provide sufficient discharge for them to flow into dryland areas. Not all of them reach the sea, as natural loss or human exploitation depletes the flow until it disappears into its dry river gravels or swamps, like the Okavango in Botswana, or an internal lake basin such as the Jordan in Israel.

The use of rivers as a source of water goes back to biblical times or even earlier. Systems appear to have been developed independently in at least three locations – south-west Asia by 5500 BC, Peru by

1200 BC and China in 350 BC. The amount of water in the river would determine how much could be used for agriculture. Variations in levels would be reflected in variations in crop yield. Figure 27.8 shows how this relationship was appreciated by the ancient Egyptians. If the river flood level was at twelve ells there was insufficient water for crops, and hunger or famine prevailed. As water levels rose the harvest was likely to improve, so that with a sixteen-ell flood there was abundance. However, the diagram also demonstrates the delicate balance between too little and too much water. With flood levels above eighteen ells the river would burst its banks and cause disaster.

The Egyptians were relatively lucky in that they utilized a perennially flowing river, even if its level did fluctuate. In some locations with seasonal rivers the annual flooding might not reach the lower parts of the course every year, and so irrigation potential was more limited.

The methods of extracting water from the river varied. As the river level would normally be lower than the land on which the crops were being grown, there had to be some method of lifting water. It might not be necessary in mountainous areas where water could be extracted at higher levels, then allowed to flow naturally down channels to the

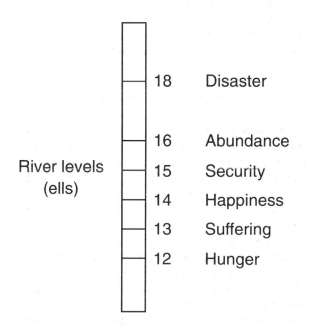

Figure 27.8 *The Nile gauge, or nileometer. The gauge measured the height of the annual flood and indicated the probable impact of the various levels on Egyptian society. One ell = 45 in., or 1.1 m.*

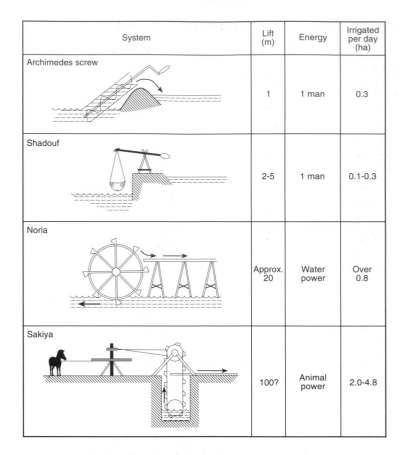

System	Lift (m)	Energy	Irrigated per day (ha)
Archimedes screw	1	1 man	0.3
Shadouf	2-5	1 man	0.1-0.3
Noria	Approx. 20	Water power	Over 0.8
Sakiya	100?	Animal power	2.0-4.8

Figure 27.9 *Traditional water-lifting systems. For each method the lift in metres, energy source and approximate area irrigated per day in hectares are indicated. Source: After Heathcote (1983).*

irrigated areas, but in lowland areas like Egypt, Iraq and Niger devices to lift water had to be invented.

There are four traditional methods of lifting water (Figure 27.9). Although useful, they are still small in comparison with the amount of water and the height of lift which can be achieved by a diesel engine. On the other hand they are simple to operate, cheap to run and require little maintenance – important considerations in countries where technical skills are scarce.

Dams

In the above examples of water use the supply of water was dependent upon the river level; levels too high or too low could cause disaster for different reasons. The ideal would be control of the water level through damming so that it became more stable and the annual fluctuations were removed. During wet years surplus water could be stored until the capacity of the dam was reached and in dry years the surplus could be drained off to sustain the river at the optimum level.

Dams have been constructed for at least 5000 years in order to control river flow. Recently dams have acquired a dual purpose, with power generation being an important feature. As technology has improved so the scale of dam building has increased and now there are many examples of vast lakes impounded by large dams. The Volta dam in Ghana can store water with an area the size of Lebanon and Lake Kariba on the Zambezi can cover over half a million hectares (Table 27.1). These artificial lakes have become a feature of Earth's surface. There is no disputing that the control of river systems by dams can have a positive effect on the local area and the national economy. If we take Egypt as an example, we see that construction of the Aswan High Dam in 1970 has stabilized the flow of the Nile (Figure 27.10). It has also allowed the generation of about 20 per cent of Egypt's electricity, which saves on import costs of fossil fuel. The area of perennially irrigated cropland has been increased, which is of vital importance in a country with high natural increases of population. Less obviously the control of water level on the Nile has had a beneficial impact on tourism. Tourists can visit the Nile temples more easily with improved navigation and assist the country's economy.

Unfortunately not all the results of big dam construction have been benefits. There have been a number of disadvantages, some natural, some unexpected and some induced through subsequent human activities.

Siltation: Like all natural lakes, reservoirs are prone to silting. Dams built in dryland environments are likely to have large areas of partial vegetation cover, especially during the dry season. When the rains start, or floods occur, there is likely to be major movement of sediment which would normally be transported in the river. When a river flows into a new lake, its velocity will fall and so the suspended sediments will be deposited in the lake. Some of the dams on the Huang Ho in China have the unenviable global reputation of being filled most rapidly by sediment. The catchment area of the Huang Ho includes the loess plateau of western China, which is easily eroded during the rainy season. The river used to be known as the Yellow River because this sediment load coloured the water. The Sanmenxia dam on the Huang Ho began impounding water in 1960 but within seven and a half years the reservoir had lost 35 per cent of its capacity; it was estimated that 3391 million m^3 had been deposited in that time! Eventually the lake will be filled with sediment

Table 27.1 *Hydropower generated per hectare inundated, and number of people displaced for selected big dam projects.*

Project and country	Approx. rated capacity (MW)	Normal area of reservoir (ha)	Kilowatts per hectare	People relocated
Pehuenche (Chile)	500	400	1250	
Guavio (Colombia)	1600	1500	1067	
Itaipu (Brazil and Paraguay)	12600	135000	93	8000 families
Sayanogorsk (Russia)	6400	80000	80	
Churchill Falls (Canada)	5225	66500	79	
Tarbela (Pakistan)	1750	24300	72	86000
Grand Coulee (USA)	2025	32400	63	
Tucurui (Brazil)	6480	216000	30	30000
Keban (Turkey)	1360	67500	20	30000
Three Gorges (China)*	13000	110000	12	1200000
Batang Ai (Sarawak)	92	8500	11	3000
Cahora Bassa (Mozambique)	2075	266000	8	25000
Aswan High Dam (Egypt)	2100	400000	5	100000
BHA (Panama)	150	35000	4	4000
Kariba (Zimbabwe/Zambia)	1500	510000	3	50000
Volta or Akosombo (Ghana)	833	848200	0.9	80000
Brokopondo (Surinam)	30	150000	0.2	5000

* Under construction; 50,000 people already relocated.
Source: After Middleton (1995).

and the value of the dam will disappear unless material is removed. All dams suffer this problem to varying degrees. The life of the dam is also important in economic terms. If power is generated only for, say forty years, instead of the predicted 100 years, it will be much more expensive per kilowatt of electricity generated.

Evaporation: When water is stored at the surface it will evaporate. In dryland areas this is a particular problem, as rates of evaporation are high because of the dry atmosphere, high inputs of energy and temperature levels. In Lake Nasser, behind the Aswan High dam, it is estimated, about 10 billion m^3 of water are lost each year through evaporation.

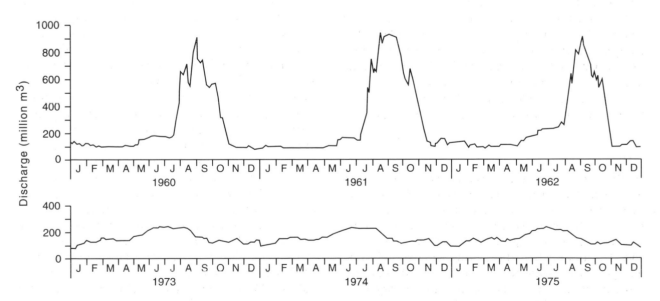

Figure 27.10 *Daily discharge régime of the river Nile at Aswan before and after construction of the High Dam.*
Source: After Beaumont et al. (1988).

Although much of this evaporated water may have previously run off direct to the sea, it is nevertheless an important loss of water resources. Unfortunately there is little that can be done to reduce it. The water surface can be covered with chemicals or artificial skins but that would be uneconomic for large lakes.

Environmental and ecological changes: The removal of sediment from the river into the lake means that the water released from the dam is relatively sediment-free. As we saw in Chapter 14, if a river loses sediment it has more energy available for erosion. Construction of the Danjiangkou dam on a tributary of the Yangtze led to degradation of the river bed and banks up to 500 km below the dam. Similar effects have been noticed on the Nile. The lack of annual sediment accumulation has caused rapid retreat of the delta coastline, with consequences for the coastal fishing industry. The loss of silt reduces the natural fertilizing effect of flooding, so to sustain yields, artificial fertilizers have to be used.

Water tables are affected by dam construction, leading to waterlogging of the immediate surrounds and the potential for salinization if there is insufficient downward movement of water. Along the coast, reduced freshwater flow can lead to an incursion of salt water into the water table.

The new lake will cover and destroy all existing vegetation. Trees may be left to decay, habitats will be lost. Such environmental impact is increasingly causing concern over major dam construction such as the Pergau dam in Malaysia and the Three Gorges project in China.

An interesting biological consequence of the increase of freshwater surface has been the encroachment of water weeds. The water hyacinth is a major problem. Within two years of construction, 50 per cent of the surface area of a lake in Surinam (South America) was covered by this plant, and Lake Kariba has experienced water-fern encroachment.

Pests and diseases: Bodies of still water provide an attractive environment for many pests and diseases in dryland areas. Malaria can be an increased problem, though in smaller dams changing water levels may strand larvae. Bilharzia is another water-related disease which has increased near major dam projects. People contract the disease through bathing, fishing, washing clothes or collecting water from areas infected by the parasitic larvae. After the Volta dam in Ghana was constructed the incidence of bilharzia in children under ten had risen to 90 per cent. Positive effects can occur. River blindness

is caused by a fly which breeds in fast-flowing sections of rivers. As some of these habitats have disappeared under the reservoirs, so the incidence of the disease has declined.

Management: Many dam schemes have been less successful than expected because of poor management and maintenance following construction. Frequently too much attention is paid to the design and construction of the project, little training being given to sustain the management of the water. Bureaucracy and complex administration can exacerbate the difficulties.

Major problems may arise where water released from dams flows into other countries. Although some co-operation may occur over the supply and use of water resources the country possessing the dam will have most control over how much water is released and when. Turkey has built a number of dams in the headwaters of the Euphrates, which eventually flows through Syria and into Iraq. The South-eastern Anatolian Project could reduce the flow of the Euphrates by as much as 60 per cent. This would severely jeopardize Syrian and Iraqi agriculture downstream. Some countries have reached agreement over shared water resources. Mexico and the United States agreed to an equal allocation of the annual average flow of the Rio Grande.

Resettlement: As well as environmental consequences, the construction of a large reservoir will also have settlement implications. Few areas of the world are unpopulated, so inundation may require considerable movement of population (Table 27.1). The Three Gorges project on the Yangtze may involve the displacement of up to 1.2 million people. In many projects, resettlement has caused much hardship. The movement of 57,000 Tongans from the area of the Kariba dam caused a major culture shock as they were moved to a very different community and environment. Experience developed through time was no longer relevant to the new area and food supplies became a problem. Often little is heard of these problems, as governments are keen to publicize the positive aspects of the project. One of the few examples of publicity was the relocation of the Egyptian temples to prevent their disappearance into Lake Nasser.

Aqueducts

Where no rivers are suitable it is possible for water to be diverted into artificial channels often called aqueducts. The Romans were great builders of aqueducts to transport water from source areas to drier locations but many parts of the world still use similar methods. At its simplest an aqueduct consists of a channel allowing water to flow by gravity to where it is needed. On Madeira and parts of the Canary Islands rock and concrete-lined channels carry water from the wetter parts of the mountains to the agricultural areas near the coast. More sophisticated aqueducts carry water in pipes and pumping may be required to enable it to cross a watershed.

Los Angeles, situated in a dry part of southern California, depends heavily on water brought by an aqueduct from the Owens valley in the mountains and from the Colorado valley. The former flows by gravity for 360 km but the latter has to be pumped along parts of this 390 km long system. As a result, costs are much higher. Los Angeles competes with the state of Arizona for the water available from the Colorado river. Other schemes now link the Colorado aqueduct with the Sacramento valley in central California. As more pumping is required to extract water from longer distances beyond the natural watershed of the Los Angeles basin so the cost increases. A cost of $20 per acre ft in the Sacramento valley rises to almost $300 per acre ft by the time the water has been pumped over the watershed to Los Angeles. Nevertheless, where water commands a high price, as drinking water, aqueducts are an efficient way of carrying it from surplus to deficit areas.

Irrigation

The usual reason for developing the water resources of an area is to provide water for irrigation. The need for irrigation may be to compensate for rainfall variability or it may be to provide a regular water supply when and where rainfall is low. Whatever the reason for the provision of water, irrigation allows crops to be produced whenever temperature conditions allow.

The amount of water required by an irrigation system will depend upon many factors, such as the type and stage of development of the crop, temperature and rainfall levels and the nature of the soil. For the most efficient use of the water these factors have to be taken into account; too much water or too little water does not produce maximum yields.

The methods of application of the irrigation water vary widely. In developing countries surface methods are most common. These can range from simple traditional 'flow diversion' techniques to large and sophisticated 'centre pivot' schemes (Plate 27.7).

Unfortunately surface and sprinkler irrigation methods in the tropics lead to large evaporation losses. To reduce them, attempts have been made to use underground water transport in plastic tubing, but pumping is required to maintain flow. Similarly, trickle irrigation can be utilized whereby water is released through small nozzles near the plants or trees. The amounts of irrigation water needed are less than with the traditional methods, as losses through seepage and evaporation are reduced. Weed competition is minimized as most of the plots are dry. Enough water can be passed through the root zone to prevent the build-up of salt. This method does have potential, especially for tree crops, where the water can be directed straight to the roots.

Although irrigation holds great potential for food production by providing the necessary water in dryland areas, there are many problems. Unless the soil and water quality are good, and unless the scheme is efficiently managed, salinization can quickly develop. Care has to be taken over disease prevention, as many harmful insects find irrigated fields to their liking. The world is full of examples of major irrigation schemes which failed for lack of understanding or management.

Ground water

Aquifers: The other major source of water in dryland areas is from groundwater. In some parts of

Plate 27.7 *Central-pivot scheme of irrigation.*
Photo: Ken Atkinson.

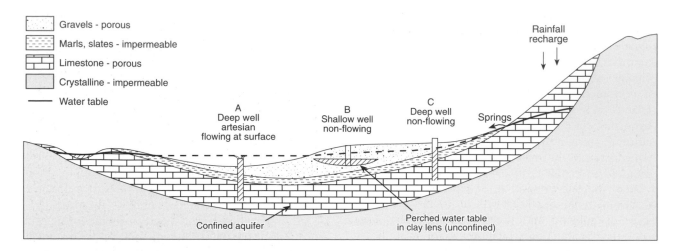

Figure 27.11   *Groundwater levels in an artesian basin.*

the world huge aquifers lie under the ground surface. These are water-bearing strata all or part of which is saturated with water and is able to yield significant quantities. Originally the water pressure in some aquifers was such that drilling for water would result in a free flow of water at the surface. Such artesian wells used to be common in Australia, where the Great Artesian Basin under much of eastern Australia allowed ready access to groundwater. Water falling on the Eastern Highlands of Australia sank into the water table and helped to sustain groundwater levels.

Aquifers can be found at a variety of levels (Figure 27.11) in many parts of the world. They achieve their greatest significance in dry areas where alternative sources are limited. Some deep aquifers are regarded as 'fossil', since they are no longer being significantly recharged. Much of the groundwater under the Sahara and the Middle East is fossil. The rate of exploitation of this non-renewable resource means that water table levels will decline rapidly. At the rates of exploitation by Saudi Arabia in the 1990s, and assuming that 80 per cent of the groundwater can be extracted, the supply will be exhausted in about fifty years. Much of this water is pumped to the surface, using abundant and cheap local fuel supplies, and used for irrigating crops of wheat. By heavily subsidizing land, equipment and the irrigation water, and by buying the wheat at several times the world price, the Saudi government has encouraged large-scale wheat farming in the desert based on fossil groundwater. Ironically the wheat could have been purchased on the world market at a quarter of the price.

On a much smaller scale many dryland countries have drilled small boreholes to the groundwater level to supply drinking water or water for animals. In theory this provides a more stable supply of water, but in practice it generates overgrazing of pastures around the borehole as farmers congregate to water their animals.

Water quality: Unfortunately not all these supplies of water are of perfect quality. Impurities in water can take the form of solids or salts. The World Health Organization sets an upper limit of 500 mg litre^{-1} for the solid content of drinking water, though often levels much higher than this will be consumed by humans and livestock. Even more severe is the problem of salt content. Domestic consumption needs water quality within the range of 500–1000 parts per million by volume (ppm). If 1 metre of water of, say 500 ppm, were added to a field of 1 ha in area as irrigation water, 5000 kg of salt would be deposited. To remove the salt, larger quantities of irrigation water would be required to flush the salt into drainage water. As the salt concentration of the water increases, so the frequency with which soil leaching is required also increases and the greater is the proportion of drainage water.

In some cases, salty irrigation water can react with the soil. On an experimental farm in Queensland water from the Great Artesian Basin was used for crop irrigation. After two years the experiment was abandoned, as the highly alkaline water had reacted chemically with the clay soils to produce a crust so hard it had to be broken up by dynamite to allow seeds to be planted! The water was used only for livestock subsequently. Similar, though less extreme, reactions have occurred in some of the calcareous soils of the Middle East. It has been estimated that within fourteen years of irrigation the salts deposited in the soil will have reached levels that are toxic to many plants.

Soil erosion

The problem

Soil erosion is not unique to dryland areas but its effects may be more apparent there than elsewhere. Erosion can take place as a general deflation of surface material together with nutrients or it can occur as gullying and sheet erosion, where large amounts of material may be removed following heavy rains (Plate 27.8). Figure 27.12 illustrates the factors affecting the types of soil erosion by water. The volumes of soil lost are difficult to estimate, especially for wind-borne material, but studies suggest values of up to 300 tonnes per hectare in the Ethiopian Highlands, where rainfall erosivity is high, compared with less than five tonnes on grazing land and less than one tonne in forested areas. Although these figures may sound severe, they need to be balanced against the rates of soil formation, as it is the net loss which is of greatest significance. Even then interpretation is not straightforward, as the erosion may take place in narrow channels which can rapidly expand, whereas soil formation will take place slowly over the whole catchment.

Estimates of the extent of soil degradation in susceptible dry lands are shown in Table 27.2. We can see that Africa and Asia have the largest areas affected by moderate or severe degradation. Interestingly, the relatively small area of dry land in Europe (largely in the Mediterranean basin) has a high proportion degraded, whilst the arid continent of Australia with its low density of population and generally low slopes has a very small proportion degraded.

Plate 27.8 *Soil erosion by gullying and sheet wash, Natal, South Africa.*
Photo: Peter Smithson.

Remedies

There are a number of ways in which cultivated soils can be protected from erosion. These can be sub-divided into three groups: (1) agronomic measures which protect the soil surface; (2) soil management techniques which improve soil structure; and (3) mechanical methods which modify surface topography to control wind and water movements. Where properly conducted, such techniques can prevent soil erosion from susceptible areas or can be used to help restore damaged areas if the erosion is not too severe. What must not be forgotten is why the problem

developed in the first place. Normally, even in dry lands, the vegetation cover is of sufficient density to prevent wind and soil erosion. It can become a problem when human activities put increased stress on the environment. The driving forces are social, economic and political factors such as population increase, unequal distribution of resources, land tenure methods, government attitudes to agriculture and the terms of trade. These factors may limit the options open to the poorer strata of society, who may have to degrade the soil resources in order to survive. Tenant farmers may have a short-term view of the land's value, trying to maximize yields rather

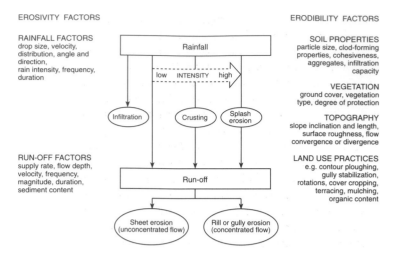

EROSIVITY FACTORS

RAINFALL FACTORS
drop size, velocity,
distribution, angle and
direction,
rain intensity, frequency,
duration

RUN-OFF FACTORS
supply rate, flow depth,
velocity, frequency,
magnitude, duration,
sediment content

ERODIBILITY FACTORS

SOIL PROPERTIES
particle size, clod-forming
properties, cohesiveness,
aggregates, infiltration
capacity

VEGETATION
ground cover, vegetation
type, degree of protection

TOPOGRAPHY
slope inclination and length,
surface roughness, flow
convergence or divergence

LAND USE PRACTICES
e.g. contour ploughing,
gully stabilization,
rotations, cover cropping,
terracing, mulching,
organic content

Figure 27.12   The main factors affecting types of soil erosion by water. Source: After Cooke and Doornkamp (1990).

than taking a long-term view of soil improvement. Here we have the ethical and practical question of who should pay: the individual farmer or society as a whole?

Desertification

A UN Conference on Desertification was convened in 1977, as, at the time, desertification was seen as a threat affecting dry lands throughout the world. It was blamed partly on declining precipitation levels, as dramatically demonstrated in the Sahel (Figure 11.7), and partly on overexploitation of a limited natural resource by increasing populations. The term is still used to denote the spread of desert-like conditions into wetter areas, but what does it really mean?

The most recent (1990) UN definition of desertification is 'land degradation in arid, semiarid and dry subhumid areas resulting mainly from adverse human impact on the environment'. The idea of desertification as purely a natural phenomenon associated with declining precipitation has been superseded. However, we must not forget the natural variability of precipitation in dry lands. It is characterized by high variability in space and time. Much of the annual precipitation falls during a few events in the rainy season. A higher frequency of events in one year would lead to a higher annual total and vice versa.

Superimposed on the high year-to-year variability, short-term trends may occur towards wetter or drier conditions. In practice it is then difficult to distinguish between adverse effects generated by human action and the dryland response to the natural climatic variability. An example of this is the increase in dust storms in the Sahel following the decline in precipitation. In Nouakchott, on the coast of Mauritania, dust storms blew, on average, ten days per year in the relatively moist 1960s. In the mid-1980s, after over twenty years of below-average rainfall, the average had risen to eighty days per year (Middleton 1995). Much of this dust would be soil particles blown into the atmosphere as a result of the drought and human activities; the relative role of these factors is difficult to assess.

Areas of desertification

The area of the world affected by desertification is not known with certainty. Despite the continuous monitoring of Earth's surface by satellite, the values quoted are still based on intelligent estimates rather than scientific data (Table 27.3). What Table 27.3

Table 27.2 *Soil degradation in susceptible dry lands, by process and continent, excluding degradation in the light category (million ha).*

Process	Africa	Asia	Australia	Europe	North America	South America	Total
Water	90.6	107.9	2.1	41.7	28.1	21.9	292.3
Wind	81.8	72.7	0.1	37.3	35.2	8.1	235.2
Chemical	16.3	28.0	0.6	2.6	1.9	6.9	66.3
Physical	12.7	5.2	1.0	4.4	0.8	0.4	23.9
Total	201.4	213.8	3.8	86.0	66.0	37.3	617.7
Area of susceptible dry land	1286.0	1671.8	663.3	299.7	732.4	516.0	5169.2
% degraded	15.6	12.8	0.6	28.6	9.0	7.2	11.9

Source: After Thomas and Middleton (1994).

does show is that a large proportion of the dry lands is affected by desertification. What it does *not* mean is that large areas of the dry lands are being engulfed by sand dunes blown in from the desert. That image may be appropriate in a few areas, but the idea is oversimplified and not an accurate picture of the way desertification works. The process of desertification is much more complex and is more likely to involve vegetational degradation on the desert fringe. Maps have been prepared to show the location of areas affected by desertification, or the risk of desertification, but they have been strongly criticized.

Causes of desertification

The development of vegetation degradation through inappropriate land use is rarely investigated through long-term scientific monitoring; most comments appear to be based on subjective judgement. We can identify three main factors which, it is argued, are likely to give rise to desertification: overgrazing, over-cultivation and deforestation. Salinization of irrigated cropland is often viewed as an additional factor in some areas. We will examine these potential causes in turn.

Overgrazing: Excessive damage to vegetation caused by too many animals being allowed to graze an area is considered by many to be the main factor behind desertification. Edible and more nutritious species are removed, leading to an invasion of coarser vegetation or even bare ground. Trampling, especially around sources of water, can compound the problem by damaging soil structure. A lower soil porosity can reduce the ability of the soil to retain moisture and support plant growth. Both soil and

vegetation factors acting together can reduce the number of animals which can be supported.

Why do the pastoralists get into this sort of situation, which is clearly harmful? As with many problems, competition for resources is important. In many countries the cultivated area has increased to cope with increasing population and higher economic expectations, pushing grazers into more marginal land. Even the American musical *Oklahoma* reminds us that 'the farmer and the cowman can't be friends'. The sinking of boreholes to provide reliable water supplies has increased grazing in surrounding areas. This increase of commercialism in pastoral farming has replaced the natural system of nomadism where the grazers moved the herds in response to vegetation growth, irrespective of political or economic controls.

Over-cultivation: Intensification of farming to produce more food can result in fewer and shorter fallow periods, leading to nutrient decline and a decrease in yield. If organic matter is not returned to the soil in sufficient quantities, soil structure can be affected. Mechanized farming has advanced into several dryland regions, with deep ploughing disturbing soil structure and increasing susceptibility to erosion. Classic examples of this are the Great Plains of the United States and the Virgin Lands campaign of central Asia in the former Soviet Union. In both cases major wind erosion resulted.

Deforestation: Clearance of forest or shrubland to increase the proportion of grasses for grazing leaves the soil exposed to erosion and may reduce the water table. Fuel wood is in great demand in many energy-poor countries, which places further demand on forest

Table 27.3 *United Nations Environment Programme (UNEP) estimates of types of dry land deemed susceptible to desertification, proportion affected and actual extent.*

Measure	1977	1984	1992
Climatic zones susceptible to desertification	Arid, semi-arid and subhumid	Arid, semi-arid and subhumid	Arid, semiarid and dry subhumid
Total dryland area susceptible to desertification (m ha)	5281	4409	5172
Proportion of susceptible drylands affected by desertification (%)	75	79	70
Total of susceptible drylands affected by desertification (m ha)	3970	3475	3592

Source: After Thomas and Middleton (1994).

resources. For example, few trees survive in the wild within about 90 km of the Sudanese capital, Khartoum. Often dried animal manure is burnt instead rather than being recycled back to the land, where it would improve soil structure.

Recent ideas

Although the above factors are usually cited as causes of desertification, recent research has questioned the validity of some of the ideas.

The state of vegetation in a dryland environment is a response to grazing pressures and the recent levels of moisture availability and fire incidence. There have been few long-term studies of vegetation in affected areas; most of the assessment has been on the basis of short-term visits, perhaps separated by a number of years. If the vegetation does decline by either process, it is the animals who are the first to be affected as food supply deteriorates. In areas of this type, supplies of fodder cannot easily be bought or brought in to offset the loss of local supplies. As the animals die, so the pressure on vegetation should decline.

In many areas affected by desertification the political, social and transport infrastructures are poorly developed. A study of the relative importance of the factors behind the Sudanese famine of 1984/5 found that the lack of rain in the 1984 wet season triggered major speculation in food. People bought stocks of cereals because they believed the price would rise. The price of food then rose beyond the reach of most rural people in the drought-affected areas. Food was available on a national level, but the mechanisms for distributing it were inadequate. Whilst deterioration of natural vegetation may have played a minor part, it was not a significant factor. In the Ethiopian drought of the same period, political factors were important in preventing the distribution of food.

Whilst there may be some evidence of a decline in vegetation and soil quality along the drier margins of the savanna belt, the idea of an encroaching desert is not based on hard scientific evidence. Nevertheless such areas are marginal, and a better understanding of the environmental and economic processes influencing life in these regions is still needed.

Conclusion

The dryland areas of the world cover an appreciable proportion of Earth's land surface. They exhibit considerable diversity in natural environment, from the moister, wooded areas of the savanna to the hyper-arid desert areas where rainfall is minimal. Many of the developing countries of the world occupy this zone; most of them are experiencing rapid population growth, which puts additional pressure on the national resources. As a result the future of the dry lands is giving cause for concern. Even potential solutions may differ, depending upon such factors as the wealth of the nation, its political system and the relative importance of the dry areas in the overall economy. These days we must not forget the attitudes of international organizations and the media. The sight of starvation on television screens brings a vivid perspective to the problems of marginal agriculture in ways which were impossible when communication was achieved on foot or by animal. Unfortunately the response to such problems is more likely to be short-term food aid with no consideration of the long-term problems which allowed the famine to develop in the first place.

We must not consider all dry lands as areas of great hardship and stress. Many economies survive adequately with the resources available, supplemented where possible by additional water for irrigation. With care and understanding dryland areas can make an effective contribution to the national economy. A good example would be the drier parts of the United States, but it does help if there is financial and technological support.

Key points

1 The dry lands are characterized by a moisture deficit caused by relatively low precipitation and high inputs of energy which lead to high levels of evaporation. Natural levels of vegetation are controlled by moisture availability. There is a strong seasonal cycle of growth associated with the wet season and generally a low density of vegetation. Where water is available crop growth can be good, but it requires the supply of adequate volumes of good-quality water.

2 Supplying water to dry lands can cause a wide range of problems. The most frequent method is dam construction, but the resulting lake has many harmful effects and incorrect use of the water can produce major problems of salinization. Under these conditions the land can become sterile, as in parts of the Indus valley.

3 Soil erosion can be another major problem in dryland areas, where bare ground is easily eroded by heavy rainstorms. It is made worse by deep ploughing and attempts at intensive commercial agriculture.

4 Desertification has been threatening the drier margins of the dryland areas, though a strict definition of the problem is difficult. As a result of overgrazing, coupled with fluctuating rainfall levels, the vegetation of the drier margins may experience stress and degrade in quality. This does not mean that the desert is advancing; it is a reflection of recent pressures on the vegetation.

Further reading

Agnew, C. and Anderson, E. (1992) *Water Resources in the Arid Realm*, London: Routledge.
A modern text concerned with the availability and use of water resources in a dryland environment. Emphasis on the Middle East and Africa.

Beaumont, P. (1993) *Drylands: Environmental Management and Development*, London: Routledge.
A full discussion of environmental management in dryland areas. Emphasis on regional examples.

Thomas, D.S.G. and Middleton, N.J. (1994) *Desertification: Exploding the myth*, Chichester: Wiley.
This book sets out to analyse the range of scientific, social and political issues surrounding desertification. Puts forward various interesting ideas about the factors, especially political ones, involved in desertification and their validity.

CHAPTER 28
Humid tropical environments

The humid tropics are those parts of the world within the tropical belt where, on balance, precipitation is greater than evapotranspiration. Its definition is somewhat arbitrary, as there is a continuous gradient from areas with rain throughout the year, through the areas of seasonal rainfall, or monsoon areas, to the deserts dominated by subtropical high-pressure cells as discussed in Chapter 27. On the eastern side of the continents the gradient is less strong, with a moist climate dominating, despite a gradual decrease in mean annual temperature, as found from Malaysia through Vietnam to Hong Kong and China and also in Brazil. Not all equatorial areas are humid, as parts of Kenya, north-east Brazil and the Galapagos Islands are all close to the equator yet for different reasons experience very dry climates.

It is an area of the world which has come into prominence recently through concern about the destruction of the tropical rain forest and its possible implications for the global environment. Commercial exploitation and population pressure have led to stresses in such areas, with a loss of resources, such as soil. We will devote a considerable part of this chapter to the problem, as the use and exploitation of the forests of the humid tropics is a major international dilemma.

Climate

The tropics have been described as the firebox of our atmospheric engine. Most of the sun's energy is absorbed here – energy which is transferred eventually into cooler, energy-poor latitudes.

The zones of the humid tropics

A simple approach to climate in the humid tropics is to distinguish three main zones: (1) the equatorial trough zone (or Inter-tropical Convergence Zone -ITCZ); (2) the monsoon areas; and (3) the trade wind zone. These are shown in Figure 28.1 together with the subtropical high-pressure belts which act as the source areas of the trade wind flows and the dominant feature of the tropical deserts. The monsoon area is really a modification of the trade wind zone brought about by the effects of the continents. Strictly, it is only the equatorial trough zone which meets the popular idea of a humid tropical climate; the other two regions are linked with the trough zone through their atmospheric circulation.

Equatorial trough

The traditional idea of the equatorial climate involves the daytime build-up of convectional clouds into massive cumulonimbus displays. Rainfall is frequent and abundant, temperatures and humidity are high, acting together to give us the tropical rain forests. At night the air is humid and still. Condensation takes place on to the forest trees and the sound of moisture dripping to the forest floor competes with that of the wildlife.

The structure of the atmosphere, though, is not as simple as this model may suggest. The multitude of names which have been used for the area give some idea of its variety – the doldrums, the intertropical front, the intertropical convergence zone, the intertropical trough, the equatorial trough or the intertropical confluence zone. For simplicity, we shall refer to it as the equatorial trough, although it does extend towards the subtropics, and is quite variable in character.

The equatorial trough has many different forms. It represents the area of low pressure somewhere near the equator towards which the trade winds blow. The precise form it takes will depend upon the stability of the trades, their moisture content and the degree of convergence and uplift. Much of the trough is over the oceans and it is only recently that satellite photographs have shown us more about the detail of cloud forms and the structure of the equatorial trough.

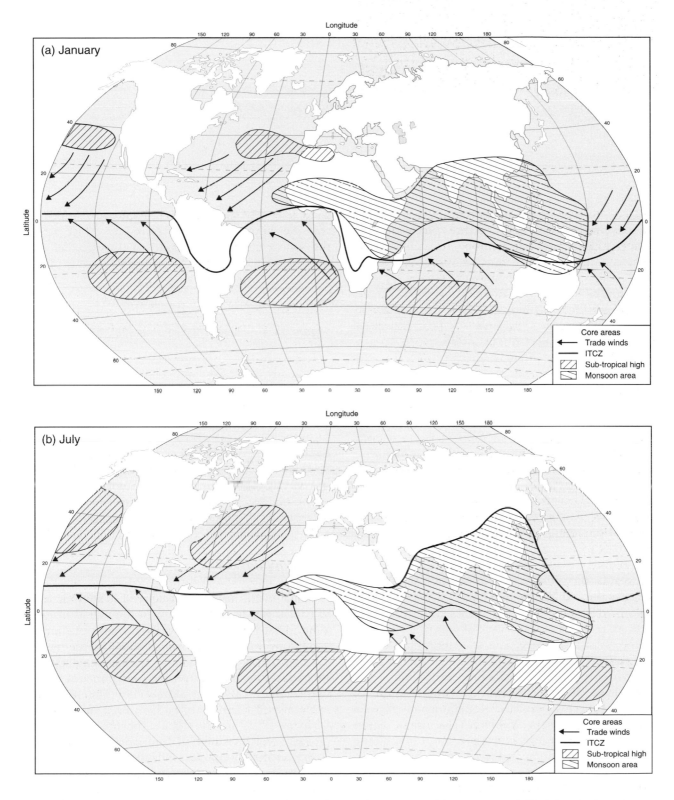

Figure 28.1 *Areas affected by the four main climatic zones in the tropics in (a) January and (b) July. The boundaries are really transitional zones and may show considerable shifts in position from year to year.*

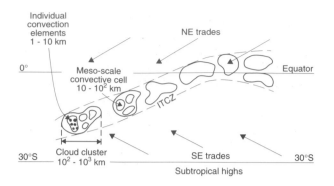

Figure 28.2 *Schematic representation of the equatorial trough or Inter-tropical Convergence Zone (ITCZ) showing the organization of associated convective cloud systems in plan view. Source: After Mason (1970).*

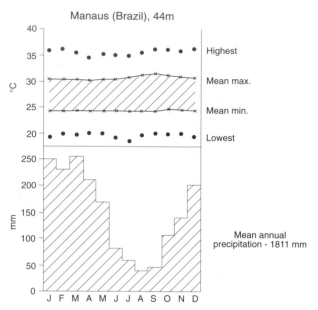

Figure 28.3 *Climatic data for Manaus, Brazil.*

The structure of the trough is variable (Figure 28.2) and careful scrutiny can show that a hierarchy of cloud is present. The larger element is a cloud cluster, perhaps 100 km to 1000 km in length. Within the cluster there are convective cells, and embedded in the cells are individual convective elements which can give the heavy rain characteristic of the equatorial trough. Over the continents the convective area expands considerably, as seen over central Africa in Plate 8.3.

What is the climate of the equatorial trough like? Figure 28.3 gives an example of mean monthly temperature and rainfall for Manaus in Brazilian Amazonia. The mean monthly maximum temperature varies by 2.8° C over the year and the mean monthly minimum by only 0.6° C. Extremes are rare and insignificant by temperate-latitude standards. The diurnal variation of temperature is usually more noticeable than the annual variation. At Manaus, mean annual rainfall is high, at 1811 mm, though even in this zone there is a drier period when rain days are fewer. It is at this time of year when burning of the forest takes place. This somewhat drier season is experienced in most of the equatorial trough zone, though its intensity and duration vary. Only a few areas have no drier season. Padang in Sumatra receives an average rainfall of 4427 mm and only one month has less than 250 mm. The driest season occurs when the trough moves polewards in response to continental heating in the southern hemisphere. As one moves farther away from the equatorial trough zone so the dry season lengthens and we reach the monsoon areas.

The humid tropics cover a wide range of climatic types and hence show considerable environmental diversity. Temperatures are fairly stable throughout the tropical world, tending to be somewhat higher during any dry season and cooler and with a smaller diurnal cycle during the wet season. The main variable is the quantity of rain and its seasonal distribution. In this section on climate, for completeness, we have included some areas which have a relatively short dry season because the origins of their climate are similar to those of more humid areas. For the rest of the chapter, we will concentrate on the more humid areas, where the dry season is shorter than the wet season and there is an annual surplus of moisture.

Geomorphology

Studies do not support the idea of a uniform level of weathering or of landforms in the humid tropical zone. What we would expect on the basis of climatic conditions is deep weathering, caused by the higher rates of chemical processes at warmer temperatures where moisture is available; the wetter an area is, the deeper the weathering should be. The processes operating on Earth's surface have been identified in Chapters 12–17. In the tropics, the intensity of these processes will be different from what it is in other environments, though it is difficult to prove that they generate land forms or landscapes that are specifically tropical.

Many parts of the tropics are based on relatively stable continental plates. Such areas have not been exposed to glaciation and therefore weathered material has been able to accumulate rather than being dispersed as happened in many parts of the northern lands. In extreme cases the weathering horizon may be as deep as 30 m in the more humid tropics, though its depth decreases where moisture is less available, and normal depths are about 3 m. How long such weathering horizons have been developing is difficult to decipher but many are believed to be long established. In the more tectonically active areas, such as Indonesia, volcanism can bury existing weathering horizons. The new lava is then rapidly weathered in turn. Erosion of the weathered material may take place through fluvial processes or by mass movement. Where slopes are steep, this may be significant, but on gentler slopes the density of biomass produces a protected zone on which slope movement is slow. One of the most important processes of the tropics is the run-off. Most of the material eroded from the humid forest is carried by run-off in the form of solution. Little coarse sediment moves as bed load, but the density of vegetation prevents all except the finest particles being carried by overland flow. Only bank erosion will produce a sudden increase in the amount of sediment transport.

In steeper areas associated with the recent volcanic chains, such as eastern Asia, Indonesia, New Guinea and the Andes, mass movement may become more significant. Slides and flows can strip away vegetation and weathered material to expose regolith and even bedrock. The effects of these processes can be seen from satellite images of the Amazon delta, where muddy waters from the Andean foothills can be distinguished as they gradually mix with the clean sea water.

In calcareous regions, solution is a highly significant and rapid process which produces unusual land form types such as the cockpit country in China and the Caribbean. However, detailed studies in karst areas indicate that the principal climatic factor affecting the erosion rate is the mean annual run-off. Whilst temperature has some influence, it appears to operate through the degree and type of soil and vegetation cover rather than as a direct control of solution. Until the tools of process studies are powerful enough to unravel the interaction of lithology, erosion rate and time, it will not be possible to determine with certainty whether latitudinal variations of climate have had more than a coincidental effect upon the development of karstic landforms.

Soils

Within the humid tropics there are a group of distinctive soil characteristics which are rarely found outside that zone. These are the processes of rapid weathering and strong leaching, the properties of a deep and highly weathered regolith and the importance of organic matter in soil fertility and management. The predominant minerals are kaolinitic clays, and hydrous oxides of iron and aluminium (sesquioxides) which give the soils their strikingly red colour. The tropical climate is important in the operation of these processes and provides the framework in which the soils develop. In addition, lithological variations and relief play an important part in the actual differentiation of soils.

Because organic matter is rapidly decomposed, the main problem of soil utilization in the tropics is its maintenance at suitable levels. The quantity of organic matter lost from the soil during one year of cultivation in the lowland tropics is of the order of two tonnes per hectare. To replace it is a major problem. Research has shown that there is no practicable means of maintaining organic matter under the cultivation of annual crops in the rain forest zone other than by an extended fallowing system. It is almost impossible to replace organic matter by fertilizers; they either become leached (N, Ca) or fixed (P). Nutrient cycling in tropical rain forests was discussed in Chapter 22.

Tropical forests

The natural vegetation of much of the humid tropics is forest. In the moister areas we find the true tropical rain forest, which covers about 13.2 per cent of the land surface of Earth, about 17×10^6 km^2. These complex and variable forests grow in lowland areas with over 1700 mm of annual rainfall and no distinct seasonality. When the monthly rainfall drops below about 120 mm for longer than one month the rain forest tends to be replaced by tropical moist forest. When there is a strong seasonality the trees become deciduous and so we find tropical deciduous forest replacing tropical rain forest.

Soils, vegetation and species

The soils associated with the forests are characterized by intense and perhaps prolonged weathering, with active leaching. Decomposition is so rapid that, despite high inputs of plant debris, the soils rarely develop a distinct organic surface layer. Moreover, the intense weathering and leaching mean that the more soluble constituents are totally removed. Iron and even silica may be mobilized. These soils have a low cation exchange capacity and a limited supply of bases such as calcium and potassium (i.e. a low base status). They possess high iron oxide contents which give red soil colours. In the FAO–UNESCO soil classification scheme these soils are mainly ferralsols and acrisols. They also include the ferricretes and laterites, which are frequently defined as tropical forest soils.

Tropical rain forest vegetation is typically diverse both in species composition and in structure. Whereas temperate forests may contain only three or four tree species per hectare, tropical forests often include as many as 100 (Figure 28.4) and studies in Amazonian Ecuador have found 473 tree species within one hectare. In general, there is a positive relationship between mean annual rainfall and the number of species. This has advantages and disadvantages when exploitation of tropical forests takes place. Most species are evergreen, and those that are leafless for any period shed their leaves at irregular intervals. There is no autumn in the sense of the leaf-fall period of deciduous forests. Intense competition and the diversity of plants lead to complex structuring of the forest, with five or more strata recognizable and a significant epiphytic component (Figure 28.5). Species diversity or richness is high, but species' distributions are often restricted to small areas (Plate 28.1).

Coupled with diversity, studies of biological interactions between species have demonstrated the complexity and fragility of the system. The classic example of this situation is the Brazil nut tree (*Bertholletia excelsa*). The flowers of this tree need to be cross-pollinated with other individuals of the

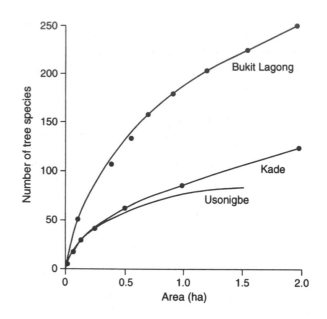

Figure 28.4 *Species/area curves for tree species in tropical forests: Bukit Lagong, Western Malaysia; Kade, Ghana and Usonigbe, Nigeria. Source: After Longman and Jenik (1987).*

same species. They have a complex physical structure which means that they are pollinated largely by certain types of bee during the one-month flowering period in November. The bees also pollinate other tree and orchid species which provide nectar at other times. If only Brazil nuts were grown, as on a plantation, there would be no suitable insects for

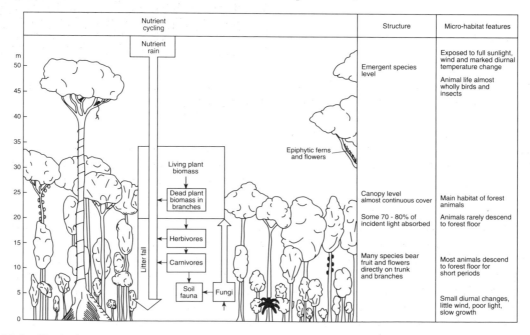

Figure 28.5 *Vertical stratification of a tropical rain forest. Source: After Collinson (1977).*

pollination, as there would be no food supply at other times of the year. It appears that the Brazil nut will not set fruit without these other tree and plant species to provide bee food when the Brazil nut is not in flower. A further complication for dispersal of the Brazil nut is the shell case. After mature nut cases have fallen to the forest floor, they are opened only by agoutis (a small guinea-pig-like rodent) which break through the shells and bury the nuts as a food store. Inevitably some are forgotten and eventually germinate. Survival of the Brazil nut therefore needs the bees, food for the bees at other times of the year and the agouti to break into the shell cases and disperse the nuts. Destruction of the forest can break these interactions and consequently the whole structure of the ecosystem.

Many different types of animals are present, taking advantage of the diverse niches provided by the vegetation. A large majority of the animals are arboreal (living in trees). Those at the canopy level rarely descend to the forest floor, but those at lower levels in the trees – the middle-zone fauna – come to the ground more frequently. In general, the high habitat and niche diversity result in high species richness. Many animals and plants are ultimately dependent on a few plant species for their existence. Because these diverse and complex ecosystems can change frequently over short distances, relatively minor disturbances, like logging, can cause species extinction. This is one of the reasons why ecologists and conservationists are so concerned about uncontrolled exploitation of the rain forest.

Biogeochemical cycling

One of the main features of tropical forests is the huge concentration of nutrients stored within the vegetation (Figure 28.6). The soil has a low nutrient storage capacity, and nutrients are retained mainly in the dense, lush vegetation. This has considerable significance, for when the forests are cleared, by artificial or natural means, most of the nutrient store is lost. The remaining soil is very infertile and therefore regeneration is slower and less luxuriant than might be expected. If crops are planted, yields decline rapidly without massive injections of fertilizer.

Under undisturbed conditions, large quantities of nutrients and energy cycle rapidly through the tropical forest, despite the fact that food chains are long and extremely complex (see Chapter 22). It has been estimated that annual biomass production (not net primary productivity) in a central Amazon rain forest near Manāus is in the order of 1100 t ha^{-1}. A high

Plate 28.1 *Amazonian rain forest near Iquitos, Peru. Note the wide height range of species and their variety.*
Photo: Peter Smithson.

proportion of this annual production is associated with the plants – the leaves, fruits and flowers of the trees, ground flora and epiphytes – which typically account for 4–9 per cent of the total biomass (compared with 1–2 per cent in temperate forests). In contrast, animals account for only 0.02 per cent of the biomass at this Amazon site. The annual litter fall is also high, but with rapid decomposition and leaching there is little surface accumulation (Plate 28.2). Most of the decomposition is the work of fungi, and soil animals, instead of feeding on the organic matter itself, tend to feed on the fungi. Earthworms are confined to the upper rooting zone in the soil and there is little mixing

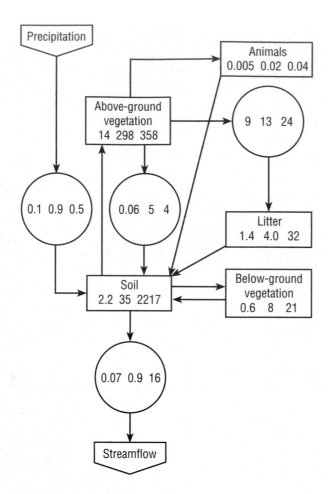

Figure 28.6 *Nutrient flow of a tropical rain forest, Panama. Figures in boxes represent total stores of phosphorus, potassium and calcium in g m⁻²; figures in circles represent annual flows of P, K and Ca in g m⁻² yr⁻¹. Source: Data from Golley (1978).*

of the soil by animals. Consequently, the soil horizons are distinct.

The nutrient budgets of tropical rain forests are interesting in that the annual turnover is much higher than that of almost any other biome. The rate of cycling is perhaps three to four times that of temperate forests, and large quantities of nutrients are returned each year to the soil (Figure 28.6). However, leaching is active, and the loss of the nutrients from the soil to surface water bodies can be intense.

In the seasonal forest, biomass production is considerably slower, averaging just over 200 t ha⁻¹. Most of the processes described above operate, though more slowly, so leaf litter accumulates at the surface, and more specialized relationships between litter and soil organisms are found. Fire is also an important factor in these areas. Fire releases nutrients to the soil and reduces the storage of nutrients

in the vegetation. However, leaching is less marked and more nutrients are retained in the soil.

Environmental change

The history of the rain forests appears to be more complex than was once thought. The diversity of species in tropical rain forest was for long believed to be the result of environmental stability. It was thought that the impact of the Pleistocene glacial periods, which had been dominant in temperate and polar regions, had not extended into the equatorial areas. Indeed, there is still controversy over the amounts of change which have occurred. Some scientists believe that the wet tropical regions of Earth have been stable for at least the last 40 million years. In that case, the rain forest development should have taken place over a long period of time. The idea is that this period of uninterrupted development has allowed rain forest plants and animals to evolve and adapt. Hence the richness and diversity of forest species to exploit the resulting ecological niches.

Other experts believe that the forests should no longer be considered ancient, almost unchanged biomes that have survived since the Tertiary. Instead it is argued that they *did* experience dramatic changes during the Quaternary Period, and probably owe much of their present diversity to the periods of isolation they experienced at that time. This isolation led to the development of many endemic species, each found exclusively in the area in which it speciated.

It is also clear that distinct variations related mainly to climatic and geological factors occur within these biomes. An idealized picture of the patterns is shown in Figure 28.7.

The humid tropics are one of the few areas of the world where the enhanced greenhouse effect is believed not to have a major impact. General circulation model predictions suggest an increase both in mean annual temperature and in mean annual rainfall, but the change is small compared with other parts of the globe. It would seem that natural change of the tropical forest environment in the near future should not be large.

Human impact on the forest

Although forests do provide food, it was only through agriculture and crop production that food supplies became sufficient to sustain a settled and concentrated population in villages, towns and eventually cities. As populations have expanded so there has always been pressure on forest reserves to increase the area of agricultural land. The European deciduous forests were decimated during this millennium; North American forests suffered similarly, though not on such a vast scale. Even the boreal forests of Canada, Scandinavia and the former Soviet Union are being utilized, though in this instance for paper and timber products rather than replacement with agriculture. It is not surprising therefore that the tropical rain forests have begun to suffer severe depredation. The scale of damage varies. In the small, fragmented states of West Africa each country has used its timber resources as a source of foreign exchange without being able to preserve large areas. For example, Ivory Coast produced 5.5 million m³ of industrial roundwood in the late 1970s. By 1991 the figure had fallen below 3 million m³. In the same time, population grew from about 6 million to almost 13 million and plantation crops increased in area under government incentives. In Brazil pressure on land has, until recently, been much less and so vast areas of rain forest remained undisturbed.

Cutting down of tropical rain forest has now reached such proportions that major concerns are being raised about the consequences. What we will consider here is, first, why are the rain forests being cut down; second, what are the impacts; third, can or should anything be done about it?

Why are the rain forests being felled?

Surprisingly, the answer is not simple. The reason why trees are felled varies. Table 28.1 lists the main factors which have been put forward as reasons for

Plate 28.2 *Ground level in the rain forest. Some recent leaf litter survives on the forest floor, but amounts are small compared with the annual fall.*
Photo: Peter Smithson.

Table 28.1 *Important factors influencing deforestation in the tropics, by major world regions.*

Region	Main factors
Latin America	Cattle ranching, resettlement and spontaneous migration, agricultural expansion, road networks, population pressure, inequitable social structures
Africa	Fuelwood collection, logging, agricultural expansion, population pressure
South Asia	Population pressure, agricultural expansion, corruption, fodder collection, fuelwood collection
South East Asia	Corruption, agricultural expansion, logging, population pressure

Source: After Kummer (1991).

deforestation by region. Common factors do appear in each region but with different emphasis (Figure 28.8). The actual felling may be done by agriculturists, by loggers or by fuelwood collectors but the key factors are access and transport. Without a method of transporting the wood or the subsequent agricultural production out of the area the only possibility of use is for subsistence agriculture. Many of the countries containing rain forest are economically poor, with large population growth rates. Poverty, low agricultural productivity and an unequal distribution of land may drive people to move

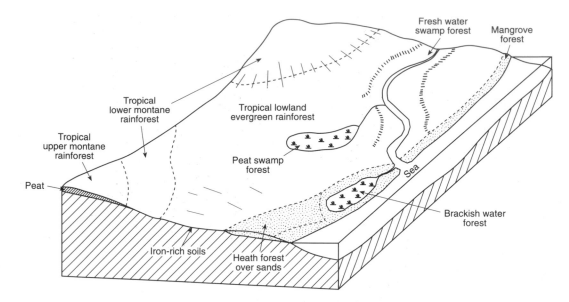

Figure 28.7 *Relations between tropical rainforest formation and environmental conditions. Source: Based on Collinson (1977).*

into forested areas with or without government permission. Equally, pressure to raise money for government can permit and encourage commercial exploitation for timber or for subsequent agricultural development, as in ranching schemes in parts of Brazil. Some agreements are with multinational corporations which provide money for investment in the hope of greater returns.

Brazil provides a good example of the way in which government policy has a major impact on forest clearance. Much of Brazil's population is concentrated on the east coast. Population has been rising faster than employment opportunities and many landless peasants flocked to the cities, where

sprawling suburbs and *favellas* or shanty towns were constructed. Drought frequently affected the agricultural north-east, causing further problems. The aim of the government became that of exploiting the vast area of forest in Amazonia as well as reducing pressure on the eastern cities. It was believed to be an empty land, rich in mineral, agricultural and water resources.

New roads were constructed to provide access to the forest lands of Rondonia and Pará (Figure 28.9). Government subsidies were provided to encourage cattle ranching and the resettlement of peasants displaced by changes in agricultural systems in the south-east of the country (Plate 28.3). Satellite monitoring of forest burning during the drier season indicated losses of nearly 80,000 km² of primary forest in Brazilian Amazonia in 1988. The rate of clearance does appear to have declined since then, for a variety of reasons. It has been realized that the fertility of the land is low unless it is carefully tended; the productivity of crops declined rapidly once the initial nutrient content was exhausted. Artificial fertilizers are expensive because of high transport costs in such locations, so peasant farmers could not afford them and hence the soil fertility declined. Unsuitable crops which did not grow well were even supplied to the settlers.

For a variety of reasons we find that there is great pressure on the tropical rain forests of the world. Their area is declining rapidly. Many ecologists believe that early in the twenty-first century only two significant areas of tropical rain forest will remain – in western Amazonia and in central Zaire.

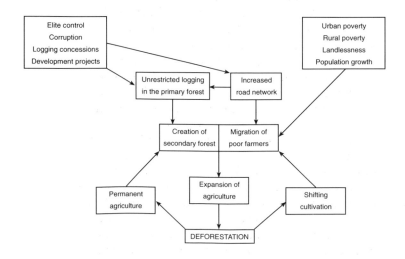

Figure 28.8 *Factors affecting deforestation in the Philippines. Source: Kummer (1991).*

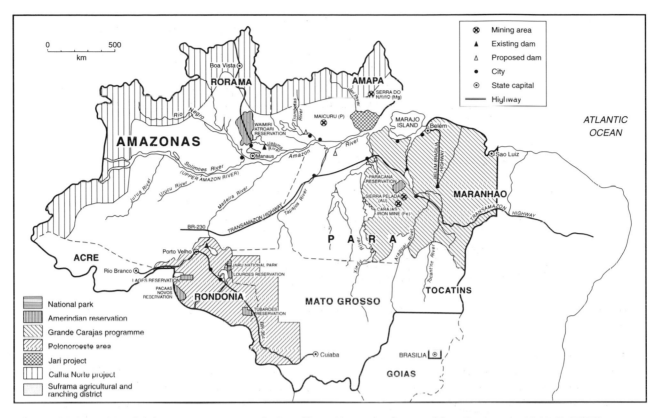

Figure 28.9 *Major development programmes in Brazilian Amazonia. Source: After Goodman and Hall (1990).*

Impacts of deforestation

When rain forests are cleared we are losing more than a collection of timber because of the implications which total clearance would have on the world environment. Here we will concentrate on the main impacts, which can be summarized as (1) loss of species diversity; (2) loss of natural resources; (3) environmental consequences; and (4) possible changes of climate on a local, regional and global scale.

Loss of biodiversity

As we have seen, rain forests are incredibly diverse in terms of the number of species which grow there and those which are dependent upon the forest for their survival. As habitats are removed through clearance, so the number of individuals will decline. There will be greater pressure on the surviving habitats and the remaining species will be forced to live in a smaller area. As we have no clear idea of the number of species in the rain forest it is hard to be precise about the rate of species loss. Estimates range from one species becoming extinct every half-hour as a result of the destruction of rain forest to between one and fifty species per day world-wide. Whatever the true number, we are definitely increasing the rate of extinctions which would occur naturally.

Loss of natural resources

The rain forests act as a reservoir of natural resources in the form of fruits, food, timber, raw materials and medicines. Species extinction through clearance would lead to the loss of any of these products. Greater concern has been expressed about the possible loss of resources which have yet to be discovered. How many species living in the rain forest might be commercially exploited? There may be much genetic diversity which is needed for improved plant breeding to sustain our increasing population. An example of this followed the blight which developed on the American maize crop in 1970 and halved production in an area which acts as the main cereal surplus area of the world. A search for other varieties of maize which were immune to this form of blight led to suitable stock being found in Mexico. Later, a new species of maize was found in the Mexican rain forest which was immune to at least seven major diseases and which could be grown in cooler, damper environments. World maize production should increase in areas previously unsuitable. Ironically the

Plate 28.3 *Agriculture on the banks of the Amazon near Iquitos, Peru. Cattle are grazing near by.*
Photo: Peter Smithson.

maize was found in an area undergoing clearance and only a few thousand stalks remained.

It is also believed that the rain forest has potential for materials used as medicines. Almost a quarter of the prescribed drugs used in the United States are derived from tropical rain forest plants, and at least 2000 rain forest plants have been identified by the US National Cancer Institute as having anti-cancer properties.

Environmental factors

Besides their biological role, the rain forests interact with their environment and affect the soils, hydrology and climate. Let us have a look at these aspects in turn.

Soils: Clearance of trees removes the main supply of nutrients to the forest soils and at the same time allows rainwater to reach the floor unmodified by the canopy. As it is very difficult to replace nutrients by the addition of fertilizer the soils will rapidly become poorer through the leaching effects of heavy rainfall (Fig 28.10).

With a complete forest cover the soils are protected from erosion by root mats, dense ground and decomposing vegetation acting as a sponge and the canopy provides a shield from intense rainfall. Once the cover is removed, run-off increases and erosion rates rise dramatically, especially on slopes. Severe erosion can strip off topsoil down to the impermeable lateritic hardpan, which makes recolonization difficult. Additionally, the run-off may become concentrated into gullies which greatly increase sediment yield into rivers. Logging activity often generates erosion along the access tracks. In Thailand in 1988 forty people were killed in mudslides which were blamed on illegal logging activities.

Run-off: The more rapid run-off associated with deforestation can give rise to flooding downstream of the affected area. An increase in flooding in the Philippines following typhoons is believed to be the result of the widespread deforestation of the islands.

Conversely some of the sediment is dropped in the river channels, in reservoirs or even on farmland if floods have occurred. Previously navigable rivers have become silted up in Madagascar, where deforestation and soil erosion are major problems. During the wet season the coastal waters change colour as vast quantities of sediment are transported to the sea.

Climate

Deforestation can affect climate on the local, the regional and perhaps even the global scales.

The local effect is obvious, as surface characteristics, such as the albedo, are being changed. The dark green of the rain forest is replaced by the lighter greens of growing crops or light browns of bare soil. Solar radiation is able to reach the surface without much interruption and moisture evaporates more easily from the exposed surface. As less moisture is stored in the sponge-like forest floor, there is a loss of water via surface run-off rather than as evaporation back to the atmosphere. It is believed that this reduction of moisture can reduce the rainfall régime, leading to drier conditions. Soil temperatures which were previously relatively stable will fluctuate more widely. Heating of the forest soils can speed up the processes of hardpan formation and nutrient leaching which quickly render them useless for agriculture.

On a regional scale, extensive deforestation can change surface temperatures through the albedo and alter the hydrological properties. Increased run-off, increased evaporation, decreased soil water storage and increased temperatures can lead to a cycle of drying. Cloud cover may be reduced as less moisture is returned to the atmosphere, giving a positive feedback effect to higher surface temperatures.

Less obviously, tropical forests may have an impact on the gaseous composition of the atmosphere, which in turn can affect surface climate. As plants photosynthesize they use carbon dioxide from the atmosphere and therefore contribute to the balance of this gas. If the forests are cleared, the subsequent vegetation growth is likely to have lower biomass and so extract less carbon dioxide. Burning of the forest will directly add carbon dioxide too. It has been estimated that if all the world's rain forests were burned between 1986 and 2000 the carbon dioxide concentration in the atmosphere could rise by up to 20 per cent.

Methane is another greenhouse gas which would increase through forest clearance. The main sources of methane are rice growing, biomass burning and cattle ruminating (see Chapter 11). All these activities may increase as forest is removed, and methane

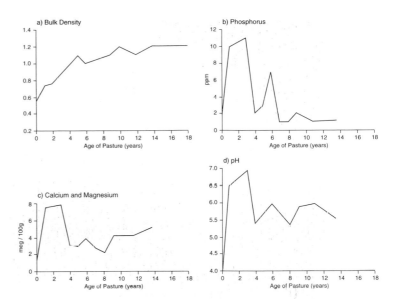

Figure 28.10 *Changes in soil properties after conversion of forest to pasture. Source: After Park (1992).*

is a more efficient greenhouse gas than carbon dioxide.

These effects lead us to the largest scale of climate impact, as the addition of greenhouse gases would add to global warming. Enhanced rates of clearance would exacerbate the problem. General circulation models have been used to predict what the impact of total forest clearance would be at both the regional and the global scales. The consensus view is that the direct effects of deforestation on regional climate may be large but the impact on global climate should be relatively small, perhaps warming Earth by about 0.3° C.

Forest management

We have seen that tropical rain forests are rich ecosystems which are being threatened by extensive clearance. At a local scale, the change in land use may benefit a small number of individuals, but on the world scale we are facing a major crisis of the wholesale extinction of species and habitats, which in turn may affect global climate. Can or should anything be done about it?

Much deforestation takes place far away from the centres of national government. In order to address the problem we have got to examine why clearance is taking place and what the alternatives may be. Passing legislation which cannot be enforced is meaningless. We have also got to be economically realistic. Brazil has one of the largest areas of this natural resource. For Brazil to cease developing it for the

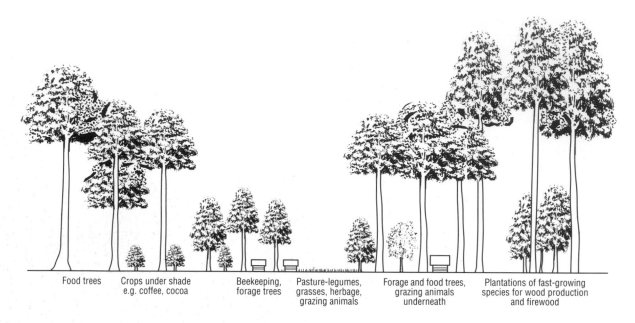

Food trees | Crops under shade e.g. coffee, cocoa | Beekeeping, forage trees | Pasture-legumes, grasses, herbage, grazing animals | Forage and food trees, grazing animals underneath | Plantations of fast-growing species for wood production and firewood

Figure 28.11 *Tropical forest management for environmental protection and sustained yields. Source: After Simmons (1989).*

Table 28.2 *Economic return from different uses of 1 ha of lowland tropical rain forest in eastern Peru.*

Use	Annual income (after labour and transport) (US$)	Net present value (twenty-year discounted) (US$)	Comments
Extractive uses			
Latex harvest	22	440	Does not include other forest products or tourism
Edible fruit harvest	400	6000	
Total (harvesting)	422	6440	
Sustainable selective logging	15	490	
Total (harvesting + sustainable logging)	437	6930	
Conventional 'development' uses			
One-time removal of marketable timber		1000	Cutting destroys extractive resources
Reforestation with *Gmelina arborea*	159	3184	Not sustainable
Total (forestry)		4184	
Intensive cattle ranching on ideal pasture	148	under 2960	Not sustainable
Total (ranching)		under 2960	

Source: After Peters *et al.* (1989).

sake of the global good is not feasible. It is often pointed out that many developed countries cleared their own forests centuries ago. Why should developing countries not benefit from the use of available forest resources? Most of them are heavily over-burdened by debt to the developed countries or the World Bank, so that much of the revenue earned from forest exploitation goes into interest payments.

One suggestion for conserving forests has been the designation of national parks or nature reserves. By 1990 there were about 550 tropical forest parks, which account for about 4 per cent of all tropical forests. There has also been a scheme to offset debt to developed countries in exchange for retaining forest land. In 1991, Mexico agreed a debt-for-nature swap with Conservation International, which agreed to purchase and write off US$4 million worth of Mexican debt from foreign creditors. In return the Mexican government agreed to invest US$2.6 million in rain forest conservation.

There is considerable scientific debate about the area of reserved land which is needed to sustain suitable habitats for animals as well as plants. A few very large areas are seen as more appropriate than many smaller sites. Unfortunately, even those sites which have been agreed are not unaffected by exploitation. Many reserves exist on paper only, with no policing or support through lack of financial resources. It is not unknown for logging to continue even in areas designated as parks. On the Indonesian island of Siberut, off Sumatra, plans were advanced to log 1500 km² of virgin forest in a reserve and replace it with an oil palm plantation.

Ideally what is needed for the rain forest is the maintenance of as much as possible of the present variety of species and habitats and the restoration of damaged areas. At the same time, the forests must be used to generate revenue at a greater rate than what could be obtained by clearance and replacement by some other use (Figure 28.11). The land must be seen to be earning its keep, otherwise, in a world where economic pressures dominate, the forests will disappear. Is such an approach possible?

Estimates have been made of the economic returns of different types of land use in a forested area of eastern Peru (Table 28.2). Low-intensity exploitation of nuts, fruit, rubber and other products together with minor logging could generate greater income than that from conventional methods of forest clearance and ranching. Even so, care must be taken over transport and marketing. The products need to have some international value to compensate for the loss of hard currency obtained from the sale of tropical hardwood. Ironically the decrease in the sale of such

Plate 28.4 *Tourist lodge set in the forest close to the Amazon to aid communication. Local styles of construction are used to imitate the natural environment, and disturbance is kept to a minimum. Photo: Peter Smithson.*

timber could increase its value, unless demand declined, placing even greater pressure on the forests.

Recently efforts have been made to develop the idea of ecotourism to provide additional income for forested areas. Tourists are encouraged to visit an area to view the beauty of virgin tropical rain forest (Plate 28.4). Although still on a small scale, and not without its own problems, it does represent an additional source of hard currency if correctly operated. The benefit here is that income is generated at the local level through guides, transport and accommodation, with no serious disturbance of the forest.

Conclusion

The humid tropics are a sensitive environment. They occupy the hotter parts of the world where solar energy is absorbed and transferred to deficit parts of the globe. Precipitation is generally high, with a dry season of variable length. Most of the soils forming under rain forest conditions are relatively poor, with nutrients rapidly recycled and stored in the biomass rather than in the soil.

The natural vegetation is a biologically rich forest with characteristic layers of growth, from the upper canopy of the tallest trees down to the dense vegetation of the forest floor. These conditions provide a variety of habitats for plants and animals. The number of species found in the average rain forest is far greater than anywhere else on Earth.

For a variety of reasons, these forests have suffered major clearance, especially in West Africa, southern India and parts of eastern Asia. The surviving areas, dominated by Latin America and Central Africa, are still experiencing great pressures for land clearance, leading to drastic losses of biodiversity and habitat.

In theory it should be possible to maintain forest cover, given sufficient investment in agriculture and forest management and a well enforced network of protected areas. Inevitably there will be further losses of forested areas, owing to social and economic difficulties, international disputes and problems of education. What needs to be stressed is that such a valuable global resource must not be destroyed. A practical form of sustainable management and control is needed.

Key points

1 The humid tropics cover the warmest parts of the globe, where there is, on average, a water surplus. Rain falls throughout the year, associated with the equatorial trough, though there may be a drier season when the active trough is farthest away. Seasonal temperature variations are small and diurnal control is strongest.

2 Because of the humid conditions the soils are the product of rapid weathering and strong leaching. Organic matter decomposes rapidly, so maintaining it at suitable levels is one of the key features of tropical soil utilization.

3 Tropical rain forest is diverse in species composition and structure. Yet at the same time it exhibits a fragile system. There are complex relationships within the ecosystem which can easily be disturbed or broken. The tropical forests of this area have been severely degraded in many parts of the world. Commercial and population pressures have led to large areas being converted to agricultural use, though it is difficult to sustain because of poor soil fertility. Besides the loss of timber and species, the clearance of forest has many environmental impacts at the local, regional and perhaps even global scales. The problem of managing the tropical forests in those areas where it survives is becoming a global concern. Countries want to maximize the utilization of their natural resources, but the resources need to be sustained for the future. How it can best be done has yet to be resolved.

Further reading

Goodman, D. and Hall, A., eds (1990) *The Future of Amazonia: Destruction or sustainable development?* London: Macmillan.
A thorough survey of Amazonian development edited by two economists. Fifteen authors cover current development of the forest, problems of environmental destruction and social conflict and how sustainable development may be achieved.

Longman, K.A. and Jenik, J. (1987) *Tropical Forest and its Environments*, second edition, Harlow: Longman.
A biological text on the nature of tropical forest, with some indication of the practical implications for anyone using or managing tropical forest land.

Park, C.C. (1992) *Tropical Rain Forests*, London: Routledge.
A recent review of the problems and prospects of the tropical rain forest ecosystem. Consequences of clearance are examined at the local, regional and global scales.

Classification is perhaps one of the most vexed and confused aspects of soil science. Although classification of soils is as necessary as for rocks, plants and animals, agreement on a universally acceptable scheme has been elusive. This is partly because soils are inherently difficult to classify, as they are a continuum, with all shades and ranges of properties. It is also partly due to the emotional forces at work, as different international and national soil survey organizations have sought to promote their own schemes. At the international level there are two main competing schemes: *Soil Taxonomy*, developed by the Soil Conservation Service of the US Department of Agriculture (USDA 1975), and the *FAO–UNESCO Soil Map of the World* classification. The latter is the scheme used in this volume for international correlation.

The FAO classification was published in 1974 as the legend for the 1 : 5,000,000 *Soil Map of the World*. It has been used increasingly for international communication ever since. With some modifications it was used in 1985 for the 1 : 1,000,000 *Soil Map of the EEC countries*. The FAO classification consists of an amalgam of traditional names (podzol, rendzina), newly coined names (luvisol, acrisol) and borrowings from the US *Soil Taxonomy* (histosol, vertisol). There are twenty-six higher classes or *orders*, conveniently lettered A to Z, and 106 lower classes or *units*. These are listed in Table A.1. together with correlations with traditional names and features.

Table A.1 Soil units of the FAO classification.

A	Acrisols	Argillic B of low base status
B	Cambisols	Brown earth with weathered B
C	Chernozems	Soils with black humic A
D	Podzoluvisols	Argillic B with irregular top
E	Rendzinas	Dark humic A over limestone
F	Ferralsols	B horizon of sesquioxides
G	Gleysols	Wet soils with gleyed horizons
H	Phaeozems	Humic A over varied horizons
I	Lithosols	Shallow over hard rock
J	Fluvisols	Weakly formed in recent alluvium
K	Kastanozems	Humic A with calcic or gypsic B
L	Luvisols	Argillic B of medium/high base status
M	Greyzems	Humic A with argillic B
N	Nitosols	Other soils with deep argillic B
O	Histosols	Soils of organic material
P	Podzols	Soils with podzolic B
Q	Arenosols	Coarse textured without horizons
R	Regosols	Soils of unconsolidated materials
S	Solonetz	Soils with alkali B
T	Andosols	Soils on volcanic ash
U	Rankers	Humic A over non-calcareous rock
V	Vertisols	Black deeply cracking clays
W	Planosols	Heavy texture in B
X	Xerosols	Saline soils of arid regions
Y	Yermosols	Weakly saline soils of arid regions
Z	Solonchaks	Salt-enriched soils

Table A.2   *1985 revisions of the FAO classification.*

E	Rendzinas	} Amalgamated into new group I Leptosols
I	Lithosols	
U	Rankers	
X	Xerosols	} Deleted. Soils classed on intrinsic properties
Y	Yermosols	
L	Luvisols	Split into high-activity clays L Luvisols and low-activity clays E Lixisols
A	Acrisols	Split into high-clay activity A Acrisols and low clay activity U Alisols
C	Calcisols	Dominated by accumulation of calcium carbonate and/or gypsum
X	Plinthosols	Soils with plinthite
*	Anthrosols	Strong human influences

The 1985 revision will be used for future work in updating old maps and producing new maps. The main changes are detailed in Table A.2. Some orders have been eliminated, others split in two, to give a new total of twenty-seven groups.

In addition to defining soil classes, soil classifications also define *diagnostic horizons* which enable the orders and units to be identified. The FAO–UNESCO scheme for naming horizons is as follows:

H	Peat (Histic)	
O	Surface organic other than peat	
A	Organo–mineral topsoil	
Ah	Mollic A	Humic. Base saturation over 50%
Ae	Umbric A	Base saturation below 50%
A	Ochric A	Light-coloured
E	Eluvial (leached and bleached)	
B	Subsoil horizon formed by weathering and/or illuviation (in-washing)	
Bw	Cambic B	Weathered
Bt	Luvic B	Clay in-washed

Bn	Natric B	Sodium-dominated
Box	Oxic	Fe and Al oxides
Bh	Humic B	Humus in-washed
Bs	Spodic B	Fe and Al in-washed
C	Unconsolidated parent material	
R	Hard rock	

Other horizon designations which can be used with A, B or C horizons are:

k	Calcic horizon	Secondary lime (e.g. Bk)
g	Gleyic horizon	Waterlogging (e.g. Ag)
y	Gypsic horizon	Secondary gypsum (e.g. Cy)
z	Salic horizon	Soluble salts (e.g. Bz)
x	Fragipan	Compact and brittle (e.g. Cx)

Other terms commonly used to describe horizons, but for which there are no horizon symbols, are:

gelic	Permanently frozen (permafrost)
plinthite	Soft iron-rich clay which hardens irreversibly on drying (laterite)
vertic	Expanding clay with signs of swelling and shrinking (cracks, hummocks)

Bibliography

Ackerman, B. (1985), 'Temporal march of the Chicago heat island', *J. Clim. Appl. Meteorol.* 24, 547–54

Addison, K. (1983), *Classic Glacial Landforms of Snowdonia*, Landform Guides 3, Sheffield: Geographical Association

Allen, J.R.L. (1968), *Current Ripples: their relations to patterns of water and sediment motion*, Amsterdam: North Holland

Andrews, J.T. (1975), *Glacial Systems: An approach to glaciers and their environments*, North Scituate, Mass.: Duxbury Press

Atkinson, T.C. (1971), 'Hydrology and Erosion in a Limestone Terrain', unpublished Ph.D. thesis, University of Bristol

Avery, B.W. (1990), *Soils of the British Isles*, Wallingford: CAB International

Ballantyne, C.K. and Harris, C. (1994), *The Periglaciation of Great Britain*, Cambridge: Cambridge University Press

Barazangi, M. and Dorman, J. (1969), 'World seismicity map of ESSA coast and geodetic survey epicenter data for 1961–67', *Bulletin of Seismological Society of America* 59, 369–80

Barry, R.G. (1969), 'Evaporation and transpiration', in R.G. Chorley (ed.), *Water, Earth and Man*, London: Methuen, 169–84.

Barry, R.G. (1971), 'Evaporation and precipitation', in R.J. Chorley (ed.), *Introduction to Physical Hydrology*, London: Methuen, 83–97

Barry, R.G. (1992), *Mountain Weather and Climate*, second edition, London and New York: Routledge

Barry, R.G. and Chorley, R.J. (1992), *Atmosphere, Weather and Climate*, sixth edition, London: Methuen

Beaumont, P., Blake, G.H., and Wagstaff, J.M. (1988), *The Middle East: A geographical study*, London: Fulton

Bennett, R.J. and Chorley, R.J. (1978), *Environmental Systems*, London: Methuen

Billings, W.D. (1974), 'Arctic and alpine vegetation: plant adaptations to cold summer climates', in J.D. Ives and R.G. Barry (eds.), *Arctic and Alpine Environments*, London: Methuen, 403–43

Bojkov, R.D. (1995), *The Changing Ozone Layer*, Geneva: World Meteorological Organization

Bolt, B.A., Horn, W.L., Macdonald, G.A. and Scott, R.F. (1975), *Geological Hazards*, Berlin: Springer-Verlag

Boorman, L.A., Goss-Custard, J.D. and McGrorty, S. (1989), *Climate Change, Rising Sea Level and the British Coast*, London: HMSO

Boulton, G.S. (1992), 'Quaternary', in P.McL.D. Duff and A.J. Smith (eds), *Geology of England and Wales*, London: Geological Society, 413–44

Boulton, G.S., Jones, A.S., Clayton, K.M. and Kenning, M.J. (1977), 'A British ice-sheet model and patterns of glacial erosion and deposition in Britain', in F.W. Shotton (ed.), *British Quaternary Studies: Recent Advances*, Oxford: Clarendon Press

Bradshaw, M. and Weaver, R. (1993), *Physical Geography: An introduction to earth environments*, St Louis: Mosby

Briggs, D. (1977), *Sediments*, London, Butterworths

Browning, K. (1985), 'Conceptual models of precipitation systems', *Meteorol. Mag.* 114, 293–319

Bruce, J.P. and Clark, R.H. (1966), *Introduction to Hydrometeorology*, Oxford: Pergamon

Budyko, M.I., Yefimova, N.A., Aubenok, L.I. and Strokhina, L.A. (1962), 'The heat balance of the surface of the earth', *Soviet Geogr.* 3, 3–16

Burt, S.D. and Mansfield, D.A. (1988), 'The great storm of 15–16 October 1987', *Weather* 43, 90–108

Butzer, K.W. (1976), *Geomorphology from the Earth*, New York: Harper and Row

Carter, R.W.G. (1988), *Coastal Environments*, London: Academic Press

Changnon, S.A. (1971), 'A note on hailstone size distributions', *J. Appl. Meteorol.* 10, 168–70

Clark, J.A., Farrell, W.E. and Peltier, W.R. (1978), 'Global changes in postglacial sea level: A numerical calculation', *Quaternary Research* 9, 265–87

CLIMAP Project Members (1976), 'The surface of the ice age earth', *Science* 191, 1131–7

Colinvaux, P. (1973), *Introduction to Ecology*, Chichester: Wiley

Collinson, A.S. (1977), *Introduction to World Vegetation*, London: Allen and Unwin

Collinson, J.D. (1986), 'Deserts', in H.G. Reading (ed.), *Sedimentary Environments and Facies*, second edition, Oxford: Blackwell, 95–112

Cooke, R.U. and Doornkamp, J.C. (1990), *Geomorphology in Environmental Management: A new introduction*, Oxford: Clarendon Press

Critchfield, H.J. (1983), *General Climatology* fourth edition, Englewood Cliffs, N.J.: Prentice-Hall

Crowe, P.R. (1971), *Concepts in Climatology*, London: Longman

Dalrymple, J.B., Blong, R.J. and Conacher, A.J. (1968), 'A hypothetical nine-unit landsurface model', *Zeitschrift für Geomorphologie* 12, 60–76

Davenport, A.G. (1965), 'Relationship of wind structure to wind loading', *Proc. Conf. Wind Effects on Structures*, Symp.16, vol. 1, London: HMSO, 53–102

Davies, J.L. (1972), *Geographical Variation in Coastline Development*, Edinburgh: Oliver and Boyd

Davies, T.A. and Gorsline, D.S. (1976), 'Oceanic sediments and sedimentary processes', in J.P. Riley and R. Chester (eds), *Chemical Oceanography*, second edition, London: Academic Press, 1–80

Davis, R.A. (1996), *Coasts*, Upper Saddle River, N.J.: Prentice Hall

Dearman, W.R. (1974), 'Weathering classification in the characterisation of rock for engineering purposes in British practice', *Bulletin of the International Association of Engineering Geology* 13, 123–27

Department of Environment (Water Data Unit) (1983) *Surface Water: United Kingdom, 1977–80*, HMSO

Duff, P.McL.D. (1993), *Holmes' Principles of Physical Geology*, fourth edition, London: Chapman and Hall

Elliot, T. (1986), 'Deltas', in H.G. Reading (ed.), *Sedimentary Environments and Facies*, second edition, Oxford: Blackwell, 113–54

Emerson, W.W. (1959), 'The structure of soil crumbs', *Journal of Soil Science* 10, 235–44

Eyre, S.R. (1968), *Vegetation and Soils: A world picture*, second edition, London: Arnold

Fleagle, R.G. and Businger, J.A. (1963), *An Introduction to Atmospheric Physics*, Int. Geophysics ser. 5, New York: Academic Press

Fookes, P.G., Sweeney, H., Manby, C.N.D. and Martin, R.P. (1985), 'Geological and geotechnical aspects of low-cost roads in mountainous terrain', *Engineering Geology* 21, 1–152

Fowler, C.M.R. (1990), *The Solid Earth*, Cambridge: Cambridge University Press

Fuh, Baw-puh (1962), 'The influence of slope orientation on micro-climate', *Acta Meteorologica Sinica* 32, 71–86

Galloway, W.E. (1975), 'Process framework for describing the morphologic and stratigraphic evolution of deltaic depositional systems', in M.L. Broussard (ed.), *Deltas: Models of exploration*, Houston, Tx: Houston Geological Society

Gay, L.W., Knoerr, K.N. and Braaten, M.O. (1971), 'Solar radiation variability on the floor of a pine plantation', *Agricultural Meteorology* 8, 39–50

Gilmour, D. and Bonell, M. (1979), 'Six minute rainfall intensity data for an exceptionally heavy tropical storm', *Weather* 34, 148–58

Golley, F. (1978), 'Decomposition and biogeochemical cycles', in *Tropical Forest Ecosystems*, Paris: UNESCO, 270–85

Goltsberg, I.A. (1969), *Microclimates of the U.S.S.R.*, Jerusalem: Israel Program for Scientific Translation

Goodess, C.M., Palutikof, J.P. and Davies, T.D (1992), *Nature and Causes of Climate Change*, London: Belhaven

Goodman, D. and Hall, A. (1990), *The Future of Amazonia: Destruction or sustainable development?* London: Macmillan

Goudie, A.S. (1984), *The Nature of the Environment*, Oxford: Blackwell

Gregory, K.J. and Walling, D.E. (1973), *Drainage Basin Form and Process: A geomorphological approach*, London: Edward Arnold

Gross, M.G. (1990), *Oceanography*, sixth edition, New York: Macmillan

Harland, W.B., Cox, A.V., Llewellyn, P.G., Pickton, C.A.G., Smith, A.G. and Walters, R. (1982), *A Geologic Time Scale*, Cambridge: Cambridge University Press

Hartmann, D.L. (1994), *Global Physical Climatology*, San Diego: Academic Press

Hastenrath, S. and Lamb, P. (1978), *Heat Budget Atlas of the Tropical Atlantic and Eastern Pacific Oceans*, Madison, Wis.: University of Wisconsin Press

Hayes, M.O. (1976), 'Morphology of sand accumulation in estuaries: an introduction to the symposium', in L.E. Cronin (ed.) *Estuarine Research* II, *Geology and Engineering*, 3–22. London: Academic Press

Heathcote, R.L. (1983), *The Arid Lands*, London: Longman

Hoek, E. and Bray, J. (1977), *Rock Slope Engineering*, second edition, London: Institute of Mining and Metallurgy

Hoffman, R.S. (1974), 'Terrestrial vertebrates', in J.D. Ives and R.G. Barry (eds), *Arctic and Alpine Environments* London: Methuen, 475–568

Houghton, J.L. (1984), *The Global Climate*, Cambridge: Cambridge University Press

Howell, D.G. (1995), *Principles of Terrane Analysis: New applications for global tectonics*, second edition, London: Chapman and Hall

Howells, M.F., Leveridge, B.E. and Reedman, A.J. (1981), *Snowdonia* London: Unwin Paperbacks

Howells, M.F., Reedman, A.J. and Campbell, S.D.G. (1991), *Ordovician (Caradoc) Marginal Basin Volcanism in Snowdonia (North-west Wales)*, London: HMSO for the British Geological Survey

Huff, F.A. and Shipp, W.L. (1969), 'Spatial correlations of storms, monthly and season precipitation', *J. Appl. Meteorol.* 8, 542–50

Huggett, R.J. (1995), *Geoecology: An evolutionary approach*, London: Routledge

Hjulström, F. (1935), 'Studies of the morphological activity of rivers as illustrated by the river Fyris', *Bulletin of the Geological Institute of the University of Uppsala* 25, 221–527

Hurni, H. (1993), 'Land degradation, famine and land resource scenarios in Ethiopia', in D. Pimental (ed.), *World Soil Erosion and Conservation*, Cambridge: Cambridge University Press, 27–61

Imbrie, J. and Imbrie, K.P. (1979), *Ice Ages: Solving the mystery*, London: Macmillan

Ince, M. (ed.) (1990) *The Rising Seas*, Conference Proceedings of Cities on Water, Venice

Inman, D.L. and Nordstrom, K.F. (1971), 'On the tectonic and morphological classification of coasts', *Journal of Geology* 79, 1–21

Institute of Hydrology (1980) *Low Flood Studies*, Wallingford

Intergovernmental Panel on Climate Change (1990), *Climate Change: The IPCC scientific assessment*, J.T. Houghton, G.J. Jenkins and J.J. Ephraums (eds), Cambridge: Cambridge University Press

Jackson, I.J. (1977), *Climate, Water and Agriculture in the Tropics*, Harlow: Longman

James, D.E. (1973), 'The evolution of the Andes', *Scientific American* 229, 60–9

Jenny, H. (1941), *Factors of Soil Formation*, New York: McGraw-Hill

Johnson, D.B. (1985), 'Urban modification of diurnal temperature cycles in Birmingham', *J. Climatol.* 5, 221–5

Jones, R.L. and Keen, D.H. (1993), *Pleistocene Environments in the British Isles*, London: Chapman and Hall

Karig, D.E. (1977), 'Growth patterns in the upper trench', in M. Talwani and W.C. Pitman (eds), *Island Arcs, Deep Sea Trenches and Back Arc Basins*, Washington, DC: American Geophysical Union, 175–85

Kearey, P. and Vine, F. (1996), *Global Tectonics*, second edition, Oxford: Blackwell

Kellogg, W.W. and Schnieder, S.H. (1974), 'Climate stabilization: For better or for worse?' *Science* 186, 1163–72

Kemp, D.D. (1994), *Global Environmental Issues: A climatological approach*, London: Routledge

King, C.A.M. (1962), *Oceanography for Geographers*, London: Edward Arnold

Kirkby, M.J., Baird, A.J., Diamond, S.M., Lockwood, J.G., McMahon, M.D., Mitchell, P.L., Shao, J., Shechy, J.E., Thornes, J.B. and Woodward, F.I. (1997), 'The Medalus slope catena model: A physically based process model for hydrology, ecology and land degradation interactions' in C.J. Brandt and J.B. Thornes (eds) *Mediterranean Desertification and Land Use*, Chichester: Wiley

Knighton, A.D. (1984), *Fluvial Forms and Processes*, London: Edward Arnold

Kummer, D.M. (1991), *Deforestation in the post-war Philippines*, Chicago: University of Chicago Press

Lapworth, C.F. (1965), 'Evaporation from a reservoir near London', *J. Inst. Water Engineering* 19, 163–81

Livingstone, A. and Warren, A. (1996), *Aeolian Geomorphology: An introduction*, Harlow: Addison Wesley Longman

Longman, K.A. and Jenik, J. (1987), *Tropical Forest and its Environment*, second edition, London: Longman

Lovell, J.P.B. (1977), *The British Isles through Geological Time*, London: Allen and Unwin

L'vovich, M.I. (1973), 'The global water balance', *Transactions of the American Geophysical Union* 54, 28–42

L'vovich, M.I. (1979), *World Water Resources and the Future*, Chelsea, Mich.: American Geophysical Union

MacArthur, R.H. (1965), 'Patterns of species diversity', *Biological Reviews* 40, 510–33

Mackay, G.A. and Gray, D.M. (1981), 'The distribution of snow cover', in D.M. Gray and D.H. Male (eds), *Handbook of Snow: principles, processes, management and use*, Toronto: Pergamon Press, 153–90

Malkus, J.S. (1958), 'Tropical weather disturbances – why do so few become hurricanes?' *Weather* 13, 75–89

Mason, B.J. (1970), 'Future developments in meteorology: An outlook to the year 2000', *Quart. J. Roy. Meteorol. Soc.* 96, 349–68

Mason, B.J. (1976), 'Towards the understanding and prediction of climatic variations', *Quart. J. Roy. Meteorol. Soc.* 102, 473–98

Mather, J.R. (1974), *Climatology: Fundamentals and applications*, New York: McGraw-Hill.

Meyer, L.D. and Wischmeier, W.H. (1969), 'Mathematical simulation of the process of soil erosion by water', *Trans. Amer. Soc. Agricult. Engineering* 12, 754–8

Middleton, N. (1995), *The Global Casino*, London: Arnold

Miller, A. (1966), *Meteorology*, Columbus, Oh: Merrill

Miller, D.H. (1977), *Water at the Surface of the Earth*, New York: Academic Press

Mitchell, J.M. (1968), 'Concluding remarks', in *Causes of Climatic Change*, Meteorological Monographs 8, No. 30, Boston, Mass.: American Meteorological Society, 155–9

Moffitt, B.J. and Ratcliffe, R.A.S. (1972), 'Northern hemisphere monthly mean 500 mb and 1000-500 mb thickness charts and some derived statistics', *Geophysical Mem.* 117

Natural Environment Research Council (1976), *Flood Studies Report* 2, Meteorological Studies, NERC

Neiburger, M., Edinger, J.G. and Bonner, W.D. (1982), *Understanding our Atmospheric Environment*, second edition, San Francisco: Freeman

Nelder, G.J. (1985), *Chester's Climate: Past and present*, privately published

Newell, R.E., Vincent, D.G., Dopplick, T.G., Ferruza, D. and Kidson, J.W. (1969), 'The energy balance of the global atmosphere', in G.A. Corby (ed.), *The Global Circulation of the Atmosphere*, London: Royal Meteorological Society, 42–90

Newson, M.D. (1981), 'Mountain streams', in J. Lewin (ed.), *British Rivers*, London: Allen and Unwin, 59–89

Newton, C. and Laporte, L. (1989), *Ancient Environments*, third edition, Englewood Cliffs, N.J.: Prentice Hall

Oke, T.R. (1987), *Boundary Layer Climates*, London: Routledge

Open University Oceanography Course Team (1992), *The Ocean Basins: Their structure and evolution*, Milton Keynes: Open University, and Oxford: Pergamon Press

Palmén, E. and Newton, C.W. (1969), *Atmospheric Circulation Systems*, Int. Geophysics Ser. 15, New York: Academic Press

Park, C.C. (1992), *Tropical Rain Forests*, London: Routledge

Park, R.G. (1988), *Geological Structures and Moving Plates*, Blackie

Peltier, L. (1950), 'The geographic cycle in periglacial regions as it is related to climatic geomorphology', *Annals of the Association of American Geographers*, 40, 214–36

Peters, D.M., Gentry, A.H. and Mendelsohn, R.O. (1989), 'Valuation of an Amazonian rain forest', *Nature*, 339, 655 6

Pethick, J. (1984), *An Introduction to Coastal Geomorphology*, London: Edward Arnold

Pike, W.S. (1992), 'Three motorway traffic accidents in hail showers on 29 March 1986: a case study', *Meteorol. Mag.* 121, 84–8

Pimm, S.L. and Redfearn, A. (1988), 'The variability of population densities', *Nature* 334, 613–14

Polunin, D., and Huxley, A. (1965), *Flowers of the Mediterranean*, London: Chatto & Windus

Reading, H.G. (ed.) (1996), *Sedimentary Environments and Facies*, third edition, Oxford: Blackwell

Sawyer, J.S. (1956), 'Rainfall of depressions which pass eastward over or near the British Isles', *Professional Notes* 7 (118), London: HMSO

Scarth, A. (1994), *Volcanoes*, London: UCL Press

Schmidt, W. (1930), 'Der tiefsten minimum temperatur in Mitteleuropa', *Naturwissenschaft* 18, 367–9

Scotese, C.R., Gahagan, L.M. and Larson, R.L. (1988), 'Plate tectonic reconstructions of the Cretaceous and cenozoic ocean basins', *Tectonophysics* 155, 27–48

Selby, M.J. (1985), *Earth's Changing Surface: An introduction to geomorphology*, Oxford: Clarendon Press

Selby, M.J. (1993), *Hillslope Materials and Processes*, second edition, Oxford: Oxford University Press

Sellers, W.D. (1965), *Physical Climatology*, Chicago: University of Chicago Press

Shellard, H.C. (1976), 'Wind', in T.J.Chandler and S.Gregory (eds), *The Climate of the British Isles*, Harlow: Longman, 39–73

Simmons, I.G. (1989), *Changing the Face of the Earth: Culture, environment, history*, Oxford: Blackwell

Skinner, B.J. and Porter, S.C. (1995), The Dynamic Earth, sixth edition, New York: Wiley

Smith, C. (1992), *Late Stone Age Hunters of the British Isles*, London: Routledge

Smith, D.G. (ed.) (1982), *The Cambridge Encyclopedia of Earth Sciences*, Cambridge: Cambridge University Press

Smith, K. (1975), *Principles of Applied Climatology*, New York: McGraw-Hill

Strahler, A.N. and Strahler, A.H. (1978), *Modern Physical Geography*, New York: Wiley

Stride, A.H. (1982), 'Sand transport', in A.H. Stride (ed.), *Offshore Tidal Sands: Processes and deposits*, London: Chapman and Hall, 58–94

Sugden, D.E. and John, B.S. (1976), *Glaciers and Landscape: A geomorphological approach*, London: Edward Arnold

Summerhayes, C.P. and Thorpe, S.A. (1996), *Oceanography: an illustrated guide*, London: Manson

Thomas, D.S.G. and Middleton, N.J. (1994), *Desertification: Exploding the myth*, Chichester: Wiley

Tricart, J. and Cailleux, A. (1972), *Introduction to Climatic Geomorphology*, London: Longman

Twitchell, K. (1991), 'The not-so-pristine Arctic', *Canadian Geographic* 111 (1), 53–60

USDA (1975), *Soil Taxonomy*, Washington DC: Soil Conservation Service

Varnes, D.J. (1958), *Landslide types and processes*, Special Publication, Highway Research Board: Washington DC, 29, 20–47

Walling, D.E. and Webb, B.W. (1983), 'Patterns of sediment yield', in K.J. Gregory (ed.), *Background to Palaeohydrology*, Chichester: Wiley

Walter, H. (1976), *Vegetation of the Earth in relation to Climate and Ecophysiological Conditions*, London: Springer

Ward, R.C. (1975), *Principles of Hydrology*, second edition, London: McGraw-Hill

Warren, A. (1979), 'Aeolian processes', in C. Embleton and J. Thornes (eds), *Process in Geomorphology*, London: Edward Arnold, 325–51

Warrick, R.A., Barrow, E.M. and Wigley, T.M.L. (1990), *The Greenhouse Effect and its implications for the European Community*, Luxembourg: Commission of the European Communities

Welch, H.E., Bergman, M.A., Siferd, T.D., Martin, K.A., Curtis, M.F., Crawford, R.E., Conover, R.J. and Hop, H. (1992), 'Energy flow through the marine ecosystem of the Lancaster Sound region, Arctic Canada', *Arctic* 45 (4), 343–57

Whittow, J.B. (1992), *Geology and Scenery in Britain*, London: Chapman and Hall

Wilson, E.O. (1992), *The Diversity of Life*, Harmondsworth: Penguin

Windley, B.F. (1984; 1995), *The Evolving Continents*, second and third edition, Chichester: Wiley

Woodwell, G.M. (1967), 'Toxic substances and ecological cycles', *Scientific American* 216 (3)

Wyllie, P.J. (1971), *The Dynamic Earth*, New York: John Wiley & Sons Ltd

Yoshino, M.M. (1975), *Climate in a Small Area*, Tokyo: University of Tokyo Press

Zirin, H. (1988), *Astrophysics of the Sun*, Cambridge: Cambridge University Press

Glossary

Abandoned meander core Ground encircled by the course of a former river meander but now isolated by cut-off across the meander neck by subsequent channel straightening; the core may be a substantial hill in an incised valley.

Ablation till A supraglacial coarse-grained sediment or **till**, accumulating as the subjacent ice melts and drains away and finally let down on to the exhumed subglacial surface.

Abnormal soils A class of soils in Dokuchaiev's classification denoting young or **azonal** soils, e.g. peat, alluvial soils, raw sands.

Abrasion Mechanical wear and tear brought about by the movement of harder rock fragments, ice pellets or organic debris against softer rocks or rock fragments.

Absolute zero The temperature at which atoms and molecules possess the minimum amount of energy and no thermal motion. It corresponds to $-273.15°$ on the Celsius scale.

Absorption The conversion of **radiation** to another form of energy.

Abyssal plain The profound, almost level, largest single component of the deep ocean, lying 4–6 km deep between mid-ocean ridges and trenches.

Accretionary prism A wedge-shaped rockmass of sediment and **ophiolite** transferred during subduction from descending oceanic plate to the adjacent continental plate.

Acid A substance capable of liberating hydrogen ions in water, measured by a **pH** of less than 7.0; acids have corrosive properties and are important agents of **rock weathering**.

Active layer The superficial layer of a **permafrost** land-surface which experiences seasonal melt and refreezing and is, consequently, the site and cause of considerable ground disturbance.

Actual evapotranspiration The amount of moisture evaporated from the ground surface and transpired by plants into the atmosphere.

Adret A mountain slope whose orientation maximizes the receipt of sunlight.

Adsorption The accumulation of ions at the surfaces of clay minerals and humic colloids in soils.

Aeolian Said of the processes, earth materials and land-forms involving the role of the wind.

Aerenchyma Air-filled spaces in the roots and stems of hydrophytic plants.

Aggradation A rise in ground level caused by the accumulation of sediments.

Aggregates Soil structural units of various shapes, composed of mineral and organic material; formed by natural processes, and having a range of stabilities.

Aiguille A steep, frost-shattered rock pinnacle.

Air capacity The percentage of soil volume occupied by air spaces or pores.

Aklé A wavy- or cuspate-edged transverse dune with the points of each cusp pointing downwind.

Albedo An index of the reflecting power of a surface. It is usually used of **short-wave radiation**. Light-coloured surfaces such as ice have a high albedo.

Alkali Said of a substance capable of liberating hydroxide ions in water, measured by a **pH** of more than 7.0, and possessing caustic properties; it can neutralize hydrogen ions, with which it reacts to form a salt and water, and is an important agent in **rock weathering**.

Allogenic successions Plant successions affected by material originating from elsewhere (seeds, sediment).

Alluvial fan A fan-shaped spread of **alluvium** deposited where a tributary stream loses power on entering a more gently sloping valley.

Alluvial toeslope The lowest component of a hill slope bordering the valley floor, where slope and channel processes interact.

Alluvium A general term for unconsolidated, granular sediments deposited by rivers.

Alp A high-altitude bench overlooking a glacial trough or the **alpine meadow** growing there; sometimes used to describe a glaciated mountain in or resembling the European Alps.

Alpha diversity Diversity within a habitat measured by the number of species of a specified taxonomic group in a specified area.

Alpine Said of a rugged, steep and high mountain or mountainous region resembling the European Alps; the region is likely to be a young **orogen** high enough to support **alpine glaciers**.

Alpine glacier A temperate, warm-based mountain glacier characterized by vigorous **mass balance**, high flow velocities and confinement to rock-walled channels.

Alpine heath A vegetation community with dominant dwarf woody shrubs adapted to high-altitude conditions.

Alpine landsystem A distinctive landsystem with integrated glacial, **cryonival**, mountain slope and **ecosystem** elements dominated by cold climate processes on steep slopes.

Alpine meadow A vegetation community characterized by grasses and flowering plants adapted to high-altitude conditions.

Alteration compounds Colloids (clays, iron oxides) formed by the weathering of rock minerals.

Altiplanation The combined processes of **frost shattering** and **solifluction** which progressively level mountain summits, leaving residual rock platforms and **blockfields**.

Alumino-silicates Silicate minerals in which aluminium substitutes for one or more silicate cation.

Ammonification The production of ammonia and ammonium-nitrogen through the decomposition of organic nitrogen compounds in soil organic matter.

Ampferer or *A-subduction* The downward displacement of continental crust, named after an Alpine geologist; buoyancy prevents the granitic crust from being recycled into the mantle but it may melt locally.

Amphidromic point An ocean hub of zero tidal range, around which co-tidal lines radiate towards adjacent coasts, caused by the impact of Earth's rotation on the tidal bulge.

Amygdaloidal The texture of an igneous rock created by the crystallization of secondary minerals in former gas vesicles or voids.

Anabatic Said of the inflowing and ascending portion of a local thermal circulation cell, driven by the warming phase of diurnal heating in a mountainous region.

Anabranching Said of the individual river channels which form a braided channel reach where intervening bars are large and stable.

Anaerobism, anaerobic The absence of free oxygen, or organisms active in the absence of free oxygen.

Anatexis The melting of adjacent (country) rock by an igneous intrusion which swells the size of the developing **pluton**.

Andesite line An imaginary line dividing the Pacific Ocean and rim into two volcanic provinces based on their magma petrology and volcanic morphology; intraplate, basaltic shield volcanoes are distinguished from subduction zone, andesitic strato-volcanoes.

Anion An atom which has gained one or more negatively charged electrons and is thus itself negatively charged.

Annual A plant that completes its life cycle in one growing season.

Antarctic Circumpolar Current The east-flowing, clockwise cold surface current which encircles the Antarctic continent in the absence of land barriers and feeds cold water into the anticlockwise **gyres** of adjacent oceans.

Antarctic convergence The boundary between the north-flowing cold, dense Antarctic water mass and less-dense, south-flowing water masses in the adjacent oceans.

Anthropogenic Said of a process or material originating from human activity.

Aphelion The point on the orbit of Earth when it is farthest from the sun.

Aquiclude A rockmass which absorbs underground water but impedes its onward transfer.

Aquifer Rocks and sediments capable of storing groundwater.

Aquifuge An impermeable rockmass which arrests underground water transfer.

Aquitard A rockmass which retards but does not arrest underground water transfer.

Arch A rockmass spanning a gap weathered or eroded through its core

Arctic-alpine Said of the flora or geomorphology of high-altitude and high-latitude regions, showing common adaptation to, or reliance on, cold climates and **cryospheric** processes; they vary according to differences in daylight

régime and general slope conditions between the two zones. See also **cryonival** and **cryophyte**.

Arctic tundra The distinctive treeless plant community or broader **geoecological** environment of the Arctic basin polewards of the **timberline**.

Area effect That part of island biogeography theory which relates the number of plant colonizers to the size of the island being colonized.

Arête A narrow, precipitous and frost-shattered mountain ridge forming the remnant divide between two glacial **cirques**.

Armouring A crust formed on soil surfaces by stones and chemical cements (iron oxides, lime or silica) which restricts infiltration in the soils of arid and semi-arid regions.

Ash Unconsolidated fine-grained **pyroclastic** material, less than 2 mm in diameter, ejected into the atmosphere by volcanic eruption.

Ash fall tuff A lithified volcanic rock formed by **ash** which has fallen out of a volcanic cloud.

Ash-flow Volcanic **ash** suspended in hot gas and capable of long-distance gravity flow over a land or sea surface as an incandescent cloud or **nuée ardente**; it forms an ash-flow tuff on cooling.

Atmometer An instrument used to measure water evaporated from porous surfaces. Atmometers are inexpensive but they are sensitive to wind speed and are not a good indicator of evaporation from an open-water surface.

Atmosphere 1 Earth's envelope of gases, representing the lightest, volatile products of geological and biological fractionation retained by gravity. 2 A unit of pressure; one atmosphere will support a column of mercury measuring 760 mm in height at sea level.

Attrition The mutual wear and tear of particles in turbulent contact with each other during transport.

Aulacogen A continental rift, frequently found at plate triple junctions, which has failed to develop full sea-floor spreading; it may form a major topographic depression guiding river basin development.

Aurora australis Light produced in the upper atmosphere of the southern hemisphere by the interaction of the solar wind and the magnetosphere. The gases emit visible radiation which causes the sky to glow like a neon light.

Aurora borealis Similar phenomenon to the *Aurora australis* but in the northern hemisphere.

Autecology The study of the ecology of individual species, in contrast to the study of whole communities.

Autocatalysis The positive feedbacks from ice sheet growth which reinforce **icehouse** conditions, including the extension of surfaces with high **albedo**, **cold air drainage**, falling sea level and seaward extension of **grounding lines**.

Autogamy A plant adaptation permitting self-pollination and therefore not dependent on insect or other pollinators.

Autogenic successions Plant successions driven by internal processes and internally induced habitat changes.

Autotrophs Plants and micro-organisms capable of synthesizing organic compounds from inorganic materials by either **photosynthesis** or oxidation reactions.

Available nutrients That proportion of the total nutrient content of soils which plants can absorb and utilize.

Available water The amount of water in a soil available for plant growth after excess water has drained away under the influence of gravity.

Average surface lowering   A rate of denudation extrapolated from sediment yields to give the average thickness of a landsurface layer removed per unit area per unit time.

Avulsion   An abrupt rerouting of stream flow into a new or abandoned channel due to **aggradation** of the floodplain.

Azonal   Soils still in a raw, immature state (young soils).

Back-arc   A basin created by crustal extension or stretching on the further side of a volcanic arc from a B-subduction zone, floored by oceanic crust and usually flooded by the sea.

Backshore   The upper part of a **beach**, lying between the ordinary high-tide mark and the **coastline**, which is waveswept only at exceptionally high tides and may be a source of landward **aeolian** sand transport.

Backwash   A seaward return pulse of water from breaking waves.

Bajada   A continuous alluvial apron in an arid environment, composed of a coalescence of **Piedmont fans**.

Bank caving   The slumping, sliding or toppling of fluvial sediments into an active river channel by current turbulence or at low flow stages when lateral support is absent.

Bankfull discharge   River discharge into a channel at maximum capacity.

Bar   A ridge of coarse, granular fluvial sediment deposited where and when stream velocity falls, especially in midstream and on the inside of meanders.

Bar   The unit of pressure under the metric (cgs) system; one bar = 0.987 atmosphere.

Barchan   An individual sand dune with a crescentic plan form, pointing downwind, with a gentle windward slope and steep leeward slope.

Barchanoid   Said of a coalescence of individual **barchan** dunes into a transverse dune.

Barrier island   A coast-parallel low ridge of coarse granular debris (sand, gravel) sheltering a lagoon on its landward side; common on wide-shelf coasts and may eventually accrete on to the land through storm washover and migration.

Basal shear stress   The shear stress exerted by earth materials, particularly glacier ice, moving over their bed.

Basal sliding   The sliding of a glacier past its bed and sides, greatly facilitated by the presence of subglacial water.

Basalt   A basic fine-grained **extrusive** igneous rock.

Base cations   Metallic cations (e.g. potassium, calcium, magnesium, sodium) that are plant nutrients and take part in **cation exchange** reactions.

Base exchange   The process whereby a basic ion in the soil solution exchanges with a basic cation adsorbed on a soil colloid.

Baseflow   A more enduring component of streamflow contributed by **groundwater** transfers.

Base saturation   The condition when the entire **cation exchange capacity** is occupied by base cations.

Basin and range   A zone of continental crustal extension marked by elevated mountain blocks (**horsts**) separated by **graben** fault basins.

Batholith   A large mass of intrusive (usually granitic) igneous rock emplaced at depth in the core of an **orogen**; eventual exposure at Earth's surface indicates removal of many kilometres of overlying rock.

Beach   A gently-sloping, concave sand and gravel accumulation occupying the **foreshore** and **backshore**, the product of net onshore sediment movement.

Beach cusp   The crescentic component of a transient, sinuous line cut in a beach by breaking waves and caused by turbulence in the **swash** zone.

Bedding plane   A planar boundary separating two layers of sediment and marking a break in the continuity of sedimentation or materials; this discontinuity is likely to be exploited during subsequent **weathering** and erosion.

Bedform   A feature developed in soft sediment by fluid motion across its surface; it involves the entrainment or deposition of sediment and is therefore representative of fluid velocity, flow conditions and sediment particle size.

Benioff or *B-subduction*   The downward displacement of oceanic crust into the mantle beneath continental (or other oceanic) crust by virtue of its greater density, leading to metamorphism and melting; named after geologists Wadati and Benioff and less commonly known as **Wadati–Benioff** subduction.

Benthic   Of the sea- or lake-bed environment.

Berm   A sand or shingle bank with a steep seaward face and flat top marking the upper limit of the **swash** zone on a beach.

Beta diversity   The diversity within a defined area which reflects changes between habitats; measured by the degree of change in a community index along a transect in the field.

Bifurcation ratio   The ratio of the number of streams of one order to those of the next highest order, providing a measure of the connectivity of the stream network; see also **stream order**.

Biocomplexity   Biodiversity studies which include social and economic considerations.

Biodiversity   'The variability among living organisms and the ecological complexes of which they are a part; this includes diversity within species, between species and of ecosystems' (United Nations Conference on the Environment and Development, 1992).

Biogenic sediment   Sediment produced by the biological activity of living organisms and consisting wholly or partly of their remains or derivatives.

Bioherm   An alternative name for a **reef** which stresses its biogenic origin.

Biomass   The total weight of living biological organisms within a specified unit (area, community, population).

Biome   A major ecological community extending over large areas; the dominant plants have a similar physiognomy.

Biosphere   The zone occupied by living organisms at the common boundary of Earth's lithosphere, hydrosphere and atmosphere and depending for its raw materials on geological fractionation and **photosynthesis**.

Biotic climax   The interacting complex of plants, soils and animals which develops in a specified region in response to climate, environmental factors and time.

Bioturbation structure   A sedimentary structure, such as a burrow or cast, produced by the motion or behaviour of living organisms.

Black body   An ideal radiating substance which emits and absorbs all the radiation appropriate to its absolute temperature.

Black box   A system which is treated as a unit without any understanding of its internal relationships. Only the inputs and outputs are identified.

Blockfield   Mountain-top debris dominated by large angular blocks and representing a residual product of **frost shattering**, from which fine debris has been flushed out by wind or water; also known as **felsenmeer**.

Blocky   A type of soil structure which is cube-like and consists of sides with angular or sub-angular corners.

Blow-out   A **deflation** depression, eroded by wind from the face of a vegetated dune.

Bomb   Unconsolidated, blocky **pyroclastic** material, larger than 64 mm in diameter, ejected by volcanic activity.

Bora   A dry, cold air drainage which penetrates the Adriatic basin from the European mainland to the north, especially in winter.

Bore   The leading edge of a **tidal wave** which rises as it moves landwards in constricted estuaries.

Boundary shear stress   The shear stress exerted by the movement of water over a stream bed.

BP   Years **B**efore the **P**resent, based on radiometric dating of a past event and counting back from the base year of AD 1950.

Breaker coefficient   A relationship between wave height, wavelength and beach slope which determines the inshore limit of breaking waves.

Breaking wave   A sea surface wave, whose oversteepened crest outruns its base once it begins to **shoal**.

Brickearth   A lithified deposit of reworked **loess**.

Brine   Sea-water, distinguished from fresh water by its relatively high concentration of dissolved salts.

Brittle failure   The deformation of rock mass by fracture, including **faulting**, which usually occurs abruptly when the strain rate exceeds the ability of the rock to deform plastically.

Bulk density   The weight of soil per unit volume, including all air spaces.

Buried channel   The bedrock channel of a former river infilled now with sediment.

Buried soil   A type of fossil soil buried beneath sediment and no longer at the surface.

Butte   An isolated, **scree**-fringed rock pinnacle in a desert environment representing a remnant of formerly extensive horizontally bedded rocks.

Calcification   The accumulation of calcium carbonate in soil horizons.

Caldera   A major landsurface depression containing one or more volcanic vents and forming by large-scale subsidence as the parent **magma** chamber or **diapir** gradually cools and contracts.

Calving   The detachment of **icebergs** from a glacier or ice-shelf which terminates in water into the receiving lake or marine basin.

Cambering   Tensile fracture and gravitational sliding or sagging of strong, brittle rocks where they are undermined by the failure of underlying weaker, more ductile rocks on hill-slopes.

Capillary   Said of the connected pores or fine 'tubes' in soil which are capable of retaining and moving water against gravity by surface tension or suction, and said of the water itself.

Capillary water   Water held within the capillary pores of soils; mostly available to plants.

Carbonation   The solution of carbon dioxide in water, forming weak carbonic acid which enhances its 'aggressivity' or solution potential; a key preliminary stage in the solution of limestones.

Carnivores   Organisms which are flesh-eating and therefore occupy the third or higher **trophic level** in **ecosystems**.

Cascading system   A system composed of a chain of subsystems which have both spatial magnitude and geographical location, and which are linked by a cascade of mass or energy.

Catchment   A three-dimensional landsystem or **drainage basin** which converts precipitation and groundwater inputs to stream flow and whose components are assessed in terms of their influence on these processes.

Catena   The sequence of soils which occupy a slope transect, from the topographic divide to the bottom of the adjacent valley.

Cation   An atom which has lost one or more negatively charged electrons and is thus itself positively charged.

Cation exchange   The process whereby cations in the soil solution exchange with those adsorbed on soil colloids.

Cation exchange capacity (CEC)   The total amount of exchangeable cations which a soil can adsorb on its colloidal surfaces.

Cavitation   The implosion of bubbles or cavities against a channel wall during rapid, turbulent streamflow and its enhancement of fluid shear stress.

Chamaephytes   Plants which hug the ground surface and where buds are located on the ground surface.

Channel flow   The confinement and concentration of surface water movement in a fluvial channel.

Channel network   The pattern and connectivity of all channels draining a **catchment**.

Channel segment   A short length of fluvial channel selected for the purpose of assessing or modelling relations between channel geometry, stream discharge and sediment transfer.

Chelate   A complex organic compound containing a central metallic ion (e.g. iron, calcium, copper) surrounded by organic chemical groups.

Chelating agent   An organic substance capable of weathering metallic ions from rock.

Chemical energy   A form of energy bound up within the chemical structure of a substance.

Chemical sediment   A non-**clastic** and often crystalline sediment, derived from mineral or organic sources and formed by precipitation from a solution or suspension.

Chemical weathering   The disaggregation of rock mass caused by chemical alteration of some or all of its constituent minerals in the conditions prevailing at or near the landsurface.

Chinook   A dry downflow in the lee of the Rocky Mountains, warming adiabatically on descent and warmer in absolute terms at any given altitude than on its windward ascent.

Chionophilous   Able to survive very long winter seasons completely covered by snow; snow-loving.

Chlorite   A 2:1:1 clay mineral with 2:1 mica units held together by a magnesium (brucite) sheet.

Chlorosis   Yellowing of green leaves caused by lack of an essential plant nutrient or by toxic amounts of acid rain.

Chute   A narrow channel containing a fast-flowing stream in a braided river; or a steep, rock-lined channel between rock pinnacles which funnels debris on to lower slopes.

Cinder   A vesicular **pyroclastic** fragment ejected during volcanic eruption.

Circular sliding   A slope failure whose failure surface is along the arc of a circle.

Cirque   A rock basin excavated into a mountainside by the erosive power of a **cirque glacier** and possessing some or all of the following: steep retaining rock walls, a gently inclined floor or rock basin and barrier, abundant signs of glacial scour and a terminal moraine.

Cirque glacier   A small mountain glacier which excavates and occupies a **cirque**; its diminutive size renders it particularly sensitive to local climate and climatic change.

Clast   A rock fragment derived by weathering and erosion from existing rock mass; individual clasts more than 2 mm in diameter are often distinguished from smaller fragments, which form a rock matrix.

Clastic sediment   A sediment composed of rock fragments, regardless of their individual size, rather than chemical precipitates or biogenic material; it may become lithified by the precipitation of a chemical cement.

Clay minerals   Crystalline colloids smaller than 2 μm in diameter; mostly new minerals formed by **weathering** and **soil formation processes**, and very important in determining the properties of soils.

Cleavage   A rock texture in fine-grained materials with parallel microplanes or fractures, dependent on individual **platy** crystal structure or the alignment of platy minerals in rock mass; the term is also used to describe the tendency for such rocks to split along these planes.

Climax community   The **plant community** which marks the end point of a **succession**; it is relatively stable and in equilibrium with prevailing environmental conditions.

Climax pattern   The pattern of climax communities which develop in a defined area.

Coastal cell   A discrete unit of coastline identified for management purposes, recognizing the integration of coast-parallel as well as coast-normal water and sediment transfers and multiple-use by human socio-economic activity.

Coastal plain   A gently sloping landsurface which forms a continuum with the **continental shelf** and is susceptible to small sea level changes; it is likely to be wide on **trailing-edge** (**passive margin**) coasts and narrow on **leading-edge** (**convergent margin**) coasts.

Coastline   The boundary between land and sea or, more precisely, a permanently exposed landsurface and the highest occurrence of storm waves.

Cohesion   The intermolecular bonding of constituents of earth materials by chemical, magnetic and electrostatic forces.

Cohesive strength   The portion of total rock mass strength dependent on the extent of **cohesion** developed between particles or crystals by intermolecular forces.

Cold air drainage   The gravity flow of cold air by virtue of its greater density.

Cold glacier   A glacier whose mass is of predominantly cold, polythermal ice with thin surface and basal layers periodically at **pressure melting point**; typical of the greater part of large ice sheets.

Cold sea-water weathering   The chemical alteration of rock at the sea bed by cold water, as distinct from hydrothermal circulation, primarily by hydration, although some oxidation may also occur.

Cold stage   A period of Earth history in which global climates are significantly colder than at present: snow, ice and **permafrost** cover large portions of continental landsurfaces and polar oceans, with prevailing **cryospheric** and **icehouse** conditions. Up to twenty cold stages have dominated the Quaternary period and the term is more appropriate than 'glacial period' or 'ice age' – glaciers were absent from many areas for much of the time.

Colluvium   Granular debris accumulating towards the base of slopes as a result of mass wasting of bedrock and older slope deposits.

Columnar   A type of soil structure consisting of vertical units with rounded tops; usually in the subsoil of alkaline clays.

Competition   Between organisms with similar growth requirements, and which compete for them.

Components   The pathways by which energy and matter flow between elements of the system.

Compound translation failure   A slope failure event involving two or more processes, such as rotation and sliding, or sliding and flow, as material properties and combinations change during initial motion.

Compressive flow   A deceleration in glacier flow which creates compressive stress, longitudinal crevasses and thickening.

Compressive strength   The maximum compression an earth material may resist before failure occurs.

Conceptual model   A general model based on conceptual and qualitative ideas of relationships.

Conduction   The process of heat transfer through matter without movement of the matter itself. It is the process whereby heat travels through solids.

Cone sheet   A **dyke** intruded upwards and radially outwards from a **magma** reservoir, giving it a cone shape.

Conformable sequence   An unbroken rock-forming sequence in which each successive layer forms an undisturbed contact with its predecessor, with little or no time separation.

Connectance   The number of functioning ecological links which connect the members of a defined community.

Constructive margin   The boundary between two diverging crustal plates where new oceanic crust is formed.

Contact metamorphism   Thermal **metamorphism**, involving chemical alteration or recrystallization of rock in direct or close contact with a **magma**.

Continental drift   The relative movement of continents over Earth's surface summarized by Alfred Wegener in 1912 and now known to be part of the wider processes of plate tectonics and sea-floor spreading.

Continental rise   The lowermost section of a submerged continental margin, which rises gently from an **abyssal plain** before steepening into the **continental slope**.

Continental shelf   The area of continental crust which lies below sea level and extends beyond the coast as a shallow, gently sloping plain as far as the **continental shelf break**; with an average width of 70 km, it is more extensive on passive than convergent plate margins.

Continental slope   The submerged continental margin seaward of the **continental shelf break** which steepens and extends down to the **continental rise**.

Continuum   A range of properties in which change occurs continuously and more or less smoothly, rather than in an abrupt, stepwise manner or discontinuum.

Convection   A process of heat transfer in a fluid, involving the movement of substantial volumes of the fluid concerned. Convection is very important in the atmosphere and, to a lesser extent, in the oceans.

Cordillera   Mountain chains or ranges, usually long (10^3 km) but narrow (10^2 km) in extent; subduction orogens typically form **cordilleran** mountain systems.

Core   Earth's innermost, high-temperature and dense nickel–iron sphere.

Core-stone   The residual, relatively unweathered core of a larger joint-bound block whose outer parts are severely weathered or disintegrated.

Co-tidal line   A line linking points on a map at which high tide occurs simultaneously, usually measured in hours before or after the high tide at a suitable reference point.

Coversand   An extensive sand sheet, generally thin and lacking bedforms, covering a landsurface adjacent to a current or former ice sheet from whose fine-grained debris it was deflated.

Craton   An area of continental crust, generally stable throughout the Phanerozoic, comprising a crystalline core or **shield** and marginal **platform** of metamorphic and sedimentary rocks.

Creep   Slow and continuous non-recoverable **plastic** deformation of rock mass, soil or ice under gravitational stress accomplished by intergranular motion.

Crevasse   A brittle fracture caused by **extending flow** in a moving glacier; also used to define a channel breaching a river bank or **levée**.

Crevasse splay   A fan of coarse alluvium spread over the flood plain from a channel **crevasse**.

Crust   Earth's outermost solid sphere, representing the upper part of the lithosphere and differentiated into lighter, thicker continental crust and denser, thinner oceanic crust.

Crustal extension   Crustal stretching and consequential thinning, achieved by faulting and rifting, in response to a number of tectonic and geomorphic processes; it is associated particularly with divergent plate boundaries.

Crustal shortening   Crustal compression and consequential thickening, achieved by folding and thrusting, and associated particularly with convergent plate boundaries.

Crustal thickening   An increase in crustal thickness caused by intracontinental thrusting in an **A-subduction** zone, where the buoyancy of light continental crust precludes **B-subduction** and **resorption**; isostatic balance is maintained by the displacement of denser crust below the thickened pile.

Crusting   The reduction of soil infiltration capacity by surface compaction during heavy rainfall, the dispersal of fine grains into pores and subsequent drying.

Crustose lichens   Lichens with flat, crust-like growth form; lichens are symbiotic associations between a fungus and an alga.

Cryofracture   An alternative term for **frost shattering**.

Cryonival   Said of processes related to the combination or close proximity of snow and ice, such as frost weathering at the margins of a snow patch and sourced by its diurnal thaw–freeze.

Cryopedology   Soil-forming processes occurring in low-temperature conditions, usually in the presence of **permafrost**.

Cryophyte   A cold-tolerant plant of the **arctic–alpine** community with life forms and cycles adapted to persistent snow and ground ice.

Cryosphere   The portion of Earth's hydrosphere where water is perennially frozen as snows, **glacier** or **ground ice**.

Cushion mosses   Mosses with erect stems.

Cut and fill   The erosion and subsequent sedimentation of new fluvial channels cut through existing fluvial sediments, and the sedimentary structures so formed.

Cyclic climax theory   The theory of climax vegetation which emphasizes cyclical rather than unidirectional change; mostly discussed in relation to tropical forests.

Darcy's Law   A definition of the relationship between the transmission of water through porous earth material, water viscosity and the height of the free water surface above the point of measurement.

Debris cone   A cone-shaped mass of rounded boulders deposited where a mountain stream meets a flat valley floor.

Debris flow   A mass wasting process in which fluidized **colluvium** or other loose, granular debris moves downhill in a series of rapid turbulent pulses.

Decalcification   The removal of carbonates (mostly calcium carbonate) from a soil horizon.

Decollément   A zone along which overlying rocks or other earth materials have become detached from and have moved over underlying materials, normally as the respective materials respond to shear stress.

Decomposers   Organisms that feed on dead or decaying organic matter.

Deflation   The removal by wind of fine granular materials, especially sand, silt and snow.

Deflation hollow   A surface depression in a material which has undergone **deflation** or been abraded by wind-blown sand.

Deformation   The alteration of earth material from an initial shape and internal structure, involving compression, extension, **folding**, **faulting**, shear, etc.

Delayed flow   An estimated component of stream flow which is delayed in reaching the channel after a precipitation event but does not form part of the **baseflow**.

Denitrification   Bacterial processes occurring in soils, in the absence of free oxygen, to break down nitrates and nitrites with the evolution of free nitrogen.

Density   The mass of a unit volume of a substance, at a specified temperature and pressure. It is normally measured in kg m^{-3} or g cm^{-3}.

Denudation   The combined processes of **weathering**, mass wasting and erosion which cause the disaggregation of rock mass and its removal to lower-lying ground; it may proceed far enough to cause the virtual destruction and levelling of continental crust down to **sea level**.

Denudation chronology   A scheme of landsurface development which saw many variations developed over almost a century up to the 1960s, based on belief in a **denudation** cycle; modern **plate tectonics** provides more realistic mechanisms but more complex geomorphological histories.

Denudation cycle   A simple scheme of continental **denudation** which involved the cyclical stimulation of uplift of a new landsurface, followed by progressive denudation down to a low residual plain and its renewed uplift or 'rejuvenation'.

Depletion curve The expression on a **hydrograph** of the gradual decline in the **baseflow** component of stream discharge due to dwindling groundwater discharge.

Deposition Any means by which sediments may accumulate from a condition of motion or activity such as mechanical fall-out from suspension, chemical precipitation or the assemblage of biogenic debris.

Depression storage The short-term storage of precipitation in small surface depressions, from which it may evaporate or percolate underground, and its capacity to delay the start of **overland flow**.

Desertification 'Land degradation in arid, semi-arid and dry subhumid areas resulting mainly from adverse human impacts' (United Nations Environment Programme, 1992).

Desert pavement A loose, gravelly surface layer protecting underlying rock or fine sediments and probably representing a residual material from which fines have been deflated or washed out.

Destructive margin The boundary between two convergent crustal plates, where ocean crust is consumed by subduction; however, it also triggers igneous activity which forms new crust at volcanic arcs.

Detrivores Organisms that feed on dead or decaying organic matter.

Devensian stage The most recent global **cold stage**, known as the Weichselian in Europe and Wisconsinan in North America, extending from c. 115 ka to 10 ka BP.

Diagenesis Minor, non-destructive changes in the mechanical or chemical properties of rock shortly after initial emplacement, associated with the final stages of **lithification**.

Diamicton A non-sorted, terrigenous sediment containing a very wide range of particle sizes; the term is descriptive, not generic, and a prefix would identify its origin, e.g. a **glacial** diamicton.

Diapir A bulbous intrusion of less dense igneous or sedimentary rock which forms a dome or broad fold in denser surrounding rock through which it rises.

Diffuse double layer A model to explain the distribution of ions concentrated near colloid surfaces in soils; the inner layer at the colloid surface consists of positively charged cations, whereas the outer layer in the soil solution contains equal amounts of cations and anions.

Dilatancy A property of fine-grained sediments which, unusually, causes expansion and stiffening when compressed through the rearrangement of grains into a larger volume and consequent intake of water into the voids.

Dilation Strictly speaking, a deformation involving an increase in the volume of an earth material without a change in shape but more often understood mistakenly as 'pressure release' in rock mass.

Dip The angle of inclination of a rock structure from the horizontal.

Direct deposition The sublimation of ice directly on to a cold surface from water vapour.

Discharge A volume of river flow per unit of time expressed in cubic metres per second or litres per second.

Disclimax A **climax community** maintained by human activities; the US term for 'plagioclimax'.

Discontinuous rockmass strength The lower resistance to shear offered by fractured rock mass, compared with the **intact strength** of rock between the fractures; in essence, the continuity of **cohesion** and **friction** is destroyed or reduced by the fractures, which also permit water to enter the rock mass, with further destabilizing effects.

Dispersion tectonics Tectonic processes which lead to the large-scale separation and spatial dispersion of crustal fragments.

Displaced terrane A **terrane** or crustal fragment with a distinctive suite of rocks which has been displaced tectonically away from where it formed.

Diversity The variety and relative abundance of species in a defined area.

Downwelling A convergence and subsidence of ocean surface water.

Draa A large **aeolian** sand dune complex composed of megadunes on which smaller dunes may be superimposed.

Drainage basin A specific geographical area, bounded by a **watershed** and drained by a discrete drainage network.

Drainage density The total stream channel length per unit landsurface area, normally calculated for an entire **drainage basin**.

Drainage network More or less synonymous with the **channel network** but may also include rills, gullies and larger underground pipes not considered part of a permanent surface channel network.

Drainage pattern The geometric configuration or plan of a drainage network which usually reflects catchment geology, tectonic and denudation history.

Draw-down The process and extent of gravitational or artificial withdrawal of water from **drainage basin** stores.

Dry valley A surface valley showing evidence of erosion by fluvial processes but rarely or never occupied by a modern stream.

Drumlin A large, subglacial **bedform** composed mostly of **till** and streamlined in the iceflow direction; it is indicative of active iceflow and probable deformation in the sediment body.

Dune A mobile sand-wave **bedform** shaped by fluid motion, found in a wide range of wavelengths (from cm to km) and environments, ranging from stream channel beds to coasts and deserts.

Duricrust A hard, crystalline crust found on tropical landsurfaces formed by **evaporite** minerals brought there in solution by **capillary** action from underlying soil and rock.

Dyke A columnar igneous **intrusion** which cuts discordantly through existing rock structures.

Dynamic equilibrium A form of self-regulation in a system which maintains a similar type of system.

Dynamic viscosity The resistance to flow of a fluid.

Eccentricity of the orbit The changing shape of Earth's orbit around the sun from a more circular to a more elliptical path. It varies over a cycle of almost 100,000 years.

Ecological amplitude The overall range in which an organism can function; the sum total of all individual tolerance ranges for a particular species.

Ecological niche The position of an individual species within an ecosystem, in terms of function, space and time.

Ecological optimum That part of the range of a plant species where the plant's vigour is greatest.

Ecological status The position of a **plant community** within the hierarchy of seral and climax communities.

Ecosphere The biologically inhabited part of the Earth, oceans and atmosphere.

Ecosystem Open system comprising plants, animals and their environments which is involved in the flow of energy and the circulation of matter.

Ecosystem stability The behaviour of the entire system in response to an external perturbation; can be defined by several indicators, including resistance and resilience.

Ecotone A zone of transition, and hence competition, between two contiguous **plant communities**.

Edaphic climax Climax vegetation maintained by soil conditions (e.g. wetness, chemistry).

Eddy diffusion The mixing of atmospheric matter and properties which is brought about by eddies.

Eddy viscosity The resistance to flow of a fluid caused by friction between individual strands of the flow.

Edge wave A wave moving approximately at right angles to the shore and breaking waves, as a result of the need to drain water being pushed onshore which cannot easily escape through the **swash**.

Eemian stage The penultimate global **temperate stage**, known as the Ipswichian in Britain and as the Sangamon in North America, extending from *c*. 135 ka to 115 ka BP.

Effusive Extrusive igneous activity characterized by a steady outflow of basaltic material from a fissure (cf. **explosive**).

Elastic A material condition in which strain (deformation, or change of shape) is wholly and immediately recoverable upon the removal of stress.

Elastic strain release The restoration of the original shape of a material which has experienced elastic deformation, when the stress which caused it is released; thus rock which deformed elastically when compressed will expand on 'pressure release' and may fracture if the recovery is imperfect.

Electrical conductivity (EC) The ability of a soil to conduct an electric current; used as an index of soil salinity.

El Niño The appearance of unusually warm water off the South American Pacific coast when the westward-driven equatorial ocean current periodically falters, owing to a reduction in trade wind strength in the equatorial Pacific Ocean. This, in turn, suppresses **upwelling** cold, deep water induced by these atmospheric and ocean currents; ocean–atmosphere coupling further disturbs pressure and precipitation systems throughout the region (the 'Southern Oscillation'). The effect occurs every few years, commencing around Christmas time – hence its Hispanic allusion to the Christ child.

Eluviation The removal of suspended solids or mineral colloids from a higher to a lower soil horizon by water percolation.

Emerson model A model of soil aggregate formation involving clay domains, organic linkages and quartz particles.

Endogenetic Energy supplied by Earth. It is mainly derived from the hot interior of Earth.

Energy pyramids The pyramidal structure of all **ecosystems** when measured by the flow of energy.

Entisol A soil order in the USDA classification characterized by shallow soils lacking distinct horizons.

Epeirogenesis The elevation or depression of large areas of crust without major deformation, in contrast with **orogenesis**, resulting from either thermal or mechanical processes.

Ephemeral stream flow Intermittent stream flow through all or part of a channel generated only by precipitation events and common in arid and semi-arid zones.

Epicontinental sea A partially enclosed marine basin on continental crust linked to an adjacent ocean (cf. **marginal sea**).

Equatorial counter-current An eastwards, equatorial gravity flow driven by the slight westerly rise in the ocean surface stacked up by more pronounced westerly currents.

Equilibrium The state of a system which over time tends to maintain its general structure and character in sympathy with the processes acting upon it.

Equilibrium line altitude (ELA) The altitude which marks the surface boundary between the upper accumulation and lower ablation zones of a glacier; it may be taken as the position at the end of an ablation (summer) season or the average position over several years.

Equitability A measure of ecological diversity based on information theory.

Erg A sand desert or large sand sea.

Erodibility The susceptibility of earth materials to erosion.

Erosion Any dynamic process which causes the removal of earth materials, distinguished here from **weathering**, **denudation** and **mass wasting**.

Erosivity The erosive power of a stream or other agent.

Erratic A rock fragment transported away from its source and recognizable as such after it has crossed the outcrop of its parent lithology.

Esker The alluvial bed of an en- or sub-glacial stream which may survive as a long, sinuous partially-collapsed debris ridge.

Etch-front The boundary separating weathered from unweathered rock, at the landscape rather than individual rock scale.

Eustatic Relating to global sea-level oscillation caused by absolute changes in sea-water volume.

Evaporation pan An open water tank used to measure evaporation; sizes vary between different countries.

Evaporite A mineral or sedimentary rock precipitated from a saline solution as a result of evaporation.

Evorsion The corrosion of stream channel **potholes** by pebbles swirled around by vortices and eddies.

Exchangeable cations Cations which are held on exchange sites on soil colloids and which are able to exchange with cations in the soil solution.

Exchangeable sodium percentage (ESP) The percentage of the **cation exchange capacity** occupied by exchangeable sodium ions; used as an index of soil alkalinity.

Exfoliation Mechanical or physical weathering which proceeds by the disintegration and removal of successive layers of rock mass.

Exogenetic Energy derived from outside Earth. The vast majority of such energy is from the Sun.

Explosive Extrusive igneous activity characterized by violent eruptions of **felsic** material through volcanic vents.

Extending flow A zone of accelerated flow within a glacier which creates tensile stress, lateral crevassing and thinning.

Extrusive A description of molten igneous material which erupts at Earth's surface before cooling.

Facilitation model A model of plant **succession** in which a habitat is modified by a species in such a manner as to favour its replacement by other species.

Factor of safety A measure of the balance between shear stress and shear strength in a slope; a state of **limiting equilibrium** exists when shearing forces equal resisting forces in a slope and F = 1.

Facultative relationships Relationships which exist under various conditions.

Fall velocity The specific velocity below which a moving fluid is unable to sustain a given particle size in suspension; or the rate at which suspended particles settle through a fluid.

Fall wind Any wind characterized by its descent of leeward mountain slopes, irrespective of its thermal character and origins.

Falling limb The expression on a **hydrograph** of the subsiding **quickflow** component of stream discharge.

Fatigue The progressive weakening of a material through cyclic application and removal of non-critical stress which leads to its eventual failure.

Fault A line or zone along which **faulting** has occurred in rock mass.

Fault breccia Angular rock rubble lining a fault and formed by shearing and crushing of a wider zone of rock mass bounding the fault during movement.

Faulting The process of fracture or brittle failure of rock with displacement of adjacent parts on either side of the **fault**.

Feedback The property of a system such that, when change is introduced via one of the variables in the system, its transmission through the system leads back to a change in the original variable.

Felsenmeer A German term for **blockfield**.

Felsic A mnemonic from *feldspar* and *silicate*, which identifies the silicate-rich igneous rocks characterized by their light-coloured, acidic minerals.

Fetch The extent of open water over which a dominant wind develops a wave system.

Field capacity The maximum volume of water held in the voids of a soil when gravitational drainage is complete, comprised of **capillary** and **hygroscopic** water.

Firn A stage in the transformation of snowpack towards glacial ice, with a density of 0.4–0.5 kg m^{-3}.

Fissure A wide fault or tension crack in the landsurface often associated with a linear volcanic eruption.

Fixation The transformation in soil of a plant nutrient from an **available** to an unavailable state.

Fjord A long, deep rock basin excavated by an **outlet** or valley glacier between high rock walls and flooded by the sea during deglaciation.

Flandrian stage The current global **temperate stage**, more or less synonymous with the Holocene, which commenced *c.* 10 ka BP.

Flexural isostasy Localized isostatic adjustment peripheral to, and in an opposite direction from, an area of crustal loading or unloading; due to flexure or **creep** in the lithosphere.

Flocculation The aggregation of individual suspended clay particles into larger masses, with consequential implications for their sediment dynamics.

Flood basalt An extensive basalt flow extruded from continental rifts and fissures which forms distinctive terrestrial landsurfaces.

Flood plain A lowland landsurface prone to episodic river floods and associated alluvial sedimentation.

Flow régime The range of styles of streamflow and their related **bedforms** and modes of sediment transport.

Fluid stressing The erosive power of a stream contributed by fluid shear stress at the stream bed.

Flysch Submarine **turbidite** sediments eroded from an orogenic belt during uplift and therefore located in adjacent **trench** or **back-arc** zones.

Föhn An equivalent of the **chinook** wind in the European Alps.

Fold A compressional or gravitational **plastic** deformation of earth materials which has bent and shortened a previously planar mass.

Foliation A close-spaced planar texture in rock acquired by the alignment of platy minerals during **metamorphism**.

Force Mass × acceleration measured in newtons, N (SI units).

Fore-arc A zone lying between and parallel to the **trench** and **volcanic arc** of a **B-subduction** zone, consisting of an outer ridge and an inner basin.

Foreshore That part of a beach lying between the low-water mark and maximum high-tide mark.

Formation A world vegetation type dominated by plants of similar life form and identified in a geographically distinct area.

Fossil soil A soil formed in the past under environmental conditions which no longer exist.

Fractional crystallization Separation of a magma during cooling into distinctive, usually mineral specific parts.

Fractionation The separation of a mixture into its component elements and minerals.

Fragipan A dense pan-like subsoil horizon, indurated by physical and/or chemical processes, and with a high bulk density.

Free-living fixation Fixation of atmospheric nitrogen into organic nitrogen compounds by soil micro-organisms which live freely in the soil rather than existing in association or symbiosis with a plant.

Friction strength The portion of rock or soil strength dependent on the frictional resistance between constituent particles.

Frost shattering The fracture of rock mass attributed to internal stress generated by expansion on freezing, pore water migration to a freezing plane or hydration in the **permafrost** environment; it is probable that some form of **fatigue** is involved over many freeze–thaw cycles.

Froude number An index of the type of flow in a stream.

Fundamental niche The maximum area, in terms of space, time and function, which a species would be capable of occupying in the absence of competition from other species.

Gabbro A basic, coarse-grained igneous rock; the **intrusive** near equivalent of **basalt**.

Geochemical cycle The movement of rock minerals which accompanies the **rock cycle**, characterized by aggregation, disaggregation, fractionation, refinement, changes of state and the formation of new species as rockmass itself is cycled.

Geoecology An integrated discipline which studies interactions between geological, geomorphic, ecological and meteorological components of the landscape.

Geologic process A process usually involving or associated with Earth's near surface or interior rocks, and distinct from – or combined with – **geomorphological** process.

Geomorphic   Pertaining to Earth's surface landforms and their study.

Geomorphological process   A process involved in the formation and alteration of the landforms at Earth's landsurface.

Geophytes   A class of plants which reproduce from bulbs, corms, rhizomes or tubers.

Geothermal heat flow   The heat loss from Earth's interior to space, measured at an average Earth surface flux of 82 mW m^{-2} but locally varying according to the proximity of hot-spots, volcanoes, etc.

Gibber   A desert rock surface covered with **lag gravels**.

Gilbert-type delta   A 'classic' fan delta with successive overlapping topset, foreset and bottomset beds on the **prograding** surface, advancing front and distal slope respectively.

Glacier   A large accumulation of terrestrial ice and superficial snow, metamorphosed from annual snowfall and other precipitation and capable of deformation and flow under its own mass.

Glacio-eustatic   The change in ocean water volume and global sea level in response to the growth and decay of ice sheets and glaciers, which has dominated Quaternary sea-level change.

Glacio-isostatic rebound   The **isostatic** uplift of a land-surface formerly supporting an ice sheet, due to the removal of the weight of ice and eroded rock.

Glaciomarine sediment   Rock debris released into the sea from **tidewater glaciers**, floating **ice shelves** and **icebergs**.

Glaciotectonic   Said of the deformation of ice, bedrock and sediment by glacier iceflow and consequent fold, thrust and shear structures.

Gleying   Soil processes characteristic of wet or waterlogged soils; usually denoted by bluish-grey colours and reddish mottles, produced by a complex series of oxidation and reduction reactions.

Global hydrological cycle   The global stores and transfers of water in its liquid, solid and gas phases.

Global Ocean Conveyor   A slow, three-dimensional ocean current system transferring warmer surface and intermediate water northwards, with a return southwards flow of deep, cold water (see also **thermohaline circulation**).

Gouge   Soft, fine-grained debris lining a **fault** or mineral vein.

Graben   A down-faulted rockmass flanked by parallel faults and often forming a structural valley.

Grade   A property of soil structure which describes the strength or stability of soil aggregate development.

Gradients   Changes in environmental conditions along a gradient.

Grain ballistics   Collisions between moving and stationary particles in a fluid boundary layer in which energy transferred to stationary particles moves them horizontally or entrains them in the flow.

Granite   A coarse-grained **intrusive** igneous rock.

Gravity   The force exerted on any body by Earth's mass and axial rotation; it is an important endogenetic source of energy for geological and geomorphic processes at or near Earth's surface.

Gravitational energy   The potential energy acquired by virtue of an object's distance, or further displacement away, from Earth's centre; it plays an important role in tectonic processes in the mantle and crust and in geomorphic processes at the surface.

Gravitational water   The class of soil moisture which drains from a saturated soil under the influence of gravity; it is approximately equivalent to water held in pores larger than 50 μm diameter.

Greenhouse effect   The condition in which Earth's average global temperatures are normally higher than predicted by radiation laws by virtue of the presence of substances in the lower atmosphere capable of absorbing outgoing long-wave radiation.

Grey box   A partially understood system, in which interest is centred on a restricted number of subsystems and the remainder are ignored.

Grey dune zone   A stage in a **psammosere** on sand dunes when mosses and lichens colonize the ground surface, and darken the colour of the yellow sand.

Gross primary productivity (GPP)   The total amount of solar energy fixed in **photosynthesis** by **autotrophs** per unit area per unit time.

Ground heave   Small-scale ground expansion and uplift in unconsolidated materials through hydration, ice formation and influent water seepage, with consequent disturbance of any incipient structure and strength.

Ground ice   Any form of frozen water below the land-surface, irrespective of its origin and whether it is **interstitial** or **segregated ice**.

Grounding line   The water depth at which an **ice shelf** or **tidewater glacier** begins to float by virtue of its lower density than water.

Groundwater   The portion of all subsurface water stored in saturated rock below the water table; sometimes extended to include water in the overlying unsaturated layer.

Gulf Stream   The warm, north-east-flowing current of the clockwise **gyre** in the North Atlantic Ocean, flowing from the Florida coast into the Arctic basin past Britain and Norway; also known as the North Atlantic Drift.

Gullies   A modest, steep-sided channel eroded by intermittent streamflow with a frequency and vigour capable of keeping the channel open.

Guyot   A flat-topped submarine mountain or **seamount**.

Gyre   A system of surface, wind-driven ocean currents forming a closed or partially closed circulation which transfers heat from warmer to colder surface waters.

Halocline   A zone of marked change in salinity with ocean depth.

Halophyte   A plant adapted to growth in saline environments.

Halosere   The sequence of **plant communities** which, successively, occupy a salt marsh.

Hamada   An upland desert landsurface of wind-scoured bare rock with patches of **lag gravels**.

Hardware model   A model of a system composed of real objects. A flume is a hardware model of a river; a wind tunnel is a *hardware* model of airflow near the ground.

Headward retreat   The upslope migration of the point of initiation of channel flow as it continues to attract ever more surface or saturated **overland flow**.

Heinrich event   A surge of marine-based portions of the former Laurentide ice-sheet in Canada which sent a pulse of icebergs, ice-rafted debris and cold fresh water into the north-west Atlantic Ocean.

Helical flow   A spiral motion superimposed on the general direction of streamflow or airflow which causes lateral transport of energy and entrained materials.

Heliophyte   A sun-loving plant, adapted to high exposure to sunlight in its mature form.

Hemicryptophytes   Tussock plants whose buds are located at or just below the surface of the soil.

Herbivores   Organisms which eat plants and therefore occupy the second **trophic level** in ecosystems.

Hydration   The incorporation of water into the chemical composition of a mineral, converting it from an anhydrous to a hydrous form; the term is also applied to a form of weathering in which hydration swelling creates tensile stress within a rockmass.

Hydraulic conductivity   The rate at which water is able to move through a soil.

Hydraulic drop   An abrupt step over which a stream surface falls as the stream enters a steeper segment.

Hydraulic efficiency   The conservation of potential energy in streamflow, achieved by minimizing friction losses against the channel, etc.

Hydraulic jump   An abrupt step or standing wave over which a stream surface rises, as the stream enters a less steep segment or encounters a submerged obstacle.

Hydraulic radius   The relationship between the wetted perimeter to the cross-sectional area of a stream; the higher the value, the more efficient the channel.

Hydrogenic   The description of rocks formed at the sea bed by the **hydrothermal circulation** of sea water through mid-ocean ridges.

Hydrogeology   The study of the terrestrial part of the hydrological cycle and the association between its vegetation, earth materials and streamflow.

Hydrograph   A plot of the variation of stream discharge with time at a selected point in a catchment and capable of separation into estimates of the various components of flow.

Hydrologic sequence   The topographic sequence of soils, from ridge crest to adjoining valley bottom, which reflects the changing soil–water régimes downslope.

Hydrolysis   A form of chemical weathering in which the H^+ and OH^- ions of water react with a mineral, with consequent loss of strength.

Hydrometeorology   The study of the atmospheric part of the hydrological cycle and the association between precipitation, evapotranspiration and the **drainage basin**.

Hydrophobic   A soil structure or soil constituent which repels water.

Hydrophyte   A plant adapted to growth in wet or water-logged conditions.

Hydrosere   The sequence of **plant communities** which, successively, occupy a silting-up freshwater lake.

Hydrosphere   Earth's outer, liquid envelope, concentrated almost entirely within the oceans and its liquid, gaseous and solid derivatives on the landsurface and lower atmosphere.

Hydrothermal alteration   The chemical **weathering** of minerals induced by exposure to a different thermal and moisture environment from that in which they formed.

Hydrothermal circulation   Sea-water circulation through mid-ocean ridges, entering through extension faults and pumped back out through axial vents, and its associated **hydrothermal plume** of new minerals formed by chemical reactions between the heated water, oceanic crust and magma.

Hydrothermal plume   The efflux of sea water and dissolved minerals from axial vents in mid-ocean ridges.

Hygroscopic   Water retained by or attracted to soil or dust particles and not evaporated at ordinary temperatures and pressures.

Hypsithermal   The 'climatic optimum' or period of highest global mean annual temperature, between 8000 and 5000 years ago, during the Holocene or current temperate stage.

Iceberg   A block of ice which has become detached from a floating glacier or **ice shelf** terminus.

Icefall   A steep, rapidly flowing and heavily crevassed glacier segment moving by **extending** or **surging** flow.

Icehouse   An uncommon condition in which Earth's average global temperature may be nearer that predicted by radiation laws by virtue of the presence and **autocatalytic** consequences of large ice sheets and frozen ocean surfaces.

Ice sheet   A large, subcontinental- or continental scale glacier of tabular shape which buries all or most of the landsurface.

Ice shelf   The floating portion of the margins of an **ice sheet** which, owing to the absence of basal shear stress, spreads and thins over the sea surface.

Ice wedge   A mass of **ground ice** forming a vertical wedge in desiccation/contraction cracks.

Igneous   Of molten, partly molten or **magmatic** nature and origin.

Ignimbrite   The cooled and lithified product of a volcanic **ash flow**, also known as an **ash-flow tuff**.

Inceptisol   A soil order (USDA classification) characterized by the alteration or removal of minerals other than carbonates or amorphous silica.

Incised meanders   The entrenched bedrock channels of old river meanders after rejuvenation.

Incision   The erosion of a narrow, bedrock river channel by vigorous fluvial downcutting.

Indentation tectonics   Crustal deformation caused by the penetration of one continental plate by another and involving compression, extension and rotation in the affected crust.

Individualistic communities   The concept that **plant communities** represent an assemblage of plant species with overlapping environmental requirements which arise because of random propagule availability.

Infiltration capacity   The maximum rate at which water may infiltrate soil or rock.

Infiltration rate   The rate at which water added to the surface can enter the soil.

Information theoretic index   A measure of ecological diversity derived from the theory of information (e.g. the Shannon index).

Inhibition model   The model of succession in which changes in floristic composition are prevented until an established species dies out.

Inputs   Flows of energy and matter into a system.

Inselberg   A residual hill in massive, resistant rock which has survived the **weathering** and stripping of adjacent rockmass through its superior strength.

Inshore   The shallow-water coastal zone below the low-water mark in which waves **shoal**.

Insolation   A contraction of *incoming solar radiation*. It refers to the short-wave part of the solar energy input.

Insolation weathering A form of mechanical weathering in which rockmass disintegration is attributed to diurnal thermal expansion and contraction; this form of fatigue failure is apparently most effective in a moist environment.

Intact rockmass strength The peak strength of a rockmass capable of resisting shear; **Mohr–Coulomb criteria** define it as comprising internal cohesion, friction strength and normal stress.

Interception The process which catches and stores precipitation in a vegetation layer, where it may be used, evaporated or transmitted on towards the ground.

Internal deformation The change of shape and volume of a mass of earth material due to a change in the nature or arrangement of its internal properties; a process by which material moves under its own mass.

Interspecific competition Competition between distinct species.

Interstitial ice Individual or fused ice crystals occupying the voids of a soil or rock.

Intertidal zone The zone lying between low-water and high-water marks which fluctuates in width and height range with the monthly tidal cycle.

Intraspecific competition Competition between individuals of the same species.

Intrazonal A class of soils whose profiles are dominated by local factors (e.g. geology, topography).

Intrusion An igneous rockmass of **intrusive** origin forming a sub-surface **batholith**, **dyke**, **pluton** or **sill** and exposed at the landsurface only by subsequent erosion.

Intrusive A description of molten igneous material which penetrates surrounding rock and cools and solidifies before reaching Earth's surface.

Ion An atom which has lost or gained one or more negatively-charged electrons.

Ion antagonism The blocking of the uptake of one cation by the presence in excess of another (e.g. excess calcium inhibiting iron uptake).

Ionic substitution The replacement of one or more ions in a crystal structure by ions of similar size and charge, without altering the crystal structure.

Isomorphous substitution The replacement of one atom by another of similar size in the crystal structure of clay minerals, without disrupting the structure.

Isostasy The equilibrium condition in which lighter crust 'floats' on denser mantle and whose relative proportions from one place to another maintain Earth's shape.

Isostatic adjustment The vertical and lateral displacement of crust and lithosphere in order to maintain or restore isostatic equilibrium.

Joint A fracture between the constituent parts of a rockmass, usually caused by its contraction on cooling or drying.

Jökulhlaup A flash flood of glacier meltwater from a sub-glacial or glacier-margin lake due to failure of an ice dam.

Kame A steep, isolated mound of glaciofluvial sand and gravel deposited in contact with glacier ice.

Kame moraine An irregular ridge of glaciofluvial sediments marking a glacier terminus, formed either by the wholesale meltwater reworking of a **moraine** or the coalescence of **kames** and **kame terrace** fragments.

Kame terrace A valley-side bench of glaciofluvial sediment marking the course of an ice-marginal meltwater stream.

Kaolinite A 1:1 type of clay mineral of high stability but little reactivity, composed of a silica sheet fused with one alumina sheet.

Karst geomorphology A landsystem uniquely developed on carbonate rocks by the predominance of solution and the progressive development of underground drainage.

Katabat The outflowing, descending portion of a local thermal circulation system during the cooling phase of diurnal heating in a mountainous region, or any other more general cold air drainage current.

Katamorphism Intense and rapid **weathering** of rocks by hydrolysis, hydration and oxidation under humid tropical conditions.

Kind A property of soil structure referring to the shape of the aggregates.

Kinetic energy The energy possessed by a body because of its movement. Its magnitude is equal to $\frac{1}{2} mv^2$, where m is the mass of the body and v is its velocity.

Klippe A fragment of a **nappe** dispersed away from its source into the mass of a collision **orogen**.

Knick point A step in the long profile of a stream marking the rejuvenation of fluvial incision after uplift.

Knock-and-lochan A highly abraded, rocky landsurface characterized by streamlined ridges and intervening basins, usually of glacial origin.

Krummholz A woodland of dwarf trees marginal to the **timberline**, whose individuals are severely stunted by cold and wind pruning.

Kuroshio The warm, north-east-flowing current of the clockwise **gyre** in the north Pacific Ocean, flowing from the Philippine coast and deflected southwards in the eastern Pacific by the Alaskan coast.

Laccolith A lens-shaped igneous rock body of moderate size (10^{3-4} m long and 10^{1-2} m thick) formed by accordant (**sill**-like) **intrusion** into existing rocks.

Lag deposit A residual accumulation of coarse rock fragments, too large to move in a particular force field, after the removal of fines.

Lag gravel A surface layer of loose, coarse granular debris left after the deflation or surface wash of fines.

Lag time The time lapse between a stimulus and its effect such as that between the peak of a precipitation event and **peak discharge** response of a stream.

Lahar A volcanic mudflow of liquefied volcanic and other debris, after the melt and disruption of summit glaciers or crater lakes.

Laminar flow Fluid flow in which the direction of each individual flow strand remains discrete and unidirectional, although strands may shear past each other as the channel walls are approached.

Lapilli Unconsolidated coarse-grained **pyroclastic** material 2–64 mm in diameter, ejected into the atmosphere by volcanic eruption.

Latent heat The quantity of heat absorbed or emitted during a change of state of a substance. In climatology it usually refers to the change of state of water from solid to liquid to vapour or vice versa.

Lateral convection Convectional processes in the oceans brought about by horizontal differences in density.

Laterite A reddish tropical clay composed of the sesqui-oxides of iron and aluminium, and **kaolinite**; it hardens irreversibly on drying, often with a concretionary structure.

Lava An extrusive flow of molten **magma** and the rock into which it solidifies.

Law of the minimum The law which states that the productivity of an **ecosystem** is controlled by, or is proportional to, the growth factor which is operating at a minimum (i.e. in shortest supply).

Leaching The washing out of materials in solution or suspension from a soil horizon or profile.

Leading edge The advancing edge of a continental plate, marked by a coastal orogen, narrow continental shelf and deep offshore trench which influence the nature of coastline development (cf. **trailing edge**).

Leaf area index A measure of the density of vegetation surfaces capable of intercepting sunlight and precipitation, given as the total area of leaves in the layered canopy covering a unit area of ground.

Leaf drip The concentration and onward transfer, as large drops, of precipitation intercepted by a leaf.

Leaf leaching The removal by rainfall of chemicals, including nutrients, from the surface and interior of a plant leaf.

Leaf uptake The uptake of nutrients via stomata on leaves.

Lee wave cloud A lens-shaped cloud forming in a standing wave of turbulent air in the lee of a mountain barrier, with continuous condensation as air rises and cools at its leading edge and evaporation as air falls and warms at its trailing edge.

Levée A bank of coarse debris flanking a floodplain river, formed by the concentration of suspended sediment during **overbank discharge**; boulder levées also flank **debris flows** as a result of collision and ejection during their turbulent flow.

Limiting equilibrium A state of balance between shear stress and shear strength – or eroding and resisting forces – in earth materials, defined by **Mohr–Coulomb failure criteria**.

Lineament A large-scale, linear feature of tectonic or other structural origin, visible at the landsurface; it may be (or represent the trace of) a **fault**, **suture**, fracture zone, etc.

Liquid limit The critical water content of a granular solid beyond which it develops liquid behaviour.

Lithification The transformation of unconsolidated sediments into a cohesive rockmass through **syngenetic** and **diagenetic** dewatering and cementation, compaction and crystallization.

Lithology The macroscopic character of rockmass determined by its geochemical (mineral) and mechanical (particulate) components and related structures.

Lithosere The sequence of **plant communities** which, successively, occupy a bare rock surface.

Little Ice Age The period from about 1500 to 1800 when climatic conditions in Europe were much colder than before or since. Many glaciers advanced in the Alps and Scandinavia.

Load The total mass of mineral and organic sediment transported by a stream by bed traction, suspension and solution.

Load casts A protrusion from the base of one sedimentary lamina into the underlying surface of another by **syngenetic**

deformation due to unequal settling, water content or mass.

Loch Lomond stadial The period between 10,800 and 10,000 BP when glaciers reappeared in many of the mountain areas of the British Isles. Other parts of the world suffered a decline of temperature but not so marked.

Lodgement till A subglacial **till** deposit formed by direct plastering-on of debris to the substrate by a moving glacier, where the deforming boundary lies along the till–ice contact.

Loess An accumulation of wind-blown dust which may have undergone mild **diagenesis**.

Longshore current A coast-parallel or oblique current in the **surf zone** comprising water pushed by **edge waves** and the **swash** from **refracted waves** which periodically drains seawards as **rip currents**.

Longshore drift Sediment transfer along the coast by **longshore currents**.

Long-wave radiation Electromagnetic radiation between about 3 μm and 100 μm. Earth only radiates in this waveband, so it is sometimes called *terrestrial radiation*.

Lower limit of tolerance The lower threshold of the tolerance range of a plant species, ie the point at which the species will not survive.

Lysimeter Instrument used to obtain evapotranspiration. It incorporates growing vegetation and compares changes in the water content of the soil column beneath the vegetation.

Macropore An intergranular pore in earth material which is too large (greater than 50 μm in diameter) to hold **capillary** water and acts instead as a conduit for gravity drainage.

Mafic A mnemonic from *magnesium* and *ferric* which identifies the ferromagnesian-rich igneous rocks, characterized by their dark-coloured, basic minerals.

Magma Partially molten rock material, usually with solid minerals and/or gas pockets suspended in a liquid silicate mass, from which igneous rocks solidify.

Magnetic polarity Earth's magnetic field, characterized by two poles of opposite tendency, with the dipole normally arranged so that its south pole lies in the northern hemisphere (corresponding to the north magnetic pole) and *vice versa*; this is periodically reversed on timescales of 10^{3-6} years and provides a valuable dating tool.

Major nutrient A plant nutrient needed in relatively large quantities (e.g. nitrogen, potassium, phosphorus).

Managed retreat A pragmatic approach to the coastal management of rising sea levels which rejects the use of hard defences in favour of allowing the coastline to retreat at favourable sites with the landward migration of **salt marsh** or dune barriers as soft defences.

Mangrove swamp A tropical intertidal **ecosystem** on low-energy coasts capable of high productivity, wide diversity and coastline protection against erosion.

Manning equation An equation which calculates the velocity of uniform streamflow in relation to channel slope, **hydraulic radius** and bed roughness.

Mantle Earth's internal sphere, sandwiched between the **core** and the crust, whose outer **lithosphere** and **asthenosphere** are instrumental in **plate tectonic** processes.

Mantle plume A rising limb of slow flow/creep of hot rock driven by convection in the mantle; marked initially by surface hot-spots, persistent motion leads eventually to surface rifting and sea-floor spreading.

Marginal sea   A marine basin impounded on oceanic crust, associated usually with the development of a **back arc** and partially separated from an adjacent ocean by an island **volcanic arc**.

Mass balance   The mass input (accumulation), storage and output (ablation) of ice, snow and water which constitutes the mass budget of a **glacier** per unit time, normally a 'budget' year commencing at the end of the summer ablation season.

Mass movement   An alternative term for **mass wasting** in common use but passed over here because of its inaccurate implication of the coherent movement of material *en masse*.

Mass wasting   The downslope movement of earth materials solely under the influence of gravity and without the active aid of other moving materials such as water, ice and air.

Mathematical model   A model in which all the components of the system are represented by mathematical symbols and the relations between them by equations.

Matric force   A soil suction force due to **adsorption** and capillarity in the soil matrix which resists gravity drainage.

Matrix   The finer-grained component of earth material which surrounds and infills pores between larger **clasts** and crystals; together, they create a biomodal texture and reduce **porosity**.

Mechanical weathering   The disaggregation of rockmass caused by the development of internal, tensile stress through thermal expansion, hydration and ice growth in the conditions prevailing at or near the landsurface.

Medieval Warm Epoch   A period of northern hemispherical, and possibly global, climatic warming for some three centuries between AD 800 and AD 1300 but reaching a peak at different times in different places; summer temperatures were *c.* 1° C warmer than they are today, enough to trigger substantial climatic, environmental and socio-economic change.

Megaripple   A large sand wave, with wavelengths 1–100 m and wave heights between 0.1–1 m, formed by high energy flow in shallow waters such as tidal estuaries.

Mélange   A chaotic mixture of rock material from a variety of sources, commonly associated with **subduction zones**, where it consists of subducted oceanic crust, ocean-floor sediment and, maybe, adjacent continental crust.

Melt-out   The subglacial release of ice-transported debris during a temporary or permanent melting phase.

Metamorphic aureole   The zone of rock surrounding an igneous **intrusion** which is altered by **contact metamorphism**.

Metamorphism   The mineralogical and structural alteration of rock in response to thermal and pressure conditions substantially different from those in which it formed, whilst remaining in solid state; it lies between mild **diagenesis** and the presence of a liquid phase required by **metasomatism** and **migmatization**.

Metasomatism   The alteration of existing minerals and formation of new species by fluids and gases circulating through rockmass in metamorphic belts and mid-ocean ridges.

Metastable equilibrium   A condition whereby small changes in system variables can have a major effect once they reach a certain value.

Microclimate   The climate of the landsurface, extending no more than a few metres above ground and strongly influenced by its material, morphological and organic components.

Mid-ocean ridge   A broad, linear ridge emerging from the ocean floor along rising **mantle plumes** and the focus of rifting and sea-floor spreading; basalt effusion forms new oceanic crust at this **constructive margin**.

Migmatization   The mineralogical and structural alteration of rock at extreme ranges of temperature and pressure, causing significant remelt.

Mineralization   The decomposition of organic compounds, which results in the production of mineral nutrients in ionic form.

Minerogenic sediment   Sediment (or soil) derived solely from inorganic, mineral sources.

Minor nutrient   A plant nutrient needed in relatively small amounts (e.g. calcium, sulphur, magnesium).

Misfit stream   A stream which appears to underfit its valley, indicated by its diminutive size and meander wavelengths much shorter than those of the valley itself; attributed to a climatically related reduction in stream discharge.

Moder   Surface organic matter, intermediate in form between mull and mor, and consisting mostly of partly humified plant remains.

Mohr–Coulomb criteria   The failure criteria for earth materials which defines a state of **limiting equilibrium** when the **shear stress** acting on the material exactly equals the material's internal **shear strength** comprising cohesion, internal friction and normal stress.

Molasse   The sedimentary product of syn-tectonic or early post-tectonic erosion of a new **orogen**, consisting of coarse **clastic**, and mostly terrestrial sediments.

Mollisols   A soil order (USDA classification) characterized by base-rich soils with a dark, organic-rich surface horizon.

Monadnock   An isolated mountain in a **peneplain**, representing a residual feature of extensive denudation.

Monoclimax theory   The theory of climax vegetation which emphasizes that only one type of climax will ultimately develop in a specified climatic region; this will be in stable equilibrium with climate and soil.

Montane   Said of the mountain forest belt or used more generally to denote a characteristic of mountainous terrain.

Montane forest   A cool, mountain forest community.

Montmorillonite   An expanding 2:1 type of clay mineral; isomorphous replacement of aluminium by iron and magnesium is common in the alumina sheet.

Mor   An acid organic matter horizon consisting of litter (leaves, twigs, wood) overlying partly decomposed, fermenting plant remains.

Moraine   Ridge-like accumulations of glacier debris, carried as glacier surface medial and lateral moraines or deposited at the ice margin by a variety of passive release or active push processes; it is mostly composed of glacial **till** with admixed glaciofluvial sediment and is also known as ground moraine when deposited in extensive, amorphous sheets.

Morphogenesis   The conversion of weathered regolith into a soil profile by processes of soil formation.

Morphogenetic   Said of a geographical region or process where climate is believed to have created a distinctive suite of landforms.

Morphological system A type of system in which the morphological expression is examined rather than the dynamics or interactions and flows.

Morphotectonic landform A landform created by tectonic processes, such as an **island arc**, **orogen**, **rift valley** or **passive margin** coast or any of their principal components.

Morphotectonics The construction of large-scale surface landforms by tectonic processes.

Mosaic theory The theory of climax vegetation which emphasizes the mosaic patterning of the climax cover.

Moulin A cylindrical, supraglacial 'pothole' marking the englacial transfer of a surface meltwater stream.

Mountain circulation winds A general regional pattern of wind circulation determined or modified by the insolational and mechanical character of mountainous terrain.

Mudflow The moderate to fast downslope movement of a fluidized mass of very fine debris, or its resultant landform.

Mull Well decomposed and well humified organic matter, thoroughly mixed with mineral soil by earthworms.

Nappe A **fold** which has experienced such intense deformation as to have become recumbent (horizontal) and sheared along its axis.

Natural fire A fire ignited by natural means (e.g. lightning).

Neap tide The twice-monthly tidal period when the gravitational pull of Sun and Moon are opposed (at right-angles to each other) and minimize tidal range.

Nearshore The zone of shoreline–wave interaction, subdivided landwards into **breaking wave**, **surf** and **swash** zones.

Negative feedback A feedback effect in which the initial change in the system is damped down.

Neotectonic Plate tectonic activity during the late Cenozoic era.

Net ecosystem production (NEP) The change in the **biomass** of an **ecosystem** per unit time; equivalent to **net primary productivity** minus losses due to grazing by herbivores.

Net primary productivity (NPP) The amount of energy fixed by plant **photosynthesis**, taking losses by respiration into account; it represents growth by the plant or ecosystem, and is measured per unit area per unit time.

Net radiation budget The difference between the total incoming and outgoing radiation terms. A positive value would indicate greater incoming than outgoing energy and so a warming; a negative value would indicate the reverse.

Net radiation deficit The situation in which Earth is losing more **radiant energy** than it is gaining.

Nitrate vulnerable zone (NVZ) An area where there are restrictions on the use of nitrogen fertilizers.

Nitrification The conversion of organic nitrogen compounds in soil organic matter into nitrates by soil micro-organisms.

Nitrogen fixation The conversion of atmospheric nitrogen into organic nitrogen compounds by soil micro-organisms, either free-living or in nodules of plant roots.

Nivation The erosion of surface depressions by the combined processes of rock weathering and mass wasting associated with the growth and decay of snowpack.

Nordenskjöld line The limit of the Arctic and Antarctic as defined by both the temperature of the warmest month and the temperature of the coldest month.

Normal stress The portion of rock or soil strength dependent on the anchoring effect of the mass of a particle or intact block, normal (at right-angles) to a surface on which it rests; this is at a maximum if the surface is horizontal but diminishes as slope angle increases.

Nuée ardente An incandescent (fiery) cloud of **ash** and volcanic gas developing as near-surface gravity flow after volcanic eruption, capable of destroying anything in its path.

Nunatak An isolated mountain or hill protruding through, and completely surrounded by, glacier ice.

Obligate relationships Relationships which are restricted to specified conditions only.

Obliquity of the ecliptic The tilt of Earth's axis of rotation relative to the plane of its orbit. It varies between 21.8° and 24.4° over a period of about 40,000 years.

Occupational fire A fire started by humans in order to manage an **ecosystem** for economic gain (e.g. grazing, land clearance or herding wild animals).

Offshore A zone of deeper water lying on the inner margins of the continental shelf, beyond the **nearshore** zone.

Ooze Fine-grained, marine sediment comprised of more than 30 per cent skeletal remains of **pelagic** organisms and clay minerals.

Open systems Systems which are characterized by the exchange of both matter and energy with their surroundings. The majority of natural systems are open.

Ophiolite A sliver of oceanic crust caught up in an **accretionary prism** and found out of place in a subsequent **orogen**.

Organismic community The concept which regards the **plant community** as a 'super-organism', with properties not present in its individual constituent organisms.

Orogen A linear continental mountain range elevated mechanically or thermally by plate collision, crustal shortening and uplift.

Orogenesis The formative processes of an orogen.

Oroshi A cold air drainage current blowing from the mountains of central Japan.

Orthogonal Said of a system with right-angle relationships between its components.

Outlet glacier A steep, fast-flowing glacier discharging large ice volumes from inland portions of ice sheets through a confining rock-walled channel.

Outputs The flow of energy and matter out of a system.

Outwash plain An extensive landsurface covered by glaciofluvial sediments and braided meltwater streams released from a glacier terminus, especially in unconfined piedmont zones.

Overbank discharge That portion of stream discharge not confined to the channel during a flood.

Overburden pressure A compressive stress exerted on earth material by the mass of overlying rock, soil, water or ice.

Overland flow Non-channelled surface water flow where precipitation intensity exceeds infiltration capacity, taking two forms: Horton overland flow develops as **sheetflow** on unvegetated surfaces before infiltrating or concentrating in channels, whereas saturated overland flow emerges towards valleys floors on vegetated slopes.

Oxidation A chemical weathering process involving the combination of oxygen with a mineral accompanied by a positive shift in its valency.

Palaeocurrent A former current of moving material (water, ice, air, debris, etc.) whose direction and strength may be inferred from rock structures which it created.

Palsa A small mound of peat and ice formed by **segregated ice** growth in a peat bog.

Pangaea Earth's most recent supercontinent, formed by the coalescence of most continental plates *c.* 300 Ma ago and rifted apart *c.* 200 Ma ago.

Parabolic dune A tight crescent-shaped dune with elongated arms pointing upwind, often developing from a **blow-out**.

Particle sorting The segregation of debris particles according to their size and the competence or power of a moving medium; also, the range of particle sizes in a particular sediment sample expressed by its standard deviation.

Passive margin A tectonically passive continental margin associated with divergent plate boundaries and marking the zone of initial rifting.

Patterned ground A collective term for a variety of plan-form patterns on a **permafrost** landsurface formed by turbulent heaving, sifting and collapse in the **active layer**; symmetrical patterns develop in more homogenous earth materials and flat surfaces, becoming irregular elsewhere.

Peak discharge value The highest water discharge in a stream channel stimulated by a precipitation event and appearing as a peak on its **hydrograph**.

Peat Dark organic material composed of plant residues accumulating under wet or waterlogged conditions.

Ped A natural soil **aggregate** consisting of primary particles and colloidal material.

Pediment A concave erosion surface sloping gently down to a lowland plain from a rather more abrupt contact with a mountain front.

Pediplain The coalescence of one or more **pediments** to create a more extensive lowland, considered by proponents of **denudation cycles** to develop through parallel slope retreat in semi-arid climates.

Pedological process Any process associated with the formation and development of soil.

Pegmatitic An igneous rock texture characterized by very large crystals representing the final **magma** fraction.

Pelagic Of the open ocean environment, as opposed to the ocean margin and coastline.

Pelagic sediments Sediments associated with the pelagic zone and excluding terrigenous material; they consist of the remains of marine organisms and **red clays**.

Peneplain A lowland plain on which erosion of whatever nature has progressively obliterated structural and morphological features; considered to be the final stage of a humid, fluvial **denudation cycle**.

Percolation Water transfer through the voids of unsaturated soil or rock.

Peridotite The coarse-grained, olivine-rich **ultramafic** rock which forms the asthenosphere and is the raw material for oceanic crust.

Periglacial A term formerly used to describe the environment and processes around the margins of a glacier or ice sheet and strongly influenced by its proximity; this use is too restrictive and may be misleading and a definition which emphasizes the predominance of **cryospheric** processes – without the need for glaciers – is preferred.

Perihelion The point on Earth's orbit when it is closest to the Sun.

Permafrost The enduring and continuous presence of freezing temperatures at and below the landsurface in which all soil and groundwater is frozen except for a thin surface **active layer** of summer melting.

Permafrost soil processes Soil processes occurring in frozen ground (permafrost).

Permeability The capacity of earth materials to circulate and transmit fluids (water, solutions, air, etc.) through their pores and fractures and measured as the fluid volume passing through a unit cross-section area.

P-form 'Plastically sculptured' sinuous bedrock grooves of uncertain origin but believed to be formed by high-pressure subglacial meltwater rather than glacier abrasion.

pH The measure of acidity or alkalinity of a substance, measured by the number of hydrogen ions per litre, on a logarithmic scale where neutrality = 7.0; acid and alkaline substances have a pH of less than 7.0 and more than 7.0 respectively.

Photic zone The thin, surface layer of a water body penetrated by sunlight.

Photosynthesis The synthesis of organic compounds from water and carbon dioxide, using energy absorbed by chlorophyll from the radiant energy of the Sun.

Phreatic water Water in the saturated zone of an **aquifer**.

Phreatophytes Plants which survive aridity by developing deep root systems to exploit soil water reserves.

Phytoplankton The plant (primary producer) form of marine micro-organisms which form the base of the marine food web.

Piedmont fan A fan-shaped lobe of alluvium or other debris accumulated at the break of slope along a mountain front.

Piedmont glacier A glacier which fans out across the unconfined surface of a piedmont zone as it leaves the confined channel of an **outlet** or valley glacier.

Piedmont lobe The lobate terminal zone of a **piedmont glacier**.

Piezometric surface An imaginary surface defined by the level to which water rises in a well and representing the static 'head' of water.

Pingo A large, ice-cored mound elevated by hydrostatic pressure and **segregated ice** growth on a flat **permafrost** landsurface which is waterlogged in summer.

Pioneer community The first set of plant species to colonize a newly available site which was previously unvegetated.

Pipe A narrow water conduit in soil formed through the connection of **macropores** or the removal of **swelling clays**.

Placer A terrestrial or shallow marine deposit of heavy minerals in a body of **clastic** sediments, sourced by erosion, transport and gravity deposition from a parent mineral body.

Plagioclimax A vegetational state where burning or grazing modifies the natural state of the vegetation.

Planar discontinuity A plane surface in rockmass or other earth materials (such as a fold, fracture, fault, thrust, joint, lamination, etc.) at which the continuous or homogenous properties of material on either side of the plane – providing its **intact strength** – are momentarily interrupted or lost;

Planar slide   A sliding failure along a single **planar discontinuity**, inclined at an angle less than that of the slope on which it occurs but greater than the internal friction angle of the material.

Plant community   A group of plants which form a distinct combination of species in the landscape and which interact with each other.

Plant uptake   The amount of a nutrient absorbed by a plant, or the process of the absorption.

Plastic   A condition in which material is capable of continuous and permanent deformation without fracturing.

Plastic limit   The threshold water content of a solid sediment at the point at which its behaviour changes to a plastic state.

Plate   A large and rigid 'raft' of Earth's lithosphere which is mobilized by mantle convection currents and whose boundaries are marked by the formation or destruction of oceanic crust and creation of new continental crust.

Plate tectonics   The global-scale movement and deformation of Earth's lithospheric plates, representing the surface expression of Earth's long-term geological evolution and responsible for global scale landforms.

Platy   A type of soil structure consisting of horizontal units.

Playa   An enclosed, ephemeral lake basin and its residual mud or evaporite floor in an arid or semi-arid environment.

Plinthite   A reddish clay in tropical and subtropical soils which hardens irreversibly on drying; consists of sesqui-oxides of iron and aluminium, and **kaolinite**.

Plutonic   Said of igneous rockmass formed at great depth in the lithosphere by slow cooling, and characterized by granitic texture and mineralogy.

Plutons   An igneous **intrusion** of **plutonic** character which has cooled and solidified below ground.

Polar glacier   See **cold glacier**.

Polyclimax theory   The theory of climax vegetation which emphasizes that in any region a variety of climaxes will develop in relation to soil and topographic conditions.

Polymerization   The formation of large framework minerals by the replication of smaller constituent minerals, involving the sharing of atoms and consequential strengthening of mineral structure.

Polymorphic   The description of a single mineral capable of assuming two crystalline forms.

Polyploidy   A genetic adaptation in plants, endowing them with more than two sets of chromosomes, which appears to make them particularly vigorous and successful colonizers of hostile environments.

Pool   A depression formed by erosive scour in a stream bed.

Porosity   The volume of voids in earth materials, measured as a percentage of their bulk volume and comprising interparticle pores and fractures.

Porphyritic   An igneous rock texture characterized by larger, slower-cooling crystals in a finer matrix.

Positive feedback   A feedback effect in which the initial change in the system is amplified to bring about even greater changes. It can result in instability.

Potential energy   The energy possessed by a body because of its position. A common example is the potential energy possessed by a boulder on a mountainside, which has potential energy relative to the valley bottom and eventually to sea level.

Potential evapotranspiration   The amount of moisture which would be evaporated and transpired from a short vegetation surface with no moisture deficit.

Pothole   A cylindrical hole developed in a rocky stream bed by **evorsion** or a shaft taking surface water underground.

Power threshold   The minimum power required to overcome frictional resistance to movement.

Precession of the equinoxes   The movement of the timing of the equinoxes around Earth's orbit. Currently Earth is nearest the sun in January. In about 10,000 years it will be nearest the Sun in July.

Precipitation   All types of falling and direct deposition of water or ice from the atmosphere at Earth's surface; also, the process whereby dissolved solids are deposited from a fluid.

Pressure-melting point   The temperature at which ice can melt at a given pressure; this is central to a modern understanding of glacier behaviour, since the **overburden pressure** at the base of a glacier is frequently high enough to melt basal ice at sub-zero temperatures and form supercooled meltwater.

Primary stage   The first stage in a plant **succession**.

Primary succession   The sequence of **plant communities** which, successively, occupy a natural area previously devoid of vegetation.

Primary (P) wave   The fastest of several earthquake vibrating waves, capable of passing through any Earth material by alternating compression and expansion (see also **Secondary wave**).

Principal stress   A stress which acts perpendicular to each of the three pairs of faces of a cube in a rockmass.

Principle of competitive replacement   The principle which states that in successions plant species tend to make conditions more favourable for a competing species which will replace them.

Principle of uniformitarianism   The principle that present-day analogues are used as a basis for the interpretation of observed features within the past geological record.

Prisere   A **primary succession**.

Prismatic   A type of soil structure consisting of vertical units with straight tops, usually in the subsoils of clay soils.

Process-response system   A combination of morphological and cascading systems so that the system demonstrates the manner in which form is related to process.

Producers   Autotrophic organisms capable of **photosynthesis**, i.e. organisms in the first **trophic level** of an ecological pyramid.

Progradation   The seaward extension of river floodplains by downstream sediment transfer into estuarine and delta environments.

Prostrate mosses   Mosses which grow horizontally and hug the ground.

Protalus rampart   A steep-sided ridge formed at the foot of a permanent snowbed by the accumulation of coarse, angular frost-weathered debris which has slid over, or been washed under, the snowbed.

Psammosere   A series of **plant communities** which, successively, occupy and stabilize an area of unconsolidated sand.

Pumice   A low-density, highly porous volcanic rock material formed by explosive degassing of its parent **magma**.

Pycnocline   A zone of marked density change with ocean depth.

Pyroclast   Molten or solid explosive volcanic products in the form of **ash**, **lapilli** or **tephra**; volcanic rocks are formed by their deposition and, frequently, hot welding.

Quarrying   The large-scale excavation of large bedrock blocks and fragments by moving glacier ice and high-pressure basal meltwater; replaces the now defunct term 'plucking'.

Quick clay   A clay whose shear strength collapses on disturbance.

Quickflow   The component of stream discharge contributed by water reaching the channel rapidly by direct channel precipitation, **overland flow** and subsurface **throughflow** and forming a transient **peak discharge**.

Radiant energy   Energy transmitted in the form of electromagnetic waves. These waves do not need molecules to transmit them and in a vacuum they travel at the speed of light.

Radiation   Another term for **radiant energy**.

Raised beach   A former beach abandoned by isostatic uplift or a eustatic fall in sea level and retaining mineral and organic sedimentary evidence of its origin.

Raised platform   A former marine, wave-cut rock platform abandoned by a change in sea level (see **raised beach**).

Range of tolerance   The intervening range between the upper and lower thresholds of survival for a species.

Reagent   A substance which is capable of causing a chemical reaction.

Realized niche   That proportion of the fundamental niche which is actually occupied by a plant species.

Recurrence interval   The predicted frequency or return period of a particular value of stream discharge, measured in years.

Red clay   Open ocean (**pelagic**) sea-bed sediments derived from long-distance **aeolian** or ocean current transport of clay minerals from volcanic, meteoric and ice-rafted sources and rain-out of marine organic debris; their red colour comes from ferric oxide coating.

Reduction   A chemical reaction involving the dissociation of oxygen from a mineral and accompanied by a negative shift in valency.

Reef   A shallow-water marine bench or mound constructed mostly of the carbonate-rich skeletal secretions and remains of organisms; also known as a **bioherm**.

Reflected wave   A wave which has rebounded from coastal features in its path into an incoming wave.

Refracted wave   A wave front which has been diverted from its original path as it encounters shallow water or a coastal current.

Refugia   Isolated geographical locations whose distinctive environments permit the survival of formerly widespread plant and animal species during periods of adverse environmental change; they may act as centres of dispersal in any subsequent amelioration.

Reg   A stony desert surface or **desert pavement**.

Regelation   The refreezing of supercooled water by the reduction of glacier **overburden pressure**, or of surface meltwater which has percolated into colder ice.

Regional metamorphism   The chemical and textural alteration of rock by widespread compression and heating or burial in, for example, a subduction zone.

Regolith   A general term for the superficial layer of disaggregated earth material at the landsurface, irrespective of its specific origins.

Regression   A seaward retreat of the coastline caused by a relative fall in sea level and its stratigraphic expression in the advance of terrestrial sedimentation.

Rejuvenation   The stimulation of denudation processes to renewed activity, normally by the increase in potential energy caused by tectonic or isostatic uplift; also regarded as the impetus for the first stage of a new **denudation cycle**.

Release surface   A **planar discontinuity** in earth materials along which slope failure has occurred; it may act singly, in the case of **planar slides**, or with other release surfaces in more complex failure.

Relic soil   A type of fossil soil, currently at the surface, and therefore showing properties formed by past and present **soil-forming processes**.

Remnant arc   An extinct **volcanic arc** abandoned by the migration of a **subduction zone**.

Residual soil   A soil whose parent material is solid bedrock.

Resistance   The sum of forces in earth materials mobilized to resist shearing or other forces.

Resistant residue   Soil minerals relatively resistant to **weathering** which therefore tend to accumulate in soils (e.g. heavy minerals, quartz, feldspars).

Resorption   The process of crustal recycling whereby oceanic and other material is partially or completely melted in the higher temperatures and pressures of a **subduction zone**.

Respiration   The breakdown of organic compounds, using oxygen to obtain energy for metabolic processes.

Reverse weathering   A reversal of sea-floor chemical weathering which precipitates solid phases of minerals previously taken into solution; includes the important biochemical precipitation of calcium carbonate and silica.

Reynold's number   A value which distinguishes between **laminar** and **turbulent** streamflow, dependent on the relative values of **hydraulic radius**, water velocity and viscosity.

Rheologic property   The ability of an essentially solid material to deform and flow under stress.

Rhourd   A pyramid sand dune formed in a variable wind field.

Ria   A marine inlet formed by the flooding of a coastal river valley by eustatic rise in sea level or isostatic depression.

Ridge push   A component of **sea-floor spreading** driven by gravity-sliding away from the elevated **mid-ocean ridge**.

Riegel   An abraded, cross-valley rock barrier in a glaciated valley.

Riffle   A stream bed accumulation of coarse alluvium linked to the scour of an upstream **pool**.

Rift valley   A valley formed by crustal downfaulting, normally between two parallel faults; it gives a topographic expression to a **graben**.

Rill   The smallest and most transient of stream channels, eroded during intermittent surface flow and liable to collapse or infill between precipitation events.

Rip current A narrow and intermittent current, fed by longshore currents and draining seaward through the **nearshore** zone, where it may be vigorous enough to cut a rip channel in the seabed.

Rising limb The component of a **hydrograph** which marks the increase in stream discharge during the **time of rise** to the **peak discharge**, in response to a precipitation event.

Roche moutonnée A valley floor, glaciated bedrock hump with a streamlined, abraded uphill face inclined gently up-valley and a steep, quarried downhill face.

Rock cycle The global geological cycling of lithospheric and crustal rocks from their igneous origins through all or any stages of alteration, deformation, resorption and reformation.

Rock debris The initial **blocky** product of **weathering** of a rock face.

Rockfall A free fall of **rock debris**.

Rock flour Fine debris produced by subglacial abrasion and usually flushed out as suspended sediment in meltwater.

Rock glacier A slow-moving mass of angular rock debris with sufficient interstitial or subjacent ice for it to flow like a glacier, usually found in arid cold climates.

Rock platform A wave-cut platform across a rock surface in the intertidal zone.

Rock weathering See **weathering**.

Rotor A small, overturning turbulent eddy in the airstream downwind of a mountain range; it may generate a rotor cloud if the rising and falling limbs pass through a condensation level.

Sabkha A salt-encrusted plain marked by the accumulation of **evaporite** rocks, usually on tidal flats; also used to describe an inland salt pan or **playa**.

Salina A general term used for a surface depression of periodic flooding and evaporation which leads to the accumulation of **evaporite** rocks.

Salinity The mass of total dissolved salts present in sea water, measured in g kg^{-1} or parts per thousand.

Saltation The movement of sediment particles by turbulent entrainment in water, wind or by **grain ballistics**, followed by short jumps or bounces along the bed.

Salt efflorescence A precipitation and growth of salt crystals from a fluid in rock or soil voids.

Salt weathering The granular disintegration of rock caused by **salt efflorescence** which acts as an important mechanical weathering agent through its generation of high tensile stress.

Saltmarsh A **halophytic** plant community occupying intertidal mudflats, exhibiting a weak progression in diversity and productivity shorewards as the frequency of tidal inundation falls; its surface is flooded and drained through a series of tidal creeks.

Sandar An **outwash plain** forming the superficial land-surface in piedmont or coastal zones beyond glacier margins.

Sand wave A large wave- or dune-like sand **bedform** formed by fluid motion normal to its axis.

Saprolite A soft *in situ* residue of chemically decomposed rock.

Scattering The process whereby radiation is dispersed in all directions by particles. The particles can be from the size of molecules upwards. Meteorologically, radiation which has been scattered is known as *diffuse* radiation.

Schistosity A rock texture in coarse-grained material with close, sub-parallel planes formed by the arrangement of **platy** minerals; a rather coarser form of **cleavage**.

Scree Angular rock debris or **talus** which accumulates on slopes below the rockwall from which it was weathered; it may be partially gravity-sorted and maintains a slope angle dependent on its friction strength.

Sea-floor spreading The lateral spread of the sea floor generated by mantle convection and consequential formation of new ocean crust at a mid-ocean ridge; the driving mechanism of plate tectonics.

Sea ice Floating ice formed by the freezing of sea water; not to be confused with floating glacier ice.

Sea level The mean surface elevation of the sea, normally excluding transient changes induced by tides, atmospheric pressure, upwelling and water influx.

Seamount A wholly submerged, submarine mountain; many are former sea-floor volcanoes.

Secondary (S) wave A slower-moving earthquake wave which oscillates at right angles to its direction of travel and can only pass through rock (see also **Primary wave**).

Second stage community A community occupying the second stage of a **succession**.

Sedimentary The description of a major group of both unconsolidated and **lithified** rocks formed by the eventual accumulation of rock and organic debris after a period of transport, suspension or solution.

Sedimentary basin A geographical area in which sediments accumulate, generally in the form of a continental or marine depression which acts as a gravitational 'sump'.

Sedimentary environment A general location in which groups of genetically related **sedimentary facies** are deposited, such as a fluvial or marine environment.

Sedimentary facies A parcel of sediment with distinct internal characteristics reflecting a particular depositional event or location within a broader **sedimentary environment**.

Sediment balance The volumetric input, storage and output of sediment which constitutes the sediment budget of a **drainage basin** per unit time, usually a calendar year.

Segregated ice lens A type of **ground ice** formed by the migration of pore water to a freezing plane and displacing unconsolidated soil particles to form a discrete ice lens.

Seif dune A large, sinuous linear dune drawn out parallel to the wind.

Seismic activity The sudden release of accumulated stress in rocks subjected to tectonic and other deformation and the earth shock waves or earthquakes which it propagates.

Seismic sea wave A travelling ocean surface wave caused by **seismo-volcanic** activity which grows in height as it enters shallow coastal waters and frequently causes damage and loss of life; often mistakenly confused with a **tidal wave**.

Seismo-volcanic The close association between volcanic and earthquake activity whereby either may trigger the other, and their mutual concentration at plate boundaries.

Sensible heat The heat we can feel and measure with a thermometer.

Sensitive clay A clay whose undisturbed strength is four to eight times greater than its disturbed strength and is therefore particularly prone to failure.

Seral stage A stage in a **sere** when the area is occupied by a community which creates conditions more and more favourable for a succeeding community.

Sere An alternative term for **succession**.

Set-down The local, minor fall in water level experienced in the **surf** zone when **edge waves** and incoming waves are out of phase.

Set-up The local, minor rise in water level experienced in the **surf** zone when **edge waves** and incoming waves coincide.

Severn bore The **bore** which marks the leading edge of incoming tides at the head of the Bristol Channel, Britain's largest marine inlet experiencing one of Earth's largest tidal ranges.

Shape of soil structure The morphology of the individual **peds** which make up the soil structure.

Shear strength The sum of internal forces in a rock or soil capable of resisting shear which determines the maximum shear stress the material can endure without failing.

Shear stress The tangential stress which acts on a unit of earth material and may lead to shearing.

Sheetflow Surface water flowing as a thin, continuous film rather than concentrated in a channel.

Sheeting structure A fracture or other **planar discontinuity** formed or exhumed in rockmass by the removal of overlying material accompanied by **elastic strain release**.

Shield The crystalline or high-grade metamorphic core of stable continental crust or **craton**.

Shoaling The alteration of wave height, form and velocity through sea-bed friction as it enters shallow water, transforming it into a **breaking wave**.

Shoaling wave An incoming wave experiencing sea-bed effects or **shoaling**.

Shore-normal Moving at right-angles to the shore, i.e. directly landwards or seawards.

Short-wave radiation The **radiant energy** emitted by the Sun at wavelengths below 3 μm.

Sichelwanne A crescent-shaped scar on a glaciated bedrock surface marking the removal of a flake by an abrading rock or high-pressure meltwater.

Sierra A high mountain range or **cordillera** characterized by jagged, saw-toothed peaks.

Silicates A group of minerals built around the **silicate tetrahedron**, including the olivine, pyroxene, amphibole, mica, feldspar and quartz group, and which form 95% of Earth's crustal rocks.

Silicate tetrahedron The highly stable silicate **anion**, SiO_4^{4-} which forms the basic building block of a wide range of **silicate** minerals.

Sill A tabular igneous **intrusion** which is accordant with existing rock structures.

Slab pull A gravitational component of sea-floor spreading drawing cold, dense crust into a **subduction zone**.

Slaking The disintegration of earth materials on exposure to the air or hydration.

Slickensides A rock surface polished by shearing and abrasive removal of surface roughness during movement along a **fault**.

Sliding resistance The sum of forces capable of resisting sliding on a slope, which normally consists of the proportion of the **normal stress** of a rock or soil mass acting at right angles to the slope plus friction.

Slumping A translation failure involving shearing at the upper boundary of a moving soil or rock mass and its downward rotation along a curved failure surface.

Sodium adsorption ratio (SAR) A measure of soil alkalinity, calculated by dividing the content of exchangeable sodium by the square root of the sum of exchangeable calcium and magnesium.

Soil An assemblage of loose and normally stratified, granular minerogenic and biogenic debris at the landsurface; it is the supporting medium for the growth of plants.

Soil catena The sequence of soils which occupy a slope from the topographic divide to the bottom of the adjacent valley.

Soil separate A particle-size fraction of the mineral material in soil, i.e. sand, silt or clay.

Soil Taxonomy The Comprehensive Soil Classification System devised and used by the US Department of Agriculture.

Soil texture The relative proportions of sand (2.0–0.05 mm diameter), silt (0.05–0.002 mm diameter) and clay (less than 0.002 mm diameter) mineral material in soils.

Soil-forming process Any process working to produce a soil from parent material.

Soil zonality The concept which views the distribution of soils in world-wide zones corresponding to climatic regions.

Solar wind The outflow of charged particles from the Sun that escapes the Sun's outer atmosphere at high speed. It may interact with Earth to produce the aurora.

Solid solution A single crystalline mineral phase in which one element may substitute for another without change of phase.

Solid-state recrystallization The reformation of less dense mineral species into denser forms in **regional metamorphism** without melting or other change of phase.

Solifluction A form of mass wasting involving the slow to intermediate flow of loose, granular materials above their **liquid** or **plastic limit**; often applied more narrowly to such behaviour in the **active layer** of a **permafrost** environment and – incorrectly – to soil **creep**.

Solod The original Russian term for **solodic planosol**.

Solodic planosol An acid soil with an illuvial Bt horizon which results from the degradation of **solonetz**.

Solonetz An alkaline soil with an illuvial Bt horizon with a distinct columnar structure.

Solution The change of state of a solid or gas into a liquid by mixture with a **solvent** which forms an important chemical weathering process.

Solvent A fluid capable of forming a **solution** with a solid or gaseous substance.

Species richness The number of species of a defined taxonomic group in a specified area.

Specific retention The volume of water retained by an **aquifer** after gravity flow, sometimes measured as a ratio of that volume to the volume of rock.

Specific susceptibility The susceptibility of rock to a specific weathering process determined by its specific lithological (chemical and structural) characteristics.

Specific yield The volume of water released from an **aquifer** by gravity flow, sometimes measured as a ratio of that volume to the volume of rock.

Spit A narrow, coarse-grained sediment bar extended across a bay or estuary by **longshore drift** from a headland and often curved at its free end in response to estuarine cross-currents.

Spring sapping The undercutting or **headward retreat** of the slope immediately above a spring or point of initiation of a stream channel by the concentration of erosive power.

Spring tide The twice-monthly tidal period when the gravitational pull of the Sun and that of the Moon are in line and maximize tidal range.

Stability of soil structure The ability of the soil structural units to remain coherent under an applied stress, e.g. raindrop impact, waterlogging or ploughing.

Stack A residual rock pinnacle which marks coastal cliff retreat and/or the landward advance of a **rock platform**.

Stage The height of the water surface above a specific location in a fluvial channel, usually the deepest point.

Standard deviation A measure of the variability within a data set.

Standing crop The total weight of living organic material per unit area at any one time.

Standing wave A water wave which oscillates vertically between two points without propagating horizontally.

Steady-state equilibrium The state of a system where the steady output of matter and energy is equal to the input over a particular period of time.

Stemflow That portion of intercepted precipitation which is concentrated and transferred towards the ground by plant stems and trunks.

Stomata Microscopic pores on plant leaves through which most water vapour and other gaseous exchanges take place.

Storage Locations where energy and matter may be stored for certain periods of time.

Storage capacity The amount of **available water** which a soil can hold.

Storm surge An abnormal rise in sea level driven against the coast by extreme weather events, most severe and liable to cause coastal damage when coinciding with high **spring tides**.

Stratified scree Scree material showing apparent bedding planes dipping downslope, suggesting episodic **solifluction** or **sheetflow** typical of a **permafrost** environment; this may be so, or the structures may reflect post-depositional removal or settling of fines or **platy** debris.

Stratigraphy The branch of geology which studies the form (*litho*-stratigraphy), fossils (*bio*-stratigraphy) and age (*chrono*-stratigraphy) of a layered or *stratified* sequence of rocks.

Strato-volcano A composite volcano built up by the accumulation of successive, stratified layers of **lava** and **pyroclasts** during its eruptive history.

Stream competence A measure of the ability of streamflow to maintain particles in motion, in terms of their size and current velocity.

Stream gauging The manual or automatic measurement of water velocity, depth and wetted channel cross-sectional area from which **discharge** is calculated, in cubic metres per second.

Stream order The designation of the position of a stream channel, by a value from 1 to *n*, in a network, indicative directly or indirectly of the number of tributary channels contributing to a channel of a particular number.

Stream power The rate of energy supply available for work at the stream bed, measured in W m^{-2}.

Striation A scratch on a rock surface made by the abrasive contact of a moving material of greater hardness and aligned in the direction of motion.

Strike The orientation of a geological structure (e.g. fold or fault) in the horizontal plane.

Strike-slip That part of the displacement of rocks on opposite sides of a fault which is parallel to its orientation and thus in the horizontal plane.

Stromatolite A dome-shaped calcareous mat of algae and trapped, fine-grained sediment accumulating in shallow-water lagoons.

Structural basin A crustal depression defined by large-scale geological structures rather than the product of surface erosion.

Structural control The influence of structures or **planar discontinuities** on denudation and geomorphic development of the landsurface through the defects which they create in rockmass strength and their geometric arrangement.

Structureless soil A soil showing no aggregation into **peds**, i.e. single grain or massive.

Subduction zone A linear boundary between convergent plates at which one plate is drawn or forced down under the other, usually by virtue of its greater density; the zone is marked by intense crustal deformation and **resorption**, accompanied by **seismo-volcanic** activity.

Submarine canyon A steep-sided submarine valley cut across the **continental shelf** and **slope**, frequently the continuation of, or originating from, terrestrial valleys and swept by **turbidity currents**.

Subsystems A series of smaller systems linked together by a series of flows of energy and matter.

Succession The process of change in **plant communities** which, successively, occupy a given area and culminate in climax vegetation.

Successional community A community in a succession which is creating conditions more and more favourable for a succeeding community.

Succulent A plant with fleshy, water-storing tissues, characteristic of arid or saline areas.

Suction The force with which water is held in soil, or the force required to remove water from soil.

Supercontinental cycle The periodic coalescence and rifting apart of supercontinents, driven by **plate tectonics** over *c.* 500 Ma time scales caused by the inevitability of plate collisions in one area as a result of **sea-floor spreading** in another; also known as the **Wilson cycle** (after J.T. Wilson).

Surf zone The turbulent water from breaking waves which lies between the breaker and **swash** zones in the nearshore environment.

Surging glacier A glacier enjoying a transient phase of very rapid, extending and thinning flow which exceeds the glacier's **mass balance** capacity and either leads to a runaway collapse of the glacier or a quiescent period of recovery.

Suspect terrane A crustal fragment or **terrane** amongst a collage of such fragments forming continental crust, whose external and even internal common geographic origins are in doubt or 'suspect', by virtue of the endless mobility which characterizes plate boundaries.

Suture The linear, convergent boundary at which two **continental plates** are welding together and marked by compression structures and remnants of the ocean which formerly separated them.

Swash A forward pulse of water released by a **breaking wave** after it has broken, capable of moving sand up a beach.

Swelling clay A clay capable of absorbing very large quantities of water and greatly increasing in volume which may contribute to mechanical weathering by creating tensile stress in a rockmass.

Symbiotic fixation Conversion of atmospheric nitrogen into organic nitrogen compounds by *Rhizobium* bacteria living in nodules on the roots of leguminous plants.

Syndepositional sedimentary structure A structure formed during the final settlement of a soft sediment, reflecting, for example, slumping or faulting on contraction as water is expelled, or flowage of more saturated components.

Syngenesis Minor, non-destructive changes in the mechanical or chemical properties of rock during its emplacement, which act as early stages in its **lithification**.

Talik A pocket of unfrozen ground, other than the **active layer**, within the boundary of an otherwise-frozen **permafrost** environment.

Talus An alternative name for **scree**.

Talus cone A cone of coarse, angular rock debris or **scree** accumulated at the base of a rock **chute**.

Tectonic cycle Cycles of ocean and continental expansion and contraction, synonymous with a **supercontinental** or **Wilson cycle**.

Temperate glacier A warm-based glacier, at or near **pressure-melting point** throughout and generally typical of alpine valley glaciers or the lower portions of **outlet glaciers**.

Tensile stress A stress tending to stretch, pull apart and fracture rockmass.

Tephra The collective term for all **pyroclastic** rock material thrown into the atmosphere by volcanic eruption, from fine ash to large rocks.

Tephrochronology A dating tool dependent on the recognition of individual **ash falls** or **tephra** layers from specific volcanic eruptions; it provides either a relative age in a stratigraphic sequence or an absolute age, by isotopic dating or direct knowledge of the date of eruption.

Terrace A topographic bench or hillside step, cut in bedrock or formed by sediment **aggradation**, at the margins of a river, glacier, lake or sea; it slopes steeply from a level or gently sloping upper surface.

Terrane A crustal fragment with a coherent lithological and structural identity and geological history, quite distinct from its neighbours.

Terrestrial sediment Sediments eroded from and deposited on a landsurface.

Terrigenous sediments Sediments eroded from a landsurface and deposited at continental margins.

Thalassostatic Said of river terraces cut by **incision** whilst sea level is low or falling and **aggradation** whilst sea level is high or rising.

Thalweg An imaginary line connecting the lowest points along a stream bed or valley floor.

Thermal bulge A section of crust elevated over a mantle convection current by thermal expansion.

Thermal energy The energy of a substance which is stored in the form of **sensible heat** and/or **latent heat**.

Thermal welt Crustal thickening over a rising **mantle plume** caused by the intrusion of **magma** (see also **thermal bulge**).

Thermocline A zone of marked change in temperature with ocean depth.

Thermogenesis The production of heat.

Thermohaline circulation A global, density-driven ocean circulation system controlled by differences in temperature and salinity (see also **Global Ocean Conveyor**).

Third stage community A community occupying the third stage of a succession.

Throughfall Net precipitation at the ground after passing through a vegetation canopy.

Throughflow The shallow subsurface transmission of water through soil, developing lateral movement as the onward infiltration rate is reduced and emerging as **saturated overland flow** towards valley floors; sometimes also used to identify the portion of stream **discharge** attributable to such transmission.

Thrusting The action of overriding of one geological unit by another caused by low-angled shear.

Tidal bulge The rise in ocean water surface caused by the gravitational attraction of the Moon and Sun, which moves around the global ocean following their motion relative to Earth.

Tidal current The horizontal ebb and flow of semi-diurnal and diurnal tides around the coastline.

Tidal flat An extensive, low-lying surface occupying an **intertidal zone** and commonly covered in sand, mud or saltmarsh.

Tidal pass A natural breach through a coastal barrier or **barrier island** through which tides flood and drain a landward lagoon.

Tidal wave The semi-diurnal rise and fall of the ocean surface as the **tidal bulge** sweeps around the global ocean.

Tide The regular horizontal and vertical motion of the ocean surface in response to the gravitational attraction of Sun and Moon, most noticeable at the coastline where its effects are usually amplified.

Tidewater glacier A glacier which terminates in the sea, into which **glaciomarine** environment it discharges sediment, meltwater and **icebergs**.

Till A coarse, generally unsorted and unstratified sediment deposited by glacier ice; its biomodal character, with large **clasts** and a fine-grained matrix – described in the now-defunct term 'boulder clay' – reflects the indiscriminate power of glaciers.

Timberline The upper altitudinal or latitudinal limit beyond which trees cannot normally grow; locally, microclimate may sustain pockets of trees beyond the regional timberline.

Time of rise The time elapsed from the point at which discharge increases in response to a precipitation event and the point of **peak discharge**.

Tolerance model The model of succession in which modification of a habitat by an established species has little effect on other species, as changes in the composition of a community are controlled by the life cycle of the plants.

Tombolo A sand or gravel bar connecting an island with another land mass.

Top carnivore The carnivore which occupies the highest position in a food web or **energy pyramid**.

Topoclimate A local mesoclimate extending up to 250 m above a landsurface, in which regional climate is modified by topographic and slope factors such as aspect, shade, exposure, etc.

Topographic climax Climax vegetation maintained by topographic conditions (exposure, soil hydrology or aspect).

Toppling failure A rockfall involving a column of rock (or cohesive soil) whose centre of gravity overhangs a pivot point; the column rotates outwards at its top before overturning.

Tor A residual rock pinnacle or pile on an elevated site, best developed in massive crystalline rocks, and exposed by the weathering and mass wasting of surrounding rock mass.

Trace element An essential nutrient for plant growth but only needed in very small quantities.

Trailing edge The receding **passive margin** of continental crust with a distal **orogen** or proximal rift escarpment, wide continental plain and shelf which influence coastal development (cf. **leading edge**).

Transform fault A large-scale fault between or within crustal plates, with displacement wholly or mostly in the horizontal plane (see **strike-slip**).

Transform margin A plate margin which coincides with and is guided by a **transform fault**.

Transgression The submergence of low-lying coastal land by rising sea level or land subsidence and consequential landward shifts in each component of the littoral zone.

Transitional soils An early term for intrazonal soils, i.e. soils whose profiles are controlled by local factors of geology and topography.

Translation The movement of a shallow rock or soil mass along a failure surface or **planar discontinuity**.

Transmission pore Pores in soil larger than 0.05 mm diameter which allow water to drain away under the force of gravity.

Transport(ation) A process or stage in the **rock cycle** whereby all forms of rock and organic debris are carried by moving water, ice or air from their point of origin to a point of temporary or permanent deposition.

Transported soil A soil whose parent material is a transported sediment.

Trench A narrow, linear and deep depression in the sea bed caused by the subduction of oceanic crust at a destructive plate margin.

Trench suction force The force which draws some of the upper plate down into a **subduction zone** with the descending plate.

Trimline An abrupt line on a valley side separating weathered, vegetated upper slopes from newly eroded, unvegetated lower slopes and marking the upper limit of a contemporary or recent valley glacier.

Triple junction A three-way rift developed at the common divergent margins of three adjacent crustal plates.

Trophic level The energy or feeding level of a particular organism in a food chain or food web.

Trophic relationships The food web or feeding links in a community or **ecosystem**.

Trough A narrow, deep and steep-sided rockwalled valley typical of intense erosion by a valley or **outlet glacier**.

Tsunami A seismic surface wave, triggered at sea by **seismo-volcanic activity**, which rises in elevation in shallow coastal waters and is capable of inflicting considerable shoreline damage.

Turbidite The sedimentary unit formed as a **turbidity current** comes to rest; generally moderately sorted, with a fining-upwards or graded sedimentary structure.

Turbidity current A turbulent, gravity-induced density current of suspended sediment, usually in a marine or lacustrine environment, which eventually forms a **turbidite**.

Turbulent flow Fluid motion in which individual flow strands are confused and multidirectional, with eddies developing within the general forward motion.

Ubac A mountain slope whose orientation shades it from the sun.

Ultramafic Said of igneous rock crystallized at high temperature and composed mostly of magnesium–iron (**mafic**) minerals.

Unconformity A contact between two rock units indicative of a break in a continuous sequence of rock formation; it may mark an episode of inactivity, weathering or erosion.

Unit hydrograph The model plot of river discharge over time generated in a particular catchment by a specified unit of precipitation.

Upwelling The rise of cold water to an ocean surface induced by divergent surface currents.

Vadose water Water in the unsaturated zone of an **aquifer**.

Valley train A valley-wide, braided glacial meltwater stream and sediment system emerging from a glacier terminus.

Vapour pressure deficit The extent of the vapour pressure gradient between subsurface soil and a dry atmosphere.

Vegetation formations Units of natural vegetation occupying distinct geographical regions of large size, and possessing a uniform physiognomy.

Velocity The rate of movement of an object or material, measured typically for rapid environmental processes in metres per second or in slow processes in millimetres per year.

Ventifact A loose stone shaped by wind abrasion on an arid landsurface.

Vermiculite An expanding 2:1 clay mineral with isomorphous substitution in both the silica and the alumina sheets, which makes it very reactive.

Vesicular An igneous rock texture characterized by the empty vesicles or voids vacated by gas bubbles in the original **magma**.

Viscous drag A coupling of rigid lithosphere and partially melted, moving asthenosphere once thought to be the principal mechanism of **continental drift**.

Vivipary Plant reproduction by the formation of bulblets on the parent.

Volcanic arc An arcuate line of explosive, andesitic island volcanoes erupted through oceanic crust on the landward side of a **B-subduction zone** as a result of the hydration and partial melt of downgoing crust; the arc may eventually migrate and weld on to adjacent continental crust.

Volcanic breccia A **pyroclastic** rock formed of consolidated, angular volcanic rock rubble.

Wadati–Benioff subduction The full title of what is more commonly known as **Benioff** or **B-subduction**.

Water balance The volumetric input, storage and output of water which constitutes the water budget of a **drainage basin** per unit time, normally a calendar year.

Water-holding capacity The amount of water held in soil after free drainage under the influence of gravity.

Water-layer weathering Rock weathering in the **intertidal zone** by those processes of slaking, hydration and salt weathering which are enhanced by regular cyclic hydration and drying.

Watershed The delimiting boundary of a **drainage basin**, normally at the landsurface but taking into account any lateral underground transfers determined by geological conditions; alternatively, the **drainage basin** delimited by such a boundary.

Wave An oscillatory rise and fall in a water surface, marking the horizontal transmission of wind-driven energy through the orbital motion of water particles.

Wave base The water depth at which the orbital motion of surface waves ceases.

Wave-cut notch A notch or indentation cut into the base of a cliff by wave action.

Wavelength The distance between crests of adjacent **waves**.

Wave period The time taken for consecutive wave crests to pass a fixed point.

Wave train A regular procession of water surface waves characterized by their **wave length** and **wave period**.

Weathering The progressive alteration and eventual chemical or mechanical disintegration of rockmass at the landsurface; exposure to a different thermal and moisture environment from that in which it formed renders it unstable and susceptible to weathering.

Weathering rind An outer crust of discoloured and texturally altered rock representing the initial stage in rock weathering.

Weathering solution Solution containing metallic ions and silica which is produced during weathering.

Wedge slide Sliding failure of a rock wedge bounded by two release surfaces where two **planar discontinuities** intersect; the angle of intersection must be greater than the internal friction angle of the mass but less than the parent slope angle.

Wetted perimeter The length of a river channel covered by water at a particular point and water depth; an important component of the measure of **hydraulic radius** and **efficiency**.

White alkali soils A popular term for solonchaks.

White box A system in which a detailed knowledge of all the internal structure is identified; it is rarely achieved, except in the most simple of systems.

Wildfire A fire started by humans either by accident or for malicious reasons.

Wilson cycle The alternative name for the **Supercontinental** cycle, named after its proponent, J. Tuzo Wilson, during the 1960s.

Wilting point The moisture content of soil at which there is insufficient water to maintain the turgor of the plant.

Windstorm A violent cold air gravity flow off the eastern Rocky Mountains in winter, stimulated by temperature inversion over snow.

Wind stress The force exerted by wind per unit area on an adjacent surface.

Xerosere A **primary succession** which starts on a surface suffering from drought.

Xerophyte A plant adapted to grow in arid habitats.

Yardang An exposed rock surface abraded into a streamlined shape by windblown sand.

Younger Dryas A European term for the climatic deterioration and mountain glaciation *c.* 11,000–10,000 BP, known in Britain as the **Loch Lomond stadial**.

Zeolite The lowest grade of metamorphic rock, altered at relatively low lithospheric pressures and temperatures.

Zonal Soils occurring in extensive zones on a world scale.

Zone of soil formation The upper part of the zone of **weathering** where **soil-forming processes** are most active.

Zone of weathering The upper part of the lithosphere where **weathering** processes are operating at a maximum.